AF566929

Principles of Seed Pathology

Second Edition

V. K. Agarwal
James B. Sinclair

FIRST INDIAN REPRINT, 2012

Acquiring Editor: Neil Levine
Project Editor: Les Kaplan
Marketing Manager: Greg Daurelle
Direct Marketing Manager: Arline Massey
Cover Design: Denise Craig
Typesetter: Pamela Morrell
Prepress: Kevin Luong
Manufacturing: Sheri Schwartz

Library of Congress Cataloging-in-Publication Data

Agarwal, V. K.
Principles of seed pathology / Vijendra K. Agarwal, James B. Sinclair. — 2nd ed.
p. cm.
Includes bibliographical references and index.
ISBN 0-87371-670-1
1. Seed pathology. I. Sinclair, J. B. (James Burton), 1927– . II. Title.
SB732.8.A35 1996
632'.3—dc20 96-14116
CIP

International Standard Book Number 0-87371-670-1
Library of Congress Card Number 96-14116
Printed and bound in India by Nutech Photolithographers

FOR SALE IN SOUTH ASIA ONLY.

PREFACE

The science of seed pathology is relatively young, having had its beginnings in seed health testing and control of seedborne pathogens. Since the late 1970s there has been a worldwide increase in research, outreach, and training activities related to seed pathology. Seedborne pathogens have had special consideration in seed production areas and in plant quarantine activities. Recognition of the increased interest and importance of this branch of plant pathology was given by the creation of the Danish Government Institute of Seed Pathology for Developing Countries (DGISP), Copenhagen, in 1967, of the Seed Pathology Committee by the American Phytopathological Society in 1976, and of the International Society of Plant Pathology in 1977.

This book was written to serve those interested in seed pathology. It is designed to serve as a textbook as well as a reference book for students, teachers, and researchers, and for personnel involved in seed health testing, seed production, and plant quarantine. It is to be used as a guide to the literature. Much of the illustrative material has come from the authors' files used for teaching or from their own research. Teachers will want to supplement this book with examples from their own experience and research or with information and data from other seed pathology programs.

The authors hope that this book, in addition to being of value to seed and plant pathologists, will be useful to agriculturalists interested in crop production. It was written in part to stimulate research in seed pathology and in the importance of seed pathology to the role of seedborne inocula in the epidemiology and control of plant diseases.

The scientific names of many plant pathogens and some of their hosts have been changed since the first edition. The revised scientific names of fungal pathogens (Appendix A) and anamorphs and teleomorphs (Appendix B) are provided. For this second edition, the Latin names for the hosts and pathogens have been used as provided in the following publications:

1. Farr, F. F. , Bills, G. F., Chamuris, G. P., and Rossman, A. Y., *Fungi on Plants and Plant Products in the United States*, APS Press, St. Paul, MN, 1989, 1252.
2. Hansen, E. M. and Maxwell, D. P., Species of the *Phytophthora megasperma* complex, *Mycologia*, 83, 376, 1991.
3. Sneath, P. H. A., Mair, N. S., and Sharpe, N. E., Eds., *Bergey's Manual of Systematic Bacteriology*, Vol. 2, 9th ed., Williams & Wilkins, Baltimore, 1986.
4. Publications from the Commonwealth Mycological Institute, Kew, Surrey, U. K.

The authors wish to thank G.B. Pant University of Agriculture and Technology (GBPUAT), the Institut National de la Recherche Agronomique (INRA), Versailles, France, and the University of Illinois at Urbana-Champaign (UIUC) for the use of library facilities in preparation of the manuscript. Thanks are given to M. L. Verma (GBPUAT) for typing the initial draft of the manuscript and to Danielle Clark, Nancy David, Barbara Goschy, Susan Mittelsteadt, and Susan Schmall-Ross (UIUC), who typed supplemental drafts of all or portions of the manuscript. We thank Pam Purcell Avenius (Urbana), who did portions of the art work. Particular thanks go to Richard D. McClary (UIUC) for his many trips to the libraries of the University of Illinois at Urbana-Champaign to search for articles and references.

Special thanks to students, colleagues, and friends for their patience during the preparation of the manuscript for the second edition.

Professor Agarwal is grateful to GBPUAT for permission to take up the task of preparing the material for this book and is especially grateful to Y. L. Nene, formerly from GBPUAT and now at the International Crops Research Institute for the Semi-Arid Tropics, Andhra Pradesh, India, for his guidance, inspiration, and encouragement throughout the author's career. He is grateful to D. Spire, Madame Monique Lemattre, and Y. Maury of INRA for their permission for the use of the library facilities and for their encouragement and to Irene Fiala, Gisele Lacaze, Valerie Molinero, and Roselyne Corbiere of INRA for their help in the literature search. He owes special gratitude to his parents for their guidance and inspiration, and to his wife, Kiran, and his daughters, Priyanka and Sheelu, for their patience and cooperation during the preparation of the manuscript. Dr. Agarwal expresses sincere and deepest regard to his grandfather, Babu Ram Prasad, for encouraging him to obtain an education.

Vijendra K. Agarwal
James B. Sinclair

THE AUTHORS

Vijendra K. Agarwal, Ph.D., FISST, FPSI, is a professor in the Department of Plant Pathology, College of Agriculture, Govind Ballabh Pant University of Agriculture and Technology (GBPUAT), Pantnagar (Nainital), U.P., India.

Professor Agarwal received his basic education in a government school in his native village, Jhalu (Bijnor), Uttar Pradesh. He completed his B.Sc. (Honors) in Agriculture and Animal Husbandry in 1965, his M.Sc. in Plant Pathology in 1967, and his Ph.D. in Plant Pathology in 1975 under the guidance of Y. L. Nene from GBPUAT. He joined the Department of Plant Pathology in July 1967, and has served as a senior research assistant, assistant professor, associate professor, and professor. He worked as Research Scholar at the Danish Government Institute of Seed Pathology (DGISP) for Developing Countries, Copenhagen, Denmark, from January 1969 to February 1970 with Paul Neergaard and S. B. Mathur. He was Postdoctoral Fellow at the Institut National de la Recherche Agronomique, Versailles, France, on a French Government Fellowship from April to November 1991 with D. Spire. During this period he also worked with R. Champion, Station Nationale d'Essais de Semences, La Miniere. He was a visiting scientist at the DGISP, Copenhagen, during June to July 1991.

Professor Agarwal has taught basic courses in plant pathology and seed pathology at the graduate and postgraduate levels. He has developed research and teaching programs in seed pathology. His research has concentrated on standardization of techniques for the detection, infection, transmission, and control of seedborne pathogens and on seed certification. He has worked on seedborne diseases, such as loose smut, black point and karnal bunt of wheat, brown discoloration of paddy, purple stain of soybean, grain mold of sorghum and pearl millet, anthracnose of pepper, and seedborne pathogens of forest tree seeds. The sodium hydroxide seed soak method, which he and his colleagues developed, is being used widely for the detection of the karnal and rice bunt fungus in non-treated and chemically treated wheat and rice seeds. He has published over 75 refereed research papers, 2 books/bulletins, and 70 research abstracts, along with reviews and technical articles; he has also contributed chapters to several books. Professor Agarwal has served professional societies in different capacities, including a term as Vice President of the Indian Society of Seed Technology. He has attended and presented invitational lectures/ scientific papers at several national and international conferences seminars and workshops.

James B. Sinclair, Ph.D., is a professor of plant pathology in the Department of Crop Sciences, College of Agricultural, Consumer and Environmental Sciences, University of Illinois at Urbana-Champaign, Urbana (UIUC). Professor Sinclair received his B.Sc. degree from Lawrence University, Appleton, Wisconsin, in 1951 and his Ph.D. in plant pathology from the University of Wisconsin, Madison, in 1955 under J. C. Walker, with whom he continued to work with postdoctoral appointment until 1956, when he accepted a position in the Department of Plant Pathology, Louisiana State University (LSU), Baton Rouge. At

LSU he served as an assistant professor, associate professor, and professor until 1968. Also, he was administrative assistant to the Chancellor from 1966 to 1968. He joined the Department of Plant Pathology, UIUC, in 1968 as a professor of international plant pathology. He was Campus and then All University Coordinator for the Illinois-Tehran Research Unit, 1974–78.. He was named Interim Director, National Soybean Research Laboratory, UIUC, in 1992.

Professor Sinclair has taught five graduate courses in plant pathology; he has also planned and participated in and given invitational lectures at numerous national and international conferences and workshops. He has worked in over 40 countries professionally and has directed the research of 71 graduate students, of whom 12 completed a portion of their thesis research at an overseas institution. He is a member of many national and international professional organizations and served from 1979 to 1983 as Chairman, Seed Pathology Committee, International Society of Plant Pathology.

Professor Sinclair's research has been primarily on seedborne and soilborne pathogens of soybeans and other crops and their control, and on the uptake and translocation of systematic fungicides in various crop plants. He has published over 234 refereed research papers and 259 research abstracts. He has authored, edited, or co-edited 38 books and monographs and 215 other articles. He has received the following awards: ICI/American Soybean Association Research Recognition Award, 1983; UIUC Paul A. Funk Award, 1984; U.S. Department of Agriculture Award for Distinguished Services, 1988; American Soybean Association Production Research Award, 1989; Honorary Member, Illinois Crop Improvement Association, 1990; North Central Division, American Phytopathological Society Distinguished Service Award, 1991; Land of Lincoln Soybean Association Research Award, 1991; the UIUC College of Agriculture Senior Faculty Award for Excellence in Research, 1992; Fellow, American Phytopathological Society, 1993; and Fellow, National Academy of Sciences, India, 1995.

CONTENTS

Chapter 7
Seed Transmission and Inoculation

Chapter 8
Factors Affecting Seed Transmission

Chapter 12
Mycotoxins and Mycotoxicoses

Chapter 1

Introduction

The term *seed pathology* was first used by Paul Neergaard and Mary Noble in the 1940s. Although the science of seed pathology has existed for over 100 years, the term initially was used for seed testing technology. At that time standardized procedures of testing for seed germination and purity had been established, but techniques for detection of seedborne pathogens were needed, although the fact that seed pathology was more than a testing technology was well understood.[1] Seed pathology may be defined as the study of seedborne diseases and pathogens. It includes studies on the mechanisms of infection, seed transmission, the role of seedborne inoculum in disease development, techniques for detection of seedborne pathogens and nonpathogens, seed certification standards, deterioration due to storage fungi, mycotoxins and mycotoxicoses, and control of seedborne inoculum. Seed pathology includes the study of diseases and deterioration caused by bacteria, fungi, nematodes, viroids and viruses, and physiological and mechanical disorders of seeds.

Paul Neergaard (1907–1987) is considered the father of seed pathology (Figure 1-1). His monograph on Danish species of *Alternaria* and *Stemphylium* is a standard for information on these genera.[2] He also described about 100 plant diseases and pathogens, either new to Denmark or the world. He established the seedborne nature of several hundred host–pathogen combinations, and in collaboration with colleagues at the Danish Government Institute of Seed Pathology for Developing Countries (DGISP), Copenhagen, demonstrated the etiology of seedborne *Colletotrichum*, *Bipolaris, Drechslera*, *Fusarium*, *Myrothecium*, and other fungi. He initiated the 2,4-D blotter method for detection of seedborne fungi. As Chairman of the Plant Disease Committee (PDC) of the International Seed Testing Association (ISTA) from 1956 to 1974, he helped standardize methods for detection of seedborne fungi through seed health testing workshops. His primary contribution was the writing of *Seed Pathology*, a reference and textbook that has helped in the teaching and practice of seed pathology the world over.[3] Neergaard's efforts led to the understanding of the importance of seed health testing, which is an integral part of most seed testing laboratories.[4]

Figure 1-1 Paul Neergaard. (From *Annu. Rev. Phytopathol.*, 24, 1, 1986. With permission.)

In 1967, the DGIPS was established by the Danish International Development Assistance of the Ministry of Foreign Affairs. The primary objectives of the Institute were to train plant pathologists from developing countries in seed pathology and to assist developing countries in programs of education in seed health certification and quarantine.[5] Neergaard was its first Director, from 1967 to 1982. The Institute now is affiliated with the Royal Veterinary and Agricultural University, Copenhagen, with S. B. Mathur as Director. Mary Noble is an internationally known seed pathologist specializing in seedborne diseases of cereal and clover crops. Her work on blind seed disease of perennial ryegrass (*Gloeotinia granigena*) is considered classic.[6]

The science of seed pathology has gained new appreciation since 1980 and is now recognized as an important discipline within plant pathology. In addition to developing new technology for pathogen detection, there is increased interest in and an understanding of the role of seedborne inoculum in plant disease epidemiology, seed certification, and plant quarantine. Since many seedborne pathogens perpetuate through various means, the role of seedborne inoculum in disease development cannot be studied in isolation. Similarly, control measures of seedborne pathogens may be part of an integrated disease management program.

I. TERMINOLOGY

Seed pathology is an aspect of general plant pathology that includes the relationship of plant pathogens to all types of propagative materials, including commercial and weed seeds, and other plant propagating material, such as cuttings, scions, tubers, and meristem propagules, which are of importance in trade and in the exchange of germ plasm.[1] The term *seed* as used in this book includes true seeds and dry, one-seeded fruits. A true seed is a fertilized mature ovule consisting of an embryonic plant, stored food material, and a protective seed coat. Asexually or vegetatively propagated crops in which no true seeds are involved are not considered.

All the principles that apply to the study of diseased seedlings and plants can be applied to seed pathology. The seed is a miniature plant. Seed disease is the result of an interaction, over time, between a susceptible host, a pathogen, the environment, and a transmitting agent, resulting in signs or symptoms of such effects. Any infectious agent associated with seeds that has the potential of causing a disease of a seedling or plant should be termed a *seedborne pathogen*. This term includes all plant–pathogenic bacteria, fungi, nematodes and other microorganisms, and viruses, all of which can be carried in, on, or with the seeds. The term also is used to indicate the mode of transmission and perpetuation of a plant pathogen. Often reference is made in the literature to a seedborne disease, when the correct reference should be to a seedborne pathogen. Technically, a seedborne disease is one showing symptoms on infected seeds. Many seedborne pathogens are asymptomatic.[7] Because of long-standing practices, for the sake of brevity, and assuming that the readers will understand, plant diseases may be referred to as being seedborne. The pathogenic fungus *Ustilago segetum* var. *tritici* is carried in the embryo of barley seeds and *U. segetum* var. *segetum* on the surface of barley seeds, while the nematode *Anguina tritici* is carried as an admixture in ear cockle galls with wheat seeds. The pathogens cause loose and covered smut of barley and ear cockle of wheat, respectively. These may be referred to as seedborne diseases.

Seedborne microflora is a generalized term indicating the association of bacteria, fungi, nematodes, and other microflora or, incorrectly, viruses associated with seeds, which may or may not have the potential of causing diseases of the seed or plant. It includes pathogenic and saprophytic microflora.

The terms *externally* and *internally seedborne* refer to the location of the pathogen in relation to the seed. If a pathogen is located on the outside of the functional part of the seed, it is externally seedborne, and if inside the seed, it is internally seedborne. Pathogens such as *U. segetum* var. *avenae* and *U. segetum* var. *segetum*, which survive as teliospores on the seed surface of oats and barley, respectively, are externally seedborne pathogens. Barley stripe mosaic virus and *U. segetum* var. *tritici*, which are internally seedborne pathogens, are deep seated and localized inside the barley seed and are internally seedborne. A pathogen may be both internally and externally seedborne. For example, *Sclerospora graminicola*, which causes downy mildew or green ear of pearl millet, is carried internally as hyphae in the scutellum and on the seed surface as oospores. Both inocula are capable of causing disease.

The term *seed transmission* refers to the passage of a seedborne pathogen from seeds to seedlings and plants. The rate of seed transmission depends upon the host, pathogen, environment, vectors, and their interaction over time.

II. HISTORICAL DEVELOPMENT

Documentation for transmission of plant pathogens through seeds came relatively late in plant pathology history. The French botanist du Tillet[8] showed, in 1755, that stinking or hill bunt of wheat was caused by a "poisonous substance" contained in the dust sticking on seed surfaces. In 1807, Prevost[9] proved that stinking bunt was caused by a parasitic fungus, *Tilletia caries*. The internally seedborne nature of a fungus was demonstrated by Frank[10] in 1883, describing *Colletotrichum lindemuthianum* in bean (*Phaseolus vulgaris*) seeds. In 1892, Beach[11] proved the seedborne nature of *Xanthomonas campestris* pv. *phaseoli* in bean seeds.

Rolfs[12] in 1915 showed the internal transmission of *X. c.* pv. *malvacearum* in cottonseeds and its association with lint. Another early demonstration of internal seed infection by bacteria was by Clayton[13] in 1929 studying *X. c.* pv. *campestris* on cauliflower. External transmission of a bacterium was shown first by Stewart in 1897 studying *Erwinia stewartii* on maize.[14] Virus seed transmission was studied by Mayer,[15] who showed in 1886 that tobacco seeds from tobacco mosaic-infected plants yielded diseased seedlings. McClintock[16] suggested in 1916 that cucumber mosaic (CMV) was seed transmitted. Conclusive evidence for virus seed transmission was presented by Stewart and Reddick[17] in 1917, who showed that bean seeds from bean common mosaic virus-infected plants produced infected seedlings. Doolittle and Gilbert[18] in 1919 demonstrated seed transmission of CMV in wild cucumber (*Echinocystis lobata*). They noted that wild cucumber was susceptible to CMV and that symptoms appeared 3 to 4 weeks before those in domesticated plants in the field. They demonstrated that a portion of seeds from infected plants produced diseased plants the following year and concluded that seedborne inoculum was important in the overwintering of the virus and was a source of primary inoculum.[19] The first reference to an association of plant parasitic nematodes with seeds was in Act IV, Scene 3 of Shakespeare's play *Love's Labour's Lost* (1594) in the line "sowed cockle, reaped no corn." In 1743, Needham[20] observed nematodes in cockled wheat seeds and showed that *Anguina tritici* caused wheat cockles. Since these early reports, about 3000 microorganisms and viruses have been shown to be seedborne.[21,22]

Dorogin[23] in 1923 published systems for detection of seedborne pathogens associated with crop seeds in the U.S.S.R, and in 1924 an analysis of crop seeds for plant pathogens was made compulsory. Chen[24] published a monograph on internal fungal parasites of agricultural seeds in 1920. Orton[25] in 1931 and Porter[26] in 1949 published a list of seedborne pathogens from the United States and the damage caused by them. In 1931 Alcock[27] published a list of seedborne mycoflora of forage crops, ornamental plants, and vegetables in Scotland. A list of the fungi found on barley, oats, and wheat seeds in Canada was compiled by Machacek et al.[28] *An Annotated List of Seedborne Diseases* was published by Noble et al.[29] in 1958, updated in 1968 by Noble and Richardson,[30] and revised in 1979 and 1990 by Richardson.[21,22]

III. DEVELOPMENT OF SEED HEALTH TESTING

Testing seeds for germination and purity as a measure of seed quality has been a universal practice for more than a century. The first official seed testing station was established in 1869 by Nobbe in Tharandt (Germany) with its major function to test seeds for germination and purity.[31] In 1876 another such laboratory was established in the United States by the Connecticut Agricultural Experiment Station.[32] In 1884 E. Schribaux founded the Station Nationale d'Essais de Semences in France.[31] Today most countries have one or more federal and/or state seed laboratories testing for seed germination, purity, freedom from noxious weed seeds, moisture content, and general seed health. Techniques for testing seeds for germination and purity have been standardized by the Association of Official Seed Analysts (AOSA) in the United States, and by the International Seed Testing Association (ISTA). The first step toward international cooperation in seed testing was taken at the First International Seed Testing Congress (ISTC) in Hamburg in 1906. At the Third ISTC in Copenhagen in 1921, the European Seed Testing Association was founded. During the Fourth ISTC in Cambridge in 1924, activities of the association were extended to all countries, and ISTC was reconstituted under its present name, the International Seed Testing Association (ISTA). The primary purpose of ISTA is to develop, adopt, and publish standard procedures for sampling and testing seeds and to promote uniform application of these procedures for evaluation of seeds in international trade. Secondary purposes are to promote research in all areas of seed science and technology, to encourage cultivar certification, to participate in conferences and training courses aimed at furthering these objectives, and to establish and maintain liaisons with other organizations having common or related interests in seeds. ISTA cooperates with the Association of Official Seed Analysts (AOSA) of North America, the European Economic Community (EEC), the European Plant Protection Organization (EPPO), the Food and Agricultural Organization (FAO) of the United Nations, the International Seed Trade Federation (FIS), the International Institute for Beet Research (IIRB), the International Organization for Standardization (ISO), the International Union of Biological Sciences (IUBS), and the International Union of Forestry Research Organizations (IUFRO). It publishes procedures and techniques used in seed testing known as the ISTA Rules. The rules are amended and approved on recommendations of technical committees. The 1990 edition of the rules consisted of (i) the rules describing principles and definitions and (ii) annexes, describing in detail approved methods. Authorized official seed testing stations within ISTA issue international certificates that certify the results of tests conducted in accordance with the international rules. More than 100,000 international certificates are issued each year.[33]

Nobbe's publication of "Samenkunde" in 1876 mentioned the occurrence of sclerotia and smut balls in connection with seed production and distribution. However, neither he nor Hartz, another author of that time, described any method to detect pathogens in seed samples except those visible to the unaided eye.[34]

The association of plant pathogens with seeds was verified by Bessey in 1886,[35] when he published a list of fungi detected in seeds in Iowa. Later, A. L. Smith[36] published illustrated notes on fungi found on germinated agricultural seeds in the United Kingdom. The first seed health testing laboratory was established in 1918 at the Government Seed Testing Station, Wageningen, The Netherlands, with L.C. Doyer as the first official seed pathologist.[37] L. C. Doyer was the first chairperson of the ISTA Plant Disease Committee for Seed Health Committee. Her contribution, published in the 1930 ISTA Proceedings, was a systematic survey of diseases and pests on weeds involving bacteria, fungi, insects, viruses, and other injurious organisms. This work was the basis for her presentation at the 1931 Wageningen Congress, titled "Proposals for Recording the Sanitary Conditions of Seed on the International Rules of Seed Testing." She continued as Chairperson of the Plant Disease Committee to 1949. Her *Manual for Determination of Seedborne Diseases* is used today. She was succeeded by W. F. Crosier (1949–1953), A. J. Skolko (1953–1956), and then Paul Neergaard (1956–1974). She published a manual in 1938 for determining seedborne pathogens during germination tests based on symptoms produced on seedlings.[37,38] Hiltner[39] in 1917 developed a test for the estimation of *Fusarium* infection of cereals, especially rye, based on a seedling symptoms. Seedling infection through seeds by *Microdochium nivale*, the snow mold fungus, failed to emerge from a layer of damp, crushed brick stone. Laboratory results, thus, were related closely to field performance. At the 1921 ISTA Congress in Copenhagen, Dorph-Petersen reported on field plot tests at StatsfrøKontrollen by J. Holmgaard for cultivar purity and detection of seedborne diseases in cereals. At the 1924 ISTA Congress in Cambridge, Genter presented a paper on "The Determination of Plant Diseases Transmitted by Seed." The methods used at that time, in addition to Hiltner's brick test method, were observation of seeds and seedlings during germination and purity tests. The first *International Rules for Seed Testing* was published by ISTA in 1928. It included a chapter on the determination of sanitary conditions and how to report such tests. Special attention was given to *Claviceps purpurea*, *Fusarium*, *Tilletia*, and *U. segetum* on cereals; *Ascochyta pisi* on peas; *Colletotrichum lindemuthianum* on beans; and *Botrytis*, *Colletotrichum linicola* and *Aureabasidium lini* on flax.[34]

Routine seed health testing now is carried out in most countries for seed certification and plant quarantine. Plant protection services of seed-exporting countries can issue an International Phytosanitary Certificate of the Food and Agricultural Organization based on a seed health test.

The methods used for detecting seedborne pathogens vary among laboratories. Due to germ plasm exchange, the use of uniform testing procedures was recommended by the first working group on seedborne diseases of the European Plant Protection Organization, held in Paris in 1954. In 1957 the PDC of ISTA established a comparative seed health testing program to standardize techniques for detection of seedborne pathogens. The first PDC workshop was held at the Seed Testing Station, Cambridge, U.K. in 1958 (Figure 1-2).[37] The basic prin-

Figure 1-2 Delegates to the first Seed Pathology Workshop held at the Seed Testing Station, Cambridge, England, 1958. Front row, L to R: Maria Kreitreiber (Austria), Maria de Lourdes de Florencio (Portugal), A. E. Muskett (England), Paul Neergaard (Denmark), Gillian Marshall (England), Mary Noble (Scotland), and W. F. Crosier (U.S.). Back row, L to R: Ponchet (France), J. Malone (England), (unidentified), E. R. Wallace (England), (unidentified), J. de Tempe (Netherlands), J. Baker (England), J. Mullin (Ireland), and Mme. Florencio's husband (Portugal). (From Yorinori, J. T., Sinclair, J. B., Mehta, Y. R., and Mohan, S. K., Eds., Seed Pathology — Progress and Problems, Fundacao Instituto Agronomico do Parana (IAPAR), Londrina, Brazil, 1979, 6. With permission.)

ciple of comparative seed health testing involves the distribution of referee samples to a number of scientists working independently, followed by annual workshops organized for comparing results. The main objective is to develop and standardize simple methods suitable for international application. The PDC was reorganized during the 1981 International Workshop on Seed Pathology, and referee seed health testing groups were formed for seeds of beet, crucifers, legumes, temperate cereals and grasses, and tropical and subtropical crops, and for viruses. At the 1992 PDC meeting at Angers, France, pathogen working groups were established for bacteria, fungi, nematodes, and viruses to coordinate the work of these subgroups. These groups focus on the development and standardization of seed health testing methods (one host–pathogen combination per group) or on special problems related to that work. The former PDC seminars were replaced by 3-day symposia held every 3 years. The first PDC symposium was held in August 1993 in Ottawa.[40] The latest seed testing procedures and rules, which were adopted at the 23rd ISTA Congress in 1992, became effective July 1, 1993.[41]

IV. SIGNIFICANCE

A. Reduction in Crop Yields

Seeds are the basic input for production. About 90% of the world's food crops is sown using seeds. The major world food crops are barley, beans, maize, millet, peanut, pulses, rice, sorghum, soybean, sugarbeet, and wheat. These crops are attacked by a number of pathogens, a majority of which are seedborne (Table 1-1).[21,22] It is recognized that plant diseases cause significant yield losses, but the data are variable because of differences in crop loss assessment methods. The global loss due to plant diseases is estimated to be 12% of potential production, which is equivalent to a monetary loss of $50 billion at the producer level; in terms of quantity, this is approximately 550 million tons. This loss occurs despite the use of fungicides at a cost of $1 billion.[42] The loss to cereals is 135 million tons.[42] This loss is three times that of the total food deficit of agriculturally developing countries in 1975. Losses due to disease in different regions of the world vary from 30% in the agriculturally developing countries of Asia to 25% in Europe and to 15% in North America.[43] Losses caused by seedborne inoculum or diseases have not been quantified.[44]

Analyses of the economic losses due to seedborne pathogens involve consideration of the impact on the seed production and distribution industry. Losses associated with restrictions imposed on seed production and distribution in view of the risk of the spread of disease have led to losses, including (1) those incurred by seed producers because of the minimum allowable infestations in certified seed, (2) those caused by extra costs to obtain seeds from pathogen-free areas, and (3) those caused by the cost of seed treatment because of the disease. For example, losses due to karnal bunt of wheat in northwestern Mexico are estimated to average $7.02 million per year. The major components of these costs are quality loss of infected seeds (36.2% of total costs), costs from planting restrictions (28.6%), loss of seed exports (15.7%), additional costs of transporting seeds (8.8%), and yield losses (6.4%). The direct yield and quality losses account for 42.6% of the total.[45]

Maize, rice, and wheat account for about two-thirds of the world food production from cereals. Important seed-transmitted diseases of wheat are the bunts, Bipolaris seedling blights and leaf spots, Fusarium blights, Stagonospora glume blotch, and smuts. Yield losses of 100% due to loose smut were reported in Georgia.[46] In Canada, loss due to common root rot of wheat caused by *Bipolaris sorokiniana* and *Fusarium* between 1969 and 1971 was 5.7%, equivalent to $42 million.[47,48] *Stagonospora nodorum* infection in barley and wheat resulted in losses of 1.5% and 1.0% in yields in 1974 and 1975, valued at $10 million and $5 million, respectively, in England and Wales.[49,50]

Rice is the most widely grown crop in agriculturally developing countries. The crop is attacked severely by seedborne pathogens, such as *Pyricularia oryzae* (blast), *Bipolaris oryzae* (brown spot), and *X. oryzae* pv. *oryzae* (blight). Blast was responsible for a famine in Japan during the 1930s. In the Philippines, losses due to blast may be more than 50%.[51] In 1942, one of the major factors contrib-

Table 1-1 Economically Important Seedborne Disease of Major Crop[22]

Crop	Pathogen	Disease
Agropyron (wheat grasses, quackgrass)	*Claviceps purpurea*	Ergot
	Anguina	Ear cockle, seed galls
Allium (chives, garlic, leek, onion, shallot)	*Alternaria porri*	Purple`blotch
	Botrytis acalda	Damping-off, gray mold
	Sclerotium cepivorum	White rot
	Ditylenchus dipsaci	Bloat, eelworm rot
Apium graveolens (celery)	*Septoria apiicola*	Leaf spot, late blight
Arachis hypogaea (peanut)	*Aspergillus niger*	Collar rot, crown rot
	Macrophomina phaseolina	Root rot, stem rot
	Peanut stunt virus	
	Peanut mottle virus	
	Peanut stripe virus	
Arrhenatherum (oat grasses, French ryegrass)	*Tilletia controversa*	Dwarf bunt
	Ustilago segetum var. *segetum*	Loose smut
Avena sativa (oats)	*Cochliobolus sativus*	Foot rot, seedling blight
	C. victoriae	Leaf blight
	Gibberella zeae	Scab
	Monographella nivalis	Brown foot rot, snow mold
	Phaeosphaeria avenaria	Black stem, kernel blight, leaf blotch
	P. syringae pv. *coronafaciens*	Halo blight
	Pyrenophora avenae	Leaf spot, pre-emergence blight
	Ustilago segetum var. *avenae*	Loose smut
	U. segetum var. *segetum*	Covered smut
Beta vulgaris (fodder beet, mangold, red beet, sugar beet)	*Cercospora beticola*	Leaf spot
	Curtobacterium flaccumfaciens pv. *betae*	Silvering of red beet
	Peronospora farinosa f. sp. *betae*	Downy mildew
	Pleospora bjoerlingii	Blackleg, damping-off, leaf spot
	Pseudomonas syringae pv. *aptata*	Bacterial blight
Brassica (crucifers)	*Albugo candida*	White blister
	Alternaria brassicae	Grey leaf spot
	A. brassicicola	Black spot
	Botrytis cinerea	Grey mold
	Peronospora parasitica	Downy mildew
	Phaeosphaeria maculans	Blackleg, black rot, dry rot
	Sclerotinia sclerotiorum	Drop, watery soft rot, white blight
	X. campestris pv. *campestris*	Black rot
Capsicum (chilli, pepper)	*Colletotrichum capsici*	Ripe rot, anthracnose
	X. campestris pv. *vesicatoria*	Bacterial spot of stem, leaf, fruit and seedling blight

Table 1-1 Economically Important Seedborne Disease of Major Crop[22] (continued)

Crop	Pathogen	Disease
Carthamus tinctorius (safflower)	*Alternaria carthami*	Leaf spot
	F. oxysporum f. sp. *carthami*	Wilt
	Puccinia calcitrapae var. *centaureae*	Rust
	Verticillium dahliae	Wilt
Cicer arietinum (chickpea, gram)	*Botrytis cinerea*	Grey mold
	Colletotrichum dematium	Blight
	F. oxysporum f. sp. *ciceri*	Wilt
	Phoma rabiei	Anthracnose, blight
Citrullus lanatus (watermelon)	*F. oxysporum* f. sp. *niveum*	Wilt
	Glomerella lagenaria	Anthracnose
	Pseudomonas pseudoalcaligenes subsp. *citrulli*	Blight
Citrus (lemon, citron)	*Phytophthora nicotianae* var. *parasitica*	Brown rot
	X. campestris pv. *citri*	Citrus canker
	Exocortis viroid	Exocortis
Corchorus (jute)	*Colletotrichum corchori*	Anthracnose
	M. phaseolina	Damping-off, stem rot
Coriandrum sativum (coriander)	*Protomyces macrosporus*	Stem gall
Cucumis melo (melon, cantaloupe, muskmelon)	*Glomerella lagenaria*	Anthracnose
	M. phaseolina	Charcoal rot
	Cucumber mosaic virus	Mosaic
	Melon necrotic spot virus	Necrotic spot
	Squash mosaic virus	Mosaic
Cucumis sativus (cucumber)	*Colletotrichum orbiculare*	Anthracnose
	Didymella bryoniae	Leaf spot, black rot
	P. syringae pv. *lachrymans*	Angular leaf spot
	Cucumber green mottle mosaic virus	Mosaic
	Cucumber mosaic virus	Mosaic
Cucurbita (pumpkin, vegetable marrow, squash)	*Didymella bryoniae*	Leaf spot
	F. solani f. sp. *cucurbitae*	Fusarium foot rot
	Cucumber mosaic virus	Mosaic
	Squash mosaic virus	Mosaic
Cuminum cyminum (cumin)	*F. oxysporum* f. sp. *cumini*	Wilt
Cyamopsis tetragonolobus (guar, cluster bean)	*Alternaria cyamopsidis*	Alternaria blight
	X. campestris pv. *cyamopsidis*	Blight
Daucus carota (carrot)	*Alternaria dauci*	Leaf blight
	A. radicina	Seedling blight
	X. campestris pv. *carotae*	Bacterial blight
Festuca (fescue, bunch grass)	*Cochliobolus sativus*	Blight
	Epichloe typhina	Cat's tail disease, choke
Glycine max (soybean)	*Bacillus subtilis*	Seed decay
	Cercospora kikuchii	Leaf spot, purple seed stain
	Cercospora sojina	Frogeye leaf spot
	Colletotrichum truncatum	Anthracnose, seedling blight

Table 1-1 Economically Important Seedborne Disease of Major Crop[22] (continued)

Crop	Pathogen	Disease
	D. phaseolorum var. *caulivora*	Stem canker
	D. phaseolorum f. sp. *meridionalis*	Stem canker
	D. phaseolorum var. *sojae*	Pod and stem blight
	Heterodera glycines	Cyst nematode
	M. phaseolina	Charcoal rot
	Peronospora manshurica	Downy mildew
	Phomopsis longicolla	Seed decay
	Phytophthora sojae	Root and stem rot
	P. syringae pv. *glycinea*	Bacterial blight
	Rhizoctonia solani	Damping-off, foot and basal stem rot
	X. campestris pv. *glycines*	Bacterial pustule
	Sclerotinia sclerotiorum	Sclerotinia stem rot
	Soybean mosaic virus	Mosaic
	Tobacco ringspot virus	Bud blight
Gossypium (cotton)	*Alternaria macrospora*	Leaf spot
	F. oxysporum f. sp. *vasinfectum*	Fusarium wilt
	Glomerella gossypii	Anthracnose, seedling blight, boll rot
	M. phaseolina	Root rot, stem blight
	Rhizoctonia solani	Damping-off, seedling stem canker
	Verticillium albo-atrum	Wilt
	X. campestris pv. *malvacearum*	Angular leafspot, black arm, bacterial blight, gummosis
Helianthus annuus (sunflower)	*Alternaria alternata*	Leaf spot
	A. helianthi	Leaf spot
	Botrytis cinerea	Gray mold
	M. phaseolina	Charcoal rot
	Plasmopara halstedii	Downy mildew
	Sclerotinia sclerotiorum	Wilt, white rot
Hibiscus (kenaf)	*Colletotrichum gloeosporioides*	Anthracnose
Hordeum vulgare (barley)	Barley stripe mosaic virus	False stripe
	Claviceps purpurea	Ergot
	Cochliobolus sativus	Black point, common root rot, foot rot, seedling blight
	Phaeosphaeria nodorum	Glume blotch
	Monographella nivalis	Brown foot rot, snow mold
	Pyrenophora graminea	Leaf stripe
	P. teres	Net blotch
	Rhynchosporium secalis	Scald
	Ustilago segetum var. *avenae*	Black smut, false loose smut
	U. segetum var. *segetum*	Covered smut
	U. segetum var. *tritici*	Loose smut

Table 1-1 Economically Important Seedborne Disease of Major Crop[22] (continued)

Crop	Pathogen	Disease
Juglans (walnut)	*X. campestris* pv. *juglandis*	Blight
	Cherry leaf roll virus	Leaf roll
Lactuca sativa (lettuce)	*Bremia lactucae*	Downy mildew
	Lettuce mosaic virus	Mosaic
Lens culinaris (lentil)	*Ascochyta fabae* f. sp. *lentis*	Blight
	F. oxysporum	Wilt
	Pea seedborne mosaic virus	Mosaic
Linum usitatissimum (flax, linseed)	*Alternaria linicola*	Blight
	Botrytis cinerea	Gray mold
	Colletotrichum linicola	Anthracnose, seedling blight, stem canker
	F. oxysporum f. sp. *lini*	Wilt
	Guignardia fulvida	Browning, stem break
	Melampsora lini	Rust
	Mycosphaerella linicola	Blotch, pasmo, rust
	Phoma exigua var. *linicola*	Foot rot
	Sclerotinia sclerotiorum	Wilt
Lolium temulentum (darnel)	*Claviceps purpurea*	Ergot
	Gloeotinia granigena	Blind seed disease
Lupinus (lupins)	*Botrytis cinerea*	Gray mold
	Diaporthe woodii	Phomopsis stem blight
	Glomerella cingulata	Anthracnose
	Pleiochaeta setosa	Brown spot
	Bean yellow mosaic virus	Mosaic
Lycopersicon esculentum (tomato)	*Alternaria solani*	Early blight
	C. michiganensis subsp. *michiganensis*	Canker
	Didymella lycoperisci	Stem canker, stem rot
	F. oxysporum f. sp. *lycopersici*	Wilt
	Glomerella cingulata	Anthracnose, ripe rot
	Phytophthora infestans	Late blight, fruit rot
	P. syringae pv. tomato	Speck, bacterial leaf spot
	Tobacco mosaic virus	Mosaic
	X. campestris pv. *vesicatoria*	Black spot
Malus (apples)	Tomato bushy stunt virus	
Manihot esculenta (cassava)	*X. campestris* pv. *manihotis*	Leaf spot
Medicago sativa (alfalfa)	Alfalfa mosaic virus	Mosaic
	C. michiganensis subsp. *insidiosus*	Wilt
	Colletotrichum trifolii	Anthracnose
	Ditylenchus dipsaci	Stem nematode
	P. medicaginis	Spring black stem
	Verticillium albo-atrum	Wilt
	Lucerne latent (Australian) virus	
Nicotiana tabacum (tobacco)	*Erwinia carotovora* subsp. *carotovora*	Hollow stalk
	Peronospora tabacina	Blue mold, downy mildew
	P. syringae pv. *tabaci*	Wildfire

Table 1-1 Economically Important Seedborne Disease of Major Crop[22] (continued)

Crop	Pathogen	Disease
Oryza sativa (rice)	*Alternaria padwickii*	Leaf spot, pink kernel, seedling blight, stackburn
	Aphelenchoides besseyi	White tip
	Cercospora janseana	Narrow brown leaf spot
	Cochliobolus miyabeanus	Black sheath rot, brown spot, seedling blight
	Curvularia	Black kernel
	D. angustus	"Ufra"
	Ephelis oryzae	Udbatta
	Gibberella fujikuroi	Bakanae disease, foot rot
	G. zeae	Head blight, node rot, scab
	Microdochium orzae	Leaf scald
	Pseudomonas avenae	Stripe
	Pseudomonas fuscovaginae	Sheath brown rot
	Pseudomonas glumae	Grain rot
	Pyricularia oryzae	Blast, rotten neck
	Rhizoctonia solani	Sheath blight
	Sarocladium oryzae	Sheath rot
	Tilletia barclayana	Black smut, covered smut, kernel bunt, kernel smut
	Ustilaginoidea virens	False smut
	X. oryzae pv. *oryzae*	Leaf blight
	X. oryzae pv. *oryzicola*	Leaf streak
Panicum (panic grasses, proso)	*Sporisorium destruens*	Head smut
Pennisetum glaucum (pearl millet)	*Claviceps fusiformis*	Ergot
	Sclerospora graminicola	Downy mildew, green ear
	Maeziomyces bullatus	Smut
Persea americana (avocado)	Avocado sunblotch viroid	Sunblotch
Phaseolus vulgaris (bean)	Bean common mosaic virus	Mosaic
	Colletotrichum lindemuthianum	Anthracnose
	Cucumber mosaic virus	Mosaic
	Curtobacterium flaccumfaciens pv. *flaccumfaciens*	Wilt
	Diaporthe phaseolorum	Blight
	D. phaseolorum var. *sojae*	Pod and stem blight
	F. oxysporum f. sp. *phaseoli*	Yellows, wilt
	M. phaseolina	Ashy stem blight, charcoal rot
	Phaeoisariopsis griseola	Angular leaf spot
	Phomopsis	Seed decay
	Phytophthora phaseoli	Collar rot, downy mildew
	P. syringae pv. *phaseolicola*	Grease spot, halo blight
	Rhizoctonia solani	Damping off, stem canker
	Sclerotinia sclerotiorum	Wilt, stem rot, watery soft rot
	Southern bean mosaic virus	Mosaic
	X. campestris pv. *phaseoli*	Common bacterial blight
	P. syringae pv. *syringae*	Brown spot

Table 1-1 Economically Important Seedborne Disease of Major Crop[22] (continued)

Crop	Pathogen	Disease
Pisum sativum (pea)	Ascochyta pisi	Leaf and pod spot
	F. oxysporum f. sp. *pisi*	Near wilt
	Mycosphaerella pinodes	Black spot, blight, foot rot, leaf spot
	Pea seedborne mosaic virus	Mosaic
	Pea early browning virus	Browning
	Phoma medicaginis var. *pinodella*	Leaf and pod spot
	P. syringae pv. *pisi*	Blight
Prunus (cherries, plum, peach)	Prune dwarf virus	Dwarfing
	Prunus necrotic ringspot virus	Necrosis
Raphanus sativus (radish)	*Alternaria brassicae*	Gray leaf spot
	A. brassicicola	Black leaf spot
	A. raphani	Leaf spot
	Colletotrichum higginsianum	Anthracnose, leaf spot
	Rhizoctonia solani	Damping-off, canker
Ricinus communis (castor bean)	*Alternaria ricini*	Capsule mold, seedling blight
	X. campestris pv. *ricini*	Leaf spot
Rubus idaeus (raspberry)	Raspberry bushy dwarf virus	Stunting
Secale cereale (rye)	*Claviceps purpurea*	Ergot
	Gloeotinia granigena	Blind seed disease
	Monographella nivalis	Brown foot rot, snow mold
	Tilletia controversa	Dwarf bunt
	Urocytis occulta	Stalk smut, stem smut
Sesamum indicum (gingily, sesame, til)	*Alternaria sesami*	Blight
	A. sesamicola	Leaf spot
	M. phaseolina	Charcoal rot
	Mycosphaerella sesami	Leaf spot
	Mycosphaerella sesamicola	Brown leaf spot, angular leaf spot
	X. campestris pv. *sesami*	Leaf spot
Solanum melongena (brinjal, eggplant)	*Phomopsis vexans*	Fruit rot
	Verticillium albo-atrum	Wilt
	Eggplant mosaic virus	Mosaic
Solanum tuberosum (potato)	Potato virus T	Mosaic
	Potato virus X	Mosaic
	Tobacco ringspot virus	Ringspot
	Potato spindle tuber viroid	
Sorghum biolor (milo, sorghum)	*Cochliobolus lunatus*	Leaf spot
	Colletotrichum graminicola	Red leaf, stalk rot
	Gibberella fujikuroi	Seed rot, stalk rot
	Gloeocercospora sorghi	Zonate leaf spot
	Peronosclerospora sorghi	Downy mildew
	Sporisorium cruentum	Loose smut
	Tolyposporium ehrenbergii	Long smut
Spinacia oleracea (spinach)	*F. oxysporum* f. sp. *spinaciae*	Wilt
	Peronospora farinosa f. sp. *spinaciae*	Downy mildew
	Spinach latent virus	Mosaic

Table 1-1 Economically Important Seedborne Disease of Major Crop[22] (continued)

Crop	Pathogen	Disease
Theobroma cacao (cacao)	*Crinipellis perniciosa*	Witches' broom disease
Trifolium (clovers)	*Botrytis anthophila*	Anther mold
	Ditylenchus dipsaci	Stem nematode
	Sclerotinia trifoliorum	Clover rot
Triticum aestivum (wheat)	*Alternaria triticina*	Leaf blight
	Anguina tritici	Ear cockle
	Barley stripe mosaic virus	False stripe
	Cochliobolus sativus	Foot rot, seedling blight, spot blotch
	C. tritici	Tundu, yellow ear rot, yellow slime
	Gibberella zeae	Head blight, scab, glume blotch
	Hymenella cerealis	Cephalosporium stripe
	Phaeosphaeria nodorum	Glume blotch, leafspot
	Monographella nivalis	Snow mold, brown foot rot
	Pyricularia oryzae	Blast
	Pyrenophora tritici-repentis	Yellow leaf spot
	Tilletia caries	Bunt
	T. controversa	Dwarf bunt
	Tilletia laevis	Bunt
	T. indica	Karnal bunt
	Urocystis agropyri	Flag smut
	Ustilago segetum var. *tritici*	Loose smut
	X. campestris pv. *translucens*	Black chaff
Vicia faba (fava bean)	*Ascochyta fabae*	Leaf and pod spot
	Botrytis fabae	Chocolate spot
	Bean yellow mosaic virus	Mosaic
	Broadbean stain virus	Staining
	Broadbean true mosaic virus	Mosaic
	Ditylenchus dipsaci	Stem eelworm
Vigna unguiculata (cowpea)	Blackeye cowpea mosaic virus	Mosaic
	Cowpea aphid-borne mosaic virus	Mosaic
	Cowpea banding mosaic virus	Mosaic
	Cowpea mosaic virus	Mosaic
	M. phaseolina	Ashey stem blight
	X. campestris pv. *vignicola*	Blight, canker spot
Vigna mungo (urd bean)	Bean common mosaic virus	Mosaic
	Blackgram leaf crinkle virus	Leaf crinkle
Zea mays (maize)	*Acremonium strictum*	Kernel rot
	Clavibacter michiganensis subsp. *nebraskensis*	Wilt and leaf freckles
	Cochliobolus heterostrophus	Southern leafspot or blight
	Diplodia maydis	Dry or white ear rot, root rot, seedling blight, stalk rot

Table 1-1 Economically Important Seedborne Disease of Major Crop[22] (continued)

Crop	Pathogen	Disease
	Erwinia stewartii	Leaf blight wilt
	Gibberella fujikuroi	Ear rot, kernel rot, seedling blight stalk rot
	G. fujikuroi var. *subglutinans*	Black bundle
	G. zeae	Cob rot, seedling blight
	Peronosclerospora sorghi	Downy mildew
	U. zeae	Smut
Zinnia elegans (zinnia)	*Alternaria zinniae*	Blight
	X. campestris pv. *zinniae*	Blight

uting to the Bengal famine in India was the failure of the rice crop because of brown spot. Yield loss in rice due to *F. moniliforme* has been reported to be 15% in India, 20 to 50% in Japan, and 3.7 to 14.6% in Thailand.[51] Sheath blight (*R. solani*) is considered of major economic importance in the People's Republic of China, Japan, Sri Lanka, Taiwan, and the United States.[52]

Loss in yield due to seedborne infection has varied. *Bipolaris oryzae*, *Microdochium oryzae,* and *Phyllosticta* caused reduction in seedling size, production of fewer seeds per panicle, and empty glumes, and incompletely filled seeds in rice.[53] The relationship between seedborne *Pyrenophora graminea* and yield loss varies in barley: 1:1,[54] 1:09,[55,56] 1:075,[57] 1:06.[58] *Ascochyta fabae* f. sp. *lentis*-infected lentil seeds yield plants with fewer branches, smaller roots and shoots, reduced, vigor, and lower seed yields.[59] Ascochyta blight in combination with anthracnose (*C. truncatum*) has caused yield reductions in lentil of 50% in western Canada.[60]

Bacterial leaf blight is the most serious disease of rice in southeast Asia, particularly in India, Indonesia, Japan, and the Philippines. Yield losses in Japan generally are 20 to 30%, occasionally reaching 50%.[52,61]

Cucumber mosaic virus (CMV)-infected narrow-leafed lupin (*Lupinus angustifolius*) seeds failed to produce plants and plots sown with 5% and 0.5% infected seeds produced 1.5 to 2.9% and 0.2 to 0.3% infected plants, respectively. However, the rate of virus spread by aphids was faster and resulted in greater infection at maturity in plots sown with 5% than with 0.5% infected seeds. Sowing 5% infected seeds has resulted in yield losses of 34 to 53% and CMV-infected seeds of 6 to 13%.[62] Lettuce mosaic virus results in a delay in the lettuce flower stalk growth, a reduction in seed set, and empty seeds.[63] Pea seedborne mosaic virus is the most important virus of pea in Europe, and the incidence observed after direct assay from plots sown with three cultivars of pea showed reduced growth, delayed maturity, and a yield loss of 15% due to reduced seed size.[64]

The cowpea aphid-borne mosaic virus reduced cowpea seed yield. The loss differed according to plant growth stage at the time of infection and cultivar.

Losses reached 70% for plants inoculated at the seedling stage, compared to 40% for those inoculated at flowering. The cultivar Azmerly showed a higher yield reduction than those of Fetriat and Buff.[65] Yield losses in soybeans caused by the soybean mosaic virus were related linearly to the logarithm of the percentage seed transmission (r = 0.06) and among maturity groups.[66]

Wheat plants completely affected with *Anguina tritici* (ear cockle) or *Clavibacter tritici* showed a 100% loss. Even partially cockled ears resulted in 52.4% seed loss and a reduction in 1000-seed weight.[67]

Viroid diseases have become important in several vegetatively propagated crops. Potato spindle tuber viroid is widespread in temperate regions of Canada and the United States, and occurs in the Union of South Africa and the former U.S.S.R.[68] It has been a problem to potato growers in North America since the early twentieth century and 5% yield losses have been reported in experimental trials.[69] The chrysanthemum stunt viroid nearly destroyed the entire U.S. chrysanthemum industry in the late 1940s.[70] The cadang-cadang disease of coconuts has resulted in the death of over 30 million coconut palms in the Philippines since its discovery in 1930.[71] The disease is still uncontrolled and continues to spread.[72]

Virus infection may cause plants to produce abnormal pollen. In *Pelargonium* anthers infected with tomato ringspot virus, pollen grains were shrivelled and deficient in cytoplasm.[73] Many pollen grains from *Chenopodium quinoa* plants infected with sowbane mosaic virus collapsed, were grooved, and had sunken opercula. Similar but less obvious symptoms were found in pollen of *Plantago lanceolata* infected with either the ribgrass mosaic or broadbean wilt virus.[74] Reduction in germination was observed in alfalfa pollen grains from alfalfa mosaic virus-infected plants of all clones, and within each clone, pollen from infected plants produced shorter germ tubes than pollen from virus-free plants (Figure 1-3).[75] Lychnis ringspot virus drastically reduced pollen production in *Beta vulgaris* and *Silene noctiflora*. In the latter, most stamens on diseased plants failed to elongate, and anthers were sterile.[76] Barley stripe mosaic virus induced sterility of ovules and pollen and was responsible for the yield losses in barley. Infected plants in highly susceptible cultivars often had shorter spikes and sterile florets; the absence of plump seed at each rachis node was evident on short spikes from infected plants.[77] Pollen collected from potato spindle-tuber viroid infected cv. Monona was less viable than that collected from healthy plants. Pollen collected from infected plants significantly reduced seed sets but not fruit development or seed set in all cultivars tested. For some cultivars, infected maternal plants increased the frequency of reduced fruit development and seed weight.[78]

In maize, important seedborne pathogens include *Diplodia* (ear rot, stalk rot, seedling blight), *Bipolaris maydis* (southern leaf spot or blight), *Fusarium* (stalk rot, ear rot, seedling blight), and *Erwinia stewartii* (bacterial wilt). In 1970, a new race of *B. maydis*, highly virulent on maize with T-type cytoplasm, reduced United States yields approximately 710 million bushels at an estimated loss of $1 billion.[79] The barley stripe mosaic virus (BSMV) resulted in yield losses up

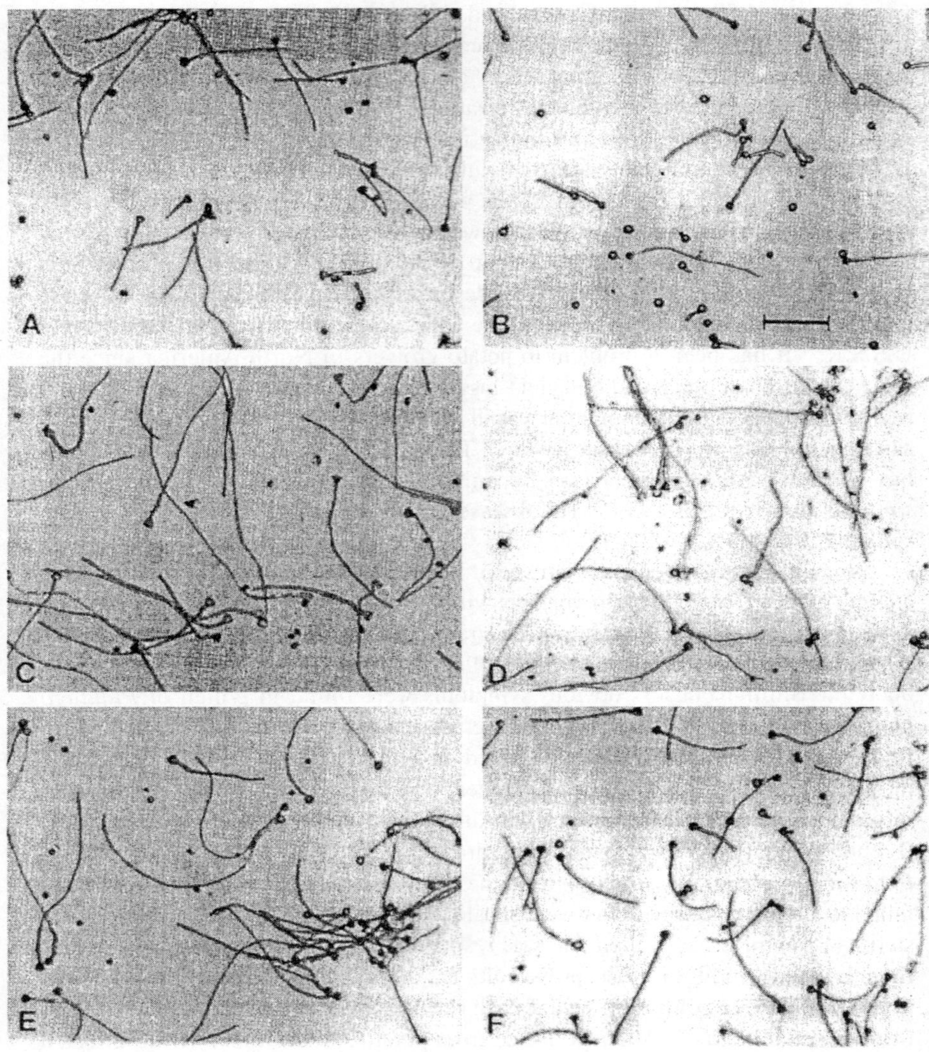

Figure 1-3 Alfalfa (*Medicago sativa*) pollen grains were germinated for 2 hr on Bacto-agar (1.5 g) and sucrose (20 g/100 ml). Pollen from alfalfa mosaic virus-free plants of clones B-19 (**A**), B-24 (**C**), and B-39 (**E**); pollen from infected plants of the same clones, (**B**), (**D**), and (**F**), respectively. The bar represents 250 μm. (From Pesic, Z. and Hiruki, C., *Can. J. Plant Pathol.*, 10, 6, 1988. With permission.)

to 64% and 75% in barley and wheat, respectively.[80] The total loss in barley exceeded $30 million during 1953 to 1970.[81] Outbreaks of black rot (*X. c.* pv. *campestris*) and black leg (*Phoma lingam*) of cabbage in the U.S. during 1972 and 1973 resulted in a crop loss of about $25 million.[82]

B. Loss in Germination and Vigor

Many seedborne pathogens become active when seeds are sown, which may result in seed decay and/or pre- or postemergence damping-off. This in turn results in a poor plant stand in the field. The loss in germination depends upon many factors, including cultivar specificity; type, amount, and location of inoculum; environmental conditions; and other factors. The effect of seedborne fungi on germination was reviewed by Halfon-Meiri.[83] In soybean, *Colletotrichum truncatum*, *D. phaseolorum* var. *sojae*, and *Sclerotinia sclerotiorum* are internally seedborne and may inhibit field emergence.[84,85] Soybean seedling emergence was reduced 12% by *Cercospora kikuchii* and 59% by *Macrophomina phaseolina*.[86,87] Soybean seed germination decreased proportionally to the increase of *D. phaseolorum* var. *sojae* and *Phomopsis* seed infection.[88,89] Tomato seeds infected with *Phytophthora nicotianae* var. *parasitica* either failed to germinate, or if they germinated, seedlings were killed by the fungus.[90] *Alternaria padwickii* caused decay of rice seeds, roots, and coleoptiles, resulting in the death of young seedlings.[91] Severe infection of *A. zinniae* has caused preemergence death, and superficial seed infection has caused disease in plants after emergence.[92]

Tilletia indica-infected wheat seeds (karnal bunt) have a significantly lower survival rate than noninfected seeds and therefore have a lower survival rate in storage.[93] Germination of *Bipolaris oryzae*-infected rice seeds was lower than healthy seeds.[94] *Fusarium moniliforme* caused seed rot, seedlings blight, and brown discoloration in the coleoptile and in primary and secondary leaves of rice.[95] Barley seeds severely infected with *Bipolaris sorokiniana* did not germinate, or if they germinated, seedlings became infected.[96] Maize seed germination was reduced by *Diplodia* but not by *Fusarium*.[97] Soybean seeds with severe symptoms caused by *F. oxysporum* failed to germinate.[98] Infection by *Phomopsis* reduced emergence and seedling establishment when soybean seeds were incubated in dry soil because low soil water potential inhibited seedling growth more than did *Phomopsis*.[99] Soybean seeds infected with *Cercospora kikuchii* always had less viability compared to healthy seeds. The loss in viability increased with an increase in seed coat discoloration. The maximum loss (more than 30%) in seed viability and germination was observed in seeds having 100% purple discoloration on the seed coat.[100] Sorghum seeds severely infected with *Colletotrichum graminicola* did not germinate, whereas mildly infected seeds exhibited poor germination and low vigor.[101] *Gloeocercospora sorghi* reduced germination and resulted in diseased seedlings in sorghum.[102] The incidence of seedborne *F. moniliforme* and *Curvularia lunata* in sorghum was correlated negatively (P = 0.01) with seed germination (r = –0.97 and –0.63, respectively) and correlated positively (P = 0.01) with seed damage (r = 0.92 and 0.84, respectively).[103] Sugarbeet seeds infected with *C. dematium* f. sp. *spinaciae* generally failed to mature and died, but if seeds matured, seed weight and germination were low.[104] *Eleusine coracana* seeds infected with *B. undulosa* and *P. grisea* germinated poorly and were unfit for sowing.[105] The percentage of seed germination was negatively correlated to the incidence of *Alternaria* in surface-sterilized Brassica seeds in 1978, 1981, and 1982 but not in 1979.[106] Poor seed germination in soybeans was associated with *Phomopsis longicolla* infection.[107] Soybean seed germination declined with increasing seed infection by *Phomopsis longicolla, Diaporthe phaseolorum* var. *sojae*, and *Diaporthe phaseolorum* var. *caulivora*.[108] *Microdochium nivale*-infected wheat seeds either did not germinate or gave rise to abnormal seedlings or visibly diseased plantlets.[109] .

C. m. subsp. *michiganensis* in pepper[110] and *Bacillus subtilis* in chickpea and soybean[111,112] reduced field emergence of naturally infected seeds. *P. s.* pv. *glycinea* from soybean seeds significantly reduced emergence of cultivar Amsoy at 20, 25, 30, or 35°C, with a greater reduction at 35°C.[113] At 40°C all soybean seeds were decayed by *P. s.* pv. *glycinea*.[114] Soybean seeds of different cultivars responded differently depending upon their resistance and susceptibility to losses in germination.[115] Losses in germination due to *B. subtilis* at higher temperature were reported in chickpeas and soybeans.[111,112,116]

Virus-infected seeds also may fail to germinate. A 31 to 35% reduction in germination in alfalfa seeds was reported due to infection by the alfalfa mosaic virus.[117] Loss in germination or vigor also may be genetic, physiological, cytological, or mechanical.[118] Spinach latent ringspot virus infection in *Nicotiana xanthi* and *N. rustica* caused a reduction in seed germination.[119] The vigor of seedlings infected by strawberry latent ringspot virus through seeds was reduced in *Chenopodium quinoa* but not in parsnip (*Pastinaca sativa*).[120]

C. Development of Plant Diseases

Seeds are the most important means for perpetuation of plant pathogens. For certain pathogens, such as those that cause loose smut of wheat, covered smut of barley, and barley stripe mosaic virus in barley and wheat, seeds are the exclusive means of survival. The role of such inocula in disease establishment in the field is well documented. Transmission of a pathogen through seeds is considered more important than other survival means. Pathogens remain viable longer in seeds than in vegetative plant parts or in soil. The host–parasite relationship within seeds also favors the earliest possible infection in the field. Since pathogens are in direct contact with the seeds, the chances of seedling infection are enhanced. Seedborne infection can provide a focus for inoculum, which may spread under favorable conditions and cause an epidemic. As few as two *X. c.* pv. *campestris*-infected cabbage seeds per 10,000 can cause an epidemic of black rot.[121] Similarly, 0.5% seed infection of *X. c.* pv. *phaseoli*[122] and 0.02% of *P. s.* pv. *phaseolicola* in bean,[123] and less than 1% of *X. c.* pv. *vesicatoria* in tomato[124] can result in an epidemic.

Teliospores of *Tilletia controversa, T. laevis*, and *T. caries* can be disseminated widely by importation through the use of contaminated seeds, animal-derived fertilizers, and natural movement of seed-eating migratory animals. It is common practice to feed severely contaminated seed to animals. The mechanical application of manure from feedlots as fertilizer could deposit teliospores on the soil surface and provide a source of infection for wheat.[125]

A pathogen can be distributed to new areas through seeds. Plant pathogens have spread with planting material since crop cultivation began. Before modern times, the movement of pathogens was restricted to certain localized areas because of limited means of travel. However, today, with expanded and rapid modes of transport, chances of spread through seeds have increased to include large areas within a country and from one country to another. The exchange of plant material between countries has enhanced the risks of introducing new pathogens with seeds. Thus, new strains or physiologic races of a pathogen may be introduced with new germ plasm from other countries.

However, severely restrictive quarantine regulations on germ plasm can do more harm than good to the economy of a country by restricting the development of much needed cultivar improvement programs. Restrictive quarantines on germ plasm can defeat the purpose, if the seeds are of an important commodity or if the commodity is freely exchanged through local markets with neighboring countries where the disease occurs. For example, if large quantities of beans, cereals, maize, peanuts, or soybeans are being imported, the chances of introducing new pathogens or strains of established pathogens is much greater than through carefully grown and monitored germ plasm seeds.

Sclerotinia has been spread from field to field and from one geographic area to another by several means, including windborne ascospores and soil adhered to seedlings, farm equipment, animals, or humans. On farms where diseased plants are used as cattle feed or bedding, the spreading of manure can introduce the pathogen to uncontaminated fields. Viable sclerotia of *S. sclerotiorum* passing through the digestive tract of a ruminant can be an important source for spread of the pathogen from infested to noninfested areas within a field or from infested fields to noninfested fields.[126]

D. Discoloration and Shriveling

Discoloration on soybean seeds, caused by various microorganisms, which has long been used for diagnostic purposes, can indicate seed quality.[127] Discoloration can indicate undesirable physical qualities, the presence of toxic metabolites, or other unfavorable seed characteristics. Soybean seeds infected with *Phomopsis longicolla* are elongated, smaller than normal, deeply fissured, and covered with whitish mycelium. Dull gray to deep brown patches and scattered, dark sunken areas on a soybean seed are due to infection by *Alternaria alternata* and *A. tenuissima*.[128] *Nematospora coryli* results in sunken, cream-colored spots with dark borders. Salmon or pink to red discoloration anywhere on a soybean seed coat may be associated with infection by *Fusarium graminearum* and *F. sporotrichioides*. Irregular brown or gray areas with black specks develop on soybean seed coats infected with *Colletotrichum truncatum*. Purple stain caused by *Cercospora kikuchii* can vary from violet to pale purple to dark purple, and the color tends to bleed from the hilum (Figure 1-4).[127] Seeds from a soybean plant with downy mildew become encrusted with oospores of *Peronospora manshurica*. Seeds from soybean plants with soybean mosaic virus are mottled with

Figure 1-4 Soybean (*Glycine max*) seed lots showing symptoms produced by seedborne pathogens. Clockwise from upper right: seeds with a light brown hilum ring from soybean mosaic virus infected plants; seeds infected with *Cercospora sojina*; seeds infected with *C. kikuchii*; seeds infected with *Phomopsis longicolla*; seeds with a black hilum ring from soybean mosaic virus infected plants; and asymptomatic seeds. (From Sinclair, J. B. and Backman, P. A., Eds., *Compendium of Soybean Diseases*, APS Press, St. Paul, MN, 1989. With permission.)

black or brown, depending on the hilum color.[129] Soybean seeds infected with *Cercospora sojina* develop conspicuous light to dark gray or brown areas that vary from minute specks to large blotches covering the entire seedcoat.[130] Some lesions show alternating bands of light and dark brown. Occasionally, brown and gray lesions diffuse into each other. Normally the seed coat cracks. The symptoms are distinct from those produced on soybean seeds by *C. kikuchii*, *Colletotrichum truncatum*, the *Diaporthe-Phomopsis* complex, and *Fusarium*.[127,129] Symptoms on soybean seeds infected by *Macrophomina phaseolina* appear as indefinite black spots or blemishes on the seedcoat.[131] Soybean seeds colonized with *F. oxysporum* appear shrunken, slightly irregular in shape, often with cracks in the seedcoat with light to dark pink discolored areas over most of the seed surface (Figure 1-5).[98]

Some pathogens that cause discoloration in soybean seeds affect only seed coat color, causing "cosmetic" or "superficial" damage, while other pathogens damage tissue in the seed coat and embryo. The relationship of seed coat color to infection by fungi and viruses and to chemical and physical properties of seeds provides a potential means for automatic determination of seed quality by color image classification techniques.[131-133] Computer image analysis shows promise for assisting in fast, accurate determination of pathological factors affecting

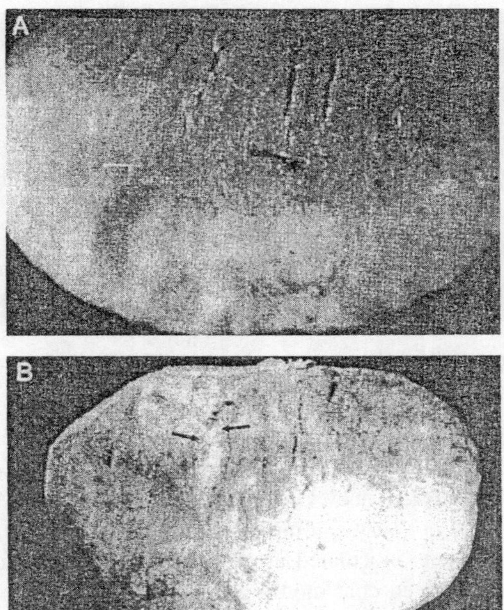

Figure 1-5 Soybean (*Glycine max*) seeds infected with *Fusarium oxysporum*: (**A**) A seed from the field showing pink discoloration and cracking of the seed coat near the hilum, and (**B**) a similar seed stained with aniline blue for 2 to 3 min showing fungus growth in hilum and production of conidiophores (*arrows*) in seed coat crack. (From Velicheti, R. K. and Sinclair, J. B., *Seed Sci. Technol.*, 19, 445, 1991. With permission.)

soybean seed quality and grade. A computer vision algorithm developed to detect and classify soybean seeds damaged by fungi differentiates damaged from undamaged seeds on the basis of color, with an accuracy of 98%.[131,132] Diseases such as Phomopsis seed decay, Alternaria seed decay, and Fusarium seed stain are differentiated.[133]

Chickpea seeds infected with *Phoma rabiei* are small and wrinkled, with dark brown lesions of various shapes and sizes with pycnidia forming in deep lesions.[134] Pea seeds infected with *M. pinodes* have dark brown lesions (Figure 1-6). A number of fungi, such as *Alternaria alternata*, *A. padwickii*, *Curvularia lunata*, and *Bipolaris oryzae*, induce a brown discoloration on rice seeds.[61] *Ascochyta fabae* causes various degrees of discoloration on fava bean, depending on infection levels.[135] *Pseudomonas glumae* causes seed discoloration and formation of brown bands across the endosperm in rice (Figure 1-7).[136]

Bean seeds infected with *Colletotrichum lindemuthianum* showed brown to light chocolate-colored, sunken cankers on seed coats.[137] Lesions may extend into the cotyledons following early infection. Less severely infected seeds showed yellowish to brown sunken lesions, which were not always distinguishable from those caused by other organisms.[137] *Ascochyta fabae* f. sp. *lentis*-infected lentil seeds were shriveled and discolored, and seed quality was reduced.[138] Reduced seed size was correlated significantly to the level of seedborne inoculum.[139]

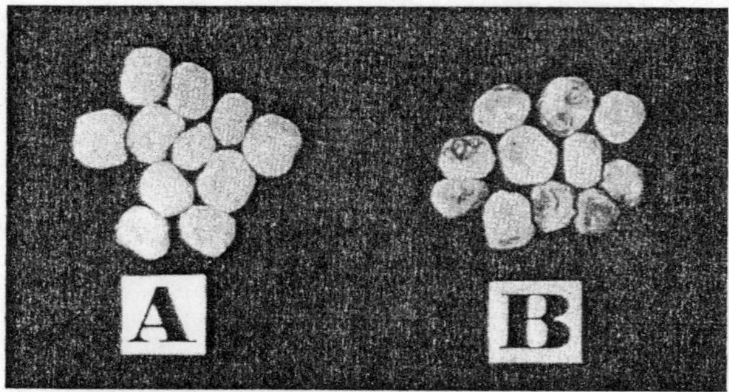

Figure 1-6 Noninfected pea (*Pisum sativum*) seeds (A) and seeds infected with *M. pinodes* (B). (From INRA, Versailles. With permission.)

Seed coat mottling in soybean had no relationship with the presence of soybean mosaic virus (SbMV). The virus, which caused necrotic symptoms on the foliage, was not detected in cv. Gwanggyo seeds.[140] The SbMV antigen content in soybean seeds did not vary significantly between plants inoculated at growth stages V1 or R2. Plants inoculated at growth stage R2 produced seeds with a greater percentage of seed coat mottling than those inoculated at growth stage V1. The percentage of mottled seed was not correlated with virus antigen content of seeds in most cultivars; however, the correlation was high in some cultivars.[141] In French bean, a significantly high percentage of deformed seeds carried the French bean mosaic virus.[142]

Ditylenchus destructor-infected peanut seeds are shrunken, and the testas and embryos have a yellow to brown or black discoloration.[143]

E. Biochemical Changes in Seeds

Many seedborne fungi induce qualitative changes in the physico–chemical properties of seeds, such as color, odor, oil content, iodine and saponification

Figure 1-7 Discoloration and formation of brown bands across the endosperm of rice (*Oryza sativa*) seeds, caused by *Pseudomonas glumae*. (Courtesy of S. Mogi, from Agarwal, P. C., Mortensen, C., and Mathur, S. B., Seedborne Diseases and Seed Health Testing of Rice, Tech. Bull. No. 3, Danish Govt. Inst. Seed Pathol. Developing Countries, Copenhagen, 1989, 106.)

value, refractive index, and protein content, thereby affecting their commercial value.

1. *Protein*

Soybean seeds infected by *Fusarium* or *Phomopsis* have lower quality oil, higher amounts of free-fatty acids, poor meal color, and other reduced quality factors when compared to noninfected seeds.[144] For example, soybean seeds contaminated with *F. solani* and stored at 80% relative humidity produced rancid, turbid oil, high in free-fatty acids.[145] Similarly, the level of free-fatty acids in crude oil derived from soybean seeds infected with *Alternaria* was higher than in that derived from noninfected seeds.[146] *Phomopsis* was more important than *Fusarium* in soybean seed deterioration; flour and oil derived from seeds infected with *Phomopsis* were unmarketable.[147] Generally, *C. kikuchii* colonizing the entire soybean seed surface tends to lower oil and increase protein content.[128] In soybean seeds encrusted with oospores of *P. manshurica*, the protein content and free-fatty acid content was higher and oil content lower than in healthy seeds.[148]

The degradation of structural and functional proteins in soybean seeds infected with *C. kikuchii* and *Phomopsis longicolla* affects both seed quality and viability. Soybean seeds contain a variety of proteins, including storage globulins, lectin, lipoxygenase, and β-amylase. Storage globulins consist of two major classes of proteins, glycinins (11S) and conglycinins (7S). β-conglycinins constitute 30% of the total protein and are composed of four subunits, denoted α, α^1, β, and β^1.[149] Glycinins also include several acidic and basic subunits.[150] Structural and quantitative variations in these protein groups influence N and S content and functional properties of soybean foods, such as solubility, coagulation temperature, and flavor. The lipoxygenases are monomeric globular proteins with molecular weight of approximately 100 KDa,[151] and are present in soybean cotyledons constituting 1 to 2% of the total seed protein. Lipoxygenases may contribute to pathogen resistance.[152] Soybean seed lectin is a 120-KDa tetrameric glycoprotein, with 30-KDa units or 15-KDa subunits, that is found primarily in cotyledons.[153] Higher levels of lectin found in seeds of cultivars resistant to *Phytophthora sojae* than in those of susceptible cultivars suggested a possible role of lectin in disease resistance.[154] Soybean seeds infected with *C. kikuchii* showed degradation of seed coat proteins, but not cotyledonary proteins, which were degraded by *P. longicolla*. Lipoxygenase was degraded only in *C. kikuchii*-infected seed coats and only in cotyledons infected with *P. longicolla*. β-amylase was partially degraded in seed coats and cotyledons infected with *P. longicolla* (Figure 1-8). Soybean seed lectins were found in seed coats infected with *P. longicolla* but not in *C. kikuchii*-infected seeds (Figure 1-9).[155]

Infection by *Alternaria padwickii* resulted in a reduction of total carbohydrates in rice. Infected seeds had 65.7% carbohydrates, compared to 76.7% in healthy seeds.[156] Infection of peanut seeds by *Aspergillus flavus*, *A. niger*, and *R. solani* increased crude fat content but reduced starch content, probably due to amylase production by these fungi.[157] *Macrophomina phaseolina* infection in peanut

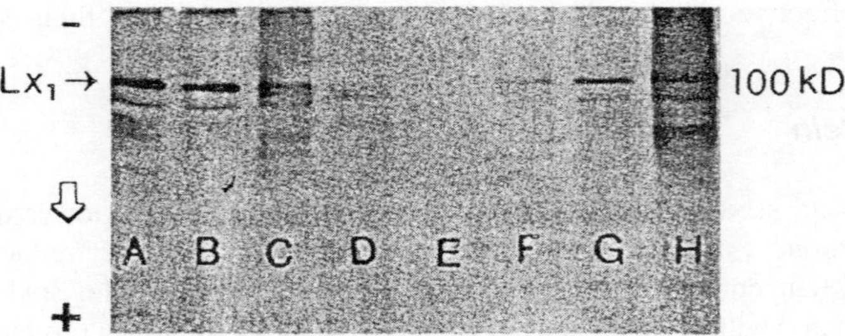

Figure 1-8 Immunoblot analysis of lipoxygenase (Lx_1) in seed coats and cotyledons of soybean (*Glycine max*) cv. Hack. (**A**) Cotyledons infected with *Cercospora kikuchii*; (**B**) noninfected cotyledons; (**C**) cotyledons infected with *Phomopsis longicolla*; (**D**) mycelia of P. longicolla; (**E**) mycelia of *C. kikuchii*; (**F**) seed coats infected with *C. kikuchii*; (**G**) seed coats of noninfected seeds; and (**H**) seed coats infected with *P. longicolla*. Molecular weight of lipoxygenase indicated on right. (From Velicheti, R. K. et al., *Plant Dis.*, 76, 779, 1992. With permission.)

caused both qualitative and quantitative damage, including discoloration of pods and seeds, and a reduction in pod and seed yield and oil content.[158]

Seeds from soybean plants infected with soybean mosaic virus had higher levels of amino acids and protein and a lower level of oil than those from noninfected plants.[159] In barley seeds, protein increased as the proportion of seed infected by barley stripe mosaic virus increased. High protein content in malting barley is undesirable; seeds are heavily discounted if it is above 13.55%.[160] Southern bean mosaic virus (SBMV) and cowpea mosaic virus (CPMV) reduced the carbohydrate fraction of cowpea seeds. SBMV infection increased total N, protein, and total amino acids, and decreased inorganic phosphorus, and nitrate

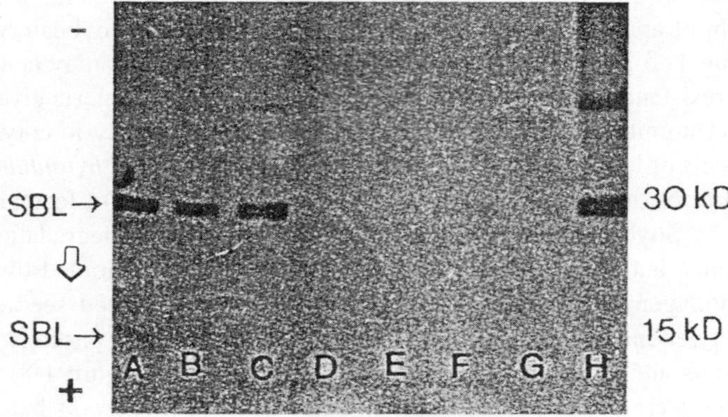

Figure 1-9 Immunoblot analysis of soybean (*Glycine max*) seed lectins (SBL) in seed coats and cotyledons of soybeans cv. Hack. (**A**) Cotyledons infected with *Cercospora kikuchii*; (**B**) noninfected cotyledons; (**C**) cotyledons infected with *Phomopis longicolla*; (**D**) mycelia of *P. longicolla*; (**E**) mycelia of *C. kikuchii*; (**F**) seed coats infected with *C. kikuchii*; (**G**) noninfected seed coats; and (**H**) seed coats infected with *P. longicolla*. Molecular weights of SBL monomer and subunits indicated on the right. (From Velicheti, R. K. et al., *Plant Dis.*, 76, 779, 1992. With permission.)

N. CPMV increased total N and protein in the seed coat but reduced it in other seed parts and increased all phosphorus fractions.[161] A significant reduction in total N content of cowpea aphid-borne mosaic-infected cowpea seeds was recorded.[65] It was higher in plants inoculated at the seedling stage. Also, a significant increase in total carbohydrate content was recorded for seeds of plants infected at the seedling stage.

2. *Oil*

Oil is synthesized in seeds by condensation of molecules of glycerine or fatty acids, or both. The glycerine and fatty acids are synthesized from carbohydrate, especially reducing sugars.[162] Naturally occurring mixtures of glycerides in oil contain a small amount of free-fatty acids. The increase in free-fatty acids shows that fungi have a high lipase activity, which can bring about fat hydrolysis. Ward[163] reported that an increase in free-fatty acids in peanuts was due to fungal invasion. Wilson[164] reported that microorganisms produced free-fatty acids during the lipolysis of triglycerides. These chemical changes resulting in increased unsaturated fatty acids are undesirable and can cause obesity and cardiovascular disorders in humans. Hence, such seeds are not fit for oil extraction and human consumption or for derived oil cake as animal feed.[165] The saponification value increased but the I-value decreased in oil extracted from *Fusarium*-infected seeds of *Brassica campestris* var. *dichotoma*.[166] This indicated that a large number of short-chained fatty acids and their glycerides were formed during oil hydolysis by lipase, secreted by the fungi.[166] The oil content in sunflower seeds is reduced due to infection of *M. phaseolina*. The oil color from inoculated seeds changes from light yellow to yellowish brown, and the free-fatty acid content increases. The oil from inoculated seeds has a greater saponification value than that from naturally infected seeds, while the I-number decreases in both types.[167] The I-number indicates the quantity of unsaturated acids present in oil. Vegetable oils are high in unsaturated fatty acids. The decreased I-value observed in *Fusarium*-infected canola oil was indicative of decreased unsaturation created by the metabolic activity of the microbes.[166,168] Decreased I-value due to fungal infection also has been reported by Raj and Saxena[169] in Indian mustard oil, by Mall and Pateria[170] in peanut oil, by Singh and Prasad[171] in sunflower oil, and by Sharma[172] in sesame oil.

The oil content of infected seeds may be reduced. The oil content of peanut seeds was reduced by infection of *Aspergillus flavus*, *A. tamarii*, *Lasiodiplodia theobromae*, *Cladosporium herbarum*, *Macrophomina phaseolina* and *Penicillium citrinum*.[173] *A. niger*, *A. tamarii*, and *M. phaseolina* reduced the oil content in sesame more than *A. flavus*, *P. citrinum*, or *Cladosporium herbarum*.[172] In sunflower seeds, *A. flavus* and *Alternaria alternata* decreased oil content.[174] Three *Fusarium* spp. reduced the percentage of oil content in canola.[166] The maximum reduction was by *F. oxysporum*, followed by *F. moniliforme*, and then *F. pallidoroseum*.[165,166] Bilgrami and Verma[175] reported that fat utilization was common among fungi. The free-fatty acid content of oil increased in *Fusarium*-infected *B. campestris* var. *dichotoma* seeds.[166]

Oil extracted from infected and stored seeds may emit a rancid odor. Oil extracted from safflower seeds infected with *A. flavus* and *A. alternata* emitted a mild, rancid odor within 2 weeks, and as the incubation period increased, the odor became more intense. However, in the case of *Alternaria carthami*, rancidity was observed after 4 weeks.[176] The oil extracted from canola seeds infected with *F. oxysporum*, *F. moniliforme* var. *dichotoma* emitted a moldy odor, and the reactive index increased.[166] The unpleasant odor was due to an increase in free-fatty acids. A high percentage of free-fatty acids showed that fungi caused a hydrolytic type of deterioration in oil, which results in oil rancidity.[177]

Wilson[164] reported a rancid taste in peanut oil due to free-fatty acids. A shift in the refractive index of oil extracted from *Fusarium*-infected seeds indicated the presence of saturated fatty acids.[163,164] Oil from *Fusarium*-infected seeds also showed a change in color, which may be due to pigments synthesized by invading fungi.[178] The oil extracted from healthy kernels was barium-yellow, and oil from diseased kernels was amber-yellow. Such infected kernels contained lower quantities of behenic and stearic acids, but higher quantities of arachidic, linoleic, and palmitic acids than noninfected ones. Four fatty acids, eicosanoic, lauric, myristic, and palmitoleic, were absent in diseased kernels.[157] Oil content was reduced due to infection by *Macrophomina phaseolina* in sunflower seeds. The color of the oil from inoculated seeds changed from light yellow to yellow-brown, and the oil had a higher free-fatty acid content and a greater saponification value than that from naturally infected seeds. The iodine number decreased in both types of infected seeds.[167]

F. Alteration in Physical Properties of Seeds

Economically important physical properties in soybean seeds, such as density, shape, size, surface area, volume, and weight, were affected by species of *Alternaria*, *Fusarium*, and *Phomopsis* but not by soybean mosaic virus. For example, *Phomopsis* reduced seed density by 4% and volume and weight by 13%, with a resultant potential of seed breakage 20 times greater than that for asymptomatic seeds. Effects were similar but less dramatic for seeds infected with *Alternaria* or *Fusarium*.[179] Soybean seed quality determined by biological, chemical, and physical characteristics is often defined poorly in commerce and categorized on a basis of visual observation, such as by comparing seeds of other color, or by noting seeds that are weathered, moldy, bicolored, discolored, and mottled. These visual classifications lower the grade or acceptance rate. For example, seeds showing more than 10% discoloration are assessed as damaged by the U.S. Department of Agriculture Federal Grain Inspection Service (Table 1-2).[180] All types of fungal damage are treated equally, regardless of source or type.[180]

Infection by *Curvularia lunata*, *Fusarium moniliforme*, and *P. sorghina* significantly reduced the starch granule size in infected sorghum seeds compared to noninfected seeds. The reduction in size increased with increase in infection severity. The infection of seeds by *F. moniliforme* also resulted in pitting on starch granule surfaces.[181] Internally seedborne *Sclerotinia sclerotiorum* was observed

Table 1-2 United States Department of Agriculture Soybean (*Glycine max*) Seed Grades and Grade Requirements[a]

		Damaged seeds		Maximum limits		
Grade[b]	Minimum test weight/bu (lb)	Heat (%)	Total (%)	Foreign material (%)	Splits (%)	Soybean of other colors (%)
No. 1	56	0.2	2	1	10	1
No. 2	54	0.5	3	2	20	2
No. 3	52	1.0	5	3	30	5
No. 4	49	3.0	8	5	40	10
Sample	—	—	—	—	—	—

[a] From *Official United States Standards for Grains* (181).
[b] Purple mottled or stained soybeans are graded not higher than No. 3, and materially weathered soybeans are graded not higher than No. 4. Sample grade soybeans (i) do not meet the requirements of No. 1, 2, 3, or 4; (ii) contain eight or more stones with an aggregate weight in excess of 0.2% of the sample weight, two or more pieces of glass, three or more seeds of *Crotalaria*, two or more castor beans (*Ricinus communis* L.), four or more particles of an unknown foreign substance(s), or a commonly recognized harmful or toxic quantity of other animal filth per 1000 soybeans; (iii) have a musty, sour, or commercially objectionable foreign odor (except garlic odor); or (iv) are heating or are otherwise of distinctly low quality.

in soybean seeds; infected seeds showed an abnormal color, and the seed coat could be separated easily from the embryo.[182]

REFERENCES

1. Neergaard, P., Opening addresses, *Seed Sci. Technol.*, 11, 471, 1983.
2. Neergaard, P., *Danish Species of Alternaria and Stemphylium*, Einar Munksgaard, Copenhagen, 1945, 560.
3. Neergaard, P., *Seed Pathology*, Vols. 1 and 2, Macmillan, London, 1977, 1187.
4. Anonymous, A biographical note on Paul Neergaard, *Seed Sci. Technol.*, 11, 459, 1983.
5. Neergaard, P., Screening for plant health, *Annu. Rev. Phytopathol.*, 24, 1, 1986.
6. Wilson, M., Noble, M., and Gray, E. G., The blind seed disease of rye-grass and its causal fungus, *Trans. R. Soc. Edinburgh*, 61, 327, 1945.
7. Sinclair, J. B., Latent infection of soybean plants and seeds by fungi, *Plant Dis.*, 75, 220, 1991.
8. du Tillet, M., Dissertation sur la cause qui corrompt et noircit les grains de bled dans les épis; et sur les moyens de preévenir cesaccidens (English transl. by H. B. Humphrey), *Phytopathological Classics*, No. 5, APS Press, St. Paul, MN, 1937, 91.
9. Prevost, B., Mémoire sur la cause immédiate de la carie ou charbon des bles, et de plusieurs autres maladies des plantes, et sur les préservatifs de la carie (English transl. by G. W. Keitt), *Phytopathological Classics*, No. 6, APS Press, St. Paul, MN, 1939, 94.
10. Frank, B., *Die Krankheiten der Pflanzen*, Breslau, 1883, 844.
11. Beach, S. A., Some bean diseases, *N.Y. State Agric. Exp. Sta. Bull.*, 48, 333, 1892.
12. Rolfs, F. M., Angular leaf spot of cotton, *S.C. Agric. Exp. Stn. Bull.*, 184, 1915.

13. Clayton, E. E., Studies of the black rot or blight disease of cauliflower, *N.Y. State Agric. Exp. Stn. Bull.* 576, 1929, 1.
14. Schuster, M. L. and Coyne, D. P., Survival mechanisms of phytopathogenic bacteria, *Annu. Rev. Phytopathol.*, 12, 199, 1974.
15. Mayer, A., Uber die Mosaikkrankheit des Tabaks [*Concerning the Mosaic Disease of Tobacco*] (English transl. by J. Johnson), *Phytopathological Classics*, No. 7, APS Press, St. Paul, MN, 1942, 11.
16. McClintock, J. A., Is cucumber mosaic carried by seed?, *Science*, 44, 786, 1916.
17. Stewart, V. B. and Reddick, D., Bean mosaic, *Phytopathology*, 7, 61, 1917.
18. Doolittle, S. P. and Gilbert, W. W., Seed transmission of curcurbit mosaic by the wild cucumber, *Phytopathology*, 9, 326, 1919.
19. Stace-Smith, R. and Hamilton, R. I., Inoculum thresholds of seedborne pathogens, Viruses, *Phytopathology*, 78, 875, 1988.
20. Needham, T., A letter concerning certain chalky tubulous concretions, called malm; with some microscopal observations on the farina of the red lilly, and of worms discovered in smutty corn, *Philos. Trans. R. Soc. London*, 42, 634, 1743.
21. Richardson, M. J., An Annotated List of Seedborne Diseases, 3rd ed., Commonwealth Mycological Institute, Kew, Surrey, U.K., 1990, 320.
22. Richardson, M. J., *An Annotated List of Seedborne Diseases*, ISTA Secretariat, Zurich, 1990.
23. Dorogin, G. N., Instructions for testing seeds to determine the presence of fungus diseases at seed control stations, *Zashch. Rast.*, 2, 1923.
24. Chen, C. C., Internal Fungous Parasites of Agricultural Seeds, *Univ. Md. Agric. Bull.*, 240, 81, 1920.
25. Orton, C. R., Seed-borne parasites — a bibliography, *W. Va. Agric. Exp. Stn. Bull.*, 245, 47, 1931.
26. Porter, R. H., Recent developments in seed technology, *Bot. Rev.*, 15, 221, 1949.
27. Alcock, N. L., Notes on common diseases sometimes seedborne, *Trans. Bot. Soc. Edinburgh*, 30, 332, 1931.
28. Machacek, J. E., Cherewick, W. J., Mead, H. W., and Broadfoot, W. C., A study of some seedborne diseases of cereals in Canada, II. Kinds of fungi and prevalence of diseases in cereal seed, *Sci. Agric.*, 31, 193, 1951.
29. Noble, M., de Tempe, J., and Neergaard, P., An Annotated List of Seed-borne Diseases, Commonwealth Mycological Institute, Kew, Surrey, U.K., 1958.
30. Noble, M. and Richardson, M. J., An Annotated List of Seed-borne Diseases, CMI, *Phytopathol. Pap.*, 8, 191, 1968.
31. One hundred years of seed testing in France, 1884–1984, *Seed Pathol. News*, 17, 11, 1986.
32. Schaad, N. W., Correlation of laboratory assays for seedborne bacteria with disease development, *Seed Sci. Technol.*, 11, 877, 1983.
33. International Seed Testing Association, Zurich, 1991, 7.
34. Wold, A., Opening address, *Seed Sci. Technol.*, 11, 464, 1983.
35. Bessey, C. E., Iowa State University, 1886 (cited in Malone and Muskett, *Pest Articles News Summaries*, Sect. B, 13, 266, 1967).
36. Smith, A. L., The fungi of germinating farm seeds, 1896-1901, *Trans. Br. Mycol. Soc.*, 1, 182, 1903.
37. Yorinori, J. T., Sinclair, J. B., Mehta, Y. R., and Mohan, S. K., Eds., Seed Pathology, Progress and Problems, Fundacâo Instituto Agronômico do Paraná, Londrina, Brazil, 1979, 274.

38. Doyer, L. C., Manual for the Determination of Seed-borne Diseases, International Seed Testing Association, Wageningen, The Netherlands, 1938, 60.
39. Hiltner, L., *Techn. Vorschr. fur die Prufung von Saatgu gultig vom 1 Juli 1916 an B. Besonderer Teil. I. Getreide, Landwirtsch*, Versuchs-Stationen, 89, 379, 1917.
40. Langerak, C. J., Meeting of the Plant Disease Committee, *ISTA News Bull.*, 102, 8, 1992.
41. Anonymous, International Seed Testing Association (ISTA), International Rules for Seed Testing, 1993, *Seed Sci. Technol.*, Suppl. 21, 296, 1993.
42. James, W. C., The cost of disease to world agriculture, *Seed Sci. Technol.*, 9, 679, 1981.
43. Cramer, H. H., *Plant Protection and World Crop Protection*, Bayer, Leverkusen, Germany, 1967, 524.
44. James, W. C. and Teng, P. S., The quantification of production constraints associated with plant diseases, *Rev. Appl. Biol.*, 4, 201, 1979.
45. Brennan, J. P., Warham, E. J., Byerlee, D., and Hernandez-Estrada, J., Evaluating the economic impact of quality reducing seedborne diseases: lessons from karnal bunt of wheat, *Agric. Econ.*, 6, 345, 1992.
46. Persons, T. D., Destructive outbreak of loose smut in a Georgia wheat field, *Plant Dis. Rep.*, 38, 422, 1954.
47. Ledingham, R. J., Atkinson, T. G., Horricks, J. S., Mills, J. T., Piening, L. J., and Tinline, R. D., Wheat losses due to common root rot in the Prairie Provinces of Canada, 1969–71, *Can. Plant Dis. Surv.*, 53, 113, 1973.
48. Piening, L. J., Atkinson, T. G., Horricks, J. S., Ledingham, R. J., Mills, J. T., and Tinline, R. D., Barley losses due to common root rot in the Prairie Provinces of Canada, 1970–72, *Can. Plant Dis. Surv.*, 56, 41, 1976.
49. Cock, L. J., The control of cereal diseases in the U.K., in Proc. 8th Br. Insecticide and Fungicide Conf., Brighton, U.K., 1975, 859.
50. King, J. E., Surveys of diseases of winter wheat in England and Wales, 1970–75, *Plant Pathol.*, 26, 8, 1977.
51. Ou, S. H., *Rice Diseases*, 2nd ed., CAB International Mycological Institute, Kew, Surrey, U.K., 1985.
52. Lee, F. N. and Rush, M. C., Rice sheath blight — a major rice disease, *Plant Dis.*, 67, 829, 1983.
53. Jairo Castano, Z., Rice grain discoloration in Colombia: etiology, damage and control, *Phytopathology*, 75, 1176, 1985.
54. Richardson, M. J., Whittle, A. M., and Jacks, M., Yield loss relationships in cereals, *Plant Pathol.*, 25, 21, 1976.
55. Mathur, R. S., Mathur, S. C., and Bajpai, G. K., An attempt to estimate loss caused by stripe disease of barley, *Plant Dis. Rep.*, 48, 708, 1964.
56. Porta Puglia, A., Delogu, G., and Vannacci, G., *Pyrenophora graminea* on winter barley seed: effect on disease incidence and yield losses, *J. Phytopathol.*, 117, 26, 1986.
57. Sunesom, C. A., Effect of barley stripe, *Helminthosporium gramineareum* Rab. on yield, *J. Am. Soc. Agron.*, 38, 954, 1946.
58. Tekauz, A., Reaction of Canadian barley cultivars to *Pyrenophora graminea*, the incitant of leaf stripe, *Can. J. Plant Pathol.*, 5, 294, 1983.
59. Kaiser, W. J. and Hannan, R. M., Seed treatment fungicides for control of seedborne *Ascochyta lentis* on lentil, *Plant Dis.*, 71, 58, 1987.
60. Morrall, R. A. A., Significance of seedborne inoculum of lentil pathogens in western Canada, Proc. First Int. Conf. Grain Legumes, France, 1992, 313.

61. Agarwal, P. C., Mortensen, C., and Mathur, S. B., Seedborne Diseases and Seed Health Testing of Rice, Tech. Bull. No. 3, Danish Govt. Institute of Seed Pathology for Developing Countries, Copenhagen, 1989, 106.
62. Jones, R. A. C. and Proudlove, W., Further studies on cucumber mosaic virus infection of narrow-leafed lupin (*Lupinus angustifolius*): seedborne infection, aphid transmission, spread and effects on grain yield, *Ann. Appl. Biol.*, 118, 319, 1991.
63. Tobias, I. and Pusztal, L., Incidence of lettuce mosaic virus on lettuce, *Zold Segtermesztesi Kutate Intezet Bull.*, 19, 117, 1986.
64. Masmoudi, K., Khetarpal, R. K., and Maury, Y., Seed transmission of pea seed-borne mosaic virus, Proc. First Eur. Conf. Grain Legumes, France, 1992, 317.
65. Allan, E. K., Olfat, E. H., and Nagwa, A. M., Effect of virus infection on seed production and virus seed transmission of legumes II, Cowpea aphid-borne mosaic virus and its effect on cowpea seed, *19th Int. Seed Testing Assn. Congr.*, Vienna, 1980.
66. Goodman, R. M. and Oard, J. H., Seed transmission and yield losses in tropical soybean infected by soybean mosaic virus, *Plant Dis.*, 64, 913, 1980.
67. Paruthi, I. J., Singh, M., and Gupta, D. C., Quantitative and qualitative losses in wheat grains due to earcockle and tundu, *Seed Res.*, 15, 83, 1987.
68. Diener, T. O. and Raymer, W. B., Potato spindle tuber virus, in Descriptions of Plant Viruses, No. 66, Commonwealth Mycological Institute and Association of Applied Biologists, Kew, Surrey, U.K., 1971.
69. Diener, T. O., *Viroids and Viroid Diseases*, John Wiley & Sons, New York, 1979, 252.
70. Lawson, R. H., Controlling virus diseases in major international flower and bulb crops, *Plant Dis.*, 65, 780, 1981.
71. Zelazny, B., Randles, J. W., Boccardo, G., and Imperial, J. S., The viroid nature of cadang-cadang disease of coconut palm, *Sci. Filip.*, 2, 45, 1982.
72. Keese, P. and Symons, R. H., Physical-chemical properties: molecular structure (primary and secondary), in *The Viroids*, Diener, T.O., Ed., Plenum Press, New York, 1987, 37.
73. Murdock, D. J., Nelson, P. E., and Smith, S. S., Histopathological examination of *Pelargonium* infected with tomato ringspot virus, *Phytopathology*, 66, 844, 1976.
74. Haight, E. and Gibbs, A., Effect of viruses on pollen morphology, *Plant Pathol.*, 32, 369, 1983.
75. Pesic, Z. and Hiruki, C., Effect of alfalfa mosaic virus on germination and tube growth of alfalfa pollen, *Can. J. Plant Pathol.*, 10, 6, 1988.
76. Bennet, C. W., *Lychnis* ringspot, *Phytopathology*, 49, 706, 1959.
77. Carroll, T. W., Barley stripe mosaic virus: its economic importance and control in Montana, *Plant Dis.*, 64, 136, 1980.
78. Grasmick, M. E. and Slack, S. A., Effect of potato spindle tuber viroid on sexual reproduction and viroid transmission on true potato seed, *Can. J. Plant Pathol.*, 64, 336, 1986.
79. Tatum, L. A., The southern corn leaf blight epidemic, *Science*, 171, 113, 1971.
80. Hagborg, W. A. L., Dwarfing of wheat and barley by the barley stripe mosaic (false stripe) virus, *Can. J. Bot.*, 32, 24, 1954.
81. Carroll, T. W., Barley stripe mosaic virus: its importance and control in Montana, *Plant Dis.*, 64, 136, 1980.
82. Neergaard, P., Report of the committee on plant diseases 1971-74, *Seed Sci. Technol.*, 3, 201, 1975.

83. Halfon-Meiri, A., The effect of seed-borne fungi on germination of seeds, in *Crop Physiology*, Gupta, U.S., Ed., Oxford and IBH, New Delhi, 1978.
84. Nicholson, J. F., Dhingra, O. D., and Sinclair, J. B., Internal seedborne nature of *Sclerotinia sclerotiorum* and *Phomopsis* spp. and their effects on soybean seed quality, *Phytopathology*, 62, 1261, 1972.
85. Wallen, V. R. and Seaman, W. L., Seedborne aspects of *Diaporthe phaseolorum* in soybean, *Phytopathology*, 52, 756, 1962.
86. Agarwal, V. K. and Joshi, A. B., A preliminary note on the purple stain disease of soybean, *Indian Phytopathol.*, 24, 810, 1971.
87. Gangopadhyay, S., Wyllie, T. D., and Luedders, V. D., Charcoal rot disease of soybean transmitted by seeds, *Plant Dis. Rep.*, 54, 1088, 1971.
88. Kmetz, K., Ellett, C. W., and Schmitthenner, A. F., Effect of crop history on the incidence of *Phomopsis* and *Diaporthe* in soybean plants and seed, Proc. Am. Phytopathol Soc., 1, 41, 1974.
89. Wallen, V. R. and Cuddy, T. F., Relation of seedborne *Diaporthe phaseolorum* to the germination of soybean, Proc. Assoc. Off. Seed Anal., 50, 137, 1960.
90. Sharma, S. L. and Sohi, H. S., Seedborne nature of *Phytophthora parasitica* causing buckeye rot of tomato, *Indian Phytopathol.*, 28, 130, 1975.
91. Mathur, S. B., Mallya, J. I., and Neergaard, P., Seedborne infection of *Trichoconis padwickii* in rice, distribution and damage to seeds and seedlings, Proc. Int. Seed Test Assoc., 37, 803, 1972.
92. Gambogi, P., Triolo, E., and Vannacci, G., Experiments on the behavior of the seedborne fungus *Alternaria zinniae*, *Seed Sci. Technol.*, 4, 333, 1976.
93. Warham, E. J., Effect of *Tilletia indica* infection on viability, germination and vigor of wheat seed, *Plant Dis.*, 74, 130, 1990.
94. Prabhu, A. S. and de A. Vieira, N. H., Sementas de arroz infectadas por *Drechslera oryzae*: germinocao, tranmsissao e controle, *EMBRADA CNPAF Boletina de Pesquisa*, 7, 1, 1989.
95. Kim, W. G., Oh, I. S., Yu, S. H., and Park, J. S., *Fusarium moniliforme* detected in seeds of corn and its pathological significanee, *Korean J. Mycol.*, 12, 105, 1984.
96. Kurppa, A., *Bipolaris sorokiniana* on barley seed in Finland, *J. Agric. Sci. Finland*, 56, 174, 1984.
97. Rheeder, J. P., Marasas, W. F. O., and VanWyk, P. S., Fungal association in corn kernels and effect on germination, *Phytopathology*, 80, 131, 1990.
98. Velicheti, R. K. and Sinclair, J. B., Histopathology of soybean seeds colonized by *Fusarium oxysporum*, *Seed Sci. Technol.*, 19, 445, 1991.
99. Gleason, M. L. and Ferriss, R. S., Influence of soil water potential on performance of soybean seeds infected by *Phomopsis* sp., *Phytopathology*, 75, 1236, 1985.
100. Singh, D. P. and Agarwal, V. K., Purple stain of soybean and seed viability, *Seed Res.*, 14, 126, 1986.
101. Saifulla, M. and Rangnathaiah, K. G., Seed health testing of sorghum with special reference to *Colletotrichum graminicola*, *Indian Phytopathol.*, 42, 73, 1989.
102. Mathur, K., Siradhana, B. S., and Lodha, B. C., Studies on seedling blight of sorghum caused by *Gloeocercospora sorghi*, *Seed Sci. Technol.*, 15, 851, 1987.
103. Hepperly, P. R., Feliciano, C., and Sotomayor, A., Chemical control of seedborne fungi of sorghum and their association with seed quality and germination in Puerto Rico, *Plant Dis.*, 66, 902, 1982.
104. Chikuo, Y. and Sugimoto, T., Infection of sugarbeet seed by *Colletotrichum dematuim* f. *spinaciae*, *Ann. Phytopathol. Soc. Jpn.*, 50, 249, 1984.

105. Pande, S., Mukuru, S. Z., Odhiambo, R. O., and Karunakar, R. I., Seedborne infection of *Eleusine coracana* by *Bipolaris nodulosa* and *Pyricularia grisea* in Uganda and Kenya, *Plant Dis.*, 78, 60, 1994.
106. Badoost, M., Gabrielson, R. L., Olson, S. A., and Mulanax, M. W., Control of *Alternaria* diseases of Brassica seed crops caused by *Alternaria brassicae* and *Alternaria brassicicola* with ground and aerial fungicide application, *Seed Sci. Technol.*, 21, 1, 1993.
107. Mayhew, W. L. and Caviness, C. E., Seed quality and yield of early planted short-season soybean genotypes, *Agron. J.*, 86, 16, 1994.
108. Zorrilla, G., Knapp, A. D., and McGee, D. C., Severity of *Phomopsis* seed decay, seed quality evaluation and field performance of soybean, *Crop Sci.*, 34, 172, 1994.
109. Cristani, C., Seedborne *Microdochium nivale* (Ces. ex Sacc.) Samuels (= *Fusarium nivale* (Fr.) Ces.) in naturally infected seeds of wheat and triticale in Italy, *Seed Sci. Technol.*, 20, 603, 1992.
110. Lai, M., Bacterial canker of bell pepper caused by *Corynebacterium michiganense*, *Plant Dis. Rep.*, 60, 339, 1976.
111. Ellis, M. A., Tenne, F. D., and Sinclair, J. B., Effect of antibiotics and high temperature storage on decay of soybean seeds by *Bacillus subtilis*, *Seed Sci. Technol.*, 5, 753, 1977.
112. Mengistu, A. and Sinclair, J. B., Seedborne microorganisms of Ethiopian-grown soybean and chickpea seeds, *Plant Dis. Rep.*, 63, 616, 1979.
113. White, J. C., Nicholson, J. F., and Sinclair, J. B., Effect of soil temperature and *Pseudomonas glycinea* on emergence and growth of soybean seedlings, *Phytopathology*, 62, 296, 1972.
114. Tenne, F. D. and Sinclair, J. B., Detection of internally-borne *Pseudomonas glycinea* in soybean seeds at high temperatures, *Proc. Am. Phytopathol. Soc.*, 1, 130, 1974.
115. Laurence, J. A. and Kennedy, B. W., Population changes of *Pseudomonas glycinea* on germinating soybean seeds, *Phytopathology*, 64, 1470, 1974.
116. Tenne, F. D., Foor, S. D., and Sinclair, J. B., Association of *Bacillus subtilis* with soybean seeds, *Seed Sci. Technol.*, 5, 763, 1973.
117. Hemmati, K. and McLean, D. L., Gamete seed transmission of alfalfa mosaic virus and its effect on seed germination and yield in alfalfa plants, *Phytopathology*, 67, 576, 1977.
118. Heydecker, W., The vigor of seeds — a review, in Fifteenth Int. Seed Testing Congr., Christchurch, New Zealand, 1968, 15 (preprint).
119. Walkey, D. G. A., Brocklehurst, P. A., and Parker, J. E., Seed transmission of viruses, Thirty-third Annu. Rep. Natl. Vegetable Res. Stn., Wellsbourne, Warwick, U.K., 82, 1983.
120. Hicks, R. G. T., Smith, T. J., and Edwards, R. P., Effects of strawberry latent ringspot virus on the development of seeds and seedlings of *Chenopodium quinoa* and *Pastinaca sativa* (parsnip), *Seed Sci. Technol.*, 14, 409, 1986.
121. Schaad, N. W., Sitterly, W. R., and Humaydan, H., Relationship of levels of seedborne *Xanthomonas campestris* as determined by laboratory assays to the development of black rot of cabbage, Third Int. Congr. Plant Pathology, P. Parey, Berlin, 1978, 66.
122. Wallen, V. R. and Sutton, M. D., *Xanthomonas phaseoli* var. *fuscans* (Burkh.) Starr and Burkh. on field bean in Ontario, *Can. J. Bot.*, 43, 437, 1965.
123. Walker, J. C. and Patel, P. N., Splash dispersal and wind as factors in epidemiology of halo blight of bean, *Phytopathology*, 54, 140, 1964.

124. Cox, R. S., The role of bacterial spot in tomato production in South Florida, *Plant Dis. Rep.*, 50, 699, 1966.
125. Fischer, G. W. and Holton, C. S., *Biology and Control of Smut Fungi*, Ronald Press, New York, 1957, 622.
126. Melouk, H. A., Singleton, L. L., Owens, F. N., and Aken, C. N., Viability of sclerotia of *Sclerotinia minor* after passage through the digestive tract of a crossbred heifer, *Plant Dis.*, 73, 65, 1989.
127. Sinclair, J. B. and Backman, P. A., *Compendium of Soybean Diseases*, 3rd ed., APS Press, St. Paul, MN, 1989.
128. Kunwar, I. K., Manandhar, J. B., and Sinclair, J. B., Histopathology of soybean seeds infected with *Alternaria*, *Phytopathology*, 76, 543, 1986.
129. Sinclair, J. B., Discoloration of soybean seeds — an indicator of quality, *Plant Dis.*, 76, 1087, 1992.
130. Singh, T. and Sinclair, J. B., Histopathology of *Cercospora sojina* in soybean seeds, *Phytopathology*, 75, 185, 1985.
131. Casady, W. W., Paulsen, M. R., Reid, J. F., and Sinclair, J. B., A trainable algorithm for inspection of soybean quality, *Am. Soc. Agric. Eng. Pap.*, 90-7522, 10, 1990.
132. Paulsen, M. R., Wigger, W. D., Litchfield, J. B., and Sinclair, J. B., Computer image analyses for detection of maize and soybean kernel quality factors, *J. Agric. Eng. Res.*, 43, 93, 1989.
133. Casady, W. W., Paulsen, M. R. and Sinclair, J. B., Optical properties of damaged soybean seeds, *Am. Soc. Agric. Eng.*, 36, 943, 1993.
134. Haware, M. P., Nene, Y. L., and Mathur, S. B., Seedborne Diseases of Chickpea, DGISPDC Tech. Bull. 1, Copenhagen, 1986, 32.
135. Madeira, A. C., Clark, J. A., Rossall, S., and McArthur, A. J., A classification system for seeds of favabean infected by *Ascochyta fabae*, *Fabis Newsl.*, 30, 48, 1992.
136. Wakimoto, S., Akaki, M., and Tsuchiya, K., Serological specificity of *Pseudomonas glumae*, the pathogenic bacterium of grain rot disease of rice, *Ann. Phytopathol. Soc. Jpn.*, 53, 150, 1987.
137. Tu, J. C., Bean anthracnose, in *Plant Diseases of International Importance II, Diseases of Vegetables and Oil Seed Crops*, Chaube, H. S., Singh, U. S., Mukhopadhyay, A. N., and Kumar, J., Eds., Prentice Hall, New York, 1992, 1.
138. Kaiser, W. J., Testing and Production of Healthy Plant Germ Plasm, DGISPDC Tech. Bull. 2, 1987, 30.
139. Beauchamp, C. J., Morrall, R. A. A., and Slinkard, A. E., The potential of control of Ascochyta blight of lentil with foliar-applied fungicides, *Can. J. Plant Pathol.*, 8, 254, 1986.
140. La, Y. J., Bak, W. C., and Oh, J. H., Immunochemical detection of soybean mosaic virus infections in the seeds of soybean cultivars in Korea, *Korean J. Plant Prot.*, 22, 26, 1983.
141. Bryant, G. R., Hill, J. H., Bailey, T. B., Tachibana, H., Durand, D. P., and Benner, H. I., Detection of soybean mosaic virus in seed by solid-phase radioimmunoassay, *Plant Dis.*, 66, 693, 1982.
142. Capoor, S. P., Garg, D. G., and Sawant, D. M., Seed transmission of French bean mosaic virus, *Indian Phytopathol.*, 39, 343, 1986.
143. Waele, D. De., Jones, B. L., Bolton, C., and Van Der Berg, E., *Ditylenchus destructor* in hulls and seeds of peanut, *J. Nematol.*, 21, 10, 1989.

144. Clear, R. M., Nowicki, T. W., and Daun, J. K., Soybean seed discolorations by *Alternaria* spp. and *Fusarium* spp., effects on quality and production of fusari-otoxins, *Can. J. Plant Pathol.*, 11, 308, 1989.
145. Anahosur, K. H., Shivasmurthy, S. C., and Hiremath, R. V., Effect of seed micro-flora on the quality of soybean oil, *Curr. Res.*, 4, 83, 1975.
146. Cavanaugh, K. J. and Sinclair, J. B., Mycotoxin production by four isolates of *Alternaria alternata* and soybean seed quality amylases of asymptomatic and symptomatic seeds colonized by *A. alternata*, *Phytopathology*, 77, 1698, 1987.
147. Hepperly, P. R. and Sinclair, J. B., Quality losses in *Phomopsis*-infected soybean seeds, *Phytopathology*, 68, 1684, 1978.
148. Marcinkowska, J., Schollenberger, M., and Boros, L., Influence of infection of soybean seeds with *Peronospora manshurica* and *Pseudomonas syringae* pv. *glycinea* on protein, oil and fatty-acid content, *Acto Agrobot.*, 37, 157, 1984.
149. Coates, J. B., Medeiros, J. S., Thanh, V. H., and Neilson, N. C., Characterization of subunits of β-conglycinin, *Arch. Biochem. Biophys.*, 263, 186, 1985.
150. Murphy, P. A., Structural characteristics of soybean glycinin and conglycinin, in *Proc. World Soybean Res. Conf.* III, Shibles, R., Ed., Westview Press, Boulder, CO, 1985, 143.
151. Christopher, J. P., Pisturious, E. K., and Axelrod, B., Isolation of a third enzyme of soybean lipoxygenase, *Biochem. Biophys. Acta*, 284, 54, 1972.
152. Peever, T. L. and Higgins, V. J., Electrolyte leakage, lipoxygenase and lipid peroxidation induced in tomato plants by specific and non-specific elicitors from *Cladosporium fulvum*, *Plant Physiol.*, 90, 867, 1989.
153. Etzler, M. E., Plant pectins: molecular and biological aspects, *Annu. Rev. Plant Physiol.*, 36, 209, 1985.
154. Gibson, D. M., Stack, S., Krell, K., and House, J., A comparison of soybean agglutinin in cultivars resistant and susceptible to *Phytophthora megasperma* var. *sojae* (race 1), *Plant Physiol.*, 70, 560, 1982.
155. Velicheti, R. K., Kollipara, K. P., Sinclair, J. B., and Hymowitz, T., Selective degradation of protein by *Cercospora kikuchii* and *Phomopsis longicolla* in soybean seed coats and cotyledons, *Plant Dis.*, 76, 779, 1992.
156. Singh, S. N. and Khare, M. N., Influence of seedborne infection of *Trichoconiella padwickii* on biomass and qualitative changes in paddy rice, *Seed Res.*, 15, 233, 1987.
157. Kamble, B. R. and Gangawane, L. V., Biochemical changes in groundnut influenced by fungi, *Seed Res.*, 15, 106, 1987.
158. Sharma, M. C. and Bhowmik, T. P., Effect of *Macrophomina phaseolina* infection on the physico-chemical components of groundnut seeds, *J. Phytopathol.*, 118, 181, 1987.
159. El-Amrety, A. S., El-Said, H. M., and Salem, D. E., Effect of soybean mosaic virus infection on quality of soybean seeds, *Agric. Res. Rev.*, 63, 155, 1985.
160. Nutter, F. W., Jr., Pederson, V. D., and Timian, R. G., Relationship between seed infection by barley stripe mosaic virus and yield lost, *Phytopathology*, 74, 363, 1984.
161. Singh, A. K. and Singh, A. K., Chemical composition of cowpea seeds as influenced by southern bean mosaic virus and cowpea virus, *Phyton. Austria*, 26, 165, 1987.
162. Sahasrabuddhe, D. L. and Kale, N. P., A biochemical study on the formation of oil in niger seed (*Guizotia abyssinca*), *Indian J. Agric. Sci.*, 3, 57, 1933.

163. Ward, H. S., The effect of moisture and temperature during storage on the germination, respiration and free fatty acids on Dixie Runner peanuts, *J. Ala. Acad. Sci.*, 27, 96, 1955.
164. Wilson, C., Concealed damage of peanut in Alabama, *Phytopathology*, 37, 657, 1947.
165. Robinson, C. H., *Fundamentals of Normal Nutrition*, Macmillan, New York, 1978.
166. Ashraf, S. S. and Basuchaudhary, K. C., Effect of seedborne *Fusarium* species on the physico-chemical properties of rapeseed oil, *J. Phytopathol.*, 117, 107, 1986.
167. Lakshmidevi, N., Prakash, H. S., and Shetty, H. S., Effect of seedborne *Macrophomina phaseolina* (Tassi) Goid on physiochemical properties of sunflower oil, *Int. J. Trop. Plant Dis.*, 10, 79, 1992.
168. Sankaram, A., *A Laboratory Manual for Agricultural Chemistry*, Asia Publishing House, Bombay, 1966, 340.
169. Raj, J. N. and Saxena, A., Effect of some seed-borne fungi on the physico-chemical properties of the oil of the Indian mustard, *Indian J. Agric. Res.*, 48, 769, 1978.
170. Mall, O. P. and Pateria, H. M., Effect of temperature on biodeterioration and aflatoxin production in groundnut seed, Proc. Symp. Mycotoxin in Food and Feed, 1983, 271.
171. Singh, B. K. and Prasad, T., Changes in oil properties of sunflower seeds, (*Helianthus annuus* L.) due to some seed-borne fungi, *Proc. Nat. Acad. Sci. India*, 51, 45, 1981.
172. Sharma, K. D., Biodeterioration of sesamum oil *in situ* by fungi, *Indian Phytopathol.*, 34, 50, 1981.
173. Lalithakumari, D., Govindaswamy, C. V., and Vidhyasekaran, P., Effects of seedborne fungi on physico-chemical properties of groundnut seed, *Indian Phytopathol.*, 24, 283, 1971.
174. Prasad, T. and Singh, B. K., Effect of relative humidity on oil properties of fungal infected sunflower seeds, *Biol. Bull.*, 5, 85, 1983.
175. Bilgrami, K. S. and Verma, R. N., *Physiology of Fungi*, Vikas Publishing House, Pvt., Ltd., New Delhi, 1978, 507.
176. Basuchaudhary, K. C. and Prasad, P. K., Qualitative changes in oilseeds due to seedborne fungi, Proc. Int. Conf. Seed Sci. Technol., Indian Soc. Seed Technol., New Delhi, 1990, 109.
177. Ward, H. S. and Diener, U. L., Biochemical changes in shelled peanuts caused by storage fungi, I. Effect of *Aspergillus tamarri*, four species of *A. glaucus* group and *Penicillium citrinum*, *Phytopathology*, 51, 244, 1961.
178. Booth, C., The Genus *Fusarium*, Commonwealth Mycological Institute, Kew, Surrey, U.K., 1971, 237.
179. Mbuvi, S. W., Litchfield, L. B. and Sinclair, J. B., Physical properties of soybean seeds damaged by fungi and a virus, *Trans. Am. Soc. Agric. Eng.*, 32, 2093, 1989.
180. Official United States Standards for Grains, Subpart I — U.S. Standards for Soybeans: Grade and Grade Requirements, U.S. Department of Agriculture, Federal Grain Inspection Service, 1990.
181. Singh, D. P. and Agarwal, V. K., Effect of grain mold infection on starch granules in sorghum, *Indian J. Plant Pathol.*, 5, 26, 1987.
182. Stovold, G. E. and Priest, M. J., A note on the incidence of internally-borne *Sclerotinia sclerotiorum* in soybean seed harvested in New South Wales, *Aust. Plant Pathol.*, 15, 83, 1986.

CHAPTER 2

Seedborne Pathogens

I. FUNGI

Fungi form a major group of pathogens that can be seedborne or transmitted through seeds. Fungi are nonvascular, heterotrophic, lack chlorophyll or other photosynthetic pigments, reproduce by means of sexual and/or asexual spores, and have assimilative bodies, which may be amoeboid or unicellular but typically are made up of multicellular branching filaments called hyphae. Hyphae grow up to 100 μm wide and branch to form the vegetative body or mycelium. Mycelium may be an interlacing tangle of hyphae, a loose wooly mass, or a compact body. In general, fungi live on dead saprobes or as parasites on living organisms.[1] Fungi belong to the Kingdom Mycetae. The classification of plant pathogenic fungi after Agrios[2] is presented in Table 2-1.

Most plant diseases are caused by fungi. Of over 100,000 described fungal species, more than 8000 are plant pathogens. Fungi are adapted to survive in seeds as well as in air, soil, and water, and in or on living or dead organic matter. Fungi acquire nutrients in several ways. A large percentage of plant-pathogenic fungi are necrotrophs — saprophytes capable of using nutrients either released from decaying plant tissue or of infected living tissues. Other fungi are obligate parasites, biotrophs that grow and reproduce in intimate association with a narrow range of living plant hosts. Both types are associated with seeds.

Typically, fungi reproduce, spread, and survive by sexual and asexual spores. Fungal taxonomy and nomenclature are based on reproduction method and spore morphology and development. In general, the asexual cycle is more important to plant and seed pathologists because large numbers of spores usually are produced, and the cycle may be repeated several times during a growing season. Seeds are the most important means of perpetuation of plant-pathogenic fungi.

Some fungi survive unfavorable conditions by producing resting (dormant) spores, such as chlamydospores and oospores or structures such as sclerotia, all of which are spread by seeds and man, as well as by animals, water, and wind.[1]

Table 2-1 General Classification of Plant Pathogenic Fungi

KINGDOM: Mycetae

DIVISION I: **Myxomycota** Produce plasmodia
- CLASS 1: **Myxomycetes** Lack mycelium (slime molds)
  - ORDER: **Physarales** Saprophytic plasmodium

- CLASS 2: **Plasmodiophoromycetes**
  - ORDER: **Plasmodiophorales** Plasmodia produced within plant cells

DIVISION II: **Eumycota** Produce mycelium

- SUBDIVISION I: **Mastigomycotina** Produce zoospores
  - CLASS 1: **Chytridiomycetes** Mycelium lacks cross walls
    - ORDER: **Chytridiales** Cell wall but no true mycelium
  - CLASS 2: **Oomycetes** Have elongated mycelium, produce zoospores
    - ORDER: **Saprolegniales** Well-developed mycelium
    - ORDER: **Peronosporales** Produce sporangia and oospores
      - FAMILY: **Pythiaceae** Sporangia on somatic hyphae or on sporangiophores of indeterminate growth
      - FAMILY: **Albuginaceae** Sporangia borne in chains
      - FAMILY: **Peronosporaceae** Sporangia borne on sporangiophores of determinate growth

- SUBDIVISION 2: **Zygomycotina**
  - CLASS: **Zygomycetes** Produce nonmotile asexual spores in sporangia
    - ORDER: **Mucorales** Spores formed in terminal sporangia
    - ORDER: **Endogonales** Mycorrhizal fungi, produce spores singly or in sporocarps

- SUBDIVISION 3: **Ascomycotina** Produce sexual spores in an ascus
  - CLASS 1: **Hemiascomycetes** Asci naked
    - ORDER: **Endomycetales** The yeasts
    - ORDER: **Taphrinales** Asci arising from binucleate cells
  - CLASS 2: **Pyrenomycetes** Asci produced in fruiting bodies
    - ORDER: **Erysiphales** Mycelium and cleistothecia on host surface
    - ORDER: **Sphaeriales** Perithecia with dark-colored, firm walls
    - ORDER: **Hypocreales** Perithecia light-colored, red or blue
  - CLASS 3: **Loculoascomycetes** Produce pseudothecia
    - ORDER: **Myriangiales** Cavities at various levels, single ascus
    - ORDER: **Dothideales** Cavities in basal layer with many asci; no pseudoparaphyses
    - ORDER: **Pleosporales** Cavities in basal layer with many asci; pseudoparaphyses present
  - CLASS 4: **Discomycetes** Asci produced at surface of cup-shaped apothecia
    - ORDER: **Phacidiales** Apothecia in stroma
    - ORDER: **Helotiales** Apothecia not in a stroma, asci released through an apical perforation

- SUBDIVISION 4: **Deuteromycotina** Imperfect fungi
  - CLASS 1: **Coelomycetes** Conidia in pycnidium or acervulus
    - ORDER: **Sphaeropsidales** Spores in a pycnidium
    - ORDER: **Melanconiales** Spores in an acervulus
  - CLASS 2: **Hyphomycetes**
    - ORDER: **Hyphales** Spores on or in hyphae
  - CLASS 3: **Agonomycetes** (Mycelia sterilia)
    - ORDER: **Agonomycetales** No spore production

Table 2-1 General Classification of Plant Pathogenic Fungi (continued)

SUBDIVISION 5: **Basidiomycotina** Sexual spores produced on a one- or four-celled basidium

CLASS 1: **Hemibasidiomycetes** Basidium with cross walls or the promycelium of a teliospore. Teliospores single or united remaining in host tissue

ORDER: **Ustilaginales** Fertilization by union of compatible spores, etc.; only basidiospores and teliospores produced

ORDER: **Uredinales** Spermatia fertilize receptive hyphae in spermogonia. Produce aeciospores, uredospores, teliospores, and basidospores

CLASS 2: **Hymenomycetes** Basidium without cross walls; basidiocarp lacking or present

ORDER: **Exobasidiales** Basidiocarp lacking, basidia produced on surface of tissue

ORDER: **Aphyllochorales** Hymenium lining surfaces of pores or tubes

ORDER: **Tulasnellales** Basidiocarps weblike, often waxy

ORDER: **Agaricales** Hymenium on radiating gills or lamellae

From Agrios, G. N., Plant Pathology, Academic Press, New York, 1988, 803. With permission.

Some fungi enter plants and seeds through natural openings, such as seed hila, hydathodes, lenticles, micropyles, and stomatal openings, and through wounds made by hail, blowing rain, or sand, animals, insects, man, or other microorganisms. Other fungi use mechanical pressure, enzymatic action or both to penetrate plant and seed tissues directly. Still others may penetrate directly and/or indirectly, depending upon environmental factors. Fungi overseason in or on seeds, on living or dead plants, in soil, and occasionally in insects.[1]

Symptoms or signs of disease produced by pathogens on seeds and host tissues assist in the diagnosis of diseases caused by fungi and identification of the causal agent.[2] Some pathogens are asymptomatic during portions of their cycle on plants.[3]

Physiologic races or strains of fungi occur and are morphologically identical but differ in such characteristics as the species and cultivars of host plants they parasitize, whether they are seedborne or not, to what extent of being seedborne, and in their interaction with the environment as it influences infection and disease development.[2] Seedborne fungi known to cause important plant diseases are presented in Table 2-2.

II. BACTERIA

Bacteria are procaryotic organisms that lack chlorophyll and multiply by fission. Bacteria have a cell membrane, a rigid cell wall and, often, one or more flagella. They may be comma shaped, ellipsoidal, filamentous, rod-shaped, spherical, or spiral. Most plant pathogenic bacteria are necrotrophs.[2]

The international standards for mapping pathovars of phytopathogenic bacteria and a list of pathovar names and pathotype strains was proposed by the ISPP Committee on Taxonomy of Phytopathogenic Bacteria in 1980.[443] However, some pathotype strains were found unsuitable, and the names of many new plant pathogens have been published. Revisions of the international standards and a list of names were published (Table 2-3).[444] At present the definition of a phyto-

Table 2-2 Seedborne Fungal Pathogens That Cause Important Disease of Major Crops[a,b]

Pathogen	Crop	Disease	Ref.
Acremonium coenophialum	*Festuca arundinacea* (tall fescue)	Leaf spot	4
	F. pratensis (meadow fescue)	Leaf spot	5
A. strictum	*A. sativa* (oats)	Stripe	6
	H. vulgare (barley)	Stripe	7
	T. aestivum (wheat)	Stripe	8, 9
	Z. mays (maize)	Black bundle, kernel rot	10
Albugo candida	*Brassica juncea* (brown mustard)	White rust	11
	B. rapa (field mustard)	White rust	11
Alternaria alternata	*Antirrhinum majus* (snapdragon)	Seedling malformation	12
	Cajanus cajan (pigeon pea)	Blight	13
	Glycine max (soybean)	Seed rot	14
	Helianthus annuus (sunflower)	Blight	15
	Lobelia (lobelia)	Blight	16
	Phaseolus vulgaris (bean)	Blight	17
	T. aestivum (wheat)	Black point	18
A. alternata f.sp. *lycopersici*	*Lycopersicon esculentum* (tomato)	Wilt	19
A. brassicae	*Brassica* (crucifers)	Gray leaf spot	20
A. brassicicola	*Brassica* (crucifers)	Black spot	21
	Crambe (kale)	Black spot	22
A. burnsii	*Cuminum cyminum* (cumin)	Leaf spot	23
A. carthami	*Carthamus tinctorius* (safflower)	Blight	24
A. cheiranthi	*Cheiranthus cheiri* (wallflower)	Black mold	25
A. cyamopsidis	*Cyamopsis tetragonoloba* (cluster bean, guar)	Blight, leaf spot	26
A. dauci	*Daucus carota* (carrot)	Leaf blight	27
A. dianthi	*D. caryophyllus* (carnation)	Leaf spot	28
A. dianthicola	*D. caryophyllus* (carnation)	Black mold, petal blight	29
A. helianthi	*H. annuus* (sunflower)	Blight	30
A. lini	*Linum usitatissimum* (flax)	Blight	31
A. linicola	*L. usitatissimum* (flax)	Blight	27
A. longipes	*Nicotiana tabacum* (tobacco)	Brown spot	32
A. longissima	*Sesamum indicum* (sesame)	Zonate leaf spot	33
A. macrospora	*Gossypium* (cotton)	Blight	34

[a] Does not necessarily refer to first report.
[b] See Appendix A for revised scientific names.

Table 2-2 Seedborne Fungal Pathogens That Cause Important Disease of Major Crops[a,b] (continued)

Pathogen	Crop	Disease	Ref.
A. padwickii	*Oryza sativa* (rice)	Seedling blight, stackburn	35
A. porri	*Allium cepa* (onion)	Purple blotch	29
A. porri f. sp. *calendulae*	*Calendula officinalis* (marigold)	Leaf spot	29
A. radicina	*Daucus carota* (carrot)	Black root rot, seedling blight	27
A. raphani	*Raphanus sativus* (radish)	Leaf spot	29
A. ricini	*Ricinus communis* (castor bean)	Capsule mold, seedling blight	36
A. sesami	*S. indicum* (sesame)	Blight, seed rot	37, 38
A. sesamicola	*S. indicum* (sesame)	Blight, seed rot	33
A. solani	*L. esculentum* (tomato)	Early blight	39
A. tenuissima	*G. max* (soybean)	Seed rot	40
A. tagetica	*Tagetes erecta* (African marigold)	Blight	41
A. triticina	*T. aestivum* (wheat)	Leaf blight	42
A. zinniae	*H. annuus* (sunflower)	Blight	43
Ascochyta fabae	*Vicia faba* (fava bean)	Leaf and pod spot	44, 45
A. fabae f. sp. *lentis*	*Lens culinaris* (lentil)	Leaf spot, blight	46
A. gossypii	*Gossypium* (cotton)	Leaf spot, seedling blight	47
A. linicola	*L. usitatissimum* (flax)	Foot rot	48
A. paspali	*Paspalum dilatatum* (Dallis grass)	Leaf spot, blight	49
A. phaseolorum	*P. lunatus* (lima bean)	Leaf spot	25
A. pinodes	*Pisum sativum* (pea)	Blight, foot rot, leaf spot	50
A. pisi	*P. sativum* (pea)	Foot rot, leaf and pod spot	50
A. rabiei	*C. arietinum* (chickpea)	Collar rot, foot rot	51
	T. alexandrinum (berseem)	Leaf spot	52
Aureobasidium caulivora	*Trifolium incarnatum* (crimson clover)	Northern anthracnose, scorch	53
A. zeae	*Z. mays* (maize)	Eyespot	54
Aspergillus flavus	*A. hypogaea* (peanut)	Alfaroot	55
	Gossypium (cotton)	Blight	56
	G. max (soybean)	Seedling blight	57
	H. annuus (sunflower)	Blight	58
	Shorea robusta (shorea)	Blight	59
	Z. mays (maize)	Blight	60
A. flavus var. *columnaris*	*Z. mays* (maize)	Seedling blight	61
A. niger	*Allium cepa* (onion)	Black mold	62
	A. hypogaea (peanut)	Collar rot, crown rot	63
A. parasiticus	*Z. mays* (maize)	Blight	64
A. quercinus	*G. max* (soybean)	Seed rot	65
Atkinsonella hypoxylon	*Danthonia spicata* (poverty oatgrass)	Blight	66

Table 2-2 Seedborne Fungal Pathogens That Cause Important Disease of Major Crops[a,b] (continued)

Pathogen	Crop	Disease	Ref.
Balansia-oryzae sativae	*O. sativa* (rice)	Blackring, sterility disease, Udbatta	67
Bipolaris maydis	Z. mays	Blight, southern leaf blight	68
B. oryzae	*O. sativa* (rice)	Brown spot	69
B. sorokiniana	*A. sativa* (oats)	Blight, leaf scorch, stem rot	69
	H. vulgare (barley)	Ear blight, foot rot, seedling blight	69
B. victoriae	*A. sativa* (oats)	Leaf blight	70
B. zeae	*Pennisetum clandestinum* (kikuya grass)	Brown spot	71
Botryodiplodia palmarum	*Capsicum* (pepper)	Leaf blight	72
B. theobromae	*A. hypogaea* (peanut)	Collar rot	73
	Cucurbita (cucurbit)	Leaf spot	74
	Dolichos biflorus	Leaf spot	75
	Gossypium (cotton)	Dry rot	76
	L. siceraria (bottle gourd)		77
	P. caribaea (Caribbean pine)		78
	Z. mays (maize)		79
Botrytis allii	*Allium cepa* (onion)	Damping-off, gray mold, neck rot	80
B. anthophila	*Trifolium pratense* (red clover)	Anther mold	81
B. cinerea	*Brassica* (crucifer)	Gray mold	82
	Callistephus chinensis (China aster)	Gray mold	83
	Carthamus tinctorius (safflower)	Gray mold	84
	Cicer arietinum (chickpea)	Gray mold	85
	H. annuus (sunflower)	Gray mold	86
	L. usitatissimum (flax)	Gray mold	48
	Lupinus (lupin)	Gray mold	87
	Phaseolus lunatus (lima bean)	Gray mold	88
	Pisum sativum (pea)	Gray mold	88
	V. faba (fava bean)	Gray mold	45
B. fabae	*V. faba* (fava bean)	Chocolate spot	89
Bremia lactucae	*Lactuca sativa* (lettuce)	Downy mildew	90
Caloscypha fulgens	*Picea sitchensis* (Sitka spruce)	Needle blight	91
Cephalosporium gramineum	*T. aestivum* (wheat)	Stripe	92
C. maydis	*Z. mays* (maize)	Late blight, slow wilt	93
Ceratocystis paradoxa	*Elaeis quineensis* (oil palm)	Black rot	94
Cercospora arachidicola	*Arachis hypogaea* (peanut)	Tikka disease	95
C. beticola	*Beta vulgaris* (red beet)	Leaf spot	96
C. canescens	*P. vulgaris* (bean)	Leaf spot, blotch	97

Table 2-2 Seedborne Fungal Pathogens That Cause Important Disease of Major Crops[a,b] (continued)

Pathogen	Crop	Disease	Ref.
C. capsici	*Capsicum annuum* (pepper)	Leaf spot	98
C. carotae	*D. carota* (carrot)	Blight, leaf spot	99
C. carthami	*C. tinctorius* (safflower)	Leaf spot	100
C. corchori	*Corchorus capsularis* (jute)	Leaf spot	101
C. kikuchii	*Cyamopsis tetragonoloba* (cluster bean, guar)	Purple seed stain	102
	G. max (soybean)	Purple seed stain	103
C. nicotianae	*N. tabacum* (tobacco)	Frogeye or green leaf spot	104
C. oryzae	*O. sativa* (rice)	Glume spot, narrow brown leaf spot	105
C. personata	*A. hypogaea* (peanut)	Tikka	95
C. ricinella	*R. communis* (castor bean)	White leaf spot	106
C. sesami	*Sesamum indicum* (sesame)	Leaf spot	107, 108
C. sojina	*G. max* (soybean)	Frogeye leaf spot	109
Cercosporella antirrhini	*A. majus* (snapdragon)	Leaf spot, shothole	110
Chalara elegans	*A. hypogaea* (peanut)	Black hull	111
Choanephora cucurbitarum	*Abelmoschus esculentus* (okra)	Fruit rot	112
Cladosporium herbarum	*Gossypium* (cotton)	Boll rot	113
	V. faba (fava bean)	Leaf mold	114
Claviceps fusiformis	*Pennisetum glaucum* (pearl millet)	Ergot	115
C. gigantea	*Z. mays* (maize)	Ergot	116
C. maximensis	*Panicum maximum* (guinea grass)	Ergot	117
C. paspali	*P. anceps* (panigrass)	Ergot	118
	Paspalum paspalodes (dallis grass)	Ergot	119
C. purpurea	*Agropyron repens* (quackgrass)	Ergot	120
	Agrostis tenuis (African daisy)	Ergot	121
	A. sativa (oat)	Ergot	122
	F. arundinacea (tall fescue)	Ergot	121
	F. rubra subsp. *commutata* (shade fescue)	Ergot	121
	H. vulgare (barley)	Ergot	123
	Lolium multiflorum (annual ryegrass)	Ergot	121
	L. perenne (perennial ryegrass)	Ergot	121
	P. quadrifarium (paspalum)	Ergot	124
	Poa pratensis (Kentucky bluegrass)	Ergot	121

Table 2-2 Seedborne Fungal Pathogens That Cause Important Disease of Major Crops[a,b] (continued)

Pathogen	Crop	Disease	Ref.
	Secale cereale (rye)	Ergot	125
	Sorghum bicolor (sorghum)	Ergot	126, 127
	T. aestivum (wheat)	Ergot	128
	Triticale	Ergot	129
C. queenslandica	*Paspalum quadrifarium* (paspalm)	Ergot	130
Colletotrichum capsici	*Capsicum* (pepper)	Anthracnose, ripe rot	131
	Dolichos uniflorus (horse gram)	Anthracnose	132
	Vigna unguiculata (cowpea)	Anthracnose	133
C. circinans	*Allium cepa* (onion)	Smudge, damping-off	134
C. corchori	*Corchorus* (jute)	Anthracnose, damping-off	135
C. curvatum	*Capsicum* (pepper)	Anthracnose, ripe spot	136
	Crotalaria juncea (sunnhemp)	Stem break	137
	Hibiscus cannabinus (kenaf)	Tip blight	138
C. dematium	*Capsicum* (pepper)	Anthracnose	139
C. dematium f. sp. *spinaciae*	*B. vulgaris* (beet)	Anthracnose	140
C. gloeosporioides	*Aeschynomene virginica* (joint vetch)	Anthracnose	141
	G. max (soybean)	Anthracnose	142
	Protea compacta (protea)	Anthracnose	143
C. gossypii	*Gossypium* (cotton)	Anthracnose, pink boll rot, seedling disease	143a
C. g. var. *cephalosporioides*	*Gossypium* (cotton)	Anthracnose	144
C. graminicola	*G. max* (soybean)	Seed rot	142
	S. bicolor (sorghum)	Anthracnose, red leaf	145
	Z. mays (maize)	Anthracnose	146
C. higginsianum	*Raphanus sativus* (radish)	Anthracnose	147
C. indicum	*Gossypium* (cotton)	Anthracnose, boll rot, seedling blight	148
C. lagenarium	*Citrullus lanatus* (watermelon)	Anthracnose	149
	Cucumis melo (cantaloupe)	Anthracnose	150, 151
	C. sativus (cucumber)	Anthracnose	152
C. lindemuthianum	*Phaseolus lunatus* (lima bean)	Anthracnose	153
	P. vulgaris (bean)	Anthracnose	153

Table 2-2 Seedborne Fungal Pathogens That Cause Important Disease of Major Crops[a,b] (continued)

Pathogen	Crop	Disease	Ref.
C. linicola	*L. usitatissiumum* (flax)	Anthracnose, seedling blight, stem canker	154
C. malvarum	*Althaea* (hollyhock)	Anthracnose	155
C. trifolii	*Trifolium* (red clover)	Anthracnose	156
C. truncatum	*Cyamopsis tetragonoloba* (cluster bean, guar)	Anthracnose	25
	G. max (soybean)	Anthracnose	157
	P. lunatus (lima bean)	Anthracnose	158
	Vicia faba (fava bean)	Anthracnose	159
	Vigna mungo (urd bean)	Anthracnose	160
	V. radiata (mung bean)	Anthracnose	160
Coniothyrium fuckelii	*Rosa alba* (white rose)	Anthracnose	161
	R. rugosa (rugose rose)	Anthracnose	161
Corynespora cassiicola	*G. max* (soybean)	Target spot	162
Curvularia lunata	*Gossypium* (cotton)	Seedling blight	163
	Setaria italica (millet)	Ear rot, seedling blight	164
	S. bicolor (sorghum)	Grain mold	165
Cylindrocladium crotalariae	*A. hypogaea* (peanut)	Black mold	166
Diaporthe p. var. *caulivora*	*G. max* (soybean)	Stem canker	167
D. p. f. sp. *meridionalis*	*G. max* (soybean)	Stem canker	168
	P. vulgaris (bean)	Pod and stem blight	168
	V. unguiculata (cowpea)	Pod and stem blight	168
D. p. var. *sojae*	*G. max* (soybean)	Pod and stem blight	167, 169, 170
	P. lunatus (lima bean)	Pod and stem blight	168
	P. vulgaris (bean)	Pod and stem blight	168
	V. unguiculata (cowpea)	Pod and stem blight	168
D. woodi	*Lupinus* (lupin)	Stem blight	171
Didymella bryoniae	*Citrullus lanatus* (watermelon)	Black rot, leaf spot	172
	Cucumis sativus (cucumber)	Black rot, leaf spot	173
	Cucurbita pepo (cantaloupe)	Leaf spot	25
D. lycopersici	*L. esculentum* (tomato)	Stem canker, stem rot	174
Diplodia maydis	*Z. mays* (maize)	Ear rot, seedling blight, stalk rot	175, 176
Drechslera avenae	*A. sativa* (oats)	Leaf spot, seedling blight	177
D. carbonum	*Z. mays* (maize)	Leaf spot	178
D. graminae	*H. vulgare* (barley)	Leaf stripe	179
D. sesami	*S. indicum* (sesami)	Blight, leaf blotch, stem rot	180
D. setariae	*Pennisetum glaucum* (bulrush millet)	Blight	181
D. teres	*H. vulgare* (barley)	Net blotch	182

Table 2-2 Seedborne Fungal Pathogens That Cause Important Disease of Major Crops[a,b] (continued)

Pathogen	Crop	Disease	Ref.
D. tritici-repents	*T. aestivum* (wheat)	Leaf spot, yellow spot	183
Elsinoe phaseoli	*P. lunatus* (lima bean)	Scab	184
	P. vulgaris (bean)	Scab	25
Epichloe typhina	*Festuca* (fescue)	Cat's tail	185
Epicoccum nigrum	*Z. mays* (maize)	Red kernel	186
	O. sativa (rice)	Pink color	187
Erysiphe betae	*B. vulgaris* (beet)	Powdery mildew	188
E. cichoriacearum	*Z. elegans* (zinnia)	Powdery mildew	189
E. pisi	*P. sativum* (pea)	Powdery mildew	190
Fusarium avenaceum	*Lupinus* (lupin)	Seedling blight	191
	Medicago truncatula (clover)	Seedling blight	192
	T. aestivum (wheat)	Seedling blight	193
F. culmorum	*T. aestivum* (wheat)	Seedling blight	193
F. equiseti	*V. unguiculata* (cowpea)	Seedling blight	194
F. graminearum	*A. sativa* (oats)	Head blight, scab	183
	H. vulgare (barley)	Head blight, scab	195
	O. sativa (rice)	Head blight, scab	196
	T. aestivum (wheat)	Head blight, scab	183
	Z. mays (maize)	Head blight, scab	198
F. lateritium f. sp. *cajani*	*Cajanus cajan* (pigeon pea)	Wilt	199
F. moniliforme	*Asparagus officinalis* (asparagus)	Seed rot	200
	O. sativa (rice)	Bakanae disease	197
	S. bicolor (sorghum)	Seed rot	201, 202
	T. aestivum (wheat)	Seed rot	203
F. m. var. *clichotona*	*B. napus* (canola)	Seedling blight	204
F. m. var. *subglutinans*	*A. cepa* (onion)	Seedling blight	205
	P. elliottii (slash pine)	Seedling blight	206
	P. taeda (loblolly pine)	Seedling blight	206
F. oxysporum	*Capsicum* (pepper)	Wilt	207
	G. max (soybean)	Seedling blight	208
	M. sativa (alfalfa)	Root rot	209
	Pseudotsuga menziesii (Douglas fir)	Root rot	210
	V. faba (fava bean)	Root rot	211
F. o. f. sp. *apii*	*Apium graveolens* (celeriac, celery)	Wilt	212
F. o. f. sp. *asparagi*	*Asparagus officinalis* (asparagus)	Root rot	200
F. o. f. sp. *betae*	*B. vulgaris* (fodder beet, red beet, sugarbeet)	Wilt	213
F. o. f. sp. *carthami*	*Carthamus tinctorius* (safflower)	Wilt	214
F. o. f. sp. *callistephi*	*Callistephus chinensis* (China aster)	Wilt	215
F. o. f. sp. *ciceri*	*Cicer arietinum* (chickpea)	Wilt	216
F. o. f. sp. *conglutinans*	*B. oleracea* var. *botrytis* (brussels sprouts)	Wilt	217

Table 2-2 Seedborne Fungal Pathogens That Cause Important Disease of Major Crops[a,b] (continued)

Pathogen	Crop	Disease	Ref.
F. o. f. sp. *crotalariae*	*C. juncea* (sunnhemp)	Wilt	218
F. o. f. sp. *cucumerinum*	*C. sativus* (cucumber)	Wilt	219
F. o. f. sp. *cumini*	*Cuminum cyminum* (cumin)	Wilt	220
F. o. f. sp. *koae*	*Acacia confusa* (acacia)	Wilt	221
	A. koa (koa)	Wilt	221
	A. koaia (acacia)	Wilt	221
F. o. f. sp. *lagenarium*	*Lagenaria siceraria* (bottle gourd)	Wilt	222
F. o. f. sp. *lini*	*L. usitatissimum* (flax)	Wilt	223
F. o. f. sp. *lycopersici*	*L. esculentum* (tomato)	Wilt	224
F. o. f. sp. *niveum*	*Citrullus lanatus* (watermelon)	Wilt	225
F. o. f. sp. *perniciosum*	*Albizia* (mimosa)	Wilt	226
F. o. f. sp. *phaseoli*	*P. vulgaris* (bean)	Wilt	227
F. o. f. sp. *pisi*	*P. sativum* (pea)	Wilt	228
F. o. f. sp. *spinaciae*	*Spinacia oleracae* (spinach)	Wilt	229
F. o. f. sp. *solani*	*G. max* (soybean)	Wilt	230
F. o. f. sp. *tracheiphilum*	*V. unguiculata* (cowpea)	Wilt	231
F. o. f. sp. *vasinfectum*	*Abelmoschus esculentus* (okra)	Wilt	232
	Gossypium (cotton)	Wilt	233
F. solani f. sp. *cucurbitae*	*Cucurbita* (cucurbits)	Foot rot	234
F. s. f. sp. *hibisci*	*Abelmoschus esculentus* (okra)	Wilt	235
F. s. f. sp. *phaseoli*	*P. vulgaris* (bean)	Root rot	236
F. sporotrichioides	*G. max* (soybean)	Root rot	237
F. udum	*Cajanus cajan* (pigeon pea)	Wilt	238
F. u. f. sp. *crotalariae*	*C. juncea* (sunnhemp)	Wilt	239
Gibberella cyanogena	*A. sativa* (oats)	Foot rot	240
G. fujikuroi	*Z. mays* (corn)	Bakanae	241
Gloeocercospora sorghi	*S. bicolor* (sorghum)	Zonate leaf spot	242
Gloeotinia granigena	*Lolium temulentum* (ryegrass)	Blind seed disease	243, 244
Glomerella cingulata	*G. max* (soybean)	Anthracnose	245
	L. esculentum (tomato)	Anthracnose, ripe rot	246
G. c. var. *orbiculare*	*Citrullus lanatus* (watermelon)	Anthracnose	247
Graphium aeruginosum	*Terminalia myriocarpa* (Rangoon creeper)	Seed rot	248
Helminthosporium papaveracea	*Papaver somniferum* (opium poppy)	Leaf stem spot	249
Humicola lanuginosa	*O. sativa* (rice)	Leaf spot	250

Table 2-2 Seedborne Fungal Pathogens That Cause Important Disease of Major Crops[a,b] (continued)

Pathogen	Crop	Disease	Ref.
Hymenella cerealis	*A. sativa* (oats)	Stripe disease	251
	H. vulgare (barley)	Stripe disease	252
	T. aestivum (wheat)	Stripe disease	253, 254
	Z. mays (maize)	Black bundle, kernel rot	255
Kabatiella linidguitii	*L. usitatissimum* (flax)	Brown stem	183
Lasiodiplodia theobromae	*A. hypogaea* (peanut)	Collar rot	256
	Cucurbita (cucubit)	Leaf spot	257
	Dolichos biflorus	Leaf spot	258
	Gossypium (cotton)	Dry rot	259
	Lagenaria siceraria (bottle gourd)	Leaf spot	257
	Pinus caribaea (Caribbean pine)	Leaf spot	260, 261
	Z. mays (maize)	Pre- and postemergence damping-off, kernel rot, stalk rot	118, 262
Leptosphaeria lindguistii	*H. annuus* (sunflower)	Leaf spot	263
Lirula macrospora	*Abies* (fir)	Cone cast	264
Macrophomina phaseolina	*A. hypogaea* (peanut)	Root rot, stem rot	265
	Corchorus (jute)	Charcoal rot	266
	Crotalaria juncea (sunnhemp)	Charcoal rot	267
	Cucurbita (cucurbit)	Charcoal rot	268
	Cucumis melo (melon)	Charcoal rot	268
	G. max (soybean)	Charcoal rot	269
	Gossypium (cotton)	Root rot, stem blight	270
	H. annuus (sunflower)	Charcoal rot	271
	Lagenaria siceraria (bottle gourd)	Charcoal rot	268
	P. vulgaris (bean)	Ashy stem blight, charcoal rot	272
	S. indicum (sesame)	Charcoal rot	273
	V. mungo (urd bean)	Charcoal rot	274
	V. unguiculata (cowpea)	Charcoal rot	275
Magnaporthe grisea	*Digitaria sanguinalis* (hairy crabgrass)	Blast	276
Marssonnia brunnea	*Populus* (poplar)	Leaf spot	277
Melampsora lini	*L. usitatissimum* (flax)	Rust	278
Microdochium nivale	*A. sativa* (oats)	Snow mold	183
	H. vulgare (barley)	Snow mold	183
	Secale cereale (rye)	Snow mold	279
	Triticale (triticale)	Snow mold	280
	T. aestivum (wheat)	Snow mold	281
M. oryzae	*O. sativa* (rice)	Snow mold	282
Maeziomyces bullatus	*Pennisetum glaucum* (pearl millet)	Smut	283, 284
Mycosphaerella brassicicola	*Brassica* (crucifers)	Black ringspot	285
M. linicola	*L. usitatissimum* (flax)	Blotch	286

Table 2-2 Seedborne Fungal Pathogens That Cause Important Disease of Major Crops[a,b] (continued)

Pathogen	Crop	Disease	Ref.
M. sesami	*S. indicum* (sesame)	Leaf spot	287
Nematospora coryli	*G. max* (soybean)	Yeast spot	288
	Gossypium (cotton)	Boll rot	183
Nigrospora oxyzae	*Z. mays* (maize)	Cob and stalk rot	289
Penicillium oxalicum	*Z. mays* (maize)	Seedling blight	290
Periconia circinata	*S. bicolor* (sorghum)	Crown rot, root rot	183
Peronospora arborescens	*Meconopsis* (Chinese poppy)	Downey mildew	291
	Papaver (poppy)	Downy mildew	292
P. arthuri	*Clarkia elegans* (clarkia)	Downy mildew	25
P. destructor	*Allium cepa* (onion)	Downy mildew	293, 294
P. ducometi	*Fagopyrum esculentum* (buckwheat)	Downy mildew	295
P. farinosa farinosa f. sp. *betae*	*B. vulgaris* (red beet, sugar beet)	Downy mildew	296
P. f. sp. *spinaciae*	*Spinacia oleracea* (spinach)	Downy mildew	297
P. manshurica	*G. max* (soybean)	Downy mildew	298
P. parasitica	*Brassica* (crucifers)	Downy mildew	299
	Raphanus sativus (radish)	Downy mildew	300
P. tabacina	*N. tabacum* (tobacco)	Blue mold, downy mildew	301
P. valerianellae	*Valerianella locusta* (corn salad)	Downy mildew	302
P. viciae	*P. sativum* (pea)	Downy mildew	303
Peronosclerospora sacchari	*Z. mays* (maize)	Downy mildew	304
P. sorghi	*S. bicolor* (sorghum)	Downy mildew	305
	Z. mays (maize)	Downy mildew	306
Phaeoisariopsis griseola	*P. vulgaris* (bean)	Angular leaf spot	307, 308
Phoma apiicola	*Apium graveolens* (celery)	Black neck, root rot, scab, seedling canker	309
P. betae	*B. vulgaris* (beet)	Black leg, damping-off	310
	B. v. var. *cicla* (Swiss chard)	Leaf spot	311
	B. v. var. *rapacea* (beet)	Leaf spot	311
P. exigua	*Cyclamen persicum* (cyclamen)	Seed rot	312
	P. lunatus (lima bean)	Leaf spot	25, 313
P. exigua var. *diversispora*	*P. vulgaris* (bean)	Black leg, damping-off	314
P. lingam	*Brassica oleracea* (cabbage)	Black leg	315
P. macdonaldii	*H. annuus* (sunflower)	Black spot	316
P. medicaginis	*M. sativa* (alfalfa)	Leaf spot, black stem	317
P. m. var. *sojae*	*G. max* (soybean)	Leaf spot	318
P. pinodella	*C. arietinum* (chickpea)	Collar rot, foot rot, blight	50, 319
P. rabiei	*Pisum sativum* (pea)	Collar rot, foot rot	320

Table 2-2 Seedborne Fungal Pathogens That Cause Important Disease of Major Crops[a,b] (continued)

Pathogen	Crop	Disease	Ref.
	T. alexandrinum (berseem, clover)	Leaf spot	321
P. rostrupii	*Daucus carota* (carrot)	Root rot	322
P. sorghina	*O. sativa* (rice)	Seedling blight	323
	Sorghum bicolor (sorghum)	Grain spot, red leaf	324
Phomopsis longicolla	*G. max* (soybean)	Seed decay	325, 326, 327
P. phaseoli	*G. max* (soybean)	Pod and stem blight	170, 328
P. vexans	*Solanum melongena* (eggplant)	Fruit rot	329
Phyllosticta antirrhini	*A. majus* (snapdragon)	Leaf spot, stem rot	330
Phytophthora capsici	*C. annuum* (pepper)	Blight, stem and fruit blight	331
P. cinnamomi	*Persea americana* (avocado)	Root rot, seedling blight	25
P. citrophthora	*Citrus* (citrus)	Foot rot, gummosis	332
P. infestans	*L. esculentum* (tomato)	Fruit rot, late blight	333
P. nicotianae var. *parasitica*	*Capsicum* (pepper)	Blight	334
	Citrus (citrus)	Gummosis, fruit rot	25
	L. esculentum (tomato)	Black eyespot	335
	S. indicum (sesame)	Blight	336
P. sojae	*G. max* (soybean)	Root and stem rot	337
P. phaseoli	*P. lunatus* (lima bean)	Downy mildew	338
Plasmodiophora brassicae	*Brassica* (crucifers)	Clubroot	339
Plasmopara halstedii	*H. annuus* (sunflower)	Downy mildew	340
P. nivea	*Impatiens balsamina* (balsam)	Downy mildew	341
P. obducens	*Petroselinum crispum* (parsley)	Downy mildew	342
Pleiochaeta setosa	*Lupinus albus* (lupin)	Brown spot	342a
Pleospora herbarum	*Allium* (garlic, onion)	Black stalk rot, leaf spot, leaf mold	183
	M. sativa (alfalfa)	Ringspot	343
Protomyces macrosporus	*Coriandrum sativum* (coriander)	Stem gall	344
Pseudocercosporella capsellae	*Brassica* (crucifer)	White leaf spot	345
P. trichachnicola	*Trichachne insularis* (trichachne)	Leaf spot	346
Puccinia antirrhini	*A. majus* (snapdragon)	Rust	215
P. arachidis	*A. hypogaea* (peanut)	Rust	347
P. calatropae var. *centaureae*	*Carthamus tinctorius* (safflower)	Rust	348
P. malvacearum	*Althaea* (hollyhock)	Rust	215
Pyricularia grisea	*Eleusine coracana* (finger millet)	Blast	349
P. oryzae	*O. sativa* (rice)	Blast, rotten neck	350
	T. aestivum (wheat)	Blast	351

Table 2-2 Seedborne Fungal Pathogens That Cause Important Disease of Major Crops[a,b] (continued)

Pathogen	Crop	Disease	Ref.
Pythium aphanidermatum	*Cucurbita pepo* (squash)	Damping-off	352
	L. esculentum (tomato)	Damping-off	353
P. debaryanum	*B. vulgaris* (beet, sugar beet)	Damping-off	354
Rhizoctonia leguminicola	*G. max* (soybean)	Black patch	355
	Trifolium (clover)	Black patch	356
R. solani	*A. hypogaea* (peanut)	Seedling blight	357
	Araucaria angustifolia (Brazilian pine)	Damping-off	358
	A. cunninghamii (pine)	Damping-off	358
	A. heterophylla (Norfolk island pine)	Damping-off	358
	C. frutescens (pepper)	Damping-off	359
	G. max (soybean)	Web blight	360
	Gossypium (cotton)	Damping-off, seedling blight	361
	L. usitatissimum (flax)	Seedling blight, wilt	362
	L. esculentum (tomato)	Seedling blight	359
	O. sativa (rice)	Sheath blight	363
	P. lunatus (lima bean)	Damping-off, stem canker	364
	P. vulgaris (bean)	Damping-off, stem canker	364
	S. melongena (eggplant)	Seedling blight	359
	V. unguiculata (cowpea)	Blight	365
Rhynchosporium secalis	*H. vulgare* (barley)	Scald	366
Sarocladium oxyzae	*O. sativa* (rice)	Sheath rot	367
Schizophyllum commune	*E. quipeensis* (oil palm)	Blight	368
Sclerophthora macrospora	*A. sativa* (oats)	Downy mildew	369
	Eleusine coracana (finger millet)	Downy mildew	370
	T. aestivum (wheat)	Downy mildew	371
Sclerospora graminicola	*Pennisetum glaucum* (pearl millet)	Downy mildew, green ear	372
Sclerotinia minor	*A. hypogaea* (peanut)	Root rot	373
S. sclerotiorum	*Brassica* (crucifers)	Watery soft rot, white blight	374
	C. tinctorius (safflower)	Blight	375
	D. carota (carrot)	Wilt	376
	G. max (soybean)	Sclerotial wilt	183
	H. annuus (sunflower)	White rot, wilt, stem rot	183
	H. vulgare (barley)	Blight	377
	L. usitatissimum (flax)	Blight	378
	P. lunatus (lima bean)	Sclerotial wilt, stem rot	183
	P. vulgaris (bean)	Sclerotial wilt	379
	P. sativum (pea)	Stem rot	376
	S. bicolor (sorghum)	Blight	377
	Trifolium (clover)	Clover rot	380

Table 2-2 Seedborne Fungal Pathogens That Cause Important Disease of Major Crops[a,b] (continued)

Pathogen	Crop	Disease	Ref.
	T. aestivum (wheat)	Blight	377
S. trifoliorum	*Astragalus sinicus* (astragalus)	Blight	381
Sclerotium cepivorum	*Allium* (Chives, garlic, onion)	White rot	382
S. rolfsii	*A. hypogaea* (peanut)	Sclerotial rot, wilt	383
	H. annuus (sunflower)	Blight	384
	L. sativa (lettuce)	Sclerotial rot, wilt	385
	P. vulgaris (bean)	Sclerotial rot, wilt	386
	T. aestivum (wheat)	Sclerotial rot, wilt	387
Septoria apiicola	*Apium graveolens* (celery)	Late blight	388
S. glycines	*G. max* (soybean)	Leaf spot	389
S. helianthi	*H. annuus* (sunflower)	Blight	390
S. lactucae	*L. sativa* (lettuce)	Leaf spot	391
S. linicola	*L. usitatissimum* (flax)	Blotch	286
S. trifoliorum	*Trifolium* (clover)	Clover rot	392
S. tritici	*T. aestivum* (wheat)	Speckled leaf spot	393
Sirococcus strobilinus	*Picea* (spruce)	Blight	394
Sphacelotheca cordobensis	*Paspalum* (Dallis grass)	Smut	395
Sporisorium cruentum	*S. bicolor* (sorghum)	Loose smut	396
S. destruens	*Panicum miliaceum* (millet)	Head smut	397, 398
	Setaria italica (millet)	Head smut	399
S. sorghi	*S. bicolor* (sorghum)	Covered smut, grain smut	400
Stagonospora avenae	*A. sativa* (oats)	Leaf blotch, stem breaking	401
S. nodorum	*A. sativa* (oats)	Glume blotch	402
	H. vulgare (barley)	Glume blotch	403
	T. aestivum (wheat)	Glume blotch	404
Tilletia ayresii	*Panicum coloratum* (millet)	Bunt	405
T. barclayana	*O. sativa* (rice)	Bunt	406
T. caries	*T. aestivum* (wheat)	Rough-spored bunt	407
T. controversa	*Aegilops* (bentgrass)	Dwarf bunt	408
	Agropyron intermedium (wheatgrass)	Dwarf bunt	409
	Arrhenatherum (oatgrass)	Dwarf bunt	410
	S. cereale (rye)	Dwarf bunt	411
	T. aestivum (wheat)	Dwarf bunt	412
T. fusca	*Festuca* (fescue)	Kernel bunt	395
T. indica	*S. cereale* (rye)	Karnal bunt	413
	T. aestivum (wheat)	Karnal bunt	414a
	Triticale (triticale)	Karnal bunt	415
T. laevis	*S. cereale* (rye)	Bunt	414
	T. aestivum (wheat)	Smooth-spored bunt	415
T. rugispora	*Paspalum* (dallis grass)	Bunt	395

Table 2-2 Seedborne Fungal Pathogens That Cause Important Disease of Major Crops[a,b] (continued)

Pathogen	Crop	Disease	Ref.
T. texana	*F. octoflora* (fescue)	Bunt	395
T. tritici	*A. cristatum* (crested wheatgrass)	Bunt	416
	A. subsecundum (bearded wheatgrass)	Bunt	416
	A. trachycaulum (slender wheatgrass)	Bunt	416
	Festuca (fescue)	Bunt	395
	S. cereale (rye)	Bunt	407
	T. aestivum (wheat)	Bunt	407
Trichothecium roseum	*Z. mays* (maize)	Seed rot	417
Urocystis agropyri	*T. aestivum* (wheat)	Flag smut	418
U. occulta	*S. cereale* (rye)	Stalk smut, stem smut	419
Uromyces betae	*B. vulgaris* (red beet, sugar beet)	Rust	420
U. viciae fabae	Lens culinaris (lentil)	Rust	421
Ustilago bullata	*Agropyron* (quack grass)	Head smut, loose smut	395
	Bromus catharticus (rescue grass)	Head smut	422
	F. idahoensis (bluebunch fescue)	Head smut	395
	F. viridula (green-leaf fescue)	Head smut	395
U. coicis	*Coix lacrymajobi* (Job's tears)	Smut	423
U. crameri	*P. mileaceum* (common millet)	Kernel smut	424
	S. italica (fox tail millet)	Smut	425
U. heterogen	*P. dichotomiflorum* (fall panicgrass)	Smut	395
U. mulfordiana	*Festuca* (fescue)	Head smut	395
U. segetum var. *avenae*	*Arrhenatherum elatius* (tall oatgrass)	Loose smut	426
	A. sativa (oats)	Loose smut	427
	H. vulgare (barley)	Black smut, false loose smut	428
U. segetum var. *segetum*	*A. sativa* (oats)	Covered smut	429
	H. vulgare (barley)	Covered smut	430
U. segetum var. *triciti*	*Agropyron* (grass)	Loose smut	395
	H. vulgare (barley)	Loose smut	431
	S. cereale (rye)	Loose smut	432
	Triticale	Loose smut	433
	T. aestivum (wheat)	Loose smut	434
U. sphaerocarpa	*F. amplissima* (fescue)	Head smut	395
U. zeae	*Z. mays* (maize)	Blunt smut, loose smut	435
Verticillium albo-atrum	*C. tinctorius* (safflower)	Wilt	436
	Gossypium (cotton)	Wilt	437
	H. annuus (sunflower)	Wilt	438

Table 2-2 Seedborne Fungal Pathogens That Cause Important Disease of Major Crops[a,b] (continued)

Pathogen	Crop	Disease	Ref.
	M. sativa (alfalfa)	Wilt	439
	S. melongena (eggplant)	Wilt	440
V. dahliae	*C. tinctorius* (safflower)	Wilt	441
	H. annuus (sunflower)	Wilt	442
	L. usitatissimum (flax)	Wilt	183

Table 2-3 Classification of Higher Ranks of Procaryotes and Affiliation of Plant Pathogenic Bacteria

KINGDOM: Procaryotae

DIVISION: **Gracilicutes**
- CLASS: **Proteobacteria** — nonphotosynthetic
  - FAMILY: **Enterobacteriaceae**
    - Genus: *Erwinia*
  - FAMILY: **Pseudomonadaceae**
    - Genus: *Acidovorax*, *Pseudomonas*, *Rhizobacter*, *Rhizomonas*, *Xanthomonas*, *Xylophilus*
  - FAMILY: **Rhlzobaceae**
    - Genus: *Agrobacterium*
  - FAMILY: **Undefined**
    - Genus: *Xylella*
- CLASS: **Oxyphotobacteria** — photosynthesis with O_2 generation

DIVISION: **Firmicutes**
- CLASS: **Firmibacteria** — simple, Gram +
  - Genus: *Bacillus*, *Clostridium*
- CLASS: **Thallobacteria** — branching Gram +
  - Genus: *Arthrobacter*, *Clavibacter*, *Curtobacterium*, *Rhodococcus*, *Streptomyces*

DIVISION: **Tenericutes**
- CLASS: **Mollicutes** — wall-less
  - FAMILY: **Spiroplasmataceae**
    - Genus: *Spiroplasma*

DIVISION: **Mendosicutes**
- CLASS: **Archaeobacteria** — unusual walls

From Goto, M., *Fundamentals of Plant Pathology,* Academic Press, New York, 1992. With permission.

bacterial pathogen is a strain or set of strains with the same or similar characteristics, differentiated at the intrasubspecific level from other strains of the same species or subspecies on the basis of distinctive pathogenicity to one or more plant hosts. Classification of a taxon as a pathovar does not exclude recognition of differences in biochemical, serological, or other nonpathogenic characteristics between that and other pathovars of the same species or subspecies, but implies that other differences at the intrasubspecific level have less taxonomic significance in comparison to differences in pathogenicity. Usually pathovars are distinguished in terms of proved differences in host range. However, clear differences in symptomatology on the same plant species (e.g., *X. oryzae* pv. *oryzae* and *X. oryzae* pv. *oryzicola*) can warrant different pathovar designations.[444,445] This definition allows further division of pathogenic species or pathovars into races. A bacterial race is a collection of strains that differ from others within a species or pathovar in their host specialization to cultivar or other germ plasm. Races are identified using host differentials, which may be cultivars or other identifiable germ plasm. Races have no nomenclatural standing. A race should be designated with letters or numbers. The term *forma specialis* is not a substitute for pathovar or race for phytopathogenic bacteria. It is applied only to pathogens that are specific for a particular host.[434]

Bacterial taxonomy comprises three distinct activities: (1) classification or grouping, which is the arrangement of strains into natural groups (taxa); (2) nomenclature, or naming, which involves the allocation of names to circumscribed groups, using the International Code of Nomenclature of Bacteria (the code) for taxa at the level of subspecies and above, and by the allocation of names or identifying terms to groups at intrasubspecifc levels; and (3) identification or diagnosis, which involves the process by which unidentified isolates are referred to known taxa.[444] A large number of bacteria occur in nature, with about 200 known to cause plant diseases. The majority of plant-pathogenic bacteria are rod shaped, either motile and Gram negative (*Acidovorax*, *Agrobacterium*, *Erwinia*, *Pseudomonas*, and *Xanthomonas*) or Gram positive (*Bacillus*, *Clavibacter*, *Curtobacterium*, *Rhodococcus*, and *Streptomyces*). Bacteria cause symptoms on all plant parts, including seeds. However, many seedborne bacteria do not cause macroscopic symptoms. Bacteria are differentiated by cultural characteristics, morphology, growth rate, and carbohydrate metabolism.[2] Bacteria are found in air, seeds, soil, and water and in or on all plants and animals, including man. They vary in characteristics and adaptability and commonly exist in mixed populations.

Most plant-pathogenic bacteria do not form spores. *Bacillus* is an exception. Many species have one to several flagella. Common types divide by binary fission. In a warm, moist environment, large numbers of new cells can be produced within a few hours.

Plant-pathogenic bacteria are disseminated by seeds or animals (including humans) and are transported in infected plant materials by cultural practices, wind-blown sand, soil and water, flowing water; or insects, mites, and nematodes. Bacteria enter seed and plant tissue through wounds or natural openings. Water-soaked tissues often predispose plant tissue to invasion. Free moisture and moderate to warm temperatures generally are required for bacterial invasion, colonization, and disease development.[2]

When conditions are unfavorable for growth and multiplication, bacteria remain dormant in or on seeds, living or dead plant tissue, soil equipment, implements, tools, etc., or in or on bodies of insects and other animals. Most plant-pathogenic species die after 10 min at 50°C, in dry conditions, or in sunlight.

Transmission through seeds is an important means of bacterial survival and in establishing epidemics.

Seedborne bacterial plant pathogens known to cause important plant diseases are presented in Table 2-4.

III. MYCOPLASMALIKE ORGANISMS

A number of plant diseases once thought to be caused by plant viruses now are associated with mycoplasmalike organisms (MLO). Doi et al.[571] first found MLOs in 1967 in the ultrathin sections of phloem sieve tubes of plants with symptoms of aster yellows and mulberry dwarf. MLOs lack a cell wall, are bounded by a triple-layered unit membrane, and have cytoplasmic ribosomes and strands of nuclear material. Their shape usually is spheroidal to ovoid or irregularly tubular to filamentous. They generally are present in the sap of phloem sieve tubes. Most plant MLOs are transmitted by psyllids and plant hoppers. They are sensitive to antibiotics, particularly of the tetracycline group.[2] They are not visible by bright-field microscopy. MLO cells are 150 to 300 nm in diameter with a bilayer membrane, and contain ribosomes and DNA. They replicate by binary fission, and some that infect vertebrates can be grown *in vitro*.[571] None of the plant diseases caused by MLO are known to be seed transmitted.

IV. FASTIDIOUS VASCULAR BACTERIA

Fastidious vascular bacteria (FVB) are parasitic bacteria that cannot be grown on simple culture media in the absence of host cells.[2] Windsor and Black[572] first reported such organisms in phloem of clover plants with club-leaf disease. FVB are transmitted by leafhoppers and are present exclusively or primarily in the phloem or xylem of plants.[2] Fastidious vascular bacteria generally are rod-shaped, 0.2 to 0.5 μm in diameter by 1 to 4 μm in length and bounded by a membrane and cell wall. However, phloem-inhabiting FVB have a cell wall that appears more as a secondary membrane than as a cell wall. They have no flagella. Several such xylem-limited bacteria have been placed in the genus *Xylella*.[2] None of the FVB that cause plant disease are seed transmitted.

V. SPIROPLASMAS

Spiroplasmas are helical mycoplasmas. They are pleomorphic cells that vary in shape from spherical or slightly ovoid, 100 to 250 nm in diameter, to helical

Table 2-4 Seedborne Bacterial Pathogens that Cause Important Disease of Major Crops[a]

Bacterium	Crop	Disease	Ref.
Acidovorax avenae subsp. *citrulli*	*Citrullus lanatus* (watermelon)	Fruit blotch	446
Agrobacterium tumefaciens	*Humulus lupulus* (hop)	Crown gall	447
	Prunus (cherry, peach, plum)	Crown gall	448
Bacillus macerans	*L. usitatissimum* (flax)	Rot	449
B. megaterium pv. *cerealis*	*T. aestivum* (wheat)	White blotch	450
B. mesentericus	*Tritifolium pratense* (clover)	Seed rot	451
B. subtilis	*C. annuum* (pepper)	Seed rot	452
	Cicer arietinum (chickpea)	Seed rot	453
	G. max (soybean)	Seed rot	454
Clavibacter iranicum	*T. aestivum* (wheat)	Yellow slime	455
C. rathayi	*Dactylis glomerata* (orchard grass)	Yellow slime	456
C. tritici	*T. aestivum* (wheat)	Tundu, yellow ear rot, yellow slime	457
C. michiganensis subsp. *insidiosus*	*Trifolium pratense* (red clover)	Wilt	458
C. michiganensis subsp. *michiganensis*	*L. esculentum* (tomato)	Canker	459
C. michiganensis subsp. *nebraskensis*	*Z. mays* (maize)	Nebraska leaf freckle, wilt	460
C. michiganensis subsp. *sepedonicus*	*L. esculentum* (tomato)	Ring rot	461
C. michiganensis subsp. *tessellarius*	*T. aestivum* (wheat)	Bacterial mosaic	462
Curtobacterium flaccumfaciens subsp. *betae*	*B. vulgaris* (beet)	Silvering	463
C. f. subsp. *flaccumfaciens*	*G. max* (soybean)	Wilt	464
	P. vulgaris (bean)	Wilt	465–467
	Z. glabra (zornia)	Seed rot	468
Erwinia carotovora subsp. *atroseptica*	*Delphinium ajacis* (larkspur)	Black leg, soft rot	469
	V. faba (fava bean)	Soft rot	470
E. c. subsp. *carotovora*	*Apium graveolens* (celery)	Crater rot, soft rot	471
	M. sativa (alfalfa)	Soft rot	451
	N. tabacum (tobacco)	Hollow stalk	472
E. c. pv. *zeae*	*Z. mays* (maize)	Soft rot	473
E. herbicola	*O. sativa* (rice)	Palea browning	474
E. rhapontici	*P. sativum* (pea)	Pink seed	475
	T. aestivum (wheat)	Pink seed	476
E. stewartii	*Z. mays* (maize)	Leaf blight, Stewart's wilt	477
Pseudomonas avenae	*O. sativa* (rice)	Stripe	478
P. cichorii	*Lactuca sativa* (lettuce)	Leaf blight	479
P. corrugata	*L. esculentum* (tomato)	Blight	480

[a] Does not necessarily refer to first report.

Table 2-4 Seedborne Bacterial Pathogens that Cause Important Disease of Major Crops[a] (continued)

Bacterium	Crop	Disease	Ref.
P. fabae	*Vicia faba* (fava bean)	Blight	481
P. fluorescens	*Onobrychis viciifolia* (sainfoin)	Seed rot	449
P. fuscovaginae	*O. sativa* (rice)	Sheath rot	482
	T. aestivum (wheat)	Sheath rot	483
P. glumae	*O. sativa* (rice)	Grain rot	484, 485
P. plantarii	*O. sativa* (rice)	Seedling blight	486
P. rubrilineans	*Z. mays* (maize)	Leaf stripe	487
Burkholderia solanacearum	A. hypogaea (peanut)	Wilt	488
	Capsicum frutescens (pepper)	Brown rot	489
	G. max (soybean)	Brown rot	490
	L. esculentum (tomato)	Brown rot	489
	M. sativa (alfalfa)	Brown rot	449
P. syringae pv. *antirrhini*	*A. majus* (snapdragon)	Leaf spot	491
P. s. pv. *apii*	*A. graveolens* (celeriac, celery)	Blight	492
P. s. pv. *aptata*	*B. vulgaris* (sugar beet)	Blight, spot	493
P. s. pv. *atrofaciens*	*H. vulgare* (barley)	Basal glume rot	494
	T. aestivum (wheat)	Basal glume rot	495
P. s. pv. *atropurpurea*	*Agropyron repens* (brome grass)	Halo blight	496
P. s. pv. *coronafaciens*	*A. sativa* (oats)	Halo blight	497
P. s. pv. *glycinea*	*G. max* (soybean)	Blight	498
P. s. pv. *hibisci*	*A. esculentus* (okra)	Leaf spot, blight	499
P. s. pv. *japonica*	*H. vulgare* (barley)	Black node	500
	T. aestivum (wheat)	Black node	500
P. s. pv. *lachrymans*	*Cucumis sativus* (cucumber)	Angular leaf spot	501
P. s. pv. *lapsa*	*Z. mays* (maize)	Stalk rot, Udbatta	502
P. s. pv. *maculicola*	*Brassica* (crucifers)	Leaf spot	503
P. s. pv. *mellea*	*N. tabacum* (tobacco)	Wisconsin leaf spot	471
P. s. pv. *phaseolicola*	*P. vulgaris* (bean)	Halo blight	504
P. s. pv. *pisi*	*P. sativum* (pea)	Blight	505
P. pseudoalcaligenes subsp. *citrulli*	*Citrullus lanatus* (watermelon)	Seedling blight	506
P. s. pv. *sesami*	*Sesamum indicum* (sesame)	Leaf spot	507
P. s. pv. *striafaciens*	*A. sativa* (oats)	Stripe	497
P. s. pv. *syringae*	*Abelmoschus esculentus* (okra)	Blight	508
	Hordeum vulgare (barley)	Kernel spot	509
	O. sativa (rice)	Spot	510
	Phaseolus lunatus (lima bean)	Brown spot	511, 512
	P. sativum (pea)	Brown spot	513
	P. vulgaris (bean)	Brown spot	514
	Sorghum bicolor (sorghum)	Leaf spot	515, 516
	T. aestivum (wheat)	Leaf spot	517, 518
	V. radiata (mungbean)	Leaf spot	519
	V. unguiculata (cowpea)	Leaf spot	520
	Zea mays (maize)	Bacterial spot	521
P. s. pv. *tabaci*	*G. max* (soybean)	Wildfire	498
P. s. pv. *tagetis*	*H. annuus* (sunflower)	Blight	522

Table 2-4 Seedborne Bacterial Pathogens that Cause Important Disease of Major Crops[a] (continued)

Bacterium	Crop	Disease	Ref.
	T. erecta (marigold)	Leaf spot	523
	N. tabacum (tobacco)	Wildfire	524
P. s. pv. *tomato*	*L. esculentum* (tomato)	Leaf spot, speck	525
P. viridiflava	*Pastinaca sativa* (parsnip)	Blight	526
	R. sativus (radish)	Seedling blight	527
Rhodococcus fascians	*Lathyrus odoratus* (sweet pea)	Fasciation, witches' broom	528
	N. tabacum (tobacco)	Fasciation	529
	T. majus (nasturtium)	Fasciation	530
Xanthomonas axonopodis	*Axonopus affinis* (carpet grass, imperial grass)	Gummosis	531
X. campestris pv. *argemonae*	*Argemone mexicana* (Mexican prickly poppy)	Blight	532
X. c. pv. *campestris*	*Brassica* (crucifers)	Black rot	533
	Cheiranthus cheiri (wallflower)	Black rot	534
	Crambe abyssinica (crambe)	Black rot	535
	Matthiola incana (garden stock)	Black rot	536
X. c. pv. *cajani*	*C. cajan* (pigeon pea)	Leaf blight	537
X. c. pv. *carotae*	*Coriandrum sativum* (coriander)	Blight	538
	Daucus carota (carrot)	Blight	539
X. c. pv. *cassavae*	*Manihot esculenta* (cassava)	Leak spot	540
X. c. pv. *citri*	*Citrus* (citrus)	Citrus canker	541
X. c. pv. *cucurbitae*	*Cucurbita maxima* (winter squash)	Leaf spot	542
X. c. pv. *cyamopsidis*	*Cyamopsis tetragonoloba* (guar)	Leaf spot	543
X. c. pv. *glycines*	*G. max* (soybean)	Pustule	498
X. c. pv. *holcicola*	*S. bicolor* (sorghum)	Leaf streak	544
X. c. pv. *incanae*	*M. incana* (garden stock)	Blight, seedling wilt	545
X. c. pv. *juglandis*	*Juglans regia* (English walnut)	Blight	546
X. c. pv. *lespedezae*	*Lespedeza* (lespedeza)	Wilt	547
X. c. pv. *malvacearum*	*Gossypium* (cotton)	Angular leaf spot, blackarm	548–550
X. c. pv. *manihotis*	*Manihot esculenta* (cassava)	Blight	551
X. c. pv. *papavericola*	*Papaver somniferum* (opium poppy)	Blight	118, 552
X. c. pv. *phaseoli*	*Lablab niger* (hyacinth bean, Indian bean)	Blight	553
	Phaseolus acutifolius (tepary bean)	Common blight	554
	P. vulgaris (bean)	Leaf spot	555
X. c. pv. *pruni*	*Prunus* (plum)	Leaf spot	556
X. c. pv. *raphani*	*R. sativus* (radish)	Leaf spot	557
X. c. pv. *ricini*	*Ricinus communis* (castor bean)	Leaf spot	558
X. c. pv. *sesami*	*Sesamum indicum* (sesame)	Blight, leaf spot	559, 560
X. c. pv. *translucens*	*H. vulgare* (barley)	Black chaff, leaf blight	561
	T. aestivum (wheat)	Black chaff, leaf blight	561

Table 2-4 Seedborne Bacterial Pathogens that Cause Important Disease of Major Crops[a] (continued)

Bacterium	Crop	Disease	Ref.
X. c. pv. *vesicatoria*	*C. annuum* (pepper)	Seedling blight, spot of fruit, stem, and leaf	562
	L. esculentum (tomato)	Bacterial spot, black spot	563
X. c. pv. *vignicola*	*V. unguiculata* (cowpea)	Spot, blight	564
X. c. pv. *vitians*	*Lactuca sativa* (lettuce)	Spot	565
X. c. pv. *undulosa*	*S. cereale* (rye)	Leaf spot	566
	Triticale (triticale)	Black chaff	567
	T. aestivum (wheat)	Black chaff	567
X. c. pv. *zinniae*	*Zinnia elegans* (zinnia)	Angular spot	568
X. oryzae pv. *oryzae*	*O. sativa* (rice)	Blight	569
X. o. pv. *oryzicola*	*O. sativa* (rice)	Leaf streak	570

and branched nonhelical filaments that are about 120 nm in diameter and 2 to 4 μm long during active growth and considerably longer (up to 15 μm) in later growth stages. Unlike the MLOs, they can be cultured on nutrient media. They lack a true cell wall and are bounded by a single, triple-layered unit membrane. They multiply by fission. They are resistant to penicillin but inhibited by amphotericin, erythromycin, neomycin, and tetracycline.[2]

Spiroplasmas were described first by Davis et al.[573] associated with corn stunt disease. In 1973, Davis and Worley[574] named the causal agents "spiroplasmas" based on their structure and mobility, which were similar to MLOs in their ability to form "fried-egg" colonies on solid media and to pass through filters with 0.22 μm micropores, their resistance to penicillin, the absence of typical cell walls, and their failure to revert to walled bacterial forms. Spiroplasmas are differentiated from MLOs by morphology and motility.[573] The citrus stubborn and barley stunt pathogens may be seed transmitted. The barley stunt agent is a spiroplasma that can be disseminated by seeds. The causal organism is spherical (47.1 to 70.6 nm in diameter), spiral (612 to 800 × 24 nm) or acinus-like (235 to 353 × 23 to 27 nm), and Gram positive. These helicoids are distributed in the leaves, roots, and stems of diseased plants and in soil.[574]

VI. VIRUSES

A virus is a set of one or more nucleic acid template molecules, normally encased in a protective coat or coats of protein or lipoprotein, that organizes its replication only within suitable host cells. Within such cells, virus replication is (1) dependent on the host's protein-synthesizing machinery, (2) organized from pools of required materials rather than by binary fission, (3) located at sites that are not separated from the host cell contents by a lipoprotein bilayer membrane, and (4) continually giving rise to variants through various kinds of change in the viral nucleic acid.[575] Most plant viruses have ssRNA genomes enclosed in either a tube-shaped or an isometric shell made up of many small protein molecules. Within these small geometric units is usually one kind of protein molecule, but some have two. Some viruses have double-stranded RNA genomes while others have single-stranded or double-stranded DNA. A few plant viruses have an outer envelope of lipoprotein. Some of the larger viruses contain several viral-coded proteins, including enzymes involved in nucleic acid synthesis. The major components of all viruses are nucleic acid and protein.[575] A virus is defined as a transmissible parasite whose nuclei acid genome weighs less than 3×10^8 Da and needs ribosomes and other components of its host cells for multiplication.[576]

Classification of biological entities into taxonomic categories (taxa) is based on similarities and/or relationships, whereas nomenclature is the assignment of names to taxa according to international rules.[577] To apply these concepts to viruses, the virology division of the International Union of Microbiological Societies established the International Committee on Taxonomy of Viruses (ICTV). This committee operates through subcommittees and specialized study groups.[577] The objectives of ICTV are to (1) develop an internationally agreed taxonomy for viruses, (2) establish internationally agreed names for taxonomic groups of viruses, and (3) communicate latest results on the classification and nomenclature of viruses to virologists, by holding meetings and publishing results. The three existing families and 32 groups of plant viruses as they appeared in the Fifth ICTV report are presented in Table 2-5.[577,578]

To identify an unknown virus, its characteristics are determined and then matched with information about previously described viruses. The identification process may follow one of two strategies, the pragmatic or the logical, although in practice a mixture normally is adopted. The pragmatic approach depends on the virus having a definite host range that may be confined to one or a small number of plant families and also depends on the assumption that plants usually are easier to identify than viruses. The logical or taxonomic approach involves determining to which group a virus belongs by some of its group-specific characteristics, such as shape and size. Then more specific tests, including host range, are used to determine whether the unknown virus already has been described.

Plant disease symptoms caused by plant viruses range from asymptomatic (latent infections) to plant death. Plant viruses may induce a wide variety of symptoms that can be confused with those produced by other parasitic agents or nutritional imbalances, including chlorosis and other discoloration, stunting, wilting, mosaic patterns, and necrosis.[579]

Viruses are identified by symptomology, host specificity, particle morphology, mode(s) of transmission, and biochemical, physical, and serological properties. The types of symptoms on indicator plants aid in identification. Plant viruses are grouped taxonomically by type of nucleic acid, size and shape, mode(s) of transmission, and serological and antigen properties.

Plant viruses are transmitted to plants by pollination; through wounds created by man, animals (primarily anthropods and nematodes), or fungi, by parasitic plants, by mechanical inoculation, and by deliberate or inadvertent human activities, such as planting infected seeds and using infected propagative plant materials.[579]

Table 2-5 The Current 3 Families and 32 Groups of Plant viruses According to the Fifth Report of the International Committee on Taxonomy of Viruses

Classification	Genome properties[a]	Number of viruses[b]
Families		
Rhabdoviridae	(–)ssRNA (1)	
Subgroup A		9
Subgroup B		4
Subgroup C (nonenveloped particles)		4
Unclassified possible species		68
Bunyaviridae	(–)ssRNA (3)	
Tospovirus		1
Reoviridae	dsRNA (10-12)	
Phytoreovirus		3
Fijivirus		3
Unnamed genus		2
Groups		
Isometric particles		
Caulimovirus	dsDNA (1)	17
Cryptovirus	dsRNA (2)	
Subgroup I		26
Subgroup II		5
Carmovirus	ssRNA (1)	17
Luteovirus	ssRNA (1)	21
Maize chlorotic dwarf virus	ssRNA (1)	3
Marafivirus	ssRNA (1)	3
Necrovirus	ssRNA (1)	4
Parsnip yellow fleck virus	ssRNA (1)	3
Sobemovirus	ssRNA (1)	16
Tombusvirus	ssRNA (1)	12
Tymovirus	ssRNA (1)	19
Comovirus	ssRNA (2)	13
Dianthovirus	ssRNA (2)	3
Fabavirus	ssRNA (2)	3
Nepovirus	ssRNA (2)	36
Pea enation mosaic virus	ssRNA (2)	1
Bromovirus	ssRNA (3)	6
Cucumovirus	ssRNA (3)	4
Quasi-isometric to bacilliform particles		
Ilarvirus	ssRNA (3)	20
Alfalfa mosaic virus	ssRNA (3)	1
Geminate particles		
Geminivirus	ssDNA (1 or 2)	
Subgroup I		10
Subgroup II		5
Subgroup III		33
Bacilliform particles		
Commelina yellow mottle virus	dsDNA (1)	14
Rod-shaped rigid particles		
Tobamovirus	ssRNA (1)	14

Table 2-5 The Current 3 Families and 32 Groups of Plant viruses According to the Fifth Report of the International Committee on Taxonomy of Viruses (continued)

Classification	Genome properties[a]	Number of viruses[b]
Tobravirus	ssRNA (2)	3
Furovirus	ssRNA (2-4)	11
Hordeivirus	ssRNA (3)	4
Filamentous particles		
Capillovirus	ssRNA (1)	4
Carlavirus	ssRNA (1)	56
Closterovirus	ssRNA (1)	22
Potexvirus	ssRNA (1)	39
Potyvirus	ssRNA (1 or 2)	153
Tenuivirus	ssRNA(4)	7

[a] Number of functional species of nucleic acid in parentheses.
[b] Definitive and possible.

From Martelli, G. P., Classification and nomenclature of plant viruses: state of the art, *Plant Dis.*, 76, 436, 1992; Franckii, R. I. B., Fauquet, C. M., Knudsen, D. L., and Brown, F., Classification and nomenclature of viruses, *Arch. Virol.*, Suppl., 2, 1991. With permission.

Successful seed transmission of viruses occurs when the host is infected systemically prior to flowering and when the virus is able to invade either pollen or the ovule, survive in the gamete, survive dehydration in seeds during storage, and not be inactivated during seed development.[579] Approximately 20% of the known plant viruses are transmitted through seeds of infected plants (Table 2-6).

In 1969, Bennett[580] reported that 47 viruses were seedborne. By 1974, the number reached 85[968] and by 1981, 119.[969] However, when synonymy was taken into consideration, 46 diseases and virus names were found in the list of seed-transmitted viruses that occurred only once or a few times in the literature, and 108 viruses were found, excluding cryptic viruses, for which there is evidence of transmission through seed or pollen. All members of cryptoviruses are transmitted through seed or pollen.[581,970] Viruses that have been recognized for several years and thought not to be seed transmitted may be seedborne at a low percentage in some hosts. Seed transmission of maize chlorotic mottle spot virus was detected in 17 of 42,000 plants from 25 seed lots.[971] Most tests with maize dwarf mosaic virus demonstrated low levels of seed transmission, i.e., one of 22,189 seeds,[972] one of 11,448 seeds,[973] and two of 29,735 seeds.[974] A portion of the new viruses that are continually being discovered are seedborne. The number of known seedborne viruses will increase because (1) previously unrecognized inoculum sources will be identified through epidemiological investigations, (2) seedborne viruses will be discovered that were not previously detected, (3) viruses will be identified that are seedborne but symptomless in certain hosts, and (4) seedborne viruses will be detected through the use of increasingly suitable and sensitive detection technology.[975]

Table 2-6 Seedborne Viruses that Cause Disease of Major Crops[a]

Virus	Crop	Ref.
Abutilon variegation	*Abutilon* (abutilon)	582, 583
Alfalfa mosaic	*Amaranthus albus* (tumble weed)	584
	Capsicum annuum (pepper)	585
	C. album (lamb's quarters)	586
	C. amaranticolor (chenopodium)	587
	C. quinoa (chenopodium)	587
	D. stramonium (jimsonweed)	586
	G. max (soybean)	588
	Holesteum umbellatum var. *umbellatum* (holesteum)	586
	Medicago murex (medic)	589
	M. polymorpha (burclover)	589
	M. sativa (alfalfa)	590, 591
	M. truncatula (barral medica)	592
	Melilotus lupulina (sweet clover)	593
	Nicandra physaloides (apple-of-Peru, shoofly plant)	594
	P. vulgaris (bean)	584
	Senecio vulgaris (groundsel)	586
	Solanum brevidens (solanum)	595
	S. nigrum (black nightshade)	586
	Stellaria media (chickweed)	586
	Trifolium alexandrinum (Egyptian clover)	596
	Vigna unguiculata (cowpea)	597
Apple mosaic	*P. dulcis* (almond)	598
	V. unguiculata (cowpea)	25
Apple stem grooving[b]	*C. quinoa* (lamb's quarters)	599
Arabis mosaic	*B. vulgaris* (red beet)	600
	C. bursa-pastoris (shepherd's purse)	600
	C. album (lamb's quarters)	600
	C. quinoa (chenopodium)	601
	Fragaria ananassa (strawberry)	602
	G. max (soybean)	601
	H. lupulus (hop)	603
	L. sativa (lettuce)	604
	Lamium amplexicaule (henbit)	600
	L. esculentum (tomato)	600
	Myosotis arvensis (forget-me-not)	600
	N. clevelandii (tobacco)	605
	Petunia hybrida (garden petunia)	605
	P. violacea (violet petunia)	605
	Plantago major (broadleaf plantain)	600
	Poa annua (annual bluegrass)	605
	Polygonum persicaria (lady's thumb)	600
	Rheum rhaponticum (rhubarb)	606
	Senecio vulgaris (groundsel)	601, 602
	Stellaria media (chickweed)	601
Arracacha A	*N. clevelandii* (tobacco)	607
Arracacha B	*C. quinoa* (chenopodium)	608
	S. tuberosum (potato)	608

Table 2-6 Seedborne Viruses that Cause Disease of Major Crops[a] (continued)

Virus	Crop	Ref.
Artichoke yellow ringspot	*Anethum graveolens* (dill)	609
	Calendula arvensis (calendula)	610
	Carduus pycnocephalus (thistle)	610
	C. amaranticolor (chenopodium)	609
	C. quinoa (chenopodium)	609
	D. stramonium (jimsonweed)	609
	Foeniculum (fennel)	609
	Gomphrena globosa (globe amaranth)	610
	N. clevelandii (tobacco)	609
	N. glauca (tree tobacco)	609
	N. glutinosa (tobacco)	609
	N. rustica (tobacco)	609
	N. tabacum (tobacco)	609
	Petunia hybrida (petunia)	609
	Reseda alba (mignonette)	609
	Stellaria media (chickweed)	610
	Vicia faba var. *minor* (fava bean)	611
Asparagus virus 2	*A. officinalis* (asparagus)	612–615
	N. tabacum (tobacco)	616
	P. hybrida (petunia)	616
	Z. elegans (zinnia)	616
Avocado black streak	*Persea americana* (avocado)	617, 618
Avocado virus 1	*P. americana* (avocado)	619
Avocado virus 2	*P. americana* (avocado)	619
Avocado virus 3	*P. americana* (avocado)	619
Barley mosaic[b]	*Hordeum vulgare* (barley)	620
Barley stripe mosaic	*Aegilops* (goat grass)	621
	Agropyron elongatum (wheatgrass)	622
	Avena fatua (wild oats)	623
	A. sativa (oats)	624
	Bromus inermis (smooth brome)	622
	Commelina communis (dayflower)	622
	H. depressum (barley)	622
	H. glaucum (barley)	622
	H. vulgare (barley)	625
	Triticum aestivum (wheat)	626
Bean common mosaic	*Crotalaria juncea* (sunnhemp)	627
	Lupinus luteus (yellow lupine)	628
	Macroptilium lathyroides (one-leaf clover)	629
	P. aborigineus (phasemy bean)	629
	P. acutifolius (tepary bean)	630
	P. angustifolius (bean)	631
	P. vulgaris (bean)	632
	Vigna angularis (adzuki bean)	633
	V. mungo (urd bean)	634
	V. radiata (mung bean)	635
	V. sesquipedalis (asparagus bean)	636
	V. unguiculata (cowpea)	637
Bean pod mottle	*G. max* (soybean)	638
Bean southern mosaic	*P. vulgaris* (bean)	639
	V. unguiculata (cowpea)	640

Table 2-6 Seedborne Viruses that Cause Disease of Major Crops[a] (continued)

Virus	Crop	Ref.
Bean yellow mosaic	*L. albus* (white lupine)	641
	L. luteus (yellow lupine)	642
	L. pilosus (blue lupine)	643
	Melilotus alba (white sweetclover)	643a
	P. vulgaris (bean)	644
	Pisum sativum (pea)	645,646
	Trifolium pratense (red clover)	647
	Vicia faba (fava bean)	648, 649
	Vigna radiata (mung bean)	650
	V. unguiculata (cowpea)	651
Beet cryptic	*Beta vulgaris* (beet)	652, 653
Beet cryptic I	*B. vulgaris* (sugar beet)	654
Beet 41 yellows[b]	*B. vulgaris* (sugar beet)	655
Beet mild yellowing	*B. vulgaris* (beet)	656
Bell pepper yellow mosaic	*C. frutescens* (pepper)	657
Blackeye cowpea mosaic	*V. mungo* (urd bean)	658
	V. unguiculata (cowpea)	659
Blueberry leaf mottle	*C. quinoa* (chenopodium)	660
	V. corymbosum (highbush blueberry)	661
	Vitis labrusca (American grape)	660
Broadbean mild mosaic[b]	*Vicia faba* (fava bean)	662
Broadbean mottle	*C. arietinum* (chickpea)	663
	P. sativum (pea)	663
	V. faba (fava bean)	664, 665
	Vigna mungo (urd bean)	666
Broadbean stain	*L. culinaris* (lentil)	667
	P. sativum (pea)	646
	V. faba (fava bean)	668
	V. palaestina (vetch)	669
Broadbean true mosaic	*V. faba* (fava bean)	183
Broadbean wilt	*V. faba* (fava bean)	670
Broadbean yellow band[b]	*V. faba* (fava bean)	671
Brome mosaic	*T. aestivum* (wheat)	672
Cacao necrosis	*G. max* (soybean)	673
	P. lunatus (lima bean)	673
Capsicum mosaic	*Capsicum* (pepper)	674
Cardamom chirke	*Amomum subulatum* (greater cardamon)	675
Cardamom greater streak[b]	*A. subulatum* (greater cardamom)	676
Carnation cryptic	*Dianthus caryophyllus* (carnation)	677
	D. chinensis (rainbow pink)	677
Cassava green mottle	*N. clevelandii* (tobacco)	675
Cherry leaf roll	*Betula pendula* (European white birch)	678
	Chenopodium bonus-henricus (goosefoot)	600
	G. max (soybean)	679
	Juglans (walnut)	680
	J. nigra (black walnut)	681
	J. regia (Persian walnut)	682
	N. clevelandii (tobacco)	683
	N. megalosiphon (tobacco)	683
	Nicotiana rustica (tobacco)	684

Table 2-6 Seedborne Viruses that Cause Disease of Major Crops[a] (continued)

Virus	Crop	Ref.
	N. tabacum (tobacco)	685
	P. vulgaris (bean)	600
	Prunus serotina (black cherry)	686
	R. rhaponticum (rhubarb)	606
	Sambucus racemosa (European red elder)	687
	Ulmus americana (American elm)	688
	Viola tricolor (viola)	600
Cherry raspleaf	*C. amaranticolor* (goosefoot)	689, 690
	C. quinoa (chenopodium)	691
	Prunus (cherry)	690
	Taraxacum officinale (dandelion)	690
Chicory yellow mottle	*Cichorium intybus* (chicory)	692
Cineraria mosaic[b]	*Senecio cruentus* (cineraria)	693–695
Citrus mosaic	*Citrus sinensis* (sweet orange)	696
Citrus tatter leaf	*Chenopodium quinoa* (goosefoot)	697
	Lilium (lily)	697
Citrus veinal necrosis	*Citrus* (citrus)	698
Citrus wood pocket	*Citrus* (citrus)	699
Citrus xyloporosis	*Citrus aurantifolia* (lime)	700
Clover (red) mosaic[b]	*Trifolium pratense* (red clover)	647
	Vicia faba (fava bean)	701
Clover (red) vein mosaic	*T. pratense* (red clover)	702–705
	V. faba (fava bean)	705
Clover (white) cryptic	*Trifolium* (clover)	706
Clover (white) mosaic	Cassia occidentalis	707
	Trifolium repens (white clover)	707
	T. pratense (red clover)	708
Clover yellow mosaic	*T. pratense* (red clover)	708
Coffee ringspot	*Coffea excelsa* (coffee)	709
Corchorus leaf mosaic chlorosis[b]	*Corchorus capsularis* (jute)	710
Cowpea aphidborne mosaic	*P. vulgaris* (bean)	711
	V. unguiculata (cowpea)	712
Cowpea banding mosaic	*V. unguiculata* (cowpea)	713
Cowpea green vein banding	*V. unguiculata* (cowpea)	675
Cowpea isometric mosaic[b]	*V. unguiculata* (cowpea)	714
Cowpea little leaf	*V. unguiculata* (cowpea)	715
Cowpea mottle	*P. vulgaris* (bean)	716
	V. subterranea (cowpea)	717
	V. unguiculata (cowpea)	716
Cowpea mild mottle	*V. unguiculata* (cowpea)	718
Cowpea severe mottle	*V. unguiculata* (cowpea)	719
Cowpea mosaic	*V. unguiculata* (cowpea)	720
	V. cylindrica (catjang-pea)	721
	V. unguiculata subsp. *sesquipedalis* (asparagus bean)	722
Cowpea severe mosaic	*V. unguiculata* (cowpea)	723
	V. unguiculata subsp. *sesquipedalis* (asparagus bean)	724
Cowpea ringspot	*V. unguiculata* (cowpea)	724

Table 2-6 Seedborne Viruses that Cause Disease of Major Crops[a] (continued)

Virus	Crop	Ref.
Cowpea stunt disease[c]	*V. unguiculata* (cowpea)	725
Crimson clover latent	*C. quinoa* (chenopodium)	726
	T. incarnatum (crimson clover)	726
Cucumber green mottle mosaic	*C. lunatus* (watermelon)	727
	Cucumis sativus (cucumber)	728
	Lagenaria siceraria (bottle gourd)	729
Cucumber leaf spot[b]	*C. sativus* (cucumber)	730
Cucumber mosaic	*A. hypogaea* (peanut)	731
	Benincassa hispida (ash gourd)	732
	Carthamus tinctorius (safflower)	733
	Cerastium fontanum (chickweed)	734
	C. holosteoides (mouse-ear chickweed)	734
	Crescentia cujete (calabash)	732
	Cucumis melo (cantaloupe, muskmelon)	735
	C. sativus (cucumber)	736
	Cucurbita flexuous (cucurbit)	737
	C. maxima (winter squash)	738
	C. moschata (winter crookneck squash)	732
	C. pepo (pumpkin)	732
	Echinocystis lobata (mock cucumber)	739
	G. max (soybean)	740
	H. vulgare (barley)	741
	Laminum purpureum (henbit)	734
	Luffa acutangula (dishcloth gourd, vegetable sponge)	732
	Lupinus angustifolius (blue lupine)	742
	L. luteus (yellow lupine)	743
	L. esculentum (tomato)	744
	Micrampelis lobata (wild cucumber)	745
	P. vulgaris (bean)	746
	Psophocarpus tetragonolobus	747
	S. tuberosum (potato)	748
	Spergula arvensis (spurry)	734
	Stellaria media (chickweed)	749
	T. subterraneum (subterranean clover)	750
	V. radiata (mungbean)	751, 752
	V. unguiculata (cowpea)	753
	V. cylindrica (catjang-pea)	754
	V. unguiculata subsp. *sesquipedalis* (asparagus bean)	712
Desmodium mosaic	*Desmodium canum* (desmodium)	755
Desmodium triflorum mottle[b]	*D. triflorum* (desmodium)	756
Dodder latent mosaic[b]	*Cuscuta californica* (dodder)	757
	C. campestris (dodder)	757
Dulcamara mottle	*Solanum dulcamara* (deadly nightshade)	758
Eggplant mosaic	*N. clevelandii* (tobacco)	759
	Petunia hybrida (petunia)	759
	S. melongena (eggplant)	759
	S. tuberosum (potato)	760
Eggplant mottled dwarf[b]	*S. tuberosum* (potato)	761

Table 2-6 Seedborne Viruses that Cause Disease of Major Crops[a] (continued)

Virus	Crop	Ref.
Elm mottle[b]	*U. glabra* (Scotch elm)	762, 763
Eucharis mottle[b]	*Eucharis candida* (lily)	675
	A. sativa (oat)	764
	Avena (Clinton oat)	764
Foxtail mosaic	*Briza maxima* (quaking grass)	765, 766
French bean mosaic	*P. vulgaris* (bean)	767
Grapevine Bulgarian latent	*Vitis vinifera* (grape)	768
	Chenopodium quinoa (goosefoot)	768
Grapevine fanleaf	*C. amaranticolor* (goosefoot)	769
	C. quinoa (goosefoot)	770
	G. max (soybean)	770
	V. vinifera (grape)	770
Guar symptomless	*Cyamopsis tetragonoloba* (cluster bean, guar)	771
Hami muskmelon seedborne	*C. melo* (cantaloupe)	772
Hemp streak[b]	*Cannabis sativa* (hemp)	773
Hippeastrum mosaic	*Hippeastrum hybridum* (amaryllis)	774
Hop mosaic	*H. japonicus* (hop)	775
Hydrangea mosaic[b]	*C. quinoa* (goosefoot)	776
Johnson grass chlorotic stripe mosaic	*Sorghum bicolor* (sorghum)	776a
Lettuce mosaic	*C. quinoa* (goosefoot)	616
	Lactuca sativa (lettuce)	777, 778
	L. serriola (prickly lettuce)	779
	Senecio vulgaris (groundsel)	780
	Sinapis arvensis (charlock)	781
Lettuce yellow mosaic[b]	*L. sativa* (lettuce)	782
Lilac ring mottle	*Celosia argentea* (woolflower)	783
	C. amaranticolor (chenopodium)	783
	C. giganteum (chenopodium)	783
	C. quinoa (goosefoot)	783
Lima bean mosaic[b]	*P. lunatus* (lima bean)	784
Lucerne latent (Australian)	*C. amaranticolor* (chenopodium)	785, 786
	C. quinoa (goosefoot)	785, 786
	Medicago sativa (alfalfa)	785, 786
Lucerne symptomless (Australian)	*C. quinoa* (goosefoot)	787
Lucerne transient streak	*M. alba* (alfalfa)	788
Lychnis ring spot	*Beta vulgaris* (red beet)	789
	Capsella bursa-pastoris (shepherd's purse)	789
	Cerastium vulgatum (mouse-eared chickweed)	789
	Lychnis divaricata (lychnis)	789
	Silene gallica (French catchfly)	789
	S. noctiflora (night-blooming catchfly)	789
	Stellaria media (chickweed)	789
Maize chlorotic mottle	*Zea mays* (maize)	790
Maize dwarf mosaic	*Z. mays* (maize)	791
Maize leaf spot[b]	*Z. mays* (maize)	792

Table 2-6 Seedborne Viruses that Cause Disease of Major Crops[a] (continued)

Virus	Crop	Ref.
Maize mosaic	*Z. mays* (maize)	793
Melon necrotic spot	*Cucumis melo* (cantaloupe, musk melon)	794
Mimosa chlorotic leaf stripe	*Albizia julibrissin* (mimosa)	795
Mulberry ringspot	*G. max* (soybean)	796
Mung bean mosaic	*V. radiata* (mungbean)	797
Mung & urd bean mosaic 1[b]	*V. mungo* (urd bean)	675
Mung & urd bean mosaic 2[b]	*V. mungo* (urd bean)	675
Mung bean yellow mosaic	*V. radiata* (mungbean)	798
Muskmelon necrotic ringspot	*Cucumis melo* (cantaloupe, musk melon)	799–801
Nicotiana velutina mosaic	*N. clevelandii* (tobacco)	802
	N. debneyi (tobacco)	802
	N. glutinosa (tobacco)	802
	N. rustica (tobacco)	802
	N. velutina (tobacco)	802
Oat mosaic	*Avena sativa* (oats)	803
Onion mosaic[b]	*Allium cepa* (chives, garlic, onion)	804
Onion yellow dwarf	*A. cepa* (garlic, onion,)	805
Panicum mosaic	*Setaria italica* (foxtail millet)	806
Papaya ringspot[b]	*Asimina triloba* (papaya)	807
	C. maxima (squash)	808
Parsley latent	*Petroselinum crispum* (parsley)	809
Pea early browning	*P. sativum* (pea)	810
	V. faba (fava bean)	811
Pea enation mosaic	*Lathyrus ochros* (sweet pea)	812
	P. sativum (pea)	812
Pea seedborne mosaic	*Lathyrus annuus* (sweet pea)	813
	L. ochrus (sweet pea)	813
	L. sativus (sweet pea)	813
	Lens culinaris (lentil)	814
	P. arvense (pea)	815
	P. sativum (pea)	816–826
	Vicia articulata (vetch)	183
	V. ervilia (vetch)	817
	V. faba (fava bean)	675
	V. palaestina (vetch)	817
	V. pannonica (Hungarian vetch)	183
	V. narbonensis (vetch)	183
	V. sativa (vetch)	817
Pea seedborne symptomless	*P. sativum* (pea)	827
Peach rosette mosaic	*Chenopodium quinoa* (goosefoot)	828
	Taraxacum officinale (dandelion)	829
	Vitis labrusca (foxgrape)	829
Peanut bunchytop[b]	*Arachis hypogaea* (peanut)	830
Peanut chlorosis[b]	*A. hypogaea* (peanut)	830
Peanut clump	*A. hypogaea* (peanut)	831
	S. italica (foxtail millet)	675
Peanut marginal chlorosis[b]	*A. hypogaea* (peanut)	832
Peanut mottle	*A. hypogaea* (peanut)	833, 834
	Lupinus albus (white lupine)	835

Table 2-6 Seedborne Viruses that Cause Disease of Major Crops[a] (continued)

Virus	Crop	Ref.
	V. unguiculata (cowpea)	836
	Voandzeia subterranea (Bamarra)	837
Peanut ring spot[b]	*A. hypogaea* (peanut)	183
Peanut rosette	*A. hypogaea* (peanut)	797
Peanut stripe	*A. hypogaea* (peanut)	838
	G. max (soybean)	839
Peanut stunt	*A. hypogaea* (peanut)	840
	G. max (soybean)	588
Pelargonium zonate leaf spot	*N. glutinosa* (tobacco)	841
Potato black ringspot	*S. tuberosum* (potato)	608
Potato M	*L. esculentum* (tomato)	842
Potato S	*S. tuberosum* (potato)	843
Potato T	*Datura stramonium* (jimson weed)	844
	Nicandra physalodes (apple-of-Peru, shoofly plant)	844
	S. demissum (wild potato)	844
Potato U[b]	*C. amaranticolor* (chenopodium)	845
	C. gigantea (chenopodium)	845
	C. quinoa (chenopodium)	845
	N. debneyi (tobacco)	845
	N. tabacum (tobacco)	845
Potato X	*S. sarachoides* (potato)	846
	S. tuberosum (potato)	847
	Tribulus terrestris (burnut)	846
Potato Y	*S. nigrum* var. *judaicum* (black nightshade)	848
	S. tuberosum (potato)	830
Prune dwarf	*Prunus* (cherry)	830
	P. armeniaca (apricot)	849
	P. avium (bird cherry)	849
	P. cerasifera (cherry plum)	849
	P. cerasus (sour cherry)	850, 851
	P. mahaleb (mahaleb cherry)	183
	P. persica (peach)	850
Prunus necrotic ring spot	*Cucurbita maxima* (winter squash)	852
	P. americana (plum)	853
	P. avium (sweet cherry)	854
	P. besseyi (bessey cherry)	851
	P. cerasifera (cherry plum)	850, 851
	P. cerasus (sour cherry)	855
	P. mahaleb (mahaleb cherry)	856
	P. pennsylvanica (wild red cherry)	858
	P. persica (peach)	857
Psophocarpus ringspot mosaic	*P. tetrogonolobus* (winged bean)	858
Radish yellow edge	*B. vulgaris* (red beet)	859
	Raphanus sativus (radish)	859, 860
	Spinacia oleracea (spinach)	859
Raspberry bushy dwarf	*Chenopodium* (goosefoot)	861
	Fragaria vesca (strawberry)	862
	Malus (apple)	863

Table 2-6 Seedborne Viruses that Cause Disease of Major Crops[a] (continued)

Virus	Crop	Ref.
	R. idaeus (raspberry)	863, 864
	R. loganobaccus (loganberry)	865
	R. phoenicolasius (phoenician berry)	864
Raspberry ringspot	*B. vulgaris* (red beet)	600
	Capsella bursa-pastoris (shepherd's purse)	600
	F. ananassa (strawberry)	866–868
	G. max (soybean)	600
	G. soja (soybean)	830
	Petunia violacea (petunia)	666
	Rubus idaeus (raspberry)	866
	S. media (chickweed)	601
Rubus Chinese seedborne	*C. quinoa* (chenopodium)	869
	N. bigelowii (tobacco)	869
	R. ideaus (raspberry)	869
Runner bean mosaic[b]	*Phaseolus coccineus* (scarlet runner bean)	870
Ryegrass cryptic	*L. temulentum* (darnel)	871
Safflower mosaic[b]	*C. tinctorius* (safflower)	872
Satsuma dwarf	*P. vulgaris* (bean)	675
Sincomas mosaic[b]	*Pachyrrhizus erosus* (yam bean)	873
Solanum apical curling	*S. tuberosum* (potato)	874
Sowbane mosaic	*Atriplex pasifica* (saltbush)	875
	Chenopodium album (lamb's quarters)	875
	C. amaranticolor (goosefoot)	876
	C. murale (goosefoot)	877
	C. quinoa (goosefoot)	877
Soybean mild mosaic	*G. max* (soybean)	878
	G. soja (wild soybean)	830
Soybean mosaic	*C. macrocarpum* (centrosema)	879
	G. max (soybean)	880
	L. albus (white lupin)	881
	P. vulgaris (bean)	882
Spinach latent	*C. cristata* (celosia)	883
	C. quinoa (chenopodium)	883
	N. clevelandii (tobacco)	884
	N. megalosiphon (tobacco)	884
	N. rustica (tobacco)	883
	N. tabacum (tobacco)	883
	N. xanthi (tobacco)	885
	Spinacia oleracea (spinach)	883
Spinach latent ringspot	*N. rustica* (tobacco)	885
	N. xanthii (tobacco)	885
Spinach temperate	*S. oleracea* (spinach)	886
Squash mosaic	*Atriplex glauca* (saltbush)	887
	B. vulgaris (beet)	887
	C. murale (chenopodium)	887
	C. quinoa (chenopodium)	887
	Citrullus vulgaris (watermelon)	888, 889
	Cucumis melo (cantaloupe, muskmelon)	889, 890
	C. melo var. *inodorus* (melon)	888

Table 2-6 Seedborne Viruses that Cause Disease of Major Crops[a] (continued)

Virus	Crop	Ref.
	Cucurbita flexuosus (cucurbit)	183
	C. maxima (winter squash)	891
	C. mixta (cucurbit)	892
	C. moschata (winter crookneck squash)	183
	C. pepo (pumpkin)	892
Strawberry latent ring spot	*Amaranthus lividus* (amaranth)	183
	Apium graveolens (celeriac, celery)	893
	Capsella bursa-pastoris (shepherd's purse)	894
	Chenopodium quinoa (goosefoot)	894
	Lamium amplexicaule (henbit)	895
	Mentha arvensis (field mint)	602
	Pastinaca sativa (parsnip)	896
	Petroselinum crispum (parsley)	897, 898
	Rubus idaeus (raspberry)	895
	Senecio vulgaris (groundsel)	895
	Solanum nigrum (black nightshade)	183
	Stellaria media (chickweed)	601
Subterranean clover mottle	*T. subterraneum* (clover)	899
Sunflower mosaic	*H. annuus* (sunflower)	900
Sunflower rugose mosaic[b]	*H. annuus* (sunflower)	901
Sunnhemp mosaic	*C. juncea* (crotalaria)	902
Sweetpotato ringspot[b]	*I. batatas* (sweetpotato)	675
Telfairia mosaic	*T. occidentalis* (telfairia)	903
Tobacco ascending mosaic	*Nicotiana tabacum* (tobacco)	904
Tobacco etch	*N. tabacum* (tobacco)	905
Tobacco mosaic	*Capsicum annuum* (pepper)	906
	C. frutescens (pepper)	907
	C. juncea (crotalaria)	908
	L. esculentum (tomato)	909–911
	Malus platycarpa (apple)	912
	M. sylvestris (apple)	912
	Plantago major (plantain)	913
	Pyrus communis (pear)	912
	S. melongena (eggplant)	910
	V. unguiculata (cowpea)	666
	Vitis vinifera (grape)	914
Tobacco necrosis	*Euonymus europaeus* (euonymus)	915
	Z. mays (maize)	916
Tobacco rattle	*Capsella bursa-pastoris* (shepherd's purse)	600
	F. moschata (strawberry)	679
	Lamium amplexicaule (henbit)	600
	Myosotis arvensis (forget-me-not)	600
	Papaver rhoeas (corn poppy)	600
	P. oleracea (portulaca)	846
	S. vulgaris (groundsel)	917
	S. sarachoides (potato)	917
	Viola arvensis (pansy)	600
	X. strumarium (cocklebur)	846

Table 2-6 Seedborne Viruses that Cause Disease of Major Crops[a] (continued)

Virus	Crop	Ref.
Tobacco ringspot	*A. hybridus* (amaranthus)	918
	C. bursa-pastoris (sheperd's purse)	600
	Cicer arietinum (chickpea, garbanzo)	919
	Cucumis melo (cantaloupe, muskmelon)	920
	F. ananassa (strawberry)	605
	G. max (soybean)	921
	Gomphrena globosa (gomphrena)	922
	Lactuca sativa (lettuce)	923
	N. glutinosa (tobacco)	606
	N. tabacum (tobacco)	924
	Pelargonium hortorum (fish geranium)	925
	Petunia hybrida (petunia)	926
	P. violacea (violet petunia)	926
	Senecio vulgaris (groundsel)	927
	S. tuberosum (potato)	608
	T. officinale (dandelion)	928
	Vigna radiata (mungbean)	929
	V. unguiculata (cowpea)	605
Tobacco streak	*A. officinalis* (asparagus)	930
	C. amaranticolor (chenopodium)	931
	C. quinoa (goosefoot)	932, 933
	Cicer arietinum (chickpea)	933
	D. stramonium (jimson weed)	932
	F. vesca var. *semperflorens* (strawberry)	934
	G. globosa (gomphrena)	933
	G. max (soybean)	935
	L. esculentum (tomato)	830
	M. alba (white clover)	933
	Nicandra physalodes (shoo-fly plant)	936
	N. clevelandii (tobacco)	933
	P. vulgaris (pinto bean)	937
	R. raphanistrum (wild radish)	938
	R. occidentalis (black raspberry)	939
	V. angularis (azuki bean)	933
	V. unguiculata (cowpea)	933
Tomato aspermy	*Stellaria media* (chickweed)	940
Tomato blackring	*A. graecizans* (tumbleweed)	941
	B. vulgaris (sugarbeet)	942
	C. cajan (pigeon pea)	941
	Capsella bursa-pastoris (shepherd's purse)	600
	Cerastium vulgatum (mouse-eared chickweed)	600
	Chenopodium album (lamb's quarters)	600
	C. fasciculosum var. *muraliforme* (chenopodium)	941
	C. quinoa (chenopodium)	601
	D. stramonium (jimsonweed)	941
	F. ananassa (strawberry)	605
	Fumaria officinalis (fumatories)	600

Table 2-6 Seedborne Viruses that Cause Disease of Major Crops[a] (continued)

Virus	Crop	Ref.
	Geranium dissectum (cutleaf geranium)	943
	G. globosa (gomphrena)	941
	G. max (soybean)	183, 943
	H. annuus (sunflower)	941
	Lactuca sativa (lettuce)	944
	Lamium amplexicaule (henbit)	600
	Ligustrum vulgare (privet)	600
	L. esculentum (tomato)	600
	M. arvensis (forget-me-not)	600
	N. debneyi (nicandra)	941
	N. physalodes (shoo-fly plant)	941
	N. clevelandii (tobacco)	601
	N. rustica (tobacco)	600
	N. tabacum (tobacco)	601
	Petunia violacea (viola)	666
	P. vulgaris (bean)	941
	Poa annua (bluegrass)	600
	P. aviculare (polygonum)	943
	P. convolvulus (wild buckwheat)	943
	P. persicaria (lady's thumb)	600
	Portulaca (portulaca)	941
	R. idaeus (raspberry)	600
	Senecio vulgaris (groundsel)	605
	Spergula arvensis (corn spurrey)	600
	Stellaria media (chickweed)	183
	Tripleurospermum maritimum (tripleurospermum)	943
	Veronica agrestis (veronica)	943
	V. radiata (mung bean)	941
	V. unguiculata (cowpea)	605, 943
	Z. elegans (zinnia)	941
Tomato bushy stunt	*L. esculentum* (tomato)	945
	Malus pumila (apple)	946
Tomato mosaic	*L. esculentum* (tomato)	947
	P. angulata (physalis)	947
	P. minima (physalis)	947
Tomato ringspot	*C. amaranticolor* (goosefoot)	769
	C. giganteum (chenopodium)	947
	Fragaria vesca (wild strawberry)	948
	G. max (soybean)	949
	Gomphrena globosa (globosa)	950
	L. esculentum (tomato)	951
	N. tabacum (tobacco)	951
	Pelargonium hortorum (geranium)	951
	R. idaeus (raspberry)	952
	Sambucus canadensis (American elder)	953
	Taraxacum officinale (dandelion)	954
	Trifolium pratense (red clover)	647
Tomato spotted wilt	*Senecio cruentus* (cineraria)	695
Turnip yellow mosaic	*Alliaria petiolata* (hedge garlic)	955

Table 2-6 Seedborne Viruses that Cause Disease of Major Crops[a] (continued)

Virus	Crop	Ref.
	Brassica (crucifer)	956
	Camelina sativa (gold of pleasure)	957
Urdbean leaf crinkle	*Vigna mungo* (urd bean)	958
	V. radiata (mungbean)	959
Vicia cryptic	*Vicia faba* (fava bean)	960
Watermelon mosaic	*Cucurbita pepo* (pumpkin)	961
	Echinocystis lobata (mock cucumber)	962
	Sechium edule (chayote)	963
Wheat streak mosaic	*Z. mays* (maize)	791
Wheat striate mosaic	*T. aestivum* (wheat)	964
Zucchini yellow mosaic	*Cucurbita pepo* (squash)	965, 966
	Ranunculus sardous (ranunculus)	967

[a] Does not necessarily refer to first report.
[b] Virus or disease reported to be seed transmitted, but not reported again.
[c] Two viruses.

VII. VIROIDS

Viroids are parasites of higher plants composed of naked, single-stranded, low molecular weight, circular RNA (molecular weight = 0.8 to 1.3×10^5), which utilizes only host components for replication. They exist in solution as rodlike structures arranged in a series of short base-paired and nonbase-paired regions.[575,976] Diseases caused by viroids were linked to viruses because of their small size and because they cause virus-like symptoms.[977] In 1971, "viroid" was coined for the causal agent of potato spindle tuber.[977] Evidence that the potato spindle tuber viroid was a free RNA was based on its low sedimentation rate and sensitivity to riboneuclease.[978] Diener[979] showed that the viroid was a low molecular weight RNA with no helper virus involved in its replication that had the ability to replicate autonomously in susceptible cells. "Viroid" was proposed as a generic term for agents and other pathogenic nucleic acids with similar properties.[977,979] Diener[980] concluded that it was a single molecular species. Semancik and Weathers[981] concluded that the citrus exocortis viroid was a low molecular weight RNA.

The viroid nature of the cause of chrysanthemum stunt disease was shown by Diener and Lawson.[982] Sogo et al.[983] established the low molecular weight of spindle tuber viroid by direct measurements using electron microscopy. The viroidal nature of coconut cadang-cadang was suggested by Randles.[984] The stunt disease of hops was found to be caused by a viroid.[985]

Viroids are distinct from viruses in that viroids (1) lack mRNA activity and have no viroid-coded protein coat, (2) depend on host factors for their multiplication, in contrast to viruses, (3) are small, with 246 nucleotides, as in the case of coconut cadang-cadang, whereas maize streak virus, the smallest known DNA virus, has 2681 and 2687 nucleotides,[987] and (4) increase in replication and

Table 2-7 Seedborne Viroids that Cause Diseases or Reported on Different Hosts

Viroids	Crop	Ref.
Avocado sunblotch	*P. americana* (avocado)	989
Apple scar skin	*Malus* (apple)	990
Chrysanthemum stunt	*C. morifolium* (chrysanthemum)	991
	L. esculentum (tomato)	992
Citrus exocortis	*C. morifolium* (chrysanthemum)	993
	L. esculentum (tomato)	993
Coconut cadang-cadang	*Cocos nucifera* (coconut palm)	994
Coleus color break	*Coleus scutellarioides* (coleus)	995
Cucumber pale fruit	*L. esculentum* (tomato)	988
Dapple apple	*Malus* (apple)	990
Potato spindler tuber	*L. esculentum* (tomato)	992
	Physalis peruviana (cap gooseberry)	118
	Scopolia sinensis (scopolia)	996
	Solanum muricatum (solanaceous)	997
	S. tuberosum (potato)	

[a] Does not necessarily refer to first report.

symptom development with increase in temperature from 20 to 35°C. In the case of viruses, high temperatures inhibit multiplication and symptom expression.[987]

Kryczynski[987] suggested that viroids were seed transmitted because of their affinity to meristematic cells. Viroid RNA is extremely small and likely to be translocated through weak symplastic connections with sporophytes.[988] Seed-transmitted viroids are presented in Table 2-7.

VIII. NEMATODES

Nematodes are, in general, eel-shaped and round in cross section, with smooth, unsegmented bodies, without legs or other appendages. The females of some species, however, become swollen at maturity and have pear-shaped or spheroid bodies. They are 300 to 1000 μm with some up to 4 mm long by 15 to 35 μm wide.[2] All plant parasitic nematodes belong to the phylum Nematoda (Table 2-8). Most of the important plant parasitic genera belong to the orders Dorylaimda and Tylenchida.

All plant-parasitic nematodes possess a stylet, used to puncture host cells and withdraw nutrients. Nematodes require a film of water for activity and mobility. Activity of nematodes also is influenced by soil aeration and temperature, which affect their survival. Under dry conditions at low temperatures, nematodes may enter into a quiescent phase and remain dormant for years. Root exudates attract nematodes toward the plant rhizosphere and can serve as hatching factors. Larvae feed either ectoparasitically (on cells at or near the root surface) or enter the host and feed endoparasitically (within the host tissues). Nematodes, after feeding,

Table 2-8 General Classifications of Plant Pathogenic Nematodes

- PHYLUM: **Nematoda**
  - ORDER: **Tylenchida**
    - SUBORDER: **Tylenchina**
      - SUPERFAMILY: **Tylenchoidea**
        - FAMILY: **Tylenchidae**
          - GENUS: *Anguina, Ditylenchus*
        - FAMILY: **Tylenchorhynchidae**
          - GENUS: *Tylenchorhynchus*
        - FAMILY: **Pratylenchidae**
          - GENUS: *Pratylenchus, Radopholus*
        - FAMILY: **Hoplolaimidae**
          - GENUS: *Hoplolaimus, Rotylenchus, Helicotylenchus*
        - FAMILY: **Belonolaimidae**
          - GENUS: *Belonolaimus*
      - SUPERFAMILY: **Heteroderoidea**
        - FAMILY: **Heteroderidae**
          - GENUS: *Globodera, Heterodera, Meloidogyne*
        - FAMILY: **Nacobbidae**
          - GENUS: *Nacobbus, Rotylenchus*
      - SUPERFAMILY: **Criconematoidea**
        - FAMILY: **Criconematidae**
          - GENUS: *Criconemella, Hemicycliophora*
        - FAMILY: **Paratylenchidae**
          - GENUS: *Paratylenchus*
        - FAMILY: **Tylenchulidae**
          - GENUS: *Tylenchulus*
    - SUBORDER: **Aphelenchina**
      - SUPERFAMILY: **Aphelenchoididea**
        - FAMILY: **Aphelenchoididae**
          - GENUS: *Aphelenchoides, Bursaphelenchus, Rhadinaphelenchus*
  - ORDER: **Dorylaimida**
    - FAMILY: **Longidoridae**
      - GENUS: *Longidorus, Xiphinema*
    - FAMILY: **Trichodoridae**
      - GENUS: *Paratrichodorus, Trichodorus*

From Agrios, G. N., *Plant Pathology*, Academic Press, New York, 1988, 803. With permission.

move (migratory) or remain on root surfaces (sedentary) for feeding. Symptoms may occur on underground plant parts as galls, knots, lesions, and/or proliferations, or on above ground parts as dwarfing, yellowing, and/or foliage wilting; or as galls on influorescences.[2] Seed transmission of nematodes has been found in species of the following nematode genera: *Anguina*, *Aphelenchoides*, *Ditylenchus*, *Heterodera*, *Panagrolaimus*, *Subanguina*, and *Rhadinaphelenchus*.

Seedborne nematodes known to cause important plant diseases are presented in Table 2-9.

Table 2-9 Seedborne Plant Parasitic Nematodes that Cause Diseases of Major Crops[a]

Nematode	Common name	Crop	Ref.
Anguina agropyronifloris	Seed gall nematode	*Agropyron smithii* (wheat grass)	998
A. agrostis	Bentgrass nematode, seed gall nematode	*Agrostis* (bentgrass)	999
		Arctagrostis latifolia (manzanita)	1000
		Dactylis glomerata (orchard grass)	1001
		Dupontia fisheri (grass)	1002
		Festuca (fescue)	1002
		Lolium rigidum (ryegrass)	1003
A. funesta	Seed gall nematode	*L. rigidum* (ryegrass)	1004
A. tritici	Ear cockle, seed gall nematode	*Avena sativa* (oats)	1005
		Secale cereale (rye)	1006
		Triticum aestivum (wheat)	1007
		T. dicoccum (emmer wheat)	118
		T. spelta (spelt wheat)	118
A. amsinckia		*Amsinckia* (amsinckia)	1008
Aphelenchoides arachidis	Testa nematode	*Arachis hypogaea* (peanut)	1009
A. besseyi	White tip	*Fraxinus americana* (white ash)	1010
		Oryza sativa (rice)	1011
		Stylosanthes hamata (Caribbean stylo)	1012
A. blastophorus	Foliar nematode	*Callistephus chinensis* (China aster)	1013
A. ritzema–bosi	Foliar nematode	*C. chinensis* (China aster)	1014
Ditylenchus angustus	Stem nematode	*O. sativa* (rice)	1015
D. destructor	Stem nematode	*A. hypogaea* (peanut)	1008
D. dipsaci	Bulb and stem nematode	*Allium cepa* (chives, garlic, onion, shallots)	1016
		A. sativa (oats)	1017
		Beta vulgaris (fodder beet, red beet, sugar beet)	1018
		D. carota (carrot)	1008
		Dipsacus fullonum (teasel)	1019
		F. sagittatum (buckwheat)	1008
		Hypochoeris radicata (cat's ear)	1020
		Medicago sativa (alfalfa)	1021
		P. coccineus (scarlet runner bean)	1008
		P. sativum (pea)	1008
		Plantago (plantain)	1022
		Taraxacum officinale (dandelion)	1023
		Trifolium pratense (red clover)	1024
		Vicia faba (fava bean)	1025
		V. sativa (vetch)	1008

Table 2-9 Seedborne Plant Parasitic Nematodes that Cause Diseases of Major Crops[a] (continued)

Nematode	Common name	Crop	Ref.
Heterodera glycines	Cyst nematode	*Glycine max* (soybean)	1026
H. goettingiana	Cyst nematode	*Pisum sativum* (pea)	1027
H. schachtii	Sugar beet cyst eelworm, cyst nematode	*B. vulgaris* (fodder beet, red beet, sugar beet)	1027
Panagrolaimus		*P. glaucum* (pearl millet)	1008
Rhadinaphelenchus cocophilus	Redring disease, wilt	*Cocos nucifera* (coconut palm)	1028
Subanguina chrysopogoni		*C. fubus* (chrysopogoni)	1008

[a] Does not necessarily refer to first report.

REFERENCES

1. Talbot, P. H. B., *Principles of Fungal Taxonomy*, Macmillan, London, 1978, 274.
2. Agrios, G. N., *Plant Pathology*, Academic Press, New York, 1988, 803.
3. Sinclair, J. B., Latent infection of soybean plants and seeds by fungi, *Plant Dis.*, 75, 220, 1991.
4. Reddick, B. B., Detection of the tall fescue endophyte with emphasis on enzyme linked immunosorbent assay, *J. Prod. Agric.*, 1, 133, 1988.
5. Prochazka, J., Bumeri, J., and Mika, V., Transmission of the endophytic fungus *Acremonium* sp. in the seed generations of grasses, *Sbornik UVTIZ, Genetika a Slecthtent*, 27, 199, 1991.
6. Kiewnick, L., Utersuchungen über den Einfluss der Samenund Bodonmikroflora auf die Lebensdauer der Spelzfruchte des Flughafers (*Avena fatua* L.), *Weed Res.*, 3, 322, 1963.
7. Lutey, R. W., Chains of spores produced by *Cephalosporium gramineum*, *Mycopathol. Appl.*, 18, 117, 1962.
8. Bruehl, G. W., *Cephalosporium* stripe disease of wheat, *Phytopathology*, 47, 641, 1957.
9. Arneson, E. and Stiers, D. L., *Cephalosporium gramineum*, a seedborne pathogen, *Plant Dis. Rep.*, 61, 619, 1977.
10. Reddy, C. S. and Holbert, J. R., The black bundle disease of corn, *J. Agric. Res.*, 27, 277, 1924.
11. Verma, U. and Bhowmik, T. P., Oospores of *Albugo candida* (Pers. ex Lev.) Kunze — its germination and role as the primary source of inoculum for the white rust disease of rape seed and mustard, *Int. J. Trop. Plant Dis.*, 6, 265, 1988.
12. Harman, G. E., Heit, C. E., Pfleger, F. L., and Braverman, S.W., Snapdragon seed blight — a serious problem caused by seed-borne fungi, *Plant Dis. Rep.*, 57, 592, 1973.
13. Kumar, K. and Patnaik, P., Seedborne nature of *Alternaria alternata* in pigeonpea: its detection and control, *Indian J. Plant Pathol.*, 3, 69, 1985.
14. Kunwar, I. K., Manandhar, J. B., and Sinclair, J. B., Histopathology of soybean seeds infected with *Alternaria alternata*, *Phytopathology*, 76, 543, 1986.
15. Svetov, V.G., Alternaria blight of sunflower along the Kuban river, *Mikol. Fitopatol.*, 9, 418, 1975.

16. Hall, T. J. and Taylor, G. S., Aerated steam treatment for the control of *Alternaria tenuis* on lobelia seeds, *Ann. Appl. Biol.*, 103, 219, 1983.
17. Gomes, J. L. L. and Dhingra, O. D., *Alternaria alternata* — a serious pathogen of white colored snap bean (*Phaseolus vulgaris*) seeds, *Fitopatol. Brasileria*, 8, 173, 1983.
18. Bhowmik, T. P., *Alternaria* seed infection of wheat, *Plant Dis. Rep.*, 53, 77, 1969.
19. Franceschini, A., Corda, P., and Carta, C., Observations on pathogenesis and epidemiology of tomato wilt due to *Alternaria alternata* f. sp. *lycopersici*, Grogan, Kimble and Misaghi, *Studi Sassaresi*, 29, 61, 1982.
20. Maude, R. B. and Humpherson-Jones, F. H., Studies on the seed-borne phases of dark leaf spot (*Alternaria brassicicola*) and gray leaf spot (*Alternaria brassicae*) of Brassicas, *Ann. Appl. Biol.*, 95, 311, 1980.
21. Knox-Davies, P. S., Relationship between *Alternaria brassicicola* and *Brassica* seeds, *Trans. Br. Mycol. Soc.*, 73, 235, 1979.
22. Kilpatrick, R. A., Fungal flora of crambe seeds and virulence of *Alternaria brassicicola*, *Phytopathology*, 66, 945, 1976.
23. Patel, R. M. and Desai, M. V., Alternaria blight of *Cuminum cyminum* and its control, *Indian Phytopathol.*, 24, 16, 1971.
24. Zazzerini, A. and Buonaurio, R., Diseases of safflower: leaf spot due to *Alternaria* spp., *Informatore Fitopatol.*, 31, 7, 1981.
25. Neergaard, P., *Seed Pathology*, Vols. 1 and 2, Macmillan, London, 1977, 1187.
26. Sowell, G., Alternaria leafspot of guar, *Plant Dis. Rep.*, 49, 605, 1965.
27. Groves, J. W. and Skolko, A. J., Notes on seed-borne fungi. II. *Alternaria*, *Can. J. Res.* Sect. C, 22, 217, 1944.
28. Crosier, W. F. and Heit, C. E., Some seedborne fungi of flowers, Proc. Assoc. Off. Seed Anal., 38, 73, 1948.
29. Neergaard, P., *Danish Species of Alternaria and Stemphylium*, Einer Munksgaard, Copenhagen, 1945, 560.
30. Hiremath, P. C., Kulkarni, M. S., and Lokesh, M. S., An epiphytotic of *Alternaria* blight of sunflower in Karnataka, *Karnataka J. Agric. Sci.*, 3, 277, 1990.
31. Ayra, H. C. and Prasada, R., Alternaria blight of linseed, *Indian Phytopathol.*, 5, 33, 1953.
32. Hopkins, J. C. F., *Tobacco Diseases with Special Reference to Africa*, Commonwealth Mycological Institute, Kew, Surrey, U.K., 1956, 178.
33. Yu, S., Mathur, S. B., and Neergaard, P., Taxonomy and pathogenicity of four seedborne species of *Alternaria* from sesame, *Trans. Br. Mycol. Soc.*, 78, 447, 1982.
34. Padaganur, G. M., The seedborne nature of *Alternaria macrospora* Zimm. in cotton, *Madras Agric., J.*, 66, 325, 1979.
35. Mathur, S. B., Mallya, J. I., and Neergaard, P., Seed-borne infection of *Trichoconis padwickii* in rice; distribution, and damage to seeds and seedlings, *Proc. Int. Seed Test. Assoc.*, 37, 803, 1972.
36. Culp, T. W., Thomas, C. A., and Zimmerman, L. H., Capsule diseases of castorbeans in Mississippi, *Plant Dis. Rep.*, 50, 35, 1966.
37. Crosier, W. F. and Heit, C. E., Some seed-borne fungi of flowers, *Proc. Assoc. Off. Seed Anal.*, 38, 73, 1948.
38. Leppik, E. E. and Sowell, G., *Alternaria sesami*, a serious seedborne pathogen of world-wide distribution, *FAO Plant Prot. Bull.*, 12, 13, 1964.

39. Miller, J. H. and Crosier, W. F., Pathogenic associates of tomato seed: their prevalence, relation to field disease and elimination, *Proc. Assoc. Off. Seed Anal. North Am.*, 26, 108, 1936.
40. Shortt, B. J., Sinclair, J. B., Helm, C. G., Jeffords, M. R., and Kogan, M., Soybean seed quality losses associated with bean leaf beetles and *Alternaria tenuissima*, *Phytopathology*, 72, 615, 1982.
41. Yu, S. H. and Lee, S. K., Blight of marigold caused by *Alternaria tagetica* in Korea, *Korean J. Plant Pathol.*, 5, 354, 1989.
42. Kumar, V. R. and Ayra, H. C., Certain aspects of perpetuation and recurrence of leaf blight of wheat in Rajasthan, *Indian J. Mycol. Plant Pathol.*, 3, 93, 1973.
43. McDonald, W. C. and Martens, J. W., Leaf and stem spot of sunflowers caused by *Alternaria zinniae*, *Phytopathology*, 53, 93, 1963.
44. Madeira, A. C., Clark, J. A., Rossall, S., and McArthur, A. J., A classification system for seeds of faba bean infected by *Ascochyta fabae*, *FABIS Newsl.*, 30, 48, 1992.
45. Sundheim, L., *Botrytis fabae*, *B. cinerea* and *Ascochyta fabae* on broad bean (*Vicia faba*) in Norway, *Acta Agric. Scand.*, 23, 43, 1973.
46. Kaiser, W. J. and Hannan R., Seed treatment fungicides for control of seedborne *Ascochyta lentis* on lentil, *Plant Dis.*, 71, 58, 1987.
47. Smith, A. L., Ascochyta seedling blight of cotton in Alabama in 1950, *Plant Dis. Rep.*, 34, 233, 1950.
48. Van der Spek, J., The influence of environment on the effect of linseed disinfection, *Meded. Landb. Hogesch.* (Gent.) 22, 535, 1957.
49. Buchanan, P. K., Systemic growth of *Ascochyta papsali* in *Paspalum*, *N. Z. J. Agric. Res.*, 27, 451, 1984.
50. Crosier, W. F., Prevalence and significance of fungus associates of pea seeds, *Proc. Assoc. Off. Seed Anal. North Am.*, 26, 101, 1936.
51. Luthra, J. C. and Bedi, K. S., Some preliminary studies on gram blight with reference to its cause and mode of perennation, *Indian J. Agric. Sci.*, 2, 499, 1932.
52. Montorsi, F., di Giambattista, G., and Porta-Punglia, A., First report of *Ascochyta rabiei* on berseem clover seeds, *Plant Dis.*, 76, 538, 1992.
53. Leach, C. M., *Kabatiella caulivora*, a seedborne pathogen of *Trifolium incarnatum* in Oregon, *Phytopathology*, 52, 1184, 1962.
54. Reifschneider, F. J. B. and Arny, D. C., Seed infection of maize (*Zea mays*) by *Kabatiella zeae*, *Plant Dis. Rep.*, 63, 352, 1979.
55. Mehan, V. K., Rao, R. C. N., McDonald, D., and Whilliams, J. H., Management of drought stress to improve field screening of peanuts for resistance to *Aspergillus flavus*, *Phytopathology*, 78, 659, 1988.
56. Huizar, H. E., Bertke, C. C., Klich, M. A., and Aronson, J. M., Cytochemical localization and ultra structure of *Aspergillus flavus* in cottonseed, *Mycopathology*, 110, 43, 1990.
57. Tsay, J. G., *Aspergillus* seedling blight of soybean and factors affecting its development, *Plant Prot. Bull.*, 32, 183, 1990.
58. Suryanarayanan, T. S. and Suryanarayanan, C. S., Fungi associated with stored sunflower seed, *J. Econ. Taxon. Bot.*, 14, 174, 1990.
59. Mittal, R. K. and Sharma, M.R., Chemical control of *Aspergillus flavus* on the seeds of *Shorea robusta*, *Indian Phytopathol.*, 33, 597, 1980.
60. Payne, G. A., Thompson, D. L., Lillehoj, E. B., Zuber, M. S., and Adkins, C. R., Effect of temperature on the preharvest infection of maize kernels by *Aspergillus flavus*, *Phytopathology*, 78, 1376, 1988.

61. Mycock, D. J., Rijkenberg, F. H. J., and Berjak, P., Systematic transmission of *Aspergillus flavus* var. *columnaris* from one maize seed generation to the next, *Seed Sci. Technol.*, 20, 1, 1992.
62. Hayden, N. J. and Maude, R. B., The role of seedborne *Aspergillus niger* in transmission of black mold of onion, *Plant Pathol.*, 41, 573, 1992.
63. Vaziri, A. and Vaughan, E. K., Aspergillus crown rot of peanut in Iran (*A. niger* on groundnut), *Plant Dis. Rep.*, 60, 602, 1976.
64. Zummo, N. and Scott, G. E., Relative aggressiveness of *Aspergillus flavus* and *A. parasiticus* on maize in Mississippi, *Plant Dis.*, 74, 978, 1990.
65. Ellis, M. A., Ilyas, M. B., and Sinclair, J. B., Effect of cultivar and growing region on internally seedborne fungi and *Aspergillus melleus* pathogenicity in soybean, *Plant Dis. Rep.*, 58, 332, 1974.
66. Clay, K. and Jones, J. P., Transmission of *Atkinsonella hypoxylon* (Clavicipitaceae) by cleistogamous seeds of *Danthonia spicata* (Graminaceae), *Can. J. Bot.*, 62, 2893, 1984.
67. Mohanty, N. N., Studies on "Udbatta" disease of rice, *Indian Phytopathol.*, 17, 308, 1964.
68. Boothroyd, C. W., Transmission of *Helminthosporium maydis* race T by infected corn seed, *Phytopathology*, 61, 747, 1971.
69. Chidambaram, P., Mathur, S. B., and Neergaard, P., Identification of seed-borne *Drechslera* species, *Friesia*, 10, 165, 1973.
70. Tveit, M., Pathogenicity of species of *Helminthosporium* from Brazilian oats, *Phytopathology*, 46, 45, 1956.
71. Sisterna, M. and Wolcan, S., *Bipolaris zeae* on *Pennisetum clandestimum* in Argentina, *Summa Phytopathol.*, 16, 184, 1990.
72. Maholay, M. N. and Sohi, H. S., Seedborne infection of *Capsicum* by *Botryodiplodia palmarum* (Cook) Petrak and Syd. and Syd. and its control, *Indian J. Hort.*, 34, 316, 1977.
73. Gilman. G. A., Black and moldy groundnuts in the Gambia, *Commonw. Phytopathol. News*, 1, 1965.
74. Maholay, M. N. and Sohi, H. S., *Botryodiplodia* seed rot of bottlegourd and squash, *Indian J. Mycol. Plant Pathol.*, 12, 32, 1982.
75. Maholay, M. N. and Sohi, H. S., Studies on *Botryodiplodia* rot of *Dolichos biflorus*, *Indian J. Mycol. Plant Pathol.*, 6, 126, 1977.
76. Roncadori, R. W., McCarter, S. M., and Crawford, J. L., Influence of fungi on cotton seed deterioration prior to harvest, *Phytopathology*, 61, 1326, 1971.
77. Anonymous, Forestry Commission U.K., *Report on Forest Research for The Year Ended March 1980*, Her Majesty's Stationary Office, London, U.K., 1980, 87.
78. Rees, A. A., Infection of *Pinus caribaea* seed by *Lasiodiplodia theobromae*, *Trans. Br. Mycol. Soc.*, 90, 321, 1988.
79. Kumar, V. and Shetty, H. S., Transmission of *Botryodiplodia theobromae* in maize (*Zea mays*), *Seed Sci. Technol.*, 11, 781, 1983.
80. Maude, R. B. and Presly, A. H., Fungal diseases, in 26th Annual report for 1975, National Vegetable Res. Stn., Wellesbourne, U.K., 1976, 158.
81. Salonen, A., The occurrence of *Botrytis anthophila* in samples of commercial seed of Finnish red clover, *Maataloust. Aikakausk.*, 32, 186, 1960.
82. Anonymous, Brassica diseases, vegetable crop production and storage, in Annu. Rep. Edinburgh Sch. Agric. 1982, 1983, 147.

83. Baker, K. F., Treatment of Seed and Planting Material: A Series of Lectures Presented at the New South Wales Association of Nurserymen's Seminar, Australia, 1969.
84. Sackston, W. E., *Botrytis cinerea* and *Sclerotinia sclerotiorum* in seed of safflower and sunflower, *Plant Dis. Rep.*, 44, 664, 1960.
85. Laha, S. K. and Grewal, J. S., *Botrytis* blight of chick pea and its perpetuation through seeds, *Indian Phytopathol.*, 136, 630, 1986.
86. Anselme, C. and Champion, R., Study of the transmission of *Botrytis cinerea* by sunflower (*Helianthus annuus*) seed, *Seed Sci. Technol.*, 3, 711, 1975.
87. Dienes, J. G., Laboratory investigation on the effectiveness of different fungicides against *Ceratophorum setosum* and *Botrytis cinerea* isolated from lupin seeds, *Novenyvedelem*, 18, 267, 1982.
88. Ogilvie, L., The control of vegetable diseases, *Occas. Publ. Hortic. Educ. Assoc.*, 5, 63, 1947.
89. Harrison, J. G., Role of seed-borne infection in epidemiology of *Botrytis fabae* on field beans, *Trans. Br. Mycol. Soc.*, 70, 35, 1978.
90. Tao, C. F., A preliminary study on the primary sources of the downy mildew of lettuce caused by *Bremia lactucae* Reg. in Chengtu, *Acta Phytophylac. Sin.*, 4, 15, 1965.
91. Sutherland, J. R., Time, temperature and moisture effects on incidence of seed infected by *Caloscypha fulgens* in Sitka spruce cones, *Can. J. Forest Res.*, 11, 727, 1981.
92. Ozaki, M., Kondo, N., and Akai, J., Seeds of wheat and soil infestation with *Cephalosporium gramineum* Nis. & Ika. causal fungus of *Cephalosporium* stripe of wheat, *Bull. Hokkaido, Prefectural Agric. Exp. Stn.*, 56, 75, 1987.
93. Samra. A. S., Sabet, K. A., and Hingorani, M. K., Late wilt disease of maize caused by *Cephalosporium maydis*, *Phytopathology*, 53, 402, 1963.
94. Omamor, I. B., Black rot of sprouted seeds of oil palm (*Elaeis quineensis*), *Trans. Br. Mycol. Soc.*, 84, 159, 1985.
95. Gibbons, W. R., Mycosphaerella leaf spots of groundnuts, *FAO Plant Prot. Bull.*, 14, 25, 1966.
96. Darpoux, H., Lebrun, A., and De la Tullaye, B., Studies on the diseases of beet carried out in 1961, *Publ. Inst. Tech. Better. Irid.*, 1961, 33; 1962, 31.
97. Dhingra, O. D. and Asmus, G. L., An efficient method of detecting *Cercospora canescens* in bean seeds, *Trans. Br. Mycol. Soc.*, 81, 425, 1983.
98. Doolittle, S. P., Diseases of peppers, in *Yearbook of Agriculture,* U.S. Department of Agriculture, Washington, D.C., 1953, 466.
99. Thomas, H. R., Cercospora blight of carrot, *Phytopathology*, 33, 114, 1943.
100. Sundararaman, S. and Ramakrishnan, T. S., A leaf-spot disease of safflower (*Carthamus tinctorius*) caused by *Cercospora carthami*, nov. sp., *Agric. J. India*, 23, 383, 1928.
101. Chowdhury, S., A *Cercospora* leaf blight of jute, *J. Indian Bot. Soc.*, 26, 227, 1948.
102. Johnson, H. W. and Jones, J., Purple stain of guar, *Phytopathology*, 52, 269, 1962.
103. Ilyas, M. B., Dhingra, O. D., Ellis, M. A., and Sinclair, J. B., Location of mycelium of *Diaporthe phaseolorum* var. *sojae* and *Cercospora kikuchii* in infected soybean seed, *Plant Dis. Rep.*, 59, 17, 1975.
104. Park, M. and Fernando, M., Some studies on tobacco diseases in Ceylon, I, *Trop. Agric.*, 88, 153, 1937.
105. Cherewick, W. J., Studies on seed-borne microflora and the effect of seed treatment of rice, *Malaysia Agric. J.*, 37, 169, 1954.

106. Singh, U. B., Phytopathological examination of seeds by Ulster method, *Curr. Sci.*, 17, 266, 1948.
107. Thomas, C. A., Notes on diseases of some special crops in 1950, *Plant Dis. Rep.*, 34, 391, 1950.
108. Kurozawa, C., Nakagawa, J., Doi, J., and Melloto, E., Behavior of 13 sesame (*Sesamum indicum*) cultivars to *Cercospora sesami*, its transmissibility by seed and control, *Fitopathol. Brasileira*, 10, 123, 1985.
109. Sherwin, H. S. and Kreitlow, K. W., Discoloration of soybean seeds by the frogeye fungus, *Cercospora sojina*, *Phytopathology*, 42, 568, 1952.
110. Buddin, W. and Wakefield, E. M., Notes on some *Antirrhinum* diseases, *Gardeners' Chronicle No. 1966*, 1924, 76.
111 Labuschagne, N. and Kotze, J. M., Incidence of *Chalara elegans* in groundnut seed samples and seed transmission of black-hull, *Plant Pathol.*, 40, 639, 1991.
112. Tai Luang Huan and Musa Bin Jamil, M., Seed-borne pathogens in okra fruit rot, *Mardi Res. Bull.*, 3, 38, 1975.
113. Hillocks, R. J. and Brettell, J. H., The association between honeydew and growth of *Cladosporium herbarum* and other fungi on cotton lint, *Trop. Sci.*, 33, 121, 1993.
114. Lenti, I., Cladosporium disease of broadbean (*Vicia faba* L.), *Novenyvedelem*, 25, 481, 1990.
115. Siddiqui, M. R. and Khan, I. D., Renaming *Claviceps microcephala* ergot fungus on *Pennisetum typhoides* in India as *Claviceps fusiformis*, *Trans. Mycol. Soc. Jpn.*, 14, 195, 1973.
116. Fuentes, S., De Lourdes de la Isla, M., Ullstrup, A. J., and Rodriguez, A. E., Ergot of maize in Mexico, *Phytopathology*, 52, 733, 1962.
117. Theis, T., An undescribed species of ergot on *Panicum maxicum* Jacq. var. Common Guinea, *Mycologia*, 44, 789, 1952.
118. Richardson, M. J., *An Annotated List of Seedborne Diseases*, ISTA Secretariat, Zurich, Switzerland, 1990, 386.
119. Wells, H. D., Burton, G. W., and Jackson, J. E., Burning of dormant dallis grass shows promise of controlling ergot caused by *Claviceps paspali* Stev. and Hall, *Plant Dis. Rep.*, 42, 30, 1958.
120. Efimova, N. S., Ergot on cereals in the foothills of Zaillisky Ala-Tau, *Referat Zh. Biol.*, 15, 211, 1958.
121. Alderman, S. C., Distribution of *Gloeotinia temulenta*, *Claviceps purpurea*, and *Anguina agrostis* among grasses in the Willamette Valley of Oregon in 1988, *J. Appl. Seed Prod.*, 6, 6, 1988.
122. Graniti, A., A host of *Claviceps purpurea* (Fr.) Tul. new for Italy: *Avena sativa* L. (cult.), *Notiz. Malatt. Piante,* 29, 16, 1955.
123. Suryanarayana, D., Outbreaks of new records, ergot on barley, *FAO Plant Prot. Bull.*, 15, 19, 1967.
124. Kobel, H. and Stopp, K., A first record of a wild form of *C. purpurea* on *Paspalum*, *Naturwissenschaften*, 54, 145, 1967.
125. Niemann, E., Possibilities of the separation of ergot from rye seed, *Angew. Bot.*, 30, 65, 1956.
126. Futrell, M. C. and Webster, O. J., Ergot infection and sterility in grain sorghum, *Plant Dis. Rep.*, 49, 680, 1965.
127. Futrell, M. C. and Webster, O. J., Host range and epidemiology of the sorghum ergot organism, *Plant Dis. Rep.*, 50, 823, 1966.
128. Brady, L. R., Phylogenetic distribution of parasitism by *Claviceps* species, *Lloydia*, 25, 1, 1962.

129. Fuentes, S. F., Triticale diseases review, in Proc. Int. Symp. Triticale, El Batan, Mexico, Int. Dev. Res. Center, Mexico City, 1974, 187.
130. Langdon, R. F. N., New species of *Claviceps*, *Pap. Dept. Bot. Univ. Q.*, 3, 39, 1954.
131. Grover, R. K. and Bansal, R. D., Seed-borne nature of *Colletotrichum capsici* in chili seeds and its control by seed dressing fungicides, *Indian Phytopathol.*, 23, 664, 1970.
132. Pangtey, Y. P. S. and Sinha, S., Two new seedborne leaf-spot diseases in horsegram in Kumaun hills, *Indian J. Agric. Sci.*, 50, 502, 1980.
133. Emechebe, A. M., Brown blotch of cowpea in northern Nigeria, *Samaru J. Agric. Res.*, 1, 20, 1981.
134. Behr, L., On a total loss of germinating onions caused by *C. dematium f. circinans*, *Zbl. Bakt.*, 116. 552. 1963.
135. Ghosh, T., Anthracnose of jute, *Indian Phytopathol.*, 10, 63, 1957.
136. Smith, R. W. and Crossan, D. F., The taxonomy, etiology and control of *Colletotrichum piperatum* (E. & E.) E. & H. and *Colletotrichum capsici* (Syd.) B. & B., *Plant Dis. Rep.*, 42, 1099, 1958.
137. Whiteside, J. O. and Herd, G. W., List of diseases of economic plants in Rhodesia, *Rhod. Agric. J. Tech. Bull.*, 5, 28, 1966.
138. Wu, S. L. and Hsu, J. S., Control of kenaf anthracnose, *Acta Phytopathol. Sin.*, 2, 127, 1956.
139. Gahukar, K. B., Raut, J. G., and Deshmukh, R. N., Seed infection in relation to pericarp infection of *Colletotrichum dematium* in chilli, *PKV Res. J.*, 13. 52. 1989.
140. Chikuo, Y. and Sugimoto, T., Histopathology of sugarbeet flowers and seed balls infected with *Colletotrichum dematium* f. *spinaciae*, *Ann. Phytopathol. Soc. Jpn.*, 55, 404, 1989.
141. TeBeest, D. O. and Brumley, J. M., *Colletotrichum gloeosporoides* borne within the seed of *Aeschynomene virginica*, *Plant Dis. Rep.*, 62, 675, 1978.
142. Roy, K. W., Seedling diseases caused in soybean by species of *Colletotrichum* and *Glomerella*, *Phytopathology*, 72, 1093, 1982.
143. Benic, L. M. and Knox-Davies, P. S., Anthracnose of *Protea compacta* caused by *Colletotrichum gloeosporioides*, *Phytophylactica*, 15, 109, 1983.
143a. Arndt, C. H., Survival of *Colletotrichum gossypii* on cotton seeds in storage, *Phytopathology*, 43, 220, 1953.
144. Lima, E. F., Carvalho, J. M. F. C., and Carvalho, L. P. de., Survival of *Colletotrichum gossypii* var. *cephalosporioides* on cotton (*Gossypium hirsutum* L. r. *latifolium*) seeds, *Fitopatol. Brasileira*, 13, 247. 1988.
145. Basuchaudhary, K. C. and Mathur, S. B., Infection of sorghum seeds by *Colletotrichum graminicola*. I. Survey, location in seed and transmission of the pathogen, *Seed Sci. Technol.*, 7, 87, 1979.
146. Warren, H. L. and Nicholson, R. L., Kernel infection, seedling blight, and wilt of maize caused by *Colletotrichum graminicola*, *Phytopathology*, 65, 620, 1975.
147. Scheffer, R. P., Anthracnose leaf spot of crucifers, *Tech. Bull. N.C. Agric. Exp. Stn.*, 92, 26, 1950.
148. Bhide, V. P., Desai, M. K., and Rane, M. S., Control of seedborne infection of anthracnose of cotton in Bombay State, *Indian Cotton Growing Rev.*, 11, 496, 1957.
149. Rankin, H. W., Effectiveness of seed treatment for controlling anthracnose and gummy-stem blight of watermelon, *Phytopathology*, 44, 675, 1954.
150. Panteleev, A. A. and Bagdasaryan, A. Z., Increase in the resistance of melons to anthracnose when treated with microelements, *Khim. sel'sk Khoz.*, 9, 50, 1971.

151. Senyurek, M., Zavrak, Y., and Olcum, S. K., Research on the bioecological basis of control of anthracnose (*Colletotrichum lagenarium* (Pass) Ell, & Halst.) on melon and watermelon in the Black Sea Region, *Bitki Koruma Bulteni*, 17, 151, 1977.
152. Horn, N. L., Wilson, W. F., and Giamalva, M., Seed and insect transmission of cucumber anthracnose, *Plant Dis. Rep.*, 41, 69, 1957.
153. Medina, A. C., Viability of seeds and a bean pathogen, *Agric. Tec. Mex.*, 3, 3, 1970.
154. Pietkiewicz, T. A. and Czyzewska, S., Effect of seed treatment on the infection of flax seedlings by the fungus, *Colletotrichum linicola*, *Rocz. Nauk Roln. Ser. A*, 81, 671, 1960.
155. Pape, H., *Krankheiten und Schädlinge der Zierpflanzen und ihre Bekämpfung*, 5th ed., P. Parey, Berlin, 1964, 625.
156. Butler, F. C., Anthracnose and sooty blotch of red clover. Two diseases new to New South Wales, *Agric. Gaz. N.S.W.*, 64, 368, 1953.
157. Lehman, S. G. and Wolf, F. A., Soybean anthracnose, *J. Agric. Res.*, 33, 381, 1926.
158. Chambers, A. Y. and Andes, J. O., Stem anthracnose (*Colletotrichum truncatum*) and its control on lima beans, *Bull. Tenn. Agric. Exp. Stn.*, 338, 19, 1962.
159. Simay, E. I., *Colletotrichum dematium* (Pers. ex Fr.) Grove f. sp. *truncata* (Schw./Arx), a new seed-transmitted pathogen of *Vicia faba* L., *Fabis Newsl.*, 26, 32, 1990.
160. Bains, S. S., Kaur, L., Dhaliwal, H. S., and Gill, A. S., Outbreak of a new anthracnose of mung and mash in Punjab, *Indian Bot. Rep.*, 8, 164, 1989.
161. Muthaiyan, M. C., Sreehari, M., and Devi, R., Note on interception of *Coniothyrium fuckelii* Sacc. on fruits of *Rosa* spp., *Plant Prot. Bull.*, 44, 28, 1992.
162. Liu, Sih Tsing, Seed-borne diseases of soybean, *Bot. Bull. Acad. Sin.*, 2, 69, 1948.
163. Gadage, N. B. and Patil, B. P., Further studies on the Curvularia leaf spot of cotton in Maharashtra, *Maharashtra Agric. Univ.*, 2, 239, 1977.
164. Grewal, J. S. and Pal, M., Seed microflora. II. Seedborne fungi of *Setaria italica*, their distribution and control, *Indian Phytopathol.*, 18, 123, 1965.
165. Singh, D. P. and Agarwal, V. K., Effect of different degrees of grain mold infection on yield and quality of sorghum seed, *Indian J. Plant Pathol.*, 7, 103, 1989.
166. Porter, D. M., Wright, F. S., Taber, R. A., and Smith, D. H., Colonization of peanut seed by *Cylindrocladium crotalariae*, *Phytopathology*, 81, 896, 1991.
167. Kmetz, K. T., Schmitthenner, A. F., and Ellett, C. W., Soybean seed decay: prevalence of infection and symptom expression caused by *Phomopsis* sp., *Diaporthe phaseolorum* var. *sojae*, and *D. phaseolorum* var. *caulivora*, *Phytopathology*, 68, 836, 1978.
168. Luttrell, E. S., *Diaporthe phaseolorum* var. *sojae* on crop plants, *Phytopathology*, 37, 447, 1947.
169. Wallen, V. R. and Seamann, W. L., Seed infection of soybean by *Diaporthe phaseolorum* and its influence on host development, *Can. J. Bot.*, 41, 13, 1963.
170. Nicholson, J. F., Dhingra, O. D., and Sinclair, J. B., Internal seed-borne nature of *Sclerotinia sclerotiorum* and *Phomopsis* spp. and their effects on soybean seed quality, *Phytopathology*, 62, 1261, 1972.
171. Petterson, D. S., Peterson, J. E., Smith, L. W., Wood, P. M., and Culvenor, C. C. J., Bioassay of contamination of lupin seed by the mycotoxin phomopsin, *Aust. J. Exp. Agric.*, 25, 434, 1985.
172. Parris, G. K., Watermelon disease control, *Proc. Fla. Hortic. Soc.*, 147, 1947.
173. Brown, M. E., Howard, E. M., and Knight, B. C., Seed-borne *Mycosphaerella melonis* on cucumber, *Plant Pathol.*, 19, 198, 1970.

174. Derbyshire, D. M., Seed transmission of *Didymella lycopersici*, *Plant Pathol.*, 9, 152, 1960.
175. Nwigwe, C., Effect of *Diplodia zeae* and β-*Phomopsis* on the germination of seeds of maize (*Zea mays*), *Plant Dis. Rep.*, 58, 414, 1974.
176. Dai, K., Nagai, M., Sasaki, H., Nakamura, H., Takechi, K., and Warbi, M., Detection of *Diplodia maydis* (Berkeley) Saccardo from imported corn seed, *Res. Bull. Plant Prot. Sci. Jpn.*, 23, 1, 1987.
177. Sheridan, J. E. and Tan, P. E. T., Incidence and survival of *Pyrenophora avenae* in New Zealand seed oats, *N. Z. J. Agric. Res.*, 16, 251, 1973.
178. Levic, J. and Pencic, V., Contribution to the investigation on protection of maize seed against *Helmithosporium carbonum* by using fungicides, *Zastita Bilja*, 34, 493, 1983.
179. Verhoeven, W. B. L., The stripe disease of barley, *Tijdschr. Plantzenzieken*, 27, 105, 1921.
180. Poole, D. D., Aerial stem rot of sesame caused by *Helminthosporium sesami* in Texas, *Plant Dis. Rep.*, 40, 235, 1956.
181. Shetty, H. S., Mathur, S. B., Neergaard, P., and Safeeulla, K. M., *Drechslera setariae* in Indian pearl millet seeds, its seedborne nature, transmission and significance, *Trans. Br. Mycol. Soc.*, 78, 170, 1982.
182. de Tempe, J., *Helminthosporium* spp. in seeds of wheat, barley, oats and rye, *Proc. Int. Seed Test. Assoc.*, 29, 117, 1964.
183. Richardson, M. J., An Annotated List of Seed-borne Diseases, 3rd ed., Commonwealth Mycological Institute, Kew, Surrey, U.K., 1990, 320.
184. Sitterly, W. R. and Epps, W. M., Lima bean scab found in South Carolina, *Plant Dis. Rep.*, 42, 1309, 1958.
185. Welty, R. E., Milbrath, G. M., Faulkenberry, D., Azevedo, M. D., Meek, L., and Hall, K., Endophyte detection in tall fescue seed by staining and ELISA, *Seed Sci. Technol.*, 14, 105, 1986.
186. Wright, W. R. and Billeter, B. A., Red kernel disease of sweet corn on the retail market, *Plant Dis. Rep.*, 58, 1065, 1974.
187. Tanaka, F. and Tsuchiya, S., Factors affecting the incidence of pink coloring of rice grains, caused by *Epicoccum purpurescens*. I. Conidial dispersion and infection to the glume and kernels, *Bull. Hokkaido Prefectural Agric. Exp. Stn.*, 56, 51, 1987.
188. Ruppel, E. G. and Tomasovic, B. J., Epidemiological factors of sugarbeet powdery mildew, *Phytopathology*, 67, 619, 1977.
189. Baker, K. F. and Locke, W. F., Perithecia of powdery mildew on zinnia seed, *Phytopathology*, 36, 379, 1946.
190. Uppal, B. N., Patel, M. K., and Kamat, M. N., Pea powdery mildew in Bombay, *Bull. Dep. Agric. Bombay*, 177, 12, 1935.
191. Shcherbina, E. A. and Orlov, V. P., Effect of nitrogen on yield of narrow leaved fodder lupin and on its infection by *Fusarium* sp. on dark grey forest soil, *Referatinvyl Zhurnol*, 6, 212, 1978.
192. Lamprecht, S. C., Marasas, W. F. O., Knox-Davies, P. S., and Calitz, F. J., Overwintering of *Fusarium avenaceum* on stems, seeds, seedpods, and roots of *Medicago truncatula, Phytophylactica,* 21, 403, 1989.
193. Mishra, C. B. P., Studies on *Fusarium* species on wheat caryopses and proof of their pathogenicity as foot disease agents, *Arch. Phytopathol. Pflanzenschutz*, 9, 123, 1973.

194. Sinha, O. K. and Khare, M. N., Site of infection and further development of *Macrophomina phaseolina* and *Fusarium equiseti* in naturally infected cowpea seeds, *Seed Sci. Technol.*, 5, 721, 1977.
195. Shands, R. G., Longevity of *Gibberella saubinetii* and other fungi in barley kernels and its relation to the emetic effect, *Phytopathology*, 27, 749, 1937.
196. Mironenko, P. V., Fusarium wilt of rice in the Caspian area, Trudy *Vsesoyuznogo Nauchnoissledovatel skogo Inst. Zashchity Rastenii*, 14, 123, 1960.
197. Andersen, A. L., The development of *Gibberella zeae* head blight of wheat, *Phytopathology*, 38, 595, 1948.
198. Dungan, G. H. and Koehler, B., Age of seed corn in relation to seed infection and yielding capacity, *J. Am. Soc. Agron.*, 36, 436, 1944.
199. Jeswani, M. D. and Gemawat, P. D., Seed-borne nature of wilt of arhar, in 3rd Int. Symp. Plant Pathology, New Delhi, 1981, 162.
200. Inglis, D. A., Contamination of asparagus seed by *Fusarium oxysporum* f. sp. *asparagi* and *Fusarium moniliforme*, *Plant Dis.*, 64, 74, 1980.
201. Mathur, S. K., Mathur, S. B., and Neergaard, P., Detection of seedborne fungi in sorghum and location of *Fusarium moniliforme* in the seed, *Seed Sci. Technol.*, 3, 683, 1975.
202. Wu, W. S. and Mathur, S. B., Evaluation of methods for detecting *Fusarium moniliforme* in sorghum seeds, *Seed Sci. Technol.*, 15, 821, 1987.
203. Pidoplichko, V. M., Phytotoxicity of *Fusarium* spp. causal agents of root rot of winter wheat, *Mykrobiol. Zh.*, 32, 700, 1970.
204. Ashraf, S. S. and Basuchaudhary, K. C., Effect of seedborne *Fusarium* species on the physico-chemical properties of rapeseed oil, *J. Phytopathol.*, 117, 107, 1986.
205. Vannacci, G., Record of *Fusarium moniliforme* var. *subglutinans* Wr. Reink on onion seeds and its pathogenicity, *Phytopathol. Mediterranea*, 20, 144, 1981.
206. Anderson, R. L., New method of assessing contamination of slash and loblolly pine seeds by *Fusarium moniliforme* var. *subglutinans*, *Plant Dis.*, 70, 452, 1986.
207. Vidhyasekaran, P. and Thiagarajan, C. P., Seedborne transmission of *Fusarium oxysporum* in chilli, *Indian Phytopathol.*, 34, 211, 1981.
208. Velicheti, R. K. and Sinclair, J. B., Histopathology of soybean seeds colonised by *Fusarium oxysporum, Seed Sci. Technol.*, 19, 445, 1991.
209. Lomeli, O. C., Leyva, M. G., and Castro, A. L., Root disease of alfalfa (*Medicago sativa* L.) in Chapingo, Mexico, and their seed transmission, *Rev. Mexicana Fitopatol.*, 9, 8, 1991.
210. Graham, J. H., and Linderman, R. G., Pathogenic seedborne *Fusarium oxysporum* from Douglas fir, *Plant Dis.*, 67, 323, 1983.
211. Ibrahim, G. and Owen, H., *Fusarium oxysporum* causal agent of root rot on broadbean in the Sudan, *Phytopathol. Z.*, 101, 80, 1981.
212. Orton, T. J., Physical association of *Fusarium oxysporum* f. sp. *apii* with germinating celery seeds following artificial inoculation, *Phytopathol. Z.*, 102, 320, 1981.
213. Gross, D. C. and Leach, L. D., Stalk blight of sugarbeet seed crops caused by *Fusarium oxysporum* f. sp. *betae*, *Phytopathology*, 63, 1216, 1973.
214. Klisiewicz, J. M., Wilt-incitant *Fusarium oxysporum* f. *carthami* present in seed from infected safflower, *Phytopathology*, 53, 1046, 1963.
215. Baker, K. F., A problem of seedsmen and flower growers — seedborne parasites, *Seed World*, 70, 38, 1952.

216. Haware, M. P., Nene, Y. L., and Rajeshwari, R., Eradication of *Fusarium oxysporum* f. sp. *ciceri* transmitted in chickpea seed, *Phytopathology*, 68, 1364, 1978.
217. Tamietti, G. and Garibaldi, A., Fusarium vascular wilt of Brussels sprouts: critical period for infection, varietal resistance and seed transmissibility, *Inform. Fitopathol.*, 32, 43, 1982.
218. Mishra, C. B. P. and Som, D., Field evaluation of Bavistin in controlling wilt of sunn-hemp, *Indian J. Mycol. Plant Pathol.*, 13, 324, 1985.
219. Takeuchi, S., Ogawa, K., and Nomura, Y., Mechanism of seed transmission of Fusarium wilt of cucumber and bottle gourd and improvement of a method for the seed disinfection test, *J. Cent. Agric. Exp. Stn.*, 28, 49, 1978.
220. Mathur, B. L. and Mathur, R. L., Role of contaminated seeds in dissemination of cumin wilt fungus *Fusarium oxysporum* f. *cumini*, *Rajasthan J. Agric. Sci.*, 1, 80, 1970.
221. Gardner, D. E., *Acacia koa* seedling wilt caused by *Fusarium oxysporum* f. sp. *koae*, f. sp. nov., *Phytopathology*, 70, 594, 1980.
222. Kuniyashu, K. and Kishi, N., Seed transmission of Fusarium wilt of bottle gourd, *Lagenaria siceraria* Standle, used as root stock of watermelon. II. The seed infection course from the infected stem of bottle gourd to the fruit and seed, *Ann. Phytopathol. Soc. Jpn.*, 43, 192, 1977.
223. Miladenov, M., Flax seed as a source of infection in the transmission of Fusarium wilt, *Rastit. Zashchi.*, 22, 29, 1974.
224. Besri, M., Phases of the transmission of *Fusarium oxysporum* f. sp. *lycopersici* and *Verticillium dahliae* by seeds of some tomato varieties, *Phytopathol. Z.*, 93, 148, 1978.
225. Tÿmchenko, V. I., Character of pathological changes in Fusariosis of watermelon and the role of seed in the spread of infection in selection and seed breeding of potato, vegetables, and Cucurbitaceae, Sokol, P. F., Ed., *Kartoplya Ovochevi Bashtanni Kult. Kiew, Idzat, Vrozhai*, 1964, 141.
226. Gill, D. L., Mimosa wilt *Fusarium* carried in seed, *Plant Dis. Rep.*, 52, 949, 1968.
227. Tuset Barrachina, J. J., Studies on wilt and desiccation of herbaceous plants. I. A *Fusarium* species pathogenic to bean in eastern Spain, *An. Inst. Nac. Invest. Argar.*, (Spain) *Ser. Prot. Veg.*, 3, 73, 1973.
228. Kerling, L. C. P., Seed transmission of the American vascular disease of peas, *Tijdschr. Plantzenzieken*, 58, 236, 1952.
229. Bassi, A., Jr., Goode, M. J., and Bowers, J. L., Spinach Fusarium decline, a widespread and destructive disease in Arkansas, *Ark. Farm Res.*, 27, 13, 1978.
230. Anahosur, K. H., Shivasmurthy, S. C., and Hiremath, R. V., Effect of seed microflora on the quality of soybean oil, *Curr. Res.*, 4, 83, 1975.
231. Kendrick, J. B., Seed transmission of cowpea Fusarium wilt, *Phytopathology*, 21, 979, 1931.
232. Gangopadhyay, S. and Kapoor, K. S., Control of Fusarium wilt of okra with seed treatment, *Indian J. Mycol. Plant Pathol.*, 7, 147, 1978.
233. Gubanov, G. Y. and Sabirov, B. G., Transmission of Fusarium wilt infection by cottonseed, *Khlopkovedstvo*, 22, 20, 1972.
234. Conroy, R. J., Fusarium root rot of cucurbits, *Agric. Gaz. N.S.W.*, 64, 655, 1953.
235. de Ribeiro, R., L. D., Robbs, C. F., Akiba, F., Kimura, O., and Sudo, S., Studies on pre- and post-emergence rot of okra (*Hibiscus esculentus* L.) in the lower Carioca Eluminense region caused by a new spiral form of *Fusarium solani* (Mart.) Appel, *Arq. Univ. Fed. Rural Rio de Janeiro*, 1, 9, 1971.

236. Nash, S. M. and Snyder, W. C., Dissemination of the root rot *Fusarium* with bean seed, *Phytopathology*, 54, 880, 1964.
237. Sinclair, J. B. and Backman, P. A., Eds., *Compendium of Soybean Diseases*, 3rd ed., APS Press, Inc., 1989, 106.
238. Haware, M. P. and Kannaiyan, J., Seed transmission of *Fusarium udum* in pigeonpea and its control by seed-treatment fungicides, *Seed Sci. Technol.*, 20, 597, 1992.
239. Uppal, B. N. and Kulkarni, N. T., Studies on Fusarium wilt of sunnhemp. I. The physiology and biology of *Fusarium vasinfectum* Atk., *Indian J. Agric. Sci.*, 7, 413, 1937.
240. Roll-Hansen, F., Investigation of *Gibberella saubinetii* (Mont.) Sacc. as foot rot of oats in 1940, *Meld. St. Frokontr. As.*, 1939-40, 32, 1941.
241. Foley, D. C., Systemic infection of corn by *Fusarium moniliforme*, *Phytopathology*, 52, 870, 1962.
242. Watanabe, T. and Hasmimoto, K., Recovery of *Gloeocercospora sorghi* from sorghum seed and soil, and its significance in transmission, *Ann. Phytopathol. Soc. Jpn.*, 44, 633, 1978.
243. Wilson, M., Noble, M., and Gay, E.G., The blind seed disease of rye-grass and its causal fungus, *Trans. R. Soc. Edinburgh*, 61, 327, 1945.
244. Matthews, D., A comparison of methods for the detection of blind seed disease (*Gloeotinia temulenta*) in ryegrass seed samples, *Seed Sci. Technol.*, 8, 183, 1980.
245. Ahn, J. K. and Chung, B. K., Effects of seed disinfectants for controlling the soybean anthracnose, *J. Plant Prot. Korea*, 9, 21, 1970.
246. Calo, R. P., Anthracnose of tomato, *Philipp. Agric.*, 41, 273, 1957.
247. Goode, M. J., Watermelon anthracnose: an old problem returns, *Ark. Farm Res.*, 17, 12, 1968.
248. Datta, B. and Sharma, G. D., A report on seed rot of *Terminalia myriocarpa* by *Graphium aeruginousm, Acta Bot.,* 18, 97, 1990.
249. Bogarada, A. P., Liman, V. E., and Spiridonova, V. P., New preparations for the protection of poppy, *Zakhist Roslin,* 28, 19, 1981.
250. Dewey, F. M., Macdonald, M. M., and Phillips, S. T., Development of monoclonal antibody — ELISA-DOT-Blot and DIP-Stick immunoassays for *Humicola lanuginosa* in rice, *J. Gen. Microbiol.*, 135, 361, 1989.
251. Kiewnick, L., Untersuchungen über den Einfluss der Samen und Bodenmikroflora auf die Lebensdauer der Spelzfrüchte des Flughafers (*Avena fatua* L.), *Weed Res.*, 3, 322, 1963.
252. Lutey, R. W., Chains of spores produced by *Cephalosporium gramineum, Mycopathol., Mycol. Appl.*, 18, 117, 1962.
253. Bruehl, G. W., Cephalosporium stripe disease of wheat, *Phytopathology*, 47, 641, 1957.
254. Arneson, E. and Stiers, D. L., *Cephalosporium gramineum,* a seedborne pathogen, *Plant Dis. Rep.*, 61, 619, 1977.
255. Reddy, C. S. and Holbert, J. R., The black bundle disease of corn, *J. Agric. Res.*, 27, 177, 1924.
256. Reddy, G. R., Reddy, A. G. R., and Rao, K. C., Effect of certain seed-borne fungi on seed germination and seedling vigour of groundnut, *J. Res. APAU*, 19, 77, 1991.
257. Maholay, M. N. and Sohi, H. S., *Botryodiplodia* seed rot of bottlegourd and squash, *Indian J. Mycol. Plant Pathol.*, 12, 32, 1982.
258. Maholay, M. N. and Sohi, H. S., Studies of *Botryodiplodia* rot of *Dolichos biflorus, Indian J. Mycol. Plant Pathol.*, 6, 126, 1977.

259. Roncadori, R. W., McCarter, S. M., and Crawford, J. L., Influence of fungi on cotton seed deterioration prior to harvest, *Phytopathology*, 61, 1326, 1971.
260. Forestry Commission U.K., Report on forest research for the year ended March 1980, London, Her Majesty's Stationary Office, U.K., 1980, 87.
261. Rees, A. A., Infection of *Pinus caribaea* seed by *Lasiodiplodia theobromae, Trans. Br. Mycol. Soc.*, 90, 321, 1988.
262. Kumar, V. and Shetty, H. S., Transmission of *Botryodiplodia theobromae* in maize (*Zea mays*), *Seed Sci. Technol.*, 11, 781, 1983.
263. Pineda, J. B., Colmenares, O., and Avila, J., Evaluation of sunflower (*Helianthus annuus* L.) hybrid seed production in relation to incidence of plant pathogens, *Agron. Trop.*, 41, 215, 1991.
264. Drachkov, V. N., Fir cone cast — a dangerous disease, *Les. Khoz*, 19, 59, 1965.
265. Reddy, G. R., Reddy A. G. R., and Rao, K. C., Effect of different seed dressing fungicides against certain seedborne fungi of groundnut, *J. Oilseeds Res.*, 8, 79, 1991.
266. Grewal, J. S. and Vir, D., Efficacy of different fungicides. I. Seed disinfection in relation to stem rot of jute and stripe disease of barley, *Indian Phytopathol.*, 11, 175, 1958.
267. Basu Chaudhary, K. C. and Pal, A. K., Infection of sunnhemp (*Crotalaria juncea*) seeds by *Macrophomina phaseolina*, *Seed Sci. Technol.*, 10, 151, 1982.
268. Shakir, A. S. and Mirza, J. H., Seedborne fungi of bottlegourd from Faisalabad and their control, *Pakistan J. Phytopathol.*, 4, 54, 1992.
269. Kunwar, I. K., Singh, T., Machado, C. C., and Sinclair, J. B., Histopathology of soybean seed and seedling infection by *Macrophomina phaseolina*, *Phytopathology*, 76, 532, 1986.
270. Radha, K., The genus *Rhizoctonia* in relation to soil moisture. II. *Rhizoctonia bataticola*, cotton root rot organism, *Indian Cotton J.*, 13, 137, 1960.
271. Lakshmidevi, N., Prakash, H. S., and Shetty, H. S., Effect of seedborne *Macrophomina phaseolina* (Tassi) Goid on physicochemical properties of sunflower oil, *Int. J. Trop. Plant Dis.*, 10, 79, 1992.
272. Rio, L. del, The effect of microorganisms on quality of bean seed used by agriculturalists in Honduras, *Ceiba*, 30, 81, 1989.
273. Singh, T. and Singh, D., Incidence of *Macrophomina phaseolina* (Tassi) Goid in sesame seeds of Rajasthan, *Indian J. Mycol. Plant Pathol.*, 20, 111, 1990.
274. Reddy, M. R. S. and Subbayya, J., *Macrophomina phaseolina* on seed health of blackgram (*Phaseolus mungo L.*), *Curr. Res.*, 10, 58, 1981.
275. Sinha, O. K. and Khare, M. N., Site of infection and further development of *Macrophomina phaseolina* and *Fusarium equiseti* in naturally infected cowpea seeds, *Seed Sci. Technol.*, 5, 721, 1977.
276. Yaegashi, H. and Asaga, K., Possibility of overwintering of crabgrass fungus, *Pyricularia grisea* (Cooke) Saccardo, by infected seeds, *Annu. Rept. Soc. Plant Protection North Jpn.*, 32, 77, 1981.
277. Spiers, A. G. and Wenham, H. T., Poplar seed transmission of *Marssonina brunnea*, *Eur. J. Forest Pathol.*, 13, 305, 1983.
278. Vasudeva, R. S., Report of the Division of Mycology and Plant Pathology, 1961, Indian Agric. Res. Inst., New Delhi, 1963,87.
279. Olvang, H., Benomyl resistance in *Gerlachia nivalis*. II. Investigations of seed lots and performance of carbendazim seed treatment, *Zeitsch. Pflanzenkrank. Pflanzenschutz*, 92, 561, 1985.

280. Cristani, C., Seedborne *Microdochium nivale* (es. ex Sacc.), Samules (=*Fusarium nivale* (Fr.) Ces.) in naturally infected seeds of wheat and triticale in Italy, *Seed Sci. Technol.*, 20, 603, 1992.
281. Miller, C. S. and Colhoun, J., Fusarium diseases of cereals. VI. Epidemiology of *Fusarium nivale* on wheat, *Trans. Br. Mycol. Soc.*, 52, 195, 1969.
282. Mia, M. A. T., Safeeulla, K. M., and Shetty, H. S., Seedborne nature of *Gerlachia oryzae*, the incitant of leaf scald of rice in Karnataka, *Indian Phytopathol.*, 39, 92, 1986.
283. Girand, D. H., *List of Intercepted Plant Pests,* 1972. U.S. Department of Agriculture, Animal and Plant Health Inspection Service, Washington, D.C. 1974.
284. Wells, H. D., Burton, G. W., and Ourecky, D. K., Polyposporium smut, a new disease on pearl millet, *Pennisetum glaucum*, in the United States, *Plant Dis. Rep.*, 47, 16, 1963.
285. Huber, G. A. and Gould, C. J., Cabbage seed treatment, *Phytopathology*, 39, 869, 1949.
286. Muskett, A. E. and Colhoun, J., Seedborne diseases of flax and their control, *Ann. Appl. Biol.*, 33, 331, 1946.
287. Leppik, E. E., Seedborne diseases on introduced plants, *Rep. North Cent. Reg. Plant Introd. Stn.*, Ames, 7, 1962, 11
288. Johnson, H. W., Chamberlain, D. W., and Lehman, S. G., Diseases of soybeans and methods of control, *U.S. Dep. Agric. Circ.*, 930, 40, 1954.
289. Dumitras, L., Some aspect of biology and control of *Nigrospora oryzae* (B & Br.) Petch Studii si Cercetari de Biologie, *Biol. Vegetala*, 34, 150, 1982.
290. Halfon-Meiri, A. and Solel, Z., Factors affecting seedling blight of sweet corn caused by seedborne *Penicillium oxalicum*, *Plant Dis.*, 74, 36, 1990.
291. Alcock, N. L., Downy mildew of *Meconopsis*, *New Flora Silva*, 5, 279, 1933.
292. Behr, L., Der falsche Mehltau am Mohn [*Peronospora arborescens* (Berk.) deBy.] Untersuchungen zur biologie des Erregers, *Phytopathol. Z.*, 27, 289, 1956.
293. Stuart, W. W. and Newhall, A. G., Further evidence of the seedborne nature of *Peronospora destructor*, *Phytopathology*, 25, 35, 1935.
294. Cook, H. T., The presence of mycelium of *Peronospora schleideni* in the flowers of *Allium cepa*, *Phytopathology*, 20, 139, 1930.
295. Zimmer, R. C., McKeen, W. E., and Campbell, C. G., Location of oospores in buckwheat seed and probable roles of oospores and conidia of *Peronospora ducometi* in the disease cycle on buckwheat, *J. Phytopathol.*, 135, 217, 1992.
296. Leach. L. D., Downy mildew of the beet caused by *Peronospora schactii* Fuckel, *Hilgardia*, 6, 203, 1931.
297. Inaba, T., Seed transmission of downy mildew of spinach and soybean, *JARQ*, 19, 26, 1985.
298. Johnson, H. W. and Lefebvre, C. L., Downy mildew on soybean seeds, *Plant Dis. Rep.*, 26, 49, 1942.
299. Polyakov, I. M. and Vladimirskaya, M. E., The role of light conditions in the resistance of cabbage to false powdery mildew, *Tr. Vses. Inst. Zashch. Rast.*, 21, 18, 1964.
300. Jang, P. and Safeeulla, K. M., Modes of entry, establishment and seed transmission of *Peronospora parasitica* in radish, *Proc. Indian Acad. Sci. Plant Sci.*, 100, 369, 1990.
301. Borovskaya, M., Transmission of *Peronospora tabacina* by seeds, *Zashch. Rast. Vredit. Bolez.*, 10, 44, 1965.

302. Champion, R. and Mecheneau, H., Method for the detection of *Peronospora valerianellae*, the causal organism of mildew on seeds of lambs lettuce (*Valerianella locusta*), comparison of culture and laboratory results, *Seed Sci. Technol.*, 7, 259, 1979.
303. Snyder, W. C., *Peronospora viciae* and internal proliferation in pea pods, *Phytopathology*, 24, 1358, 1934.
304. Yao, C. L., Magill, C. W., Frederiksen, R. A., Bonde, M. R., Wang, Y., and Wu, P. S., Detection and identification of *Peronosclerospora sacchari* in maize by DNA hybridization, *Phytopathology,* 81, 901, 1991.
305. Ahmad, R. and Majumder, S. K., Seedborne nature, detection and seed transmission of sorghum downy mildew of sorghum, in *Plant Quarantine and Phytosanitary Barriers to Trade in the ASEAN,* ASEAN Plant Quarantine Centre Training Inst. Malaysia, 173, 1987.
306. Rao, B. M., Prakash, H. S., Shetty, H. S., and Safeeulla, K. M., Downy mildew inoculum of *Peronosclerospora sorghi* in maize, *Seed Sci. Technol.*, 13, 593, 1985.
307. Sohi, H. S. and Sharma, R.D., Mode of survival of *Isariopsis griseola* Sacc., the causal agent of angular leaf spot of beans, *Indian J. Hort.*, 31, 110, 1974.
308. Saettler, A. W. and Corea, F. J., Transmission of *Phaeoisariopsis griseola* by bean seed, *J. Seed Technol.*, 12, 133, 1988.
309. Hewett, P.D., *Phoma apiicola* on celery seed, *Plant Pathol.*, 20, 123, 1971.
310. Leach, L. D. and MacDonald, L. D., Seed-borne *Phoma betae* as influenced by area of sugarbeet production, seed processing and fungicidal seed treatments, *J. Am. Soc. Sugar Beet Technol.*, 19, 4, 1976.
311. Wolcan, S. M. and Perello, A. E., Susceptibility of different organs of spinach beet and fodder beet to various isolates of *Phoma betae* Frank, *Rev. Facultad Agron.*, 65, 13, 1989.
312. Lacicowa, B., Kiecanna, L., and Pieta, D., Mycoflora of *Cyclamen persicum* Mill seeds and pathogenicity of *Phoma exigua* Desm. to this plant, *Acta Mycol.*, 26, 25, 1990.
313. Sneep, J., De *Ascochyta*-vlekkenziekte van de boon (*Phaseolus*), *Tijdschr. Plantenzieken*, 51, 1, 1945.
314. Boerema, G. H., Cruger, G., Gerlagh, M., and Nirenberg, H., *Phoma exigua* var. *diversipora* and related fungi on *Phaseolus* beans, *Z. Pflanzenkr. Pflanzenschutz*, 88, 597, 1981.
315. Jacobsen, B. J. and Williams, P. H., Histology and control of *Brassica oleracea* seed infection by *Phoma lingam*, *Plant Dis. Rep.*, 55, 934, 1971.
316. El-Sayed, F. and Maric, A., Contribution to the study of the biology and epidemiology of *Phoma macdonaldii* Boerema, pathogen of black spot of sunflower, *Zast. Bilja*, 32, 13, 1981.
317. Kernkamp, M. F. and Hemerick, G. A., Alfalfa seed loss due to *Ascochyta imperfecta* PK (black-stem), *Phytopathology*, 42, 468, 1952.
318. Shukla, C. S., Prasad, K. V. V., and Khare, M. N., Efficacy of different methods in the detection of *Phoma medicaginis* var. *sojae* associated with soybean seeds, *Indian J. Mycol. Plant Pathol.*, 20, 152, 1990.
319. Bathgate, J. A., Sivasithamparam, K., and Khan, T. N., Identity and recovery of seedborne fungal pathogens of field peas in Western Australia, *N. Z. J. Crop Hort. Sci.*, 17, 97, 1989.
320. Luthra, J. C. and Bedi, K. S., Some preliminary studies on gram blight with reference to its cause and mode of perennation, *Indian J. Agric. Sci.*, 2, 499, 1932.

321. Montorsi, F., di Giambattista, G., and Porta-Punglia, A., First report of *Ascochyta rabiei* on berseem clover seeds, *Plant Dis.*, 76, 538, 1992.
322. Osnitskaya, E. A., Control measures against carrot *Phomopsis*, *Orchard Garden*, 19, 1954.
323. Sartorato, A., Peneado, M. F. P., and Menten, J. O. M., Chemical control of *Phoma sorghina* in rice (*Oryza sativa* L.), *Rev. Brasileira Sementes*, 12, 59, 1990.
324. Wallace, G. B. and Wallace, M. M., Diseases of sorghum, *Phamphl. Dep. Agric. Tanganyika*, 53, 16, 1953.
325. Hobbs, T. W., Schmitthenner, A. F., and Kuter, G. A., A new *Phomopsis* species from soybean, *Mycologia*, 77, 535, 1985.
326. Velicheti, R. K., Kollipara, K. P., Sinclair, J. B., and Hymowitz, T., Selective degradation of proteins by *Cercospora kikuchii* and *Phomopsis longicolla* in soybean seedcoats and cotyledons, *Plant Dis.*, 76, 779, 1992.
327. Sinclair, J. B., Phomopsis seed decay of soybeans — a prototype for studying seed disease, *Plant Dis.*, 77, 329, 1993.
328. Kulik, M. M. and Yaklich, R. W., Soybean seedcoat structures: relationship to weathering resistance and infection by the fungus *Phomopsis phaseoli*, *Crop Sci.*, 31, 108, 1991.
329. Porter, R. P., Seed-borne inoculum of *Phomopsis vexans* — its extent and effects, *Plant Dis. Rep.*, 27, 167, 1943.
330. Harman, G. E., Heit, C. E., and Braverman, S. W., Seed-borne fungi of fresh *Antirrhinum majus* seed, *Plant Dis. Rep.*, 55, 639, 1971.
331. Leonian, L. H., Stem and fruit blight of peppers caused by *Phytophthora capsici* sp. nov., *Phytopathology*, 12, 401, 1922.
332. Klotz, L. J., deWolle, T. A., Roistacher, C. N., Nauer, E. M., and Carpenter, J. B., Heat treatments to destroy fungi in infected citrus seeds, *Calif. Citrogr.*, 46, 63, 1960.
333. Boyd, O. C., Evidence of the seed-borne nature of late blight (*Phytophthora infestans*) of tomatoes, *Phytopathology*, 25, 7, 1935.
334. Bhardwaj, S. S., Sharma, S. L., and Tyagi, S. N. S., Perpetuation of *Phytophthora nicotianae* var. *nicotianae* in bell pepper seeds, *Indian J. Plant Pathol.*, 5, 94, 1987.
335. Sharma, S. L. and Sohi, H. S., Seedborne nature of *Phytophthora parasitica* causing buckeye rot of tomato, *Indian Phytopathol.*, 28, 130, 1975.
336. Sehgal, S. P. and Prasad, N., Studies on the perennation and survival of sesamum *Phytophthora*, *Indian Phytopathol.*, 19, 173, 1966.
337. Klein, H. H., Etiology of the Phytophthora disease of soybeans, *Phytopathology*, 49, 380, 1959.
338. Wester, R. E., Goth, R. W., and Drechsler, C., Overwintering of *Phytophthora phaseoli*, *Phytopathology*, 56, 95, 1966.
339. Warne, L. G. G., An outbreak of club root traceable to a seed-borne infection, *J.R. Hortic. Soc.*, 69, 45, 1944.
340. Zad, J., Transmission of sunflower downy mildew by seed, *Iranian J. Plant Pathol.*, 14, 1, 1978.
341. Srivastava, S. and Singh, L., Evidence of seedborne nature of downy mildew fungus of balsam, *Seed Res.*, 16, 254, 1988.
342. D'Ercole, N., Downy mildew of parsley, *Colture Protette*, 19, 117, 1990.
342a. Paulitz, T. C. and Atlin, G. First report of brown spot of lupins caused by *Pleiochaeta setosa* in Canada, *Plant Dis.*, 76, 1185, 1992.
343. Nelson, R. R., Studies on Stemphylium leafspot of alfalfa, *Phytopathology*, 45, 352, 1955.

344. Pavgi, M. S. and Mukhopadhyay, A. N., Development of coriander fruit infected by *Protomyces macrosporus* Unger, *Cytologia*, 37, 619, 1972.
345. Petrie, G. A. and Vanterpool, T. C., *Pseudocercosporella capsellae*, the cause of white leaf spot and grey stem of Cruciferae in Western Canada, *Can. Plant Dis. Surv.*, 58, 69, 1978.
346. White, J. F., Jr. and Morrow, A. C., Endophyte-host associations in forage grasses. XII. A fungal endophyte of *Trichachne insularis* belonging to *Pseudocercosporella, Mycologia*, 82, 218, 1990.
347. West, E., Peanut rust, *Plant Dis. Rep.*, 15, 5, 1931.
348. Calvert, O. H. and Thomas, C. A., Some factors affecting seed transmission of safflower rust, *Phytopathology*, 44, 609, 1954.
349. Ranganathaiah, K. G. and Mathur, S. B., Seed health testing of *Eleusine coracana* with special reference to *Drechslera nodulosa* and *Pyricularia grisea*, *Seed Sci. Technol.*, 6, 943, 1978.
350. Suzuki, H., Experimental studies on the possibility of primary infection of *Pyricularia oryzae* and *Ophiobolus miyabeanus* internal of rice seed, *Ann. Phytopathol. Soc. Jpn.*, 2, 1, 1930.
351. Goulart, A. C. P. and Paiva, F. de A., Transmission of *Pyricularia oryzae* through wheat (*Triticum aestivum*) seed, *Fitopatol. Brasileira*, 15, 359, 1990.
352. Harman, G. E., Braverman, S. W., and Waters, E. C., *Pythium aphanidermatum* seed-borne on squash, a case of seed and seedling rot, *J. Seed Technol.*, 1, 55, 1976.
353. Gattani, M. L. and Kaul, T. N., Damping-off of tomato seedlings — its causes and control, *Indian Phytopathol.*, 4, 156, 1951.
354. Von Plotho, O., Die Pilzflora der Rübenknaule, *Zucker*, 6, 291, 1953.
355. Weimer, J. L., Blackpatch of soybean and other forage legumes, *Phytopathology*, 40, 782, 1950.
356. Gough, F. J. and Elliott, E. S., Blackpatch of red clover and other legumes caused by *Rhizoctonia leguminicola* sp. nov., *W. Va. Univ. Agric. Exp. Stn. Bull.*, 387 T, 23, 1956.
357. Ashworth, L. J. and Langley, B.C., The relationship of pod damage to kernel damage by molds in Spanish peanut, *Plant Dis. Rep.*, 48, 875, 1964.
358. Kamara, A. M., El-Lakany, M. H., Badran, O. A., and Attia, Y. G., Seed pathology of *Araucaria* spp. I. Survey of seedborne fungi associated with four *Araucaria* spp., *Aust. Forest Res.*, 11, 269, 1981.
359. Baker, K. F., Seed transmission of *Rhizoctonia solani* in relation to control of seedling damping-off, *Phytopathology*, 37, 912, 1947.
360. Hepperly, P. R., Mignucci, J. S., Sinclair, J. B., Smith, R. S., and Judy, W. H., Rhizoctonia web blight of soybean in Puerto Rico, *Plant Dis.*, 66, 256, 1982.
361. Peeples, J. L. and Bain, D. C., Infection of cotton seed by *Rhizoctonia*, *Plant Dis. Rep.*, 50, 770, 1966.
362. Asthana, R. P., Incidence of wilt disease on *Linum usitatissimum* in M. P., *Nagpur Agric. Coll. Mag.*, 31, 1, 1965.
363. Roy, A. K., Source of seedborne infection of sheath blight of rice, *Oryza*, 26, 111, 1989.
364. Leach, C. M. and Pierpoint, M., Seed transmission of *Rhizoctonia solani* in *Phaseolus vulgaris* and *Phaseolus lunatus*, *Plant Dis. Rep.*, 40, 907, 1956.
365. Sharma, S., Raghunathan, A. N., and Shetty, H. S., Histopathological studies on *Rhizoctonia solani* infected seeds of *Vigna unguiculata* L. Walp., *Curr. Sci.*, 57, 945, 1988.

366. Habgood, R. M., The transmission of *Rhynchosporium secalis* by infected barley seed, *Plant Pathol.*, 20, 80, 1971.
367. Shahjahan, A. K. M., Harahap, Z., and Rush, M. C., Sheath rot of rice caused by *Acrocylindrium oryzae* in Louisiana, *Plant Dis. Rep.*, 61, 307, 1977.
368. Bezerra, J. L. and Oliveira, D. P., De., *Schizophyllum commune* as a pathogen on oil palm (*Elaeis guipeensis*) seeds in Bahia, *Revista Theobroma*, 14, 73, 1984.
369. Summers, T. E., Downy mildew (*Sclerospora macrospora*) of oats in the south, *Plant Dis. Rep.*, 36, 347, 1952.
370. Raghavendra, S. and Safeeulla, K. M., Histopathological studies on ragi (*Eleusine coracana* L. Gaetnr.) infected by *Sclerophthora macrospora* (Sacc.), *Proc. Indian Acad. Sci.*, 88, 19, 1979.
371. Bains, S. S. and Jhooty, J. S., Seed transmission of *Sclerophthora macrospora* in wheat, *Seed Res.*, 13, 154, 1985.
372. Sundaram, N. V., Sastry, D. V. R., and Nayar, S. K., Note on the seed-borne infection of downy mildew (*Sclerospora graminicola* (Sacc.) Schroet.) of pearl millet, *Indian J. Agric. Sci.*, 43, 215, 1973.
373. Wadsworth, D. F. and Melouk, H. A., Potential for transmission and spread of *Sclerotinia minor* by infected peanut seeds and debris, *Plant Dis.*, 69, 379, 1985.
374. Neergaard, P., Mycelial seed infection of certain crucifers by *Sclerotinia sclerotiorum* (Lib.) DBy, *Plant Dis. Rep.*, 42, 1105, 1958.
375. Basu Chaudhary, K. C. and Puttoo, B. L., Seedborne nature of *Sclerotinia sclerotiorum* in safflower in India, *Indian J. Mycol. Plant Pathol.*, 21, 196, 1991.
376. Gray, E. G. and Noble, M., Sclerotinia diseases, *Scott. Agric.*, 4, 1965.
377. Miclaus, D., Sin, D., Damian, V., and Guran, M., Presence and spread of *Sclerotinia sclerotiorum* (Lib.) de Bary on seeds of various cereal and industrial crops, Analele Institutlui de Cercetaria Pentru Cereale Si, *Plante Tehnice*, Fundulea, 56, 169, 1988.
378. Mitchell, S. J., Jellis, G. J., and Cox, T. W., *Sclerotinia sclerotiorum* on linseed, *Plant Pathol.*, 35, 403, 1986.
379. Steadman, J. R., Nature and epidemiological significance of infection of bean seed by *Whetzelinia sclerotiorum*, *Phytopathology*, 65, 1323, 1975.
380. Leach, C. M., Additional evidence for seed-borne mycelium of *Sclerotinia sclerotiorum* associated with clover seed, *Phytopathology*, 48, 388, 1958.
381. Aso, T. and Imai, S., Sulphuric acid treatment of seeds of *Astragalus sinicus* for eradication of sclerotia of *Sclerotinia trifoliorum* contaminating the seeds, *Ann. Phytopathol. Soc. Jpn.*, 28, 243, 1963.
382. Hickman, C. J. and Coley-Smith, J. R., Onion white rot, *NAAS Q. Rev.*, 50, 58, 1960.
383. Clinton, P. K. S., A note on a wilt of groundnuts due to *Sclerotium rolfsii* Sacc. in Tanganyika, *East Afr. Agric. J.*, 22, 137, 1957.
384. Lakshmidevi, N., Prakash, H. S., and Shetty, H. S., Seedborne nature and transmission of *Sclerotium rolfsii* Sacc. in sunflower, *Plant Dis. Res.*, 6, 63, 1991.
385. Silveira, V. D., The phytopathological examination of seeds, *Biol. Soc. Brasil. Agron.*, 10, 143, 1947.
386. Singh, D. and Mathur, S. K., *Sclerotium rolfsii* in seeds of bean from Uganda, *Seed Sci. Technol.*, 2, 481, 1974.
387. Epps, W. M., Patterson, J. C., and Freeman, I. E., Physiology and parasitism of *Sclerotium rolfsii*, *Phytopathology*, 41, 245, 1951.
388. Hewett, P. D., Viable *Septoria* spp. in celery seed samples, *Ann. Appl. Biol.*, 61, 89, 1968.

389. Macneill, B. H. and Zalosky, H., Histological study of host parasite relationships between *Septoria glycines* Hemmi and soybean leaves and pods, *Can. J. Bot.*, 35, 501, 1957.
390. Tosi, L. and Zazzerini, A., *Septoria helianthi* Ell. & Kell, a new parasite of sunflower in Italy, *Inform. Fitopatol.*, 39, 43, 1989.
391. Moore, W. C., New and interesting plant diseases, *Trans. Br. Mycol. Soc.*, 24, 345, 1940.
392. Justham, M. and Ogilvie, L., Clover rot investigations, *Ann. Appl. Biol.*, 37, 328, 1950.
393. Cooke, B. M. and Jones, D. G., The epidemiology of *Septoria tritici* and *S. nodorum*. II. Comparative studies of head infection by *Septoria tritici* and *S. nodorum* on spring wheat, *Trans. Br. Mycol. Soc.*, 54, 395, 1970.
394. Sutherland, J. R., Lock, W., and Farris, S. H., Sirococcus blight: a seedborne disease of container-grown spruce seedlings in coastal British Columbia forest nurseries, *Can. J. Bot.*, 59, 559, 1981.
395. Fischer, G. W., *Manual of the North American Smut Fungi*, Ronald Press, New York, 1953, 343.
396. Faris, J. A. and Reed, G. M., Modes of infection of sorghum by loose kernel smut, *Mycologia*, 17, 51, 1925.
397. Krylov, S. V., Chernyaev, N. G., and Zakharov, V. V., Waterproofing millet seeds, *Zashchita Rastenii*, 8, 33, 1981.
398. Kühnel, W., Testing various methods for infecting common and Italian millet with the smut fungi, *S. destruens* and *U. crameri* as well as investigations of the duration of the infectible state of the host plant, *Nachrichtenbl. Dtsch. Pflanzenschutzdienst (Berlin)*, 15, 241, 1961.
399. Dorovskaya, L. M., Biological features of millet smut in the Ural region and measures for seed disinfection, *Ref. Zh.*, 55, 1243, 1976.
400. Reed, G. M. and Faris, J. A., Influence of environmental factors on the infection of sorghums and oats by smuts, *Am. J. Bot.*, 11, 518, 1924.
401. Lund, S. and Shands, H. L., Seedling infection of oats by *Septoria avenae*, *Phytopathology*, 45, 181, 1955.
402. Bronnimann, A., Zur Kenntnis von *Septoria nodorum* Berk., dem Erreger der Spelzenbräune und einer Blattdürre des Weizens, *Phytopathol. Z.*, 61, 101, 1968.
403. Holmes, S. J. I. and Colhoun, J., *Septoria nodorum* as a seed-borne pathogen of barley, *Plant Pathol.*, 22, 194, 1973.
404. Baker, C. J., Influence of environmental factors on development of symptoms on wheat seedlings, grown from seed infected with *Leptosphaeria nodorum*, *Trans. Br. Mycol. Soc.*, 55, 443, 1971.
405. Whiteside, J. O. and Herd, G. W., List of diseases of economic plants in Rhodesia, *Rhodesia Agric. J. Tech. Bull.*, 5, 1966, 28.
406. Chowdhury, S., Studies in the bunt of rice (*Oryza sativa* L.), *Indian Phytopathol.*, 4, 25, 1951.
407. Swinburne, T. R., Infection by *Tilletia caries* (DC.) Tul., the causal organism of bunt, *Trans. Br. Mycol. Soc.*, 46, 145, 1963.
408. Dewey, W. G. and Tingey, D. C., A new grass host for dwarf bunt, *Plant Dis. Rep.*, 42, 18, 1958.
409. Hardison, J. R., New grass host records and life history observations of dwarf bunt in eastern Oregon, *Plant Dis. Rep.*, 38, 345, 1954.
410. Hardison, J. R. and Corden, M. E., A serious outbreak of dwarf bunt in tall oat grass in Oregon, *Plant Dis. Rep.*, 36, 343, 1952.

411. Tverdunova, T. A., Timely seed treatment of winter rye, *Zashchita Rastenii*, 3, 22, 1974.
412. Hoffman, J. A. and Purdy, L. H., Effect of stage of development of winter wheat on infection by *Tilletia controversa*, *Phytopathology*, 57, 410, 1967.
413. Aujla, S. S. and Sharma, I., New host records of *Neovossia indica, Indian Phytopathol.*, 40, 439, 1987.
414. Agarwal, V. K., Karnal bunt of wheat — a seedborne disease of considerable importance, *Seed Res.*, 14, 1, 1986.
414a. Agarwal, V. K., Verma, H. S., and Khetarpal, R. K., Occurrence of partial bunt on triticale, *FAO Plant Prot. Bull.*, 25, 210, 1977.
415. Bambeg, R. H., Holton, C. S., Rodenhiser, H.A., and Woodward, R.W., Wheat dwarf bunt depressed by common bunt, *Phytopathology*, 37, 556, 1947.
416. Fischer, G. W., Spontaneous occurrence of *Tilletia caries* (*T. tritici*) on seven new hosts, *Phytopathology*, 40, 9, 1950.
417. Kumar, V. and Shetty, H. S., An ear and kernel rot of maize caused by *Trichothecium roseum, Curr. Sci.*, 54, 486, 1985.
418. Purdy, L. H., Soil moisture and soil temperature, their influence on infection by the wheat flag smut fungus and control of the disease by three seed treatment fungicides, *Phytopathology*, 56, 98, 1966.
419. Ling, L., Factors affecting infection in rye smut and subsequent development of the fungus in the host, *Phytopathology*, 31, 617, 1941.
420 Emdal, P. S. and Foldo, N. E., Seedborne inoculum of *Uromyces betae*, *Seed Sci. Technol.*, 7, 93, 1979.
421 Prasada, R. and Verma, U. N., Studies on lentil rust, *Uromyces fabae* (Pers.) de Bary in India, *Indian Phytopathol.*, 1, 142, 1948.
422. Falloon, R. E., Fungicide control of *Ustilago bullata* seedling and shoot infection on *Bromus catharticus*, *N. Z. J. Exp. Agric.*, 8, 173, 1980.
423. Titatarn, S., Chiengkul, A., Unchalisangkas, D., Chamkrachang, W., Chew-chin, N., and Chandrasrikul, A., Occurrence of *Ustilago coicis* on *Coix lacryma-jobi* in Thailand, *Plant Dis.*, 67, 434, 1983.
424. Kuhnel, W., Testing various methods for infecting common and Italian millet with the smut fungi: *S. destruens* and *U. crameri* as well as investigations of the duration of the infectible state of the host plant, *Nachr. Bl. Dtsch. Pflsch.-Dienst.*, Berl., N. F., 15, 241, 1961.
425. Kulkarni, S., Siddaramaiah, A. L., Bhat, R. P., and Basavarajah, A. B., Control of navane (fox tail millet) smut by seed treatment, *Curr. Res.*, 8, 128, 1979.
426. Tomala-Bed-narek, J., Studies on the control of the smut *U. avenae* on *A. elatius* by seed treatment, *Roczn. Nauk Rol.*, Ser. A., 83, 929, 1961.
427. Mills, J. T., The development of loose smut (*Ustilago avenae*) in the oat plant with observations on spore formation, *Trans. Br. Mycol. Soc.*, 49, 651, 1966.
428. Shevchenko, I. S. and Dunenko, M. A., Black loose smut of barley, *Zashch. Rast. Mosk.*, 15, 57, 1970.
429. Brandwein, P. E., Seedling invasion of covered smut of oats, *Phytopathology*, 34, 481, 1944.
430. Tapke, V. F., Studies on the natural inoculation of seed barley with covered smut (*Ustilago hordei*), *J. Agric. Res.*, 60, 787, 1940.
431. Campbell, W. P. and Tyner, L. E., Comparison of degree and duration of susceptibility of barley to ergot and true loose smut, *Phytopathology*, 49, 348, 1959.
432. Kozhevnikov, S., Smut disease in the Krasnoyarsk region, *Referat. Zh. Biol.*, 14, 209, 1958.

433. Thapliyal, P. N., Sinha, A. P., and Agarwal, V. K., Effect of loose smut infection on the growth and morphology of triticale plants, *Indian Phytopathol.*, 34, 264, 1981.
434. Batts, C. C. V., Observations on the infection of wheat by loose smut [*Ustilago tritici* (Pers.) Rostr.)], *Trans. Br. Mycol. Soc.*, 38, 465, 1955.
435. Beke, F. and Martin, J., The influence of blister smut of maize on yield, *Mag. Mezog.*, 13, 17, 1958.
436. Schuster, M. L. and Nuland, D. S., Seed transmission of safflower Verticillium wilt fungus, *Plant Dis. Rep.*, 44, 901, 1960.
437. Evans, G., Wilhelm, S., and Snyder, W. C., Dissemination of the Verticillium wilt fungus with cotton seed, *Phytopathology*, 56, 460, 1966.
438. Sackston, W. E. and Martens, J. W., Dissemination of *Verticillium alboatrum* on seed of sunflower (*Helianthus annuus*), *Can. J. Bot.*, 37, 759, 1959.
439. Christen, A. A., Seed tranmsission of Verticillium wilt in alfalfa, *Phytopathology*, 71, 208, 1981.
440. Kadow, K. J., Seed transmission of Verticillium wilt of eggplants and tomatoes, *Phytopathology*, 24, 1265, 1934.
441. Klisiewicz, J. M., Assay of *Verticillium* in safflower seed, *Plant Dis. Rep.*, 58, 926, 1974.
442. Sedun, F. S. and Sackston, W. E., Effect of daylength, host line, and pathogen isolate on development of Verticillium wilt in sunflower, *Can. J. Plant Pathol.*, 4, 109, 1982.
443. Anonymous, Nomenclatural revisions of plant pathogenic bacteria and list of names, 1980–88, *Rev. Plant Pathol.*, 70, 211, 1991.
444. Goto, M., *Fundamentals of Bacterial Plant Pathology*, Academic Press, New York, 1992.
445. Young, J. M., Takikawa, Y., Gardan, L., and Stead, D. E., Changing concepts in the taxonomy of plant pathogenic bacteria, *Annu. Rev. Phytopathol.*, 30, 67, 1992.
446. Gitaitis, R. D., Semiselective and diagonistic media used for seedborne assay for watermelon fruit blotch, ISTA Plant Dis. Committee Symp. Seed Health Testing, Ottawa, 1993.
447. Komarova, A. A., Bacterial canker of hops, *Tr. Zhitomir. Nauchno-Issled. Sta. Khmelevod.*, 6, 216, 1959.
448. Cochran, L. C., Cooper, W. C., and Blodgett, E. C., Seeds of root stocks of fruit and nut trees, in *Yearbook of Agriculture*, U.S. Department of Agriculture, Washington, D.C., 1961, 233.
449. Naumova, N. A., *Testing of Seeds for Fungous and Bacterial Infections* (translated from Russian), 3rd ed., Israel Prog. Scientific Translation, Jerusalem, 1972, 145.
450. Hosford, R. M., Jr., White blotch incited in wheat by *Bacillus megaterium* pv. *cerealis*, *Phytopathology*, 72, 1453, 1982.
451. Nikitina, K. V., Infection of clover seeds by the pathogens of bacterial blight, *Tr. Po Prikl. Bot. Genet. Sel.*, 51, 221, 1974.
452. Van Vaerenbergh, J., Walvaert, W., and Vandevelde, R., On bacterial seed infection of *Capsicum annuum* (L.), in 33rd Symp. Phytopharmacy and Phytiatry, Part III, Ghent, Belgium, 1981, 729.
453. Mengistu, A. and Sinclair, J. B., Seedborne microorganisms of Ethiopian grown soybean and chickpea seeds, *Plant Dis. Rep.*, 63, 616, 1979.
454. Sinclair, J. B., The seed-borne nature of some soybean pathogens, the effect of *Phomopsis* spp. and *Bacillus subtilis* on germination and their occurrence in Illinois, *Seed Sci. Technol.*, 6, 957, 1978.

455. Scharif, G., *Corynebacterium iranicum* sp. nov. on wheat (*Triticum vulgare* L.) in Iran and a comparative study of it with *C. tritici* and *C. rathayi*, *Entomol. Phytopathol. Appl. Tehran*, 19, 1, 1961.
456. Williams, A. S. and Taylor, L. H., Rathay's disease of orchard grass found in Virginia, *Plant Dis. Rep.*, 41, 598, 1957.
457. Mathur, R. S. and Ahmad, Z. U., Longevity of *Corynebacterium tritici* causing tundu disease of wheat, *Proc. Natl. Acad. Sci. India Sect. B*, 34, 335, 1964.
458. Cormack, M. W., Longevity of the bacterial wilt organism in alfalfa hay, pod debris, and seed, *Phytopathology*, 51, 260, 1961.
459. Bryan, M. K., Studies on bacterial canker of tomato, *J. Agric. Res.*, 41, 825, 1930.
460. Schuster, M. L., Hoff, B., Mandel, M., and Lazar, I., Leaf freckles and wilt, a new corn disease, *Proc. Annu. Corn Sorghum Res. Conf.* (Chicago), 27, 176, 1963.
461. Larson, R. H., The ringrot bacterium in relation to tomato and eggplant, *J. Agric. Res.*, 69, 309, 1944.
462. McBeath, J. H. and Adelman, M., Detection of *Corynebacterium michiganense* subsp. *tessellarius* in seeds and wheat plants, *Phytopathology*, 76, 1099, 1986.
463. Keyworth, W. G. and Howell, J. S., Studies on silvering disease of red beet, *Ann. Appl. Biol.*, 49, 173, 1961.
464. Dunleavy, J., Stunt disease of soybean caused by *Corynebacterium* sp., *Phytopathology*, 52, 730, 1962.
465. Burkholder, W. H., The longevity of the pathogen causing the wilt of the common bean, *Phytopathology*, 35, 743, 1945.
466. Schuster, M. L. and Sayre, R. M., A coryneform bacterium induces purple-colored seed and leaf hypertrophy of *Phaseolus vulgaris* and other Leguminosae, *Phytopathology*, 57, 1064, 1967.
467. Rickard, S. F., Physiology of bacterial wilt of bean, *Diss. Abstr.*, 25, 2161, 1964.
468. Chavarro, C. A., Lopex, G. C. A., and Lenne, C. A., Characteristics and pathogenicity of *Corynebacterium flaccumfaciens* (Hedges) Dows, causal agent of bacterial wilt of *Zornia* spp. and its effect on the yield of *Z. glabra* CIAT 7847 and *Phaseolus vulgaris*, *Acta Agron.*, 35, 64, 1985.
469. Pape, H., *Krankheiten und Schadlinge der Zierpflanzen und ihre Bekämpfung*, 5th ed., Paul Parey, Berlin, 1964, 625.
470. Marcelo, A. and Fernandes, M., Identification of plant pathogenic bacteria in broadbean (*Vicia faba* L.) seeds, 8th Int. Conf. Plant Pathogenic Bacteria, Versailles, 1992.
471. Elliott, C., *Manual of Bacterial Plant Pathogens*, 2nd ed., Chemical Botanica, Waltham, MA, 1951, 186.
472. Hankin, L. and Sands, D. C., Microwave treatment of tobacco seed to eliminate bacteria on the seed surface, *Phytopathology*, 67, 794, 1977.
473. Pauer, G. D., *Erwinia carotovora* f. sp. *zeae*, the bacterium causing stalk rot of maize in the Republic of South Africa, *S. Afr. J. Agric. Sci.*, 7, 581, 1964.
474. Kim, Y. C., Kim, K. C., and Choi, B. H., Palea browning disease of rice caused by *Erwinia herbicola* and ice nucleation activity of the pathogenic bacterium, *Korean J. Plant Pathol.*, 5, 72, 1989.
475. Huang, H. C., Philippe, L. M. and Philippe, R. C., Pink seed of pea: a new disease caused by *Erwinia rhapontici*, *Can. J. Plant Pathol.*, 12, 445, 1990.
476. McMullen, M. P., Stack, R. W., Miller, J. D., Bromel, M. C., and Youngs, V. L., *Erwinia rhapontici*, a bacterium causing pink wheat kernels, *Proc. N. D. Acad. Sci.*, 38, 78, 1984.

477. Ivanoff, S. S., Stewart's wilt disease of corn, with emphasis on the life history of *Phytomonas stewartii* in relation to pathogenesis, *J. Agric. Res.*, 47, 749, 1933.
478. Shakya, D. D., Chung, H. S., and Vinther, F., Transmission of *Pseudomonas avenae,* the cause of bacterial stripe of rice, *J. Phytopathol.*, 116, 92, 1986.
479. Krustev, K., Bacterial rot on lettuce, *Nauchni Tr. Vissh. Selskostop. Inst. Vasil Kolariv*, 14, 215, 1965.
480. Kritzman, G., Tomato seed assay for seedborne pathogenic bacteria, *Hassadeh*, 69, 614, 1989.
481. Schmiedeknecht, M. and Gorlitz, H., The occurrence of a broadbean bacterial disease in the D. D. R., *Nachrichtenbl. Dtsch. Pflanzenschutdienst* (Berlin), 20 37, 1966.
482. Zeigler, R. S. and Alvarez, E., Bacterial sheath brown rot of rice caused by *Pseudomonas fuscovaginae* in Latin America, *Plant Dis.*, 71, 592, 1987.
483. Duveiller, E. and Martinez, C., Seed detection of *Pseudomonas fuscovaginae* in wheat, *Mededelingen Faculteit Landbouwwetensch*, Rijksuniversiteit, Gent., 55, 1047, 1990.
484. Tsushima, S., Tsuno, K., Mogi, S., Wakimoto, S., Saito, H., and Naito, H., The infection period and multiplication of *Pseudomonas glumae* on rice grains, *Proc. 5th Int. Cong. Plant Pathol.*, Kyoto, 1, 93, 1988 (Abstract).
485. Wakimoto, S., Akai, M., and Tsuchiya, K., Serological specificity of *Pseudomonas glumae*, the pathogenic bacterium of grain rot disease of rice, *Ann. Phytopathol. Soc. Jpn.*, 53, 150, 1987.
486. Azegami, K., Tabei, H., and Fukuda, T., Entrance into rice grains of *Pseudomonas plantarii*, the causal agent of seedling blight of rice, *Ann. Phytopathol. Soc. Jpn.*, 54, 633, 1988.
487. Dange, S. R. S., Payak, M. M., and Renfro, B. L., Seed transmission of bacterial leaf stripe of maize caused by *Pseudomonas rubrilineans*, *Indian Phytopathol.*, 31, 119, 1978.
488. Machmud, M. and Middleton, K. J., Transmission of *Pseudomonas solanacearum* through groundnut seeds, *Bacterial Wilt Newsl.*, 7, 4, 1991.
489. Moffett, M. L., Wood, B. A., and Hayword, A. C., Seed and soil sources of inoculum for the colonization of the foliage of solanaceous hosts by *Pseudomonas solanacearum*, *Ann. Appl. Biol.*, 98, 403, 1981.
490. Krasnova, M. V., The effect of some physical factors on the causal agents of bacteriosis in soybean seeds, *J. Microbiol. Kiev*, 25, 50, 1963.
491. Simpson, C. J., Jones, G. E., and Taylor, J. D., A seedling blight of Antirrhinum caused by *Pseudomonas antirrhini, Plant Pathol.*, 20, 127, 1971.
492. Chupp, C. and Sherf, A. F., *Vegetable Diseases and Their Control*, Ronald Press, New York, 1960, 693.
493. Ark, P. A. and Leach, L. D., Seed transmission of bacterial blight of sugarbeet, *Phytopathology*, 36, 549, 1946.
494. Sutton, M. D., Bacteriophages of *Pseudomonas atrofaciens* in cereal seeds, *Phytopathology*, 56, 727, 1966.
495. Toben, H., Mavridis, A., and Rudolph, K. W. E., Basal glume rot (*Pseudomonas syringae* pv. *atrofaciens*) on wheat and barley in FRG and resistance screening of wheat, *Bull. OEPP*, 19, 119, 1989.
496. Reddy, C. S. and Godkin, J., A bacterial disease of bromegrass, *Phytopathology*, 13, 75, 1923.
497. Elliott, C., Bacterial stripe blight of oats, *J. Agric. Res.*, 35, 811, 1927.

498. Graham, J. H., Overwintering of three bacterial pathogens of soybean, *Phytopathology*, 43, 189, 1953.
499. Brown, C. S. and McCarter, S. M., A leaf spot and blight of *Abelmoschus moschatus* caused by a pathovar of *Pseudomonas syringae*, *Plant Dis.*, 75, 424, 1991.
500. Fukuda, T., Azegami, K., and Tabei, H., Histological studies on bacterial black node of barley and wheat caused by *Pseudomonas syringae* pv. *japonica*, *Ann. Phytopathol. Soc. Jpn.*, 56, 252, 1990.
501. Naumann, K., Über das Auftreten von Bakterien in Gurkensamen aus Früchten, die durch *Pseudomonas lachrymans infiziert* waren, *Phytopathol. Z.*, 48, 258, 1963.
502. Rangarajan, M. and Chakravarti, B. P., A strain of *Pseudomonas lapsa* isolated from corn seeds causing bacterial stalk rot in India, *Plant Dis. Rep.*, 51, 764, 1967.
503. Bakker, M., Bacterial spot disease of cauliflower and other *Brassica* species caused by *Pseudomonas maculicola* (McCulloch) Stevens, *Tijdschr. Plantenziekten*, 57, 75, 1951.
504. Grogan, R. G. and Kimble, K. A., The role of seed contamination in the transmission of *Pseudomonas phaseolicola* in *Phaseolus vulgaris*, *Phytopathology*, 57, 28, 1967.
505. Skoric, V., Bacterial blight of pea: overwintering, dissemination, and pathological histology, *Phytopathology*, 17, 611, 1927.
506. Sowell, G., Jr. and Schaad, N. W., *Pseudomonas pseudoalcaligenes* subsp. *citrulli* on watermelon: seed transmission and resistance of plant introductions, *Plant Dis. Rep.*, 63, 437, 1979.
507. Vajavat, R. M. and Chakravarti, B. P., Survival of *Pseudomonas sesami* and effect of an antagonistic bacterium isolated from seeds on the control of the disease in seed, *Indian Phytopathol.*, 31, 286, 1978.
508. Nakada, N. and Takimoto, K., Bacterial blight of *Hibiscus*, *Ann. Phytopathol. Soc. Jpn.*, 1, 13, 1923.
509. Peters, R. A. and Timian, R. G., A bacterial kernel spot of barley caused by *Pseudomonas syringae* pv. *syringae*, *Phytopathology*, 71, 1117, 1981.
510. Koroleva, I. B., Gvozdyak, R. I., and Pasichnik, L. A., *Pseudomonas oryzicola* causal agent of bacterial disease of rice in Ukraine, *Mikrobiol. Zh.*, 47, 34, 1985.
511. Thaung, M. M. and Walker, J. C., Studies on bacterial blight of lima bean, *Phytopathology*, 47, 413, 1957.
512. Arcila, M. J. and Trujillo, G., *Pseudomonas syringae* pv. *syringae*, causal agent of brown spot of bean (*Phaseolus vulgaris* L.) infects seeds of Sieva bean (*Phaseolus lunatus* L.) *Rev. Faculta Agron. Univ. Central Venezuela,* 15, 235, 1989.
513. Jindal, K. K. and Bhardwaj, S. S., Etiology of bacterial blight of pea caused by *Pseudomonas syringae* pv. *syringae, Plant Dis. Res.*, 4, 165, 1989.
514. Hoitink, H. A., Hagedorn, D. J., and McCoy, E., Survival, transmission, and taxonomy of *Pseudomonas syringae* Van Hall, the causal organism of bacterial brown spot of bean (*Phaseolus vulgaris* L.), *Can. J. Microbiol.*, 14, 437, 1968.
515. Tarr, S. A. J., *Diseases of Sorghum, Sudan Grass and Broom Corn*, Commonwealth Mycological Institute, Kew, Surrey, U.K., 1962, 380.
516. Gaudet, D. A. and Kokko, E. G., Seedling diseases of sorghum grown in southern Alberta caused by *Pseudomonas syringae* pv. *syringae, Can. J. Plant Pathol.*, 8, 208, 1986.
517. Otta, J. D., Occurrence and characteristics of isolates of *Pseudomonas syringae* on winter wheat, *Phytopathology*, 67, 22, 1977.
518. Koleva, N., Bacteriosis of winter wheat, *Rastitelna Zaschita*, 29, 15, 1981.

519. George, M. W. and Tripepi, R. R., Bacterial leaf spot and stem collapse of mungbean caused by *Pseudomonas syringae* pv. *syringae*, *Plant Dis.*, 74, 394, 1990.
520. Gardner, M. W., Bacterial spot of cowpea and lima bean, *J. Agric. Res.*, 31, 841, 1925.
521. Pady, S. M., Bacterial leaf blight and top rot of corn in Nebraska and Kansas, *Plant Dis. Rep.*, 28, 826, 1944.
522. Atlas de Gotuzzo, E., Gally de Minaverry, T., and Gonzalez, B., Transmission of *Pseudomonas syringae* pv. *tagetis*, *Fitopatol. J.*, 23, 15, 1988.
523. Hellmers, E., Bacterial leaf spot of African marigold (*Tagetes erecta*) caused by *Pseudomonas tagetis* sp. n., *Acta Agric. Scand.*, 5, 185, 1955.
524. Dowson, W. J., *Plant Diseases Due to Bacteria*, 2nd ed., Cambridge University Press, London, 1957, 232.
525. Devash, Y., Okon, Y., and Henis, Y., Survival of *Pseudomonas tomato* in soil and seeds, *Phytopathol. Z.*, 99, 175, 1980.
526. Hunter, J. E. and Cigna, J. A., Bacterial blight incited in parsnips by *Pseudomonas marginalis* and *Pseudomonoas viridiflava*, *Phytopathology*, 71, 1238, 1981.
527. Shakya, D. D. and Vinther, F., Occurrence of *Pseudomonas viridiflava* in seedlings of radish, *J. Phytopathol.*, 124, 123, 1989.
528. Jacobs, S. E. and Dadd, A. H., Antibacterial substances in seed coats and their role in the infection of sweet peas by *Corynebacterium fascians* (Tilford) Dowson, *Ann. Appl. Biol.*, 47, 666, 1959.
529. Misra, A., Jha, V., Jha, S., and Sharma, B. P., Seed-borne bacterial tumors in tobacco, in *Proc. 5th Int. Conf. Plant Pathogenic Bacteria,* Cali, Colombia, 1982, 210.
530. Baker, K. F., Bacterial fasciation disease of ornamental plants in *California, Plant Dis.*, 34, 121, 1950.
531. Castano, J. J. and Thurston, H.D., Gummosis of imperial grass, *Phytopathology*, 54, 498, 1964.
532. Srinivasan, M. C., Patel, M. K., and Thirumalachar, M. J., A new bacterial blight disease of *Argemone mexicana*, *Proc. Natl. Inst. Sci. India Sect. B*, 27, 104, 1961.
533. Shiomi, T., Dry heat sterilization for seed stain of cabbage black rot disease, *Agric. Hort.*, 66, 1177, 1991.
534. Pirone, P. P., *Diseases and Pests of Ornamental Plants*, 4th ed., Ronald Press, New York, 1970, 546.
535. Mihail, J. D., Taylor, S. J., Verslues, P. E., and Hodge, N. C., Bacterial blight of *Crambe abyssinica* in Missouri caused by *Xanthomonas campestris*, *Plant Dis.*, 77, 569, 1993.
536. Wilson, R. D., Black rot of garden stocks, *Agric. Gaz. N.S.W.*, 53, 33, 1942.
537. Rai, M. and Singh, D. V., Survival and virulence of *Xanthomonas campestris* pv. *cajani* in debris, seed, soil and in vitro, *Indian Phytopathol.*, 39, 467, 1986.
538. Godlewska-Lipowa, W., Investigations on the etiology of bacterial necrosis of coriander inflorescences, *Acta Agrobot.*, 19, 59, 1966.
539. Burÿkhina, E. K. and Buyanova, N. D., Bacteriosis of carrot, *Zashch Rast. Mosk.*, 7, 57, 1962.
540. Anonymous, Annu. Rep. Cassava Program, Centro Internacional de Agricultura Tropical, Cali, Colombia, 1980, 93.
541. Limber, D. P. and Frink, P. R., The inspection of imported plants, in *Plant Diseases, Yearbook of Agriculture*, U.S. Department of Agriculture, Washington, D.C., 1953, 159.

542. McLean, D. M., A seed-borne bacterial cotyledon spot of squash, *Plant Dis. Rep.*, 42, 425, 1958.
543. Singh, J. and Swarup, J., Survival and virulence of *Xanthomonas campestris* pv. *cyamopsidis* in plant debris, seed, soil and *in vitro*, *Indian J. Plant Pathol.*, 6, 93, 1988.
544. Ramakrishnan, T. S., *Diseases of Millets*, Indian Council Agric. Res., New Delhi, 1963, 152.
545. Kendrick, J. B. and Baker, K. F., Bacterial blight of garden stocks and its control by hot-water seed treatment, *Calif. Agric. Exp. Stn. Bull.*, 665, 23, 1942.
546. Alexandri, A., Gheorghiu, E., Stoian, E., and Manalescu, T., Researches on bacterial blight of walnut plants, *Lucr. Stiint. Inst. Cercet. Horti-vitic*, 6, 835, 1965.
547. Ayers, T. T., Lefebvre, C. L., and Johnson, H. W., Bacterial wilt of lespedeza, *U.S. Dep. Agric. Tech. Bull.*, 704, 1939, 22.
548. Brinkerhoff, L. A. and Hunter, R. E., Internally infected seed as a source of inoculum for the primary cycle of bacterial blight of cotton, *Phytopathology*, 53, 1397, 1963.
549. Honervogt, B. and Lehmann-Danzinger, H., Comparison of thermal and chemical treatment of cotton seed to control bacterial blight (*Xanthomonas campestris* pv. *malvacearum*), *J. Phytopathol.*, 134, 103, 1992.
550. Meshram, M. K. and Sheo, R., Effect of bacterial blight infection at different stages of crop growth on intensity and seed cotton yield under rainfed conditions, *Indian J. Plant Prot.*, 20, 54, 1992.
551. Elango, F. and Lozano, J. C., Seed transmission of *Xanthomonas manihotis*, *Fitopatol. Colombiana*, 8, 15, 1979.
552. Christoff, A., A bacterial blight of the opium poppy, *Renseign. Agric.* (Sofia), 12, 27, 1931.
553. Tarr, S. A. J., Diseases of economic crops in the Sudan, *FAO Plant Prot. Bull.*, 3, 113, 1955.
554. Cafati, C. R. and Saettler, A. W., Transmission of *Xanthomonas phaseoli* in seed of resistant and susceptible *Phaseolus* genotypes, *Phytopathology*, 70, 638, 1980.
555. Burkholder, W. H., The bacterial blight of the bean: a systemic disease, *Phytopathology*, 11, 61, 1921.
556. DuPlessis, H. J., Systemic invasion of plum seeds by *Xanthomonas campestris* pv. *pruni* through stalks, *J. Phytopathol.*, 130, 37, 1990.
557. White, H. E., Bacterial spot of radish and turnip, *Phytopathology*, 20, 653, 1930.
558. Khalil, O. and Sabet, K. A., A study of certain etiological aspects of bacterial leaf-blight disease of castorbean, *Sudan Agric. J.*, 6, 26, 1971.
559. Habish, H. A. and Hammad, A. H., Survival and chemical control of *Xanthomonas sesami*, *FAO Plant Prot. Bull.*, 19, 36, 1971.
560. Rao, Y. P. and Durgapal, J. C., Seed transmission of bacterial blight disease of sesamum (*Sesamum orientale* L.) and eradication of seed infection, *Indian Phytopathol.*, 19, 402, 1966.
561. Boosalis, M. G., The epidemiology of *Xanthomonas translucens* (J.J. & R.) Dowson on cereals and grasses, *Phytopathology*, 42, 37, 1952.
562. Crossan, D. F. and Morehart, A. L., Isolation of *Xanthomonas vesicatoria* from tissues of *Capsicum annuum*, *Phytopathology*, 54, 358, 1964.
563. Cox, R. S., The role of bacterial spot in tomato production in south Florida, *Plant Dis. Rep.*, 50, 699, 1966.

564. Shekhawat, G. S. and Patel, P. N., Seed transmission and spread of bacterial blight of cowpea and leaf spot of greengram in summer and monsoon seasons, *Plant Dis. Rep.*, 61, 390, 1977.
565. Ohata, K., Serizawa, S., Azegami, K., and Shirata, A., Possibility of seed transmission of *Xanthomonas campestris* pv. *vitians*, the pathogen of bacterial spot of lettuce, *Bull. Natl. Inst. Agric. Sci.*, C, 36, 81, 1982.
566. Mehta, Y. R., Management of *Xanthomonas campestris* pv. *undulosa* and *hordei* through cereal seed testing, *Seed Sci. Technol.*, 18, 467, 1990.
567. Duveiller, E. and Bragard, C., Comparison of immunofluorescence and two assays for detection of *Xanthomonas campestris* pv. *undulosa* in seeds of small grains, *Plant Dis.*, 76, 999, 1992.
568. Dimitrov, S. and Tsoneva, R., Bacterial burn (angular spot) of zinnia, a new disease in Bulgaria, *Rast. Zasch.*, 25, 31, 1977
569. Fang, C. T., Liu, C. F., and Chu, C. L., A preliminary study on the disease cycle of the bacterial leaf blight of rice, *Acta Phytopathol. Sin.*, 2, 173, 1956.
570. Shekhawat, G. S. and Srivastava, D. N., Mode of seed infection and transmission of bacterial leaf streak of rice, *Ann. Phytopathol. Soc. Jpn.*, 38, 4, 1972.
571. Doi, Y., Teranaka, M., Yora, K., and Asuyama, H., Mycoplasma, or PLT group-like microorganisms found in phloem-elements of plants infected with mulberry dwarf, potato witches' broom, aster yellows, or Paulownia witches' broom, *Ann. Phytopathol. Soc. Jpn.*, 33, 259, 1967.
572. Windsor, I. M. and Black, L. M., Clover club leaf: a possible rickettsial disease of plants, *Phytopathology*, 62, 1112, 1972.
573. Davis, R. E., Worley, J. F., Whitcomb, R. F., Ishijima, T., and Steere, R. L., Helical filaments produced by a mycoplasma-like organism associated with corn stunt disease, *Science*, 176, 521, 1972.
574. Davis, R. E. and Worley, J. F., Spiroplasma: Motile helical microorganism associated with corn stunt disease, *Phytopathology*, 63, 402, 1973.
575. Matthews, R. E. F., *Plant Virology*, 3rd ed., Academic Press, New York, 1991.
576. Gibbs, A. and Harrison, B., *Plant Virology, The Principles*, Edward Arnold, London, 1976, 292.
577. Martelli, G. P., Classification and nomenclature of plant viruses: state of the art, *Plant Dis.*, 76, 436, 1992.
578. Francki, R. I. B., Fauquet, C. M., Knudson, D. L., and Brown, F., Classification and nomenclature of viruses, *Arch. Virol. Suppl.*, 2, 1991.
579. Smith, K. M., *A Textbook of Plant Virus Diseases*, J. & A. Churchill, London, 1937.
580. Bennett, C. W., Seed transmission of plant viruses, *Adv. Virus Res.*, 14, 221, 1969.
581. Mink, G. I., Pollen and seed-transmitted viruses and viroids, *Annu. Rev. Phytopathol.*, 31, 375, 1993.
582. Keur, J. Y., Seed transmission of the virus causing variegation of *Abutilon*, *Phytopathology*, 23, 20, 1933.
583. Keur, J. Y., Studies on the occurrence and transmission of virus diseases in the genus *Abutilon*, *Bull. Torrey Bot. Club*, 61, 53, 1934.
584. Kaiser, W. J. and Hannan, R. M., Additional hosts of alfalfa mosaic virus and its seed transmission in tumble pigweed and bean, *Plant Dis.*, 67, 1354, 1983.
585. Sutic, D., The role of cayenne pepper seed in virus transmission, *Phytopathol. Z.*, 36, 84, 1959.
586. Cali, S., Investigations of virus infection of lucerne in Central Anatolia and their transmission through weeds and seeds, *Arastirma Yayinlari Serisi Yayin*, 72, 104, 1990.

587. Pospieszny, H., Transmission of alfalfa mosaic virus through *Chenopodium quinoa* and *Chenopodium amaranticolor* seeds, Zeszyty Problemowe Postepow Nauk Rolniczych, 330, 99, 1986.
588. Iizuka, N. and Yoshida, K., Incidence of mosaic disease in soybean in Hokkaido, and seed transmission of the causal viruses, *Res. Bull. Hokkaido Nat. Agric. Exp. Stn.*, 150, 33, 1988.
589. Jones, R. A. C. and Pathipanawat, W., Seedborne alfalfa mosaic virus infecting annual medics (*Medicago* spp.) in Western Australia, *Ann. Appl. Biol.*, 115, 263, 1989.
590. Tosic, M. and Pesic, Z., Investigations of alfalfa mosaic virus transmission through alfalfa seeds, *Phytopathol. Z.*, 83, 320, 1957.
591. Belli, G., Notes and experiments on the transmission of lucerne mosaic virus through the seed and demonstration of its exclusion from clones of virus-infected vines, *Ann. Fac. Milano*, 10, 15, 1962.
592. Dall, D. J., Randles, J. W. and Francki, R. I. B., The effect of alfalfa mosaic virus on productivity of annual barral medic, *Medicago truncatula, Aust. J. Agric. Res.*, 40, 807, 1989.
593. Paliwal, Y. C., Virus diseases of alfalfa and biology of alfalfa mosaic virus in Ontario and Western Quebec, *Can. J. Plant Pathol.*, 4, 175, 1982.
594. Gallo, J. and Ciampor, F., Transmission of alfalfa mosaic virus through *Nicandra physaloides* seeds and its location in embryo cotyledons, *Acta Virol.*, 21, 344, 1977.
595. Valkonen, J. P. T., Pehu, E., and Witanabe, K., Symptom expression and seed transmission of alfalfa mosaic virus and potato yellowing virus (SB-22) in *Solanum brevidens* and *S. tuberosum*, *Potato Res.*, 35, 403, 1992.
596. Mishra, M. D., Raychaudhuri, S. P., Ghosh, A., and Wilcoxon, R. D., Berseem mosaic, a seed transmitted virus disease, *Plant Dis.*, 64, 490, 1980.
597. Nain, P. S., Rishi, N., and Bishnoi, S. S., Growth stages of cowpea (*Vigna unguiculata*) vis-a-vis seed transmission of viruses, *Indian J. Virol.*, 10, 33, 1994.
598. Barba, M., Detection of apple mosaic and prunus necrotic ringspot viruses in almond by ELISA, *Arch. Phytopathol., Pflanzensch.*, 22, 279, 1986.
599. Van der Meer, F. A., Virology, in *Annual Report, 1973,* Institute of Phytopathological Research, Krijthe, J. M. and Van Noort, A. J., Eds., Wageningen, The Netherlands, 1973.
600. Lister, R. M. and Murant, A. F., Seed transmission of nematode-borne viruses, *Ann. Appl. Biol.*, 59, 49, 1967.
601. Hanada, K. and Harrison, B. D., Effect of virus genotype and temperature on seed transmission of nepo-viruses, *Ann. Appl. Biol.*, 85, 79, 1977.
602. Taylor, C. E. and Thomas, P. R., The association of *Xiphinema diversicaudatum* (Micoletsky) with strawberry latent ringspot and arabis mosaic viruses in a raspberry plantation, *Ann. Appl. Biol.*, 62, 147, 1968.
603. Kriz, J., Transmission of curl disease and infectious sterility of hops to the progeny through the seeds, *Ann. Acad. Tchecosl. Agric.*, 32, 951, 1959.
604. Walkey, D. G. A., Seed transmission of arabis mosaic virus in lettuce (*Latuca sativa*), *Plant Dis. Rep.*, 51, 883, 1967.
605. Lister, R. M., Transmission of soil-borne viruses through seed, *Virology*, 10, 547, 1960.
606. Tomlinson, J. A. and Walkey, D. G. A., The isolation and identification of rhubarb viruses occurring in Britain, *Ann. Appl. Biol.*, 59, 415, 1967.
607. Jones, R. A. C. and Kenten, R. H., Arracacha virus A., Commonwl. Mycol. Assoc. Appl. Biol., Descriptions of Plant Viruses, No. 216, 1980, 4.

608. Jones, R. A. C., Tests for seed transmission of four potato viruses through potato true seed, *Ann. Appl. Biol.*, 100, 315, 1982.
609. Rana, G. L., Kyriakopoulou, P. E., and Martelli, G. P., Artichoke yellow ringspot virus, CMI/ABB Description Plant Viruses No. 271, 1983, 4.
610. Kyriakopoulou, P. E., Rana G. L. and Roca, F., Geographic distribution, natural host range, pollen and seed transmissibility of artichoke yellow ringspot virus, *Annu. Inst. Phytopathol. Benaki,* 14, 139, 1985.
611. Avgelis, A. D., Katis, N., and Grammatikaki, G., Broadbean wrinkly seed caused by artichoke yellow ringspot nepoviruses, *Ann. Appl. Biol.* 121, 133, 1992.
612. Bertaccini, A., Giunchedi, L., and Pollini, C. P., Survey on asparagus virus diseases in Italy, *Acta Hortic.*, 271, 279, 1990.
613. Evans, T. A. and Stephens, C. T., Association of asparagus virus II with pollen from infected asparagus (*Asparagus officinalis*), *Plant Dis.*, 72, 195, 1988.
614. Falloon, P. G., Falloon, L. M., and Grogan, R. G., Survey of California asparagus for asparagus virus I, Asparagus virus II, and tobacco streak virus, *Plant Dis.*, 70, 103, 1986.
615. Weissenfels, M. and Schmelzer, K., Investigations on the amount of losses due to viruses in asparagus (*Asparagus officinalis* L.), *Arch. Phytopathol. Pflanzensch.*, 12, 67, 1976.
616. Fujisawa, I., Goto, T., Tsuchizaki, T., and Iizuka, N., Some properties of asparagus virus II isolated from *Asparagus officinalis* in Japan, *Ann. Phytopathol. Soc. Jpn.*, 49, 683, 1983.
617. Jordan, R. L., Dodds, J. A., and Ohr, H. D., Evidence for virus-like agents in avocado, *Phytopathology,* 73, 1130, 1983.
618. Alper, M., Bar-Joseph, M., Salamon, R., and Loebenstein, G., Some characteristics of an avocado strain of tobacco mosaic virus, *Phytoparasitica*, 6, 15, 1978.
619. Jordan, R. L., Dodds, J. A., and Ohr, H. D., Detection of double-stranded RNA in avocado, *Phytopathology*, 71, 884, 1981.
620. Dhanraj, K. S. and Raychaudhuri, S. P., A note on barley mosaic in India, *Plant Dis. Rep.*, 53, 766, 1969.
621. Nitzany, F. E. and Gerechter, Z. K., Barley stripe mosaic virus host range and seed transmission tests among Gramineae in Israel, *Phytopathol. Mediterr.*, 2, 11, 1962.
622. Inouye, T., Studies on barley stripe mosaic in Japan, *Ber. Ohara Inst. Landwirtsch. Biol.*, 11, 413, 1962.
623. Chiko, A. W., Natural occurrence of barley stripe mosaic virus in wild oats (*Avena fatua*), *Can. J. Bot.*, 53, 417, 1975.
624. McKinney, H. H. and Greeley, L. W., Biological characteristics of barley stripe mosaic virus strains and their evolution, *U.S. Dept. Agric. Tech. Bull.*, 1324, 84, 1965.
625. Sutic, D. and Tosic, M., Transmission of mosaic virus by barley seeds, *Adv. Front. Plant Sci.*, 17, 205, 1966.
626. McNeal, F. H. and Afanasiev, M. M., Transmission of barley stripe mosaic through the seed in eleven varieties of spring wheat, *Plant Dis. Rep.*, 39, 460, 1955.
627. Singh, R. and Singh, A. K., Observations on mosaic disease of sunnhemp (*Crotalaria juncea* L.), *Phytopathol. Mediterr.*, 16, 132, 1977.
628. Frencel, I. and Pospieszny, H., Viruses in natural infections of yellow lupin (*Lupinus luteus* L.) in Poland. IV. Bean common mosaic virus (BCMV), *Acta Phytopathol. Acad. Sci. Hung.*, 14, 279, 1979.
629. Provvidenti, R. and Braverman, S. W., Seed transmission of bean common mosaic virus in phasemy bean, *Phytopathology*, 66, 1274, 1976.

630. Provvidenti, R. and Cobb, E. D., Seed transmission of bean common mosaic virus in tepary bean, *Plant Dis. Rep.*, 59, 966, 1975.
631. Klein, R. E., Wyatt, S. D., and Kaiser, W. J., Incidence of bean common mosaic virus in USDA *Phaseolus* germplasm collection, *Plant Dis.*, 72, 301, 1988.
632. Stewart, V. B. and Reddick, D., Bean mosaic, *Phytopathology*, 7, 61, 1917.
633. Tsuchizaki, T., Yora, K., and Asuyama, H., Seed transmission of viruses in cowpea and azuki bean plants. II. Relations between seed transmission and gamete infection, *Ann. Phytopathol. Soc. Jpn.*, 36, 237, 1970.
634. Agarwal, V. K., Nene, Y. L., and Beniwal, S. P. S., Detection of bean common mosaic virus in urdbean (*Phaseolus mungo*) seeds, *Seed Sci. Technol.*, 5, 619, 1977.
635. Kaiser, W. J. and Mossahebi, G. H., Natural infection of mungbean by bean common mosaic virus, *Phytopathology*, 64, 1209, 1974.
636. Snyder, W. C., A seedborne mosaic of asparagus bean, *Vigna sesquipedalis*, *Phytopathology*, 32, 518, 1942.
637. Sachchidananda, J., Singh, S., Prakash, N., and Verma, V. S., Bean common mosaic virus on cowpea in India, *Z. Pflanzenkr. Pflanzensch.*, 80, 88, 1973.
638. Lin, M. T. and Hill, J. H., Bean pod mottle virus: occurrence in Nebraska and seed transmission in soybeans, *Plant Dis.*, 67, 230, 1983.
639. Uyemoto, J. K. and Grogan, R. G., Southern bean mosaic virus: evidence of seed transmission in bean embryos, *Phytopathology*, 67, 1190, 1977.
640. Shepherd, R. J. and Fulton, R. W., Identity of a seed-borne virus of cowpea, *Phytopathology*, 52, 489, 1962.
641. Porembskaya, N. B., Seed transmission of virus diseases of lupins, *Tr. Vses. Inst. Zashch. Rast.*, 20, 54, 1964.
642. Blaszczak, W., Seed transmission of narrow leavedness of yellow lupin (NYL), *Genet. Polon.*, 4, 65, 1963.
643. Jones, R. A. C. and McLean, G. D., Virus diseases of lupins, *Ann. Appl. Biol.*, 114, 609, 1989.
643a. Phatak, H. C., Seed-borne plant viruses — identification and diagnosis in seed health testing, *Seed Sci. Technol.*, 2, 3, 1974.
644. Hino, T., Studies on the adzuki bean mosaic virus, *Ann. Phytopathol. Soc. Jpn.*, 27, 138, 1962.
645. Inouye, T., A seed-borne mosaic virus of pea, *Ann. Phytopathol. Soc. Jpn.*, 33, 38, 1967.
646. Haack, I., Detection of seed transmissible viruses in seeds of fava beans and peas by means of ELISA, *Arch. Phytopathol. Pflanzensch.*, 26, 337, 1990.
647. Hampton, R. O., Seed transmission of viruses in red clover, *Phytopathology*, 57, 98, 1967.
648. Kaiser, W. J. and Eskandari, F., Studies with bean yellow mosaic virus in Iran, *Iran J. Plant Pathol.*, 6, 26, 1970.
649. Aftab, M., Mughal, S. M., and Ghafoor, A., Occurrence and identification of bean yellow mosaic virus from faba bean in Pakistan, *Indian J. Virol.*, 5, 88, 1989.
650. Bengno, D. A. and Favali-Hedayat, M. A., Investigations on previously unreported or noteworthy plant viruses and virus diseases in Philippines, *FAO Plant Prot. Bull.*, 25, 78, 1977.
651. Mali, V. R., Khalikar, P. V., and Gaushal, D. H., Seed transmission of poty and cucumo viruses in cowpea in India, *Indian Phytopathol.*, 36, 343, 1983.
652. Kassanis, B., Russell, G. E., and White, R. F., Seed and pollen transmission of beet cryptic virus in sugarbeet plants, *Phytopathol. Z.*, 91, 76, 1978.

653. Stanarius, A., Meyer, U., and Kuhne, T., Effect of cultural conditions on the concentration of beet cryptic virus in sugarbeet plants (*Beta vulgaris* var. *altissima*), *Arch. Phytopathol. Pflanzensch.*, 25, 421, 1989.
654. Abdel-Salam, A. M. and Amin, A. H., An Egyptian isolate of beet curly top virus: new differential hosts, physical properties, seed transmission and serological studies, *Bull. Fac. Agric. Univ. Cairo,* 41, 843, 1990.
655. Clinch, P. E. M. and Loughnane, J. B., Seed transmission of virus yellows of sugarbeet (*Beta vulgaris* L.) and the existence of strains of this virus in Eire, *Proc. R. Soc. Dublin*, 24, 307, 1948.
656. Fritzsche, R., Proeseler, G., Karl, E., and Klienhempel, H., Detection of seed transmission of beet mild yellowing virus (BMYV), *Phytopathol. Z.*, 106, 360, 1983.
657. Capoor, S. P. and Sawant, D. M., Studies on bell pepper yellow mosaic virus, *Indian Phytopathol.*, 40, 443, 1987.
658. Provvidenti, R., Seed transmission of blackeye cowpea mosaic virus in *Vigna mungo*, *Plant Dis.*, 70, 981, 1986.
659. Zettler, F. W. and Evans, I. R., Blackeye cowpea mosaic virus in Florida host range and incidence in certified cowpea seed, *Proc. Fla. State Hortic. Soc.*, 85, 99, 1972.
660. Uyemoto, J. K., Tschenberg, E. F., and Hummer, D. K., Isolation and identification of a strain of grapevine Bulgarian latent virus in concord grapevine in New York State, *Plant Dis. Rep.*, 61, 949, 1977.
661. Childress, A. M. and Ramsdell, D. C., Detection of blueberry leaf mottle virus in highbush blueberry pollen and seed, *Phytopathology*, 76, 1333, 1986.
662. Devergne, J. C. and Cousin, R., Broad bean mosaic and seed patterning symptoms, *Ann. Epiphyt.*, 17, 147, 1966.
663. Fortass, M. and Bos, L., Broadbean mottle virus in Morocco: variability, interaction with food legume species and seed transmission in fava bean, pea and chickpea, *Neth. J. Plant Pathol.*, 98, 329, 1992.
664. Makkouk, K. M., Bos, L., Rizkallah, A., Azzam, O. I., and Katul, L., Broadbean mottle virus: identification, host range, serology, and occurrence of fava bean (*Vicia faba*) in West Asia and North Africa, *Neth. J. Plant Pathol.*, 94, 195, 1988.
665. Bawden, F. C., Chaudhuri, R. P., and Kassanis, B., Some properties of broadbean mottle virus, *Ann. Appl. Biol.*, 38, 774, 1951.
666. Phatak, H. C., Seed-borne plant viruses — identification and diagnosis in seed health testing, *Seed Sci. Technol.*, 2, 3, 1974.
667. Makkouk, K. M. and Azzam, O. I., Detection of broadbean stain virus in lentil seed groups, *LENS Newsl.*, 13, 37, 1986.
668. Cockbain, A. J., Cook, S. M., and Bowen, R., Transmission of broadbean stain virus and Echtes Ackerbohnenmosaik virus to field beans (*Vicia faba*) by weevils, *Ann. Appl. Biol.*, 81, 331, 1975.
669. Makkouk, K. M., Azzam, O. I., Katul, L., Rizkallah, A., and Koumari, S., Seed transmission of broadbean stain virus in the wild legume, *Vicia palaestina* Boiss, *Fabis Newsl.* 16, 40, 1986.
670. Putz, C. and Kuszala, M., Two new viruses attacking broadbean (*Vicia faba* L.) in France. I. Identification trials and evaluation of economic importance, *Ann. Phytopathol.*, 5, 447, 1973.
671. Russo, M., Gallitelli, D., Vovlas, C., and Savino, V., Properties of broadbean yellow band virus, a possible new tobravirus, *Ann. Appl. Biol.*, 105, 223, 1984.
672. Von Wechmar, M. B., Kaufmann, A., and Rybicki, E. P., Brome mosaic virus is transmitted through wheat seed, in 4th Int. Congr. Plant Pathol., 1983, 260.

673. Kenten, R. H., Cacao necrosis virus, CMI/AAB *Descriptions of Plant Viruses,* No. 173, 1977, 4.
674. Stijger, C. C. M. M. and Rast, A. T. B., Prospects of dry heat treatment at 70°C for disinfection of pepper seed infected with Capsicum mosaic virus, *Mededel Facul. Landbouwwerenschappen,* Rijksumiversiteit Gent, 53, 473, 1988.
675. Brunt, A., Crabtree, K., and Gibbs, A., *Viruses of Tropical Plants,* CAB International, U.K., 1990, 707.
676. Phatak, H. C. and Summanwar, A. S., Detection of plant viruses in seeds and seed stocks, *Proc. Int. Seed Test. Assoc.*, 32, 625, 1967.
677 Lisa, V., Luisoni, E., and Milne, R. G., A possible virus cryptic in carnation, *Ann. Appl. Biol.*, 98, 431, 1981.
678. Schimanski, H. H., Albrecht, H. J., and Kegler, H., Seed transmission of cherry leaf roll virus in birch (*Betula pendula* (Roth)), *Arch. Phytopathol. Pflanzensch.*, 16, 231, 1980.
679. Lister, R. M. and Murant, A. F., Seedborne viruses: viruses with nematodes as vectors, *Annu. Rep. Hortic. Res. Inst. Scotl.*, 8, 56, 1961.
680. Quacquarelli, A. and Savino, V., Cherry leaf roll virus in walnut. II. Distribution in *Apulia* and transmission through seed, *Phytopathol. Mediterr.*, 16, 154, 1977.
681. Kolber, M., Nemeth, M., and Szentivanyi, P., Routine testing of English walnut mother trees and group testing of seeds by ELISA for detection of cherry leaf roll virus infection, *Acta Hortic.*, 130, 161, 1983.
682. Topchiiska, M. L., Detection of cherry leaf roll virus (CLRV) in seeds of *Juglans regia* by ELISA, *Acta Hortic.*, 311, 290, 1993.
683. Schmelzer, K., Das Ulmenscheckungs — virus, *Phytopathol. Z.*, 64, 39, 1969.
684. Cooper, V. C. and Walkey, D. G. A., Thermal inactivation of cherry leaf roll virus in tissue cultures of *Nicotiana rustica* raised from seeds and meristem tips, *Ann. Appl. Biol.*, 88, 273, 1978.
685. Schmelzer, K., Untersuchungen an viren der Zier- und Wildgeholze, 5, Mitteilung; Virosen an *Populus* und *Sambucus*, *Phytopathol. Z.*, 55, 317, 1966.
686. Schimanski, H. H., Schmelzer, K., and Albrecht, H. J., Seed transmission of cherry leaf roll virus in black cherry *Prunus serotina* Ehrh, *Zentralbl. Bakteriol. Parasitenkd. Infektionskr. Hyg.*, 2, 131, 1976.
687. Schimanski, H. H. and Schmelzer, K., Contribution to the knowledge of the transmissibility of cherry leaf roll virus by seeds of *Sambucus racemosa* L., *Zentralbl. Bakteriol. Parasitenkd. Infektionskr. Hyg.*, 127, 673, 1972.
688. Callahan, K. L., *Prunus* host range and pollen transmission of elm mosaic virus, *Phytopathology,* 47, 5, 1957.
689. Nyland, G. B., Possible virus-induced genetic abnormalitites in tree fruits, *Science*, 137, 598, 1962.
690. Nyland, G. B., Lownsbery, F., Lowe, S. K., and Mitchell, J. F., The transmission of cherry raspleaf virus by *Xiphinema americanum*, *Phytopathology,* 59, 1111, 1969.
691. Wagon, H. K., Traylor, J., Williams, H. E., and Weiner, A. C., Investigations of cherry raspleaf disease in California, *Plant Dis. Rep.*, 52, 618, 1968.
692. Vovlas, C., Seed transmission of chicory yellow mottle virus, *Phytopathol. Mediterr.*, 12, 104, 1973.
693. Dickson, B. T., A mosaic-like disease of *Cineraria*, *Annu. Rep. Quebec Soc. Prot. Plants*, 1920, 46, 1920.
694. Dickson, B. T., A mosaic-like disease of *Cineraria*, *Annu. Rep., Quebec Soc. Prat. Plants*, 1920, 46, 1920.

695. Jones, L. K., Streak and mosaic of *Cineraria, Phytopathology*, 34, 941, 1944.
696. Dakshinamurti, V. and Subbaya, J., Studies on transmission of citrus mosaic — a new disease of *Citrus sinensis* in Andhra Pradesh, in 3rd Int. Symp. Plant Pathology, New Delhi, 1981, 40.
697. Inouye, N., Maeda, T., and Mitsuhata, K., Citrus tatter leaf virus isolated from lily, *Ann. Phytopathol. Soc. Jpn.*, 45, 712, 1979.
698. Sawant, D. M., Choudhari, K. G., and Desai, U. T., Veinal chlorosis virus disease of citrus, *Indian J. Mycol. Plant Pathol.*, 13, 346, 1983.
699. Calavan, E. C., Wood pocket disease of lemons and seedless limes, *Calif. Citrogr.*, 42, 265, 1957.
700. Childs, J. F. L., Transmission experiments and xyloporosis-cachexia relations in Florida, *Plant Dis. Rep.*, 40, 143, 1956.
701. Matsulevich, B. P., The effect of clover mosaics on the productivity of red clover, *Agrobiology* (Moscow), 2, 75, 1957.
702. Hagedorn, D. J. and Hanson, E. W., The relationship between Wisconsin pea stunt and red-clover vein mosaic, *Phytopathology*, 41, 15, 1951.
703. Hagedorn, D. J. and Hanson, E. W., A comparative study of the viruses causing Wisconsin pea stunt and red clover vein mosaic, *Phytopathology*, 41, 813, 1957.
704. Osborne, H. T., Vein mosaic of red clover, *Phytopathology*, 27, 1051, 1937.
705. Sander, E., Biological properties of red clover vein mosaic virus, *Phytopathology*, 49, 748, 1959.
706. Boccardo, G., Milne, R. G., Luicsoni, E., Lisa, V., and Accott, G. P., Three seedborne cryptic viruses containing double-stranded RNA isolated from white clover, *Virology*, 147, 29, 1985.
707. Joshi, R. D., Prakash, J. and Dubey, L. N., Detection of white clover mosaic virus in eastern Uttar Pradesh, *Indian J. Mycol. Plant Pathol.*, 11, 157, 1981.
708. Hampton, R. O., Seed transmission of white clover mosaic and clover yellow mosaic viruses in red clover, *Phytopathology*, 53, 1139, 1963.
709. Reyes, T. T., Seed transmission of coffee ring spot by excelsa coffee (*Coffea excelsa*), *Plant Dis. Rep.*, 45, 185, 1961.
710. Ghose, T. and Basak, M., Transmission of leaf mosaic (chlorosis) in jute, *Jute Bull.*, 94, 1961.
711. Yilmaz, M. A. and Ozaslan, D., Determination of cowpea aphidborne mosaic virus by enzyme linked immunosorbent assay on cowpea and bean seeds, *Doga Turk Tarim Ormancilik Dergisi*, 13, 870, 1989.
712. Anderson, C. W., Seed transmission of three viruses in cowpea, *Phytopathology*, 47, 515, 1957.
713. Sharma, S. R. and Varma, A., Transmission of cowpea banding mosaic and cowpea chlorotic mosaic spot viruses through the seeds of cowpea, *Seed Sci. Tech.*, 14, 217, 1986.
714. Chenulu, V. V., Sachchidananda, J., and Mehta, S. C., Studies on a mosaic disease of cowpea from India, *Phytopathol. Z.*, 63, 381, 1968.
715. Sharma, S. R. and Varma, A., Three sap transmissible viruses from cowpea in India, *Indian Phytopathol.*, 28, 192, 1975.
716. Shoyinka, S. A., Bozarth, R. F., Reese, J., and Rossel, H. W., Cowpea mottle virus: a seedborne virus with distinctive properties infecting cowpea in Nigeria, *Phytopathology*, 68, 693, 1978.
717. Robertson, D. G., Seed-Borne Viruses of Cowpea in Nigeria, B.Sc. Thesis, University of Oxford, England, 1966, 111.

718. Brunt, A. A. and Kenten, R. W., Cowpea mild mottle, a newly recognized virus infecting cowpea (*Vigna unguiculata*) in Ghana, *Ann. Appl. Biol.*, 74, 67, 1973.
719. Santos, A. A., Lin, M. T., and Kitajima, E. W., Characterization of two poty-viruses isolated from cowpea (*Vigna unguiculata*) in Piaui State, *Fitopathol. Brasileira*, 9, 567, 1984.
720. Diwakar, M. P. and Mali, V. R., Cowpea mosaic virus disease — a new record for Marathwada, *J. Madras Agric. Univ.*, 1, 274, 1977.
721. Capoor, S. P. and Varma, P. M., Studies on a mosaic disease of *Vigna cylindrica* Skeels, *Indian J. Agric. Sci.*, 26, 95, 1956.
722. Dale, W. T., Observations on a virus disease of cowpea in Trinidad, *Ann. Appl. Biol.*, 36, 327, 1949.
723. Shepard, R. J., Properties of a mosaic disease of cowpea and its relationship to bean pod mottle virus, *Phytopathology*, 54, 466, 1964.
724. Phatak, H. C., Diaz-Ruiz, J. R., and Hull, R., Cowpea ringspot virus: a seed transmitted cucumovirus, *Phytopathol. Z.*, 87, 132, 1976.
725. Pio-Ribeiro, G., Wyatt, S. D., and Kuhn, C. W., Cowpea stunt: a disease caused by a synergistic interaction of two viruses, *Phytopathology*, 68, 1260, 1978.
726 Kenten, R. H., Cockbain, A. J., and Woods, R. D., Crimson clover latent virus — a newly recognized seed-borne virus infecting crimson clover *(Trifolium incarnatum), Ann. Appl. Biol.*, 96, 79, 1980.
727. Lee, K. W., Lee, B. C., Park, H. C., and Lee, Y. S., Occurrence of cucumber green mottle virus disease of watermelon in Korea, *Korean J. Plant Pathol.*, 6, 250, 1990.
728. Yakovleva, N., Control of green mosaic of cucumber, *Zashch. Rast. Vredit. Bolez.*, 10, 50, 1965.
729. Komuro, Y., Tochicharer, H., Fukatsu, R., Nagai, Y., and Yoneyama, S., Cucumber green mottle mosaic virus (watermelon strain) in watermelon and its bearing on deterioration of watermelon fruit known as "konnyaku" disease, *Ann. Phytopathol. Soc. Jpn.*, 37, 34, 1971.
730. Weber, I., Cucumber leaf spot virus, Assoc. Appl. Biol., Descriptions of Plant Viruses, No. 319, 1986, 4.
731. Cai, Z. N., Xu, Z. Y., Wang, D., and Yu, S. L., Studies on the defection of peanut seedborne virus by enzyme-linked immunosorbent assay (ELISA), *Acta Phytopathol. Sin.*, 16, 23, 1986.
732. Sharma, Y. R. and Chohan, J. S., Transmission of cucumis viruses 1 and 3 through seeds of cucurbits, *Indian Phytopathol.*, 26, 596, 1973.
733. Chauhan, L. S. and Singh, D. R., Occurrence of mosaic on safflower in northern India, *Indian Phytopathol.*, 32, 301, 1979.
734. Keyworth, W. G., Plant pathology, *Rep., Natl. Veg. Res. Stn.*, 22, 75, 1972.
735. Kendrick, J. B., Cucurbit mosaic transmitted by muskmelon seed, *Phytopathology*, 24, 820, 1934.
736. Doolittle, S. P., The mosaic disease of cucurbits, *U.S. Dept. Agric. Bull.*, No. 879, 1, 1920.
737. Rader, W. E., Fitzpatrick, H. F., and Hildebrand, E. M., A seedborne virus of muskmelon, *Phytopathology*, 37, 809, 1947.
738. Mukhopadhyay, S. and Kalisanker, S., Transmission of cucumis virus (cucumber mosaic virus) through seeds of *Cucurbita maxima* L., *Sci. Cult.*, 34, 436, 1968.

739. Horvath, J. and Szirmai, J., Investigations of a virus disease of wild cucumber *(Echinocystis lobata* (Michx. Torr. & Gray), *Acta Phytopathol. Acad. Sci. Hung.*, 8, 329, 1973.

740. Shen, S. L., Wang, S. Q., and Chen, Y. F., Isolation and identification of cucumber mosaic virus transmitted through soybean seed, *Acta Phytopathol. Sin.*, 14, 251, 1984.

741. Von Wechmar, M. B., Kaufmann, A., and Rybicki, E. P., Serological detection of cucumber mosaic virus in the South African cereal crops: seedborne CMV in barley, in Barley Yellow Dwarf Proceedings, Dec. 6–8, CIMMYT, Mexico, 147, 1983.

742. Wells, H. D., Cucumber mosaic virus is seed-borne in blue lupines, *Phytopathology*, 54, 627, 1964.

743. Troll, H. J., On the question of seed transmission of the browning virus of *Lupinus luteus, Nachrichtenbl. Dtsch. Pflanzenschutzdienst* (Berlin), 11, 218, 1957.

744. Van Koot, Y., Enkele nieuwe gezichtspunten betreffende het virus van het tomatenmosaiek, *Tijdschr. Plantenziekten*, 55, 152, 1949.

745. Doolittle, S. P. and Gilbert, W. W., Seed transmission of cucurbit mosaic by the wild cucumber, *Phytopathology*, 11, 326, 1921.

746. Meiners, J. P., A strain of cucumber mosaic virus seed-borne in bean (*Phaseolus vulgaris* L.), *Proc. Am. Phytopathol. Soc.*, 1, 37, 1974.

747. Jeyanandarajoh, P., Seedborne viruses infecting three important leguminous food crops in Sri Lanka, *Seed Sci. Technol.*, 20, 629, 1992.

748. Somerville, P. A., Campbell, R. N., Hall, D. H., and Rowhani, A., Natural infection of potatoes (*Solanum tuberosum*) by a legume strain of cucumber mosaic virus, *Plant Dis.*, 71, 18, 1987.

749. Tomlinson, J. A. and Carter, A. L., Studies on the seed transmission of cucumber mosaic virus in chickweed (*Stellaria media*) in relation to the ecology of the virus, *Ann. Appl. Biol.*, 66, 381, 1970.

750. Jones, R. A. C. and McKirdy, S. J., Seedborne cucumber mosaic virus infection of sub-terranean clover in Western Australia, *Ann. Appl. Biol.*, 110, 73, 1990.

751. Purivirojkul, W., Sittiyos, P., Hsu, C. H., Poehlman, J. M., and Sehgal, O. P., Natural infection of mungbean *(Vigna radiata)* with cucumber mosaic virus, *Plant Dis. Rep.*, 62, 530, 1978.

752. Asian Vegetable Research and Development Center (AVRDC), Cucumber mosaic virus: host range, seed transmission and sources of resistance, in Progress Report 1985, Shanhua, Taiwan, 1987, 171.

753. Lin, M. T., Santos, A. A., and Kitajima, E. W., Host reactions and transmission of two seed-borne cowpea viruses from central Brazil, *Fitopatol. Brasileira*, 6, 193, 1981.

754. Brantley, B. B., Kuhn, C. W., and Sowell, G., Jr., The effect of cucumber mosaic virus on southern pea *(Vigna sinensis), Proc. Am. Soc. Hortic. Sci.*, 87, 355, 1965.

755. Edwardson, J. R., Purcifull, D. E., Zettler, F. W., Christie, R. G., and Christie, S. R., A virus isolated from *Desmodium canum*: characterization and electron microscopy, *Plant Dis. Rep.*, 54, 161, 1970.

756. Sutere, B. D. and Joshi, R. D., *Desmodium triflorum* mottle — a virus disease, *Geobios*, 5, 266, 1978.

757. Bennett, C. W., Studies on dodder transmission of plant viruses, *Phytopathology*, 34, 905, 1944.

758. Gibbs, A. J., Hecht-Poinar, E., Woods, R. D., and McKee, R. K., Some properties of three related viruses: Andean potato latent, dulcamara mottle, and ononis yellow mosaic, *J. Gen. Microbiol.*, 44, 1977, 1966.

759. Mayee, C. D., Storage of seed for pragmatic control of a virus causing mosaic disease of bringal (egg plant), *Seed Sci. Technol.*, 5, 555, 1977.
760. Jones, R. A. C. and Fribourg, C. E., Beetle, contact and potato true seed transmission of Andean potato latent virus, *Ann. Appl. Biol.*, 86, 123, 1977.
761. Danesh, D. and Lockhart, B. E. L., Eggplant mottled dwarf virus in potato in Iran, *Plant Dis.*, 73, 856, 1989.
762. Callahan, K. L., Pollen transmission of elm mosaic virus, *Phytopathology*, 47, 5, 1957.
763. Jones, A. T. and Mayo, M. A., 20th Annual Report for the Year 1973, Scottish Hortic. Res. Inst., Invegowrie, Scotland, 1974.
764. Paulsen, A. Q. and Niblett, C. L., Purification and properties of foxtail mosaic virus, *Phytopathology*, 67, 1346, 1977.
765. Paulsen, A. Q., Reactions of eleven *Chenopodium* species of seven sap-transmissible grass viruses, *Phytopathology*, 60, 1307, 1970.
766. Paulsen, A. Q. and Sill, W. H., Hosts, symptoms and characteristics of a sap-transmissible virus isolate from foxtails, *Phytopathology*, 59, 1942, 1969.
767. Capoor, S. P., Rao, D. G., and Sawant, D. M., Seed transmission of French bean mosaic virus, *Indian Phytopathol.*, 39, 343, 1986.
768 Uyemoto, J. K., Taschenberg, E. F., and Hummer, D. K., Isolation and identification of a strain of grapevine Bulgarian latent virus in concord grapevine in New York State, *Plant Dis. Rep.*, 61, 949, 1977.
769. Dias, H. F., Host range and properties of grapevine fanleaf and grapevine yellow mosaic viruses, *Ann. Appl. Biol.*, 51, 85, 1963.
770. Cory, L. and Hewitt, W. B., Some grapevine viruses in pollen and seeds, *Phytopathology*, 58, 1316, 1968.
771. Hansen, A. J. and Lesemann, D. E., Occurrence and characteristics of a seed transmitted potyvirus from Indian, African, and North American guar, *Phytopathology*, 68, 41, 1978.
772. Sun, Y., Xu, H. Y., Bi, Q., Li, G. X., and Wang, J. M., Detecting hami muskmelon seedborne virus and effect of heat treatment on seedborne virus, *Acta Phytopathol. Sin.*, 15, 62, 1985.
773. Brierly, P., Viruses described primarily on ornamental or miscellaneous plants, *Plant Dis. Rep.*, (Suppl.), 150, 1944.
774. Brants, D. and Van der Heuvel, J., Investigation of Hippeastrum mosaic virus in *Hippeastrum hybridum, Neth. J. Plant Pathol.*, 71, 145, 1965.
775. Blattny, C. and Osvald, V., Transmission of hop viruses by seed, *Preslia (Prague)*, 26, 1, 1954.
776. Thomas, B. J., Barton, R. J., and Tuszynski, A., Hydrangea mosaic virus, a new ilarvirus from *Hydrangea macrophylla* (Saxifragaceae), *Ann. Appl. Biol.*, 103, 261, 1984.
776a. Izadpanath, K., Huth, W., Lesemann, D. E., and Vetten, H. J., Properties of johnsongrass chlorotic stripe mosaic virus, *J. Phytopathol.*, 137, 105, 1993.
777. Zink, F. W., Grogan, R. G., and Welch, J. E., The effect of the percentage of seed transmission upon subsequent spread of lettuce mosaic virus, *Phytopathology*, 46, 662, 1956.
778. Hunter, D. G. and Bowyer, J. W., Cytopathology of lettuce mosaic virus-infected lettuce seeds and seedlings, *J. Phytopathol.*, 137, 61, 1993.
779. Van Hoff, H. A., Seed transmission of lettuce mosaic virus in *Lactuca serriola, Tijdschr. Plantenziekten*, 65, 44, 1959.

780. Kemper, A., On the occurrence of lettuce mosaic virus on lettuce (*Lactuca sativa* L.) and on ground sel (*Senecio vulgaris* L.), *Z. Pflanzenkr. Pflanzensch.*, 69, 11, 1962.
781. Ertunc, F., Detection of lettuce mosaic virus (LMV) from infected lettuce (*Lactuca sativa* L.) and wild mustard (*Sinapsis arvensis* L.) seeds by non-precoated indirect enzyme-linked immunosorbent assay, *Ankara Univ. Ziraat Facultesi Yayinlari*, 1253, 18, 1992.
782. Vasudeva, R. S., Raychaudhuri, S. P., and Pathanian, P. S., Yellow mosaic of lettuce, *Curr. Sci.*, 17, 244, 1948.
783. Van der Meer, F. A., Huttinga, H., and Maat, D. Z., Lilac ring mottle virus: isolation from lilac, some properties, and relation to lilac ringspot disease, *Neth. J. Plant Pathol.*, 82, 67, 1976.
784. Sawant, D. M. and Capoor, S. P., Seed transmission of lima bean mosaic virus, *Indian Phytopathol.*, 36, 659, 1983.
785. Jones, A. T., Forstev, R. L. S., and Mohmed, N. A., Purification and properties of Australian lucerne latent virus, a seed-borne virus having affinities with nepoviruses, *Ann. Appl. Biol.*, 92, 49, 1979.
786. Blackstock, J. M., Lucerne transient streak and lucerne latent, two new viruses of lucerne, *Aust. J. Agric. Res.*, 29, 291, 1978.
787. Remah, A., Jones, A. T., and Mitchell, M. J., Purification and properties of lucerne Australian symptomless virus, a new virus infecting lucerne in Australia, *Ann. Appl. Biol.*, 109, 307, 1986.
788. Paliwal, Y. C., Identification and distribution in eastern Canada of lucerne transient streak, a virus newly recognized in North America, *Can. J. Plant Pathol.*, 5, 75, 1983.
789. Bennett, C. W., Lychnis ringpot, *Phytopathology*, 49, 706, 1959.
790. Jensen, S. G., Wysong, D. S., Ball, E. M., and Higley, P. M., Seed transmission of maize chlorotic mottle virus, *Plant Dis.*, 75, 497, 1991.
791. Hill, J. H., Martinson, C. A., and Russell, W. A., Seed transmission of maize dwarf mosaic and wheat streak mosaic viruses in maize and response of inbred lines, *Crop Sci.*, 14, 232, 1974.
792. Tsventkov, D., Leaf spot of cereals, *Rastit. Zasht.*, 15, 29, 1967.
793. Tosic, M. and Sutic, D., Investigation of maize mosaic virus transmission through corn seed, *Ann. Phytopathol.*, 9, 235, 1977.
794. Gonzalez-garza, R., Gumpf, D. J., Kishaba, A. N., and Bohn, G. W., Identification, seed transmission, and host range pathogenicity of a California isolate of melon necrotic spot virus, *Phytopathology*, 69, 340, 1979.
795. Martin, E. M. and Kim, K. S., A new type of plant virus causing striped chlorosis of Mimosa, *Phytopathology*, 77, 935, 1987.
796. Tsuchizaki, T., Hibino, H., and Saito, Y., Mulberry ringspot virus isolated from mulberry showing ringspot symptoms, *Ann. Phytopathol. Soc. Jpn.*, 37, 266, 1971.
797. Kaiser, W. J., Danesh, D., Okhovat, M., and Mossahebi, H., Diseases of pulse crops (edible legumes) in Iran, *Plant Dis. Rep.*, 52, 687, 1968.
798. Benigno, D. A. and Favali-Hedayat, M. A., Investigations on previously unreported or noteworthy plant viruses and virus disease in the Philippines, *FAO Plant Prot. Bull.*, 25, 78, 1977.
799. Gonzalez-Garza, R., Gumpf, D. J., Kishaba, A. N., and Bohn, G. W., Identification, seed transmission and host range, and pathogenicity of a California isolate of melon necrotic spot virus, *Phytopathology*, 69, 340, 1979.

800. Coudriet, D. L., Kishaba, A. N., and Carroll, J. E., Transmission of muskmelon necrotic ringspot virus in muskmelons by cucumber beetles, *J. Econ. Entomol.*, 72, 560, 1979.
801. Kishi, K., Necrotic spot of melon, a new viral disease, *Ann. Phytopathol. Soc. Jpn.*, 32, 138, 1966.
802. Randles, J. W., Harrison, B. D., and Robert, I. M., *Nicotiana velutina* mosaic virus: purification, properties and affinities with other rod-shaped viruses, *Ann. Appl. Biol.*, 84, 193, 1976.
803. Carr, A. J. H., Plant pathology, in Report of the Welsh Plant Breeding Station 1970, University College of Wales, Aberysthwyth, 1971, 40.
804. Protsenko, L. V., Modes of transmission of onion mosaic virus, *Kartoplya Ovochevi Bashtanni Kul't.*, 1, 147, 1965.
805. Louie, R. and Lorbeer, J. W., Mechanical transmission of onion yellow dwarf virus, *Phytopathology*, 56, 1020, 1966.
806. Niblett, C. L., Paulsen, A. Q., and Toler, R. W., Panicum mosaic virus, CMI/AAB Descriptions Plant Viruses, No. 117, 1977, 4.
807. Bayot, R. G., Villegas, V. N., Magdalita, P. M., Jovellana, M. D., Espino, T. M., and Exconde, S. B., Seed transmissibility of papaya ringspot virus, *Philipp. J. Crop Sci.*, 15, 107, 1990.
808. Singh, S. J., Studies on a virus causing mosaic disease of pumpkin (*Cucurbita maxima* Duch.), *Phytopathol. Mediterr.*, 20, 184, 1981.
809. Bos, L., Huttinga, H., and Maat, D.Z., Parsley latent virus, a new and prevalent seed transmitted but possibly harmless virus of *Petroselinum crispum, Neth. J. Plant Pathol.*, 85, 125, 1979.
810. Box, L. and Van der Want, J. P. H., Early browning of pea, a disease caused by a soil and seedborne virus, *Tijdschr. Plantenziekten*, 68, 368, 1962.
811. Mahir, M. A. M., Fortass, M., and Bos, L., Identification and properties of a deviant isolate of the broad bean yellow band serotype of pea early-browning virus from fava bean (*Vicia faba*) in Algeria, *Neth. J. Plant Pathol.*, 98, 237, 1992.
812. Kovachevski, I., Investigations of pea enation mosaic virus, *Rastenievud. Nauki.*, 15, 108, 1978.
813. Makkouk, K. M., Kumari, S. G., and Shehadesh, A., Seed transmission of pea seedborne mosaic virus in *Lathyrus* and *Vicia* forage legume species, *Z. Pflanzenkr. Pflanzensch.*, 99, 561, 1992.
814. Hampton, R. O. and Muehlbauer, F. J., Seed transmission of the pea seed-borne mosaic virus in lentils, *Plant Dis. Rep.*, 61, 235, 1977.
815. Zimmer, R. C., and Alikhan, S. T., New seedborne virus of field peas, *Can. Agric.*, 21, 6, 1976.
816 Khetarpal, R. K., Bossennec, J. M., Burghofer, A., Cousin, R., and Maury, Y., Effect of pea seedborne mosaic virus on the pod yield of field pea, *Agronomie*, 8, 811, 1988.
817. Kohnen, P. D., Dougherty, W. G., and Hampton, R. O., Detection of pea seedborne mosaic potyvirus by sequence specific enzymatic amplification, *J. Virol. Meth.*, 37, 253, 1992.
818. Wang, D. W. and Maule, A. J., A model for seed transmission of a plant virus: genetic and structural analyses of pea embryo invasion by pea seed-borne mosaic virus, *Plant Cell*, 6, 777, 1994.
819. Musil, M., Seed transmission of pea leaf roll mosaic virus, *Tagungsber. Akad. Landwirtschaftswiss. D.D.R.*, 184, 345, 1980.

820. Musil, M., Leskova, O., and Rapi, J., The influence of some factors on transmission of pea leaf rolling mosaic virus by pea seed, *Biol. Czech.*, A 36, 889, 1981.

821. Dusi, A. N., Nagata, T., and Iizuka, N., First report of pea seedborne mosaic virus in Brazil, *Plant Dis.*, 77, 1264, 1993.

822. Stevenson, W. R. and Hagedorn, D. J., Further studies on seed transmission of pea seedborne mosaic virus in *Pisum sativum, Plant Dis. Rep.*, 57, 248, 1973.

823. Thottappilly, G. and Schmutterer, H., A sap, seed-fungal, and insect transmissible virus of pea, *Z. Pflanzenkr. Pflanzensch.*, 75, 1, 1968.

824. Wang, D. and Maule, A. J., Early embryo invasion as a determinant in pea of the seed transmission of pea seedborne mosaic virus, *J. Gen. Virol.*, 73, 1615, 1992.

825. Wang, D., Woods, R. D., Cockbain, A. J., Maule, A. J., and Biddle, A. J., The susceptibility of pea cultivars to pea seedborne mosaic virus infection and virus seed transmission in the U.K., *Plant Pathol.*, 42, 42, 1993.

826. Zimmer, R. C. and Lamb, R. J., Amplification and spread of pea seedborne mosaic virus in field grown peas, *Can. J. Plant Pathol.*, 15, 17, 1993.

827. Kowalska, C. and Beczner, L., Characteristics of a seedborne virus in *Pisum sativum,* in *Problems in Plant Virology*, H. Kleinhempel, Ed., Tagungs-bericht Akademie Landwrits-chaftswissenschaften D. D. R., 1980, 353.

828. Dias, H. F. and Cation, D., The characterization of a virus responsible for peach rosette mosaic and grape decline in Michigan, *Can. J. Bot.*, 54, 1228, 1976.

828. Ramsdell, D. C. and Myers, R. L., Epidemiology of peach rosette mosaic virus in a concord grape vineyard, *Phytopathology*, 68, 447, 1978.

830. Manadahar, C. L., *Introduction to Plant Viruses*, S. Chand, New Delhi, 1978, 333.

831. Thouvenel, J. C., Fauquet, C., and Lamy, D., Seed transmission of groundnut clump virus, *Oleagineux*, 33, 503, 1978.

832. Van Velsen, R. J., Marginal necrosis, a seedborne virus of *Arachis hypogea* variety "Schwarz 21" in New Guinea, *Papua-New Guinea Agric. J.*, 14, 38, 1961.

833. Bijaisoradat, M. and Kuhn, C. W., Detection of two viruses in peanut seeds by complementary DNA hybridization tests, *Plant Dis.*, 72, 956, 1988.

834. Kuhn, C. W., Symptomatology, host range, and effect on yield of seed-transmitted peanut virus, *Phytopathology*, 55, 880, 1965.

835. Demski, J. W., Wells, H. D., Miller, J. D., and Khan, M. A., Peanut mottle virus epidemics in lupines, *Plant Dis.*, 67, 166, 1983.

836. Demski, J. W., Alexander, A. T., Stefani, M. A., and Kuhn, C. W., Natural infection, disease reactions and epidemiological implications of peanut mottle virus in cowpea, *Plant Dis.*, 67, 267, 1983.

837. Li, R. H., Zettler, F. W., Elliott, M. S., Petersen, M. A., Still, P. E., Baker, C. A., and Mink, G. I., A strain of peanut mottle virus seedborne in Bambarra groundnut, *Plant Dis.*, 75, 130, 1991.

838. Zettler, F. W., Elliott, M. S., Purcifull, D. E., Mink, G. I., Gorbet, D. W., and Knauft, D. A., Production of peanut seed free of peanut stripe and peanut mottle viruses in Florida, *Plant Dis.*, 77, 747, 1993.

839. Green, S. K. and Lee, D. R., Occurrence of peanut stripe virus (PstV) on soybean in Taiwan — effect on yield and screening for resistance, *Trop. Pest Manage.*, 35, 123, 1989.

840. Troutman, J. L., Bailey, W. K., and Thomas, C. A., Seed transmission of peanut stunt virus, *Phytopathology*, 57, 1280, 1967.

841. Gallitelli, D., Properties of a tomato isolate of Pelargonium zonate leaf spot virus, *Ann. Appl. Biol*, 100, 457, 1982.

842 Shmyglya, V. A., Makaev, S. Sh., Yunis, S., and Osman, A., Mediator method for diagnosis of plant viruses, *Sel' Skokhozyaiatvennaya Biol.*, 5, 190, 1991.
843. Jafarpour, B. Teakle, D. S., and Thomas, J. E., Incidence of potato virus S and X and potato leaf roll virus in potatoes in Queensland, *Aust. Plant Pathol.*, 17, 4, 1988.
844. Slazar, L. F. and Harrison, B. D., Host range, purification and properties of potato virus T., *Ann. Appl. Biol.*, 89, 223, 1978.
845. Jones, R. A. C., Fribourg, C. E., and Koeing, R. A., A previously undescribed nepo virus isolated from potato in Peru, *Phytopathology*, 73, 195, 1983.
846. Allen, T. C. and Davis, J. R., Distribution of tobacco rattle virus and potato virus X in leaves, roots and fruits and/or seed of naturally infected weeds, *Am. Pot. J.*, 59, 149, 1982.
847. Darozhkin, M. A. and Grabenshchykava, S. I., Transmission of virus X by potato seeds, *Vestsi Akad. Navuk B. SSR Ser. Biyal. Navuk*, 5, 80, 1974.
848. Eskrous, J. K., Habib, H. M., Kishtah, A. A., and Ismail, M. H., A strain of potato virus Y isolated from *Solanum nigrum* var. *judaicum* in Egypt, *Phytopathol. Mediterr.*, 22, 53, 1983.
849. Schimanski, H. H. and Meyer, U., Joint occurrence of prune dwarf virus and Prunus necrotic ringspot virus in seeds of sour cherry (*Cerasus vulgaris* Mill), *Arch. Phytopathol. Pflanzensch.*, 27, 419, 1991.
850. Mink, G. I. and Aichele, M. D., Detection of Prunus necrotic ringspot and prune dwarf viruses in *Prunus* seed and seedlings by enzyme-linked immunosorbent assay, *Plant Dis.*, 68, 378, 1984.
851. Gilmer, R. M. and Way, R. D., Pollen transmission of necrotic ringspot, and prune dwarf viruses in sour cherry, *Phytopathology*, 50, 624, 1960.
852. Das, C. R., Milbrath, J. A., and Swensen, K. G., Seed and pollen transmission of Prunus ringspot virus in Buttercup squash, *Phytopathology*, 51, 64, 1961.
853. Hobart, O. F., Introduction and spread of necrotic ringspot virus in sour cherry nursery trees, *Iowa State Coll. J. Sci.*, 30, 381, 1956.
854. Ascui, M. L. and Alvareza, M., Identification of Prunus necrotic ringspot virus (NRSV) by ELISA, *Agric. Tech.* (Santiago), 48, 71, 1988.
855. Megahed, E. S. and Moore, J. D., Differential mechanism transmission of *Prunus* viruses from seed of various *Prunus* spp. and from different parts of the same seed, *Phytopathology*, 57, 821, 1967.
856. Cochran, L. C., Passage of the ringspot virus through Mazzard cherry seeds, *Science*, 104, 269, 1946.
857. Cochran, L. C., Passage of the ringspot virus through peach seeds, *Phytopathology*, 40, 964, 1950.
858. Fortuner, R., Fauquet, C., and Lourd, M., Diseases of the winged bean in Ivory Coast, *Plant Dis. Rep.*, 63, 194, 1979.
859. Natsuaki, T., Yamashita, S., Doi, Y., and Yora, K., Radish yellow edge virus, a seed-borne small spherical virus newly recognized in Japanese radish (*Raphanus sativus* L.), *Ann. Phytopathol. Soc. Jpn.*, 45, 313, 1979.
860. Natsuaki, T., Yamashita, S., Doi, Y., Okuda, S., and Teranaka, M., Radish yellow edge virus, a seedborne virus with double-stranded RNA, of a possible new group, *Ann. Phytopathol. Soc. Jpn.*, 49, 593, 1983.
861. Lister, R. M. and Cadman, C. H., Relationships and possible mode of spread of filamentous viruses infecting woody Rosaceous plants (6th Eur. Symp. Fruit Tree Virus Diseases, Belgrade, June 1 to 8, 1965), *Zast. Bilja*, 16, 233, 1965.

862. Murant, A. F., Chambers, J., and Jones, A. T., Spread of raspberry bushy dwarf virus by pollen, its association with crumbly fruit, and problems of control, *Ann. Appl. Biol.*, 77, 271, 1974.
863. Cadman, C. H., Filamentous virus infection of fruit trees and raspberry and their possible mode of spread, *Plant Dis.*, 49, 230, 1965.
864. Murant A. F., Raspberry bushy dwarf virus, CMI/AAB Descriptions of Plant Viruses No. 165, 1976, 4.
865. Ormerod, P. J., A virus associated with loganberry degeneration disease, *Annu. Rep. East Malling Res. Stn.*, 165, 1970.
866. Mellor, F. C. and Stace-Smith, R., Reaction of strawberry to a ringspot virus from raspberry, *Can. J. Bot.*, 41, 865, 1963.
867. Vaughan, E. K. and Wiedman, H. W., A disease of strawberry caused by a virus from red raspberry, *Plant Dis. Rep.*, 39, 893, 1955.
868. Vaughan, E. K., Johnson, F., Fitzpatrick, R. F., and Stace-Smith, R., Diseases observed on bramble fruits in the Pacific North West, *Plant Dis. Rep.*, 35, 34, 1951.
869. Barbara, D. J., Ashby, S. C., and McNamara, D. G., Host range, purification and some properties of Rubus Chinese seedborne virus, *Ann. Appl. Biol.*, 107, 45, 1985.
870. Mishra, M. D., Raychaudhuri, S. P., and Chandra, K. J., Detecting virus infection in seeds through embryo-culture, *Proc. Int. Seed Test. Assoc.*, 32, 617, 1967.
871. Anonymous, Rothamsted Exp. Stn. Rep., 1980, Part I, Harpenden, Herts, U.K., 1981, 314.
872. Chauhan, L. S. and Singh, D. R., Occurrence of mosaic on safflower in northern India, *Indian Phytopathol.*, 32, 301, 1980.
873. Fajardo, T. G. and Maranon, J., The mosaic disease of sincamas, *Pachyrhizus erosus* (L.) Urban, *Philipp. J. Sci.*, 68, 129, 1932.
874. Hooker, W. J. and Salazar, L. F., A new plant virus from the high jungle of the Eastern Andes: Solanum apical leaf curling virus (SALCV), *Ann. Appl. Biol.*, 103, 449, 1983.
875. Bennett, C. W. and Costa, A. S., Sowbane mosaic caused by a seed-transmitted virus, *Phytopathology*, 51, 546, 1961.
876. Kado C. I., Biological and biochemical characterization of sowbane mosaic virus, *Virology*, 31, 217, 1967.
877. Dias, H. F. and Waterworth, H. E., The identity of a seed-borne mosaic virus of *Chenopodium amaranticolor* and *C. quinoa*, *Can. J. Bot.*, 45, 1285, 1967.
878. Takahashi, K., Tanaka, T., and Tsuda, Y., Soybean mild mosaic virus, *Ann. Phytopathol. Soc. Jpn.*, 40, 103, 1974.
879. Morales, F. J., Niessen, A. I., Castano, M., and Calvert, L., Detection of a strain of soybean mosaic virus affecting tropical forage species of *Centrosema*, *Plant Dis.*, 74, 648, 1990.
880. Iizuka, N., Seed transmission of viruses in soybean, *Tohoku Natl. Agric. Exp. Stn. Bull.*, 46, 131, 1973.
881. Vroon, C. W., Pietersen, G., and Van Tonder, H. J., Seed transmission of soybean mosaic virus in *Lupinus albus* L., *Phytophylactica*, 20, 269, 1988.
882. Castano, M. and Morales, F. J., Seed transmission of soybean mosaic virus in *Phaseolus vulgaris* L., *Fitopathol. Brasileira*, 8, 103, 1983.
883. Bos, L., Huttinga, H., and Maat, D. Z., Spinach latent virus, a new ilarvirus seed-borne in *Spinacia oleracea*, *Neth. J. Plant Pathol.*, 86, 79, 1980.
884. Stefanac, Z. and Wrischer, M., Spinach latent virus: some properties and comparison of two isolates, *Acta Bot. Croat.*, 42, 1, 1983.

885. Walkey, D. G. A., Brocklehurst, P. A., and Parker, J. E., Seed transmission of viruses, Thirty-third Annual Report of the National Vegetable Research Station, Wellesbourne, Warwick, U.K., 1983.
886. Natsuaki, T., Yamashita, S., Doi, Y., Okuda, S., and Teranaka, M., Two seedborne double-stranded RNA viruses, beet temperate virus and spinach temperate virus, *Ann. Phytopathol. Soc. Jpn.*, 49, 709, 1983.
887. Lockhart, B. E. L., Seed transmission of squash mosaic virus in *Chenopodium* spp., *Plant Dis.*, 69, 946, 1985.
888. Nelson, M. R. and Knuhtsen, H. K., Squash mosaic virus variability: epidemiological consequences of differences in seed transmission frequency between strains, *Phytopathology*, 63, 918, 1973.
889. Avgelis, A. D. and Katis, N., Occurrence of squash mosaic virus in melons in Greece, *Plant Pathol.* 38, 111, 1989.
890. Tolin, S. A., Lambe, R. C., and Chevone, B. I., Identification of seedborne squash mosaic virus in muskmelon, *Phytopathology*, 75, 629, 1985.
891. Leppik, E. E., Some epiphytotic aspects of squash mosaic, *Plant Dis. Rep.*, 48, 41, 1964.
892. Grogan, R. G., Hall, D. H., and Kimble, K. A., Cucurbit mosaic virus in California, *Phytopathology*, 49, 366, 1959.
893. Walkey, D. G. A. and Whittingham-Jones, S. G., Seed transmission of strawberry latent ringspot virus in celery (*Apium graveolens* var. dulce), *Plant Dis. Rep.*, 54, 802, 1970.
894. Schmelzer, K., Das Ulmenscheckungs virus, *Phytopathol. Z.*, 64, 39, 1969.
895. Murant, A. F. and Goold, R. A., Strawberry latent ringspot virus, *Annu. Rep. Scott. Hortic. Res. Int.*, 48, 1969.
896. Hicks, R. G. T., Smith, T. J., and Edwards, R. P., Effects of strawberry latent ringspot virus on the development of seeds and seedlings of *Chenopodium quinoa* and *Pastinaca sativa* (parsnip), *Seed Sci. Technol.*, 14, 409, 1986.
897. Bellardi, M. G. and Bertaccini, A., Parsley seeds infected by strawberry latent ringspot virus (SLRV), *Phytopathol. Mediterr.*, 30, 198, 1991.
898. Hanson, C. M. and Campbell, R. N., Strawberry latent ringspot virus from "plain" parsley in California, *Plant Dis. Rep.*, 63, 142, 1979.
899. Wroth, J. M. and Jones, R. A. C., Subterranean clover mottle sobemovirus: its host range, resistance in subterranean clover and transmission through seed and by grazing animals, *Ann. Appl. Biol.*, 121, 329, 1992.
900. Gupta, K. C., Studies on the identity of sunflower mosaic virus, in 3rd Int. Symp. Plant Pathol., New Delhi, 1981, 117.
901. Singh, J. P., Mechanical and seed transmission of causal agent of sunflower rugose mosaic in Kenya, *Sunflower Newsl.*, 3, 13, 1979.
902. Capoor, S. P., Southern sunnhemp mosaic virus: a strain of tobacco mosaic virus, *Phytopathology*, 52, 393, 1962.
903. Anno-Nyako, F. O., Seed transmission of Telfairia mosaic virus in fluted pumpkin (*Telfairia occidentalis* Hook) in Nigeria, *J. Phytopathol.* 121, 85, 1988.
904. Fernandez, S. Q., A new virus recorded on tobacco, *Rev. Agric. Cuba*, 4, 81, 1971.
905. Khudyna, I. P., Plant virus diseases and their control, in Trans. Conf. Plant Virus Diseases, Acad. Sci., U.S.S.R., Moscow, 1941, 340.
906. Tosic, M., Sutic, D., and Pesic, Z., Transmission of tobacco mosaic virus through pepper (*Capsicum annuum* L.) seed, *Phytopathol. Z.*, 97, 10, 1980.
907. McKinney, H. H., Two strains of tobacco mosaic virus, one of which is seed-borne in an etch-immune pungent pepper, *Plant Dis. Rep.*, 36, 184, 1952.

908. Niazi, F. R., Raychaudhari, S. P., and Mahmood, K., Effect of seed dressing sunnhemp for the prevention of tobacco mosaic virus infection, *Trop. Pest Manage.*, 28, 238, 1982.
909. Taylor, R. H., Grogan, R. G., and Kimble, K. A., Transmission of tomato mosaic virus in tomato seed, *Phytopathology*, 51, 837, 1961.
910. Cicek, Y. and Yorganci, U., Studies on the incidence of tobacco mosaic virus on certified seed of tomato, pepper and eggplant in Agean region, *J. Turkish Phytopathol.*, 20, 57, 1991.
911. Shmyglya, V. A., Makaev, S. Sh., and Aktaa, S., Characteristics of the diagnosis of tobacco mosaic virus transmitted by tomato and tobacco seeds, *Sel'-skokhozyaistvennaya Biol.*, 11, 60, 1984.
912. Gilmer, R. M. and Wilks, J. M., Seed transmission of tobacco mosaic virus in apple and pear, *Phytopathology*, 57, 214, 1967.
913. Prochazkova, Z., Presumed role of mucilage of plantain seeds in spread of tobacco mosaic virus, *Biol. Plant.*, 19, 259, 1977.
914. Gilmer, R. M. and Kelts, L. J., Transmission of tobacco mosaic virus in grape seed, *Phytopathology*, 58, 277, 1968.
915. Bojnansky, V. and Koslijarova, V., *Euonymus* mosaic, *Biol. Plant.*, 10, 322, 1968.
916. Von Wechmar, M. B., Lindsey, S., and Buckton, P., Discovery of a new and potentially serious disease of maize, Tech. Comm. 232, Dept. Agric., South Africa, 1992, 139.
917. Cooper, J. I. and Harrison, B. D., The role of seed hosts and the distribution and activity of vector nematodes in the ecology of tobacco rattle virus, *Ann. Appl. Biol.*, 73, 53, 1973.
918. Sammons, B. and Barnett, O. W., Tobacco ringspot virus from squash grown in South Carolina and transmission of the virus through seed of smooth pigweed, *Plant Dis.*, 71, 530, 1987.
919. Dhingra, K. L., A red leaf disease of chickpea in India caused by a strain of tobacco ringspot virus, in 3rd Int. Symp. Plant Pathol., New Delhi, 1981, 91.
920. McLean, D. M., Seed transmission of tobacco ringspot virus in cantaloupe, *Phytopathology*, 52, 21, 1962.
921. Desjardins, P. R., Latterell, R. L., and Mitchell, J. E., Seed transmission of tobacco ring-spot virus in Lincoln variety of soybean, *Phytopathology*, 44, 86, 1954.
922. Kahn, R. P., Scott, H. A., and Monroe, R. L., Eucharis mottle strain of tobacco ringspot virus, *Phytopathology*, 52, 1211, 1962.
923. Grogan, R. G. and Schnathorst, W. C., Tobacco ringspot virus — the cause of lettuce calico, *Plant Dis. Rep.*, 39, 803, 1955.
924. Marcelli, E., Observations on a new virus disease of tobacco transmissible through seed, *Tobacco (Rome)*, 59, 404, 1955.
925. Scarborough, B. A. and Smith, S. H., Seed transmission of tobacco and tomato ringspot viruses in geraniums, *Phytopathology*, 65, 835, 1975.
926. Henderson, R. G., Transmission of tobacco ringspot by seed of petunia, *Phytopathology*, 21, 225, 1931.
927. Keyworth, W. G., Plant pathology, *Rep. Natl. Veg. Res. Stn.*, 21, 106, 1971.
928. Tuite, J., The natural occurrence of tobacco ringspot virus, *Phytopathology*, 50, 296, 1960.
929. Shivanathan, P., A seed-borne virus of *Phaseolus aureus* (Roxb.), symposia on virus diseases of tropical crops, *Trop. Agric. Res. Ser.*, 10, 143, 1977.
930. Phatak, H. C., Seedborne plant viruses — identification and diagnosis in seed health testing, *Seed Sci. Technol.*, 2, 3, 1974.

931. Shukla, D. D. and Gough, K. H., Tobacco streak, broadbean wilt, cucumber mosaic and alfalfa mosaic viruses associated with ringspot of *Ajuga reptans* in Australia, *Plant Dis.*, 67, 221, 1983.
932. Brunt, A. A., Research summary: departmental and sectional reports on crop protection, mycology and bacteriology, virology, *Rep. Glasshouse Crops Res. Inst.*, 75, 1969.
933. Kaiser, W. J., Wyatt, S. D., and Klein, R. E., Epidemiology and seed transmission of two tobacco streak virus pathotypes associated with seed increases of legume germ plasm in Eastern Washington, *Plant Dis.*, 75, 258, 1991.
934. Johnson, H. A., Converse, R. H., Amorao, A., Espeio, J. I., and Frazier, N. W., Seed transmission of tobacco streak virus in strawberry, *Plant Dis.*, 68, 390, 1984.
935. Ghanekar, A. M. and Schwenk, F. W., Seed transmission and distribution of tobacco streak virus in six cultivars of soybeans, *Phytopathology*, 64, 112, 1974.
936. Salazar, L. F., Abad, J. A., and Hooker, W. J., Host range and properties of a strain of tobacco streak virus from potatoes, *Phytopathology*, 72, 1550, 1982.
937. Thomas, W. D., Jr. and Graham, R. W., Seed transmission of red node virus in pintobean, *Phytopathology*, 41, 959, 1951.
938. Cupertino, F. P., Grogan, R. G., Petersen, L. J., and Kimble, K. A., Tobacco streak virus infection in tomato and some natural weed hosts in California, *Plant Dis.*, 68, 331, 1984.
939. Converse, R. H. and Lister, R. M., The occurrence and some properties of black raspberry latent virus, *Phytopathology*, 59, 325, 1969.
940. Noordam, D., Bijl, M., Overbeek, S. C., and Quiniones, S. S., Virussen uit *Campanula rapunculoides en Stellaria media* en hun relatie tot komkommermozaiek virus en tomaat-aspermy virus, *Neth. J. Plant Pathol.*, 71, 61, 1965.
941. Kaiser, W. J., Bock, K. R., Guthrie, E. J., and Meredith, G., Occurrence of tomato black ring virus in potato cultivar Anett in Kenya, *Plant Dis. Rep.*, 62, 1088, 1978.
942. Gibbs, A. J. and Harrison, B. D., Nematode transmitted viruses in sugarbeet in East Anglia, *Plant Pathol.*, 13, 144, 1964.
943. Murant, A. F. and Lister, R. M., Seed transmission in the ecology of nematode-borne viruses, *Ann. Appl. Biol.*, 59, 63, 1967.
944. Morand, J. C. and Poutier, P. C., Ringspot of lettuce, a strain of tomato black ring virus, *Ann. Phytopathol.*, 10, 101, 1978.
945. Tomlinson, J. A. and Faithfull, E. M., Studies on the occurrence of tomato stunt bushy virus in English rivers, *Ann. Appl. Biol.*, 104, 485, 1984.
946. Kegler, G. and Schimanski, H. H., Investigations on the spread and seed transmissibility of tomato bushy stunt virus in pome and stone fruit in GDR, *Arch. Phytopathol. Pflanzensch.*, 18, 105, 1982.
947. Xuan, T. H., Calilung, V. J., and Benigno, D. A., Epidemiology of tomato mosaic virus and cucumber mosaic virus in the Philippines, *Philipp. Agric.*, 71, 421, 1988.
948. Cory, L. and Hewitt, W. B., Some grape vine viruses in pollen and seeds, *Phytopathology*, 58, 1316, 1968.
949. Kahn, R. P., Seed transmission of tomato ringspot virus in the Lincoln variety of soybeans, *Phytopathology*, 46, 295, 1956.
950. Keplinger, J. A. and Braun, A. J., Seed transmission of tomato ringspot virus in *Gomphrena globosa, Plant Dis. Rep.*, 57, 433, 1973.
951. Hollings, M., Stone, O. M., and Dale, W. T., Tomato ringspot virus in *Pelargonium* in England, *Plant Pathol.*, 21, 46, 1972.
952. Braun, A. J. and Keplinger, J. A., Seed transmission of tomato ringspot virus in raspberry, *Plant Dis. Rep.*, 57, 431, 1973.

953. Uyemoto, J. K., Gilmer, R. M., and Williams, E., Sap transmissible viruses of elderberries in New York, *Plant Dis. Rep.*, 55, 913, 1971.
954. Mountain, W. L., Powell, C. A., Forer, L. B., and Stouffer, R. F., Transmission of tomato ringspot virus from dandelion via seed and dagger nematodes, *Plant Dis.*, 67, 867, 1983.
955. Pelikanova, J., Hedge garlic — a spontaneous host of turnip yellow mosaic virus, *Sbornik UVTIZ Ochrana Rostlin,* 26, 17, 1990.
956. Spak, J., Kubelkova, D., and Hnilicka, E., Seed transmission of turnip yellow mosaic virus in winter turnip and winter oilseed rapes, *Ann. Appl. Biol.*, 123, 33, 1993.
957. Hein, A., Transmission of turnip yellow mosaic virus by seeds of *Camelina sativa* (gold of pleasure), *Z. Pflanzenkr. Pflanzensch.*, 91, 549, 1984.
958. Beniwal, S. P. S., Chaubey, S. N., and Bharthan, N., Presence of urdbean leaf crinkle virus in seeds of mungbean germplasm, *Indian Phytopathol.*, 33, 360, 1980.
959. Kolte, S. J. and Nene, Y. L., Studies on symptoms and mode of transmission of the leaf crinkle virus of urdbean (*Phaseolus mungo* L.), *Indian Phytopathol.*, 25, 401, 1972.
960. Diseases of grain legumes, in Rothamsted Exp. Stn. Rep. 1978, Rothamsted, U.K., 1979.
961. Bhargava, K. S. and Joshi, D. S., Detection of watermelon mosaic virus in Uttar Pradesh, *Curr. Sci.*, 29, 443, 1960.
962. Lindberg, G. D., Hall, D. H., and Walker, J. C., A study of melon and squash mosaic viruses, *Phytopathology*, 46, 489, 1956.
963. Singh, S. J., Occurrence and epidemic of a mosaic disease of chow-chow in India, in 3rd Int. Symp. Plant Pathol., New Delhi, 1981, 60.
964. Abdel-Hak, T., Ghobrial, E., and Kamel, A. H., Studies on striate mosaic of wheat in Egypt, *Agric. Res. Rev.*, 55, 1, 1977.
965. Greber, R. S., Persley, D. M., and Herrington, M. E., Some characteristics of Australian isolates of zucchini yellow mosaic virus, *Aust. J. Agric. Res.*, 39, 1085, 1988.
966. Schrijnwerkers, C. C. F. M., Huijerts, N., and Bos, L., Zucchini yellow mosaic virus: two outbreaks in the Netherlands and seed transmissibility, *Neth. J. Plant Pathol.*, 97, 187, 1991.
967. Al-Musa, A. W., Severe mosaic caused by zucchini yellow mosaic virus in cucurbits from Jordan, *Plant Pathol.*, 38, 541, 1989.
968. Phatak, H. C., Seed-borne plant viruses — identification and diagnosis in seed health testing, *Seed Sci. Technol.*, 2, 3, 1974.
969. Mandahar, C. L., Virus transmission through seed and pollen, in *Plant Diseases and Vectors*, Maramorosch K. and Harris, K. F., Eds., Academic Press, New York, 1981, 241.
970. Stace-Smith R. and Hamilton, R. I., Inoculum thresholds of seedborne pathogens, viruses, *Phytopathology*, 78, 875, 1988.
971. Jensen, S. G., Wysong, D. S., Ball, E. M., and Higley, P. M., Seed transmission of maize chlorotic mottle virus, *Plant Dis.*, 75, 497, 1991.
972. Mikel, M. A., D'Arcy, C. J., and Ford, R. E., Seed transmission of maize dwarf mosaic virus in sweet corn, *Phytopathol. Z.*, 110, 185, 1984.
973. Hill, J. H., Martinson, C. A., and Russell, W. A., Seed transmission of maize dwarf mosaic and wheat streak mosaic viruses in maize and response of inbred lines, *Crop Sci.*, 14, 232, 1974.

974. Williams, L. E., Findley, W. R., Dollinger, E. J., and Ritter, R. M., Seed transmission of maize dwarf mosaic virus in corn, *Plant Dis. Rep.*, 52, 863, 1968.
975. Mathur, S. B., Haware, M. P., and Hampton, R. O., Identification, significance and transmission of seedborne pathogens, in *World Crops: Cool Season Food Legumes*, Summerfield, R. J., Ed., 1988, 351.
976. Keese, P. and Symons, R. H., The structure of viroids and virusoids, in *Viroids and Viroid-like Pathogens*, Semancik, J. S., Ed., CRC Press, Inc., Boca Raton, 1987.
977. Diener, T. O., Potato spindle tuber "virus". IV. A replicating, low molecular weight RNA, *Virology*, 45, 411, 1971.
978. Diener, T. O. and Raymer, W.B., Potato spindle tuber virus: a plant virus with properties of a free nucleic acid, *Science*, 158, 378, 1967.
979. Diener, T. O., *Viroids and Viroid Diseases*, John Wiley & Sons, New York, 1979.
980. Diener, T.O., Potato spindle tuber viroid. VIII. Correlation of the infectivity with a UV-absorbing component and thermal denaturation properties of RNA, *Virology*, 50, 606, 1972.
981. Semancik, J. S. and Weathers, L.G., Exocortis virus: an infectious free-nucleic acid plant virus with unusual properties, *Virology*, 47, 456, 1972.
982. Diener, T. O. and Lawson, R. H., Chrysanthemum stunt, a viroid disease, *Virology*, 51, 94, 1973.
983. Sogo, J. M., Koller, T., and Diener, T. O., Potato spindle tuber viroid. X. Visualization and size determination by electron microscopy, *Virology*, 55, 70, 1973.
984. Randles, J. W., Association of two ribonucleic acid species with cadang-cadang disease of coconut palm, *Phytopathology*, 65, 163, 1975.
985. Sasaki, M. and Shikata, E., Studies on the host range of hop stunt disease in Japan, *Proc. Jpn. Acad. Ser. B.*, 53, 103, 1977.
986. Howell, S. H., The molecular biology of plant DNA viruses, *CRC Crit. Rev. Plant Sci.*, 2, 287, 1984.
987. Kryczynski, S., Transmission of viroids and viruses by tissue implantation and transport across the callus barrier, *Phytopathol. Z.*, 106, 63, 1983.
988. Kryczynski, S., Paduch-Cichal, E., and Skerezeczkowski, L. J., Transmission of three viroids through seeds and pollen of tomato plants, *J. Phytopathol.*, 121, 51, 1988.
989. Wallace, J. M. and Drake, R. J., A high rate of seed transmission of avocado sunblotch virus from symptomless trees and the origin of such trees, *Phytopathology*, 52, 237, 1962.
990. Hadidi, A., Hansen, A. J., Parish, C. L., and Yang, X., Scar skin and dapple apple viroids are seedborne and persistent in infected apple trees, *Res. Virol.*, 142, 289, 1991.
991. Monsion, M., Bachelier, J. C., and Dunez, J., Some properties of a viroid, chrysanthemum stunt, *Ann. Phytopathol.*, 5, 467, 1973.
992. Kryczynski, S., Paduch-Cichal, E., and Skrzeczkowski, L. J., Transmission of three viroids through seed and pollen of tomato plants, *J. Phytopathol.*, 121, 51, 1988.
993. Semancik, J. S., Citrus exocortis viroid, CMI/AAB Descriptions of Plant Viruses, No. 226, 1980, 4.
994. Randles, J. W. and Imperial, J. S., Coconut cadang-cadang viroid, CMI/AAB Descriptions of Plant Viruses, No. 287, 1984.
995. Singh, R. P. and Boucher, A., High incidence of transmission and occurrence of a viroid in commercial seeds of *Coleus* in Canada, *Plant Dis.*, 75, 184, 1991.

996. Singh, R. P. and Finnie, R. E., Seed transmission of potato spindle tuber metavirus through the ovule of *Scopolia sinensis*, *Can. Plant Dis. Surv.*, 53, 153, 1973.
997. Puchta, H., Herold, T., Verhoeven, K., Roenhorst, A., Ramm, K., Schmidt-Puchta, W., and Sanger, H. L., A new strain of potato spindle tuber viriod (PSTVd-N) exhibits major sequence differences as compared to all other PSTVd strains sequenced so far, *Plant Mol. Biol.*, 15, 509, 1990.
998. Norton, D. C. and Sass, J. E., Pathological changes in *Agropyron smithii* induced by *Anguina agropyronifloris*, *Phytopathology*, 56, 769, 1966.
999. Courtney, W. D. and Howell, H. B., Investigations on the bent grass nematode, *Anguina agrostis* (Steinbuch, 1799), Filipjev 1936, *Plant Dis. Rep.*, 36, 75, 1952.
1000. Goodey, T., Eelworm galls mistaken for ergot in flowers of Canadian grasses, *Nature (London)*, 169, 456, 1952.
1001. Hardison, J. R. and Jensen, H. J., A seed nematode observed in orchard grass in Oregon, *Plant Dis. Rep.*, 37, 388, 1953.
1002. Hardison, J. R., Seed disorders of forage plants, *Plant Diseases,* in Yearbook of Agriculture, U.S. Department of Agriculture, Washington, D.C., 1953, 272.
1003. Bird, A. F., Stynes, B. A., and Thomson, W. W., A comparison of nematode and bacteria-colonized galls induced by *Anguina agrostis* in *Lolium rigidum, Phytopathology*, 70, 1104, 1980.
1004. Price, P. C., Fisher, J. M., and Kerr, A., On *Anguina funesta* n. sp. and its association with *Corynebacterium* sp. in infecting *Lolium rigidum, Nematologica*, 25, 76, 1979.
1005. Thorne, G., *Principles of Nematology*, McGraw-Hill, New York, 1961, 553.
1006. Leukel, R. W., The nematode disease of wheat and rye, *U.S. Dep. Agric. Farmers' Bull.*, 1607, 11, 1929.
1007. Raeder, J. M., A note on the longevity of the wheat nematode, *Anguina tritici* Steinbuch, *Plant Dis. Rep.*, 38, 268, 1954.
1008. Caubel, G., Nematodes as pests of quarantine significance to seed, Workshop on Quarantine for Seed in the Near East, Aleppo, 1991.
1009. Bridge, J., Box, W. S., Page, L. T., and McDonald, D., The biology and possible importance of *Aphelenchoides arachidis*, a seed-borne endo-parasitic nematode of groundnuts from northern Nigeria, *Nematologica*, 23, 253, 1977.
1010. Gokte, N., Mathur, V. K., Lal, A., and Rajan, *Fraxinus americana* — a new host record for *Aphelenchoides besseyi*, *Indian J. Nematol.*, 19, 80, 1989.
1011. Crally, E. M., White tip of rice, *Phytopathology*, 39, 5, 1949.
1012. Gokte, N., Mathur, V. K., Lal, A., and Rajan, On the occurrence of *Aphelenchoides besseyi* and some free living nematode species in *Stylosanthes hamata* seeds, *Nematol. Mediterr.*, 20, 63, 1992.
1013. Burckhardt, F., New observations on the presence of *Aphelenchoides* spp. in seeds of *Callistephus chinensis, Nachrichtenbl. Dtsch. Pflanzenschutzdienstes* (Braunschweig), 24, 132, 1972.
1014. Brown, E. B., A seed-borne attack of chrysanthemum eelworm *(Aphelenchoides ritzema-bosi)* on the annual aster *(Callistephus chinensis), J. Helminthol.*, 30, 145, 1956.
1015. Seth, L. N., Annual Report of the Mycologist for the Year Ended 31st March, 1939, Mandalay, Burma, 1939, 6.
1016. Goodey, T., *Anguillulina dipsaci* on onion seed and its control by fumigation with methyl bromide, *J. Helminthol.*, 21, 45, 1945.
1017. Robertson, D., Stem eelworm of oats, *Scottish Agric.*, 34, 209, 1955.
1018. Dunning, R. A., Beet stem eelworm, *Nematologica*, 1, 189, 1956.

1019. Goodey, J. B., The control of *Anguillulina dipsaci* on the seed of teazel and red clover by fumigation with methyl bromide, *J. Helminthol.*, 23, 171,1949.

1020. Godfrey, G. H., *Tylenchus dipsaci* on *Hypochaeris radiata* in Hawaii, *Phytopathology*, 21, 759, 1931.

1021. Brown, E. B., Lucerne stem eelworm in Great Britain, *Nematologica*, Suppl. 2, 369, 1957.

1022. Hodson, W. E. H., The occurrence of *Tylenchus dipsaci* Kuhn in wild host plants in southwest England, *J. Helminthol.*, 7, 143, 1929.

1023. Godfrey, G. H., Dissemination of the stem and bulb infesting nematode, *Tylenchus dipsaci*, in the seeds of certain composites, *J. Agric. Res.*, 28, 473, 1924.

1024. Tinnila, A., Apila-ankeroisen siemenlevinta ja kotimaisten apilakanotojen tuhonalttius, *Maatalous Koetoiminta*, 14, 228, 1960.

1025. Diercks, R. and Klewitz, R., Zur Samenubertragbarkeit einer am Ackerbohnen vorkommeden Herkunft des Stengelalchens, *Ditylenchus dipsaci* (Kuhn), *Nematologica*, 7, 155, 1962.

1026 Epps, J. M., Survival of the soybean cyst nematode in seed stocks, *Plant Dis. Rep.*, 53, 403, 1969.

1027. Sethi, C. L., Nath, R. P., Mathur, V. K., and Ahuja, S., Interceptions of plant parasitic nematodes from imported seed/plant material, *Indian J.Nematol.*, 2, 89 1972.

1028. Fenwick, D. W., Red ring disease of coconuts in Trinidad and Tobago, *Colonial Office Report*, Her Majesty's Stationery Office, London, 1957, 55.

CHAPTER 3

Location of Seedborne Inoculum

Pathogens associated with seeds are transmitted as an infection or infestation (contamination). Infection implies that the pathogen is carried internally, embedded in seed tissues (Figure 3-1). When a pathogen is carried passively, it is a contaminant or infestant. In this latter case, the pathogen is carried either on the seed surface or mixed with the seed in the form of infected plant parts, fruiting structures, or infested soil particles (concomitant contamination) (Figure 3-2).

A true seed consists of a fertilized mature ovule, an endosperm of stored nutrient (rarely missing), and a protective coat or coats.[1] Any or all seed parts can become infected; however, a pathogen may be restricted to a specific area of a seed or present in more than one seed part. For predictable seed transmission to occur, infection of a particular seed part is essential. For example, hyphae of *Sclerospora graminicola* invade all parts of seed tissues of pearl millet, but hyphae in the scutellum alone cause systemic infection in seedlings.[2] For transmission of a majority of the seedborne viruses, embryos must be infected. If a virus fails to invade the embryo, transmission rarely occurs.[3]

The location of inoculum within a seed depends on the type of pathogen, the mode and time of infection, the environmental conditions at the time of infection, the cultivar, and other factors.

A number of techniques have been used to detect internally seedborne inoculum, including (1) whole mount preparation, (2) seed component plating, (3) microtome sectioning,[4,5] and (4) embryo counting. Whole mount preparations of softened, cleared, and stained seed components often have been used to demonstrate the presence of dormant mycelium. Hand cut or microtomed sections using paraffin-embedding techniques have been used to study the location of mycelium in seeds. Polarized, fluorescence, transmission, and electron and scanning electron microscopy also have been used for detailed histopathological studies.[4] Embryo counts have been used for detection of loose smut and downy mildew fungi in cereals and millets. For histopathological studies, seed sample(s) must be carefully selected. Variability in infection should be anticipated and infected seeds separated based on an estimation of the quantity of internal inoculum. Naturally

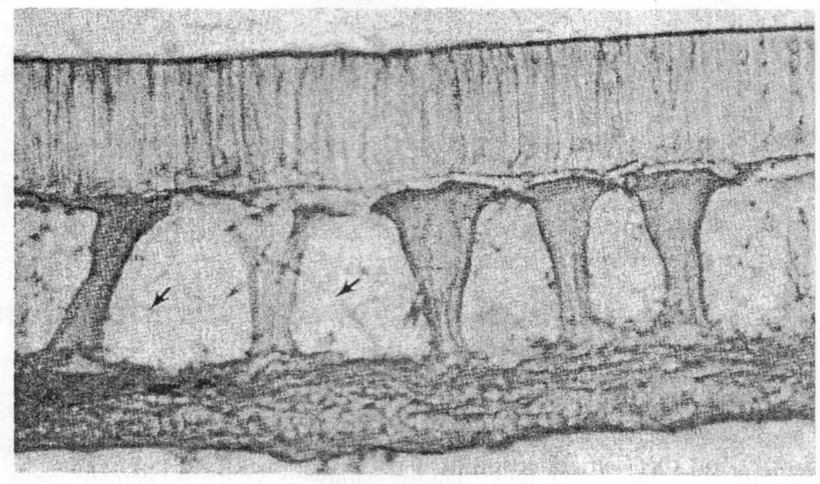

A

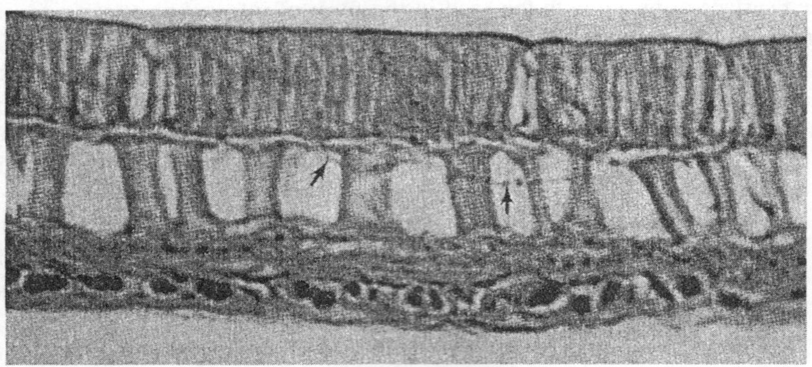

B

Figure 3-1 Location of various fungi in crop seeds. (**A**) Hyphae (*arrows*) of *Diaporthe phaseolorum* var. *sojae* (*Phomopsis*) (seed decay) in the three cell layers of a soybean (*Glycine max*) seed coat. (From Ilyas, M. B., Dhingra, O. D., Ellis, M. A., and Sinclair, J. B., *Plant Dis. Rep.*, 59, 17, 1975. With permission.) (**B**) Hypha (*arrows*) of *Cercospora kikuchii* (purple seed stain) in the three cell layers of a soybean (*G. max*) seed coat. (From Ilyas, M. B., Dhingra, O. D., and Sinclair, J. B., *Plant Dis. Rep.*, 59, 17, 1975. With permission.) (**C**) Hyphae of *Rhizoctonia solani* in tissues of bell pepper (*Capsicum frutescens*) seeds: *1*, Hyphae on the exterior and in the inner layer of the seed coat; *2*, Hyphae in the hypocotyl tissues; *3*, Hyphae in the inner layer of the seed coat and in an attached remnant of the funiculus; *4*, Hyphae in the endosperm (c = cotylendon, e = endosperm, ec = endosperm cuticle, h = hypocotyl, it = inner layer of seed coat, and ot = outer layer of seed coat.) (Adapted from Baker, K. F., *Seed Biology*, Vol. 2, Kozlowski, T. T., Ed., Academic Press, New York, 1972, 353. With permission.) (**D**) Hyphae (*arrows*) and conidia of *Alternaria sesamicola* associated with tissues of sesame (*Sesamum indicum*) seeds. (From Singh, D., Mathur, S. B., and Neergaard, P., *Seed Sci. Technol.*, 8, 84, 1980. With permission.)

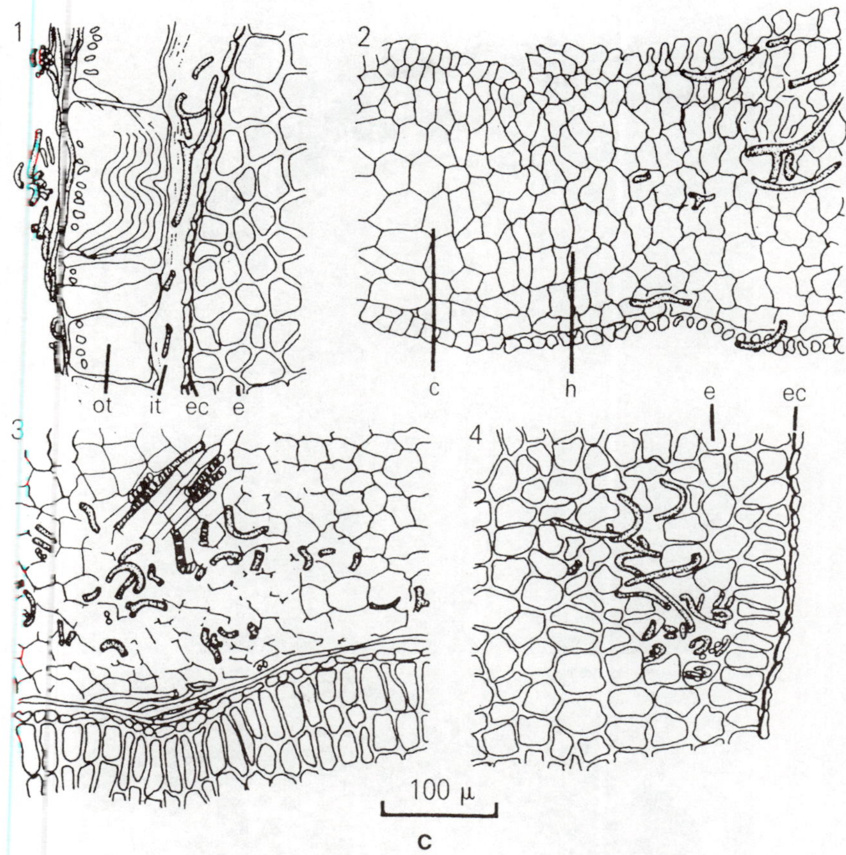

Figure 3-1 Continued.

infected seed samples should have priority over artificially inoculated ones. Seeds should not be processed so that the pathogen invades new tissues. For example, soaking seeds in water at room temperature or incubating seeds for a long period should be avoided.[4]

I. HISTOPATHOLOGY OF SOME SEEDBORNE PATHOGENS

Characteristics based on histopathological studies could serve as guidelines in distinguishing seven pathogenic fungi that colonize soybean seeds, including hyphal width, color of mature hyphae, presence or absence of hyphal aggregates, and presence of mycelial mats, sclerotia, or oil globules in the hyphae. The fungi distinguished from one another based on these characteristics were *Alternaria alternata*, *Cercospora kikuchii*, *C. sojina*, *Colletotrichum truncatum*, *Fusarium*, *Macrophomina phaseolina*, and *Phomopsis* (Table 3-1).[5,6,7] If the fungus hyphae

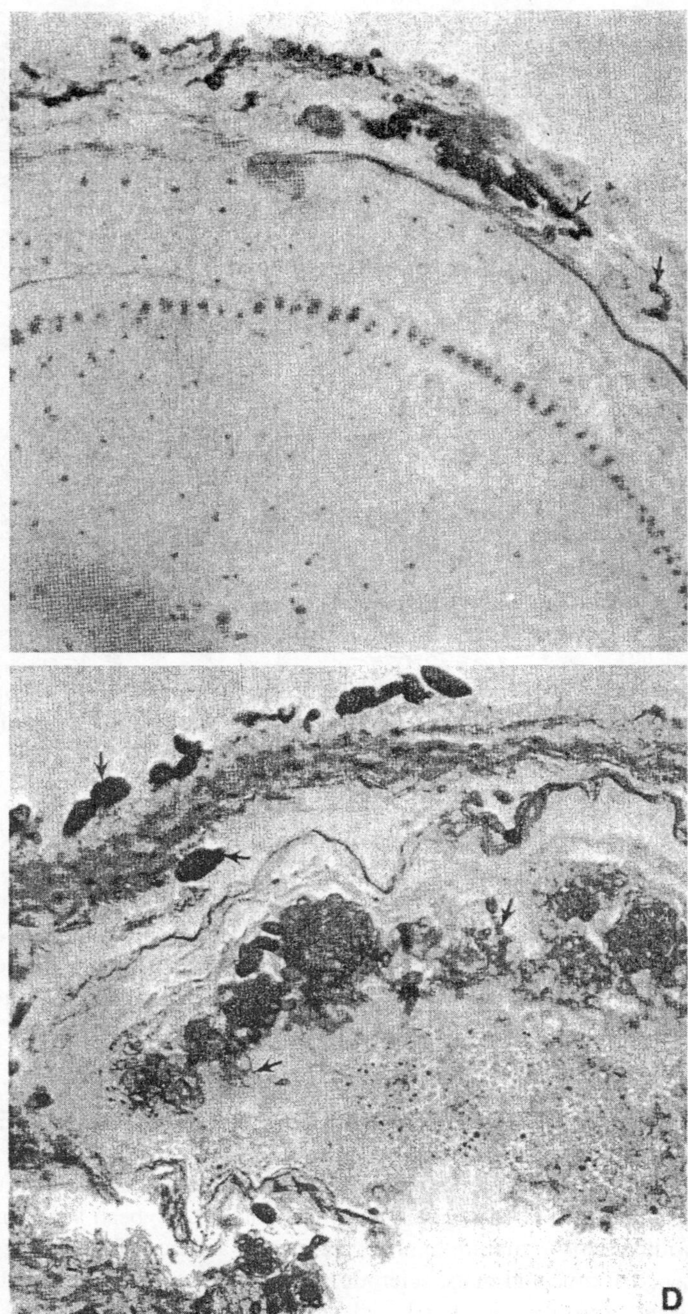

Figure 3-1 Continued.

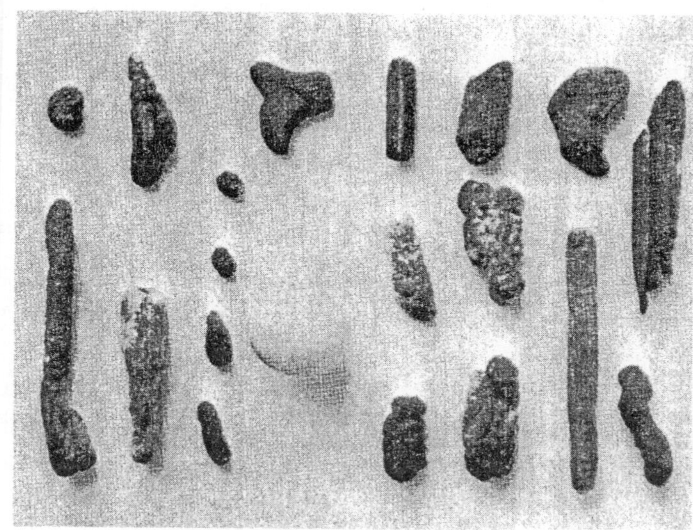

Figure 3-2 Various sizes of sclerotia of *Sclerotina sclerotiorum* (sclerotial rot) compared to a soybean (*Glycine max*) seed.

were hyaline, they stained red or green, with safranin and light green; if lightly pigmented, cell walls stained brown, and cytoplasm stained green or red. Hyaline hyphae stained blue with cotton blue; if lightly pigmented, cell walls stained brown and the cytoplasm blue. If hyphae were heavily pigmented, they would not take any stain.

A. alternata in soybean seeds was found concentrated near the hilum region on the seed coat surface and in all layers of the seed coat and endosperm (Figure 3-3).[8] Histological examination of symptomatic soybean seeds showed that hyphae and sclerotia of *M. phaseolina* were ecto- and endophytic; hyphae were inter- and intracellular in the tissues of the seed coat, embryos, and endosperm (Figure 3-4). Seed coats of heavily infected seeds were fragmented. The fungus produced sclerotia in 4% of asymptomatic seeds plated on acidified (pH 4.5) potato-dextrose agar at 25° ± 2°C. The sclerotia developed near or adjacent to the seed coat, endosperm, and hypodermis.[9] *Phomopsis* colonized seed coat and embryo tissues, while *C. sojina* and *C. truncatum* were confined to seed coat tissues. *Phomopsis* was restricted to seed coat tissues when *C. truncatum* was present (Figure 3-1A).[10] Mycelia of *C. truncatum* was present in the three layers of soybean seed coats, with acervuli formed only in the palisade layer (Figure 3-5).[11] *Fusarium oxysporum* hyphae were found over the seed coat surface, producing macro- and microconidia, and within the hilar region of soybean seeds. Hyphae of the fungus were observed in all layers of the seed coat, but not in the cotyledons or endosperm. Terminal and intercalary chlamydospores were found only in hyphae growing on the underside of seed coats (Figure 3-6).[6] Hildebrand and Koch[12] described fragments of mycelium of *Peronospora manshurica* within the palisade and hourglass cell layers of the outer seed coat and concluded that

Table 3-1 Comparison of the Appearance of Various Fungi in Soybean Seed Coat Tissues in Thin Section Prepared for Bright-Field Microscopy

Fungus	Hyphal width in μm	Unstained mature hypha	Hyphal aggregate	Mycelial mat	Sclerotia	Oil globule
Alternaria alternata	1.8–5.4	Brown	No	Yes	No	No
Cercospora kikuchii	1.3–3.0	Light brown	Yes	No	No	No
C. sojina	1.3–2.7	Brown	Yes	Yes	No	No
Colletotrichum truncatum	3–11	Brown	No	Yes	No	Yes[a]
Fusarium	1.4–3.6	Hyaline	No	No	No	No
Macrophomina phaseolina	1.8–4.5	Brown	No	No	Yes	No
Phomopsis	3.8–8.7	Hyaline	No	Yes	No	Yes

[a] Prominent oil globules.

Summarized from Kunwar, I. K., Singh, T., Machado, C. C., and Sinclair, J. B., *Phytopathology,* 76, 532, 1986; and from Singh, T. and Sinclair, J. B., *Phytopathology,* 75, 185, 1985.

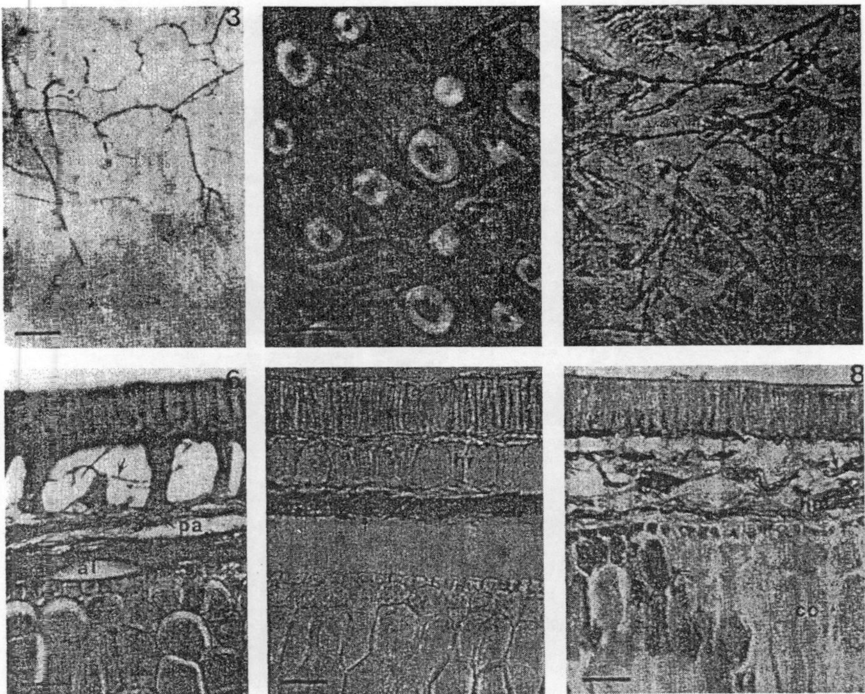

Figure 3-3 Photomicrographs of soybean seeds infected with *Alternaria alternata*. **3–5**, cleared whole mounts; **3**, hilum region showing hyphae; **4**, seed coat with hourglass layer on top in surface view showing hyphae (*arrows*) traversing among hourglass (hr) cells; and **5**, endosperm showing fanlike branching (*arrows*) of hyphae; **6–8**, traverse sections of seeds: **6**, stained section of lightly infected seed, showing hyphae (*arrows*) in hourglass layer of the seed coat (note that paenchymatous [pa] and aleurone layers [al] are distinct), **7**, unstained section of moderately infected seed showing brown discoloration and hyphal mat (*arrow*) below the hourglass layer (hr), **8**, unstained section of severely infected seed with brown discoloration (*arrows*) in seed coat, endosperm, and cotyledon (co). Note hyphal mat (hm) in indistinguishable endodermis and endosperm. Scale bar = 40 μm. (From Kunwar, I. K., Manandhar, J. B., and Sinclair, J. B., *Phytopathology*, 76, 453, 1986. With permission.)

the hyphae were senescent. Roongruangsree et al.[13] suggested that, in addition to oospores on encrustated seeds, *P. manshurica* could survive in soybean seeds as thin-walled mycelia between the hourglass cell layer and the parenchymatous tissue of the seed coat and on the seed surface as resistant, thick-walled, resting, or sclerotized hyphae.[13]

Infection in soybean seeds by *P. longicolla, D. phaseolorum* var. *sojae*, and *D. phaseolorum* var. *caulivora* was located mainly in the seed coat, with less in cotyledons.[14] Oospores of *Peronospora ducometi,* causal agent of buckwheat downy mildew, were found in calyx remnants attached to the seeds, on the outside of the seedcoat, and/or in the spermoderm between the seed coat and endosperm.[15] *Phoma rabiei* was located primarily on or in the seed coat of chickpea and

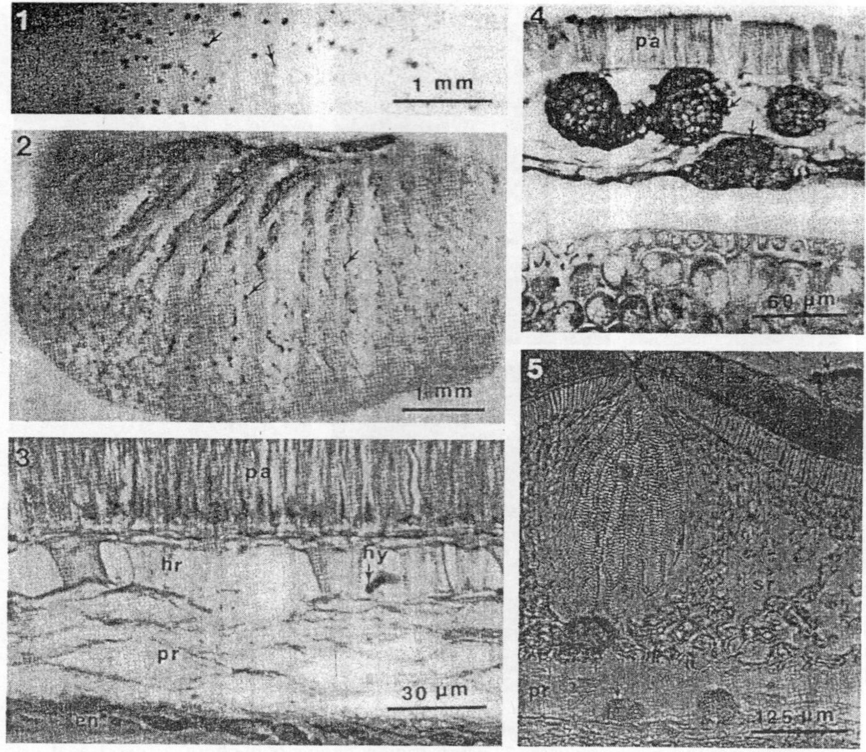

Figure 3-4 Soybean seeds naturally infected with *Macrophomina phaseolina*. (**1**) Asymptomatic seed after 2 d of incubation, showing subepidermal brown sclerotia (*arrows*). (**2**) Symptomatic seed with fragmented seed coat and black sclerotia (*arrows*) within the seed coat. (**3–5**) Photomicrographs of transverse sections of soybean seeds. (**3**) Asymptomatic seed with sparse hyphae (hy) in hourglass layer (hr). The palisade (pa), hourglass layer (hr), parenchyma (pr), and endosperm (en) of the seed coat are undamaged. (**4**) Asymptomatic seed after 2 d of incubation showing (*arrows*) in all the layers of the seed coat except the palisade (pa). (**5**) Sclerotia (*arrows*) in the hilum region above the palisade layer (pa), below the tracheid (tr), and in the sclerenchyma (sr) and parenchyma (pr) layers. (From Kunwar, I. K., Singh, T., Machado, C. C., and Sinclair, J. B., *Phytopathology*, 76, 532, 1986. With permission.)

occasionally the cotyledons, but rarely the embryo.[16] *Stagonospora nodorum* mycelium was confined mostly to pericarp and aleurone layer in shriveled seeds and to all the parts of discolored, heavily infected wheat seeds. The mycelium was thick, brown, and septate and occurred inter- and intracellularly.[17] A rice seed sample obtained from plants severely damaged by blast yielded about 65% *Pyricularia oryzae* on the hulls, including empty glumes, lemma, palea, and rachilla, whereas pericarps carried approximately 25% infection. The fungus was not detected in the embryo but in 4% of the endosperm. Sporulation was initiated at the hilum and extended to the adjacent dorsal portion when the second empty glume was close to the hilum.[18] *Lasiodiplodia theobromae*, cause of stalk and kernel rot of maize, was present in all seed parts including the embryo.[19] Using

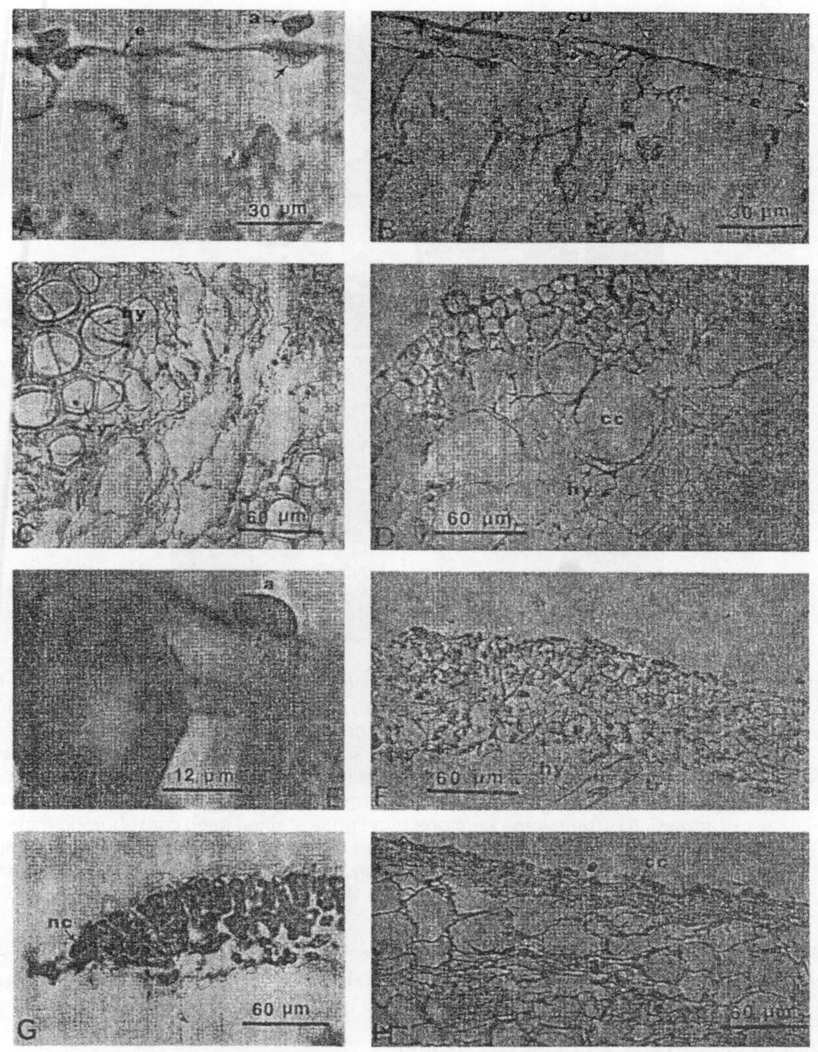

Figure 3-5 Photomicrographs of transverse sections of soybean leaf tissues stained with safranin and light green showing germination and penetration by *Colletotrichum truncatum* and *Glomerella glycines.* (**A**) At the point of contact between the appressorium (*a*) of *C. truncatum* and epidermis (*e*), upper tangential epidermal cell wall thickened, dark, and developing a lesion (*arrow*): (**B**) Hypha (hy) of *C. truncatum* between the cuticle (cu) and upper tangential and anticlinal (*arrow*) epidermal cell wall (*e*) in the cotyledon. (**C**) Hyphae (hy) of *C. truncatum* in the xylem vessels (xy) of the petiole. (**D**) Hypertrophied cortical cells (cc) of the petiole surrounded by hyphae (hy) of *C. truncatum.* (**E**) Appressorium (a) of *G. glycines* showing the thin wall (*arrows*) at point of contact with the epidermis. (**F**) Abundant hyphae (hy) of *G. glycines* in the lamina and trichome (tr). (**G**) Necrotic cells (nc) in the margin of lamina inoculated with *C. truncatum.* (**H**) Collapsed cortical cells (cc) in the petiole inoculated with *C. truncatum.* (From Manandhar, J. B., Kunwar, I. K., Singh, T., Hartman, G. L., and Sinclair, J. B., *Phytopathology,* 75, 704, 1985. With permission.)

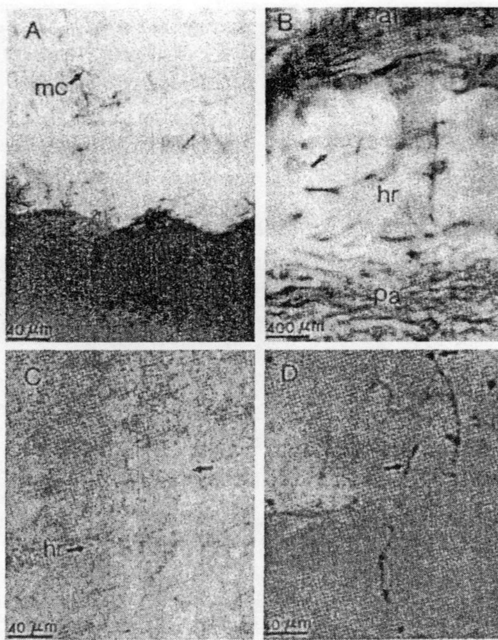

Figure 3-6 Photomicrographs of soybean seeds infected with *Fusarium oxysporum*. (**A, C, D**) Cleared whole mounts. (**A**) Hyphae, macro- and microconidia (mc) growing within the hourglass cell layer. (**C**) Transverse section of an infected seed coat with hourglass cell layer on top in surface view showing hyphae (*arrows*) transversing through the hourglass cells (hr). (**D**) The underside of an infected seed coat showing hyphae with terminal and intercalary chlamydospores (*arrows*). (**B**) Transverse sections of a lightly stained seed coat showing hyphae (*arrow*) in the hourglass cell (hr) layer of the seed coat with the parenchymatous (pa) layer below and the palisade (pal) layer above. (From Velicheti, R. K. and Sinclair, J. B., *Seed Sci. Technol.*, 19, 445, 1991. With permission.)

scanning electron microscopy, *F. moniliforme* was found in the pedicel or tip cap end of all asymptomatic maize kernels, as well as in the embryo and endosperm (Figure 3-7).[20] The mycelium and oospores of *Peronosclerospora sorghi* was found in the pericarp and embryo of maize seeds.[21,22] *Curvularia lunata* and *F. moniliforme* were found in all sorghum seed parts, including the style remnant, tip cap, pericarp, endosperm, and embryo.[23] The inoculum of *Phoma sorghina* was detected in the style remnant, tip cap, and pericarp of sorghum seeds.[23] *Colletotrichum graminicola* mycelium was found in the pericarp, endosperm, and embryo of sorghum. The endosperm and embryonic tissues were distorted. Acervuli with setae also were found in grain tissues.[24] *Gloeocercospora sorghi* was found as sclerotia in sorghum endosperm.[25]

Mycelium of *Sclerospora graminicola* has been observed in the scutellum and endosperm of pearl millet seeds,[26,27] although whether or not seed transmission occurred was uncertain.[28] The mycelium and oospores of *Sclerophthora macrospora* have been found both internally and externally in finger millet seeds.

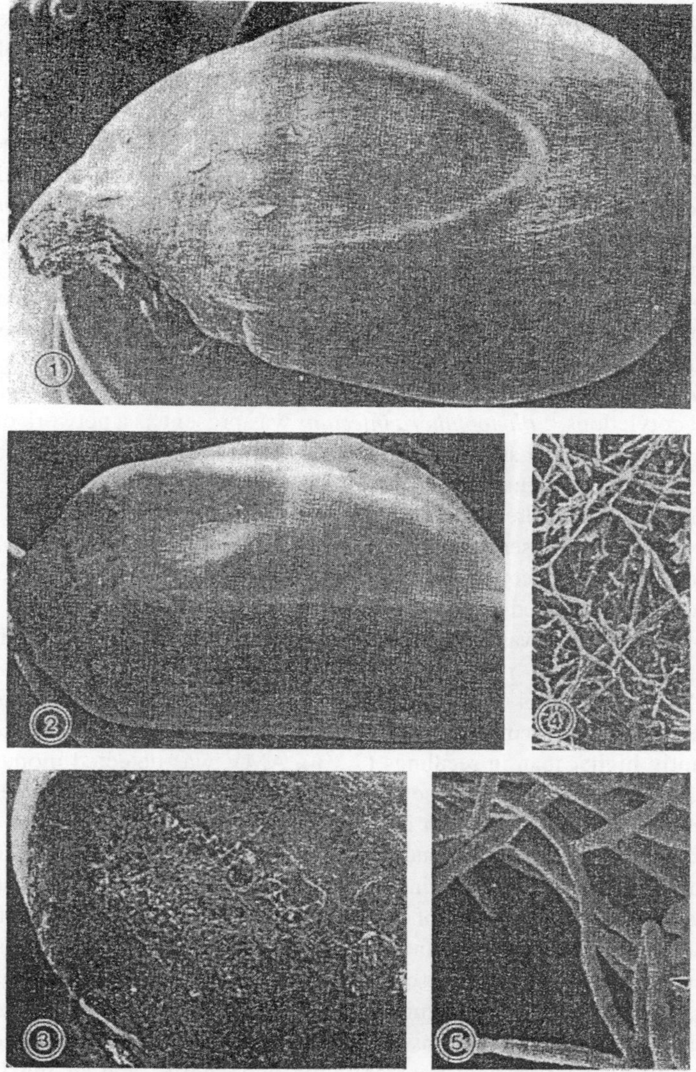

Figure 3-7 Scanning electronmicrographs of maize seeds showing external infection by *Fusarium moniliforme*. (**1**) Sound seed representing a typical nontoxic and noninfected seed (× 12). (**2**) Seed showing fungus over a damaged area just distal to the tip cap end (× 10). (**3**) The tip cap end in (**2**) showing external hyphae (× 19). (**4**) and (**5**) Further enlargements of the same area of (**3**) showing the hyphae in (**4**) (× 367) and a conidium at the tip of a conidiophore (*arrow*) in (**5**) (× 2743). (Adapted from Bacon, C. W., Bennett, R. M., Hinton, D. M., and Voss, K. A., *Plant Dis.*, 76, 144, 1992.)

The oospores were detected in the pericarp, and mycelium in the pericarp, endosperm, and embryo.[29]

Cleared whole mounts and microtome sections showed broad, hyaline coenocytic mycelium of *Albugo candida* in the seed coat and endosperm of white-encrusted canola and mustard seeds. In discolored seeds with white mycelium and oospores and in shriveled discolored seeds, mycelium was observed in the seed coat, palisade layer, endosperm, and cotyledons. The oospores generally were confined to the seed coat.[30] *F. o.* f. sp. *ciceri* was found as chlamydospore-like structures in the hilum region of chickpea seeds.[31] Oospores of *Peronospora parasitica* were found in radish pericarp and embryo tissues.[32] *Plasmopara halstedii* mycelium was found in the testa and inner layer of the pericarp of infected mature sunflower seeds. Embryos were not infected. Hyphae that grew from tissues of the hilum region into the ovary caused seed infection. The mean ratio of infected seeds among mature seeds of systemically infected plants was 28%.[33]

Sclerotinia sclerotiorum survived in bean seeds as dormant mycelium in the testa and cotyledons.[34] *Phytophthora infestans* was present as a network of coenocytic mycelia in the gel surrounding infected tomato seeds, in the seed coat and remnants of the funiculus, and between the endosperm and seed coat tissues. Occasionally spores were present on the seed surface.[35] Histopathological preparations showed the presence of mycelium and oospores of *Plasmopara obducens* in balsam seed. Seeds collected from diseased capsules and grown in sterilized soil invariably developed downy mildew symptoms.[36]

P. s. pv. *syringae* was found in large numbers in proximity to embryos of sorghum seeds.[37]

Alfalfa mosaic virus (AMV) was detected by ELISA in alfalfa seeds, seedlings, seed coats, and embryos. The incidence (20.6%) of AMV in seeds was significantly higher than in seedlings (7.3%). AMV was detected more often in seed coats than in embryos. The difference in detection levels in seeds and seedlings was related to the high incidence of AMV in seed coats and the low incidence in embryos.[38] The French bean mosaic virus was found in the testa, cotyledons, and embryos of dry infected bean seeds and also in the testa, cotyledons, embryos, and plumules of germinating seeds.[39] Virus concentration was high in embryos of dry (90%) and germinated (90%) seeds but highest in the plumules (100%) of germinated seeds. Virus concentration in the testa and cotyledons of dry, mature, and germinated seeds was low.[39] Lettuce mosaic virus aggregates and cytoplasmic pinwheel inclusions were observed throughout infected embryonic tissue, in nonembryonic endosperm, and in the first true leaves.[40]

Ripe bean pods of fava bean infested by *Ditylenchus dipsaci* contained many dried, resistant, fourth-stage larvae attached to the seeds, especially in the hilum slit. Sometimes these nematodes were massed together to form "eelworm wool." When the testa was removed from heavily infested seeds, necrotic patches containing *D. dipsaci* could be seen on either side of the radicle in cotyledonary depressions. More than 10,000 larvae have been removed from individual seeds. Small, shriveled seeds were more frequently and more heavily infested than large seeds.[41] Fourth-stage larvae and occasionally adults were present both on the seed coat and inside the seed on or in the cotyledons, and in debris mixed with the fava bean seeds.[42] Coconut cadang-cadang viroid was detected in the embryos and husks of the nuts.[43,44]

II. SEED INFECTION

All the seed parts can be colonized. Some important pathogens known to infect different seed parts are the following.

A. Embryo

Examples of some fungi found in the embryo are *Alternaria alternata* in sunflower,[45] *A. brassicicola* in cabbage,[46] *A. helianthi* in sunflower,[47] *A. padwickii* in rice,[48,49] *A. sesamicola* in sesame,[50] *A. triticina* in wheat,[51] *A. zinniae* in zinnia,[52] *Ascochyta fabae* f. sp. *lentis* in lentil,[53] *Bipolaris sorokiniana* in wheat,[54] *Botrytis fabae* in fava bean,[55] *Colletotrichum graminicola* in sorghum,[24,56] *C. lindemuthianum* in cowpea,[57] *Curvularia lunata* in sorghum,[23] *Didymella bryomiae* in watermelon,[58] *Fusarium moniliforme* in maize,[59,60] and sorghum,[23] *F. oxysporum* f. sp. *cumini* in cumin,[61] *F. solani* f. sp. *curcurbitae* in cucurbits,[62] *Gloeotinia granigena* in *Lolium perenne*,[63] *Lasiodiplodia theobromae* in maize,[19] *Macrophomina phaseolina* in soybean,[9] *Peronosclerospora sorghi* in maize,[21] *Phoma rabiei* in chick pea,[64] *Phomopsis* in soybean,[10] *Phytophthora nicotianae* var. *parasitica* in sesame,[65] *Plasmopara halstedii* in sunflower,[66] *Pyricularia oryzae* in rice,[67] *Sclerophthora macrospora* in finger millet,[29] *Sclerospora graminicola* in the scutellum of pearl millet,[68] *S. maydis* in maize,[69] *Stagonospora nodorum* in wheat,[17] and *Ustilago segetum* var. *tritici* in the scutellum of wheat (Figure 3-8).[70]

Examples of some phytobacteria found in the embryo are *Clavibacter m.* subsp. *michiganensis* in tomato,[71] *C. m.* pv. *nebraskensis* in maize, *Pseudomonas s.* pv. *lachrymans* in cucumber,[72] *P.s.* pv. *phaseolicola* in beans (Figure 3-9),[73] *Xanthomonas c.* pv. *phaseoli* in beans,[74] and *X. c.* pv. *manihotis* in cassava.[75]

Most seed-transmitted plant viruses are found in the embryo in nature. Examples are alfalfa mosaic virus in alfalfa,[38] barley stripe mosaic virus (BSMV) in barley,[76] blackeye cowpea mosaic virus in *Vigna mungo*,[77] bean common mosaic virus in bean,[78-80] *Phaseolus acutifolius* var. *latifolius*, and urdbean,[81] cowpea aphid-borne mosaic virus in cowpea,[82] French bean mosaic virus in bean,[39] lettuce mosaic virus in lettuce,[83] pea seedborne mosaic virus in pea,[84] peanut mottle virus in peanut,[85] peanut stripe mosaic virus in peanut,[86] prune dwarf virus in sweet cherry,[87] southern bean mosaic virus in bean,[88] soybean mosaic virus in soybean,[89] squash mosaic virus in cantaloupe,[90] tobacco mosaic virus in apple and pear,[91] and tobacco ringspot virus in soybean (Figure 3-10).[92]

The coconut cadang-cadang viroid also is transmitted through coconut embryos.[43] Barley embryos infected by the BSMV had virus particles in the cytoplasm of the embryonic shield and plumule.[93] In general, the percentage seed transmission of seedborne viruses is related closely to the percentage of embryo infection.

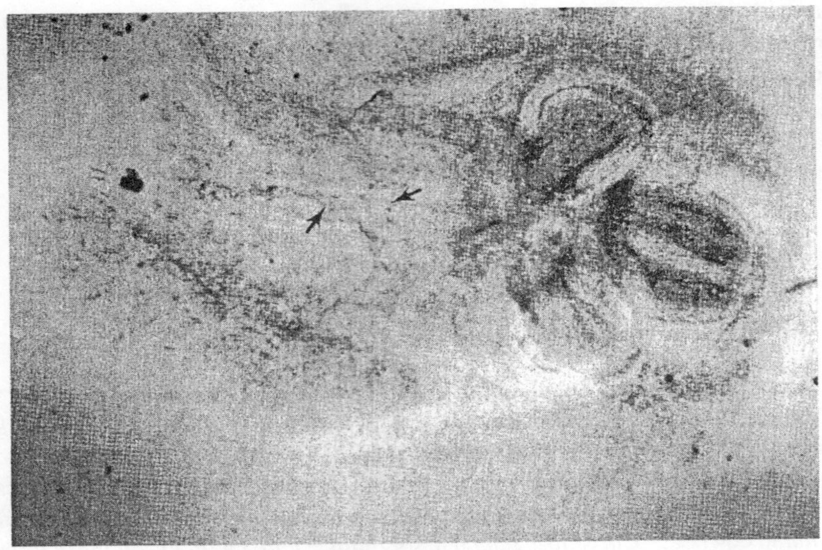

Figure 3-8 Hyphae (*arrow*) of *Ustilago segetum* var. *tritici* (loose smut of wheat) scattered in the scutellum of a wheat (*Tricium aestivum*) seed.

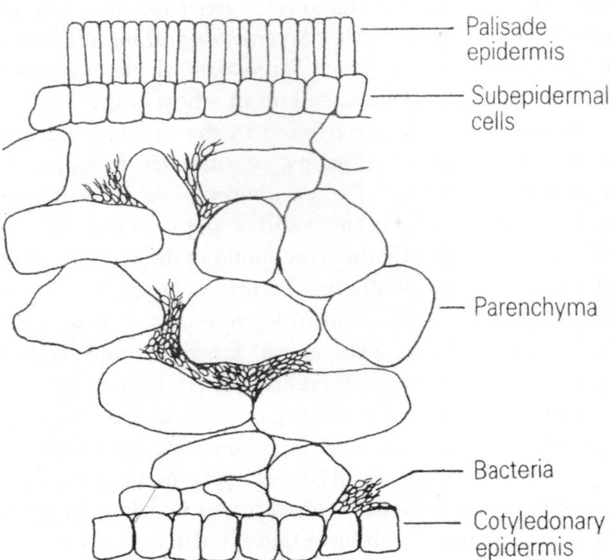

Figure 3-9 Cells of *Pseudomonas syringae* pv. *phaseolicola* (bacterial halo blight) in the embryo of a bean (*Phaseolus vulgaris*) seed. (Adapted from Taylor, J. D., Dudley, C. L., and Presly, L., *Ann. Appl. Biol.*, 93, 267, 1979.)

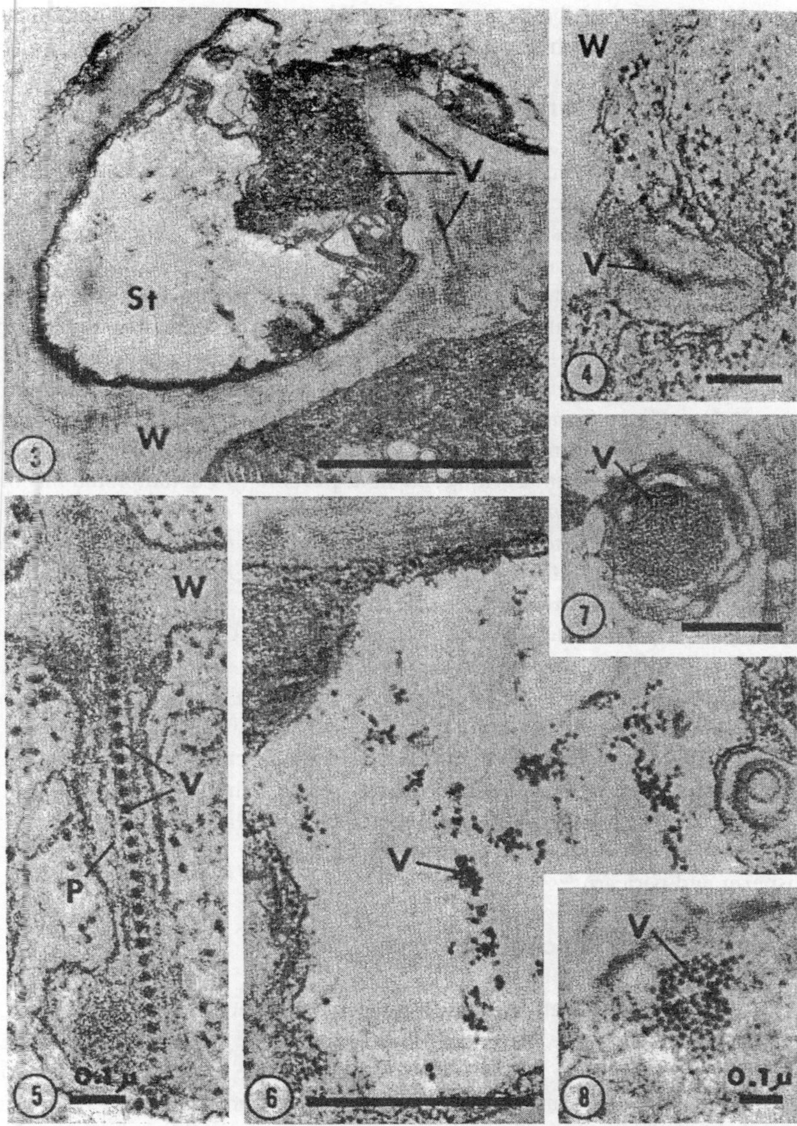

Figure 3-10 Ovules of soybean (*Glycine max*) infected with tobacco ringspot virus (bud blight). An aggregate of viruslike particles in the lumen of a sieve tube in the integument is shown in (**3**). There are two fragments of a tubule containing viruslike particles embedded in the protruded wall. In the nucellus, viruslike particles are frequently found in tubules embedded in cell wall protusions (**4**) or associated with plasmodesmata (**5**); occasionally, clusters of viruslike particles are found in the vacuole of a nucellar cell (**6**). An aggregate of viruslike particles located within layers of the membrane in the vacuole of a nucellar cell is shown in (**7**), and (**8**) shows a viruslike crystal present in a degenerated nucellar cell. P = plasmodesma, St = sieve tube, V = viruslike particles, W = cell wall. Bars = 0.5 μm unless otherwise indicated. (From Yang, A. F. and Hamilton, R. I., *Virology*, 62, 26, 1974. With permission.)

An example of a seedborne nematode is *Ditylenchus dipsaci*, found in cotyledonary lesions near the hypocotyl of *Vicia faba* seeds.[94]

B. Endosperm and Perisperm

In the Poaceae, the endosperm has an outer aleurone layer, and the cotyledons are small. In the Fabaceae, the endosperm is a thin cell layer surrounding the embryo. The endosperm is a common site of infection by seedborne fungi, for example *Alternaria alternata* in soybean[8] and sunflower,[45] *A. brassicicola* in cabbage,[30,46] *A. helianthi* in sunflower,[47] *A. padwickii* in rice,[97] *A. sesamicola* in sesame,[50] *A. zinniae* in zinnia,[52] *Albugo candida* in canola and mustard,[30] *Ascochyta fabae* f. sp. *lentis* in lentil,[53] *Bipolaris nodulosa* in *Eleusine coracana*,[96] *B. oryzae* in rice,[97] *Colletotrichum graminicola* in sorghum,[56] *C. lindemuthianum* in cowpea,[57] *Curvularia lunata* in sorghum,[23] *Didymella bryoniae* in cucumber,[98] *Fusarium moniliforme* in maize[20] and sorghum,[23] *Gloeocercospora sorghi* sclerotia in sorghum,[25] *Gloeotinia granigena* in *L. perenne*,[63] *Lasiodiplodia theobromae* in maize,[19] *Macrophomina phaseolina* in soybean,[9] *Microdochium oryzae* in rice,[48] *Peronosclerospora sorghi* in sorghum,[101] *Pyricularia oryzae* in rice,[18] *Sclerophthora macrospora* in finger millet,[29] *Sclerospora graminicola* in pearl millet,[100] and *Stagonospora nodorum* in wheat.[17]

Examples of phytobacteria found in endosperm are *Clavibacter michiganensis* subsp. *nebraskensis* in maize,[72] *C. michiganensis* subsp. *insidiosus* in the aleurone layer in alfalfa,[101] *C. m.* subsp. *sepedonicus* in tomato,[102] *Erwinia stewartii* in maize,[103] *Pseudomonas* s. pv. *lachrymans* in cucumber,[74] and *X. oryzae* pv. *oryzae* in rice.[104] Most viruses found in the embryo are located in the endosperm. An exception is tobacco mosaic virus, which is not found in the embryo but in the endosperm of tomato seeds.[105,106] Maize dwarf mosaic virus is not found in the embryo of mature maize kernels but occasionally in the endosperm and pericarp.[107]

C. Seed Coat and Pericarp

The testa and seed coat consist of external structures such as the hilum, a micropyle, a chalaza, raphe, hairs, and hair outgrowths. In many seeds, a testa is lacking. Examples of some important fungi found in the seed coat are *Alternaria alternata* in sunflower[108] and soybean,[8] *A. brassicae* and *A. brassicicola* in cabbage and canola,[109] *A. helianthi* in sunflower,[47] *A. padwickii* in rice,[49] *A. sesamicola* in sesame,[50] *A. triticina* in wheat,[51] *A. zinniae* in zinnia,[52] *Albugo candida* in canola and mustard,[30] *Aschochyta fabae* f. sp. *lentis* in lentil,[53] *Phoma rabiei* in chick pea,[64] *Bipolaris nodulosa* in *Eleusine coracana*,[98] *B. oryzae* in rice,[110] *Botrytis cinerea* in linseed,[111] *B. fabae* in fava beans,[55] *Cercospora sojina* in soybean,[10] *Cochliobolus sativus* in barley,[112] *Colletotrichum dematium* f. sp. *spinaciae* in sugarbeet,[113] *C. graminicola* in sorghum,[56] *C. lindemuthianum* in cowpea,[57] *C. truncatum* in soybean,[10,11] *Curvularia lunata* in sorghum,[23] *Diaporthe phaseolorum* var. *sojae* in soybeans,[114] *Didymella bryoniae* in cucumber,[98] *D. lycopersici* in tomato,[115] *Fusarium avenaceum* in parenchyma cells of funicular scar tissue of subterranean clover,[116] *F. moniliforme* in maize,[20,60] *F. oxysporum* f. sp. *asparagi* in asparagus,[117] *F. o.* f. sp. *ciceri* as chlamydospore-like structures in hilum of chick pea,[118] *F. o.* f. sp. *cucumerinum* in bottle gourd and cucumber,[119] *F. o.* f. sp. *lagenarium* in bottle gourd,[120] *F. o.* f. sp. *lini* in flax,[121] *F. o.* f. sp. *vasinfectum* in cotton,[122] *F. solani* f. sp. *cucurbitae* in cucurbits,[62] *Lasiodiplodia theobromae* in maize,[19] *Macrophomina phaseolina* in sesame[123] and soybeans,[9,124] *Peronospora farinosa* in sugarbeet,[125] *P. tabacina* in tobacco,[126] *Peronosclerospora sorghi* in maize,[21] *Phoma sorghina* in sorghum,[23] *Phomopsis* in soybean,[10] *Phytophthora infestans* in tomato,[35,127] *Plasmopara halstedii* in sunflower,[66] *P. manshurica* in soybeans,[13] *Pyricularia grisea* in *E. coracana*,[98] *P. oryzae* in rice,[18] *Sarocladium oryzae* in rice,[128] *Sclerophthora macrospora* in finger millet,[29] *Sclerospora graminicola* in pearl millet,[129] *Sclerotinia sclerotiorum* in bean[34] and sunflower,[130] *Stagonospora nodorum* in wheat,[17] *Ustilago segetum* var. *avenae* in oats,[131] *Verticillium alboatrum* in alfalfa (Figure 3-11),[132] and *V. dahliae* in sunflower.[133,134]

Examples of some phytobacteria found in seed coats are *Bacillus subtilis* in parenchyma and hourglass cell layers of soybean (Figure 3-12),[135] *Clavibacter michiganensis* subsp. *sepedonicus* in tomato,[104] *C. m.* subsp. *nebraskensis* in maize,[72] *Pseudomonas avenae* in rice,[136] *P. s.* pv. *pisi* in pea,[137] *P. s.* pv. *lachrymans* in cucumber,[138] and *X. c.* pv. *malvacearum* in cotton.[139]

In certain crops the seed coat contains vascular elements, such as bean, cotton, cucurbits, and tomato. The raphe of these seeds provides a favorable penetration site for phytobacteria, namely, *P. syringae* pv. *pisi* in pea[137] and *P. s.* pv. *phaseolicola* and *X. c.* pv. *phaseoli* in bean.[76] Phytobacteria also can survive as a dry film on the seed surface as does *P. s.* pv. *pisi* in pea.[137]

Seed coat transmission of plant viruses is uncommon. Some examples of nonembryonic seed transmission are alfalfa mosaic virus in alfalfa,[38] cucumber green mottle mosaic virus in cucumber,[140] French bean mosaic virus in French bean,[39] oat mosaic virus in glumes and testas of oats,[141] onion mosaic virus in onion flower residues,[142] prune dwarf virus in sweet cherry,[89] tobacco mosaic virus in apple, pear, pepper, and tomato,[93,106,143] and tomato spotted wilt virus in tomato.[144]

TMV is held loosely on tomato seed surfaces during seed extraction processes and remains viable for years, depending on the strain and storage conditions. Infested seeds serve as an inoculum source for epidemics of tobacco mosaic on tomato in the U.S.[145-148] Some viruses are carried in both embryo and seed coat — bean common mosaic virus in bean[149] and urd bean,[83] southern bean mosaic virus in bean,[150] and cowpea mosaic virus in cowpea.[151]

D. Glume Infection

Glume infection was reported in the following fungi and bacteria: *Alternaria padwickii* in rice,[49] *Cochliobolus sativus* within the parenchyma and sclerenchyma cells of the lemma and palea of barley,[112] *Drechslera graminea* and *D.*

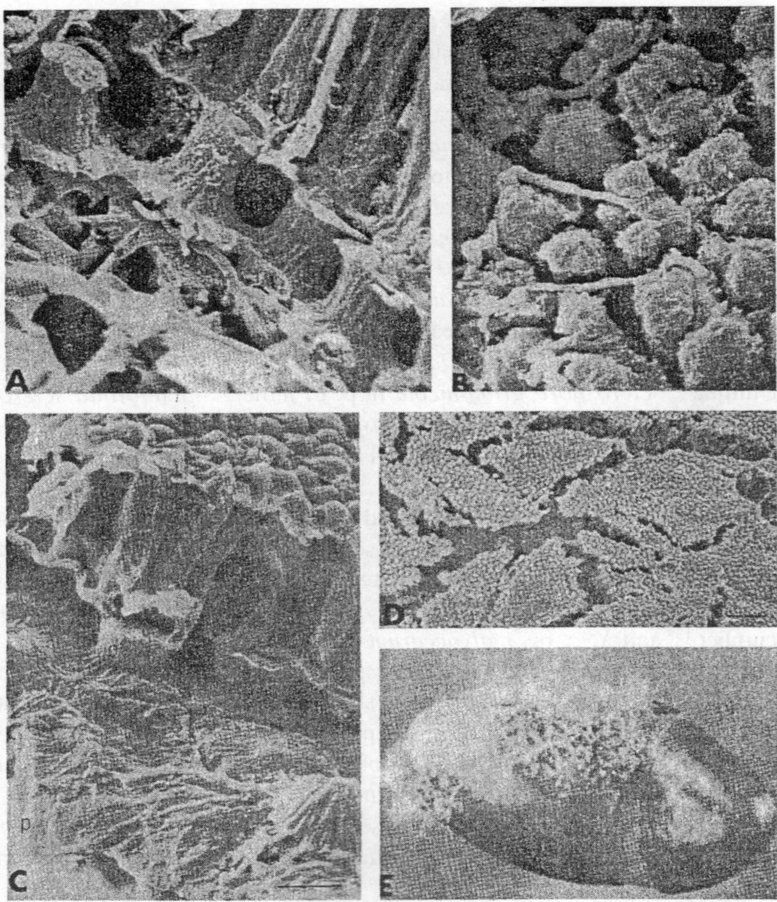

Figure 3-11 Scanning electron and bright-field micrographs showing alfalfa (*Medicago sativa*) seeds without and with hyphae of *Verticillium alboatrum* (root rot and wilt). (**A**) Freeze fracture of a seed revealing hyphae within osteoclerid cells (*arrow*) (× 3210) and (**B**) between osteosclerid cells (× 2195). (**C**) Anatomy of the outer seed coat: cuticle (c) and outer integument, which consists of malphighian (palisade) cells (m), osteosclerid cells (o), and parenchyma (p) cells (bar = 7 μm). (**D**) Surface view of an imbibed severely infected seed (bar = 40 μm). (**E**) Conidiophores and mycelial growth from a seed. (From Christen, A. A., *Phytopathology*, 72, 412, 1982. With permission.)

teres in barley,[152,153] *D. avenae* in oats,[154] *Pyricularia oryzae* in rice,[67] *Peronosclerospora sorghi* as oospores in sorghum[99]; *Sarocladium oryzae* in rice, and *Pseudomonas avenae* in rice,[128] and *X. oryzae* pv. *oryzae* in rice husks.[155]

A pathogen may be associated with different parts of a seed simultaneously; *Phoma rabiei* occurred as a spore contamination on the seed surface on chickpea, as hyphae in the seed coat and/or cotyledonary cells, and as hyphae with subepidermal pycnidia in the seed coat.[64] In eggplant, pepper (Figure 3-1C), tomato and zinnia, *R. solani* occurred as hyphae and sclerotia on the seed surface or as hyphae

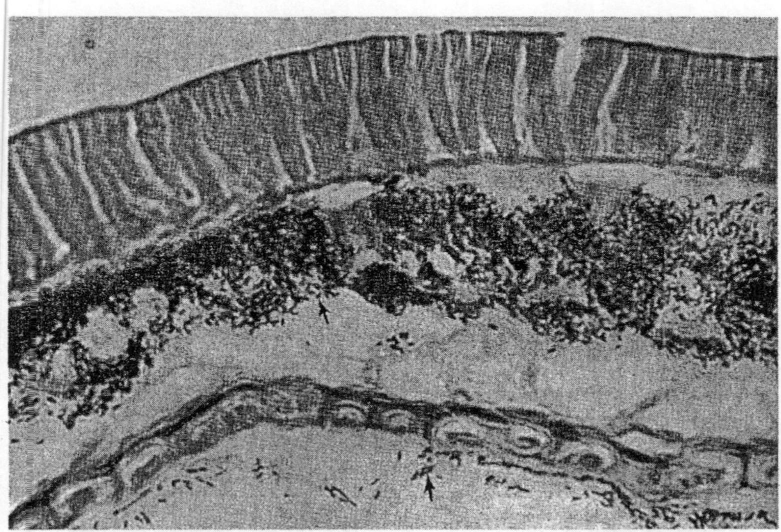

Figure 3-12 Cells of the bacterium of *Bacillus subtilis* (seed decay) (*arrow*) within the cell layers of a soybean (*Glycine max*) seed coat. (From Tenne, F. D., Foor, S. R., and Sinclair, J. B., *Seed Sci. Technol.*, 5, 763, 1977. With permission.)

in the attached funiculus remnants (inner layers of the seed coat, endosperm, and embryo), particularly at the radicle tip.[156] *Cercospora kikuchii* hyphae occurred in the seed coat, cotyledons, and embryo of soybean seeds and were equally abundant in the hourglass and parenchyma cell layers, with a high concentration in the hilum region of the hourglass cell layer but a sparse concentration in the palisade cell layers. The hourglass cell layer consists, in part, of an amorphous, starch-rich matrix that can provide a nutrient source for microorganisms.[114] Similarly, *D. phaseolorum* var. *sojae* was found as hyphae in seed coats, cotyledons, and embryos of soybean seeds, being abundant in the hourglass cell layer, less so in the parenchyma cell layer, and least in the palisade cell layer of the seed coat.[114] In sesame, *Alternaria sesamicola* occurred in subepidermal layers of the seed coat and occasionally in the endosperm and embryo. In severely infected seeds, it invaded all seed parts and sporulated within seeds.[50] *P. s.* pv. *phaseolicola* in bean was confined largely to the parenchyma cell layer of the seed coat, but bacterial masses were found surrounding the cotyledons and embryo.[75] In tomato seed, *C. m.* subsp. *sepedonicus* was found between the integuments and endosperm, and in the micropylar region between the embryo and endosperm.[102]

III. SEED INFESTATION OR CONTAMINATION

Infestation is an important means of transmitting pathogens on seeds without an active relationship between pathogen and seed. Such pathogens are carried as (1) an infestation or contamination of the seed coat or pericarp surface or (2) as propagules, infected plant debris, or infested soil mixed with seeds; for example,

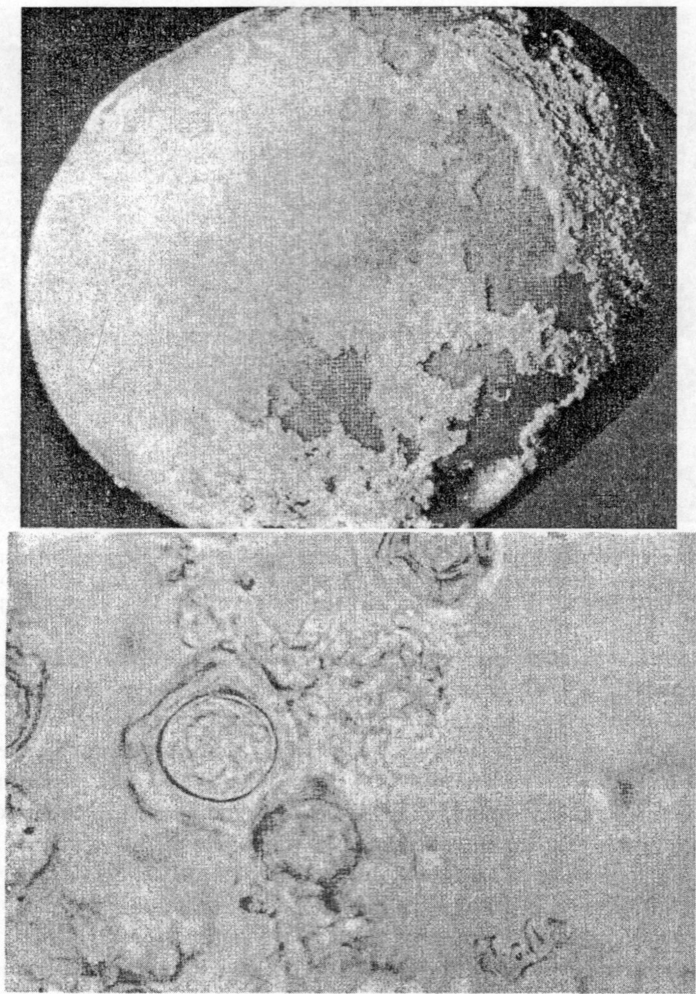

Figure 3-13 A soybean (*Glycine max*) seed encrusted with oospores of *Peronospora manshurica* (downy mildew) (*above*) and mature oospores under bright-field microscopy (*below*). (From Sinclair, J. B., Ed., *Compendium of Soybean Disease,* 2nd ed., APS Press, St. Paul, MN, 1982, 16. With permission.)

A. brassicicola has been found in cabbage,[30] *Peronospora manshurica* in soybean (Figure 3-13),[13] *Albugo candida* in canola and mustard,[30] and *Maesziomyces bullatus* in pearl millet[157] (see Chapter 7). Seed contamination occurred during harvesting, threshing, and/or post-threshing operations. *Cercospora kikuchii*, *C. sojina*, and *Phomopsis* were recovered at a significantly ($P = 0.05$) higher rate from seeds aseptically removed from pods than from threshed soybean seeds.[158,159]

REFERENCES

1. Kozlowski, T. T. and Gunn, C. R., Importance and characteristics of seeds, in *Seed Biology*, Vol. 1, Kozlowski, T. T., Ed., Academic Press, New York, 1972, 1.
2. Shetty, H. S., Mathur, S. B., and Neergaard, P., *Sclerospora graminicola* in pearl-millet seeds and its transmission, *Trans. Br. Mycol. Soc.*, 74, 127, 1980.
3. Bennett, C. W., Seed transmission of plant viruses, *Adv. Virus Res.*, 14, 221, 1969.
4. Singh, D., Histopathology of some seed-borne infection: a review of recent investigations, *Seed Sci. Technol.*, 11, 651, 1983.
5. Sinclair, J. B., Latent infection of soybean plants and seeds by fungi, *Plant Dis.*, 75, 220, 1991.
6. Velicheti, R. K. and Sinclair, J. B., Histopathology of soybean seed colonized by *Fusarium oxysporum, Seed Sci. Technol.*, 19, 445, 1991.
7. Sinclair, J. B., Multiple fungal infections of soybean seeds in preharvest and postharvest deterioration, in Physiological–Pathological Interactions Affecting Seed Deterioration, West, S. H., Ed., Crop Science Society of America, Madison, WI, 1986, 65.
8. Kunwar, I. K., Manandhar, J. B., and Sinclair, J. B., Histopathology of soybean seeds infected with *Alternaria alternata*, *Phytopathology*, 76, 543, 1986.
9. Kunwar, I. K., Singh, T., Machado, C. C., and Sinclair, J. B., Histopathology of soybean seed and seedlings infection by *Macrophomina phaseolina*, *Phytopathology*, 76, 532, 1986.
10. Kunwar, I. K., Singh, T., and Sinclair, J. B., Histopathology of mixed infection by *Colletotrichum truncatum* and *Phomopsis* spp. or *Cercospora sojina* in soybean seeds, *Phytopathology*, 75, 489, 1985.
11. Nik, W. Z. and Lim, T. K., Occurrence and site of infection of *Colletotrichum dematium* f. sp. *truncatum* in naturally infected soybean seeds, *J. Plant Prot. Trop.*, 1, 87, 1984.
12. Hildebrand, A. A. and Koch, L. W., A study of systemic infection by downy mildew with special reference to symptomatology, economic significance, and control, *Sci. Agric.*, 31, 505, 1951.
13. Roongruangsree, U-Tai, Olson, L. W., and Lange, L., The seedborne inoculum of *Peronospora manshurica* causal agent of soybean downy mildew, *J. Phytopathol.*, 123, 233, 1988.
14. Zorrilla, G., Knapp, A. D., and McGee, D. C., Severity of Phomopsis seed decay, seed quality evaluation, and field performance of soybean, *Crop Sci.*, 34, 172, 1994.
15. Zimmer, R. C., McKeen, W. E., and Campbell, C. G., Location of oospores in buckwheat seed and probable role of oospores and conidia of *Peronospora ducometi* in the disease cycle on buckwheat, *J. Phytopathol.*, 135, 217, 1992.
16. Dey, S. K. and Singh, G., Seedborne infection of *Ascochyta rabiei* in chickpea and its transmission to aerial plant parts, *Phytoparasitica*, 22, 31, 1994.
17. Agrawal, K., Singh, T., Singh, D., and Mathur, S. B., Studies on glume blotch disease of wheat. I. Location of *Septoria nodorum* in seed, *Phytomorphology*, 35, 87, 1985.
18. Chung, H. S. and Lee, C. U., Detection and transmission of *Pyricularia oryzae* in germinating rice seed, *Seed Sci. Technol.*, 11, 625, 1983.
19. Kumar, V. and Shetty, H. S., Seedborne nature and transmission of *Botryodiplodia theobromae* in maize (*Zea mays*), *Seed Sci. Technol.*, 11, 781, 1983.

20. Bacon, C. W., Bennett, R. M., Hinton, D. M., and Voss, K. A., Scanning electron microscopy of *Fusarium moniliforme* within asymptomatic corn kernels and kernels associated with equine leukoencephalomalacia, *Plant Dis.*, 76, 144, 1992.
21. Rao, B. M., Prakash, H. S., Shetty, H. S., and Safeeulla, K. M., Techniques to detect seedborne inoculum of *Peronosclerospora sorghi* in maize, *Seed Sci. Technol.*, 12, 593, 1984.
22. Rao, B. M., Shetty, H. S., and Safeeulla, K. M., Establishment of different isolates of *Peronosclerospora sorghi* in maize seed, *Indian Phytopathol.*, 40, 348, 1987.
23. Singh, D. P., Agarwal, V. K., and Khetarpal, R. K., Etiology and host–pathogen relationship of grain mould of sorghum, *Indian Phytopathol.*, 41, 389, 1988.
24. Prasad, K. V. V., Khare, M. N., and Jain, A. C., Site of infection and further development of *Colletotrichum graminicola* (Ces.) Wilson in naturally infected sorghum grains, *Seed Sci. Technol.*, 13, 37, 1985.
25. Mathur, K., Siradhana, B. S., and Lodha, B. C., Studies on seedling blight of sorghum caused by *Gloeocercospora sorghi*, *Seed Sci. Technol.*, 15, 851, 1987.
26. Shetty, H. S., Mathur, S. B., and Neergaard, P., *Sclerospora graminicola* in pearl-millet seeds and its transmission, *Trans. Br. Mycol. Soc.*, 74, 127, 1980.
27. Suryanarayana, D., Occurrence of an unknown fungal mycelium inside the sound grains produced on partly formed green ears of bajra plants, *Sci. Cult.*, 28, 536, 1962.
28. Williams, R. J., Downy mildews of tropical cereals, *Adv. Plant Pathol*, 2, 1, 1984.
29. Raghavendra, S. and Safeeulla, K. M., Seed transmission of the downy mildew of finger millet (Ragi), *Seeds Farms*, 2(4), 9, 1978.
30. Sharma, J., Studies on Seedborne Mycoflora and Some Important Seedborne Diseases of Rape and Mustard Grown in Rajasthan, Ph.D. Thesis, Univ. Rajasthan, Jaipur, 1989, 209 pp.
31. Haware, M. P., Nene, Y. L., and Rajeshwari, R., Eradication of *Fusarium oxysporum* f. sp. *ciceri* transmitted in chickpea seed, *Phytopathology*, 68, 1364, 1978.
32. Jang, P. and Safeeulla, K. M., Production and viability of *Peronospora parasitica* in radish, *Proc. Indian Acad. Sci.*, 100, 117, 1990.
33. Doken, M. T., *Plasmopara halstedii* (Farl.) Berl. et de Toni in sunflower seeds and the role of infected seeds in producing plants with systemic symptoms, *J. Phytopathol.*, 124, 23, 1989.
34. Tu, J. C., The role of white mold-infected white bean (*Phaseolus vulgaris* L.) seeds in the dissemination of *Sclerotinia sclerotiorum* (Lib.) de Bary, *J. Phytopathol.*, 121, 40, 1988.
35. Vartanian, V. G. and Endo, R. M., Survival of *Phytophthora infestans* in seeds extracted from tomato fruits, *Phytopathology*, 75, 375, 1985.
36. Srivastava, S. and Singh, L., Evidence of seedborne nature of downy mildew fungus of balsam, *Seed Res.*, 16, 254, 1988.
37. Gaudet, D. A. and Kokko, E. G., Seedling disease of sorghum grown in southern Alberta caused by *Pseudomonas syringae* pv. *syringae*, *Can. J. Plant Pathol.*, 8, 208, 1986.
38. Pesic, Z. and Hiruki, C., Difference in the incidence of alfalfa mosaic virus in seed coat and embryo of alfalfa seed, *Can. J. Plant Pathol.*, 8, 39, 1986.
39. Capoor, S. P., Rao, D. G., and Sawant, D. M., Seed transmission of French bean mosaic virus, *Indian Phytopathol.*, 39, 343, 1986.
40. Hunter, D. G. and Bowyer, J. W., Cytopathology of lettuce mosaic virus infected lettuce seeds and seedlings, *J. Phytopathol.*, 137, 61, 1993.

41. Hooper, D. J., Stem eelworm (*Ditylenchus dipsaci*), a seed and soilborne pathogen of field bean (*Vicia faba*), *Plant Pathol.*, 20, 25, 1971.
42. Caubel, G., ISTA Handbook on Seed Health Testing Working Sheet No. 57, *Vicia faba, Ditylenchus dipsaci* (Kühn) Filipjev. International Seed Testing Association, Zurich, 1987.
43. Hanold, D. and Randles, J. W., Coconut cadang-cadang disease and its viroid agent, *Plant Dis.*, 75, 330, 1991.
44. Randles, J. W. and Imperial, J. S., Coconut cadang-cadang viroid, in Descriptions of Plant Viruses, No. 287, Commonw. Mycol. Inst., Assoc. Appl. Biol., Kew, Surrey, U.K., 1984.
45. Svetov, V. G., Alternaria blight of sunflower along the Kuban river, *Mikol. Fitopatol.*, 9, 418, 1975.
46. Knox-Davies, P. S., Relationship between *Alternaria brassicicola* and *Brassica* seeds, *Trans. Br. Mycol. Soc.*, 73, 235, 1979.
47. Raut, J. G., Location of *Alternaria helianthi* in sunflower seeds and its transmission from seed to plant, *Indian Phytopathol.*, 38, 522, 1985.
48. Cheeran, A. and Raj, J. S., A technique for the separation of the whole embryo of rice grains for the detection of fungal mycelium, *Agric. Res. J. Kerala*, 10, 125, 1972.
49. Srinivasaiah, S. M., Ranganathaiah, K. G., and Nanjegowda, D., Studies on transmission and location of *Trichoconiella padwickii* in rice seeds, *Indian Phytopathol.*, 37, 31, 1984.
50. Singh, D., Mathur, S. B., and Neergaard, P., Histological studies of *Alternaria sesamicola* penetration in sesame seed, *Seed Sci. Technol.*, 8, 85, 1980.
51. Kumar, V. R. and Arya, H. C., Certain aspects of perpetuation and recurrence of leaf blight of wheat in Rajasthan, *Indian J. Mycol. Plant Pathol.*, 3, 93, 1973.
52. Srivastava, R. N. and Gupta, J. S., Seed transmission of *Alternaria zinniae*, its location in the seed and control, *Indian Phytopathol.*, 37, 83, 1984.
53. Morrall, R. A. A. and Beauchamp, C. J., Detection of *Ascochyta fabae* f. sp. *lentis* in lentil seed, *Seed Sci. Technol.*, 6, 383, 1988.
54. Weniger, W., Pathological morphology of durum wheat grains affected with black point, *Phytopathology*, 13, 48, 1923.
55. Harrison, J. G., Role of seed-borne infection in epidemiology of *Botrytis fabae* on field beans, *Trans. Br. Mycol. Soc.*, 70, 35, 1978.
56. Basuchaudhary, K. C. and Mathur, S. B., Infection of sorghum seeds by *Colletotrichum graminicola*. I. Survey, location in seed and transmission of the pathogen, *Seed Sci. Technol.*, 7, 87, 1979.
57. Prasanna, K. P. R., Seed health testing of cowpea with special reference to anthracnose caused by *Colletotrichum lindemuthianum*, *Seed Sci. Technol.*, 13, 821, 1985.
58. Rankin, H. W., Effectiveness of seed treatment for controlling anthracnose and gummy stem blight of watermelon, *Phytopathology*, 44, 675, 1954.
59. Lawrence, E. B., Nelson, P. E., and Ayers, J. E., Histopathology of sweet corn seed and plants infected with *Fusarium moniliforme* and *F. oxysporum*, *Phytopathology*, 71, 379, 1981.
60. Kim, W. G., Oh, I. S., Yu, S. H., and Park, J. S., *Fusarium moniliforme* detected in seeds of corn and its pathological significance, *Korean J. Mycol.*, 12, 105, 1984.
61. Singh, R. D., Choudhary, S. L., and Patel, K. G., Seed transmission and control of Fusarium wilt of cumin, *Phytopathol. Mediterr.*, 1, 19, 1972.

62. Tousson, T. A. and Synder, W. C., The pathogenicity, distribution, and control of two races of *Fusarium* (Hypomyces) *solani* f. *cucurbitae*, *Phytopathology*, 51, 17, 1961.
63. Neill, J. C. and Hyde, E. O. C., Blind seed diseases of ryegrass, *N.Z. J. Sci. Technol. Sect. A*, 20, 281, 1939.
64. Maden, S., Singh, D., Mathur, S. B., and Neergaard, P., Detection and location of seed-borne inoculum of *Ascochyta rabiei* and its transmission in chickpea (*Cicer arietinum*), *Seed Sci. Technol.*, 3, 667, 1975.
65. Sehgal, S. P. and Prasad, N., Studies on the perennation and survival of sesamum *Phytophthora*, *Indian Phytopathol.*, 19, 173, 1966.
66. Cohen, Y. and Sackston, W. E., Sunflower seed infection with *Plasmopara halstedii*, in 2nd International Congress of Plant Pathology, University of Minnesota, St. Paul, 1974, 446.
67. Suzuki, H., Studies on the influence of some environmental factors on the susceptibility of the rice plant to blast and Helminthosporium disease and on the anatomical characters of the plant. I. Influence of differences in soil moisture, *J. Coll. Agric. Tokyo*, 13, 45, 1934.
68. Gan-Bobo, M. S. and Dostaler, D., Recensement et incidience de la mycoflore des semences du millet perle au Niger, *Seed Sci. Technol.*, 18, 567, 1990.
69. Purakusumah, H., Seed-borne primary infection in downy mildew, *Sclerospora maydis* (Racib.) Butler, *Nature (London)*, 207, 1312, 1965.
70. Khanzada, A. K., Rennie, W. J., Mathur, S. B., and Neergaard, P., Evaluation of two routine embryo test producers for assessing the incidence of loose smut infection in seed samples of wheat (*Triticum aestivum*), *Seed Sci. Technol.*, 8, 363, 1980.
71. Bryan, M. K., Studies on bacterial canker of tomatoes, *J. Agric. Res.*, 41, 825, 1930.
72. Naumann, K., Über das Auftreten von Bakterien in Gurkensamen' aus Früchten, die durch *Pseudomonas lachrymans infiziertwaren*, *Phytopathol. Z.*, 48, 258, 1963.
73. Taylor, J. D., Dudley, C. L., and Presly, L., Studies on halo-blight seed infection and disease transmission in dwarfbeans, *Ann. Appl. Biol.*, 93, 267, 1979.
74. Zaumeyer, W. J. and Thomas, H. R., A monographic study of bean diseases and methods of their control, *U.S. Dep. Agric. Tech. Bull.*, 868, 255, 1957.
75. Elango, F. and Lozano, J. C., Transmission of *Xanthomonas manihotis* in seed of cassava (*Manihotis esculanta*), *Plant Dis.*, 64, 784, 1980.
76. Gold, A. H., Suneson, C. A., Houston, B. R., and Oswald, J. W., Electron microscopy and seed and pollen transmission of rod-shaped particles associated with the false stripe virus disease of barley, *Phytopathology*, 44, 115, 1954.
77. Provvidenti, R., Seed transmission of blackeye cowpea mosaic virus in *Vigna mungo*, *Plant Dis.*, 70, 981, 1986.
78. Ekpo, E. J. A. and Saettler, A. W., Distribution pattern of bean common mosaic virus in developing bean seed, *Phytopathology*, 64, 269, 1974.
79. Hagita, T. and Tamada, T., Detection of bean common mosaic virus in fresh bean seeds by immune electron microscopy, *Bull. Hokkaido Exp. Stn.*, 51, 83, 1984.
80. Provvidenti, R. and Cobb, E. D., Seed transmission of bean common mosaic virus in Tepary bean, *Plant Dis. Rep.*, 59, 966, 1975.
81. Agarwal, V. K., Nene, Y. L., and Beniwal, S. P. S., Location of bean common mosaic virus in urdbean seed, *Seed Sci. Technol.*, 7, 455, 1979.
82. Phatak, H. C., Seed-borne plant viruses — identification and diagnosis in seed health testing, *Seed Sci. Technol.*, 2, 3, 1974.

83. Ryder, E. J., Transmission of common lettuce mosaic virus through the gametes of the lettuce plant, *Plant Dis. Rep.*, 48, 522, 1964.
84. Stevenson, W. R. and Hagedorn, D. J., Further studies on seed transmission of pea seedborne mosaic virus in *Pisum sativum*, *Plant Dis. Rep.*, 57, 248, 1973.
85. Adams, D. B. and Kuhn, C. W., Seed transmission of peanut mottle virus, *Phytopathology*, 67, 1126, 1977.
86. Demski, J. W. and Lovell, G. R., Peanut stripe virus and the distribution in peanut seed, *Plant Dis.*, 69, 734, 1985.
87. Kelley, R. D. and Cameron, H. R., Location of prune dwarf and prunus necrotic ring spot viruses associated with sweet cherry pollen and seed, *Phytopathology*, 76, 317, 1986.
88. Uyemoto, J. K. and Grogan, R. G., Southern bean mosaic virus: evidence for seed transmission in bean embryos, *Phytopathology*, 67, 1190, 1977.
89. Porto, M. D. and Hagedorn, D. J., Seed transmission of a Brazilian isolate of soybean mosaic virus, *Phytopathology*, 65, 713, 1975.
90. Alvarez, M. and Campbell, R. N., Transmission and distribution of squash mosaic virus in seeds of cantaloupe, *Phytopathology*, 68, 257, 1978.
91. Gilmer, R. M. and Wilks, J. M., Seed transmission of tobacco mosaic virus in apple and pear, *Phytopathology*, 57, 214, 1967.
92. Owusu, G. K., Crowley, N. C., and Francki, R. I. B., Studies on the seed-transmission of tobacco ringspot virus, *Ann. Appl. Biol.*, 61, 195, 1968.
93. Razvyazkina, G. M. and Cheremushkina, N. P., Localization of barley stripe mosaic virus in the embryo, *S. Kh. Biologya*, 7, 633, 1972.
94. Diercks, R. and Klewitz, R., Zur Samenübertragbarkeit einer am Ackerbohnen vorkommenden Herkunft des Stengelälchens, *Ditylenchus dipsaci* (Kühn), *Nematologica*, 7, 155, 1962.
95. Ganguly, D., Studies on the stackburn disease of rice and identity of the causal organism, *J. Indian Bot. Soc.*, 26, 233, 1947.
96. Ranganathaiah, K. G. and Mathur, S. B., Seed health testing of *Eleusine coracana* with special reference to *Drechslera nodulosa* and *Pyricularia grisea*, *Seed Sci. Technol.*, 6, 943, 1978.
97. Agarwal, V. K. and Singh, O. V., Relative percentage incidence of seed-borne fungi associated with different varieties of rice seeds, *Seed Res.*, 2, 23, 1974.
98. Lee, Du-Hyung, Mathur, S. B., and Neergaard, P., Detection and location of seed-borne inoculum of *Didymella bryoniae* and its transmission in seedlings of cucumber and pumpkin, *Phytopathol. Z.*, 109, 301, 1984.
99. Ahmad, R. and Majumdar, S. K., 1987, Seedborne nature detection and seed transmission of sorghum downy mildew on sorghum, in Plant Quarantine and Phytosanitary Barriers to Seed in the ASEAN, 9–10 Dec., Selangor Malaysia ASEAN Plant Quarantine Center Institute, 173, 1987.
100. Shetty, H. S., Khanzada, A. K., Mathur, S. B., and Neergaard, P., Procedures for detecting seedborne inoculum of *Sclerospora graminicola* (Sacc.) Schroet. in pearlmillet (*Pennisetum typhoides* Burm.) Stapf & Hubb., *Seed Sci. Technol.*, 6, 943, 1978.
101. Cormack, M. W. and Moffatt, J. E., Occurrence of the bacterial wilt organism in alfalfa seed, *Phytopathology*, 46, 407, 1956.
102. Larson, R. H., The ring rot bacterium in relation to tomato and eggplant, *J. Agric. Res.*, 69, 309, 1944.
103. Ivanoff, S. S., Stewart's wilt disease of corn, with emphasis on the life history of *Phytomonas stewartii* in relation to pathogenesis, *J. Agric. Res.*, 47, 749, 1933.

104. Srivastava, D. N. and Rao, Y. P., Seed transmission and epidemiology of the bacterial blight disease of rice in North India, *Indian Phytopathol.*, 17, 77, 1964.
105. Broadbent, L., The epidemiology of tomato mosaic. XI. Seed transmission of TMV, *Ann. Appl. Biol.*, 56, 177, 1965.
106. Taylor, R .H., Grogan, R. G., and Kimble, K. A., Transmission of tobacco mosaic virus in tomato seed, *Phytopathology*, 51, 837, 1961.
107. Mikel, M. A., D'Arcy, C. J., Rhodes, A. M., and Ford, R. E., Seed transmission of maize dwarf mosaic virus in sweet corn, *Phytopathology*, 72, 1138, 1982.
108. Singh, D., Mathur, S. B., and Neergaard, P., Histopathology of sunflower seeds infected by *Alternaria tenuis*, *Seed Sci. Technol.*, 5, 579, 1977.
109. Domsch, K. H., Die Raps — und Kohlschotenschwärze, *Z. Pflkrank. Pflanzensch.*, 64, 65, 1957.
110. Fazli, S. F. I. and Schroeder, H. W., Kernel infection of Blue Bonnet 50 rice by *Helminthosporium oryzae*, *Phytopathology*, 56, 507, 1966.
111. Van der Spek, J., The influence of environment on the effect of linseed disinfection, *Meded. Landb. Hogesch. Gent.*, 22, 535, 1957.
112. Stevenson, I. L., Timing and nature of seed infection of barley by *Cochliobolus sativus*, *Can. J. Plant Pathol.*, 3, 76, 1981.
113. Chikuo, Y. and Sugimoto, T., Infection of sugarbeet seed by *Colletotrichum dematium* f. *spinaciae*, *Ann. Phytopathol. Soc. Jpn.*, 50, 249, 1984.
114. Ilyas, M. B., Dhingra. O. D., Ellis, M. A., and Sinclair, J. B., Location of mycelium of *Diaporthe phaseolorum* var. *sojae* and *Cercospora kikuchii* in infected soybean seed, *Plant Dis. Rep.*, 59, 17, 1975.
115. Derbyshire, D. M., Seed transmission of *Didymella lycopersici*, *Plant Pathol.*, 9, 152, 1960.
116. Kellock, A. W., Stubbs, L. L., and Parbery, D. G., Infection of seed of subterranean clover and allied species by species of *Fusarium*, *Aust. J. Agric. Res.*, 31, 297, 1980.
117. Inglis, D. A., Contamination of asparagus seed by *Fusarium oxysporum* f. sp. *asparagi* and *Fusarium moniliforme*, *Plant Dis.*, 64, 74, 1980.
118. Haware, M. P., Nene, Y. L., and Rajeshwari, R., Eradication of *Fusarium oxysporum* f. sp. *ciceri* transmitted in chickpea seed, *Phytopathology*, 68, 1364, 1978.
119. Takeuchi, S., Ogawa, K., and Nomura, Y., Mechanism of seed transmission of Fusarium wilt of cucumber and bottle gourd, and improvement of a method for the seed-disinfection test, *J. Cent. Agric. Exp. Stn.*, 28, 49, 1978.
120. Kuniyashu, K. and Kishi, K., Seed transmission of Fusarium wilt of bottlegourd, *Lagenaria siceraria* Standle, used as root stock of watermelon. II. The seed infection course from the infected stem of bottlegourd to the fruit and seed, *Ann. Phytopathol. Soc. Jpn.*, 43, 192, 1977.
121. Tu, C. C. and Cheng, Y. H., Studies on seed transmission of flax Fusarium wilt and its control with seed treatment, *J. Agric. Res. China*, 25, 134, 1976.
122. Bakry, M. A. and Rizk, R. H., Seed transmission of *Fusarium oxysporum* f. *vasinfectum*, the causal agent of cotton wilt in the United Arab Republic, *Agric. Res. Rev. Cairo*, 45, 1, 1967.
123. Abdou, Y. A., El Hassan, S. A., and Abbas, H. K., Seed transmission and pycnidial formation in sesame wilt disease caused by *Macrophomina phaseoli* (Maubl.) Ashby, *Agric. Res. Rev.*, 57, 63, 1979.
124. Gangopadhyay, S., Wyllie, T. D., and Luedders, V. D., Charcoal rot disease of soybean transmitted by seeds, *Plant Dis. Rep.*, 54, 1088, 1971.

125. Leach, L. D., Downy mildew of the beet, caused by *Peronospora schachtii* Fuckel, *Hilgardia*, 6, 203, 1931.
126. Borovskaya, M., Transmission of *Peronospora tabacina* by seeds, *Russ. Zascher Rash. Med. Boleznei*, 3, 44, 1965.
127. Boyd, O. C., Evidence of the seed-borne nature of late blight (*Phytophthora infestans*) of tomatoes, *Phytopathology*, 25, 7, 1935.
128. Mukerjee, P., Histopathological study of sheath rot of rice, Proc. Int. Symp. Rice Research: New Frontiers, Directorate of Rice Hyderabad, India, 1990, 241.
129. Shetty, H. S., Mathur, S. B., and Neergaard, P., Occurrence of *Sclerospora graminicola* (Sacc.) Schroet. inoculum in pearlmillet (*Pennisetum typhoides* (Burm.) Stapf & Hubb.) seeds and its transmission, in *3rd Int. Congr. Plant Pathology*, P. Parey, Berlin, 1978, 120.
130. Tollenaar, H. and Bleiholder, H., Distribution of the mycelium of *Sclerotinia sclerotiorum* in sunflower seed, *Agric. Technol.*, 31, 44, 1971.
131. Zade, A., Neure Untersuchungen über die Lebensweise und Bekämpfung des Haferflugbrandes (*Ustilago avenae* (Pers) Jens.), *Angew. Bot.*, 6, 113, 1924.
132. Christen, A. A., Demonstration of *Verticillium alboatrum* within alfalfa seed, *Phytopathology*, 72, 412, 1982.
133. Klisiewicz, J. M., Role of infected sunflower seed in survival and dissemination of *Verticillium dahliae*, *Proc. Am. Phytopathol. Soc.*, 1, 39, 1974.
134. Klisiewicz, J. M., Assay of *Verticillium* in sunflower seed, *Plant Dis. Rep.*, 58, 926, 1974.
135. Tenne, F. D., Foor, S. R., and Sinclair, J. B., Association of *Bacillus subtilis* with soybean seeds, *Seed Sci. Technol.*, 5, 763, 1977.
136. Shakya, D. D., Chung, H. S., and Vinther, F., Transmission of *Pseudomonas avenae* the cause of bacterial stripe of rice, *J. Phytopathol.*, 116, 92, 1986.
137. Skoric, V., Bacterial blight of pea: overwintering, dissemination, and pathological histology, *Phytopathology*, 17, 611, 1927.
138. Wiles, A. B., Studies on *Pseudomonas lachrymans* in cucumber, *Phytopathology*, 41, 38, 1951.
139. Brinkerhoff, L. A. and Hunter, R. E., Internally infected seed as a source of inoculum for the primary cycle of bacterial blight of cotton, *Phytopathology*, 53, 1397, 1963.
140. Inouye, T., Inouye, N., Asatani, M., and Mitsuhata, K., Studies on cucumber green mottle mosaic virus in Japan, *Ber. Ohara Inst. Landwirtsch. Biol. Okayama Univ.*, 14, 49, 1967.
141. Carr, A. J. H., Plant pathology, in Report of the Welsh Plant Breeding Station, 1970, University College of Wales, Aberysthwuth, 1971, 40.
142. Protsenko, L. V., Modes of transmission of onion mosaic virus, *Kartoplya Ovochevi Bashtanni Kult*, 1, 147, 1965.
143. Demski, J. W., Tobacco mosaic virus is seed borne in pimento peppers, *Plant Dis.*, 65, 723, 1981.
144. Phatak, H. C., Seed-transmitted plant viruses — general aspects, in *Seed Pathology Problems and Progress*, Yorinori, J. T., Sinclair, J. B., Mehta, Y. R., and Mohan, S. K., Eds., Fundacao Instituto Agronomico do Parana IAPAR, Brazil, 1979, 141.
145. Alexander, L. J., Inactivation of tobacco mosaic virus from tomato seed, *Phytopathology*, 50, 627, 1960.
146. Gooding, G. V. and Suggs, E. G., Seedborne tobacco mosaic virus in commercial source of tomato seed, *Plant Dis. Rep.*, 60, 441, 1976.

147. John, C. A. and Sova, C., Incidence of tobacco mosaic virus on tomato seed, *Phytopathology*, 45, 636, 1955.
148. Raychaudhuri, S. P., Studies on internal browning of tomato, *Phytopathology*, 42, 591, 1952.
149. Crowley, N. C., Studies on the seed transmission of plant virus diseases, *Aust. J. Biol. Sci.*, 10, 449, 1957.
150. McDonald, J. G. and Hamilton, R. I., Distribution of southern bean mosaic virus in the seed of *Phaseolus vulgaris*, *Phytopathology*, 62, 389, 1972.
151. Khatri, H. L. and Chohan, J. S., Studies on some factors influencing seed transmission of cowpea mosaic virus in cowpea, *Indian J. Mycol. Plant Pathol.*, 2, 40, 1972.
152. Vogt, E., Ein Beitrag zur Kenntnis von *Helminthosporium gramineum* Rbh., *Arb. Biol. Reichsanst. Land Forstwirtsch*, Berlin Dahlem, 11, 387, 1923.
153. Ravn, F. K., Nogle *Helminthosporium* Arter og de af dem fremkaldte Sygdomme hos Byg og Havre, *Bot. Tidsskr.*, 23, 101, 1900.
154. Rathschlag, H., Studien über *Helminthosporium avenae*, *Phytopathol. Z.*, 2, 469, 1930.
155. Wakimoto, S., Overwintering of *Xanthomonas oryzae* on unhulled grains of rice, *Agric. Hort.*, 30, 1501, 1955.
156. Baker, K. F., Seed transmission of *Rhizoctonia solani* in relation to control of seedling damping off, *Phytopathology*, 37, 912, 1947.
157. Thakur, R. P. and King, S. B., Smut disease of pearlmillet, Inf. Bull. No. 25, Int. Crops Res. Semi-Arid Trop., Patancheru, A.P. India, 17, 1988.
158. Yorinori, J. T. and Sinclair, J. B., Harvest and assay methods for seedborne fungi in soybeans and their pathogenicity, *Trop. Plant. Dis.*, 1, 53, 1983.
159. Singh, T. and Sinclair, J. B., Histopathology of *Cercospora sojina* in soybean seeds, *Phytopathology*, 75, 185, 1985.

CHAPTER 4

Mechanism of Seed Infection

The seed infection process is affected by environmental conditions during crop development. Successful seed infection is the exception and not the rule. Plants may be infected, but the inoculum may or may not become established in the seeds. Among the many factors that influence seed infection and the relationship between plant and seed infection are host and host genotype, pathogen and pathogen genotype, vector(s), and the environment over time.

I. SEED INFECTION

Seed infection is the establishment of a pathogen within any part of a seed, which may occur systemically, either through the vascular system or plasmodesmatic connections or directly through floral infection or penetration of the ovary wall, seed coat, or natural openings. Infection of any single seed may take place by more than one process. For example, bean seed infection by *X. c.* pv. *phaseoli* can occur through the vascular system of a systemically infected plant, by suture infection followed by invasion of the funiculus, raphe, and/or seed coat, or through the micropyle.[1] Entry into seeds from a mother plant may occur by (1) the pathogen colonizing the spikelets from an infected boot leaf while the panicle is still enclosed, (2) infected glumes providing inoculum after panicle emergence, or (3) the pathogen entering through the hilar region from infected vascular tissue.[2] Infection by *Peronospora parasitica* conidia of the stigma and of the ovary wall and establishment of infection in the ovary of radish have been reported.[3] Infection through the stigma and ovary results in embryonic infection. A direct correlation has been found between embryonic infection and seed transmission of the downy mildew fungus.[3]

Phytophthora capsici oospores have been found in seed embryos of squash seeds. The fungus penetrated the cuticle directly and crossed the integument stratum into the embryo.[4] Infection of maize seeds by *F. moniliforme* results in the induction and accumulation of the PRms proteins (pathogenesis-related proteins). The increase is due to an increase in translatable PRms mRNA.[5] PSbMV infection can be latent through the vegetative phase and reach relatively high concentrations in floral parts and seeds. Thus PSbMV may be maintained at high levels in seeds in the absence of symptoms in the plant and without transmission between plants by vectors.[6]

A. Systemic Infection through Flower, Fruit, or Seed Stalk

Seeds may become systemically infected from the flower or fruit stalk (pedicel, peduncle) or through the seed stalk (funiculus).[7,8] Some vascular-infecting bacteria, fungi, and most viruses that infect embryos pass systemically from the infected mother plant to the seeds. In bottle gourd, *F. o.* f. sp. *lagenarium* have invaded seeds from the vascular bundles of fruits.[9] A few fungi infect seeds systemically from the infected mother plant; examples are *D. p.* var. *sojae* in soybean,[10] *F. moniliforme* and *F. oxysporum* in cotton,[11] *F. o.* f. sp. *pisi* in pea,[12] *F. o.* f. sp. *lycopersici* in tomato, *F. o.* f. sp. *matthiolae* in garden stocks,[13] *Plasmopara halstedii* in sunflower,[14] *Pleospora bjorlingii* in sugarbeet,[15] *Septoria glycines* in soybean,[16] and *Verticillium dahliae* in spinach and sugarbeet.[17] *Acremonium strictum* systemically colonize sorghum plants, migrating from roots to grain through the stem and from the mother plant to its progeny through seeds.[18]

A few pathogens infect seeds through vascular elements, such as *X. c.* pv. *campestris* in *Brassica* and *X. c.* pv. *incanae* in garden stocks.[19] Seeds of rice infected by *P. avenae* were harvested from plants on which no symptoms appeared beyond the seedling stage. Thus the bacterium was transmitted to seeds from latently infected plants and survived after infection of the seedling stage.[20] *X. c.* pv. *pruni* moved into plum seeds through fruit stalks after the bacterium gained entry to the vascular system through leaves and shoots (Figure 4-1).[21] *P. s.* pv. *lachrymans* survived in the vascular system of *Cucumis sativus* and invaded plants, fruits, and seeds following its introduction through the base of the stem or flower. Plants and seeds were considered symptomless carriers.[22]

Seed infection by *X. c.* pv. *phaseoli* through the vascular system of bean was shown by the presence of a yellowish discoloration on the hilum.[23] Systemic invasion of developing bean seeds occurred through the vascular system without symptom production from an infected dorsal suture, through the funiculus, into the raphe and then the seed coat.[1] The possibility of systemic invasion of cabbage seed by *X. c.* pv. *campestris* was proposed by Monteith.[24] The bacterium was shown to invade seeds through the peduncle, silique, and funiculus.[25,26] *Erwinia stewartii* invaded maize kernels through the vascular system. Because the pedicel vascular elements terminate in the chalazal region, further progression was accomplished by dissolution of the chalazal areas, resulting in lysigenous cavities filled with bacterial ooze.[27,28] *P. s.* pv. *pisi* infection often took place through sepals, then spread to the peduncle and ultimately killed the flowers, and also caused shriveling of young pea pods. The bacterium migrated into the seeds from pod tissues through the funiculus, then the micropyle, and into the seed coat.[29]

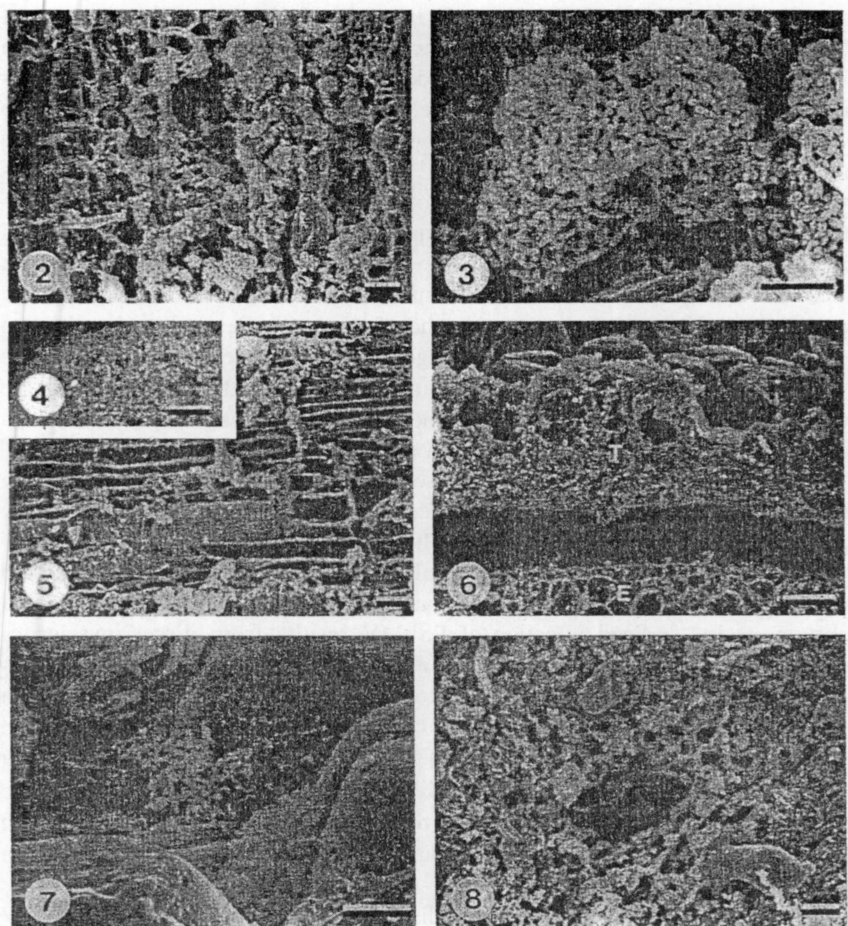

Figure 4-1 Bacteria in fruit stalks and seed of mature plum fruit 10 weeks after inoculation with *Xanthomonas campestris* pv. *pruni* through the stalks. Bacterial colonies were seen in the parenchyma (**2**) and xylem vessels (**5**) of fruit stalks. (**3 and 4**) Higher magnification of portions of (**2**) and (**3**) to show bacterial cells within aggregates. (**6**) Longitudinal section through seed showing the testa (T) and endosperm (E). (**7**) Bacterial clumps on the surface of testa. (**8**) Longitudinal section of testa revealing a vascular channel. Bars represent 6 µm in 4, 7, and 8, 10 µm in 3 and 5, and 20 µm in 2 and 6. (Adapted from Du Plessis, H. J., *J. Phytopathol.*, 130, 37, 1990. With permission.)

Systemic invasion of seeds through the vascular system was shown for *C. m.* subsp. *michiganensis* in tomato,[30] *C. m.* subsp. *sepedonicus* in tomato,[31] *X. o.* pv. *oryzae* in rice,[32] *X. c.* pv. *vesicatoria* in pepper,[33] and *P. s.* pv. *lachrymans* in cucumber.[34]

A majority of the seed-transmitted viruses have been found in the embryo. Seed infection by viruses occurs through systemic invasion of the ovule directly from the infected mother plant. Embryo infection by plant viruses occurs in three ways — systemic infection of the ovule from the mother plant, introduction of the virus into the embryo sac through infected pollen, or more rarely, penetration of the embryo during embryo and seed development.[35] Seed infection depends on the capability of the virus to invade the host plant systemically prior to development of floral primordia, differentiation of male and female gametophytes, and development of callose walls.[35,36] Callose wall formation before virus invasion has resulted in virus-free gametophytes, since it separated the gamotyphytes from parental tissues.[36] For example, BSMV was found in ovules and pollen grains in floral primordia of barley plants infected prior to development of the callose wall.[37,38] Invasion of the ovule, integuments, and nucellus occurred through plasmodesmata that connect mother plants to ovaries and ovules, through the funiculus, and into the cells of integuments and nucellus.[35,39] The callose layer lacks plasmodesmata. Hence, once the callose layer has been formed, the movement of viruses is reduced.[37,38] The tomato bushy stunt virus has been recovered from 50 to 65% of seeds from symptomless, infected tomato fruits.[40] The tobacco streak virus was present in seeds of hybrid strawberry plants where one or both parents were infected.[41] Seed transmission of BSMV in barley occurred because of embryo infection. The virus invaded juvenile primary meristems, which resulted in the infection of both eggs and sperm. Thus infected gametes resulted in infected zygotes and embryos during fertilization and embryogenesis.[39,42] Seed infection of PSbMV results from direct invasion of immature pea embryos, and resistance to seed infection in nonpermissive cultivars probably occurs at this step.[43] Potato progeny were infected by the potato spindle tuber viroid, which were infected through pollen or the ovule.[44-46] Seed transmission of the viroid could be 100% in potato cvs. Katahdin and Russet Sebago when both parents were infected.[45] The rate of seed transmission of the viroid in tomato was 6, 9, or 11% when the female parent, male parent, or both parents were infected, respectively.[46,47] The viroid also was seed transmitted in *Scopolia sinensis*.[48]

Viruses invade seeds through tissue disintegration of the mother plant. TMV has contaminated tomato seeds during the extraction process.[49,50] Some important viruses that infect ovules are the alfalfa mosaic virus in alfalfa,[51] BSMV in barley,[52] bean common mosaic virus in bean,[53] beet cryptic virus in sugar beet,[54] black raspberry latent virus in black raspberry,[55] broad bean stain virus in fava bean,[56] cowpea banding mosaic virus in cowpea,[57] cowpea aphid-borne mosaic virus in cowpea,[58] elm mosaic virus in *Ulmus americana*,[59] broadbean true mosaic virus in fava bean,[55] lettuce mosaic virus in lettuce,[60] Lychnis ringspot virus in *Lychnis divaricata* and *Silene noctiflora*,[61] pea seedborne mosaic virus in pea,[62] Prunus necrotic ringspot virus in *Cucurbita maxima*,[63] squash mosaic virus in *Cucumis melo*,[64] TMV in grapes,[65] tobacco ringspot virus in soybean,[66] tomato ringspot virus in raspberry,[67] tobacco streak virus in strawberry,[41,68] and tomato blackring virus in *Rubus*.[69]

The nematode *Ditylenchus dipsaci* may remain in the vicinity of the hilum or move through the pedicel and placenta and into ovary walls of onion seeds through the funiculus.[70]

B. Penetration through the Stigma

Certain pathogens follow the path of pollen. Fungus spores lodge on the stigma and germinate, and then the germ tube enters the style, infects the ovary, and becomes established in developing seeds. Chlamydospores of *Ustilago segetum* var. *tritici* in barley and wheat were claimed to germinate on style surfaces and with the hyphae passing through pollen tubes into the integuments and embryo.[71] However, this could not be confirmed. In pearl millet, seed infection by *Sclerospora graminicola* took place through airborne sporangia, which produced a germ tube that passed through the stigma and style and infected the ovary.[72] Infection by *Claviceps fusiformis* in pearl millet occurred primarily through stigmas.[73-75] Conidiophores and spores of *Botrytis anthophila* often enveloped anthers and stigmas of red clover flowers and hyphae ramified in the pollen grains, and later conidiophores broke through anther walls. After reaching the stigma, spores germinated with thin hyphae, which traversed the stylar canal and later increased in diameter.[76] Conidia of *A. alternata* lodged on the stigma and germinated, and hyphae entered the ovary through the style and became established in ovules and developing seeds of pepper.[77] Spraying alfalfa flowers with *V. alboatrum* spores resulted in infection of stigmas and styles and presence of hyphae in pollen grains and tubes. Infection of stigmas and styles led to infection of seeds. The fungus appeared latent in the style during all stages of seed development. It was not detected in seeds from pods with infected styles. Under humid conditions, the fungus in the remnant stylar tissue of a mature seed pod was able to colonize the pod and seed coat.[78] Alfalfa pollen was susceptible to infection by *V. alboatrum* in culture. Fungal hyphae penetrated the exine wall with a thin penetration peg without forming appressoria. Penetration occurred more rapidly through the germinative pores than through other parts of the host wall. During the early stage of hyphal invasion, pollen grains showed plasmolysis, breakage of cytoplasmic membranes, and formation of large vacuoles in the cytoplasm. Invading hyphae filled the cell lumen. Eventually, the infected pollen ruptured, leaving a cluster of compact hyphal cells encircled by remnants of the pollen cells walls.[79] Infected alfalfa pollen may sometimes lead to the spread of inoculum to healthy plants by insects. The alfalfa leaf-cutter bee has been used as a pollinator for the production of alfalfa seeds and may be the most important insect in the transmission of the fungus via alfalfa pollen.[79]

Seed infection by many viruses occurs through introduction into the embryo sac by male gametophytes (pollen). If a male gamete carrying the virus unites with an egg cell, the embryo becomes infected. If both male gametes are infected and one unites with the polar nuclei, the endosperm also may be infected.[80] The first experimental evidence of pollen transmission of a virus was demonstrated with bean common mosaic virus in bean.[81] A number of plant viruses have been reported to be pollen transmitted. Yang and Hamilton[82] found virus-like particles in the pollen wall of soybeans infected with the tobacco ringspot virus. Brlansky et al.[83] found BSMV particles in certain gametophytic and sporophytic of barley cells during critical stages of pollination, fertilization, and embryogenesis. Before fertilization, BSMV particles were seen in the pollen tube between the integument and ovary wall, and in the pollen-tube discharge within the degenerating synergid. During and after fertilization, virus particles also were found in the zygote, endosperm, persistent synerigid, and nucellus. During embryogenesis, BSMV particles were evident in the embryo integument and ovary wall. Some particles were scattered throughout the cytoplasm, while others were in mono- or multi-layer aggregates within the cytoplasm. Many particles were associated with spindle or cytoplasmic microtubules or with abnormal plastids. The prevalence of virus particles in cells associated with sexual reproduction suggested that virus nucleoprotein played a role in pollen transmission of BSMV in barley (Figure 4-2).[83] Pollen from virus-infected plants may be infected[82,84] or pollen exine may be contaminated with viruses[85,86] during pollen development, and such pollen may be involved in the spread of disease. Electron microscopy of thin sections of pollinia from *Cymbidium* and *Cattleya* infected with Odontoglossum ringspot virus (ORSV) and *Cymbidium* mosaic virus (CyMV), respectively, showed the presence of particles typical of the respective viruses. About 10% of the pollen grains in pollinia from ORSV-infected orchids were infected, while less than 1% of those from the CyMV-infected orchids contained visible particles.[87] Pollen surfaces contaminated with infectious plant viruses may serve as a vehicle by which certain mechanically transmissible viruses may spread within a plant species.[83] This route of virus spread would be more common and efficient in open- than in self-pollinated plants and could play a role in the seed transmission of some plant viruses. Asparagus is an open-pollinated plant, and asparagus plants could become infected with asparagus virus2 (AV2) by mechanical means via windborne or bee-carried pollen infested with the virus. Infectious particles of AV2 have been detected in pollen and wash solutions of pollen collected from AV2-infected plants. AV2 antigens were localized on the exine of asparagus pollen using ELISA and protein A-linked latex antiserum.[88]

Alfalfa seed infection by the alfalfa mosaic virus originating from pollen, ovules, or both was found in pods and seeds 12–15 d after pollination among healthy and AMV-infected plants; this was before maturation, which was associated with virus inactivation.[89] Prunus necrotic ringspot and prune dwarf viruses occurred in pollen of almond both externally and internally.[90] The attachment of virus particles to the pollen surface was relatively stable.[90] The prunus necrotic ringspot virus (PNRSV) was shown to be borne on the surface of pollen of almond and cherry.[91] PNRSV was pollen borne in cherry and infected trees pollinated by virus-contaminated pollen.[92] Virus particles were localized on the outer walls of some pollen grains. Hamilton et al.[90] using ELISA, showed the presence of PNRSV antigens on the exine of bees and hand-collected sweet cherry pollen.

Transmission of the prune dwarf virus (PDV) by pollen resulted in infection and caused death of peach trees.[93] Transmission electron microscopy revealed virus-like particles in cytoplasm of pollen grains from PDV-infected sweet cherry seeds.[94] The birch strain of cherry leaf roll virus (CLRV) was transmitted through pollen and the ovule to seeds of their natural hosts.[95,96] There is some evidence that pollen infected by CLRV introduced the virus into birch trees.[96,97] ELISA showed that in birch trees infected with the cherry leaf roll virus through pollen

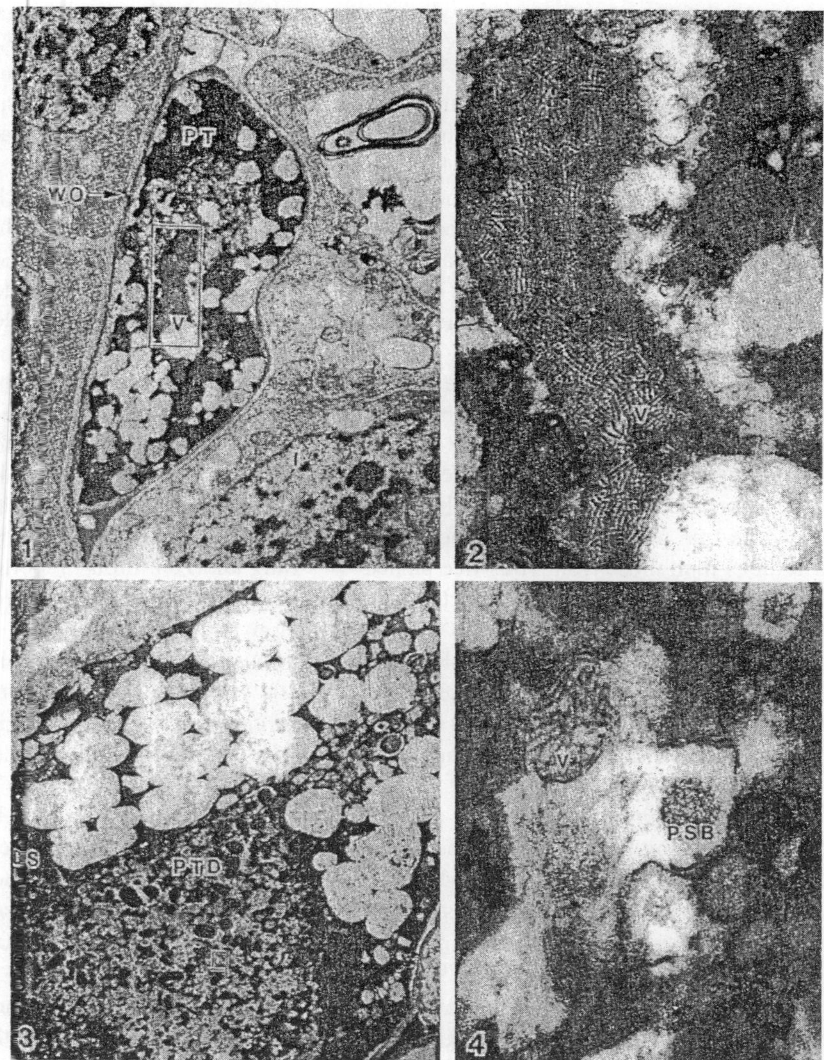

Figure 4-2 Virus particles of barley stripe mosaic in pollinated ovaries prior to fertilization. (**1**) Large aggregate of virus particles (*outlined area*) in a segment of pollen tube that is penetrating between the wall of the ovary and the integument (original magnification × 9880). (**2**) Magnified view of the large aggregate of virus particles shown in the outlined area in (1). The particles are surrounded by dense cytoplasmic material (original magnification × 46,800). (**3**) Pollen tube discharge containing virus particles (*outlined area*) within a portion of a degenerating synergid (original magnification × 3400). (**4**) A small aggregate of virus particles is seen in an enlargement of the outlined area of (**3**). The aggregate appears to be bound by a single membrane. A polysaccharide body also is evident in the contents of the pollen tube discharge (original magnification × 69,300). (Abbreviations used: *DS*, degenerating synergid; *PSB*, polysaccharide body; *PT*, pollen tube; *PTD*, pollen tube discharge; *V*, virus particle; *WO*, wall of ovary.) (From Brlansky, R. H., et al., Can. J. Bot., 64, 853, 1986. With permission.)

into which the antigen was introduced, the virus multiplied within the embryos but not the seed coat or the pericarp/wings.[96] Raspberry bushy dwarf virus was located on and probably in the pollen grains of *Rubus parviflorus*.[98] Sdoodee and Teakle[68] showed that *Thrips tabaci* transmitted tobacco streak virus by feeding on *Chenopodium amaranticolor* in the presence of pollen from infected tomato plants. Presumably, contaminated pollen fell into feeding wounds created by *T. tabaci*, and the virus was released and infected the plant. Strain O of Arracacha virus B was pollen transmitted to seeds produced on healthy potato plants but not to the plants themselves.[99] Squash mosaic virus antigen was detected on unwashed pollen of several cucurbits but not on washed or washed and damaged pollen.[100] Apple mosaic virus was shown to replicate within germinating pollen grains of birch (*Betula pendula*).[101]

The severe strain of potato spindle tuber, chrysanthemum stunt, and cucumber pale fruit viroids were transmitted through tomato seeds and pollen.[102] Viruses have been implicated in altering the development,[103] morphology,[104] or vigor[105] of pollen grains. Germination of pollen from alfalfa mosaic virus-infected alfalfa plants was significantly reduced within and among clones, compared to pollen from virus-free plants. In addition, pollen from infected plants produced shorter germ tubes than those from virus-free plants.[106,107] Yang and Hamilton[82] reported that the role of soybean pollen infected with the tobacco ringspot virus in seed transmission was negligible, due to poor germination and slow elongation of infected germ tubes. Childress and Ramsdell[108] demonstrated that pollen from highbush blueberry infected with blueberry leaf mottle virus had reduced germination and slow tube elongation, and hence could not compete successfully with virus-free pollen. The plant viroids and viruses known to infect or infest pollen are listed in Table 4-1.

Pollen-transmitted viruses also may be ovule transmitted. For example, BSMV is transmitted less through pollen than through ovules in barley. There was more seed infection when both ovules and pollen grains were from an infected plant than when only the ovule was from an infected plant.[149] Alfalfa mosaic virus in alfalfa was transmitted at a much higher frequency through pollen (0.5 to 26.5%) than through ovules (0 to 9.5%).[150]

Infection by the artichoke yellow ringspot virus through pollen was not detected in fava bean seeds, although pollen grains were contaminated by the virus.[151] Potato spindle tuber viroid (PSTVd) was detected in pollen from PSTVd-infected potato cv. Katahdin. True potato seeds from each fruit were 35 to 66% infected with the viroid.[152]

C. Penetration through the Ovary Wall and Seed Coat

Pathogens may directly penetrate the ovary wall or seed coat. This type of infection can be restricted to the seed coat or may extend into the endosperm, cotyledons, or embryo.

Maddox[153] first observed loose smut hyphae in wheat seed embryos. Since then, many workers have demonstrated that barley and wheat embryos were infected during seed development following flower infection. There is no agree-

Table 4-1 Viroids and Viruses Reported to Infect or Infest Pollen

Viroid or virus	Crop	Ref.
Alfalfa mosaic	*Medicago sativa* (alfalfa)	89
Apple mosaic	*Betula pendula* (birch)	101
Arracacha virus B	*Solanum tuberosum* (potato)	99
Asparagus virus 2	*Asparagus officinalis* (asparagus)	109
Artichoke yellow ringspot	*Apium graveolens* (celery)	110
	Chenopodium amaranticolor (chenopodium)	110
	C. quinoa (chenopodium)	110
	Datura stramonium (thornapple)	111
	Foeniculum (fennel)	110
	L. esculentum (tomato)	110
	Nicotiana clevelandii (tobacco)	111
	N. glauca (tobacco)	110
	N. glutinosa (tobacco)	111
	N. tabacum (tobacco)	111
	Petunia hybrida (petunia)	111
	Reseda alba (migonette)	110
Avocado sunblotch	*Persea americana* (avocado)	112
Barley stripe mosaic	*Hordeum vulgare* (barley)	113
Bean common mosaic	*Phaseolus vulgaris* (bean)	81
Bean yellow mosaic	*Melilotus alba* (sweet clover)	114
Beet cryptic	*Beta vulgaris* (beet)	54
Black raspberry latent	*Rubus* (black raspberry)	55
Blackeye cowpea mosaic	*V. unguiculata* (cowpea)	115
Broad bean stain	*Vicia faba* (fava bean)	56
Broad bean true mosaic	*V. faba* (fava bean)	56
Cherry leaf roll	*Prunus* (plum)	59
	Sambucus racemosa (scarlet elder)	116
Cherry raspleaf	*Prunus* (cherry)	117
Chrysanthemum stunt	*L. esculentum* (tomato)	118
Cochorus leaf mosaic chlorosis	*Corchorus capsularis* (jute)	119
Cowpea aphid-borne mosaic	*V. unguiculata* (cowpea)	58
Cowpea banding mosaic	*V. unguiculata* (cowpea)	57
Cowpea severe mosaic	*V. unguiculata* (cowpea)	115
Cucumber green mottle	*Cucumis sativus* (cucumber)	120
Cucumber mosaic	*G. max* (soybean)	121
	Stellaria media (chickweed)	122
Cucumber pale fruit	*L. esculentum* (tomato)	118
Elm mottle	*Syringa vulgaris* (lilac)	123
Grapevine fanleaf	*Vitis vinifera* (grape)	124
Lettuce mosaic	*Lactuca sativa* (lettuce)	60
Lucerne latent (Australian)	*Chenopodium quinoa* (chemopodium)	125
Lychnis ringspot	*Lychnis divaricate* (campion)	61
	Silene noctiflora (catchfly)	61
Maize whiteline mosaic	*Z. mays* (maize)	126
Mulberry ringspot	*Glycine max* (soybean)	127
Onion mosaic	*Allium cepa* (onion)	128
Onion yellow dwarf	*A. cepa* (onion)	129, 130
Pea seedborne mosaic	*Pisum sativum* (pea)	62
Pelargonium zonate spot	*Nicotiana glutinosa* (tobacco)	131
Potato spindle tuber	*Solanum tuberosum* (potato)	118
Prune dwarf	*Prunus avium* (sweet cherry)	132

Table 4-1 Viroids and Viruses Reported to Infect or Infest Pollen (continued)

Viroid or virus	Crop	Ref.
	P. cerasus (sour cherry)	133
	P. dulcis (almond)	134
	P. mahaleb (mahaleb cherry)	132
Prunus necrotic ringspot	*P. avium* (sweet cherry)	132
	P. cerasus (sour cherry)	135
	P. dulcis (almond)	134
	P. mahaleb (mahaleb cherry)	132
Radish yellow edge	*Raphanus sativus* (radish)	136
Raspberry bushy dwarf	*Rubus idaeus* (raspberry)	137
	R. loganbaccus (loganberry)	138
Raspberry ringspot	*Fragaria virginiana* (strawberry)	69
	Rubus (brambles)	69
Southern bean mosaic	*P. vulgaris* (bean)	139
Sowbane mosaic	*Chenopodium hybridum* (sowbane)	140
Spinach latent	*C. quinoa* (chenopodium)	141
	Spinacia oleracea (spinach)	141
Soybean mosaic	*G. max* (soybean)	142
Squash mosaic	*Cucumis melo* (cantaloupe, muskmelon)	64
Tobacco streak	*G. max* (soybean)	142
	Datura stramonium (jimson weed)	143
	Fragaria (strawberry)	144
	L. esculentum (tomato)	145
Tomato blackring	*B. vulgaris* (beet)	146
	Rubus (brambles)	69
Tomato bushy stunt	*Prunus avium* (sweet cherry)	147
Tomato ringspot	*Pelargonium hortorum* (fish geranium)	148

ment on the mechanism of embryo infection. Early investigators suggested that the infective diploid hyphae penetrated between stigma papillae and grew intercellularly through the style into the ovary.[71,154-157] Lang[71] reported that germinating teliospores directly penetrated the ovary wall and grew into the embryo. Many workers could not confirm the stigma as an infection path but reported that hyphae enter the ovary wall through the epidermis.[157-162] Pedersen[159] suggested that the barley embryo secretes a chemotactic substance that attracts fungal hyphae and that infection took place through the barley ovary wall and not the stigma. There was evidence that infection by hyphae of the loose smut fungus in barley and wheat occurred by direct penetration of the ovary wall.[158,162] Teliospores of *U. segetum* var. *tritici* germinated on the stigma/ovary surface and formed dicaryotic promycelia (hyphae), which penetrated the ovary wall. The hyphae, after penetration, traversed the parenchyma, passed along the integuments on the ventral side of the grain, crossed the endosperm at the base of the grain, and entered the scutellum. From the scutellum, they permeated through most parts of the embryo, being found in abundance in the hypocotyl and growing-point regions. The hyphae primarily were intracellular in the pericarp and testa, and intercellular in the aleurone endosperm, scutellum, and embryo of barley.[163]

The soybean seed coat consists of a cuticle and three cell layers, epidermis or palisade, hypodermis or hourglass, and endodermis, which is made up of the

Figure 4-3 Cross section of a soybean (*Glycine max*) seed coat infected with *Colletotrichum truncatum* (anthracnose) showing an acervulus with setae (*s*) and hyphae (*arrows*) from initial colonization (*p*) in the aleurone layer. (From Rodriguez-Marcano, A. and Sinclair, J. B., *Plant Dis. Rep.*, 62, 873, 1978. With permission.)

aleurone layer and parenchymatous cells.[164] The micropyle is distal to the peduncle, and the raphe is proximal to the peduncle. The peduncle is an area of structural weakness because the hourglass cells in this region are absent or small. The hilum is a complex structure with a tracheid bar extending either the length of the hilum, or in two parts, one near the micropyle and the other near the chalaza.[165] Routes by which pathogens invade soybean seeds vary with each pathogen. Wolf and Baker[164] suggested that breaks in the soybean seed coat, which are genetically controlled,[166] provided sites for entrance and colonization by fungi.

Colletotrichum truncatum germinate on the soybean seed surface and hyphae penetrate the palisade layer within 24 to 30 h. After 48 h, hyphae are evident throughout the palisade layer and grow down through the hypodermis and into the protein-rich aleurone layer. After 72 h, fungal hyphae form dark-colored acervuli on the epidermal surface (Figure 4-3).[167] *C. lagenarium* colonized watermelons rind splits and invaded the epidermal layer of palisade cells and sclerenchyma layers of the seed coat. Hyphal strands extended into parenchyma and other seed coat tissues.[168] Hyphae of *Acroconidiella eschscholtziae* penetrated poppy capsules at the point of attachment and ramified throughout the fruit interior. Later they either adhered to the surface or penetrated the seed coats of mature seeds.[169]

In soybean, *Alternaria* penetrated either directly through the culticle or hilum tissues or indirectly through the micropyle[170] or pits where the seed coat was weak.[171] Hyphae of *C. kikuchii* and *Phomopsis* entered soybean seeds through the hilar groove into the tracheids and aggregated in stellate parenchyma. They penetrated also through epidermal pores or cracks in the seedcoat. Both fungi invaded the embryo, endosperm, and seedcoat tissues. *Phomopsis* hyphae were commonly found in embryonic tissues, in the space between the seed coat and embryo, and between the cotyledons. In general, *C. kikuchii* colonized the seed coat, were occasionally found in the cotyledons, and were rarely in the hypocotyl–radicle axis. It formed hyphal aggregates in the seed coat and caused necrosis of cotyledonary cells and vascular elements.[172] Soybean seed infection by *C. kikuchii* began 30 d after full bloom and continued to the yellow-pod stage, but symptoms were not apparent until the yellow-pod stage, regardless of the time of infection.[173] *M. phaseolina* penetrated and colonized soybean seeds without inducing symptoms, but formed sclerotia in asymptomatic seeds when conditions were favorable for germination. After 3 to 4 d, sclerotia were formed in cotyledons of asymptomatic seeds, and after 4 to 5 d, in the hypocotyl radicle axis.[174] *Phomopsis longicolla* invaded soybean seeds indirectly through the funiculus and hilum or directly through the seedcoat.[175]

Rice seeds were infected by *Fusarium moniliforme* at flowering stage, and in the case of severe infection, kernels exhibited a reddish coloration due to conidia formation. The optimum stage for infection was at flowering; then infection decreased for 3 weeks.[176] Conidia production and ascospore discharge on diseased culms occurred in the field at flowering and with maturity, which enhanced seed infection and contamination.[177] Embryos also may carry inoculum.[178] *S. graminicola* mycelium invaded all parts of pearl millet seed tissues, but growing-on tests indicated that only mycelium present in the embryo caused systemic infection in seedlings and plants. The embryo was the only living seed part and therefore the only tissue on which an obligate parasite like *S. graminicola* could survive during storage.[179] Infection by *Curvularia lunata* and *F. moniliforme* occurred in 3 d and by *P. sorghina* in 8 d after anthesis in sorghum. *C. lunata* infection progressed to the pedicel and gynaecium at 10 d after anthesis and entered the embryo and endopserm through the placental sac and modified endosperm cells. *F. moniliforme* infected the pedicel and basal ovary at 5 to 10 d after anthesis, and mycelia produced clumps between the ovary wall and aleurone layer and directly penetrated the endosperm and embryo. *Phoma sorghina* colonized the ovary wall through the hylar region but did not colonize other seed parts.[180]

The seed ball of sugar beet consists of a seed cap and true seed with a pericarp. *C. dematium* f. sp. *spinaciae* is seedborne in sugar beet, and its flowers and seed balls are susceptible to infection at all developmental stages. When infection occurred at early-flowering, germ tubes produced appressoria and an infection peg, which penetrated the cuticle within 3 to 4 d. Mycelia spread inter- and intracellularly. Infected cells collapsed, and lesions developed within 5 d. Mycelia were found on the surface of blackened seed caps or pericarps and in the pericarp. In severe infection, mycelia were found between the seed coat and cotyledons. When infection occurred at late-flowering, mycelia were found on the seed surface and less frequently inside the seed ball. Such hyphae were found on the periphery of the apical pore or inside the seed coat (Figure 4-4).[181]

The mycelium of *A. sesami* occurred in the subepidermal seed coat layers and occasionally in the endosperm and embryo of sesame. In severely infected sesame seeds, it invaded all parts and sporulated within the seed. A heavy aggre-

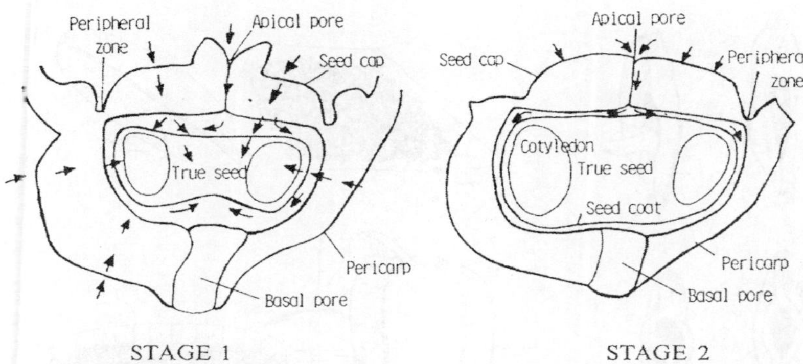

Figure 4-4 Diagram of a transverse section of sugar beet seed ball. Arrows indicate the possible routes of invasion of *Colletotrichum dematium* f. sp. *spinaciae*. (From Chikuo, Y. and Sugimoto, T., *Ann. Phytopathol. Soc. Jpn.*, 55, 404, 1989. With permission.)

gate of mycelium in the hilum region suggested that it penetrated at this point, whereas the thick inner cuticle of the seed coat and outer cuticle of endosperm appeared to resist penetration.[182] Fruits of carrot were vulnerable to *A. dauci* from early development to maturity. Fruits infected early contained seed structures completely colonized by the fungus. Tissues and spiny prickles of the vallecular ridges of fully developed fruits were the most common sites of infection, colonization, and sporulation. The fungus was confined to the outer tissues of dried pericarps and did not penetrate the seed coat and endosperm.[183] *Cylindrocladium crotalariae* hyphae ramified intra- and intercellularly throughout the testae of discolored peanut seeds. In seeds with a dark brown testa, hyphae were observed in cotyledonary tissue.[184] *Colletotrichum capsici* invaded pepper seeds by two pathways: mycelium in the inner surface of diseased fruit invaded seeds through the seed coat and colonized the outer layer of the endosperm or in the placenta remnant invaded the outer layer of the endosperm.[185]

If the fruits of cucumber, eggplant, pepper, tomato, zinnia, etc., were infected, seed infection resulted from either inoculum penetration through the funiculus or seed coat, or seeds became contaminated during the extraction process. *Rhizoctonia solani* spread through the fruits of eggplant, pepper, and zinnia, and then infected seeds from the placenta; it then penetrated the developing ovule directly or young seeds without a lignified seed coat (Figure 4-5). After seed coat lignification, the hyphae invaded through the funiculus. As the endosperm cuticle formed, infection was restricted. The point of union with the funiculus was the last to cutinize, and until it was sealed off, it provided a bridge for penetration of the endosperm.[186]

Fungi known to penetrate through the ovary wall and seed coat are *Alternaria brassicae* and *A. brassicicola* in cabbage and canola,[187] *A. alternata* in sunflower[188] and wheat,[189] *A. sesami* in *Sesamum indicum,*[190] *Aureobasidium caulivora* in *Trifolium incarnatum,*[191] *Claviceps fusiformis* in pearl millet,[74,192]

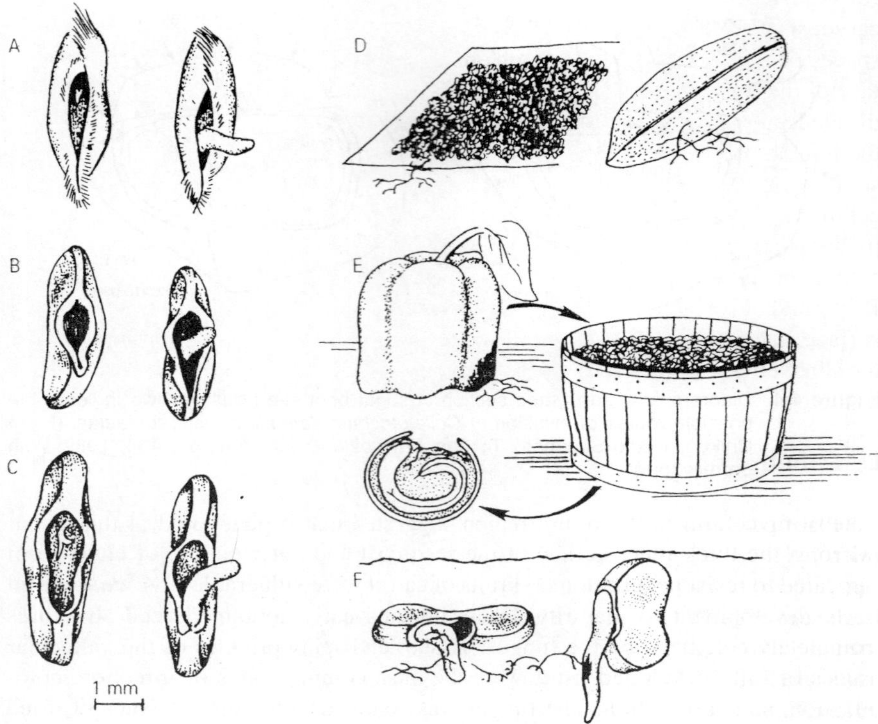

Figure 4-5 Transmission of *Rhizoctonia solani* in vegetable seeds. (**A** to **C**) *left*: Natural openings in seed coats of (**A**) pepper (*Capsicum frutescens*), (**B**) eggplant (*Solanum melongena*), and (**C**) tomato (*Lycopersicon esculentum*), through which infection occurs; and (**A** to **C**), *right*: beginning of germination of the seeds. (**D**) Invasion of zinnia (*Zinnia elegans*) flowers piled on canvas on infested soil. *Rhizoctonia solani* grows over or through the canvas into the seed. (**E**) Pepper fruits, infected from contact with infested soil, are chopped up and placed in barrels to ferment prior to separation of seeds; *Rhizoctonia* permeates the mass and penetrates the seeds. (**F**) *Rhizoctonia* grows out from an infected seed when planted and attacks neighboring seeds (or seedlings) as well as infests the soil. (Adapted from Baker, K. F., *Seed Biology*, Vol. 2, Kozlowski, T. T., Ed., Academic Press, New York, 1972, 330.)

Cochliobolus sativus in barley,[193] *Fusarium moniliforme* in sorghum,[194] *F. oxysporum* f. sp. *vasinfectum* in cotton,[195] *Phoma rabiei* in gram,[196] *S. graminicola* in pearl millet,[72] and *V. alboatrum* in spinach.[197]

Bacteria invade seeds through the seed coat or ovule wall. *X. campestris* pv. *malvacearum* entered cotton seeds through the basal end at the chalaza.[198] Similarly, infection of tomato seed by *C. m.* subsp. *michiganensis* began at the chalazal end and continued into the innermost cells of the seed coat.[199] In castor bean, *X. campestris* pv. *ricini* spread from the capsule through the pericarp into the seed integument.[200] *P. syringae* pv. *phaseolicola* penetrated from external pod lesions to underlying bean seeds.[201]

P. syringae pv. *japonica* first infected the palea and lemma, then invaded the caryopsis of rice seeds through the funiculus and multiplied in the intercellular spaces of the caryopsis parenchyma.[202] *Pseudomonas glumae* colonized the surface of the basal portion of the lodicule and inner surface of the lemma. The cells invaded intercellular spaces of the outer epidermis and spongy parenchyma of the lemma of rice seeds.[203] *Pseudomonas plantarii* entered rice glumes through stomata on the inner and outer epidermis. It multiplied and spread in the intercellular spaces of the parenchyma beneath the inner epidermis.[204] *X. campestris* pv. *vesicatoria* penetrated flowers and resulted in bacterial proliferation in warts of pepper. Bacterial multiplication was observed in warts of fruits at all ages. Bacterial cells were bound to the fruit surface by a fibrillar material. On the fruit surface, bacteria multiplied in small aggregates submerged in slime. At later stages the slime covered the entire surface in young fruits, and the bacteria were detected in the ovary of inoculated fruits that did not shed.[205]

D. Penetration through Natural Openings and Injuries

Plant pathogens gain entry through natural openings, such as the hilum and micropyle, or through threshing injuries or wounds. Histopathological and scanning electron microscope studies showed *Cercospora sojina* hyphae within soybean seed coat tissues. The fungus penetrated seeds indirectly through pores and the parechymatous seed coat tissues and cracks. Hyphal mats and aggregates were associated with the hyphae. Hyphal aggregates were abundant in the hilar region, moderately common in seed coat layers, and occasionally found on the seed surface and in the space between the seed coat and embryo. It was rarely observed in the hypocotyl–radicle axis. Fungal infection was not found in cotyledons (Figure 4-6).[206] *Phoma betae* invaded sugar beet seeds through the basal pore in the pericarp, the peripheral zone forming the slit between the seed cap and pericarp, and the apical pore in the seed cap.[207] *Alternaria sesamicola* penetrated the hilum of sesame seeds, but the endosperm cuticle resisted penetration.[208] The hilum was highly absorptive and provided entry for the bacteria.[80] *X. campestris* pv. *phaseoli* invaded bean seeds through the micropyle from injured pods.[80] Similarly, infection of cucumber seeds by *P. syringae* pv. *lachrymans* took place from the funiculus through the micropyle.[34] Seed coat cracks or injuries enhanced seed invasion by *P. syringae* pv. *phaseolicola* in beans,[80] *X. campestris* pv. *manihotis* in cassava,[209] and *X. campestris* pv. *translucens* in barley and wheat.[210]

Erwinia herbicola entered the lemma and palea of rice seeds through stomata and multiplied in the intercellular spaces of the parenchyma. Stomata were open mainly on the inner surface of the lemma and palea, with a few on the outer surface of the lemma. They were connected to each other through the intercellular spaces of parenchyma. This infection site was not specific for *E. herbicola*.[211] The bacterial seed decay caused by *P. glumae* and the bacterial seedling blight caused by *P. plantarii* showed the same infection pattern in rice seeds.[212] *P. syringae* pv. *japonica* in wheat and barley seeds first infected the lemma and palea through stomata, then invaded the caryopsis through the funiculus and

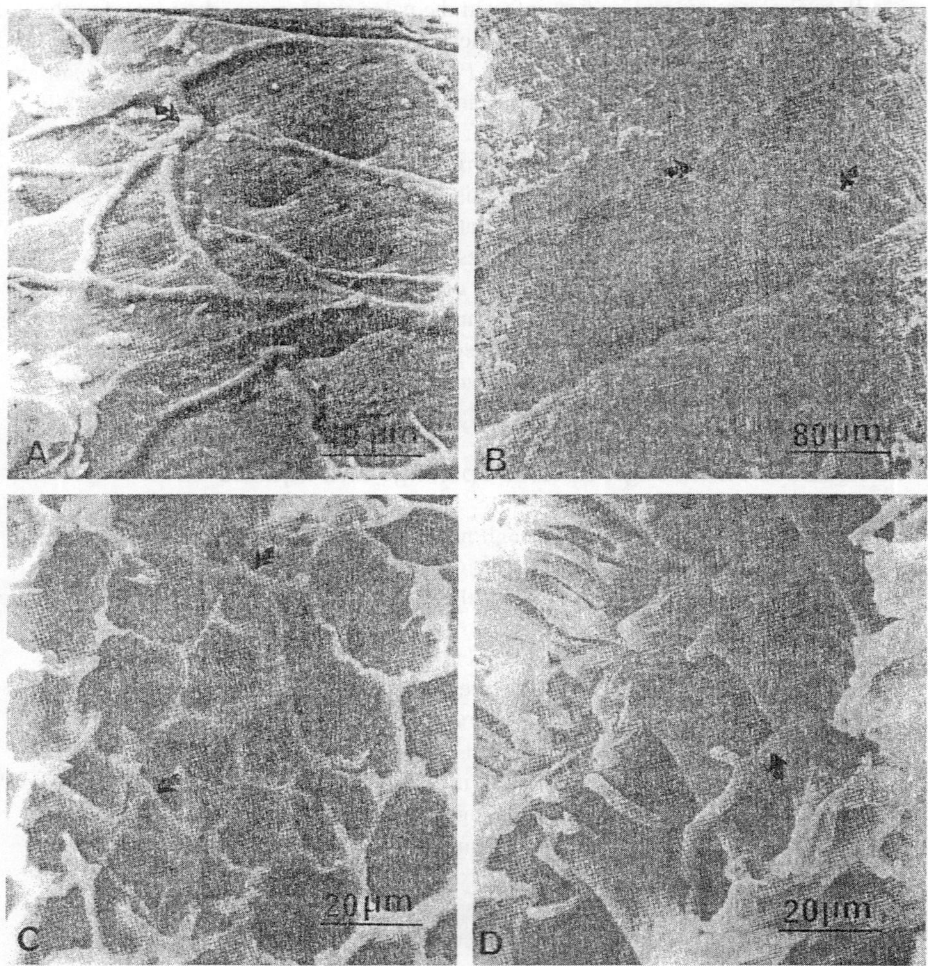

Figure 4-6 Scanning electron photomicrographs of soybean (*Glycine max*) seeds from plants inoculated with isolate LA1 of *Cercospora sojina*. (**A**) Hyphae (*arrows*) on the seed coat surface and hyphae penetrating through seed pores. (**B**) Hyphae (*arrows*) in hilar region and in hilar tracheids. (**C**) Tangential section showing hyphae (*arrows*) colonizing the honeycomblike palisade cell layer. (**D**) Cross section of a seed coat showing hyphae (*arrow*) in the hypodermal (hourglass) cell layer. (From Singh, T. and Sinclair, J. B., *Phytopathology,* 75, 185, 1985. With permission.)

multiplied in the intercellular spaces of the caryopsis. No bacteria were observed in the vascular bundles.[202]

II. SEED INFESTATION OR CONTAMINATION

Seed infestation or contamination refers to the passive association of a pathogen with seeds. The pathogens may adhere to the surface or be mixed with seeds.

Seed infestation is the exclusive means of survival of certain seedborne pathogens. Seeds may be infested at any time during harvest, extraction, threshing, or processing until packaging. For example, the number of *A. besseyi* nematode propagules in rice seeds in different cultivars varied from 0 to 900 per 100 seeds.[213] Pathogens may infest or contaminate seed in one or more of the following ways.

A. Pathogens Adhering to the Seed Surface

Infective propagules of pathogens may adhere to seed surfaces during harvest or postharvest operations. Fungal spores, such as chlamydospores, oospores, teliospores, urediospores, bacterial vegetative cells, or virus particles, have been found on seed surfaces. During mechanical harvest and transport of wheat seeds, the fragile bunt balls of *Tilletia controversa* ruptured and released millions of dust-like spores. In the field, those were carried by the wind and eventually settled on soil and became a source of primary inoculum. Uncontamined wheat seeds became contaminated with such spores during handling and transport.[214] Pearl millet seeds were contaminated with spore balls of *Maeziomyces bullatus* from ruptured sori at threshing.[215]

Spores of the following fungi can be carried on seed coat surfaces: *A. alternata* in flax and soybean, *A. brassicae* and *A. brassicicola* in crucifers, *A. longipes* in tobacco, *A. radicina* in carrot, *Phoma rabiei* in chickpea, *Bipolaris sorokiniana* and *B. oryzae* in rice, *Drechslera avenae* in oats, *F. oxysporum* f. sp. *callistephi* in China aster, *Melampsora lini* in flax, *M. bullatus* in pearl millet, teliospores of *Tilletia indica* in wheat, *Peronosclerospora sorghi* in sorghum, *Peronospora antirrhini* in snapdragon, *P. manshurica* in soybean, *P. tabacina* in tobacco, *P. valerianellae* in *Valerianella lucusta*, *Phoma medicaginis* var. *pinodella* in pea, *Phytophthora phaseoli* in bean, *Protomyces macrosporus* in coriander, *Puccinia calcitrapae* var. *centaureae* in safflower, *Pyricularia oryzae* in rice, *Rhizoctonia solani* in eggplant, pepper and tomato, *Sclerospora graminicola* in pearl millet, *Sclerophthora macrospora* in finger millet, *Sporisorium cruentum* and *S. holci-sorgi* in sorghum, *Tilletia barclayana* in rice, *T. caries, T. controversa, T. laevis* and *Urocystis agropyri* in wheat, *Uromyces betae* in sugar beet, *Ustilago segetum* var. *segetum* in barley, *U. segetum* var. *avenae* and *U. segetum* var. *segetum* in oat, and *U. zeae* in maize.

A number of bacteria, such as the following, contaminate seed surfaces: *C. flaccumfaciens* pv. *flaccumfaciens* in bean, *P. syringae* pv. *phaseolicola* in bean, *P. s.* pv. *lachrymans* in cucumber, *P. s.* pv. *tomato* in tomato, *C. m.* subsp. *michiganensis* in pepper, *X. c.* pv. *campestris* in cabbage, and *X. c.* pv. *vesicatoria* in tomato. The sticky cells of *Rhodococcus fascians* adhered to sweet pea seeds under moist conditions.[216] Contaminated bean seeds were a source of primary inoculum of *X. c.* pv. *phaseoli.* The surface population of these bacteria ranged from 0×10^4 bacteria per seed.[217] Soybean seeds became contaminated by *X. campestris* pv. *glycines* during threshing. Even though the pathogen population was small, epidemic population levels were established on seedlings when these seeds were planted in the field.[218]

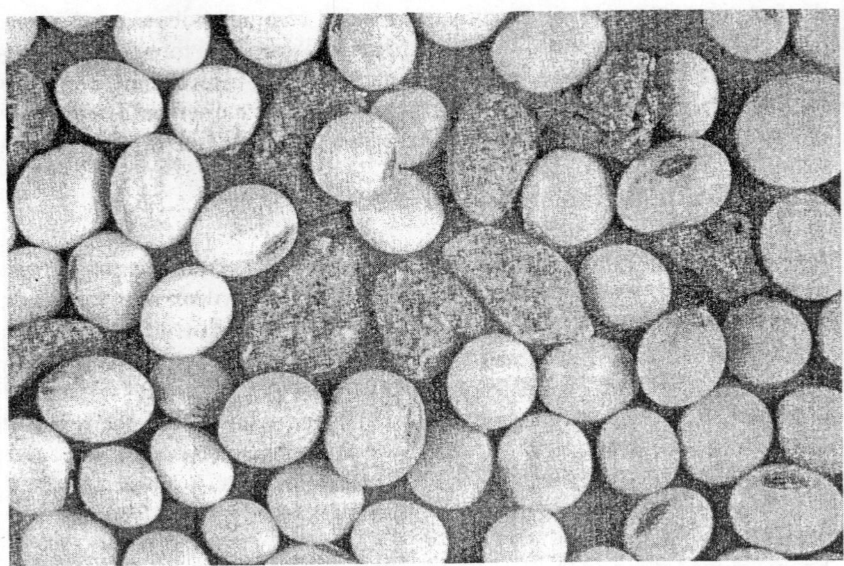

Figure 4-7 Soil peds containing cysts of *Heterodera glycines* (soybean cyst nematode) mixed with a soybean (*Glycine max*) seed lot. (From J. B. Sinclair, Ed., *Compendium of Soybean Diseases*, APS Press, St. Paul, MN, 1982, 48. With permission.)

Seed-contaminating nematodes such as *Aphelenchoides besseyi* larvae in rice seeds have been found coiled inside the palea and on the surface of lodicules.[219] Fukano[220] recorded as many as 1241 and 132 such nematodes per seed in severely and slightly affected rice cultivars, respectively, of which about 90% inhabited the husk inner surface and the remainder the kernels. *Ditylenchus dipsaci* larvae in alfalfa and onion,[7] *A. arachidis* larvae within peanut testa,[221] and the fourth-stage larvae adults of *A. ritzemabosi* and *A. blastophorus* lay between the seed coat and embryo of China aster seeds.[222] The desiccated larva of *D. dipsaci* adhered to the seed surfaces of clover, *Dipsacus fullonum,* and alfalfa.[223]

Infected or contaminated pulp may be mixed with seeds during the extraction process. Examples are *Fulvia fulva* in tomato,[224] *F. o.* f. sp. *lycopersici* in tomato,[225] *C. michiganensis* subsp. *michiganensis* in tomato,[226] and *P. syringae* pv. *lachrymans* in cucumber.[227] TMV in tomato is carried in the gelatinous matrix on tomato seed testa during seed extraction.[80]

B. Concomitant Seed Contamination

Concomitant contamination refers to the mixture of inoculum with seeds in the form of plant pathogen propagules, namely sclerotia, nematode galls, or infected plant debris, or infested soil clods/particles mixed with seeds (Figure 4-7).

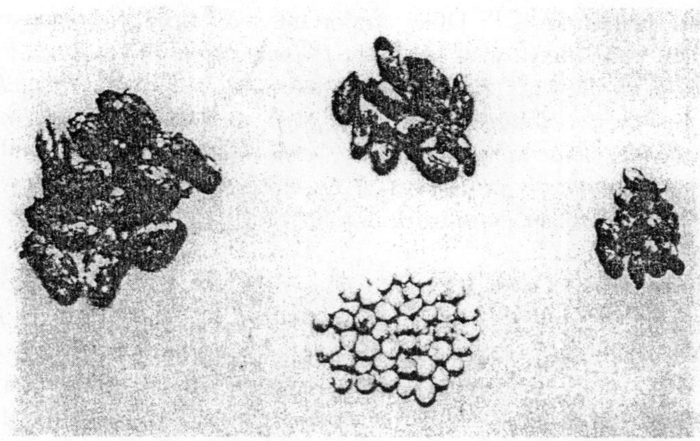

Figure 4-8 Various sizes of sclerotia of *Claviceps fusiformis* compared to pearl millet seeds (*top*). (From Thakur, R. P., et al., *Phytopathology*, 74, 201, 1984. With permission.)

1. Pathogenic Structures

Fungal fruiting structures, such as sclerotia, can be mixed with seeds during threshing. Examples are *Claviceps fusiformis*, cause of ergot of pearl millet; *C. purpurea,* cause of ergot of rye; *Sclerotinia sclerotiorum,* cause of rots of crucifers and legumes; and *S. trifoliorum* of clovers. At threshing *C. fusiformis* sclerotia were mixed with pearl millet seeds. Sclerotia also survived in soil (Figure 4-8).[228] *Anguina tritici* nematode galls can be mixed with wheat seeds. The outer wall of the galls was undifferentiated and compressed with lignification of disorganized cells, compared to the well-formed and fully expanded outer wall of healthy seeds. In addition, the inner cells of the galls were destroyed and the space filled with larvae.[229] The mean number of adults per gall ranged from 10 to 80, but as many as 283 from a single large-size gall have been detected.[230] Another example was the galls of *A. agropyronifloris* and *A. agrostis* mixed with grass seeds. *C. tritici* was carried as a contaminant on the surface of galls of *A. tritici* in wheat.

2. Mixed with Infected Plant Parts

Straw or pieces of infected plant parts mixed during harvesting and threshing can carry pathogens in a seed lot. For example, the teliospores of *Melampsora lini* were carried in infected chaff mixed with flax seeds[231] and those of *Puccinia malvacearum* on bits of bracts mixed with hollyhock seeds. *Botrytis cinerea, B. fabae, Fusarium oxysporum, Phoma pinodella*, and *Stemphylium botryosum* were mixed with plant residues attached to fava bean seeds.[232] The peduncle and the vascular tissues of tomato fruit and seeds were colonized by *F. oxysporum* f. sp. *lycopersici.* The percentage colonization of lower fruit clusters was higher than that of upper ones. The pathogen was carried on the seed surface in pulp residues adhering to seed surfaces.[233] Other important seedborne pathogens carried in infected plant parts mixed with seeds are *Phoma rabiei* in chickpea,[234] *Colletotrichum trifolii* in clover,[235] *F. o.* f. sp. *lycopersici* in tomato, *Rhynchosporium secalis* in barley,[236] *Sclerophthora macrospora* in oats,[237] *Septoria apiicola* in flax,[238] *Uromyces viciae-fabae* in lentil,[239] *Verticillium alboatrum* in alfalfa,[240] *X. campestris* pv. *manihotis* in cassava,[209] *C. michiganensis* subsp. *insidiosus* in alfalfa,[7] *Erwinia c.* subsp. *carotovora* in tobacco,[241] and *P. syringae* pv. *phaseolicola* in bean.[242]

Nematodes carried with seeds are *D. dipsaci* in alfalfa and clover[243] and *Aphelenchoides ritzemabosi* and *A. blastophorus* in China aster.[222] *D. dipsaci* occurs in a large number of tissues of many plants with inflorescences or hollow-flowering stems and remains in a viable condition mixed with onion seeds. As these parts dry out, a large number of nematodes coil in upon the dry tissues, which forms small dry fragments that accompany the seeds as they fall out of the seed capsules. A seed lot may carry the nematode within seeds and in the debris accompanying.[80]

3. Soil

Contaminated soil may be mixed with seeds. Examples are *Plasmodiophora brassicae* in turnip,[244] *F. solani* f. sp. *phaseoli* in bean,[245] *F. oxysporum* f. sp. *cumini* in cumin,[80] *Macrophomina phaseolina* in soil in kidney beans,[246] *Phoma pinodella* in pea,[80] and *R. fascians* in nasturtium and sweet pea.[7] Small clumps of soil carried *M. phaseolina, Pythium ultimum*, and *R. solani* along with chickpea seeds.[247]

Heterodera glycines cysts were carried in seed bags either loose with soybeans or in soil peds, which are small irregular bits of soil formed during harvest (Figure 4-7). The number of cysts per soil ped varied from one to four.[248,249]

REFERENCES

1. Zaumeyer, W. J., Seed infection by *Bacterium phaseoli*, *Phytopathology*, 19, 96, 1929.
2. Baker, K. F. and Smith, S. H., Dynamics of seed transmission of plant pathogens, *Annu. Rev. Plant Pathol.*, 4, 311, 1966.
3. Jang, P. and Safeeulla, K. M., Mode of entry, establishment and seed transmission of *Peronospora parasitica* in radish, *Proc. Indian Acad. Sci., Plant Sci.*, 100, 369, 1990.
4. Melo, M. B. de, Matsuoka, K., and Ansari, C. V., Histopathological examination of natural and induced infections by *Phytophthora capsici* on squash seeds, *Fitopatol. Brasileira,* 17, 96, 1992.
5. Cordero, M. J., Raventos, D., and Sequndo, B. San., Induction of PR proteins in germinating maize seeds infected with the fungus *Fusarium moniliforme, Physiol. Mol. Plant Pathol.,* 41, 189, 1992.

6. Ligat, J. S. and Randles, J. W., An eclipse of pea seedborne mosaic virus in vegetative tissue of pea following repeated transmission through the seed, *Ann. Appl. Biol.*, 122, 39, 1993.
7. Neergaard, P., *Seed Pathology,* Vols. 1 and 2, Macmillan, London, 1977, 1187.
8. Wallen, V. R., Host–parasite relations and environmental influences in seed-borne diseases, *Symp. Soc. Gen. Microbiol.,* 14, 187, 1964.
9. Kuniyasu, K. and Kishi, K., Seed transmission of Fusarium wilt of bottle gourd, *Lagenaria siceraria* Standle, used as rootstock of watermelon. II. The seed infection course from the infected stem of bottle gourd to the fruit and seed, *Ann. Phytopathol. Soc. Jpn.,* 43, 192, 1977.
10. Crittenden, H. W., Increase of *Diaporthe phaseolorum* var. *sojae* on soybean pods due to corn earworm, *Phytopathology*, 58, 883, 1968.
11. Rudolph, B. A. and Harrison, G. J., The invasion of the internal structure of cotton seed by certain *Fusaria, Phytopathology,* 35, 542, 1945.
12. Macneill, B. H. and Howard, H., A pea wilt disease new to Ontario, *Proc. Can. Phytopathol. Soc.,* 26, 13, 1959.
13. Baker, K. F., A problem of seedsmen and flower growers — seedborne parasites, *Seed World*, 70, 38, 1952.
14. Cohen, Y. and Sackston, W. E., Seed infection and latent infection of sunflowers by *Plasmopara halstedii, Can. J. Bot.*, 52, 231, 1974.
15. Osinska, B., The importance of *Pleospora björlingii* Byford (*Phoma betae* Frank) in the occurrence of black leg of sugarbeet, *Hodowia Rosl. Aklim Naslinn.*, 23, 133, 1979.
16. MacNeal, B. and Zalasky, H., Histopathological study of host–parasite relationships between *Septoria glycines* Hemmi and soybean leaves and pods, *Can. J. Bot.,* 35, 501, 1957.
17. Van der Spek, J., Internal carriage of *Verticillium dahliae* by seeds and its consequences, *Meded. Fac. Landbouwwet Rijksuniv.* (Gent.), 37, 567, 1972.
18. Bandopadhyay, R., Mughogho, L. K., and Satyanarayana, M. V., Systemic infection of sorghum by *Acremonium strictum* and its transmission through seed, *Plant Dis.*, 71, 647, 1987.
19. Schuster, M. L. and Coyne, D. P., Survival mechanisms of phytopathogenic bacteria, *Annu. Rev. Phytopathol.*, 12, 199, 1974.
20. Shakya, D. D., Chung, H. S., and Vinther, F., Transmission of *Pseudomonas avenae* the cause of bacterial stripe of rice, *J. Seed Technol.,* 116, 92, 1986.
21. Du Plessis, H. J., Systemic invasion of plum seed and fruit by *Xanthomonas campestris* pv. *pruni* through stalks, *J. Phytopathol.*, 130, 37, 1990.
22. Kritzman, G. and Zutra, D., Systemic movement of *Pseudomonas syringae* pv. *lachrymans* in the stem, leaves, fruits and seeds of cucumber, *Can. J. Plant Pathol.,* 5, 273, 1983.
23. Burkholder, W. H., The bacterial blight of the bean: a systemic disease, *Phytopathology,* 11, 61, 1921.
24. Monteith, J., Seed transmission and overwintering of cabbage black rot, *Phytopathology,* 11, 53, 1921.
25. Cook, A. A., Larson, R. H., and Walker, J. C., Relation of the black rot pathogen to cabbage seed, *Phytopathology,* 42, 316, 1952.
26. Walker, J. C., The mode of seed infection by the cabbage blackrot organism, *Phytopathology*, 40, 30, 1950.

27. Ivanoff, S. S., Stewart's wilt disease of corn, with emphasis on the life history of *Phytomonas stewarti* in relation to pathogenesis, *J. Agric. Res.,* 47, 749, 1933.
28. Schuster, M. L., Hoff, B., Mandel, M., and Lazar, I., Leaf freckles and wilt, a new corn disease, *Proc. Annu. Corn Sorghum Res. Conf.* (Chicago), 27, 176, 1973.
29. Skoric, V., Bacterial blight of pea: overwintering, dissemination, and pathological histology, *Phytopathology,* 17, 611, 1927.
30. Bryan, M. K., Studies on bacterial canker of tomato, *J. Agric. Res.,* 41, 825, 1930.
31. Larson, R. H., The ring rot bacterium in relation to tomato and eggplant, *J. Agric. Res.,* 69, 309, 1944.
32. Durgapal, J. C., Singh, B., and Pandey, R. K., Mode of infection of rice seeds by *Xanthomonas oryzae, Indian J. Agric. Sci.,* 50, 624, 1980.
33. Crossan, D. F. and Morehart, A. L., Isolation of *Xanthomonas vesicatoria* from tissues of *Capsicum annuum, Phytopathology,* 54, 358, 1964.
34. Naumann, K., On the occurrence of bacteria in cucumber seeds from fruits infected with *Pseudomonas lachrymans, Phytopathol. Z.,* 48, 258, 1963.
35. Bennett, C. W., Seed transmission of plant viruses, *Adv. Virus Res.,* 14, 221, 1969.
36. Mandahar, C. L., Virus transmission through seed and pollen, in *Plant Diseases and Vectors,* Maramorosch, K. and Harris, K. F., Eds., Academic Press, New York, 1981, 241.
37. Carroll, T. W. and Mayhew, D. E., Anther and pollen infection in relation to the pollen and seed transmissibility of two strains of barley stripe mosiac virus, *Can. J. Bot.,* 54, 1604, 1976.
38. Carroll, T. W. and Mayhew, D. E., Occurrence of virions in developing ovules and embryo sacs of barley in relation to the seed transmissibility of barley stripe mosaic virus, *Can. J. Bot.,* 54, 2497, 1976.
39. Carroll, T. W., Seed-borne viruses: virus–host interactions, in *Plant Diseases and Vectors,* Maramorosch, K. and Harris, K. F., Eds., Academic Press, New York, 1981, 293.
40. Tomlinson, J. A. and Faithfull, E. M., Studies on the occurrence of tomato bushy stunt virus in English rivers, *Ann. Appl. Biol.,* 104, 485, 1984.
41. Johnson, H. A., Jr., Converse, R. H., Amorao, A., Espeio, J. I., and Frazier, N. W., Seed transmission of tobacco streak virus in strawberry, *Plant Dis.,* 68, 390, 1984.
42. Carroll, T. W., Hordeiviruses biology and pathology, in *The Plant Viruses,* Vol. 2, Van Regenmortel, M. H. V. and Fraenkel Corfat, H., Eds., Plenum Press, New York, 1986, 373.
43. Wang, D. and Maule, A. J., Early embryo invasion as a determinant in pea of the seed transmission of pea seed-borne mosaic virus, *J. Gen. Virol.,* 73, 1615, 1992.
44. Fernow, K. H., Peterson, L. C., and Plaisted, R. L., Potato spindle tuber virus in seeds and pollen of infected potato plants, *Am. Potato J.,* 47, 75, 1970.
45. Hunter, D. E., Darling, H. M., and Beale, W. L., Seed transmission of potato spindle tuber virus, *Am. Potato J.*, 46, 247, 1969.
46. Singh, R. P., Seed transmission of potato spindle tuber virus in tomato and potato, *Am. Potato J.*, 47, 225, 1970.
47. Leont'eva, Y. A., Tomatoes and potato spindle tuber viroid, *Zaschch. Rast.*, 8, 22, 1980.
48. Singh, R. P. and Finnie, R. E., Seed transmission of potato spindle tuber metavirus through the ovule of *Scopolia sinensis, Can. Plant Dis. Surv.,* 53, 153, 1973.
49. Broadbent, L., The epidemiology of tomato mosaic. XI. Seed transmission of TMV, *Ann. Appl. Biol.,* 56, 177 1965.

50. Taylor, R. H., Grogan, R. G., and Kimble, K. A., Transmission of tobacco mosaic virus in tomato seed, *Phytopathology*, 51, 837, 1961.
51. Hemmati, K. and McLean, D. L., Gamete seed transmission of alfalfa mosaic virus and its effect on seed germination and yield in alfalfa plants, *Phytopathology*, 67, 576, 1977.
52. Timian, R. G., Seed transmission of barley stripe mosaic virus in relation to the length of time parental plants are infected, *Phytopathology*, 60, 1317, 1970.
53. Medina, A. C. and Grogan, R. G., Seed transmission of bean mosaic viruses, *Phytopathology*, 51, 452, 1961.
54. Kassanis, B., Russell, G. E., and White, R. F., Seed and pollen transmission of beet cryptic virus in sugarbeet plants, *Phytopathol. Z.*, 91, 76, 1978.
55. Converse, R. H. and Lister, R. M., The occurrence and some properties of black raspberry latent virus, *Phytopathology*, 59, 325, 1969.
56. Vorra-Urai, S. and Cockbain, A. J., Further studies on seed transmission of broad bean stain virus and Echtes Ackerbohnenmosaik virus in field beans *(Vicia faba)*, *Ann. Appl. Biol.*, 87, 365, 1977.
57. Sharma, S. R. and Varma, A., Transmission of cowpea banding mosaic and cowpea chlorotic spot viruses through the seeds of cowpea, *Seed Sci. Technol.*, 14, 217, 1986.
58. Tsuchizaki, T., Yora, K., and Asuyama, H., Seed transmission of viruses in cowpea and Azuki bean plants. II. Relations between seed transmission and gamete infection, *Ann. Phytopathol. Soc. Jpn.*, 36, 237, 1970.
59. Callahan, K. L., *Prunus* host range and pollen transmission of elm mosaic virus, *Diss. Abstr.*, 17, 1861, 1957.
60. Ryder, E. J., Transmission of common lettuce mosaic virus through the gametes of the lettuce plant, *Plant Dis. Rep.*, 48, 522, 1964.
61. Bennett, C. W., Lychnis ringspot, *Phytopathology*, 49, 706, 1959.
62. Stevenson, W. R. and Hagedorn, D. J., Further studies on seed transmission of pea seedborne mosaic virus in *Pisum sativum*, *Plant Dis. Rep.*, 57, 248, 1973.
63. Das, C. R., Milbrath, J. A., and Swenson, K. G., Seed and pollen transmission of prunus ringspot virus in Buttercup squash, *Phytopathology*, 51, 64, 1961.
64. Alvarez, M. and Campbell, R. N., Transmission and distribution of squash mosaic virus in seeds of cantaloupe, *Phytopathology*, 68, 257, 1978.
65. Gilmer, R. M. and Kelts, L. J., Transmission of tobacco mosaic virus in grape seeds, *Phytopathology*, 58, 277, 1968.
66. Yang, A. F. and Hamilton, R. I., The mechanism of seed transmission of tobacco ringspot virus in soybean, *Virology*, 62, 26, 1974.
67. Braun, A. J. and Keplinger, J. A., Seed transmission of tomato ringspot virus in raspberry, *Plant Dis. Rep.*, 57, 431, 1973.
68. Sdoodee, R. and Teakle, D. S., Transmission of tobacco streak virus by *Trips tabaci:* a new method of plant virus transmission, *Plant Pathol.*, 36, 377, 1987.
69. Lister, R. M. and Murant, A. F., Seed transmission of nematode-borne viruses, *Ann. Appl. Biol.*, 59, 49, 1967.
70. Goodey, T., Further observations on *Anguillulina dipsaci* infestation of the onion scape and inflorescence, *J. Helminthol.*, 21, 60, 1945.
71. Lang, W., Zur Ansteckung der Gerste durch *Ustilago nuda*, *Ber. Dtsch. Bot. Ges.*, 35, 4, 1917.
72. Safeeulla, K. M., Biology and Control of the Downy Mildews of Pearlmillet, Sorghum, and Fingermillet, Downy Mildew Res. Lab., Mysore University, Mysore, India, 1976, 304.

73. Luttrell, E. S., The disease cycle and fungus–host relationship in Dallis grass ergot, *Phytopathology,* 67, 1461, 1977.
74. Sundaram, N. V., Ergot of bajra, in Advances in Mycology and Plant Pathology, Prof. R. N. Tandon's Birthday Celebration Comm., New Delhi, 1975, 155.
75. Thakur, R. P. and Williams, R. J., Pollination effects on pearlmillet ergot, *Phytopathology,* 70, 80, 1980.
76. Silow, R. A., A systemic disease of red clover caused by *Botrytis anthophila* Bond., *Trans. Br. Mycol. Soc.,* 18, 239, 1933.
77. Halfon-Meiri, A. and Rilsky, J., Internal mold caused in sweet pepper by *Alternaria alternata:* fungus ingress, *Phytopathology,* 73, 67, 1983.
78. Huang, H. C., Hanna, M. R., and Kokko, E. G., Mechanism of seed contamination by *Verticillium alboatrum* in alfalfa, *Phytopathology,* 75, 482, 1985.
79. Huang, H. C. and Kokko, E. G., Infection of alfalfa pollen by *Verticillium alboatrum, Phytopathology,* 85, 859, 1985.
80. Baker, K. F., Seed pathology, in *Seed Biology,* Vol. 2, Kozlowski, T. T., Ed., Academic Press, New York, 1972, 317.
81. Reddick, D. and Stewart, V. B., Transmission of the virus of bean mosaic in seed and observations on thermal death-point of seed and virus, *Phytopathology,* 9, 445, 1919.
82. Yang, A. F. and Hamilton, R. I., The mechanism of seed transmission of tobacco ringspot virus in soybean, *Virology,* 62, 26, 1974.
83. Brlansky, R. H., Carroll, T. W., and Zaske, S. K., Some ultra-structure aspects of the pollen transmission of barley stripe mosaic virus in barley, *Can. J. Bot.,* 64, 853, 1986.
84. Carroll, T. W. and Mayhew, D. E., Occurrence of virions in developing ovules and embryo sacs of barley in relation to the seed transmissibility of barley stripe mosaic virus, *Can. J. Bot.,* 54, 2497, 1976.
85. Hamilton, R. I., Leung, E., and Nichols, C., Surface contamination of pollen by viruses, *Phytopathology,* 67, 395, 1977.
86. Cole, A., Mink, G. I.,and Regev, S., Location of prunus necrotic ringspot virus on pollen grains from infected almond and cherry trees, *Phytopathology,* 72, 1542, 1982.
87. Hamilton, R. I. and Valentine, B., Infection of orchid pollen by Odontoglossum ringspot and cymbidium mosaic viruses, *Can. J. Plant Pathol.,* 6, 185, 1984.
88. Evans, T. A. and Stephens, C. T., Association of asparagus virus II with pollen from infected asparagus *(Asparagus officinalis), Plant Dis.,* 72, 195, 1988.
89. Bailiss, K. W. and Offei, S. K., Alfalfa mosaic virus in lucerne seed during seed maturation and storage, and in seedlings, *Plant Pathol.,* 39, 539, 1990.
90. Hamilton, R. I., Nichols, C., and Valentine, B., Survey for prunus necrotic ringspot and other viruses contaminating the exine of pollen collected by bees, *Can. J. Plant Pathol.,* 6, 196, 1984.
91. Digiaro, M., Di Terlizzi, B., and Savino, V., Ilar-viruses and almond pollen, in Proc. 8th Congress of the Mediterranean Phytopathol. Union — Agadir, Morocco, 1990, 81.
92. George, J. A. and Davidson, T. R., Pollen transmission of necrotic ringspot and sour cherry yellows viruses from tree to tree, *Can. J. Plant Sci.,* 43, 276, 1963.
93. Topchliska, M., Effect of prune dwarf virus on vegetative phenomenon in peach, *Gradinarska Lozarska Nauka.,* 20, 33, 1983.

94. Kelley, R. D. and Cameron, H. R., Location of prune dwarf and prunus necrotic ring spot viruses associated with sweet cherry pollen and seed, *Phytopathology,* 76, 317, 1986.
95. Callahan, K. L., Prunus host range and pollen transmission of elm mosaic virus, *Phytopathology,* 47, 5, 1957.
96. Cooper, J. I., Massalski, P. R., and Edwards, M., Cherry leaf roll virus in the female gametophyte and seed of birch and its relevance to vertical virus transmission, *Ann. Appl. Biol.,* 105, 55, 1984.
97. Cooper, J. L., The possible epidemiological significance of pollen and seed transmission in the cherry leaf roll virus/*Betula* spp. complex, *Mitt. Biol. Bundesanst. Land. Forstwirtsch.,* 170, 17, 1976.
98. Bulger, M. A., Stace-Smith, R., and Martin, R. R., Transmission and field spread of raspberry bushy dwarf virus, *Plant Dis.,* 74, 514, 1990.
99. Jones, R.A.C., Tests for transmission of four potato viruses through potato true seed, *Ann. Appl. Biol.,* 100, 315, 1982.
100. Nolan, P. A. and Campbell, R. N., Squash mosaic virus detection in individual seeds and seed lots of cucurbits by enzyme-linked immunosorbent assay, *Plant Dis.,* 68, 971, 1984.
101. Gotlieb, A. R., Isolation, Characterization and Transmission of Apple Mosaic Virus in White Birch and Isolation and Characterization of a Latent Virus in Yellow Birch, Ph.D. Thesis, University of Wisconsin, 1974, 94.
102. Kryczynski, S., Paduch-Cichal, E., and Skrzeczkowski, L. J., Transmission of three viroids through seed and pollen of tomato plants, *J. Phytopathol.,* 121, 51, 1988.
103 Huang, B., Hills, G. J., and Sunderland, N., Virus infected pollen grains in *Paeonia emodi, J. Expl. Bot.,* 34, 1392, 198.
104. Haight, E. and Gibbs, A., Effect of viruses on pollen morphology, *Plant Pathol.,* 32, 369, 1983.
105. Childress, A. M. and Ramsdell, D. C., Bee-mediated transmission of blueberry leaf mottle virus in infected pollen in highbush blueberry, *Phytopathology,* 77, 167, 1987.
106. Pesic, Z. and Hiruki, C., Effect of alfalfa mosaic virus on germination and tube growth of alfalfa pollen, *Can. J. Plant Pathol.,* 10, 6, 1988.
107. Pesic, Z., Hiruki, C., and Chen, M. H., Detection of viral antigen by immunogold cytochemistry in ovules, pollen and anthers of alfalfa infected with alfalfa mosaic virus, *Phytopathology,* 78, 1027, 1988.
108. Childress, A. M. and Ramsdell, D. C., Detection of blueberry leaf mottle virus in highbush blueberry pollen and seed, *Phytopathology,* 76, 1333, 1986.
109. Falloon, P. G., Falloon, L. M., and Grogan, R. G., Survey of California asparagus for asparagus virus I, asparagus virus II, and tobacco streak virus, *Plant Dis.,* 70, 103, 1986.
110. Kyriakopoulou, P.E., Rana, G.L., and Roca, F., Geographic distribution, natural host range, pollen and seed transmissibility of artichoke yellow ringspot virus, *Ann. de l'Institut Phytopathol.,* Benaki, 14, 139, 1985.
111. Rana, G. L., Kyriakopoulou, P. E., and Martelli, G. P., Artichoke yellow ringspot virus, Descriptions of Plant Viruses, No. 271, Commonw. Mycol. Inst. Assoc. Appl. Biologists, 1983, 4.
112. Desjardins, P. R., Drake, R. J., Atkins, E. L., and Bergh, B. O., Pollen transmission of avocado sunblotch virus experimentally demonstrated, *Calif. Agric.,* 33, 14, 1979.

113. Gardner, W. S., Electron microscopy of barley stripe mosaic virus: comparative cytology of tissues infected during different stages of maturity, *Phytopathology,* 57, 1315, 1967.
114. Frandsen, N. O., Untersuchungen zur virusresistenzzuchtung bei *Phaseolus vulgaris* L. I. Phytopathol. Untersuch., *Z. Pflanz.,* 31, 381, 1952.
115. Brunt, A., Crabtree, K., and Gibbs, A., *Viruses of Tropical Plants,* CAB International, Kew, Survey, U.K., 1990, 707.
116. Schimanski, H. H. and Schmelzer, K., Zur Kenntnis der Ubertragbarkeit des Kirschenblattroll-virus (cherry leaf-roll virus) durch den samen von *Sambucus racemosa* L., *Zentralbl. Bakteriol. Parasitenkd. Inf.,* 127, 673, 1972.
117. Hanson, A. G., Nyland, G., McElroy, F. D., and Stace-Smith, R., Origin, cause, host range and spread of cherry raspleaf disease in North America, *Phytopathology,* 64, 721, 1974.
118. Kryczynski, S., Paduch-Cichal, E., and Skrzecz-Kowski, L. J., Transmission of three viroids through seed and pollen of tomato plants, *J. Phytopathol.,* 121, 51, 1988.
119. Ghose, T. and Basak, M., Transmission of leaf mosaic (chlorosis) in jute, *Jute Bull.,* 24, 94, 1961.
120. Hollings, M., Komuro, Y., and Tochihara, H., Cucumber green mottle virus, Descriptions of Plant Viruses, No. 154, Commonwl. Mycol. Inst. Assoc. Appl. Biol., 1975, 4.
121. Iizuka, N., Seed transmission of viruses in soybean, *Bull. Tohoku Natl. Agric. Exp. Stn.,* 46, 131, 1973.
122. Tomlinson, J. A. and Carter, A. L., Studies on the seed transmission of cucumber mosaic virus in chickweed *(Stellaria media)* in relation to the ecology of the virus, *Ann. Appl. Biol.,* 66, 381, 1970.
123. Schmelzer, K., Untersuchungen an viren der zier-und Wildgeholze, 5, Mitteilung: virosen an Populus und Sambucus, *Phytopathol. Z.,* 55, 317, 1966.
124. Cory, L. and Hewitt, W. B., Some grapevine viruses in pollen and seeds, *Phytopathology,* 58, 1316, 1968.
125. Blackstock, J. McK., Lucerne transient streak and lucerne latent, two new viruses of lucerne, *Aust. J. Agric. Res.,* 29, 291, 1978.
126. Lourie, R., Gordon, D. T., Knobe, J. K., Gingery, R. E., Bradfute, O. E., and Lipps, P. E., Maize white line mosaic virus in Ohio, *Plant Dis.,* 66, 167, 1982.
127. Tsuchizaki, T., Hibino, H., and Saito, Y., Mulberry ringspot virus isolated from mulberry showing ring spot symptom, *Ann. Phytopathol. Soc. Jpn.,* 37, 266, 1971.
128. Cheremushkina, N. P., The spread of onion mosaic virus seeds, Tr. Selektsii i Semenovod, *Ovoschch. Kult.,* 6, 52, 1979.
129. Louie, R. and Lorbeer, J. W., Mechanical transmission of onion yellow dwarf virus, *Phytopathology,* 56, 1020, 1966.
130. Hardtl, H., Die Ubertragung der Zwiebelgelbstreifigkeit durch den samen, *Z. Pflanzenkr. Pflanzensch.,* 79, 694, 1972.
131. Gallitelli, D., Quacquarelli, A., and Martelli, G. P., Pelargonium zonate leaf spot, Descriptions of Plant Viruses, No. 272, Commonwl. Mycol. Inst. Assoc. Appl. Biol., 1983, 4.
132. Schimanski, H. H. and Schade, C., Nachwies des nekrotischen Ringflecken-virus der Kirsche im Pollen der Vogelkirsche *(Prunus Avium* L.) und des chlorotischen ringflecken-virus der Kirsche im pollen der Steinweischel *(Prunus mahaleb* L.), *Arch. Phytopathol. Pflanzenschutz,* 10, 3, 1974.

133. Gilmer, R. M. and Way, R. D., Pollen transmission of necrotic ringspot and prune dwarf viruses in sour cherry, *Phytopathology,* 50, 624, 1960.
134. Digiaro, M. and Savino, V., Role of pollen and seeds in the spread of ilarviruses in almond, *Adv. Hort. Sci.,* 6, 134, 1992.
135. Davidson, T. R., Field spread of prunus necrotic ringspot in sour cherries in Ontario, *Plant Dis. Rep.,* 60, 1080, 1976.
136. Natsuaki, T., Radish yellow edge virus, Description of Plant Viruses, No. 298, Assoc. Appl. Biol., 1985, 4.
137. Cadman, C. H., Filamentous viruses infecting fruit trees and raspberry and their possible mode of spread, *Plant Dis. Rep.,* 49, 230, 1965.
138. Ormerod, P. J., A virus associated with loganberry degeneration disease, *Annu. Rep. East Malling Res. Stn.,* 1969, 165, 1970.
139. Crowley, N. C., Studies on the time of embryo infection by seed-transmitted viruses, *Virology,* 8, 116, 1959.
140. Francki, R. I. B. and Miles, R., Mechanical transmission of sowbane mosaic virus carried on pollen from infected plants, *Plant Pathol.,* 34, 11, 1985.
141. Bos, L., Huttinga, H., and Maat, D. Z., Spinach latent virus, a new ilarvirus seedborne in *Spinacia oleracea, Neth. J. Plant Pathol.,* 86, 79, 1980.
142. Ghanekar, A. M. and Schwenk, F. W., Seed transmission and distribution of tobacco streak virus in six cultivars of soybeans, *Phytopathology,* 64, 112, 1974.
143. Blakeslee, A. F., A graft-infectious disease of *Datura* resembling a vegative mutation, *J. Genet.,* 11, 17, 1921.
144. Johnson, H. A., Converse, R. H., Amorao, A., Espejo, J. I., and Frazier, N. W., Seed transmission of tobacco streak virus in strawberry, *Plant Dis.,* 68, 390, 1984.
145. Sdoodee, R. and Teakle, D. S., Seed and pollen transmission of tobacco streak virus in tomato *(Lycopersicon esculentum* cv. Grosse Lisse), *Aust. J. Agric. Res.,* 39, 469, 1988.
146. Gibbs, A. J. and Harrison, B. D., Nematode transmitted viruses in sugarbeet in East Anglia, *Plant Pathol.,* 13, 144, 1964.
147. Kegler, G. and Kegler, H., Research on natural transmission of tomato bushy stunt virus in fruit trees, in *Problems in Plant Virology No. 184*, Kleinhempel, H., Ed., Tagungsbericht Akademie Landwirtschaftswissenschafter Deutschen Demokrat. Repub., 1980, 297.
148. Scarborough, B. A. and Smith, S. H., Seed transmission of tobacco and tomato ring-spot viruses in geraniums, *Phytopathology,* 65, 835, 1975.
149. Timian, R. G., Barley stripe mosaic virus seed transmission and barley yield as influenced by time of infection, *Phytopathology,* 57, 1375, 1967.
150. Frosheiser, F. I., Alfalfa mosaic virus transmission to seed through alfalfa gametes and longevity in alfalfa seed, *Phytopathology,* 64, 102, 1974.
151. Avgelis, A. D., Katis, N., and Grammatikaki, G., Broadbean wrinkly seed caused by artichoke yellow ringspot nepovirus, *Ann. Appl. Biol.,* 121, 133, 1992.
152. Singh, R. P., Boucher, A., and Somerville, T. H., Detection of potato spindle tuber viroid in the pollen and various parts of potato plant pollinated with viroid-infected pollen, *Plant Dis.,* 76, 951, 1992.
153. Maddox, F., Eastfield experiments, smut and bunt, *Agric. Gaz. J. Counc. Agric. Tasmania,* 4, 92,1896.
154. Brefeld, O., Neue Untersuchungen und Ergebnisse uber die naturliche Infektion und verbreitung der Brandkrankheiten des Getreides, *Klub Landw. Berlin Nachr.,* 466, 4224, 1903.

155. Lang, W., Die Bluteninfektion beim Weizemflugbrand, *Zentralbl. Bakteriol. Parasitenkd.*, 25, 86, 1910 (Abstract 2).
156. McAlpin, D., The Rust of Australia, Their Structure, Life History, Treatment and Classification, University of Melbourne, 1910, 288.
157. Batts, C. C. V., Observations on the infection of wheat by loose smut [*Ustilago tritici* (Pers.) Rostr.], *Trans. Br. Mycol. Soc.*, 38, 465, 1955.
158. Kozera, W., The course of early stage of infection of barley by *Ustilago nuda, Hodowla Rosl. Aklim. Nasienn.*, 12, 269, 1968.
159. Pedersen, P. N., Infection of barley by loose smut, *Ustilago nuda* (Jens.) Rostr., *Friesia,* 5, 341, 1956.
160. Ruttle, M. L., Studies on barley smuts and on loose smut of wheat, *N.Y. Agric. Exp. Stn. Technol. Bull.*, 221, 39, 1934.
161. Vanderwalle, R., Note sur la biologie d'*Ustilago nuda tritici* Schaf., *Bull. Inst. Agric. Stn. Rech. de Gembloux,* 11, 103. 1942.
162. Shinohara, M., Anatomical studies on barley loose smut *(Ustilago nuda* (Jens.) Rostrup). I. Path of embryo infection in the developing caryopsis, *Bull. Coll. Agric. Vet. Med. Nihon Univ.*, 29, 84, 1972.
163. Malik, M. M. S., and Batts, C. C. V., The infection of barley by loose smut [*Ustilago nuda* (Jens.) Rostr.], *Trans. Br. Mycol. Soc.*, 43, 117, 1960.
164. Wolf, W. J. and Baker, F. L., Scanning electron microscopy of soybean, *Cereal Sci. Today,* 17, 125, 1972.
165. Kondo, M., Der anatomische Bau einiger auslandischer Hulsenfruchte, die jetzl viel in den Handel Kommen. Zeitschrift fur der nahrungs und genuss mitted, *Gebraruchsgegenstaede*, 25, 1, 1913.
166. Liu, H. L., Inheritance of defective seedcoat in soybeans, *J. Hered.,* 40, 317, 1949.
167. Rodriguez-Marcano, A. and Sinclair, J. B., Fruiting structures of *Colletotrichum dematium* var. *truncata* and *Phomopsis sojae* formed in soybean seed, *Plant Dis. Rep.,* 62, 873, 1978.
168. Rankin, H. W., Effectiveness of seed treatment for controlling anthracnose and gummy-stem blight of watermelon, *Phytopathology,* 44, 675, 1954.
169. Davis, L. H., The heterosporium disease of California poppy, *Mycologia*, 44, 366, 1952.
170. Kunwar,I. K., Manandhar, J. B., and Sinclair, J. B., Histopathology of soybean seeds infected with *Alternaria alternata, Phytopathology,* 76, 543, 1986.
171. Vaughan, D. A., Kunwar, I. K., Sinclair, J. B., and Bernard, R. L., Routes of entry of *Alternaria* sp. into soybean seed coats, *Seed Sci. Technol.*, 16, 725, 1988.
172. Singh, T. and Sinclair, J. B., Further studies on the colonization of soybean seeds by *Cercospora kikuchii* and *Phomopsis* sp., *Seed Sci. Technol.,* 14, 71, 1986.
173. Sakai, Y. and Ogawa, M., Chemical control of soybean seed disease caused by *Cercospora kikuchii* Matsumoto et Tomoyasu, *Bull. Hiroshima Perfect. Agric. Exp. Stn.,* 46, 33, 1983.
174. Kunwar, I. K., Singh, T., and Sinclair, J. B., Histopathology of soybean seed and seedling by *Macrophomina phaseolina, Phytopathology,* 76, 532, 1986.
175. Roy, K. W. and Abney, T. S., Colonization of pods and infection of seeds by *Phomopsis longicolla* in susceptible and resistant soybean lines inoculated in the greenhouse, *Can. J. Plant Pathol.,* 10, 317, 1988.
176. Seto, F., Studies on the "bakanae" disease of the rice plant. V. On the mode of infection of rice by *Gibberella fujikuroi* (Saw.) Wr. in the flowering period and its relation to the occurrence of the so-called "bakanae" seedlings., *Forsch. Pflkr. Kyoto,* 3, 43, 1937.

177. Ou, S. H., *Rice Diseases,* CAB Int. Mycol. Inst., Kew, Surrey, U.K., 1985, 380.
178. Hino, T. and Furuta, T., Studies on the control of bakanae disease of rice plants, caused by *Gibberella fujikuroi.* II. Influence on flowering season on rice plants and seed transmissibility through flower infection, *Bull. Chugoku Agric. Exp. Stn.,* E-2, 97, 1968.
179. Shetty, H. S., Mathur, S. B., and Neergaard, P., *Sclerospora graminicola* in pearl-millet seeds and its transmission, *Trans. Br. Mycol. Soc.*, 74, 127, 1980.
180. Singh, D. P., Agarwal, V. K., and Khetarpal, R. K., Etiology and host-pathogen relationship of grain mould of sorghum, *Indian Phytopathol.*, 41, 389, 1988.
181. Chikuo, Y. and Sugimoto, T., Histopathology of sugarbeet flowers and seed balls infected with *Colletotrichum dematium* f. *spinaciae, Ann. Phytopathol. Soc. Jpn.,* 55, 404, 1989.
182. Singh, D., Mathur, S. B., and Neergaard, P., Histopathological studies of *Alternaria sesamicola* penetration in sesame seed, *Seed Sci. Technol.,* 8, 85, 1980.
183. Strandberg, J. O., Infection and colonization of inflorescences and mericarps of carrot by *Alternaria dauci, Plant Dis.,* 67, 1351, 1983.
184. Porter, D. M., Wright, F. S., Taber, R. A, and Smith, D. H., Colonization of peanut seeds by *Cylindrocladium crotalariae, Phytopathology,* 81, 896, 1991.
185. Sangchote, S., and Juangbhanich, P., Seed transmission of *Colletotrichum capsici* on pepper (*Capsicum* spp.), *Kasetsart J. Nat. Sci.,* 18, 7, 1984.
186. Baker, K. F., seed transmission of *Rhizoctonia solani* in relation to control of seedling damping-off, *Phytopathology,* 37, 912, 1947.
187. Domsch, K. H., Die Raps und Kohlschotenschwarze, *Z. Pflanzenkr. Pflanzensch.*, 64, 65, 1957.
188. Singh, D., Mathur, S. B., and Neergaard, P., Histopathology of sunflower seeds infected by *Alternaria tenuis, Seed Sci. Technol.,* 5, 579, 1977.
189. Bhowmik, T. P., *Alternaria* seed infection of wheat, *Plant Dis. Rep.,* 53, 77, 1969.
190. Leppik, E. E., *Alternaria sesami,* a serious seedborne pathogen of world-wide distribution, *FAO Plant Prot. Bull.,* 12, 1, 1964.
191. Leach, C. M., *Kabatiella caulivora,* a seedborne pathogen of *Trifolium incarnatum* in Oregon, *Phytopathology,* 52, 1184, 1962.
192. Reddy, K. D., Govindaswamy, C. V., and Vidhyasekaran, P., Studies on ergot disease of cumbu (*Pennisetum typhoides*), *Madras Agric. J.,* 56, 367, 1969.
193. Stevenson, I. L., Timing and nature of seed infection of barley by *Cochliobolus sativus, Can. J. Plant Pathol.,* 3, 76, 1981.
194. Castor, L. L. and Frederiksen, R. A., Histopathology of *Fusarium moniliforme* infection of sorghum kernels, *Phytopathology,* 71, 208, 1981.
195. Gubanov, G. Y. and Sabirov, B. G., Transmission of Fusarium wilt infection by cotton seed, *Khlopkovedstvo,* 22, 22, 1972.
196. Luthra, J. C. and Bedi, K. S., Some preliminary studies on gram blight with reference to its cause and mode of perennation, *Indian J. Agric. Sci.,* 2, 499, 1932.
197. Snyder, W. C. and Wilhelm, S., Seed transmission of Verticillium wilt of spinach, *Phytopathology,* 52, 365, 1962.
198. Tennyson, G., Invasion of cotton seed by *Bacterium malvacearum, Phytopathology,* 26, 1083, 1936.
199. Patino-Mendez, G., Studies on the pathogenicity of *Corynebacterium michiganense* (E. F. Sm.) Jensen and its transmission in tomato seed, *Diss. Abstr.,* 27, 1692, 1967.
200. Khalil, O. and Sabet, K. A., A study of certain aetiological aspects of bacterial leaf blight disease of castor bean, *Sudan Agric. J.,* 6, 26, 1971.

201. Taylor, J. D., Dudley, C. L., and Presley, L., Studies of halo-blight seed infection and disease transmission in dwarf beans, *Ann. Appl. Biol.*, 93, 267, 1979.
202. Fukuda, T., Azegami, K., and Tabei, H., Histopathological studies on bacterial black node of barley and wheat caused by *Pseudomonas syringae* pv. *japonica, Ann. Phytopathol. Soc. Jpn.,* 56, 252, 1990.
203. Tsushima, S., Tsuno, K., Modi, S., Wakimoto, S., and Saito, H., The multiplication of *Pseudomonas glumae* on rice grains, *Ann. Phytopathol. Soc. Jpn.*, 53, 663, 1987.
204. Azegami, K., Taei, H., and Fukuda, T., Entrance into rice grains of *Pseudomonas plantarii*, the casual agent of seedling blight of rice, *Ann. Phytopathol. Soc. Jpn.*, 54, 633, 1988.
205. Bashan, Y. and Okon, Y., Internal and external infections of fruits and seeds of pepper by *Xanthomonas campestris* pv. *vesicatoria, Can. J. Bot.*, 64, 2865, 1986.
206. Singh, T. and Sinclair, J. B., Histopathology of *Cercospora sojina* in soybean seeds, *Phytopathology*, 75, 185, 1985.
207. El-Nashaar, H. and Bugbee, W. M., Scanning electron microscopic examination of sugarbeet flowers and fruits infected with *Phoma betae, J. Am. Soc. Sugarbeet Technol.*, 21, 11, 1981.
208. Singh, D., Mathur, S. B., and Neergaard, P., Histological studies of *Alternaria sesamicola* penetration in sesame seed, *Seed Sci. Technol.*, 8, 85, 1980
209. Persly, G. J., Studies on the survival and transmission of *Xanthomonas manihotis* on cassava seed, *Ann. Appl. Biol.*, 93, 159, 1979.
210. Wallin, J. R., Seed and seedling infection of barley, bromegrass and wheat by *Xanthomonas translucens* var. *cerealis, Phytopathology,* 36, 446, 1946.
211. Tabei, H., Azegami, K., and Fukuda, T., Infection site of rice grain with *Erwinia herbicola*, the causal agent of bacterial palea browning of rice, *Ann. Phytopathol. Soc. Jpn.,* 54, 637, 1988.
212. Tabei, H., Azegami, K., Fukuda, T., and Goto, T., Stomatal infection of rice grain with *Pseudomonas glumae,* the causal agent of the bacterial grain rot of rice, *Ann. Phytopathol. Soc. Jpn.*, 55, 224, 1989.
213. Rao, J., Kauraw, L. P., and Prakash, A., Incidence of white tip nematode in freshly harvested rice seed, *Seed Res.,* 11, 74, 1983.
214. Trione, E. J., Dwarf bunt of wheat and its importance in international wheat trade, *Plant Dis.,* 66, 1083, 1982.
215. Thakur, R. P., and King, S., Smut disease of pearlmillet, *Int. Crops Res. Inst. Semi-Arid Trop. Inf. Bull. 25,* Patancheru, A. P., India, 17, 1988.
216. Baker, K. F., Bacterial fasciation disease of ornamental plants in California, *Plant Dis. Rep.,* 34, 121, 1950.
217. Weller, D. M. and Saettler, A. W., Evaluation of seed-borne *Xanthomonas phaseoli* and *X. phaseoli* var. *fuscans* as primary inocula in bean blights, *Phytopathology,* 70, 148, 1980.
218. Groth, D. E. and Braun, E. J., Survival, seed transmission and epiphytic development of *Xanthomonas campestris* pv. *glycines* in the North-Central United States, *Plant Dis.*, 73, 326, 1989.
219. Nandkumar C., Prasad, J. S., Rao, Y. S., and Rao, J., Investigations on the white tip nematode (*Aphelenchoides besseyi*), *Indian J. Nematol.*, 5, 62, 1975.
220. Fukano, H., Ecological studies on white tip disease of rice plant caused by *Aphelenchoides besseyi* Christie and its control, *Fukuoka Agric. Exp. Stn. Bull.,* 18, 108, 1962.

221. Bridge, J., Bos, W. S., Page, L. J. and McDonald, D., The biology and possible importance of *Aphelenchoides arachidis*, a seedborne endoparasitic nematode of groundnuts from northern Nigeria, *Nematologica*, 23, 253, 1977.
222. Burckhardt, F., New observations on the presence of *Aphelenichoides* spp. in seed of *Callistephus chinensis, Nachrichtenbl. Dtsch. Pflanzenschutzdienstes* (Braunschweig), 24, 132, 1974.
223. Caubel, G., Epidemiology and control of seed-borne nematodes, *Seed Sci. Technol.* 11, 989, 1983.
224. Walker, J. C., *Plant Pathology*, 3rd ed., McGraw-Hill, New York, 1969, 819.
225. Besri, M., Phases of the transmission of *Fusarium oxysporum* f. sp. *lycopersici* and *Verticillium dahliae* by seeds of some tomato varieties, *Phytopathol. Z.*, 93, 148, 1978.
226. Gardner, M. W. and Kendrick, J. B., Bacterial spot of tomato and pepper, *Phytopathology*, 13, 307, 1923.
227. Leben C., Survival of *Pseudomonas syringae* pv. *lachrymans* with cucumber seed, *Can. J. Plant Pathol.*, 3, 247, 1981.
228. Thakur, R. P., Rao, V. P., and Williams, R. J., The morphology and disease cycle of ergot caused by *Claviceps fusiformis* in pearl millet, *Phytopathology*, 74, 201, 1984.
229. Gupta, P. and Swarup, G., On the earcockle and yellow ear rot diseases of wheat, *Indian Phytopathol.,* 21, 318, 1968.
230. Midha, S. K. and Swarup, G., Studies on the wheat seed galls caused by *Anguina tritici, Indian J. Nematol.,* 4, 53, 1974.
231. Muskett, A. E., Seed-borne fungi and their significance, Presidential address, *Trans. Br. Mycol. Soc.,* 33, 1, 1950.
232. Simary, E. I., Results of seed tests. II. Occurrence of some pathogenic fungi in plant residues on faba bean seeds, *Fabis Newsl.*, 30, 42, 1992.
233. Besri, M., A study of the mechanism of seed transmission of *Fusarium oxysporum* f. sp. *lycopersici* and *Verticillium dahliae, Phytopathol. Z.,* 93, 148, 1978.
234. Sattar, A., On the occurrence, perpetuation and control of gram (*Cicer arietinum* L.) blight caused by *Ascochyta rabiei* (Pass.) Labrousse, with special reference to Indian conditions, *Ann. Appl. Biol.*, 20, 612, 1933.
235. Butler, F. C., Plant diseases anthracnose and sooty blotch of red clover: two diseases new to New South Wales, *Agric. Gaz. N.S.W.,* 64, 368, 1953.
236. Reed, H. E., Studies on *Rhynchosporium secalis* (Oud.) Davis causing scald of barley, *Diss. Abstr.,* 20, 49, 1959.
237. Summers, T. E., Downy mildew (*Sclerospora macrospora*) of oats in the south, *Plant Dis. Rep.,* 36, 347, 1952.
238. Christensen, J. J., Henderson, L., and Aragaki, M., Dissemination of *Septoria apiicola, Phytopathology,* 43, 468, 1953.
239. Prasada, R. and Verma, U. N., Studies on lentil rust — *Uromyces fabae* (Pers.l) de Bary, in India, *Indian Phytopathol.,* 1, 142, 1948.
240. Sheppard, J. W. and Needham, S. N., Verticillium wilt of alfalfa in Canada, occurrence of seedborne inoculum, *Can. J. Plant Pathol.,* 2, 159, 1980.
241. McIntyre, J. L., Sands, D. C., and Taylor, G. S., Overwintering, seed disinfestation, and pathogenicity studies of the tobacco hollow stalk pathogen, *Erwinia carotovora* var. *carotovora, Phytopathology,* 68, 435, 1978.
242. Guthrie, J. W., Routine methods for detecting and enumerating seed-borne bacterial plant pathogens, *J. Seed Technol.,* 4, 78, 1979.

243. Edwards, E. T., Stem nematode diseases of lucerne with review of literature concerning the causal organism *Tylenchus dipsaci* (Kuhn) Bast, *Agric. Gaz. N.S.W.*, 43, 305, 1932.
244. Warne, L. G. G., An outbreak of clubroot traceable to a seedborne infection, *J. R. Hortic. Soc.*, 69, 45, 1944.
245. Nash, S. M. and Snyder, W. C., Dissemination of the root rot *Fusarium* with bean seed, *Phytopathology,* 54, 880, 1964.
246. Watanabe, T., Fungi associated with commercial kidney bean seed and their pathogenicity to young seedlings of kidney bean, *Ann. Phytopathol. Soc. Jpn.,* 38, 111, 1972.
247. Raabe, R. D., Association of soil particles with seeds and three pathogens of chickpea in California, *Plant Dis.,* 69, 238, 1985.
248. Epps, J. M., Survival of soybean cyst nematodes in seed bags, *Plant Dis. Rep.*, 52, 45, 1968.
249. Epps, J. M., Survival of the soybean cyst nematode in seed stocks, *Plant Dis. Rep.*, 53, 403, 1969.

CHAPTER 5

Factors Affecting Seed Infection

Successful seed infection is the most important aspect of seed transmission for pathogens. Generally, plant pathogens can be divided on the basis of whether or not they are seed transmitted. For a pathogen to be seed transmitted, it must infect or infest seeds with inoculum capable of causing disease in plants after planting. Even if a seed carries a pathogen, if the pathogen is not capable of causing the disease, then it is considered not to be seed transmitted. An infected plant does not necessarily result in seed infection or infestation. Successful seed infection where seed transmission has been shown depends upon a number of factors, including host genotype, environment, crop management, type of inoculum, seed quality, plant growth stages, mother plant infection, insect infestation, and pathogen interaction.

I. HOST GENOTYPE

Cultivars can react differently to seed infection due to incompatibility to infection, restricted pathogen invasion of the embryo or ovary wall due to inhibitors, or impermeability of a testa to aqueous solutions, because of a waxy coating on the seed surface, compact and uniform arrangement of cells, or reduced amounts of amino acids or phenolic compounds. For example, *Cercospora kikuchii* infection in soybean seeds ranged from 3 to 40% and 30 to 85% in resistant and susceptible cultivars, respectively.[1] The greater occurrence of *Cercospora kikuchii* isolates producing high qualities of cercosporin and the correlation between cercosporin production and extent of purple stain or *C. kikuchii* colonization indicated a role for cercosporin in facilitating soybean seed coat colonization by *C. kikuchii*.[2] However, the incidence of purple seed stain among 17 soybean cultivars was unrelated to disease development on foliage.[3] In resistant cultivars of barley and wheat, *U. segetum* var. *tritici* may fail to invade the embryo[4] or ovary wall.[5] Germination of *Ustilago segetum* var. *tritici* chlamydospores was higher on stigmas of susceptible than resistant wheat cultivars.[6] Embryos of susceptible wheat cultivars became infected with well-developed hyphal clumps not found in embryos of resistant cultivars.[7] In resistant barley kernels, hyphae of the loose smut fungus (*U. segetum* var. *tritici*) were not observed in the aleurone, endosperm, or embryo, although they may persist in the chalazal region and parenchyma associated with vascular bundles. In susceptible kernels, hyphae were abundant in the chalazal region, and in integuments, aleurones, endosperms, and embryos.[8] Pathogens that colonize the pericarp, endosperm, or embryo of seeds, such as *F. o.* f. sp. *apii* in celery seeds, invaded tissues of compatible hosts but not of incompatible ones following artificial inoculation.[9]

The site of resistance to maize kernel infection by *F. moniliforme* was found in the pericarp, but genotype of the endosperm, embryo, or cytoplasm had little effect on resistance.[10] Resistant lines had silks that grew for a longer period after pollination than silks of susceptible lines.[11] Most authors considered silk colonization as the primary mode of entry of *F. moniliforme* in developing kernels.[12] Thus breeding for resistance to kernel infection by *F. moniliforme* involved the selection of inbred parents for delayed silk senescence.[13] High lysine endosperms (opaque) of maize caryopses were most susceptible to *F. moniliforme* infection.[14] Infection of maturing pearl millet stigmas resulted in a rapid dissolution and necrosis of tissue and prevented colonization by *Sclerospora graminicola*. This suggested that seeds were unlikely to become infected internally.[15] A seed infection index was developed for comparing susceptibility of bean cultivars to seed infection by fungal pathogens

$$(\text{SII}) = \text{MP} \times \text{mP} \times \text{P1}$$

where MP = maximum percentage of infected seeds in any single planting, mP = mean percentage of infected seeds in 10 plantings, and PI = proportion of plantings in which an infected seed was detected.[16]

MP and mP can vary between 0 and 100 and P1 between 0 and 1 in beans. Where SII = 0, no infection occurred under any agroclimatic conditions, and where at SII = 100, all seeds became infected under all agroclimatic conditions. The accuracy of SII depended on the number of plantings and diversity of agroclimatic conditions, both of which should be as large as possible. The incidence of pod infection of fava bean by *Ascochyta fabae* varied between genotypes following field inoculation of young plants and was confirmed in the incidence of seed infection.[17] Among soybeans, with nearly identical maturities, genotypes resistant to seedborne *Cercospora kikuchii* and *Phomopsis* had a shorter growth period between R7 to R8 growth stages and a greater rate of moisture loss than susceptible soybeans. The time between growth stages R7 and R8 was associated with major pod and seed dry down and the incidence of seed infection.[18] Resistance to *C. kikuchii* was related to hard seed coat in wild soybeans and pod resistance in domesticated ones.[19]

Ovary infection of *Trifolium pratense* (red clover) through the stigma and style was considered unlikely, as female organs contain one or more substances that inhibit growth of *Botrytis anthophila*.[20] Black-seeded accessions of chickpea generally were more resistant to *Phoma rabiei* than white-seeded types.[21] Likewise, black-seeded cultivars of *Phaseolus vulgaris* were resistant to *Rhizoctonia solani*, while most white-seeded ones were susceptible to seed infection and preemergence damping-off. Extracts of seed coats of white-seeded cultivars stimulated, while those of black-seeded cultivars containing phenolic compounds inhibited growth of *R. solani*.[22] Extracts of dwarf beans (*P. vulgaris* var. *nanus*) with colored testa inhibit *Colletotrichum lindemuthianum* in culture. Fungistatic compounds were found in ripening seeds before testas developed their mature color.[23] *Bipolaris sorokiniana* infected lemmas of resistant and susceptible barley cultivars, and the fungus was isolated more readily from the latter than the former. Growth of conidial germ tubes of *B. sorokiniana* was greater in extracts of noninoculated lemmas of susceptible barleys than of resistant ones.[24] There was no significant relationship between *Colletotrichum gossypii* var. *cephalosporioides* infection in field and cottonseed infection. The least and most resistance cultivars showed 9% and 5% seed infection, respectively.[25] Antibacterial compounds active against *Rhodococcus fascians* were located in seed coats of bean, pea, and sweet pea. An inverse relationship existed between the sensitivity of a strain of *R. fascians* to sweet pea compounds and its ability to infect sweet pea seedlings.[26] A single recessive gene controls resistance of barley stripe mosaic virus to seed transmission in barley.[27]

Seed transmission of bean common mosaic virus (BCMV) did not occur in genotypes possessing the dominant necrosis gene in bean.[28] The cv. Pinto 114 was highly resistant to seed infection by the US 2 strain of BCMV.[29] Gunnery and Datta[30] isolated a low-MW RNA from barley embryos that inhibited initiation of protein synthesis. Such molecules may play a role in inhibiting virus synthesis and thus virus seed transmission in some species.[31] In pea cultivars with a high incidence of seed infection by PSbMV, the virus invaded immature embryos, multiplied in embryonic tissues, and persisted during seed maturation. In contrast, in cultivars without seed infection, there was no evidence of invasion in immature embryos.[32]

II. ENVIRONMENT

Environmental conditions at flowering and during seed development influence seed infection and inoculum localization in seeds. Thus conditions in a growing area can influence infection levels. Seed infection of wheat by *Stagonospora nodorum* was least in the mountain region and greatest in the southern plains of Georgia.[33] Cottonseeds produced in South Carolina in 1954, generally, were only lightly infected by hyphae of *Colletotrichum gossypii*, but those produced in the coastal plain and southern Piedmont areas in 1955 were heavily infected. Most seeds produced in the northern Piedmont in 1955 were lightly infected. The degree of infection and reduction in seed viability were related directly to the amount and frequency of rainfall at boll opening.[34] Seed infection of flax by *B. cinerea* was found throughout the United Kingdom, with a tendency for it to be higher in the coastal and upland areas.[35]

It requires 4 or 5 d at 20 and 15°C, respectively, at 100% relative humidity to reach the level of soybean seed infection by *Diaporthe* and *Phomopsis* attained within 3 d at 25°C.[36] High temperature and low moisture early in the season favored soybean *Macrophomina phaseolina* seed infection, whereas high temperature and moisture from mid- to late season favored development of *C. kikuchii*.[37] Forced early maturation increased the incidence of *Phomopsis* seed infection from 5.3 to 31.5% and decreased seed germination from 89.4 to 61.4% over all soybean cultivars, which demonstrated that a low incidence of *Phomopsis* seed infection in late maturing cultivars was not due to resistance but to escapes from infection.[38] Early maturing cultivars mature seed during the warm, often humid weather of late summer and characteristically have a high incidence of seedborne *Phomopsis*. Late maturing cultivars mature seed during the cooler and often drier weather of mid-autumn and usually have good seed quality.[38]

A. Moisture

Generally, humid areas with high temperatures favor seed infection. For example, flax seed produced in eastern and southern England was relatively free of *C. linicola*, *Aureobasidium lini*, and *Phoma*, while the cooler and wetter regions north and west of the Pennines favored infection and seed transmission.[35] Relative humidity, particularly at flowering and seed development, played a major role in seed infection and the establishment of inoculum within seeds. High relative humidity at flowering was conducive to infection of barley and wheat seeds by the loose smut fungus.[39-42] When flowers of inoculated Little Club wheat were exposed for 8 d to low (11 to 30%) and high (50 to 85%) relative humidities, seeds from flowers subjected to low humidity produced 21.9%, while those from high humidity produced 93.9% smutted plants.[41] High relative humidity was essential for soybean seed infection by *F. oxysporum*.[43] The incidence of ergot in rye and pearl millet seeds was greater under high relative humidity and low temperature.[44,45] *Phoma betae* was seedborne in certain sugarbeet seedlots, especially those produced in regions with rain showers or high relative humidity during seed maturity and harvest.[46] A prolonged period of warm and moist weather in southeast and south central Nebraska during September and October 1973 was the major reason for various degrees of preharvest sorghum seed discoloration. Kernel moisture in sorghum was an important factor in determining the degree of discoloration due to *Alternaria*, *Bipolaris*, *F. moniliforme*, and *F. roseum*.[47] More seed infection by *A. alternata* was found from snapdragon plants grown at above 80% relative humidity than from those produced in dry areas.[48] Wheat infection by *F. avenaceum*, *F. culmorum*, and *Microdochium nivale* was favored by high rainfall, especially during ear emergence and maturation.[49] High humidity (90 ± 4%) coupled with high temperature (33°± 1°C day and 24°± 1°C night), or lower humidity (43°± 4°C) with high temperature (33°± 1°C day and 24°± 1°C night) resulted in 49% and 4% seed infection, respectively, by *Phomopsis* in soybeans.[50] High relative humidity coupled with low temperature and frequent rainfall at flowering and seed development favored the incidence of black point and karnal bunt on triticale and wheat.[51-53]

Soybean seeds produced under wet conditions or when wet conditions prevail between physiological and harvest maturity were most susceptible to infection by *Phomopsis*.[37] Soybean seed infection by *C. kikuchii* increased with increased pod wetness periods up to 30 h, but not less than 24 h.[54] The incidence of *Stagonospora nodorum* in wheat seeds reached 23% in 1990 during a wet season, but only 2% in 1991, an atypically dry season.[55]

The high incidence of seedborne infections in areas with high humidity and rainfall resulted in a shift of seed production areas. In the United States, a majority of bean and pea seeds were produced in the New England states, resulting in heavy losses due to *C. lindemuthianum* and *Xanthomonas* in bean and *Ascochyta* in pea. The seed production areas for these crops now are in the dry areas of California, Idaho, and Oregon. Wheat seeds free of *X. c.* pv. *undulosa* were harvested from fields in Oregon where conditions were unfavorable for disease development.[56]

B. Temperature

The role of temperature on seed infection has been studied extensively. The optimum temperature for peanut pod penetration and kernel infection ranged from 26 to 38°C, 21 to 32°C, and 16 to 32°C for *Aspergillus niger*, *A. flavus*, *Macrophomina phaseolina*, and *Rhizopus stolonifer*, respectively.[57] A maximum of 19 to 20° ± 1°C and a minimum of 8 to 10° ± 1°C followed by intermittent rainfall during wheat anthesis created favorable conditions for infection by *Tilletia indica*.[58] The optimum for infection of soybean seeds with *C. kikuchii* is 25°C; no infection was observed at either 15 or 35°.[54] High more than low air temperatures during rice flowering favored embryo infection by *Gibberella fujikuroi*.[59] Infection by the alfalfa mosaic virus through alfalfa pollen and ovules was less at a constant 29° ± 1°C or 24° ± 2°C. The infection at alternating 29 and 22°C was greater than at 29°C, but less than at a constant 18 or 24°C.[60] When cv. Starr peanut plants were kept at 21 and 35°C during flowering and pegging, 1 of 140 (0.7%) and 5 of 1,036 (0.5%) seeds, respectively, were infected by the peanut mottle virus M3.[61] High temperatures in May and wet and cool conditions during late May to early July in Denmark favored a high incidence of *Bipolaris sorokiniana* in barley seeds the following harvest.[62]

C. Wind Velocity

Spore dispersal of *Ustilago segetum* var. *tritici* was correlated positively with rain and wind speed,[63] and high wind velocity favored spore dissemination.[39] Dispersal of barley loose smut fungus spores (*U. segetum* var. *tritici*) in Montana seed fields was uniform around point sources of inoculum, indicating spread by fluctuating wind currents. There was a linear relationship between infection level and log 10 of the distance from the inoculum source. Dispersion in high-velocity winds resulted in heavy infection close to inoculum sources but diminished with distance.[64]

D. Rainfall

Heavy rainfall decreased wheat flower infection by loose smut fungus spores because they were washed off smutted earheads.[39] Cultivation and harvesting methods in sugar beet delayed ripening and increased exposure of seeds to rainfall, thus increasing the incidence of *Phoma betae*.[65] Cloudy and rainy periods during the 60 d preceding harvest favored build up of *P. betae* inoculum and its spread to seed stalks and flower parts. Deep seed penetration was associated with rainfall during the curing of cut seed stalks in the field.[66] High levels of *Stagonospora nodorum* occurred in wheat during summers with abundant rainfall.[67]

Colletotrichum gossypii infection levels in cottonseeds increased with the number of rainy days in August and September in the southern United States. Similarly, infection levels of *Alternaria*, *Fusarium equiseti*, and *F. pallidoroseum* varied between locations and in different years within a location. Changes in infection levels were attributed to precipitation and temperature variations during the growing season.[68] The prevalence of light and frequent rains (6.9 to 20.5 mm for 4 to 5 d) accompanied by mild temperatures (6.4 to 23.3°C) during flowering were conducive to loose smut fungal infection in wheat seeds.[69] When soybean plants lodged after flooding, and pods came in contact with wet soil, a high level of seed infection by *Phomopsis* was recorded.[37] Soybean seed infection by *Phomopsis* increased if wet weather occurred between physiological maturity and harvest.[70] Wheat seed infection by *Stagonospora nodorum* was favored by rainfall during grain maturation.[71]

E. Irrigation

Overhead irrigation tended to increase seed infection by bacteria and fungi. Use of "drop method" irrigation in Israel decreased plant and seed infection in several crops.[72] In spring-sown lentil, irrigation increased seedborne *Ascochyta fabae* f. sp. *lentis* compared to Autumn-sown crops in New Zealand.[73] The average incidence of infected wheat seed lots by *S. nodorum* was 23% (1 to 71% range) in 1990, a moist season, and 2% (0 to 19% range) in 1991, an atypical dry season in New York.[55]

III. CROP MANAGEMENT

A. Seed Production Area

Conditions in northern Illinois were favored by optimum soil moisture because rainfall was adequate and occurred at regular intervals, whereas in southern Illinois, soil moisture was limited because of inadequate rainfall and poor soil conditions. Weather and soil conditions in northern plots were more conducive to plant growth and high-quality soybean seed production than those in southern plots. Seed weight and percent radicle emergence were greatest from northern plots, whereas numbers of fungus-infected and noninfected nonviable seeds were greatest from southern plots. The number of seeds infected by *Aspergillus*, *Chaetomium*, and *Phomopsis* was highest in northern plots, whereas that of seeds infected by *Alternaria*, *Cercospora kikuchii*, *Cladosporium*, *Colletotrichum*, *Fusarium*, *Macrophomina phaseolina*, *Nematospora coryli*, *Penicillium*, and *Phoma* was highest in southern plots.[74] Lentil seed infection by *A. fabae* f. sp. *lentis* occurs up to 250 m from the primary inoculum source.[75] The occurrence of *A. fabae* f. sp. *lentis* was significantly higher from spring-sown seeds, compared to those from autumn-sown crops in New Zealand.[73]

B. Tillage

Tillage primarily affects yield and weed control, but other parameters in crop production such as plant pathogens also may be affected. Tillage operations affected the amount of overseasoned host debris in proximity to an emerging crop and could indirectly affect plant pathogens that depended on debris as a primary source of inoculum. The incidence of bacterial blight, bacterial pustule, wildfire, and anthracnose was greater under nontillage than conventional tillage; *Fusarium*, *Phytophthora*, and *Rhizoctonia* root rots were lower under nontillage.[76,77] Conventional tillage resulted in a higher recovery of seedborne *Alternaria* and bacteria than nontillage during isolated years; however, tillage did not affect any seed quality variable during a 3-yr period.[77] No tillage has been reported to decrease the incidence of purple seed stain in soybeans compared with conventional tillage methods.[78]

C. Plant Population

Plant population or spacing can influence the microclimate around developing seeds. A high plant population or close spacing increases relative humidity, which may be conducive for seedborne infection. In rice, close spacing (15 cm) favored the incidence of seedborne fungi, such as *Alternaria longissima*, *A. padwickii*, *Curvularia lunata*, *B. oryzae*, *Fusarium pallidoroseum*, and *Sclerotium*.[79]

A significantly higher recovery of seedborne *Alternaria* was obtained with decreased row width (76 to 25 cm) in soybeans.[77] From naturally infected plants, 20% *Diaporthe phaseolorum* var. *sojae* was recorded in soybean seeds from low plant populations (10,000 to 16,000 plants per acre), compared to 1% in seeds from high populations (185,000 to 3,000,000 plants per acre). At low populations, lateral branches were partially broken from the erect central stem and came in contact with the soil surface. About 19% of the seeds from lodged branches were infected, compared with 2% from erect central stems of the same plants.[80]

The distance between infected and healthy plants also affected the degree of seed infection. For example, seed infection by the loose smut of wheat fungus generally was 0.1%, but seed infection of a seed sample from within 30 cm of a smutted head was 1.5%.[63]

D. Fertilizers

Field applications of N can influence seed infection. High N reduced the incidence of maize kernels infected with *F. moniliforme*,[81] but high N (150 to 200 kg/ha) compared to low N (0 to 15 kg/ha) increased seedborne infection by *Alternaria padwickii*, *Curvularia lunata*, *Phoma*, and *Trichothecium* in rice.[79] Applications of 120 and 180 kg/ha N favored a higher occurrence of loose smut in wheat seeds compared to 0 or 60 kg/ha N.[69] Production inputs of supplementary ammonium nitrate (50 kg/ha) increased the incidence of *Fusarium avenaceum*, *F. graminearum*, *F. sporotrichioides*, and *F. poae* up to 125% in barley, triticale, and wheat seeds.[82] The effect of potash (K_2O) on *Phomopsis* in soybean seeds was most significant at 300 and 450 kg/ha. In the absence of K_2O, disease incidence was higher. At the highest concentration of K_2O in soil and leaves, a lower disease incidence was recorded in the absence of lime.[83]

E. Growth Regulators

Growth regulators modified soybean physiological maturity (R7 to R8 growth stages), which influenced seed infection by either increasing or decreasing the conditions favorable for seed infection. Most seed infection occurred after the R7 growth stage.[84] A longer interval between R7 and R8 growth stages increased the opportunity for seed infection. However, genotypes with a fast dry down period had a shorter time for potential seed infection and were likely to have fewer infected seeds.[38] The growth regulator chlorflurenol (methyl 2-chloro-9-hydroxyfluorene-9-carboxylate) applied to soybeans at midseason altered the maturation rate, delayed maturity, extended the length of growth stage, and increased seed infection by *C. kikuchii* and *Phomopsis*; however, ethephon (2-chloroethyl phosphonic acid) tended to hasten maturity, shorten the R7 growth stage period, and decrease the percentage of seedborne fungi. The growth stage interval consistently associated with seed infection was from R7 to R8. Modification of the R7 to R8 interval of susceptible and moderately resistant cultivars by growth regulators was directly associated with changes in levels of seed infection.[85] Application of ethephon 280 g a.i./ha at the flag leaf stage increased up to 125% the incidence of infection of *Fusarium avenaceum*, *F. graminearum*, *F. sporotrichioides*, and *F. poae* infection in barley, triticale, and wheat seeds.[82]

Growth regulators such as chlormequat and ethephon and N (0 or 140 kg/ha) did not change the susceptibility of wheat heads to *G. zeae*, but shortened plants may be subject to higher inoculum doses because they are closer to ejected ascospores. It is suggested that the dwarfed plant architecture may influence the microclimate and ascospores and perithecia production on the soil surface.[86]

F. Weeds

Infection in *Phaseolus vulgaris* seeds by *F. pallidoroseum* was 5.6, 12, 10, 21, or 32%, respectively, in plots weeded regularly, or weeded once at 35, 50, or

65 d after sowing, or not at all.[87] *Rhizoctonia solani* was slightly higher in bean seeds from plots without weeding than in plots weeded once at or before the seed-filling stage (R6 to R7 growth stage).[87]

Bentazon plus sethoxydrim herbicide application to soybeans resulted in significantly heavier seed, and seeds from nonweeded control plots had a significantly higher incidence of *Phomopsis* than seeds from herbicide-treated plots.[77] Stress caused by herbicide injury resulted in a higher incidence of *Chaetomium* and *Phomopsis* in soybean seeds.[37]

G. Irrigation

Black point, caused by *Alternaria alternata* in wheat, increased after irrigation during the milk or middough stages.[88] A high positive correlation was obtained between soybean seed infection by *Diaporthe* and *Phomopsis* and average temperature during irrigation, with a lower negative correlation between seed infection and plant age during irrigation.[36]

H. Frost

Frost resulted in damage to soybean pods and seeds, which was followed by increased numbers of seeds infected with *Alternaria*, *Aspergillus*, and *Fusarium*.[89] Frost damage predisposed soybean seeds to fungal infection.[37,90,91] Freezing of soybean plants was associated with increased seed infection by *A. alternata* and *G. zeae* and reduced infection by *P. longicolla*.[91]

I. Harvest Time

Delayed harvest favored seedborne infections. *D. phaseolorum* var. *caulivora* and *D. phaseolorum* var. *sojae* were isolated from mature but not ripened soybean seeds, but the incidence increased as plants matured and seeds ripened. *C. kikuchii* was isolated more frequently from soybean seeds of late maturing cultivars than early maturing ones in Mississippi.[92] In Indiana, delayed harvest resulted in significantly more infection by *P. sojae* but not by *C. kikuchii* in soybean seeds.[93]

Weather conditions during crop maturation affected disease development and resulted in differences among soybean maturity groups. Within a growing region, seeds of cultivars developing earliest and exposed to the highest temperatures and rainfall were more susceptible to *C. kikuchii* and *Phomopsis* than late maturing cultivars.[74]

The bean common mosaic virus in bean seed coats was inactivated during seed maturation.[94] The cowpea banding mosaic virus was present in cowpea seed coats, cotyledons, plumules, and radicles of immature seeds but was appreciably inactivated during the maturation process. The virus was not present in seed coats of mature seeds nor in mature and dry pods, although it was present in immature pods.[95] The incidence of alfalfa mosaic virus in alfalfa decreased rapidly with seed maturation, whereas the antigen incidence declined slowly and was always higher than infective virus. Infective virus and antigen incidence were higher in mature seeds of cv. Maris Kabul than cv. Europe because virus inactivation was more rapid in the latter.[96]

IV. TYPE OF INOCULUM

Boot and spray inoculations of wheat with filiform secondary sporidia of *Tilletia indica* resulted in lower levels of karnal bunt than similar inoculations with allantoid secondary sporidia. Filiform sporidia were produced on liquid cultures, whereas allantoid secondary sporidia were produced on solid media.[97] A less virulent isolate of *F. o.* f. sp. *lycopersici* caused a higher percentage of tomato seed infection than a more virulent one by reaching the fruit and therefore the seeds more easily.[98]

At full seed and yellow pod stages, both strains B (mild strain) and D (severe strain) of the soybean mosaic virus (SbMV) was detected in soybean seed coats and embryos, but only strain B was detected in cotyledons. As seeds matured, detection using both ELISA and bioassay declined on French bean cv. Top Crop, especially in seed coats at full maturity when both strains lost infectivity. The detection values of strain B were higher than those of strain D. The virus in the embryo was considered responsible for transmission, but the virus in seed coat could have had an effect on seed transmission.[99] Resistance to SbMV seed transmission was strain–specific, and the soybean line x SbMV strain interactions were found for seed transmission.[100]

Seed transmission of tobacco streak virus (TSV) in beans depended on early movement and replication in pollen-associated tissues. The antigen levels of two virus isolates were similar in bean flower petals and ovaries infected with either isolate. However, mean antigen levels of one strain in stamen tissue were higher than those of the second strain.[101]

V. SEED QUALITY

Poor quality seeds carry a higher percent of seedborne pathogens then good quality seeds. In an alfalfa seed lot, the highest level (7.5%) of internal infection of *Verticillium albo-atrum* occurred in the smallest (<1.28 mg/seed) seeds, whereas no infection was found in the largest (>2.64 mg/seed) seeds.[102] A high percentage (52%) of the cowpea aphid-borne mosaic virus was transmitted in small, shriveled cowpea seeds, whereas wrinkled seed had 25% and large, smooth seeds had no virus present.[95] Likewise seed transmission of the peanut stripe virus was higher in small rather than in large peanut seeds.[103]

VI. PLANT GROWTH STAGE

The degree of seed infection may vary depending upon plant age, stage of development, and severity of plant infection. In general, a higher level of seeds get infected if plants were infected prior to flowering.

Fusarium moniliforme in maize kernels was detected 2 weeks after midsilk and decreased each week up to 35 to 65% at 10 weeks. *Acremonium strictum* infection in maize seeds was detected 3 to 4 weeks after midsilk and infection increased weekly to 30 to 45% at 10 weeks after midsilk.[104] The number of cottonseeds and embryos infected by *Colletotrichum gossypii* var. *cephalosporioides* was higher on plants when developed bolls were infected than at any other growth stage.[105,106] In screening barley and wheat cultivars against loose smut, floret inoculation gave a higher infection level than seedling inoculation, especially for wheat, but problems of plant mortality and standardization of tissue age were greater with the former method. Floret inoculation identified barley and wheat cultivars exhibiting embryo or seedling resistance.[107]

Wheat seed infection by *Pyrenophora tritici-repentis* occurs primarily after the early dough stage by first colonizing the glume, lemma, or palea.[108] Artificial inoculation at the time of flowering resulted in infection of all parts of the caryopsis, including the embryo of barley by the two forms of *D. teres* (*teres* and *maculata*) of the pathogen.[109] Soybean seeds infected with *C. kikuchii* were observed when large and small pods were inoculated with a conidial suspension. No infection was observed when flowers in either full or postbloom stages were inoculated.[54]

The highest percentage of loose smut infection in wheat was attained by inoculating at anthesis.[110] The percentage of embryos containing large amounts of hyphae decreased when inoculations were made after anthesis.[111] Embryos of barley were most susceptible to the loose smut fungus within 4 d of heading, with susceptibility decreasing after heading.[112] Parii[113] reported that the most effective stage for infection of winter wheat by the loose smut fungus was at full flowering and of barley at the beginning of flowering. In wheat, postanthesis was the best stage for seed infection by *A. alternata*.[114] Rice in flowering and milk stages was more susceptible than in soft dough and mature stages for seed infection by *Bipolaris oryzae*.[115] Developing barley seeds were susceptible to infection by *Drechslera graminea* during all stages of development, from head emergence through the soft-dough stage.[116] Susceptibility of sorghum ovaries to ergot (*Sphacelia sorghi*) decreased after fertilization. They were susceptible for up to 5 d, after which conidia were unable to infect ovaries.[117] Heavy infection of rye grain by *Fusarium culmorum* took place during flowering, whereas if infection took place during grain maturity, the percent of infection was low.[118]

Changes occurred in rice glumes after anthesis when a barrier was produced against infection by *Pyricularia oryzae*. In early infections, conidia and hyphae remained between glumes and caryopses, enabling survival of inoculum during unfavorable conditions.[119] Floret infection by the ergot fungus in barley and wheat during and shortly after anthesis resulted in high levels of infection. However, susceptibility was lost at 15 d after anthesis initiation. In fertilized ovaries, no infection occurred 9 d after fertilization.[120] Pearl millet earheads were susceptible to *Claviceps fusiformis* from emergence to 6 d, with a maximum at 4 d.[44] *Macrophomina phaseolina*-infected sunflower seeds were susceptible at anthesis when the outer quarter of the inflorescence radius was complete. The infection progressed until just before seeds reach commercial maturity.[121] Inoculation of *C.*

kikuchii to soybeans during full flower gave maximum infection.[122] Inoculations of spring wheat with *Stagonospora nodorum* at any growth stage resulted in seed infection; however, infection increased particularly between stages 6–10.1.[123,124] Inoculation of wheat ears at boot II or awn emergence stage with a mixed suspension of secondary sporidia from different cultures of *T. indica* resulted in more than 81% seed infection.[125] The disease did not develop when plants were inoculated at panicle initiation or milk stages.[126] In contrast, pigeon pea seed infection by *Fusarium pallidoroseum*, *Lasiodiplodia theobromae*, and *Phomopsis* did not occur until after seeds matured and then increased with time past field maturity.[127]

Location of inoculum within seeds was influenced by time of plant infection. For example, if flax seeds were infected early by *Aureobasidium lini* and *Septoria linicola*, hyphae grew into the embryo and killed it. If infection took place after seed coat development, hyphae did not invade the embryo and were confined to the seed coat.[128] Soybean seeds infected with *C. kikuchii* were observed when either small or large pods were inoculated with a conidial suspension, but no infection was observed when flowers either in full or postbloom were inoculated.[54] The incidence of *Colletotrichum gossypii* was highest in cottonseeds from plants inoculated at the boll stage.[129]

The total population of *Pseudomonas glumae* in rice seeds at heading time after inoculation at the boot stage was less than 10^7 cfu/g fresh weight of seeds, but it increased to 10^8 to 10^9 cfu/g during 6 d after heading.[130]

Maximum seed transmission of the bean common mosaic virus in bean seeds was recorded for test plants inoculated at the primary leaf stage. Twelve of fourteen cultivars tested did not transmit any strain in seeds of plants inoculated 30 d after sowing.[29] Seed transmission of the soybean mosaic virus in *Lupinus albus* seeds could not be detected in cv. Kiev seeds from flowers that already had been formed at the time of inoculation. However, seed transmission was observed in seeds obtained from flowers not yet formed at that time.[131] Cowpea plants inoculated with the cowpea banding mosaic virus before flowering produced a higher proportion of infected seeds compared to plants inoculated at different times after flowering; transmission declined progressively until plants inoculated at 77 d after germination did not produce any infected seeds.[95]

The level of virus seed transmission to progeny was influenced by the time plants became infected. In general, plants must be infected systemically before or shortly after flowering for successful seed transmission. Bean plants infected with the bean common mosaic virus (BCMV) during early vegetative growth produced more infected seeds than seeds from plants infected at later stages. A low percentage of transmission occurred if the plants were infected after blooming.[132,133] Seed transmission of BCMV was 17 and 11.9% in seeds obtained from plants infected at seedling and flowering stages, respectively.[134] Seeds from urd bean plants, which developed symptoms caused by the BCMV within 30 d of planting, had 17.3% seed transmission. Seed transmission decreased to 12.1% in plants that showed symptoms 30 to 40 d after planting, and decreased further to 5.1% if plants developed symptoms 40 to 50 d after planting.[135] Seedborne BSMV in barley was greatest when inoculations were made 10 d before heading.[136] In the susceptible barley cv. Compana, seed transmission of the BSMV increased with inoculations up to boot stage, with the maximum being 63.7%. Inoculations from heading to hard dough stages resulted in decreased seed transmission from 10.1 to 14.7%.[137] No transmission of BSMV occurred in plants inoculated 7 d prior to anthesis or thereafter. It was not clear whether a mechanical barrier prevented the virus from invading seeds after fertilization or if some inactivating substance was present following fertilization.[138] Couch[139] found that lettuce plants inoculated with the lettuce mosaic virus just before flowering produced fewer virus-infected seeds than plants infected soon after planting. Plants that became infected after flowering did not transmit the virus through seeds. The tobacco ringspot virus may have been transmitted up to 100% in soybean seeds from plants infected before flowering, but the percentage decreased if plants were infected just prior to or immediately after flowering.[140] Seed transmission of the cowpea banding mosaic virus varied from 15 to 32% in different cowpea cultivars. Seeds from plants infected 31 d after sowing carried 32% virus, while those infected 33 and 35 d after sowing carried 22 and 21.5%, respectively.[141] Soybean plants inoculated with the soybean mosaic or cucumber mosaic viruses just before flowering produced fewer infected seeds than those inoculated earlier.[142] Time of inoculation affected the soybean mosaic virus transmission through soybean seeds. Transmission was 16 or 3% if inoculation was before or after onset of flowering, respectively.[143] Seed transmission of the pea seedborne mosaic virus in pea was highest from plants infected before flowering.[144]

A greater reduction in soybean plant growth and seed yield and a higher percentage of mottled seeds and seed infection by soybean mosaic virus occurred with early infection. Virus titer was higher in young than in old plants and in plants infected at an early compared to a later growth stage.[145] The percentage seed transmission of tobacco mosaic virus was higher in seeds from pepper plants inoculated at 10 d than those at 30 or 60 d.[146]

VII. MOTHER PLANT INFECTION

It has been suggested that seed infection correlates with symptom severity in the mother plant. Sorghum ears harvested from plants systemically infected with *Peronosclerospora sorghi* often contained infected seeds. Seeds from plants with severe symptoms gave a higher level of seed transmission than those with mild symptoms.[147] Infection of pepper seeds by *Colletotrichum capsici* was correlated with fruit spotting, with 0 to 35% seed infection from clean fruits and 1 to 40% infection from spotted fruits.[148] The incidence of fava bean seed infection by *Ascochyta fabae* after harvest was strongly correlated with foliar and pod infection, but the range of seed infection was narrower than that for pods.[17] Peanut seed colonization with *Cylindrocladium crotalariae* was directly correlated with disease incidence in the field. Seed colonization increased as disease incidence increased.[149] Infection of fava bean seeds by *Ascochyta fabae* increased with pod infection. Seeds from symptomless pods had a low level of infection, suggesting latent or systemic infection.[150] Moderate infection of pepper seeds with *Colletot-*

richum dematium was confined to the seed coat and superficial layers of the outer endosperm and stem region, while in heavily infected seeds, inter- and intracellular mycelium and acervuli were observed throughout. Severe infection caused disintegration of seed coat parenchymatous tissue and depletion of nutrients in the endosperm and embryo, disruption of cell division in the palisade cells, and formation of lysigenous cavities in the embryo.[151] Wheat seed infection by *Pyrenophora tritici-repentis* was positively correlated with tan spot severity on flag leaves shortly after anthesis.[108]

A significant correlation was found between the incidence of *X. c.* pv. *phaseoli* on bean plants and seeds.[152] Smith and Hewitt, [153] studying 51 bean cultivars infected with the bean common mosaic virus, reported that seed transmission ranged from less than 1% to more than 75%. Seed transmission was correlated with cultivar reaction to the virus. Symptomless carriers had seed transmission ranging from 1.1 to 6.8%; plants with mild, moderate, or severe symptoms had 9.4 to 20.8%, 20.3 to 23.3%, and 30.4 and 36.1%, respectively.[153] A similar correlation of seed transmission with disease severity was reported by Singh et al.[137] with barley cultivars infected with the barley stripe mosaic virus (BSMV). Four tolerant cultivars showed less than 15% transmission, whereas severely diseased cultivars produced up to 75% infected progeny. A correlation between symptom intensity and seed infection of BSMV in barley also was reported.[154]

Seed transmission of the peanut stunt virus varied with disease severity on plants from which the seeds were harvested and seed size. Seed transmission in 11 of 5307 peanut seeds from severely stunted plants and 1 of 13,676 seeds from mildly stunted plants was recorded.[155] The rate of seed transmission of cucumber mosaic virus in lupin depended on time of plant inoculation. The maximum rate was 25.4% for plants inoculated at the mid-vegetative growth stage (58 days postemergence).[156] Host factors as well as bean pod mottle virus (BPMV) infections that modify late season seed maturation influenced *Phomopsis* seed infection in soybeans. BPMV increased pod infection by *Phomopsis*, if susceptible soybean entries were considered, and seedborne *Phomopsis* due to increased pod infection. However, seed infection was lower than pod infection in resistant soybean entries. Major seed infection occurred during the R7 to R8 growth stages, when pod and seed moisture declined. BPMV infection extended dry down periods, allowing increased levels of seedborne *Phomopsis*.[157]

VIII. INSECT INFESTATION

Insects may help establish seed infection. In Delaware, maize earworms made holes through the exocarp, mesocarp, and endocarp of soybean pods of cvs. Dorman and Hill and exposed seeds to *Diaporthe phaseolorum*. It led to a manyfold increase in pod infection.[158] Bean leaf beetle (*Cerotoma trifurcata*) caused extensive pod damage in soybean pods, with the incidence of *Alternaria tenuissima* correlated with pod injury (r = 0.53) (Figure 5-1). The wounding and stressing of pods associated with beetle feeding predisposed seeds to infection

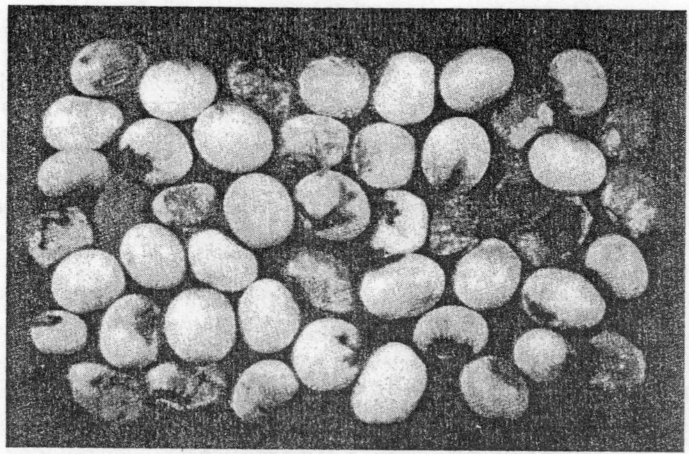

Figure 5-1 A soybean (*Glycine max*) seed sample, showing smaller than normal, shriveled, discolored seeds injured by *Cerotoma trifurcata* (bean leaf beetle) and infected with *Alternaria tenuissima*. (From Shortt, B. J., Sinclair, J. B., Helm, C. G., Jeffords, M. R., and Kogan, M., *Phytopathology*, 72, 615, 1982. With permission.)

by *A. tenuissima*.[159] The asparagus beetle produced tunnels in asparagus seeds, which resulted in the contamination of seeds by *Fusarium oxysporum* f. sp. *asparagi* and *F. moniliforme*.[160]

The number of soybean seeds infected with *A. tenuissima* was related to bean leaf beetle injury of pods and differential cultivar susceptibility to pod injury.[159] Green stinkbugs (*Acrosternum hilare*) damaged soybean pods and seeds by mechanical feeding and transmitted *Nematospora coryli*, the yeast spot disease.[161] In comparison to noninjured seeds, southern green stinkbug (*Nezara viridula*) injury resulted in higher numbers of seeds infected with *Chaetomium*, *Fusarium*, and *Penicillium*.[162] Levels of *Fusarium* and plant pathogenic bacteria increased with increased stinkbug populations and damage in soybeans. High levels of stinkbug infestation caused an increase of 65% in infected seeds.[162] A higher incidence of *Alternaria* was found associated with the pod injury, caused by bean leaf beetles (*Cerotoma trifurcata*), because of their preference for narrow rows or weeds between narrow rows in soybean.[37,77]

IX. PATHOGEN INTERACTION

Seeds can be the site of antagonism and synergism among microorganisms and viruses. *Aspergillus flavus* growing on peanut shell kernels had a depressing effect on the rate and extent of kernel infection by *Macrophomina phaseolina*.[163] *Bipolaris maydis* predisposed maize seeds to fungal invasion by *Aspergillus flavus*, *Fusarium moniliforme*, and *Penicillium* and aflatoxin production.[164] Significant negative correlations were obtained between the level of soybean seed infection by *Cercospora kikuchii* and that by *Alternaria*, *Fusarium*, and *Phomop-*

sis, with a significant positive correlation between seed infection by *Fusarium* and *Phomopsis*.[165]

The grain mold pathogens, *Curvularia lunata*, *F. moniliforme*, and *Phoma sorghina*, interacted with each other in culture and *in vivo* in sorghum. *F. moniliforme* was antagonistic to *C. lunata* and *P. sorghina*, whereas *C. lunata* was antagonistic to *P. sorghina*.[166] *F. moniliforme* produced a toxin, fusaric acid, which may be inhibitory to *C. lunata* and *Phoma sorghina*. In addition, *F. moniliforme* was found to colonize the endosperm and embryo of grain extensively, which may restrict colonization by *C. lunata* and *F. moniliforme*.[166] A negative correlation between the isolation frequency of *F. graminearum* and *F. moniliforme* from maize seeds was reported.[167,168] They attributed this to competition for substrate, production of antagonistic substances, or environmental conditions that differently influence maize ear infection by these two fungi. *F. moniliforme* was negatively correlated to *F. moniliforme* var. *subglutinans* and *Diplodia maydis* maize kernel infection.[168] A negative association in maize also has been reported between *Diplodia macrospora* and *D. maydis*.[168] Reports from South Africa indicated that *F. moniliforme* was adapted to warm, dry climates, *F. moniliforme* var. *subglutinans* in maize to cooler, more temperate climates, and *F. graminearum* to intermediate climates.[169] An antagonistic effect was observed between *Cercospora kikuchii* and *Phomopsis* infecting soybean seeds; the higher the level of *C. kikuchii* seed infection, the lower seed infection by *Phomopsis*.[170] Cercosporin is a nonspecific toxin produced by *C. kikuchii* and *C. sojina*, which was fungistatic to *Alternaria alternata*, *Diaporthe phaseolorum* f. sp. *sojae*, *Fusarium*, *Macrophomina phaseolina*, and *Phomopsis longicolla*. This inhibition suggested involvement of cercosporin in seed colonization by *Cercospora*.[171]

In Puerto Rico, *C. kikuchii* infection in soybean seeds was antagonistic to *Fusarium* and *Phomopsis* infection. These latter two fungi were recovered three to six times more often from unstained than from purple-stained seeds. In contrast, *C. kikuchii* was not antagonistic to *Phomopsis* in Illinois, except when incidence of *C. kikuchii* exceeded 10%. Recoveries of *Fusarium* and *Phomopsis* increased and germination and recoveries of *C. kikuchii* decreased when harvest was delayed in Puerto Rico.[171] The incidence of *Phomopsis* in soybean seeds increased five times in plants inoculated with the bean pod mottle virus.[172] Inoculations of soybean plants with the soybean mosaic virus before or during flowering increased the incidence of seed infection by *P. longicolla* (Figure 5-2).[173]

Multiviral infection of plants influences seed transmission. Seed transmission of the soybean mosaic virus (SbMV) in plants infected with the bean pod mottle virus (BPMV) either decreased[174] or increased, depending upon the cultivar.[175-177] Seed transmission of SbMV was 11.1 and 6%, respectively, in seeds from soybean plants infected with SbMV alone or SbMV plus the BPMV.[174] Quiniones et al.[175] found that SbMV caused mottling on 92% of seeds and, when combined with the BPMV, on 96% of seeds; SbMV was transmitted in 27% of the seeds from plants inoculated with the virus alone and in 39% of the seeds inoculated with both viruses. A mixed infection of the cowpea chlorotic mottle virus and southern bean mosaic virus (SBMV) (cowpea strain) increased seed transmission of SBMV from 12 to 20% in cowpea.[178]

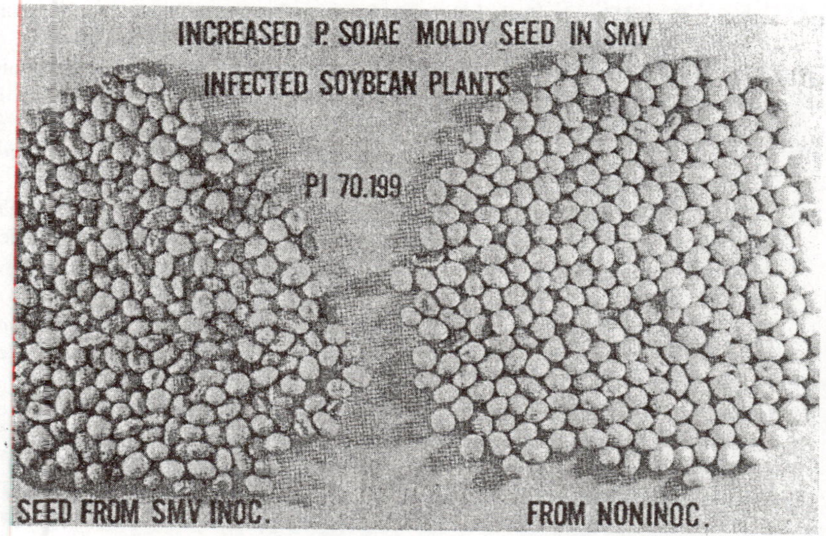

Figure 5-2 Soybean (*Glycine max*) seed lots from soybean mosaic virus (SbMV)-inoculated field plants (*left*), and noninoculated plants (*right*) showing increased incidence of Phomopsis seed decay (*Phomopsis*) from SbMV-inoculated plants. (Courtesy of P.R. Hepperly, U.S. Department of Agriculture, Agricultural Research Station, Mayaguez, Puerto Rico.)

Peanut seeds heavily infested with testa nematode, *Aphelenchoides arachidis*, had higher levels of infection by *Fusarium*, *Macrophomina phaseolina*, *Rhizoctonia solani*, and *Sclerotium rolfsii*. Nematode infestation resulted in damage to the testa, which facilitated fungal entry.[179]

REFERENCES

1. Roy, K. W. and Abney, T. S., Purple seed stain of soybeans, *Phytopathology*, 66, 1045, 1976.
2. Velicheti, R. K. and Sinclair, J. B., Production of cercosporin and colonization of soybean seed coats by *Cercospora kikuchii*, *Plant Dis.*, 78, 342, 1994.
3. Orth, C. E. and Schuh, W., Resistance of 17 soybean cultivars to foliar, latent and seed infection by *Cercospora kikuchii*, *Plant Dis.*, 78, 661, 1994.
4. Gaskin, T. A. and Schafer, J. F., Some histological and genetic relationships of resistance of wheat to loose smut, *Phytopathology*, 52, 602, 1962.
5. Batts, C. C. V., Observations on the infection of wheat by loose smut (*Ustilago tritici* (Pers.) Rostr.), *Trans. Br. Mycol. Soc.*, 38, 465, 1955.
6. Michalikova, A., Effect of the composition of substances in the stigma of some wheat varieties on chlamydospore germination of *Ustilago tritici* and on the resistance of plants to penetration of the pathogen, *Acta Inst. Bot. Acad. Sci. Slov. B*, 1, 63, 1976.
7. Krivchenko, V. I. and Yamaleev, A. M., Field and embryonic resistance of wheat species to loose smut, *Tr. Prikl. Bot. Genet. Sel.*, 53, 57, 1974.

8. Loiselle, R. and Shands, R. G., Influence of maternal tissue on loose smut infection of hybrid barley kernels, *Can. J. Bot.*, 38, 741, 1960.
9. Orton, T. J., Physical association of *Fusarium oxysporum* f. sp. *apii* with germinating celery seeds following artificial inoculation, *Phytopathol. Z.*, 102, 320, 1981.
10. Scott, G. E. and King, S. B., Site of action of factors for resistance to *Fusarium moniliforme* in maize, *Plant Dis.*, 68, 804, 1984.
11. Headrick, J. M., Pataky, J. K., and Juvik, J. A., Relationships among carbohydrate content of kernels, conditions of silks after pollination and the response of sweet corn inbred lines to infection of kernels by *Fusarium moniliforme*, *Phytopathology*, 80, 328, 1990.
12. Headrick, J. M. and Pataky, J. K., Resistance to kernel infection by *Fusarium moniliforme* in sweet corn inbred lines and the effect of infection on emergence, *Plant Dis.*, 73, 887, 1989.
13. Headrick, J. M. and Pataky, J. K., Maternal influence on the resistance of sweet corn lines to kernel infection by *Fusarium moniliforme*, *Phytopathology*, 81, 268, 1991.
14. Warren, H. L., Comparison of normal and high-lysine maize inbreds for resistance to kernel rot caused by *Fusarium moniliforme*, *Phytopathology*, 68, 1331, 1978.
15. Semisi, S. T. and Ball, S. F. L., Infection of the pearlmillet (*Pennisetum americanum*) inflorescence by the downy mildew fungus (*Sclerospora graminicola*), *Plant Pathol.*, 38, 571, 1989.
16. Asmus, G. L. and Dhingra, O. D., The use of a seed infection index for comparing the susceptibility of bean cultivars to internally seedborne pathogens, *Seed Sci. Technol.*, 13, 53, 1985.
17. Lockwood, G., Jellis, G. J., and Aubury, R. G., Genotypic influence on the incidence of infection by *Ascochyta fabae* in winter hardy faba beans (*Vicia faba*), *Plant Pathol.*, 34, 341, 1985.
18. Ploper, L. D., Abney, T. S., and Roy, K. W., Influence of soybean genotype on rate of seed maturation and its impact on seedborne fungi, *Plant Dis.*, 76, 287, 1992.
19. Fujita, Y. and Suzuki, H., Histological study on the resistance in soybean (*Glycine max*) and a wild soybean (*G. soja*) to purple seed stain caused by *Cercospora kikuchii*, *Ann. Phytopathol. Soc. Jpn.*, 54, 151, 1988.
20. Jost, J. P., Volken, P. A., and Kern, H., A disease of the anthers of *Trifolium pratense* caused by *Botrytis anthophila*, *Ber. Schweiz. Bot. Ges.*, 74, 277, 1964.
21. Kaiser, W. J., Occurrence of three fungal diseases of chickpea in Iran, *FAO Plant Prot. Bull.*, 20, 74, 1972.
22. Prasad, K. and Weigle, J. L., Association of seedcoat factors with resistance to *Rhizoctonia solani* in *Phaseolus vulgaris*, *Phytopathology*, 66, 342, 1976.
23. Presser, E., The importance of the seedcoat in the resistance of dwarf bean varieties (*Phaseolus vulgaris* var. *nanus*) to *Colletotrichum lindemuthianum*, *Zuchter*, 36, 36, 1966.
24. Anderson, W. H. and Banttari, E. E., The effect of *Bipolaris sorokiniana* on yield, kernel weight, and kernel discoloration in six-row spring barleys, *Plant Dis. Rep.*, 60, 754, 1976.
25. Pizzinatto, M. A., Cia, E., and Fuzatto, M. G., Relationship between the severity of cotton ramulosis under field conditions and the presence of *Colletotrichum gossypii* var. *cephalosporioides* in the seeds produced, *Fitopatol. Brasileira*, 19, 50, 1994.

26. Jacobs, S. E. and Dadd, A. H., Antibacterial substances in seed coats and their role in the infection of sweet peas by *Corynebacterium fascians* (Tilford) Dowson, *Ann. Appl. Biol.*, 47, 666, 1959.
27. Carroll, T. W., Gossel, P. L., and Hockett, E. A., Inheritance of resistance to seed transmission of barley stripe mosaic virus in barley, *Phytopathology*, 69, 431, 1979.
28. Drijfhout, Y. and Bos, L., The identification of two new strains of bean common mosaic virus, *Neth. J. Plant Pathol.*, 83, 13, 1977.
29. Morales, F. J. and Castano, M., Seed transmission characteristics of selected bean common mosaic virus strains in differential bean cultivars, *Plant Dis.*, 71, 51, 1987.
30. Gunnery, S. and Datta, A., An inhibitor RNA of translocation from barley embryos, *Biochem. Biophys. Res. Commun.*, 142, 383, 1987.
31. Matthews, R. E. F., *Plant Virology*, 3rd ed., Academic Press, New York, 1991, 835.
32. Wang, D., Woods, R. D., Cockbain, A. J., Maule, A. J., and Biddle, A. J., The susceptibility of pea cultivars to pea seedborne mosaic virus infection and virus seed transmission in the U.K., *Plant Pathol.*, 42, 42, 1993.
33. Cunfer, B. M., The incidence of *Septoria nodorum* in wheat seed, *Phytopathology*, 68, 832, 1978.
34. Arndt, C. H., Cotton seed produced in South Carolina in 1954 and 19[illegible]5, its viability and infestation by fungi, *Plant Dis. Rep.*, 40, 1001, 1956.
35. Muskett, A. E., Seed health in relation to flax, *Br. Agric. Bull.*, 5, 2, 1953.
36. Balducchi, A. J. and McGee, D. C., Environmental factors influencing infection of soybean seeds by *Phomopsis* and *Diaporthe* species during seed maturation, *Plant Dis.*, 71, 209, 1987.
37. Jordan, E. G., Sinclair, J. B., Manandhar, J. B., and Thapliyal, P. N., Soil type and other field conditions affecting seedborne fungi in Illinois soyabeans, *Seed Sci. Technol.*, 20, 619, 1992.
38. Wilcox, J. R., Abney, T. S., and Frankenberger, E. M., Relationships between seedborne soybean fungi and altered photoperiod, *Phytopathology*, 75, 797, 1985.
39. Atkins, I. M., Merkle, O. G., Porter, K. B., Lahr, K. A., and Weibel, D. E., The influence of environment on loose smut percentages, reinfection, and grain yields of winter wheat at four locations in Texas, *Plant Dis. Rep.*, 47, 192, 1963.
40. Ross, J. G., Semeniuk, W., Taylor, D. K., and Jenkins, B. C., Factors affecting the degree of infection of barley by loose smut (*Ustilago nuda*) (Jens.) Rostr., Sci. Agric., 28, 481, 1948.
41. Tapke, V. F., The role of humidity in the life cycle, distribution and control of loose smut fungus in wheat, *Phytopathology*, 19, 103, 1929.
42. Tapke, V. F., Influence of humidity on floral infection of wheat and barley by loose smut, *J. Agric. Res.*, 43, 503, 1931.
43. Dunleavy, J., Fusarium blight of soybean, *Proc. Iowa Acad. Sci.*, 68, 106, 1962.
44. Reddy, K. D., Govindaswamy, C. V., and Vidhyasekaran, P., Studies on ergot disease of cumbu (*Pennisetum typhoides*), *Madras Agric. J.*, 56, 367, 1969.
45. Marshall, G. M., The incidence of certain seed-borne diseases in commercial seed samples. II. Ergot, *Claviceps purpurea* (Fr.) Tul. in cereals, *Ann. Appl. Biol.*, 48, 19, 1960.
46. Leach, L. D., Seed-borne *Phoma* and its relation to the origin of sugarbeet seed lots, in Proc. 4th General Meeting — 1946, Am. Soc. Sugarbeet Technol., 1946, 381.
47. Doupnik, B., Fungi associated with preharvest discoloration in grain sorghum, *Proc. Am. Phytopathol. Soc.*, 1, 103, 1974.

48. Harman, G. E., Heit, C. E., Pfleger, F. L., and Braverman, S. W., Snapdragon seed blight — A serious problem caused by seed-borne fungi, *Plant Dis. Rep.*, 57, 592, 1973.
49. Mishra, C. B. P., Studies on *Fusarium* species on wheat cariopses and proof of their pathogenicity as foot disease agents, *Arch. Phytopathol. Pflanzench.*, 9, 123, 1973.
50. Spilker, A., Schmitthenner, A. F., and Ellett, C. W., Effects of humidity, temperature, fertility and cultivar on the reduction of soybean seed quality by *Phomopsis* species, *Phytopathology*, 71, 1027, 1981.
51. Bedi, K. S., Sikka, M. R., and Mundkur, B. B., Transmission of wheat bunt due to *Neovossia indica* (Mitra) Mundkur, *Indian Phytopathol.*, 2, 20, 1949.
52. Hansen, E. W. and Christensen, J. J., Black point disease of wheat in the United States, *Tech. Bull. Minn. Agric. Exp. Stn.*, 206, 30, 1955.
53. Khetarpal, R. K., Agarwal, V. K., and Chauhan, K. P. S., Studies on the influence of weather conditions on the incidence of black point and karnal bunt of triticale, *Seed Res.*, 9, 725, 1980.
54. Schuh, W., Effect of pod development stage, temperature, and pod wetness duration on the incidence of purple seed stain of soybeans, *Phytopathology*, 82, 446, 1992.
55. Shah, D. and Bergstrom, G. C., Assessment of seed-borne *Stagonospora nodorum* in New York soft winter wheat, *Plant Dis.*, 77, 468, 1993.
56. Duveiller, E. and Bragard, C., Comparison of immunofluorescence and two assays for detection of *Xanthomonas campestris* pv. *undulosa* in seeds of small grains, *Plant Dis.*, 76, 999, 1992.
57. Jackson, C. R., Peanut kernel infection and growth *in vitro* by four fungi at various temperatures, *Phytopathology*, 55, 46, 1965.
58. Joshi, L. M., Singh, D. V., and Srivastava, K. D., Meteorological conditions in relation to incidence of karnal bunt of wheat in India, in 3rd Int. Symp. Plant Pathol., Indian Agric. Res. Instit., New Delhi, 1981, 11.
59. Hino, T. and Furuta, T., Studies on the control of Bakanae disease of rice plants, caused by *Gibberella fujikuroi*. II. Influence on flowering season of rice plants and seed transmissibility through flower infection, *Bull. Chugoku Agric. Exp. Stn.*, E 2, 97, 1968.
60. Frosheiser, F. I., Alfalfa mosaic virus transmission to seed through alfalfa gametes and longevity in alfalfa seed, *Phytopathology*, 64, 102, 1974.
61. Adams, D. B. and Kuhn, C. W., Seed transmission of peanut mottle virus in peanuts, *Phytopathology*, 67, 1126, 1977.
62. Jorgensen, J., The incidence of *Cochliobolus sativus* on seed samples of barley and oats in Denmark during 20 years related to climatical conditions and other factors, Preprint of Report from the Danish State Seed Testing Station, 11, 107, 1986.
63. Loria, R. and Jones, A. L., Dispersal of *Ustilago tritici* teliospores and subsequent loose smut infection as influenced by environmental factors, *Phytopathology*, 71, 238, 1981.
64. Mathre, D. E. and Johnston, R. H., Dispersion of loose smut in foundation barley seed fields in a semi-arid climate, *Plant Dis. Rep.*, 59, 979, 1975.
65. Byford, W. J., Factors influencing the prevalence of *Pleospora bjorlingii* on sugarbeet seed, *Ann. Appl. Biol.*, 89, 15, 1978.

66. Leach, L. D. and MacDonald, J. D., Seed-borne *Phoma betae* as influenced by area of sugarbeet production, seed processing and fungicidal seed treatments, *J. Am. Soc. Sugar Beet Technol.*, 19, 4, 1976.
67. Hewett, P. D., A survey of seedborne fungi of wheat. I. The incidence of *Leptosphaeria nodorum* and *Griphosphaeria nivalis*, *Trans. Br. Mycol. Soc.*, 48, 59, 1965.
68. Klich, M. A., Mycoflora of cotton seed from the southern United States: a three-year study of distribution and frequency, *Mycologia*, 78, 706, 1986.
69. Verma, H. S., Singh, A., and Agarwal, V. K., Environmental factors in relation to loose smut infection in wheat seeds, *Indian Phytopathol.*, 39, 423, 1986.
70. Rupe, J. C. and Ferriss, R. S., The effect of moisture on infection of soybean seeds by *Phomopsis* sp., *Phytopathology*, 74, 632, 1984.
71. Cunfer, B. M., Epidemiology and control of seed-borne *Septoria nodorum* on wheat, *Seed Sci. Technol.*, 11, 707, 1983.
72. Halfon-Meiri, A., Personal communication, The Volcanic Center, Bet-Dagan, Israel, 1984.
73. Knight, T. L., Martin, R. J., and Harvey, I. C., Management factors affecting lentil production in mid-Canterbury, *Proc. Annu. Conf. N. Z. Agron. Soc.*, 19, 17, 1989.
74. Jordan, E. G., Manandhar, J. B., Thapliyal, P. N., and Sinclair, J. B., Soybean seed quality of 16 cultivars and four maturity groups in Illinois, *Plant Dis.*, 72, 64, 1988.
75. Pedersen, E. A., Bedi, S., and Morrall, R. A. A., Gradients of Ascochyta blight in Saskatchewan lentil crops, *Plant Dis.*, 77, 143, 1993.
76. Unger, P. W. and McCalla, T. M., Conservation tillage system, *Adv. Agron.*, 33, 1, 1990.
77. Bowman, J. E., Hartman, G. L., McClary, R. D., Sinclair, J. B., Hummel, J. W., and Wax, L. M., Effect of weed control and row spacing in conventional tillage, reduced tillage, and nontillage on soybean seed quality, *Plant Dis.*, 70, 673, 1986.
78. Tyler, D. D. and Overton, J. R., Non-tillage advantages for soybean, *Glycine max* cultivar Forrest, seed quality during drought stress, *Agron. J.*, 7, 344, 1982.
79. Agarwal, V. K., Singh, O. V., and Modgal, S. C., Influence of different doses of nitrogen and spacing on the seedborne infections of rice, *Indian Phytopathol.*, 28, 38, 1975.
80. Wilcox, J. R. and Abney, T. S., Association of pod and stem blight with stem breakage in soybeans, *Plant Dis. Rep.*, 55, 776, 1971.
81. Ooka, J. J. and Kommedahl, T., Kernels infected with *Fusarium moniliforme* in corn cultivars with opaque-2-endosperm or male-sterile cytoplasm, *Plant Dis. Rep.*, 61, 162, 1977.
82. Martin, R. A., MacLeod, J. A., and Caldwell, C., Influences of production inputs on incidence of infection by *Fusarium* species on cereal seed, *Plant Dis.*, 75, 784, 1991.
83. Ito, M. F., Mascarenhas, H. A. A., Tanaka, M. A. S., Tanaka, R. J., Gallo, P. B., and Miranda, M. A. C., Residual effect of potassium and liming on the incidence of *Phomopsis* spp. in soybean seeds, *Fitopatol. Brasileira*, 19, 44, 1994.
84. Ploper, L. D. and Abney, T. S., Influence of host genotype on rate of moisture loss during soybean seed maturation, in Proc. South Soybean Dis. Workers Co. 13th, 1986, 17.
85. Abney, T. S. and L. D. Ploper, Growth regulator effects on soybean seed maturation and seedborne fungi, *Plant Dis.*, 75, 787, 1991.

86. Fauzi, M. T. and Paulitz, T. C., The effect of plant growth regulators and nitrogen on Fusarium head blight of the spring wheat cultivar Max, *Plant Dis.*, 78, 289, 1994.
87. Chagas, D. and Dhingra, O. D., Effect of timing of weed control on the incidence of seed-borne fungi in dry bean seeds, *Fitopatol. Brasileira*, 4, 423, 1979.
88. Conner, R. L., Influence of irrigation timing on black point incidence in soft white spring wheat, *Can. J. Plant Pathol.*, 9, 301, 1987.
89. Tervet, I. W., The influence of fungi on storage on seed, viability, and seedling vigor of soybean, *Phytopathology*, 35, 3, 1945.
90. Kmetz, K. T., Schmitthenner, A. F., and Ellett, C. W., Soybean seed decay: prevalence of infection and symptom expression caused by *Phomopsis* sp., *Diaporthe phaseolorum* var. *sojae* and *D. phaseolorum* var. *caulivora*, *Phytopathology*, 68, 836, 1978.
91. Osorio, J. A. and McGee, D. C., Effects of freeze damage on soybean seed mycoflora and germination, *Plant Dis.*, 76, 879, 1992.
92. Kilpatrick, R. A., Fungi associated with soybean seeds within the pods at Stoneville, Mississippi in 1951, *Phytopathology*, 42, 285, 1952.
93. Wilcox, J. R., Laviolette, F. A., and Athow, K. L., Deterioration of soybean seed quality associated with delayed harvest, *Plant Dis. Rep.*, 58, 130, 1974.
94. Hagita, T. and Tamada, T., Detection of bean common mosaic virus in fresh bean seeds by immune electron microscopy, *Bull. Hokkaido Exp. Stn.*, 51, 83, 1984.
95. Sharma, S. R. and Varma, A., Transmission of cowpea banding mosaic and cowpea chlorotic spot viruses through the seeds of cowpea, *Seed Sci. Technol.*, 14, 217, 1986.
96. Bailiss, K. W. and Offei, S. K., Alfalfa mosaic virus in lucerne seed during seed maturation and storage, and in seedlings, *Plant Pathol.*, 39, 539, 1990.
97. Warham, E. J. and Burnett, P. A., Influence of media on pathogenicity and morphology of secondary sporidia of *Tilletia indica*, *Plant Dis.*, 74, 525, 1990.
98. Montorsi, F., Giambattista, G., and Conca, G., Transmissibility from plant to seed of two isolates of *Fusarium oxysporum* f. sp. *lycopersici* (race 1) with different virulence towards two susceptible cultivars of tomato, *Sementi Elette*, 38, 31, 1992.
99. Iwai, H., Ito, T., Sato, K. and Wakimoto, S., Distribution patterns of soybean mosaic virus strains B and D in soybean seeds at different growth stages, *Ann. Phytopathol. Soc. Jpn.*, 51, 475, 1985.
100. Bowers, G. R., Jr. and Goodman, R. M., Strain specificity of soybean mosaic virus seed transmission in soybean, *Crop Sci.*, 31, 1171, 1991.
101. Walter, M. H., Kaiser, W. J., Klein, R. E., and Wyatt, S. D., Association between tobacco streak ilarvirus seed transmission and anther tissue infection in bean, *Phytopathology*, 82, 412, 1992.
102. Gilbert, R. G. and Peaden, R. N., Dissemination of *Verticillium albo-atrum* in alfalfa by internal seed inoculum, *Can. J. Plant Pathol.*, 10, 73, 1988.
103. Xu, Z., Chen, K., Zhang, Z., and Chen, J., Seed transmission of peanut stripe virus in peanut, *Plant Dis.*, 75, 723, 1991.
104. King, S. B., Time of infection of maize kernels by *Fusarium moniliforme* and *Cephalosporium acremonium*, *Phytopathology*, 71, 796, 1981.
105. Lima, E. F., Carvalho, J. M. F. C., Carvalho, L. P. de, and Costa, J. N. da., Transport and transmissibility of *Colletotrichum gossypii* var. *cephalosporioides* through cotton seed, *Fitopatol. Brasileira*, 10, 105, 1985.

106. Tanaka, M. A. S. and Menten, J. O. M., Relationship between cotton resistance to ramulose and seed transmission of the pathogen, *Summa Phytopathol.*, 18, 227, 1992.
107. Jones, P. and Dhitaphichit, P., Comparison of responses to floret and seedling inoculation in wheat — *Ustilago tritici* and barley — *U. nuda* combinations, *Plant Pathol.*, 40, 268, 1991.
108. Schilder, A. M. C. and Bergstrom, G. C., Infection of wheat seed by *Pyrenophora tritici-repentis, Can. J. Bot.*, 72, 510, 1994.
109. Youcef-Benkada, M., Bendamane, B. S., Sy, A. A., Barrault, G., and Albertini, L., Effect of inoculation of barley inflorescences with *Drechslera teres* upon the location of seedborne inoculum and its transmission to seedlings as modified by temperature and soil moisture, *Plant Pathol.*, 43, 350, 1994.
110. Loria, R., Wiese, M. V., and Jones, A. L., Effects of free moisture, head development and embryo accessibility on infection of wheat by *Ustilago tritici*, *Phytopathology*, 72, 1270, 1982.
111. Ohms, R. E. and Bever, W. M., Effect of time of inoculation of winter wheat with *Ustilago tritici* on the percent of embryos infected and the abundance of hyphae, *Phytopathology*, 46, 157, 1956.
112. Campbell, W. P. and Tyner, L. E., Comparison of degree and duration of susceptibility of barley to ergot and true loose smut, *Phytopathology*, 49, 348, 1959.
113. Parii, I. F., Flowering phases of cereals and loose smut, *Zashch. Rast.*, 17, 43, 1974.
114. Aulakh, K. L. and Joshi, L. M., Black point of wheat, *Indian Phytopathol.*, 74, 41, 1974.
115. Fazli, I. S. F. and Schroeder, H. W., Effect of kernel infection of rice by *Helminthosporium oryzae* on yield and quality, *Phytopathology*, 56, 1003, 1966.
116. Teviotdale, B. L. and Hall, D. H., Factors affecting inoculum development and seed transmission of *Helminthosporium gramineum*, *Phytopathology*, 66, 295, 1976.
117. Puranik, S. B., Padaganur, G. M., and Hiremath, R. V., Susceptibility period of sorghum ovaries to *Sphacelia sorghi*, *Indian Phytopathol.*, 26, 586, 1973.
118. Baltzer, U., Untersuchungen über die Anfälligkeit des Roggens für Fusariosen, *Phytopathol. Z.*, 2, 377, 1930.
119. Bernaux, P., Evolution of the susceptibility of rice glumes to *Pyricularia oryzae* Cav. and *Drechslera oryzae* (Br. de Haan) Sub & Jain: consequences for disease transmission, *Agronomie*, 1, 261, 1981.
120. Puranik, S. B. and Mathre, D. E., Biology and control of ergot on male sterile wheat and barley, *Phytopathology*, 61, 1075, 1971.
121 Raut, J. G., Anthesis and seed development stages of sunflower in relation to seedborne infection of *Macrophomina phaseolina* and comparison of charcoal rot affected plants with healthy plants, in 3rd Int. Symp. Plant Pathol., Indian Agric. Res. Inst., New Delhi, 1981, 116.
122. Laviolette, F. A. and Athow, K. L., *Cercospora kikuchii* infection of soybean as affected by stage of plant development, *Phytopathology*, 62, 771, 1972.
123. Brönnimann, A., On *Septoria nodorum* Berk, the pathogen causing leaf blotch and glume blotch of wheat, *Phytopathol. Z.*, 61, 101, 1968.
124. Cooke, B. M. and Jones, D. G., The epidemiology of *Septoria tritici* and *S. nodorum*. II. Comparative studies of head infection by *Septoria tritici* and *S. nodorum* on spring wheat, *Trans. Br. Mycol. Soc.*, 54, 395, 1970.

125. Dhiman, J. S. and Bedi, P. S., A method for inoculation and detection of karnal bunt of wheat under field conditions, in 3rd Int. Symp. Plant Pathology, Indian Agric. Res. Inst., New Delhi, 1981, 230.
126. Singh, R. A. and Krishna, A., Susceptible stage for inoculation and effect of Karnal bunt on viability of wheat seed, *Indian Phytopathol.*, 35, 54, 1982.
127. Ellis, M. A., Foor, S. R., and Paschal, E. H., Effect of internally seedborne fungi on germination of pigeonpea in Puerto Rico, *Proc. Soc. Puertarriquena Ciencias Agric.*, 2, 8, 1977.
128. Johansen, G., Hørsygdomme, *Tidsskr. Planteavl.*, 48, 187, 1943.
129. Tanaka, M. A. S. and Menten, J. O. M., Relationship between cotton resistance ramulose and seed transmission of the pathogen, *Summa Phytopathol.*, 18, 227, 1992.
130. Tsushima, S., Tsuno, K., Mogi, S., Wakimoto, S. and Saito, H., The multiplication of *Pseudomonas glumae* on rice grains, *Ann. Phytopathol. Soc. Jpn.*, 53, 663, 1987.
131. Vroon, C. W., Pietersen, G., and Van Tonder, H. J., Seed transmission of soybean mosaic virus in *Lupinus albus* L., *Phytophylactica*, 20, 169, 1988.
132. Fajardo, T. G., Progress on experimental work with the transmission of bean mosaic, *Phytopathology*, 18, 155, 1928.
133. Fajardo, T. G., Studies on the mosaic disease of the bean (*Phaseolus vulgaris*), *Phytopathology*, 20, 469, 1930.
134. Allam, E. K., El-Bagoury, H., Olfat, E. H., and Nagwa, A. M., Effect of virus infection on seed production and virus seed transmission of legume. II. Bean common mosaic virus and its effect on cowpea yield, in 19th Int. Seed Testing Assoc. Congr., Vienna, 1980.
135. Agarwal, V. K., Nene, Y. L., Beniwal, S. P. S., and Verma, H. S., Transmission of bean common mosaic virus through urdbean (*Phaseolus mungo*) seeds, *Seed Sci. Technol.*, 7, 103, 1979.
136. Eslick, R. F. and Afanasiev, M. M., Influence of time of infection with barley stripe mosaic on symptoms, plant yield and seed infection of barley, *Plant Dis. Rep.*, 39, 722, 1955.
137. Singh, G. P., Arny, D. C., and Pound, G. S., Studies on the stripe mosaic of barley, including effects of temperature and age of host on disease development and seed infection, *Phytopathology*, 50, 290, 1960.
138. Timian, R. C., Barley stripe mosaic virus seed transmission and barley yield as influenced by time of infection, *Phytopathology*, 57, 1375, 1967.
139. Couch, H. B., Studies on the seed transmission of lettuce mosaic virus, *Phytopathology*, 45, 63, 1955.
140. Athow, K. L. and Bancroft, J. B., Development and transmission of tobacco ringspot virus in soybean, *Phytopathology*, 49, 697, 1959.
141. Sharma, S. R. and Varma, A., Cure of seed transmitted cowpea banding mosaic disease, *Phytopathol. Z.*, 83, 144, 1975.
142. Iizuka, N., Seed transmission of viruses in soybean, *Tohoku Natl. Agric. Exp. Stn. Bull.*, 46, 131, 1973.
143. Bowers, G. R., Jr. and Goodman, M. R., Soybean mosaic virus: infection of soybean seed parts and seed transmission, *Phytopathology*, 69, 569, 1979.
144. Musil, M., Leskova, O., and Rapi, J., The influence of some factors on transmission of pea leaf rolling mosaic virus by pea seed, *Biologia (Czechoslovakia)* A, 36, 889, 1981.
145. Tu, J. C., Symptom severity, yield, seed mottling and seed transmission of soybean mosaic virus in susceptible and resistant soybean: the influence of infection stage and growth temperature, *J. Phytopathol.*, 135, 28, 1992.

146. Kim, J. S., Lee, S. H., and Lee, M. W., Seed transmission of TMV in red pepper (*Capsicum annuum* L.), *Res. Rep. Rural Dev. Adm. Crop Prot.*, Rep. Korea, 31, 10, 1989.
147. Sommartaya, T., Pupipat, U., Intrama, S., and Renfro, B. L., Seed transmission of *Sclerospora sorghi* Weston and Uppal, the downy mildew of corn in Thailand, *Kasetsart J.*, 8, 12, 1975.
148. Siddiqui, M. R., Singh, D., and Gaur, A., Prevalence of chili anthracnose fungus on seeds and its effective control, *Seed Res.*, 5, 67, 1977.
149. Porter, D. M., Wright, F. S., Taber, R. A., and Smith, D. H., Colonization of peanut seed by *Cylindrocladium crotalariae*, *Phytopathology*, 81, 896, 1991.
150. Garber, U., Beer, H., and Motta, G., Studies on the relation between pod and seed infection with *Ascochyta fabae* Speg. in faba beans, *Nachrichtenblatt Dtsch. Pflanzenschutz-Dienste*, 44, 12, 1992.
151. Surekha, C., Singh, T., and Singh, D., Histopathology of *Colletotrichum dematium*-infected chilli seeds, *Acta Bot. Indica*, 18, 226, 1990.
152. Valarini, P. J., Menten, J. O. M., and Lollato, M. A., Incidence of common bacterial blight in the field and transmission of *Xanthomonas campestris* pv. *phaseoli* through bean seeds evaluated by different methods, *Summa Phytopathol.*, 18, 160, 1992.
153. Smith, F. L. and Hewitt, W. B., Varietal susceptibility to common bean mosaic and transmission through seed, *Calif. Agric. Exp. Stn. Bull.*, 62, 18, 1938.
154. Engsbro, B., Barley stripe mosaic virus (BSMV) comparison of symptom intensity and yield and seed infection in spring barley, *Ann. Phytopathol.*, 9, 273, 1978.
155. Troutman, J. L., Bailey, W. K., and Thomas, C. A., Seed transmission of peanut stunt virus, *Phytopathology*, 57, 1280, 1967.
156. Geering, A. D. W. and Randles, J. W., Interactions between a seed-borne strain of cucumber mosaic cucumovirus and its lupin host, *Ann. Appl. Biol.*, 124, 301, 1994.
157. Abney, T. S. and Ploper, L. D., Effects of bean pod mottle virus on soybean seed maturation and seed-borne *Phomopsis* spp, *Plant Dis.*, 78, 33, 1994.
158. Crittenden, H. W., Increase of *Diaporthe phaseolorum* var. *sojae* on soybean pods due to corn earworm, *Phytopathology*, 58, 883, 1968.
159. Shortt, B. J., Sinclair, J. B., Helm, C. G., Jeffords, M. R., and Kogan, M., Soybean seed quality losses associated with bean leaf beetle and *Alternaria tenuissima*, *Phytopathology*, 72, 615, 1982.
160. Inglis, D. A., Contamination of asparagus seed by *Fusarium oxysporum* f. sp. *asparagi* and *Fusarium moniliforme*, *Plant Dis.*, 64, 74, 1980.
161. Sinclair, J. B., Ed., *Compendium of Soybean Diseases*, 2nd ed., APS Press, St. Paul, MN., 1982, 104.
162. Russin, J. S., Layton, M. B., Orr, D. B., and Boethel, D. J., Stink bug damage to soybeans, *La. Agric.*, 32, 24, 1988.
163. Jackson, C. R., Reduction of *Sclerotium bataticola* infections of peanut kernels by *Aspergillus flavus*, *Phytopathology*, 55, 934, 1965.
164. Doupnik, B., Maize seed predisposed to fungal invasion and aflatoxin contamination by *Helminthosporium maydis* ear rot, *Phytopathology*, 62, 1367, 1972.
165. McGee, D. C., Brandt, C. L., and Burris, J. S., Seed mycoflora of soybeans relative to fungal interactions, seedling emergence, and carryover of pathogens to subsequent crops, *Phytopathology*, 70, 615, 1980.
166. Singh, D. P. and Agarwal, V. K., Interaction between grain mold pathogens of sorghum, *Indian J. Plant Pathol.*, 4, 101, 1986.

167. Blaney, B. J., Ramsey, M. D., and Tyler, A. L., Mycotoxins and toxigenic fungi in insect-damaged maize harvested during 1983 in Far North Queensland, *Aust. J. Agric. Res.*, 37, 235, 1986.
168. Rheeder, J. P., Marasas, W. F. O., and Van Wyk, P. S., Fungal associations in corn kernels and effects on germination, *Phytopathology*, 80, 131, 1990.
169. Marasas, W. F. O., Kriek, N. P. J., Wiggins, V. M., Steyn, P. S., Towers, D. K., and Hastie, T.J., Incidence, geographical distribution and toxigenicity of *Fusarium* species in South African corn, *Phytopathology*, 69, 1181, 1979.
170. Velicheti, R. K. and Sinclair, J. B., Reaction of seed-borne soybean fungal pathogens to cercosporin, *Seed Sci. Technol.*, 20, 149, 1992.
171. Hepperly, P. R. and Sinclair, J. B., Relationships among *Cercospora kikuchii*, other seed mycoflora and germination of soybeans in Puerto Rico and Illinois, *Plant Dis.*, 65, 130, 1981.
172. Stuckey, R. E., Ghabrial, S. A., and Reicosky, D. A., Increased incidence of *Phomopsis* sp. in seeds from soybean infected with bean pod mottle virus, *Plant Dis.*, 66, 826, 1982.
173. Hepperly, P. R., Bowers, G. R., Jr., Sinclair, J. B., and Goodman, R. M., Predisposition to seed infection by *Phomopsis sojae* in soybean mosaic virus, *Phytopathology*, 69, 846, 1979.
174. Ross, J. P., Interaction of the soybean mosaic and bean pod mottle viruses infecting soybeans, *Phytopathology*, 53, 887, 1963.
175. Quiniones, S. S., Dunleavy, J. M., and Fisher, J. W., Performance of three soybean varieties inoculated with soybean mosaic virus and bean pod mottle virus, *Crop Sci.*, 11, 662, 1971.
176. Ross, J. P., Effect of time and sequence of inoculation of soybeans with soybean mosaic and bean pod mottle viruses on yields and seed characters, *Phytopathology*, 59, 1404, 1969.
177. Schmitthenner, A. F. and Gordon, D. T., Effects of multiple viral infections on soybean yields, *Phytopathology*, 59, 1547, 1969.
178. Kuhn, C. W. and Dawson, W. O., Multiplication and pathogenesis of cowpea chlorotic mottle virus and southern bean mosaic virus in single and double infections in cowpea, *Phytopathology*, 63, 325, 1973.
179. McDonald, D., Bos, W. S., and Gumel, M. H., Effect of infestation of peanut (groundnut) seed by the testa nematode, *Aphelenchoides arachidis*, on seed infection by fungi and on seedling emergence, *Plant Dis. Rep.*, 63, 464, 1979.

CHAPTER 6

Longevity of Seedborne Pathogens

I. LONGEVITY

Seeds remain viable for many years. Crop seeds with low moisture content remain viable for longer time periods when stored under low temperature. Crocker[1] listed the life span of many crop seeds. The longevity of seedborne pathogens can be independent of the seeds they inhabit and depends on the capability of the pathogen to remain viable as well as virulent from one season to the next in or on seeds. The longevity period of certain seedborne fungi, bacteria, nematodes, viroids, and viruses is given in Tables 6-1 to 6-4.

Pathogens may live longer than the seeds they colonize. For example, *Colletotrichum lindemuthianum* remained active after the bean seeds it colonized lost their viability.[123] Flax seeds retained germinability for 18 to 24 months, whereas *Botrytis cinerea, Colletotrichum linicola*, and *Aureobasidium lini* survived for more than 4 years in seeds.[12] The tobacco ringspot virus remained viable in soybean seeds for 5 years at 16 to 32°C, even when most seeds failed to germinate.[114] In such situations, a pathogen would be a parasite while the seed lives and a saprophyte after the seed dies. Some bacteria die before seeds lose their viability; most survive as long as the seed is viable, and a few others longer.[124] Bacterial pathogens that infect bean and other legume seeds serve as excellent examples of the survival of bacteria in seeds beyond seed viability. *C. flaccumfaciens* pv. *flaccumfaciens* remained viable from 5 to 24 years.[46] Schuster and Sayre[126] isolated *X. c.* pv. *phaseoli* from 15-year-old beans seeds and *C. flaccumfaciens* pv. *flaccumfaciens* from 8-year-old bean seeds stored at 10°C. Internal populations of *C. m.* subsp. *nebraskensis* in maize were not affected by either moisture content at harvest (39 or 25%) or by drying at 35°C.[127] *Burkholderia solanacearum* was not isolated from peanut seeds stored for 1 year.[128]

Pathogens may not survive seed development and drying stages. In maize seeds, the hyphae of *Peronosclerospora sorghi, Sclerospora sacchari,* and *Sclerophthora rayssiae* var. *zeae* were inactivated when seeds were dried to below

Table 6-1 Longevity of Some Seedborne Fungi

Fungus	Crop	Viability (years)	Ref.
Acremonium strictum	*Zea mays* (maize)	7	2
Alternaria alternata	*Hordeum vulgare* (barley)	10	3
		6	4
	Triticum aestivum (wheat)	10	3
A. brassicicola	*Brassica oleracea* (cabbage)	7	5, 6
	B. oleracea var. *botrytis* (cauliflower)	12	6
Ascochyta pisi	*Pisum sativum* (pea)	7	7
	Vicia faba (fava bean)	9	8
Bipolaris oryzae	*Oryza sativa* (rice)	10	9
Botrytis acalda	*Allium cepa* (onion)	3.5	10
B. anthophila	*Trifolium pratense* (red clover)	6	11
B. cinerea	*Linum usitatissimum* (flax)	3.33	12
B. fabae	*Phaseolus vulgaris* (bean)	0.75	13
Bipolaris	*Secale cereale* (rye)	1.16	14
B. sorokiniana	*H. vulgare* (barley)	11	15
	Triticum aestivum (wheat)	10	3
Cercospora beticola	*Beta vulgaris* (beet)	2.5	16
C. kikuchii	*Glycine max* (soybean)	2	17
Colletotrichum gloeosporioides	*Capiscum frutescens* (pepper)	0.75	18
G. gossypii	*Gossypium* (cotton)	13.5	19
C. graminicola	*Z. mays* (maize)	3	20
Diaporthe phaseolorum	*G. max* (soybean)	2.5	21
Drechslera avenae	*Avena sativa* (oats)	10	22
D. graminea	*H. vulgare* (barley)	5	23
		10	24
D. teres	*H. vulgare* (barley)	10	3
Fusarium graminearum	*H. vulgare* (barley)	2.25	24
	Z. mays (maize)	2	2
F. moniliforme	*Z. mays* (maize)	8	2
F. oxysporum	*Trifolium pratense* (red clover)	6	10
F. solani f. sp. *cucurbitae*	*Cucurbita* (cucurbit)	2	25
Gloeotinia granigena	*Lolium* (ryegrass)	2	26
Macrophomina phaseolina	*Phaseolus vulgaris* (bean)	2.42	27
Microdochium nivale	*T. aestivum* (wheat)	3.5	28
Peronospora manshurica	*G. max* (soybean)	5	29
Phoma betae	*B. vulgaris* (beet)	5	30
P. exigua	*Phaseolus vulgaris* (bean)	2.5	31
P. lingam	*Brassica* (crucifers)	1.16	32
	B. oleracea (cabbage)	4	5
Pyricularia oryzae	*O. sativa* (rice)	2.58	33
Sclerospora graminicola	*Pennisetum glaucum* (pearl millet)	2	34
Sclerotinia sclerotiorum	*P. vulgaris* (beans)	3	35

Table 6-1 Longevity of Some Seedborne Fungi (continued)

Fungus	Crop	Viability (years)	Ref.
Septoria apiicola	*Apium graveolens* (celery)	3	36
Stagonospora nodorum	*T. aestivum* (wheat)	7, 2	3, 37
Tilletia barclayana	*O. sativa* (rice)	3	38
T. caries	*T. aestivum* (wheat)	9	39
Uromyces betae	*Beta vulgaris* (beet)	2	40
Ustilago segetum var. *tritici*	*H. vulgare* (barley)	11	41, 42
	T. aestivum (wheat)	7	43

15% moisture content.[129,130] Isolate M_2 of the peanut mottle virus was infective in immature peanut seeds, but inactivated during seed maturation.[131] The soybean mosaic virus was inactivated in immature soybean embryos during maturation in seeds of cv. Merit.[132] The International Board for Plant Genetic Resources recommended –180°C for the long-term storage of seeds.[133] The viability of seeds stored at –10 to –20°C was studied for a range of species, and viability was maintained at this temperature.[134] Likewise, a number of seedborne pathogenic fungi were found to remain viable when stored dry at –20°C for periods of up to 14 years without loss of pathogenicity (Table 6-5).[133] However, the viability of pathogenic fungi on seeds declined during storage at room temperature, the rate of decline varying among species (Table 6-1).

II. FACTORS INFLUENCING LONGEVITY

Longevity varies among pathogens. For example, Shipton et al.[135] found that two-thirds of wheat seeds originally infected by *Stagonospora nordorum* were free of the fungus 7 to 9 months after storage. Also, Von Wechmar[136] found in wheat seeds stored for 1 year that *S. nodorum* decreased from 50 to 1 to 5%. Survival of the seedborne inoculum in wheat was nearly zero after 2 year storage.[137] Earlier, the fungus was found to survive up to 7 years.[3] In contrast, it was reported that after 12 months storage, the level of seed infection by *S. nodorum* in wheat increased 12 to 14%.[138] Seedborne infection of the same pathogen increased 1 to 25% during storage for up to 24 months.[139] This increase was related to a decrease in other seedborne fungi, *Alternaria* and *Epicoccum*, which were competitive with but not antagnostic to *S. nodorum* on oxgall agar.[139]

In general, pathogens survive longer in seeds under cool and dry conditions than under high temperature and relative humidity. Viability of seedborne pathogens is affected by a number of factors.

A. Host Genotype

Fungus survival depends on the host species. *Ascochyta pisi* survived in fava bean seeds for 9 years[8] and in pea seeds for 7 years,[9] with survival varying among seed lots. Survival of *Aureobasidium lini, B. cinerea,* and *Colletotrichum linicola*

Table 6-2 Longevity of Some Seedborne Bacteria

Bacterium	Crop	Viability (years)	Ref.
Bacillus mesentericus	*Trifolium pratense* (red clover)	19	44
C. michiganensis subsp. *insidious*	*Medicago sativa* (alfalfa)	3	45
C. tritici	*Triticum aestivum* (wheat)	5	46
Curtobacterium flaccumfaciens pv. *flaccumfaciens*	*Phaseolus vulgaris* (bean)	15	47, 48
E. carotovora subsp. *atroseptica*	*T. pratense* (red clover)	19	44
E. carotovora subsp. *carotovora*	*Nicotiana* (tobacco)	0.66	49
Burkholderia solanacearum	*T. pratense* (red clover)	19	44
P. syringae pv. *apii*	*Apium graveolens* (celery)	1	50
P. syringae pv. *glycinea*	*Glycine max* (soybean)	1.33	51
P. syringae pv. *lachrymans*	*Cucumis sativus* (cucumber)	2	52, 53
P. syringae pv. *phaseolicola*	*P. vulgaris* (bean)	3	50
P. syringae pv. *pisi*	*Pisum sativum* (pea)	0.84	54
P. syringae pv. *sesami*	*Sesamum indicum* (sesame)	0.9	55
P. syringae pv. *tabaci*	*G. max* (soybean)	1.5	56
	Nicotiana (tobacco)	2	57
P. syringae pv. *tomato*	*Lycopersicon esculentum* (tomato)	20	58
Xanthomonas campestris pv. *campestris*	*Brassica oleracea* (cabbage)	3	59
X. campestris pv. *cyamopsidis*	*Cyamopsis tetrogonoloba* (guar)	1	60
X. campestris pv. *malvacearum*	*Gossypium* (cotton)	4.75	61
		7	62
X. campestris pv. *manihotis*	*Manihot esculenta* (cassava)	1.5	63
X. campestris pv. *phaseoli*	*Phaseolus vulgaris* (bean)	15	47
		3	64
X. campestris pv. *sesami*	*S. indicum* (sesame)	1.33	65
X. campestris pv. *glycines*	*G. max* (soybean)	2.5	56
X. campestris pv. *vesicatoria*	*Capsicum frutescens* (pepper)	10	58
	L. esculentum (tomato)	2.33	66
X. campestris pv. *vitians*	*Lactuca sativa* (lettuce)	1.91	67
X. campestris pv. *zinniae*	*Zinnia elegans* (zinnia)	4	68
X. oryzae pv. *oryzae*	*O. sativa* (rice)	0.16	69
		0.9	70

Table 6-3 Longevity of Some Seedborne Viroids and Viruses

Viroid/Virus	Crop	Viability (years)	Ref.
Alfalfa mosaic	*Medicago sativa* (alfalfa)	3	71
		5	72
		7	73
Barley stripe mosaic	*Hordeum vulgare* (barley)	3	74
		19	75
	Triticum aestivum (wheat)	6.5	76
Bean common mosaic	*Vigna angularis* (azuki bean)	4	77
		30	76
	P. vulgaris (bean)	3	79
		38	80
		6	81
Blackeye cowpea mosaic	*Vigna mungo* (urdbean)	3	82
Broad bean strain	*Vicia faba* (fava bean)	1	83
		4	84
Broad bean true mosaic	*V. faba* (fava bean)	1	83, 84
Brome mosaic	*T. aestivum* (wheat)	10	85
Cowpea aphid-borne mosaic	*Vigna unguiculata* (cowpea)	2	86
Cowpea banding mosaic	*V. unguiculata* (cowpea)	3.5	87
Cucumber green mottle mosaic	*Cucumis sativus* (cucumber)	0.5	88
		3	89
Cucumber mosaic	*Cucumis melo* (cantaloupe)	3	90
		5	91
	P. vulgaris (bean)	2.25	92
		0.50	93
	Stellaria media (chickweed)	1.75	94
Dodder latent mosaic	*Cuscuta campestris* (dodder)	1	95
Eggplant mosaic	*Solanum melongena* (eggplant)	0.6	96
French bean mosaic	*Phaseolus vulgaris* (bean)	6	97
Hop mosaic	*Humulus lupulus* (hop)	2	98
Lychnis ringspot	*Lychnis divaricata* (lychnis)	3.3	99
	Silene noctiflora (night-flowering catchfly)	2.15	99
Potato spindle tuber	*Solanum tuberosum* (potato)	21	100
Prune dwarf	*Prunus* (prune)	3.5	101
Prunus necrotic ringspot	*P. pennsylvanica* (wild red cherry)	6	102
Raspberry ringspot	*Capsella bursa-pastoris* (shepherd's purse)	6	103
	Stellaria media (chickweed)	6	103
Southern bean mosaic	*P. vulgaris* (bean)	0.6	104
		3	105
Sowbane mosaic	*Chenopodium murale* (goosefoot)	6.5	106
		14	107
Soybean mosaic	*Glycine max* (soybean)	2	108
		1	109
Squash mosaic	*Cucurbita pepo* (squash)	3	110
		5	91

Table 6-3 Longevity of Some Seedborne Viroids and Viruses (continued)

Viroid/Virus	Crop	Viability (years)	Ref.
Tobacco mosaic	*Lycopersicon esculentum* (tomato)	3	111
		9	112
Tobacco ringspot	*Petunia violacea* (violet petunia)	0.6	113
	G. max (soybean)	5	114
		0.75	115
	Nicotiana (tobacco)	5.5	116
Tomato blackring	*Capsella bursa-pastoris* (shepherd's purse)	6	103
	S. media (chickweed)	6	103

in flax seeds ranged from 16 to 55, 16 to 40, and 26 to 69 months, respectively, in different seed lots.[12]

B. Inoculum

Survival of seedborne pathogens depends upon the amount of inoculum per seed, the location of inoculum in the seeds, and the type of survival propagule. In celery seeds *Septoria apiicola* hyphae were short lived compared to spore inoculum.[140] *Pyricularia oryzae* conidia survived for 1 to 2 years on rice seeds, but deep-seated hyphae survived up to 4 years.[24] Hyphae of *Cercospora beticola* survived in sugar beet seeds for 2 to 2.5 years, but conidia were short lived.[15] Fischer and Holton[141] found in general that mature smut fungi spores maintain viability longer than less mature ones. The viability of surface-borne spores and hyphae of *Alternaria brassicicola* on Brassica seeds declined rapidly in storage, and they may not be important sources of inoculum after 2 years. The rate of decrease in viability of internal embryonic infection is less and constitutes a major means of survival. Internal infection persisted for 12 years at 50% relative humidity and 10°C.[142] *Aureobasidium lini* survived longer in heavily infected flax seeds than in seeds with light infection.[12] The level of seedborne *Botrytis cinerea* in chickpea decreased during storage until the next planting season, the extent depending upon cultivar and initial infection.[143] Neergaard[144] generalized that (1) hyaline fungi with the thin-walled conidia were short lived; (2) fungi with strong pigmentation and thick conidial walls were fairly long lived; (3) fungi with fruiting bodies, such as acervuli and pycnidia, remained viable for long time

Table 6-4 Longevity of Some Seedborne Nematodes

Nematode	Crop	Viability (years)	Ref.
Anguina tritici	*Triticum aestivum* (wheat)	28	117
		14	118
Aphelenchoides besseyi	*Oryza sativa* (rice)	3	119
		2	120
Ditylenchus dipsaci	*Avena sativa* (oats)	8	121
	Medicago sativa (alfalfa)	5	121
Heterodera glycines	*Glycine max* (soybean)	1.8	122

periods; (4) smut fungi were long lived; and (5) fungi with thick-walled resting mycelia survived the longest.

Bacteria either die prior to the time seeds lose viability or may survive longer than seeds. Viruses that are embryo borne may persist for long periods. Often they remain infective as long as the seeds are viable. Even if the seeds lose viability, certain viruses may persist. BCMV remained infective in bean seeds for 30 years.[78] However, viruses that are not borne in the embryo, such as TMV in tomato and cucumber green mottle mosaic virus in cucumber seeds, lose infectivity rapidly during storage. The infectivity of TMV in tomato seeds declined rapidly after 1 year in storage, although the virus was found to persist up to 3 years.[111] The infectivity of cucumber green mottle virus in cucumber seeds declined rapidly during the first 7 months of storage.[91]

Nematodes can remain viable for many years because the larvae enter a quiescent phase. Larvae of *Anguina tritici* (ear cockle of wheat) were resistant to desiccation inside nematode galls and remained viable for 28 years.[117]

C. Seed Storage Containers

Air-tight containers favor seedborne pathogen survival. *X. campestris* pv. *zinniae* remained virulent in zinnia seeds for 2 years in paper bags and 4 years in hermetically sealed packages.[69] BCMV survived for 30 years in bean seeds[78] and 38 years in seeds stored in sealed jars at room temperature.[80]

D. Storage Environment

Cool, dry conditions favor survival of seedborne inoculum. The recovery of *Sclerotinia sclerotiorum* in Lee 68 soybean seeds was less when stored for 18 months at 22° ± 3°C than at 3° ± 1°C.[145] *C. kikuchii* lost viability more rapidly in soybean seeds at 10.8 to 15% than at 6.3% moisture content, and the loss in viability was accelerated by storage for 10 months at 16 to 28°C, but there was less loss of viability at 8 or 12°C, regardless of seed moisture content.[16] The germination of *Phoma rabiei* conidia carried superficially on gram seeds after threshing was 50% after 5 months in storage at 25 to 30°C,[146] while *B. anthophila* survived in red clover seeds for 6 years at 5% relative humidity in air-tight containers.[11] *Colletotrichum graminicola* survived in maize kernels for more than 3 years at 4°C.[17] *Botrytis aclada* persisted for 4.5 years in onion seeds stored at 40% relative humidity and 22°C, or at 50% relative humidity and 10°C.[147,148] The incidence of seedborne *Alternaria brassicae* was reduced more than 50% when rape seeds were stored for 6 to 8 months at 25°C.[149] The pathogen was eliminated from seeds during storage at 29 to 35°C for 5 months, whereas reduction in seed infection was slow at 5°C compared to higher temperatures.[150] *C. gossypii* and *F. moniliforme* survived in cottonseeds for 13.5 years at 1°C.[18] *Puccinia calciptrapae* var. *centaureae* teliospores on safflower seeds lost infectivity after 18 months at room temperature, but maintained infectivity for 18 months at 3°C.[151]

Table 6-5 Viability of Seedborne Fungi after Storage at –20˚C

Fungus	Host	Years	Seed infection (%) At storage	Seed infection (%) After storage
Alternaria dauci	*Daucus carota* (carrot)	9–14	22	21
A. radicina	*D. carota* (carrot)	14	37	28
Ascochyta fabae	*Vicia faba* (fava bean)	9–13	14	14
A. pisi	*P. sativum* (pea)	8–12	23	19
Bipolaris sorokiniana	*Triticum aestivum* (wheat)	8–12	43	33
Colletotrichum lindemuthianum	*Phaseolus vulgaris* (bean)	12	99	93
Phaeosphaeria maculans	*Brassica oleracea* (cabbage)	11–13	13	12
P. nodorum	*T. aestivum* (wheat)	9–14	50	39
Pleospora bjoerlingii	*Beta vulgaris* (sugar beet)	14	30	23
Microdochium nivalis	*T. aestivum* (wheat)	9–12	19	19
Mycosphaerella pinodes	*Pisum sativum* (pea)	8–11	18	14
Pyrenophora graminea	*Hordeum vulgare* (barley)	11–12	47	44
Pyrenophora teres	*H. vulgare* (barley)	8–12	24	16

Adapted from Hewett, P. D., *Seed Sci. Technol.*, 15, 73, 1987. With permission.

The following pathogens survived in seeds for long periods of time at 5˚C: *Stagonospora nodorum* in wheat for 10 years,[152] *Phyllosticta cajani* in pigeon pea for 8 years,[153] and *Microdochium oryzae* in rice for 11 years.[154] However, *R. oryzae* in rice survived for 1 year and 2 months at 26 to 28˚C,[155] and *S. nodorum* lost viability in 3 to 4 years in seeds stored at 25˚C.[156] Storing lentil seeds infected by *Ascochyta fabae* f. sp. *lentis* for 4 years at 20˚, 5˚, –18˚, and –160 to –196˚C (liquid N) did not affect fungal pathogenicity. Germination of infected seeds was lower than that of healthy seeds at all temperatures, whether stored for 1 d, or 1, 2, 3, and 4 years. There was a significant reduction in germination of infected but not healthy seeds at each temperature over the 4-year period.[157] *Colletotrichum lindemuthianum* in bean survived at least 4 years in seeds when air dried and stored at 40˚C.[158] *A. fabae* f. sp. *lentis* survived in lentil seeds at 46˚C for 3 years.[159] Oospores of *Sclerospora graminicola* remained viable in storage for 8 years at 10˚C.[160] Under storage conditions used by seed companies in Denmark, *Alternaria brassicicola* survived for 7 years and 8 months in cabbage and cauliflower seeds.[161] *Penicillium oxalicum* lost viability in maize seeds after storage for 12 months at 20 to 25˚C.[162] The viability of seedborne fungi after storage at –20˚C is summarized in Table 6-5.

P. syringae pv. *phaseolicola* survived longer in bean seeds at 40–50% relative humidity and 7 to 10˚C than that at 10 to 27˚C.[163] *X. campestris* pv. *manihotis* survived in cassava seeds for 18 months at 60% relative humidity and at 5˚C.[63] *C. michiganensis* subsp. *insidiosus* remained viable and virulent in alfalfa seeds for 3 years at 21–26˚C.[47] Storage of bean seeds for 3–4 years reduced infection by *X. c.* pv. *phaseoli.*[164] *X. c.* pv. *glycines* survived in soybean seeds up to 9 months at 10.6 to 40˚C, for 1 year and 1 month at 6 to 10˚C.[165] *Pseudomonas avenae* in rice survived 8 years at 5˚C,[166] but the bacterium was viable for 2 years in seedlots stored at room temperature in South Korea.[167]

Bean common mosaic virus in mungbean seeds survived for more than 6 years at 2 to 4°C.[168]

Reduction in body water content of nematodes increased the rate of dehydration and reduced body food reserves, and chilling at 5 to 10°C during dehydration adversely affected the nematode's survival.[169] Storage at low temperature favored survival of viruses without loss in seed transmission. Prune dwarf virus was recovered from cherry seeds after 54 months at 5°C without loss in transmission.[104] Potato spindle tuber viroid was detected in 100% of the progeny tested after true potato seeds were stored for 12 years at 4°C. Tests on selfed true potato seeds from the viroid infected cv. Monoma showed a 100% transmission rate after subinoculation of initial bioassay plants.[170]

E. Storage Period

The percentage of seed infection may decrease as the storage period increases. The recovery of *B. cinerea* decreased over a 6-month period in sunflower seeds.[171] *Ustilago segetum* var. *tritici* hyphae remained viable in wheat up to 7 years at –2 to 0°C and in barley seeds for 11 years.[41-43] Seed transmission declined in barley as the storage period increased until all infection was eliminated. The decline in seed transmission could be due to death of the germ portion of seeds during storage.[42] *Aureobasidium caulivorum* survived up to 9 months in clover seeds, with transmission decreasing from 92.5 to 0.28%.[172] *Phaeoisariopsis griseola* declined in bean seeds from 50 to 10% after 9 months and could not be recovered after 12 months in storage.[173] One- to 2-year-old *Peronospora manshurica* oospores had 30 to 39% viability, while an 8-year-old collection had 20% viable oospores. The germ tubes from oospores 3 to 8 years old were mostly straight, while those tubes from 2-year-old oospores had a tendency to coil and branch.[174]

Internal and external pea seed infection by *P. s.* pv. *pisi* persisted for at least 3 years on seeds, although the degree of infection declined each year the seed was stored.[175] The viability of *P. syringae* pv. *phaseolicola* declined in bean seeds at 10% per year the first 3 to 4 years, with a projected extinction point after 5 to 6 years.[176] No viable bacteria were detected after 10 years, but the bacterium may survive in numbers below the threshold level for detection.[163,176]

The percentage of seed transmission of French bean mosaic virus in French bean decreased during successive years of storage over 6 years in sealed vials.[97] The rate of seed transmission of cowpea banding mosaic virus in cowpea remained high for 2 years and then declined, and no seed transmission was obtained after 3.5 years. However, by this time seed viability was reduced.[89] The level of Prunus necrotic ringspot virus in *Prunus pennsylvanica* seeds remained constant at 60 to 70% for the first 4 years at 2°C but was reduced to less than 5% by year 6.[105] The rate of seed transmission of eggplant mosaic virus in eggplant was about 41 and 2% after 30 and 210 d storage, respectively.[177] The storage of bean seeds for 3 to 4 years reduced *X. c.* pv. *phaseoli* infection.[178]

The nematode *Aphelenchoides besseyi* survived in rice seeds for 3 years, although viability decreased from 62 to 46%.[119]

F. Presence of Antagonistic Microflora

The presence or absence of other microflora on seeds influenced survival of seedborne pathogens (see Chapter 5). Bacteriophages can reduce the bacterial population of *X. c.* pv. *oryzae* on rice seeds during storage. Viable bacteria could be detected up to 2 months on infected rice seeds stored at 25 to 35°C after which no bacteria were detected.[64]

REFERENCES

1. Crocker, W., Longevity of Seeds, *J. N.Y. Bot. Gard.*, 46, 25, 1945.
2. Dungan, G. H. and Koehler, B., Age of seed corn in relation to seed infection and yielding capacity, *J. Am. Soc. Agron.*, 36, 436, 1944.
3. Machacek J. E. and Wallace, H. A. H., Longevity of some fungi in cereal seed, *Can. J. Bot.*, 30, 164, 1952.
4. Christensen, J. J., Longevity of fungi in barley kernels, *Plant Dis. Rep.*, 47, 639, 1963.
5. Neergaard, P., Kan Svampes Levetid forlaenges ved Lysbehandling Arsberetning, *J. E. Ohlsens Enkes Plantepathol. Lab.*, 6, 9, 1941.
6. Maude, R. B. and Humpherson-Jones, F. M., Studies on the seedborne phases of dark leaf spot (*Alternaria brassicicola*) and grey left spot (*Alternaria brassicae*) of Brassicas, *Ann. Appl. Biol.*, 95, 311, 1980.
7. Wallen, V. R., The effect of storage for several years on the viability of *Ascochyta pisi* in pea seed and on the germination of the seed and emergence, *Plant Dis. Rep.*, 39, 674, 1955.
8. Sprague, R., Host range and life history studies of some leguminous *Ascochytae*, *Phytopathology*, 19, 917, 1929.
9. Suzuki, H., Experimental studies on the possibility of primary infection of *Piricularia oryzae* and *Ophiobolus miyabeanus* internal of rice seed, *Ann. Phytopathol. Soc. Jpn.*, 2, 1, 1930.
10. Maude, R. B. and Presly, A. H., Neck rot (*Botrytis allii*) of bulb onions, I. Seedborne infection and its relationship to the disease in the onion crop, *Ann. Appl. Biol.*, 86, 163, 1977.
11. Narkiewicz-Jodko, M., L'observations sur l'etat de semences du trefle rouge pendant leur conservation, 17th Int. Seed Testing Assoc. Congr., Warsaw, 4, 1974.
12. Colhoun, J. and Muskett, A. E., A study of the longevity of the seed-borne parasite of flax in relation to the storage of the seed, *Ann. Appl. Biol.*, 35, 429, 1948.
13. Harrison, J. G., Role of seed-borne infection in epidemiology of *Botrytis fabae* on field beans, *Trans. Br. Mycol. Soc.*, 70, 35, 1978.
14. Luke, H. H. and Barnett, R. D., The influence of harvest condition on the germinability and mycoflora of rye seed, *Seed Sci. Technol.*, 7, 431, 1979.
15. Christensen, J. J., Studies on the parasitism of *Helminthosporium sativum, Univ. Minn. Agric. Exp. Stn. Tech. Bull.*, 11, 42, 1922.
16. Wenzl, H., The importance of seed infection of beet seed by *Cercospora beticola* and its control, *Pflanzenschutzberichte*, 23, 33, 1959.

17. Lehman, S. G., Survival of the purple seed stain fungus in soybean seeds, *Phytopathology,* 42, 285, 1952.
18. Smith, R. W. and Crossan, D. F., The taxonomy, etiology and control of *Colletotrichum piperatum* (E & E) E and H and *Colletotrichum capsici* (Syd.) B & B., *Plant Dis. Rep.,* 42, 1099, 1958.
19. Arndt, C. H., Survival of *Colletotrichum gossypii* on cotton seeds in storage, *Phytopathology,* 43, 220, 1953.
20. Warren, H. L., Survival of *Colletotrichum graminicola* in corn kernels, *Phytopathology,* 67, 160, 1977.
21. Wallen, V. R. and Seaman, W. L., Seed infection of soybean by *Diaporthe phaseolorum* and its influence on host development, *Can. J. Bot.,* 41, 13, 1963.
22. Sheridan, J. E. and Tan, P. E. T., Incidence and survival of *Pyrenophora avenae* in New Zealand seed oats, *N.Z. J. Agric. Res.*, 16, 251, 1973.
23. Leukel, R. W., Dickson, J. G., and Johnson, A. G., Effects of certain environmental factors on stripe disease of barley and on the control of the disease by seed treatment, *U.S. Dept. Agric. Tech. Bull.,* 341, 39, 1933.
24. Shands, R. G., Longevity of *Gibberella saubinettii* and other fungi in barley kernels and its relation to the emetic effect, *Phytopathology,* 27, 749, 1937.
25. Conroy, R. J., Fusarium foot rot of cucurbits, *Agric. Gaz. N.S.W.,* 64, 655, 1953.
26. Hardison, J. R., Blind disease of perennial ryegrass, *Ore. Agric. Exp. Stn. Circ.,* 177, 11, 1948.
27. Watanabe, T., *Macrophomina phaseoli* found in commmercial kidney bean seed and in soil and pathogenicity to kidney bean seedlings, *Ann. Phytopathol. Soc. Jpn.,* 38, 100, 1972.
28. Ponchet, J., Etude des Communaute's Mycopericarpiques du Caryopse de Ble, Institut National de la Recherche Agronomique, Paris, 1966, 115.
29. Naumova, E. S. and Obtemperanskaya, M. S., Modes of survival of the causal agent of Peronospora disease of soybean, *Peronospora manshurica* (Naum.) Syd. in the intervegetative period, *Mikol. Fitopatol.*, 22, 500, 1988.
30. Newton, W. and Bosher, J. E., The longevity of *Phoma betae* in garden beet seed, *Sci. Agric.*, 26, 305, 1946.
31. Sneep, J., Ascochyta spot disease of the bean (*Phaseolus*), *Tijdschr. Plantenziekten,* 51, 1, 1945.
32. Lloyd, B., The transmission of *Phoma lingam* (Tode) Desm. in the seeds of swede, turnip, chou, moellier, rape and kale, *N.Z. J. Agric. Res.,* 2, 649, 1959.
33. Oran, Y. K., Research on the taxonomy, bioecology and harmfulness of the rice blast fungus, *Piricularia oryzae* Bri. et cav. in southeast Anatolia and the resistance of rice varieties, *Bitki Koruma Bull. Suppl.,* 1, 49, 1975.
34. Shetty, H. S., Mathur, S. B., and Neergaard, P., *Sclerospora graminicola* in pearl millet seeds and its transmission, *Trans. Br. Mycol. Soc.*, 74, 127, 1980.
35. Tu, J. C., The role of white mold-infected white bean (*Phaseolus vulgaris* L.) seed in the dissemination of *Sclerotinia sclerotiorum* (Lit.) deBary., *J. Phytopathol.*, 121, 40, 1988.
36. Mujica, F., La Septoriosis del Apio, *Simiente*, Santiago, 12, 81, 1943.
37. Badadoost, M. and Hebert, T. T., Incidence of *Septoria nodorum* in wheat seeds and its effect on plant growth and grain yield, *Plant Dis.*, 68, 125, 1984.
38. Ou, S. H., Rice Diseases, Commwl. Mycol. Instit., Kew, Surrey, U.K., 1985, 380.
39. Caspar, R., Über den Einfluss aüsser Faktoren auf den Steinbrand befall des Weizens, *Kuhn-Arch.,* 23, 205, 1926.

40. Agarkov, V. A. and Assaul, B. D., Studies of sugar beet rust, in *Ways of Obtaining Good Harvests of Sugarbeet, Cereals and Bean Crops,* Sadovnik, D. D., Ed., State Publishing House for Agricultural Literature, Kiev, 1963, 164.
41. Porter, R. H., Longevity of *Ustilago nuda* in barley seed, *Phytopathology,* 45, 637, 1955.
42. Russell, R. C., The influence of aging of seed on the development of loose smut in barley, *Can. J. Bot.,* 39, 1741, 1961.
43. Tapke, V. F., Longevity of *Ustilago nuda*, *Phytopathology,* 43, 407, 1953.
44. Nikitina, K. V., Infection of clover seeds by the pathogens of bacterial blight, *Tr. Prikl. Bot. Genet. Sel.,* 51, 221, 1974.
45. Cormack, M. W., Longevity of the bacterial wilt organism in alfalfa hay, pod, debris and seed, *Phytopathology,* 51, 260, 1961.
46. Mathur, R. S. and Ahmad, Z. U., Longevity of *Corynebacterium tritici* causing tundu disease of wheat, *Proc. Natl. Acad. Sci. India Sect. B.,* 34, 335, 1964.
47. Schuster, M. L. and Sayre, R. M., A coryneform bacterium induces purple-colored seed and leaf hypertrophy of *Phaseolus vulgaris* and other leguminosae, *Phytopathology,* 57, 1064, 1967.
48. Burkholder, W. H., The longevity of the pathogen causing the wilt of the common bean, *Phytopathology,* 35, 743, 1945.
49. McIntyre, J. L., Sands, D. C., and Taylor, G. S., Overwintering, seed disinfestation and pathogenicity studies of the tobacco hollow stalk pathogen, *Erwinia carotovora* var. *carotovora, Phytopathology,* 68, 435, 1978.
50. Chupp, C. and Sherf, A. F., *Vegetable Diseases and Their Control,* Ronald Press, New York, 1960, 693.
51. Kent, G. C., A study of soybean diseases and their control, in Report on Agricultural Research for the Year Ending June 30, 1945, Iowa Agricultural Experimental Station, Ames, 1945, 221.
52. Volcani, Z., Survival ability of *Pseudomonas lachrymans*, the causal agent of leaf spot disease of cucumber, in Summaries of Research Work, 1967–1969, Volcani Instit. Agric. Res., Div. Plant Pathol., Bet Dagan, 38, 1970.
53. Walker, J. C., Chand, J. N., and Wade, E. K., Relation of seed-and soil-borne inoculum to epidemiology of angular leaf spot of cucumber in Wisconsin, *Plant Dis. Rep.,* 47, 15, 1963.
54. Skoric, V., Bacterial blight of pea: overwintering, dissemination, and pathological histology, *Phytopathology,* 17, 611, 1927.
55. Vajavat, R. M. and Chakravarti, B. P., Survival of *Pseudomonas sesami* and effect of an antagonistic bacterium isolated from seeds on the control of the disease in seed, *Indian Phytopathol.,* 31, 286, 1978.
56. Graham, J. H., Overwintering of three bacterial pathogens of soybean, *Phytopathology,* 43, 189, 1953.
57. Johnson, J. and Murwin, H. F., Experiments on the control of wildfire of tobacco, *Wis. Agric. Exp. Stn. Res. Bull.,* 62, 35, 1925.
58. Bashan, Y. Okon, Y., and Henis, Y., Long term survival of *Pseudomonas syringae* pv. *tomato* and *Xanthomonas campestris* pv. *vesicatoria* in tomato and pepper seeds, *Phytopathology,* 72, 1143, 1982.
59. Clayton, E. E., Second progress report of blackrot (*Pseudomonas campestris*) investigations on Long Island: seed infection and seasonal development, *Phytopathology,* 15, 48, 1925.
60. Gandhi, S. K. and Chand, J. N., Host range and survival of *Xanthomonas campestris* pv. *cyamopsidis, Indian Phytopathol.*, 42, 280, 1989.

61. Hunter, R. E. and Brinkerhoff, L. A., Longevity of *Xanthomonas malvacearum* on and in cotton seed, *Phytopathology,* 54, 617, 1964.
62. Schnathorst, W. C., Longevity of *Xanthomonas malvacearum* in dried cotton plants and its significance in dissemination of the pathogen on seed, *Phytopathology,* 54, 1009, 1964.
63. Persley, G. J., Studies on the survival and transmission of *Xanthomonas manihotis* on cassava seed, *Ann. Appl. Biol.,* 93, 159, 1979.
64. Basu, P. K. and Wallen, V. R., Influence of temperature on the viability, virulence, and physiologic characteristics of *Xanthomonas phaseoli* var. *fuscans in vivo* and *in vitro, Can. J. Bot.,* 44, 1239, 1966.
65. Habish, H. A. and Hammad, A. H., Survival and chemical control of *Xanthomonas sesami, FAO Plant Prot. Bull.,* 19, 36, 1971.
66. Mickail, K. Y., Bishay, F., and Farag, N. S., Tomato seed treatment and seed storage to control bacterial spot or scab, *Agric. Res. Rev. (Cairo),* 48, 23, 1970.
67. Ohata, K., Serizawa, S., Azegami, K., and Shirata, A., Possibility of seed transmission of *Xanthomonas campestris* pv. *vitians,* the pathogen of bacterial spot of lettuce, *Bull. Jpn. Natl. Inst. Agric. Sci. C.,* 36, 81, 1982.
68. Strider, D. L., Detection of *Xanthomonas nigromaculans f.* sp. *zinniae* in zinnia seed, *Plant Dis. Rep.,* 63, 869, 1979.
69. Kauffman, H. E. and Reddy, A. P. K., Seed transmission studies of *Xanthomonas oryzae* in rice, *Phytopathology,* 65, 663, 1975.
70. Singh, R. A. and Rao, M. H. S., A simple technique for detecting *Xanthomonas oryzae* in rice seeds, *Seed Sci. Technol.,* 5, 123, 1977.
71. Frosheiser, F. I., Virus-infected seeds in alfalfa seed lots, *Plant Dis. Rep.,* 54, 591, 1970.
72. Frosheiser, F. I., Alfalfa mosaic virus transmission to seed through alfalfa gametes and longevity in alfalfa seed, *Phytopathology,* 64, 102, 1974.
73. Babovic, M. V., The transmission rate of alfalfa mosaic virus by lucerne seed, *Acta Biol. Yugosl.,* 13, 83, 1976.
74. McNeal, F. H., Mills, I. K., and Berg, M. A., Variation in barley stripe mosaic virus incidence in wheat seed due to storage and continuous propagation and the effect of the disease on yield and test weight, *Agron. J.,* 53, 128, 1961.
75. McNeal, F. H., Berg, M.A., and Carroll, T. W., Barley stripe mosaic virus data from six infected spring wheat cultivars, *Plant Dis. Rep.,* 60, 730, 1976.
76. Scott, H. A., Serological detection of barley stripe mosaic virus in single seeds and dehydrated leaf tissues, *Phytopathology,* 51, 200, 1961.
77. Tsuchizaki, T., Yora, K., and Asuyama, H., Seed transmission of viruses in cowpea and azukibean plants. II. Relations between seed transmission and gamete infection, *Ann. Phytopathol. Soc. Jpn.,* 36, 237, 1970.
78. Pierce, W. H. and Hungerford, C. W., Symptomatology, transmission, detection and control of bean mosaic in Idaho, *Idaho Agric. Exp. Stn. Res. Bull.,* 7, 37, 1929.
79. Nelson, R., Investigations in the mosaic disease of bean (*Phaseolus vulgaris* L.), *Mich. Agric. Exp. Stn. Tech. Bull.,* 118, 71, 1932.
80. Walters, E. C., Jr., Thirty-eight years behind the times and still they germinate, *Assoc. Off. Seed Anal. Newsl.,* 36, 8, 1962.
81. Polak, J. and Chod, J., Possibility of selecting healthy Saxa beans from ones attacked by bean mosaic virus, *Ochr. Rost.,* 3, 125, 1967.
82. Provvidenti, R., Seed transmission of blackeye cowpea mosaic virus in *Vigna mungo, Plant Dis.,* 70, 981, 1986.

83. Cockbain, A. J., Cook, S. M., and Vorraurai, S., Diseases of field beans (*Vicia faba* L.), Rothamsted Exp. Stn. Rep., 1973, 142.
84. Vorraurai, S. and Cockbain, A. J., Further studies on seed transmission of broad-bean stain virus and Echtes Ackerbohnenmosaik virus in field beans (*Vicia faba*), *Ann. Appl. Biol.,* 87, 365, 1977.
85. Von Wechmar, M. B., Kaufmann, A., and Rybicki, E. P., Brome mosaic virus is transmitted through wheat seed, Int. Congr. Plant Pathol. Abstr., 1983, 1036.
86. Khatri, H. L. and Chohan, J. S., Studies on some factors influencing seed transmission of cowpea mosaic virus in cowpea, *Indian J. Mycol. Plant Pathol.,* 2, 40, 1972.
87. Sharma, S. R. and Varma, A., Tranmission of cowpea banding mosaic and cowpea chlorotic spot viruses through the seeds of cowpea, *Seed Sci. Technol.,* 14, 217, 1986.
88. Van Koot, Y. and Van Dorst, H. J. M., Virus diseases of the cucumber in the Netherlands, *Tijdschr. Plantenziekten,* 65, 257, 1959.
89. Yakovleva, N., Control of green mosaic of cucumber, *Zashch. Rast. Vredit Bolez,* 10, 50, 1965.
90. Rader, W. E., Fitzpatrick, H. F., and Hildebrand E. M., A seed-borne virus of muskmelon, *Phytopathology,* 37, 809, 1947.
91. Middleton, J. T. and Bohn, G. W., Cucumbers, melons, squash, in *Yearbook of Agriculture,* U.S. Department of Agriculture, Washington, D.C., 483, 1953.
92. Meiners, J. P., Waterworth, H. E., Smith, F. F., Alconero, R., and Lawson, R. H., A seed transmitted strain of cucumber mosaic virus isolated from bean, *J. Agric. Univ. P. R.,* 61, 137, 1977.
93. Bos, L. and Maat, D. Z., A strain of cucumber mosaic virus seed transmitted in beans, *Neth. J. Plant Pathol.,* 80, 113, 1974.
94. Tomlinson, J. A. and Walker, V. M., Further studies on seed transmission in the ecology of some aphid-transmitted viruses, *Ann. Appl. Biol.,* 73, 293, 1973.
95. Bennett, C. W., Latent virus of dodder and its effect on sugarbeet and other plants, *Phytopathology,* 34, 77, 1944.
96. Mayee, C. D., Storage of seed for pragmatic control of a virus causing mosaic disease of brinjal (eggplant), *Seed,Sci. Technol.,* 5, 555, 1977.
97. Capoor, S. P., Garg, D. G., and Swant, D. M., Seed transmission of French bean mosaic virus, *Indian Phytopathol.,* 39, 343, 1986.
98. Blattny, C. and Osvald, V., Prenos viros chmele (*Humulus lupulus* L.) na potomstvo semenem, *Preslia, 26, 1, 1954.*
99. Bennett, C. W., Lychnis ringspot, *Phytopathology,* 49, 706, 1959.
100. Singh, R. P., Boucher, A., and Wang, R. G., Detection, distribution, and long term persistence of potato spindle tuber viroid in true potato seed from Heilong-jiang, China, *Am. Potato J.,* 68, 65, 1991.
101. Gilmer, R. M., Longevity of sour cherry yellows virus in infected cherry seeds, *Plant Dis. Rep.,* 48, 338, 1964.
102. Fulton, R. W., Transmission of plant viruses by grafting, dodder, seed and mechanical inoculation, in *Plant Virology,* Corbett, M. K. and Sisler, H. D., Eds., University of Florida Press, Gainesville, 1964, 39.
103. Lister, R. M. and Murant, A. F., Seed transmission of nematode-borne viruses, *Ann. Appl. Biol.,* 59, 49, 1967.
104. Zaumeyer, W. J. and Harter, L. L., Two new virus diseases of beans, *J. Agric. Res.,* 67, 305, 1943.

105. Skotland, C. B. and Burke, D. W., A seed-borne bean virus of wide host range, *Phytopathology,* 51, 565, 1961.
106. Bennett, C. W. and Costa, A. S., Sowbane mosaic caused by a seed-transmitted virus, *Phytopathology,* 51, 546, 1961.
107. Bennett, C. W., Seed transmission of plant viruses, *Adv. Virus Res.,* 14, 221, 1969.
108. Kendrick, J. B. and Gardner, M. W., Soybean mosaic: seed transmission and effect on yield, *J. Agric. Res.,* 27, 91, 1924.
109. Provvidenti, R., Gonsalves, D., and Ranalli, P., Inheritance of resistance to soybean mosaic virus in *Phaseolus vulgaris, J. Hered.,* 73, 302, 1982.
110. Middleton, J. T., Seed transmission of squash mosaic virus, *Phytopathology,* 34, 405, 1944.
111. Alexander, L. J., Inactivation of tobacco mosaic virus from tomato seed, *Phytopathology,* 50, 627, 1960.
112. Broadbent, L., The epidemiology of tomato mosaic. XI. Seed transmission of TMV, *Ann. Appl. Biol.,* 56, 177, 1965.
113. Henderson, R. G., Transmission of tobacco ringspot by seed of petunia, *Phytopathology,* 21, 225, 1931.
114. Laviolette, F. A. and Athow, K. L., Longevity of tobacco ringspot virus in soybean seed, *Phytopathology,* 61, 755, 1971.
115. Athow, K. L. and Bancroft, J. B., Development and transmission of tobacco ringspot virus in soybean, *Phytopathology,* 49, 697, 1959.
116. Valleau, W. D., Symptoms of yellow ringspot and longevity of the virus in tobacco seed, *Phytopathology,* 29, 549, 1939.
117. Fielding, M. J., Observations on the length of dormancy in certain plant infecting nematodes, *Proc. Helminthol. Soc. Wash.*, 18, 110, 1951.
118. Raeder, J. M., A note on the longevity of the wheat nematode, *Anguina tritici, Plant Dis. Rep.*, 38, 268, 1954.
119. Yoshii, H. and Yamamoto, S., A rice nematode disease "Senchu Shingare Byo." II. Hibernation of *Aphelenchoides oryzae, J. Fac. Agric. Kyushu Univ.*, 9, 223, 1950.
120. Todd, E. H., Further studies on the white tip disease of rice, *Proc. Assoc. South. Agric. Wkrs.,* 49, 141, 1952.
121. Goffart, H., *Nematoden der Kulturpflanzen Europas*, Paul Parey, Berlin, 1951, 144.
122. Epps, J. M., Survival of soybean cyst nematodes in seed bags, *Plant Dis. Rep.,* 52, 45, 1968.
123. Medina, A. C., Viability of seeds and a bean pathogen, *Agric. Tec. Mex.,* 3, 3, 1970.
124. Schuster, M. L. and Coyne, D. P., Detection of bacteria in bean seed, *Annu. Rep. Bean Improvement Coop.,* 18, 71, 1975.
125. Burkholder, W. H., Bacteria as plant pathogens, *Annu. Rev. Microbiol.,* 2, 389, 1948.
126. Schuster, M. L. and Sayre, R. M., A coryneform bacterium induces purple colored seed and leaf hypertorphy of *Phaseolus vulgaris* and other legumes, *Phytopathology,* 57, 1064, 1967.
127. Biddle, J. A., McGee, D. C., and Braun, E. J., Seed transmission of *Clavibacter michiganense* subsp. *nebraskense* in corn, *Plant Dis.,* 74, 908, 1990.
128. Zeng, D. F., Tan, Y. J., and Xu, Z. Y., Survival of the *Pseudomonas solanacearum* in peanut seeds, *Bacterial Wilt Newsl.*, 10, 8, 1994.
129. Chang, S., A review of studies on downy mildew of maize in Taiwan, *Indian Phytopathol.,* 23, 270, 1970.

130. Mikoshiba, H., Studies on the control of downy mildew disease of maize in the tropical countries of Asia. IV. Infection source and disease cycle of downy mildew of maize in Indonesia, *Jpn. J. Trop. Agric.*, 23, 57, 1979.
131. Adams, D. B. and Kuhn, C. W., Seed transmission of peanut mottle virus, *Phytopathology,* 67, 1126, 1977.
132. Bowers, G. R., Jr. and Goodman, R. M., Effect of soybean mosaic virus: infection of soybean seed parts and seed transmission, *Phytopathology,* 69, 569, 1979.
133. Hewett, P. D., Pathogen viability on seed in deep freeze storage, *Seed Sci. Technol.,* 15, 73, 1987.
134. Justice, O. L. and Bass, L. N., Principles and Practice of Seed Storage, U.S. Department of Agriculture, Agricultural Handbook No. 506, Washington, D.C., 1978.
135. Shipton, W. A., Boyd, W. R. J., Rosielle, A. A., and Shearer, B. I., The common Septoria diseases of wheat, *Bot. Rev.,* 37, 231, 1971.
136. Von Wechmar, B., Investigation on the survival of *Septoria nodorum* Berk, on crop residue, *S. Afr. J. Agric. Sci*., 9, 93, 1966.
137. Kruger, J. and Hoffmann, G. M., Zur über Lebensdauer von *Septoria nodorum* und *S. avenae f.* sp. *triticea* an Sommerweizensoatgut under Einfluz der Langertemperatur, *Z. Pflanzenkr. Pflanzensch.,* 85, 413, 1978.
138. Bronnimann, A., On *Leptosphaeria nodorum,* the pathogen causing glume blotch and leaf dessication of wheat, *Phytopathol. Z.,* 61, 101, 1968.
139. Cunfer, B. M., Survival of *Septoria nodorum* in wheat seed, *Trans. Br. Mycol. Soc.,* 77, 161, 1981.
140. Krout, W. S., Treatment of celery seed for the control of Septoria blight, *J. Agric. Res.,* 21, 369, 1921.
141. Fischer, G. W. and Holton, C. S., *Biology and Control of the Smut Fungi*, Ronald Press, New York, 622, 1957.
142. Maude, R. B. and Humpherson-Jones, F. M., Studies on the seedborne phases of dark leaf spot (*Alternaria brassicicola*) and grey spot (*Alternaria brassicae*) of Brassicas, *Ann. Appl. Biol.,* 95, 311, 1980.
143. Laha, S. K. and Grewal, J. S., Botrytis blight of chickpea and its perpetuation through seed, *Indian Phytopathol.*, 36, 630, 1983.
144. Neergaard, P., *Seed Pathology,* Vols. 1 and 2, Macmillan, London, 1977, 1187.
145. Nicholson, J. F., Dhingra, O. D., and Sinclair, J. B., Internal seedborne nature of *Sclerotinia sclerotiorum* and *Phomopsis* sp. and their effects on soybean seed quality, *Phytopathology,* 62, 1261, 1972.
146. Sattar, A., On the occurrence, perpetuation and control of gram (*Cicer arietinum* L.) blight caused by *Ascochyta rabiei* (Pass.) Labousse, with special reference to Indian conditions, *Ann. Appl. Biol.,* 20, 612, 1933.
147. Narkiewicz-Jodko, M., L'observations sur l'etat de semence du trefle rouge pendant leur conservation, Proc. 17th ISTA Congr., Warsaw, 1974, 4.
148. Maude, R.B. and Presly, A. H., Fungal diseases, in 26th Annual Report for 1975, *National Veg. Res. Stat.*, Wellesbourne, U.K., 1976.
149. Petrie, G. A., Fungi associated with seeds of rape, turnip, flax, and safflower in western Canada, 1968–73, *Can. Plant Dis. Surv.,* 54, 155, 1974.
150. Chahal, A. S., Seed-borne infection of *Alternaria brassicae* in Indian mustard and its elimination during storage, *Curr. Sci.,* 50, 621, 1981.
151. Halfon-Meiri, A., Seed transmission of safflower rust (*Puccinia carthami*) in Israel, *Seed Sci. Technol.,* 11, 835, 1983.

152. Siddiqui, M. R. and Mathur, S. B., Survival of *Septoria nodorum* Berk. in wheat seed stored at 5°C, *FAO/IBPGR Genet. Newsl.*, 75/76, 7, 1989.
153. Kumar, K., Mathur, S. B., and Neergaard, P., Seedborne nature of *Phyllosticta cajani* in pigeon pea (*Cajanus cajan*), 21st Int. Seed Congr., Brisbane, 1986.
154. Mia, M. A. T., Mathur, S. B., and Neergaard, P., *Gerlachia oryzae* in rice seed, *Trans. Br. Mycol. Soc.*, 84, 337, 1985.
155. Thomas, M. S., Dry season survival to *Rhynchosporium oryzae* in rice leaves and stored seeds, *Mycologia,* 76, 1111, 1984.
156. Cunfer, B. M., Long term viability of *Septoria nodorum* in stored wheat seed, *Cereal Res. Commun.*, 19, 347, 1991.
157. Kaiser, W. J., Stanwood, P. C., and Hannan, R. M., Survival and pathogenicity of *Ascochyta fabae* f. sp. *lentis* in lentil seeds after storage for four years at 20° to –196°C, *Plant Dis.,* 73, 762, 1989.
158. Tu, J. C., Epidemiology of anthracnose caused by *Colletotrichum lindemuthianum* on white bean (*Phaseolus vulgaris*) in southern Ontario; survival of the pathogen, *Plant Dis.,* 67, 402, 1983.
159. Kaiser, W. J. and Hannan, R. M., Incidence of seedborne *Ascochyta lentis* in lentil germplasm, *Phytopathology,* 76, 355, 1986.
160. Shetty, H. S., Mathur, S. B., and Neergaard, P., *Sclerospora graminicola* in pearl-millet seeds and its transmission, *Trans. Br. Mycol. Soc.,* 74, 127, 1980.
161. Neergaard, P., *Danish Species of Alternaria and Stemphylium*, Einar Munksgaard, Copenhagen, 1945, 560.
162. Halfon-Meiri, A. and Solel, Z., Factors affecting seedling blight of sweet corn caused by seedborne *Penicillium oxalicum, Plant Dis.,* 74, 36, 1990.
163. Taylor, J. D., Dudley, C. L., and Presly, L., Studies of halo-blight seed infection and disease transmission in dwarf beans, *Ann. Appl. Biol.,* 93, 267, 1979.
164. Valarini, P. J., Menten, J. O. M., and Lollato, M. A., Incidence of common bacterial blight in the field and transmission of *Xanthomonas campestris* pv. *phaseoli* through bean seeds evaluated by different methods, *Summa Phytopathol.*, 18, 160, 1992.
165. Singh, R. B. and Jain, J. P., Survival of *Xanthomonas campestris* pv. *glycines* — the incitant of bacterial pustule of soybean, *Indian J. Mycol. Plant Pathol.,* 18, 246, 1988.
166. Shakya, D. D., Vinther, F., and Mathur, S. B., World wide distribution of bacterial stripe pathogen of rice identified as *Pseudomonas avenae, Phytopathol. Z.,* 114, 256, 1985.
167. Shakya, D. D., Etiology and Seed Transmission of Bacterial Stripe of Rice, Ph.D. Thesis, Pt. II, Seoul National University, Korea, 1981.
168. Jeyanandarajah, P. and Brunt, A. A., Occurrence of bean commom mosaic virus in mungbean in Sri Lanka, *FAO Plant Prot. Bull.*, 41, 107, 1993.
169. Huang, C. S. and Chiang, Y. C., The influence of temperature on the ability of *Aphelenchoids besseyi* to survive dehydration, *Nematologica,* 21, 351, 1975.
170. Grasmick, M. E. and Slack, S. A., Effect of potato spindle tuber viroid on sexual reproduction and viroid transmission in true potato seed, *Can. J. Bot.,* 64, 336, 1986.
171. Anselme, C. and Champion, R., Study of the transmission of *Botrytis cinerea* by sunflower (*Helianthus annuus*) seed, *Seed Sci. Technol.,* 3, 711, 1975.
172. Minyaeva, O. M., Dissemination of clover anthracnose with seeds, *Bull. Soc. Nat. Moscow NS,* 56, 91, 1951.

173. Orozco-Sarria, S. H. and Cardona-Alvarez, C., Evidence of seed transmission of angular leaf spot of bean, *Phytopathology,* 49, 159, 1959.
174. Pathak, V. K., Mathur, S. B., and Neergaard, P., Detection of *Peronospora manshurica* (Naum.) Syd. in seeds of soybean, *Glycine max, EPPO Bull.,* 8, 21, 1978.
175. Lawyer, A. S., Diseases caused by bacteria, in Compendium of Pea Diseases, Hagedorn, D. J., Ed., APS Press, St. Paul, MN. 1984.
176. Taylor, J. D. and Dudley, C. L., Bacterial diseases, in 28th Annual Report for 1977, *National Veg. Res. Sta.*, Wellesbourne, U.K., 104, 1978.
177. Mayee, C. D., Systemic lesions of seed-borne brinjal mosaic virus in some varieties of brinjal, *Curr. Sci.,* 43, 692, 1974.
178. Valarini, P. J., Menten, J. O. M., and Lollato, M. A., Incidence of common bacterial blight in the field and transmission of *Xanthomonas campestris* pv. *phaseoli* through bean seeds evaluated by different methods, *Summa Phytopathol.*, 18, 160, 1992.

CHAPTER 7

Seed Transmission and Inoculation

I. SEED TRANSMISSION

Seed transmission refers to the passage of inoculum from an infected or infested seed to a plant. The agents that are transmitted and cause disease are seed-transmitted pathogens, whereas pathogens that are associated with seeds but do not play a role in disease development are nonseed-transmitted pathogens. For example, the curly top virus is found in the perisperm of sugar beet seeds and thus is carried by seeds but the virus is not transmitted from infected seeds to seedlings.[1]

The mechanism of seed transmission varies with host and pathogen and is governed by specific host–pathogen reactions. Any part of a seed may be infected. The process is broadly classified as systemic and nonsystemic seed transmission. Systemic transmission is when infected or infested seeds on germination result in a systemic disease in the plant to a stage of symptom development. Nonsystemic seed transmission results in either pre- or postemergence infection or symptom production on the plant. All seedborne pathogens may be seed transmitted by more than one means. The rate of transmission is dependent upon the degree to which the pathogen is carried from seed to seedling and to plant.

The oospores of downy mildew fungi serve as resting propagules, which enable these fungi to survive desiccation during seed maturation and storage, since the mycelium, sporangia, or conidia may not tolerate dehydration and long-term storage.[2] However, survival of mycelium in internal seed layers has been demonstrated, such as that of *Phytophthora nicotianae* var. *parasitica* in sesame embryos,[3] *Sclerospora graminicola* in pearl millet,[4,5] *Peronosclerospora sorghi* in sorghum and maize,[6-8] *Plasmopara halstedii* in sunflower,[9,10] *Peronospora viciae* in peas,[11] *Peronospora farninosa* in beets,[12] and *Peronospora manshurica* in soybeans.[2] The internal seedborne nature of *S. graminicola* in pearl millet has been characterized as debatable.[13] Moreover, the presence of mycelia of *Peronosclerospora maydis, P. philippinensis, P. sacchari, Sclerophthora rayssiae* var.

zeae, P. sorghi, and *Sclerophthora macrospora* in maize seeds has been reported.[7,14-18] However, transmission of these pathogens through infected seeds depended on moisture content of seeds.[7,19] Survival of downy mildew pathogens as mycelium between seasons appeared less likely since dry seeds are stored, and such seeds failed to develop downy mildew on resulting seedlings.[20] However, *P. sorghi* formed oospores within maize seeds, which is an exception.[21] Seed transmission of *S. macrospora* does not appear important epidemiologically, as the small ears from infected maize plants are usually discarded during processing.[20]

Erwinia stewartii, the causal agent of Stewart's wilt, has been reported as seed transmitted in maize seeds, although direct evidence has been limited to laboratory studies. Despite the lack of evidence for seed transmission in the field, more than 50 countries have phytosanitary restrictions for seedborne *E. stewartii* in maize seeds to avoid introduction of the pathogen.[22]

Pea seedborne mosaic virus (PSbMV) was shown to have temperature-dependent latency in symptom expression; the number of plants exhibiting symptoms in different pea cultivars was higher at 29°C than at 18°C.[23] PSbMV was detected from symptomless pea plants when assayed on *Chenopodium amaranticolor.*[24] Latency was detected using ELISA: a significantly larger number of plants reacted positively when 5 weeks old compared to those at 3 weeks.[25] A seed lot of pea cv. Belinda was shown to transmit at a high frequency a latent strain of PSbMV, which was weakly detectable using ELISA after 5 weeks. This virus was present in a subliminal quantity in seeds but below the sensitivity threshold of ELISA.[26] The incidence of AMV in alfalfa seeds (24%) and seedlings (16%) was significantly different. This difference in detection rates of AMV in seed and seedlings was due to the higher incidence of AMV in seed coats.[27] Therefore, it is better to differentiate between seed infection rate and seed transmission rate. Seed infection rate is used to express both seed coat and embryo infection. Seed transmission rate refers only to embryo infection and is equivalent to detection rate for seedlings. Thus seed transmission rate is more appropriate than seed infection rate in expressing the extent of AMV transmission through seeds.

A. Systemic Seed Transmission

Systemic seedborne pathogens are transmitted either by infection of different seed parts (i.e., embryo, endosperm, or seed coat) or by contamination of the seed coat.

1. Embryo Infection

Embryo infection generally results in systemic infection. The pathogens become active as the germination process begins. Such pathogens follow the growing point of the plant and can express symptoms at different stages of plant growth.

Ustilago segetum var. *tritici* survives as dormant mycelium in the scutellum of embryos of infected wheat and barley seeds (Figure 7-1). The mycelium becomes active when seeds germinate and moves intercellularly through embryo

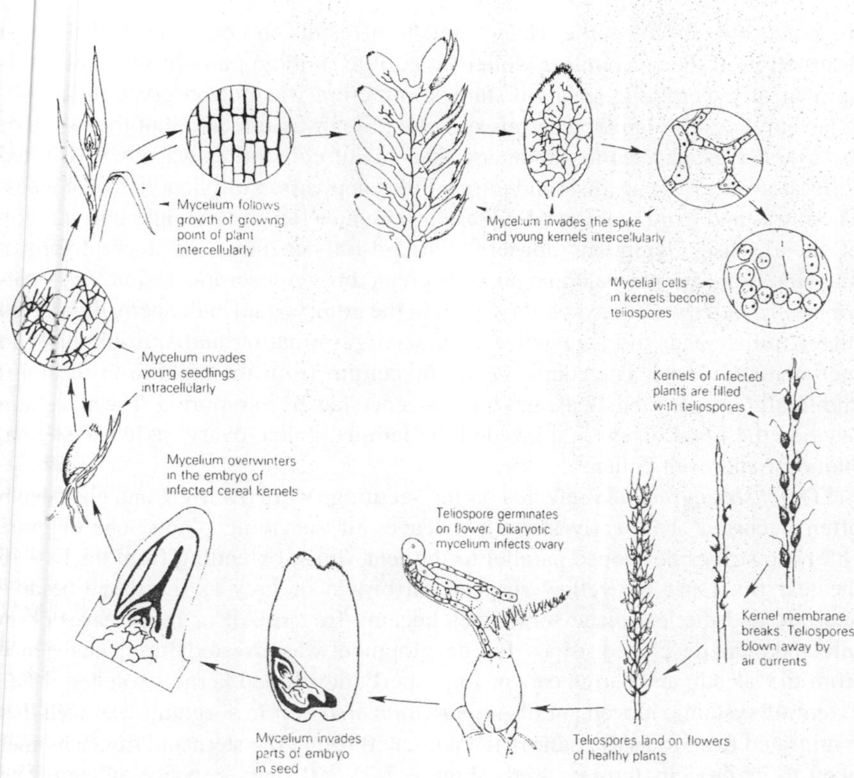

Figure 7-1 Disease cycle of *Ustilago segetum* var. *tritici* (loose smut of barley [*Hordeum vulgare*] and of wheat [*Triticum aestivum*]). (Adapted from Agrios, G. N., *Plant Pathology*, Academic Press, New York, 1988, 480. With permission.)

tissues and the young seedlings until it reaches the growing point of the plant. The mycelium follows plant growth and remains just behind the growing point. The mycelium invades all young spikelets, where it becomes intracellular and replaces most tissues of the spike except the rachis. The mycelium in the infected seed is transformed into teliospores, which are held together by the delicate membrane of host tissue. On maturity the membrane ruptures, and teliospores are released and carried to healthy plants by air currents. These teliospores fall on the flowers of healthy plants and germinate through the formation of a basidium consisting of one to four cells (haploid hyphae). The cells germinate and produce short uninucleate hyphae and sexually compatible haploid hyphae fuses, and then produce dicaryotic mycelium. This dicaryotic mycelium penetrates the flower through the stigma or young ovary walls and becomes established within the scutellum when seeds mature; it then remains dormant until seed germination.[28]

The mycelium of *Stagonospora nodorum* in wheat seed embryos during germination moves through the scutellum to the scutellar node, then the coleoptilar node, and eventually the plumule. It infects the radicle from the coleorhizal end. The extraembryonal mycelium from the pericarp penetrates either through the coleorhiza or coleoptile. Heavy infection results in cell death and eventual destruction of the coleoptile. Swollen coleoptile (knobs) parts result from development of hyperplastic areas. In studies, infection was not observed in the vascular bundles. The inner epidermis of coleoptiles was infected, but the region of the true leaves adjacent to the inner epidermis of coleoptiles was not.[29] Affected cells showed severe necrosis and intracellular deposition of phenolic compounds. The intra- and extra-embryonal infection may have led to systemic transmission of the disease. Symptoms appeared on 3-d-old seedlings as a purple-brown discoloration, later developing into discrete, brown necrotic lesions.[29,30] *Sclerophthora macrospora* mycelium, found in the embryo and endosperm of infected finger millet seeds, became active with seed germination and spread rapidly to meristematic tissue. The course of the mycelium from the embryo to seedlings and finally in root, stem, leaf, and inflorescence has been reported. The mycelium reached the floral organs and invaded the lamina, palea, ovary, style and stigma, stamens, and even pollen.[31]

Drechslera graminea-infected barley seedlings were dwarfed, and plant death often occurred at an early stage. On leaves of surviving plants, one or more chlorotic stripes developed parallel to the leaf veins, extending from the base to the leaf tip. Later the yellow stripes turn brown or grey as the tissue became necrotic and the leaf blade sometimes became frayed. All or most leaves of an infected plant developed stripes. Ear development was arrested during emergence from the sheath, and barren ears or improperly developed kernels resulted.[32] The extent of systemic movement of *Acremonium strictum* in sorghum was such that it migrated from roots to grains in the panicle through the stem and from a mother plant to its progeny through seeds (Figure 7-2).[33] *P. oryzae* in rice, in temperate and subtropical regions, overwintered as mycelium and conidia on diseased straw and seeds. The pathogen also has been reported from cultivated and wild host species.[34] The fungus was transmitted from the hilum through the pericarp to the extruded tip of the scutellum or epiblast, to the coleoptile, and then to the primary leaf. Radicle infection occurred from the infected pericarp into the extended tip of the coleorhiza.[35,36]

When hyphae of *Fusarium moniliforme* were present in the basal part of a maize embryo directly above the closing layer, it ramified throughout the seed prior to germination. Also, the pathogen grew from an infected kernel upward to the cob, where it infected other developing kernels by entering the pedicle and then the ovary. There is no direct vascular connection between the cob and ovule. If the fungus was carried on the pericarp surface or within the tip cap tissue, infection did not occur until after germination.[37]

In soybean, *Colletotrichum truncatum* hyphae in infected cotyledons were responsible for preemergence damping-off and symptomless establishment of internal hyphae. The hyphae became established in stem cortical cells without an apparent effect and remained localized in the immediate area until flowering. They then resumed growth and penetrated the lower stem, petioles, leaves, and developing seeds and pods, without symptom development. As a host matures, the fungus may fruit on stems and pods. Hyphae were found in carpel cells, ovarial locules, and cotyledons.[38]

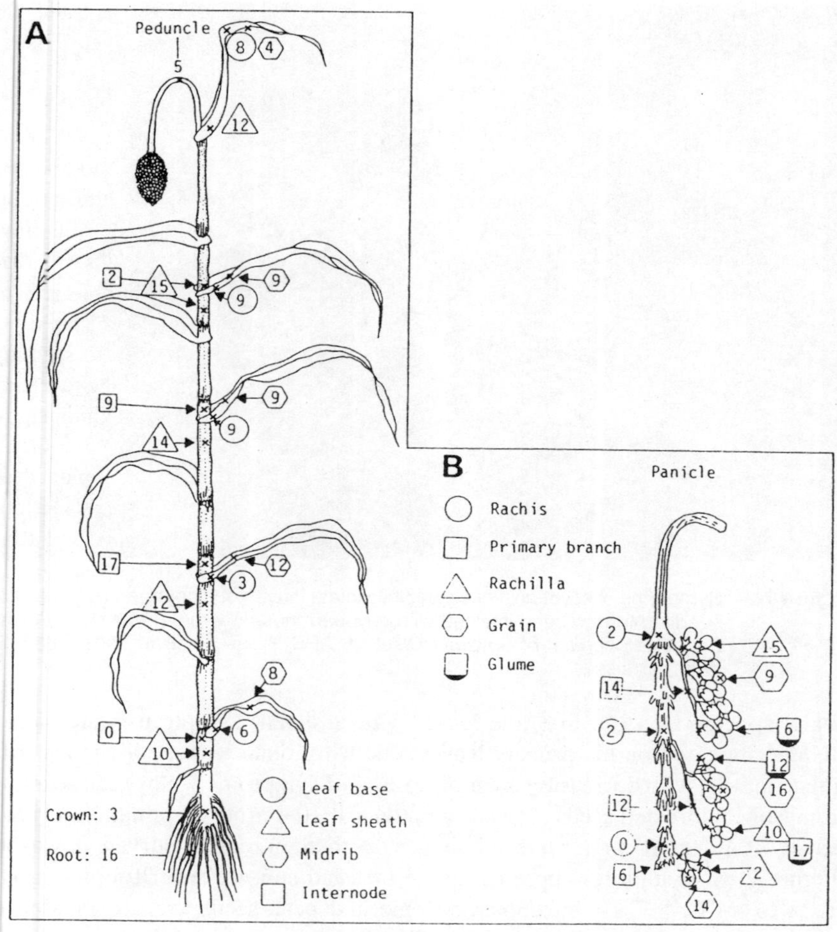

Figure 7-2 Parts of (**A**) sorghum (*Sorghum bicolor*) plant and (**B**) panicle from which *Acremonium strictum* was isolated. Numbers refer to the plant pieces from which *A. strictum* was isolated out of 20 pieces plated on potato-dextrose agar. (From Bandyopadhyay, R., et. al., *Plant Dis.*, 71, 647, 1987. With permission.)

Systemic seed transmission has been demonstrated for some seedborne bacteria: *C. flaccumfaciens* pv. *flaccumfaciens* in beans, *X. campestris* pv. *campestris* in cabbage. *X. c.* pv. *vesicatoria* in pepper, and *X. c.* pv. *phaseoli* in beans. *X. c.* pv. *phaseoli* occurred either in the embryo and seed coat or on the seed coat of beans. The bacterium entered through rifts in the cotyledon cuticle at germination, moved intercellularly, and entered the seedling vascular system. Lesions sometimes developed on leaves and stems.[39-41]

A majority of seed-transmitted viruses persist in embryos. Transmission rate depends upon virus strain, host environmental conditions, and other factors. Viruses may exhibit symptoms at any growth stage. Symptoms of BCMV in

Figure 7-3 Symptoms of soybean mosaic on unifoliolate leaves of a soybean (*Glycine max*) seedling from a seed infected with soybean mosaic virus. (From J. B. Sinclair, Ed., *Compendium of Soybean Disease,* APS Press, St. Paul, MN, 1982, 56. With permission.)

beans appeared on the first true leaves.[42] In urdbean, symptom expression of BCMV appeared on the primary leaves, first trifoliolate leaves, or second trifoliolate leaves, with decreasing order of severity of symptoms.[43] Soybean seedlings from SbMV-infected seeds showed symptoms at 10 d after germination. Primary leaves of infected plant were misshapen, with their edges curved downward (Figure 7-3). Symptoms appeared on the first and subsequent trifoliolate leaves at 22 to 24°C.[44] TMV infections in apple and pear seeds were expressed on seedings prior to transplanting or handling.[45] Some viroids and viruses did not produce symptoms on seedlings, including avocado sunblotch viroid on avocado,[46] Prunus necrotic ringspot virus in peach,[47] cherry leaf roll virus on elm,[48] raspberry ringspot virus on raspberry and strawberry, and tomato blackring virus on strawberry.[49] Pea seedborne mosaic virus under field conditions was symptomless in 5 to 10% of the pea plants carrying the virus.[50] Symptoms became masked in plants with symptoms in 6 to 8 weeks, and most infected plants appear normal at full-bloom stage.[51]

2. Nonembryonic Infection

Seed coat infection occasionally leads to systemic infection. For example, the cotyledons of onion seedlings from *Botrytis acalda*-infected seeds became infected by hyphal invasion from infected seed coats attached to the cotyledons at emergence. Symptoms developed after leaf senescence. Hyphae from initial infection by conidia invaded successive leaves by first infecting the tip and then grew downward into the leaf and invaded the onion bulk neck.[52] In coriander, chlamydospores of *Protomyces macrosporus* were carried in the pericarp. The pathogen became systemic in the host before or during preflowering, with host generative cells being destroyed.[53]

Pseudomonas avenae may be located between the glume and the pericarp or deeper in rice seeds. During seed germination the coleoptile became infected, and invasion by the bacterium took place through stomata of the coleoptile and first leaf. The rice coleoptile is known to have open-type stomata on both the abaxial and adaxial surfaces.[54] The bacterium multiplied in substomatal chambers and invaded intercellularly, reaching the lacunae, which runs the full length of leaves. The resulting lesions appeared as stripes. The bacteria were not observed in either xylem or phloem vessels. Occasionally cell wall disintegration and brown disorganized cells were seen.[55] *X. c.* pv. *phaseoli* invaded the vascular system of the bean seedlings. Bacterial colonization continued into the developing bean plant.[56]

A number of bacteria are localized in seed coat tissues and can result in a systemic infection of the host. *X. c.* pv. *malvacearum* in cottonseeds survived in the basal end of the chalaza, as a contaminant on the seed surface, or in the lint. At germination, the basal cap adhering to the cotyledon was carried above ground, where infection took place through the stomata or wounds, and symptoms developed on the cotyledons. Secondary spread was through irrigation water and splashing rain.[57] During germination, bean seeds infected with *X. c.* pv. *phaseoli* swelled rapidly with water, and rifts developed in the epidermis. The bacteria entered cotyledons through these rifts, resulting in systemic invasion of seedlings.[58] Systemic infection resulted in development of leaf lesions and stem cankers.

During germination of cabbage seeds, *X. c.* pv. *campestris* entered cotyledonary stomata. The bacterium progressed intercellularly until it reached the vascular system.[57] Similarly, *C. f.* pv. *flaccumfaciens* in bean and *C. m.* subsp. *michiganensis* in tomato were carried in the seed coat and invaded cotyledons and then the vascular system before becoming systemic.[57]

Three viruses, TMV in pepper and tomato, cucumber green mottle virus in cucumber, and tomato spotted wilt virus in cineraria and tomato, generally are found in the seed coat. Seedling infection results from injuries during uprooting and transplanting.[59-62] The incidence of TMV was highest when pepper seedlings were transplanted and lowest when seeds were planted directly.[59]

3. Seed Coat Contamination

In a few cases, seed coat contamination can lead to systemic infection (Figure 7-4). Examples are *Sporisorium sorghi, S. cruentum, Tilletia caries, T. laevis, T. controversa, U. segetum* var. *avenae, U. segetum* var. *segetum, U. zeae*, and TMV in tomato.

The dormant spores of *Tilletia controversa* germinated and produced promycelia, which bore a whorl of spindle-shaped sporidia at their tip. Compatible sporidia (+ and –) usually fused in pairs. Fused sporidia underwent plasmogamy

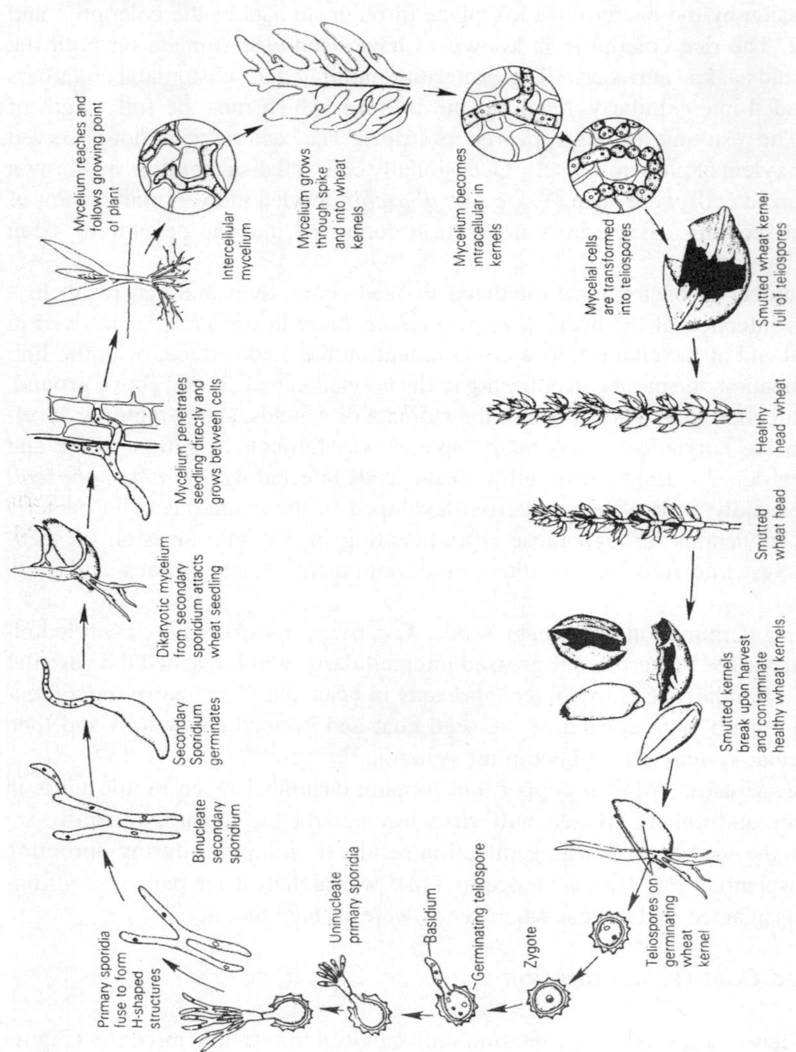

Figure 7-4 Disease cycle of *Tilletia* (covered smut or bunt of wheat [*Triticum aestivum*]). (Adapted from Agrios, G. N., *Plant Pathology*, Academic Press, New York, 1980, 484. With permission.)

(but not caryogamy), separate from the promycelium, and germinated to produce binucleate hyphae, which were pathogenic to susceptible wheat cultivars.[63] Primary sporidia that did not fuse germinated with haploid, mononucleate hyphae that grew on many media but were not pathogenic to wheat. Binucleate hyphae penetrated young wheat seedlings at multiple points and grew slowly toward apical meristem tissue, where they remained quiescent until the proper photoperiod for floral infection.[64] As the infected plant approached the heading stage and the kernel tissue enlarged, the pathogen changed from its slow-growing vegetative stage to a rapid-growing reproductive phase. Fungal teliospores replaced the contents of the kernel to form smut or bunt balls.[63] In the Pacific Northwest, United States, plant infection occurred during January through March in association with a continuous snow cover of 60 to 90 d. Presumably, snow cover provided the high moisture and cool temperatures conducive to teliospore germination on the soil surface and subsequent infection.[63]

Infection in sorghum by *Sporisorium sorghi* took place between germination and emergence of seedlings. The mature spore on germination gave rise to four-celled promycelia, giving rise to sporidia both laterally and terminally. Sometimes infection hyphae developed in place of sporidia. Infection hypha penetrated the radicle in mesocotyl region. The mycelium became systemic and remained in the meristematic tissues. The fungus kept pace with plant growth. However, in rapidly growing plants, the fungus did not reach the apical meristem and was left behind in the lower portions. Such plants resulted in healthy main tillers but infected secondary tillers.[65]

Teliospores of *T. laevis* and *T. caries* are carried on wheat seed surfaces as a contaminant. In temperate regions, teliospores remain dormant in soil for long periods. When such seeds are planted, systemic infection of the plant occurs. Teliospores germinated with seeds and produced primary and secondary sporidia. Germination of secondary sporidia resulted in penetration by dicaryotic hyphae. Pericarps became infected by *T.caries* at 4 d and coleoptiles at 7 d after sowing. The number of penetration points increased until the coleoptile shriveled. Hyphae passed inter- and intracellularly through the coleoptile to primary and secondary leaves, intercellularly through successive leaf bases into the growing point, and/or from the base of the fourth or fifth leaf into the tissues immediately beneath the growing point. Hyphae were found in the growing point at emergence of the fifth leaf, just before internode elongation. Smut developed once hyphae reach the shoot apex. If a natural barrier, such as lignification or physical distance, developed before hyphae reached the shoot apex, smut did not develop.[66,67]

In oats, teliospores of *U. segetum* var. *avenae* and *U. segetum* var. *segetum* were carried as contaminants beneath the seed lemma and palea. Dormant hyphae also were carried in the seed coat. *U. segetum* var. *avenae* penetrated coleoptiles and first internodes of seedlings directly and reached the growing point through the third or fourth leaf base at 14 to 17 d after sowing and before internode elongation. Later it was carried upward by plant growth. Spores are formed by segmentation of sporiferous hyphae.[68] Similarly, direct penetration of oat seedlings by *U. segetum* var. *segetum* can occur without hyphal fusions.[69] All the culms may not be smutted, or the oldest and youngest primary lateral culms could be smutted while one or more intermediate tiller families remain healthy. This may be the result of multiple infections.[69,70] Thus four types of infection processes by *U. segetum* var. *avenae* and *U. segetum* var. *segetum* may occur in a single host plant.[69] Sampson[71] observed a barrier of lignified cells between nodal tissues containing hyphae of *U. segetum* var. *segetum* and the apical growing point in resistant oat plants and concluded that if a plant produced this barrier before the fungus reached the growing point, culm or culms would not become infected. Fungal growth retardation might account for resistance according to this hypothesis.

Teliospores of *U. segetum* var. *segetum* carried as a seed-surface contaminant on barley gave rise to sporidia that penetrated seedlings through coleoptiles. The fungus followed plant growth and replaced the seeds with smut spores. Tillers arising from a single node of the principal culm tended to be either all smutted or healthy. When tiller families were infected differentially, older tillers frequently were not. However, the principal culm, which always is oldest, does not fit this age-frequency pattern; it is frequently infected.[72] Teliospores of *U. zeae* carried on the seed surface of maize, may send out germ tubes that penetrate young epidermal cells directly.[73] *S. cruentum* entered through the radicle, mesocotyl, or epicotyl and became systemic in sorghum.[74]

Chlamydospores of *Urocystis occulta*, carried on the surface of rye seeds, germinated, and binucleate infection hyphae directly penetrated epidermal cell walls of the coleoptile. The host cell wall was thickened and softened at the point of entry. Hyphal development in a susceptible line was rapid, with hyphae found in the coleoptilar internode after 21 d and in the meristematic region after 28 d.[75]

Teliospores of *Puccinia calcitrapae* var. *centaureae* carried on the surface of safflower seeds produced basidiospores, which directly penetrated seedling tissues during germination and before emergence.[76] Oospores of *Peronospora manshurica* on the surface of soybean seeds resulted in systemic infection (Figure 3-13). Such plants produced windborne conidia, which initiated local infections. Some systemically infected plants may die prematurely.[77,78] Similarly, oospores of *Sclerospora graminicola* carried on the seed surface resulted in systemic infection of pearl millet seedlings.

Pimento pepper seedlings may not show symptoms of tobacco mosaic virus (TMV) infection in seedling beds. Many seedlings developing from seeds of infected parents may be contaminated with TMV. About 28% of roots and 3% of shoots from infected seeds carry the virus. At transplanting, TMV associated with the pepper seedlings may be introduced into wounds on leaves, stems, or root hairs to initiate infection. Many infections may be the result of root rather than shoot infection. Thus virus particles on seed coat may carry the virus from seeds to seedlings.[79]

Aphelenchoides besseyi (nematode), which causes white tip in rice, is carried beneath rice hulls as immature, preadult larvae. Nematode larvae released with the sprouting of rice seeds were carried inside the seedling leaf sheath and later up to the growing points of leaves and stems. These finally reached the glume exterior and entered the palea. Symptoms of infection were that leaf tips turned pale yellow to white in the tillering stage and later brown and became necrotic.

Infected plants were shorter, unthrifty, and bore small panicles with distorted glumes and low fertility.[80] The larvae coiled up inside seeds and became dormant. The number of larvae inside a grain depended on the infection severity. Fukano[81] found that the number of larvae in severely and slightly affected seeds was 1241 and 132 per 100 seeds, respectively. A majority (90%) of larvae were inside the husk, with the remainder on the seed surface.[81]

Ditylenchus dipsaci and *D. angustus* are seedborne nematodes in onion and red clover, and rice, respectively. In onion, infected seeds do not germinate, or, if they germinate, seedlings are pale, twisted, and may die. In red clover, leaves, leaf stalks, and flower buds are thickened and twisted. In rice, symptoms of *D. angustus* appeared as chlorosis or streaks on upper leaves, panicles remained enclosed in the leaf sheath, peduncles turned dark brown, and normal grain formation occurred only at the panicle tips. *D. dipsaci* survived in seeds as preadult or fourth-stage larvae. Larvae became active under optimum conditions and entered the host through the stylet near the root cap, where they passed through the fourth molt and became adults. Fourth-stage larvae of the next generation congregated near the growing point of a seed stalk, invaded the inflorescence, and became established within the onion seeds.[82,83]

B. Nonsystemic Seed Transmission

Nonsystemic seed transmission is common and results from seed infection, contamination on the seed surface, or pathogens mixed with seeds.

1. Embryo Infection

Examples of embryo infection that may lead to nonsystemic seed transmission are *Ascochyta pisi*,[84] *Phoma pinodella*, and *Mycosphaerella pinodes* in peas; *Colletotrichum lindemuthianum* in beans; and *C. truncatum* in soybeans.[85] These pathogens were carried as hyphae in the embryo and seed coat. After seed germination, *A. pisi* caused lesions on the first true leaves and later on stems and petioles. Pycnidia developed on lesions under moist conditions, and pycnidiospores were carried by wind to other plants. The other pathogens followed a similar pattern of spread.

2. Seed Coat Infection

Seed coat infection generally results in nonsystemic seed transmission. Infected seeds either fail to germinate, or if the pathogen is localized in the seed coat, it infects cotyledons or young seedlings. If infected seeds do not emerge, inoculum is produced in the soil and may not spread; if seedlings and plants are infected, it is produced above ground and is a source of secondary spread. An example of the former is *Fulvia fulva* in tomato.[86] *F. oxysporum* f. sp. *lagenarium* infected emerging seedlings of bottle gourd.[87] This fungus multiplied on infected seedlings, and inoculum spread during crop growth. *Microdochium nivale* caused preemergence death when infected wheat seeds were sown and thus inoculum was introduced into the soil. Also, it multiplied and spread through a crop initially as hyphae and conidia, later by ascospores, and finally, at flowering, by ascospores and conidia produced on upper parts of mature plants.[88] Hyphae of *Pyrenophora teres* carried in the caryopsis grew into developing barley coleoptiles and carried above ground. The fungus multiplied on foliage, first producing conidia, then perithecia later in the growing season. Infection from seeds often produced streaks on coleoptiles, and the fungus penetrated underlying leaves.[89] Pycnidia of *Stagonospora nodorum* frequently were produced on diseased coleoptiles of wheat under high humidity.[90] Fava bean seeds infected by *A. fabae* produced infected seedlings, which developed dark, target-like lesions on which pycnidia was produced. Pycnidiospores spread within the crop to infect pods and seeds.[91,92] *Pyricularia oryzae* in rice seeds was transmitted from the hilum through the pericarp to the extended tip of the scutellum or epiblast, then to the coleoptile, and finally to the primary leaf. Radicle infection occurred from an infected pericarp and then to the extended tip of the coleorhiza.[93] Seedborne *Phoma lingam* in *Brassica* infected seedlings and produced pycnidia. Pycnidiospores were carried to adjacent seedlings.[94] Bean seeds heavily infected with *P. syringae* pv. *phaseolicola* often failed to produce seedlings. If seedlings developed, they showed no or only slight symptoms.[95]

3. Seed Coat Contamination

Seeds with seed coat contamination can produce healthy seedlings, but the inoculum can remain viable in the soil and cause infection at later stages of plant development. Teliospores of *Tilletia barclayana,* carried either in partially bunted rice seeds or on the surface of rice seeds, produced sporidia, which lodged on the stigma, and hyphae from the sporidia penetrated through the style, reaching the chalazal end of the ovary. Hyphae remained between the aleuone layer and seed coat, utilizing the endosperm. This provided space for sori development. Early floret infection resulted in complete transformation of a seed into a sorus; late infection resulted in partially infected seeds. The fungus caused local infection of individual florets which represented a distinct infection site.[96] Similarly, teliospores of *T. indica,* carried on the surface of wheat seeds, were introduced into the soil at planting.[97]

Teliospores of Karnal bunt normally germinated with a stout promycelium bearing 32 to 128 filiform primary sporidia, which germinated either directly, giving rise to lateral or terminal mononucleate hyphae, or indirectly to form sterigmata from which secondary, mononucleate falcate sporidia could be carried to the wheat spike by air currents. Hyphal anastomosis was observed on the glume surface of wheat. Germ tubes arose from secondary sporidia and penetrated stomatal openings of glumes, lemmae, and/or paleae. During early stages of infection, intercellular hyphae were found among chlorenchyma and parenchyma cells in the distal to mid- but not basal portions of the glume, lemma, and palea, and not the ovary, subovarian tissues, rachilla, or rachis. Later, hyphae grew intercellularly into the floret base and the subovarian tissue and entered the ovary pericarp through the funiculus. Hyphae were found in the rachis during later stages

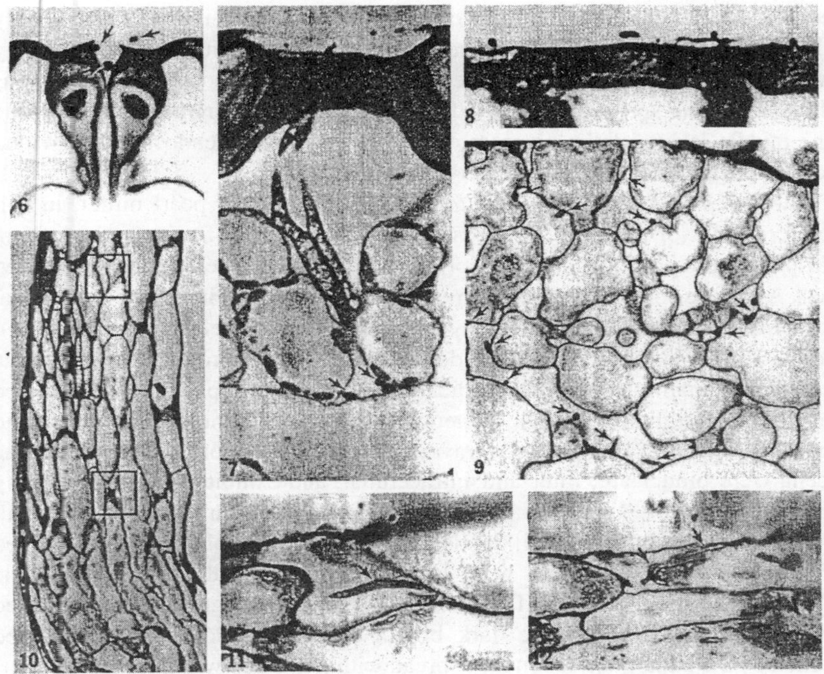

Figure 7-5 Establishment and growth of *Tilletia indica* hyphae in wheat (*Triticum aestivum*) florets. (**6**) One section of a complete series through a transversely sectioned stoma of a glume, showing transversely sectioned hyphae above and below the stomatal edges. The hyphae (*arrows*) were confined to the vestibule above the guard cells in this and all other sections of this stoma (original magnification × 1990). (**7**) Two hyphae have penetrated into the substomatal chamber of a longitudinally sectioned stoma of a glume. Other intercellular hyphae also are present (*arrows*) (original magnification × 1730). (**8**) Hyphae on the surface of a glume fixed 9 d after inoculation (original magnification × 1270). (**9**) Numerous intercellular hyphae (*arrows*) among chlorenchyma cells of a glume fixed 2 d after inoculation (original magnification × 790). (**10**) One of a series of longitudinal serial sections through the basal portion of a glume fixed 4 d after inoculation. The glume had a hyphae (in squares shown in **11** and **12**) that terminated in the subovarian tissue in adjacent sections (original magnification × 305). (**11** and **12**) High magnification of the hyphae (*arrows*) located within the squares of (**10**). The micrographs are orientated 90° counterclockwise from that shown in (**10**) (original magnification × 1400). (From Goates, B. J., *Phytopathology*, 78, 1434, 1988. With permission.)

of infection. The epidermis of the ovary was not penetrated, even after prolonged contact with germinating sporidia (Figure 7-5).[97-99] *T. indica* became systemic and moved both upward and downward in the rachis and rachilla, infecting all spikelets and florets in wheat. Infection as late as the dough stage at points away from the primary infection sites could be explained by the limited systemic nature of the pathogen.[100] Inoculations of wheat with secondary sporidia of *T. indica* at the boot stage resulted in 35% and at heading in 29% spike infection, respectively. Inoculations at the boot stage resulted in 65% of the seeds in an infected spike showing signs of infection, and inoculation at heading resulted in 52% of the seeds showing infection.[101] In Karnal bunt, atrophy of the endosperm and embryo was the only obvious host reaction to infection. The fungus intercepted nutrients that normally flow from the vascular bundle in the pericarp into the endosperm. The fungus proliferated in the space created by disintegration of the parenchymatous tissue in the middle layers of the pericarp, and shrinkage of the endosperm provided space for the development of masses of teliospores.[102]

Teliospores of *Maeziomyces bullatus,* cause of smut of pearl millet, in soil or crop residues or adhering to seeds, served as a source of primary inoculum. Teliospores were produced in smut sori from dicaryotic mycelial cells in infected florets. A mature sorus contained numerous spore balls of teliospores. Individual teliospores germinated at flowering by a four-celled promycelium, then produced sporidia. They also reproduced by budding, and the buds were infective. Sporidia of compatible mating types formed dicaryotic infectious hyphae, which infected through young stigmas. Hyphae penetrated the flower through the stigma and reached the upper ovary wall, then traversed the style without lateral spread. The mycelia were binucleate, inter-and intracellular, and exhibited slight branching with two- to four-lobed haustoria. Hyphae advanced down through the ovary wall and invaded the ovule. Before all the tissue was involved, hyphal walls gelatinized to form spore balls. Smut sori were larger than the normal seeds and became visible 2 weeks after infection. The sori initially were shiny green but later turned brown and ruptured to release dark brown teliospores. Pollination prevented infection by the pathogen. Secondary spread within the crop was minimal because of a prolonged latent period (2 weeks) by which time flowering was almost complete. However, smut sori on rupturing released teliospores, which produced sporidia. These sporidia infected the panicles of late tillers or tillers of nearby fields during the same season.[103]

The quiescent larvae of *A. besseyi*, which causes white tip of rice, are located in the space between the husk and caryopsis of dry rice seeds. The nematodes were released from seeds during sprouting and carried up to the growing point and finally to the panicles. The nematode infected rice florets from the flag leaf sheath before panicle emergence. At this time, the juvenile nematode ceased to emerge from eggs, but hatchlings continued to develop to adults. As the infected grains became mature, the nematodes become quiescent due to dehydration. The affected plants produced small, deformed, and sterile seeds.[104-106]

4. Concomitant Contamination

Seed contamination either as pathogenic propagules or mixed with seeds in infested soil or plant debris results in nonsystemic seed transmission. Pathogens are introduced in the soil along with seeds and cause infection at any plant stage. Typical examples are ergot fungi of pearl millet and sorghum, and nematode earcockle of wheat.

The sclerotia of *Sclerotinia sclerotiorum* can be introduced in the soil through seeds and upon germination produce hyphae, which directly penetrate the plant at the seedling stage. Ascospores play a major role in the disease spread and can cause infection at any stage of plant development.[57]

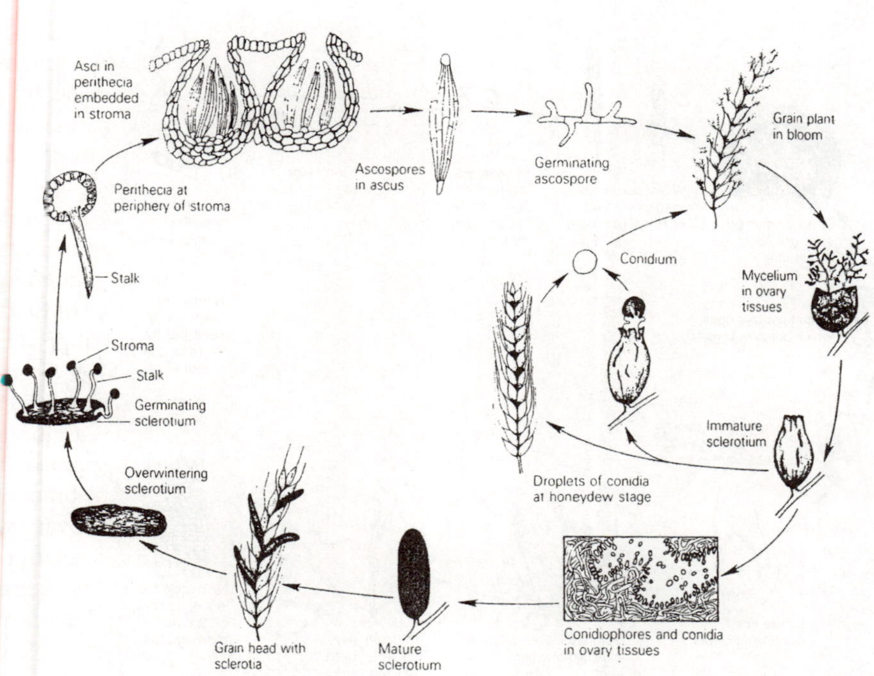

Figure 7-6 Disease cycle of *Claviceps purpurea* (ergot of small grains). (Adapted from Agrios, G. N., *Plant Pathology,* Academic Press, New York, 1988, 393. With permission.)

The primary source of *Claviceps fusiformis* inoculum was sclerotia left in the field from the previous crop and sown with seeds or in some cases infected alternative hosts. Under favorable weather, sclerotia germinated and released ascospores, which infected pearl millet inflorescences at the protogyny stage. Conidia from alternative hosts also initiated infection. Infection occurred only through stigmas, and once pollination occurred, infection was prevented. The fungus invaded florets through stigma and colonized the ovaries within 3 to 4 d. The ovaries were replaced by fungal hyphae and conidia within 10 d; sclerotia became visible within 20 to 25 d after inoculation. Pollination reduced ergot infection by inducing stylar constriction. Withered stigmas prevented infecting hyphae from entering the ovaries. Honeydew symptoms became visible within 4 to 6 d of ascosporial infection. The secondary disease cycle was initiated from macro- and microconidia in the honeydew, which were disseminated by splashing rain, wind, insects, or physical contact with healthy inflorescences. Sclerotia were produced in infected florets after the honeydew phase. Sclerotia were dark brown to black, with variable shapes and sizes.[107,108] Pearl millet flowers were susceptible to ascospore infection only after stigma emergence but before pollination and fertilization.[108]

Sclerotia of *Claviceps purpurea* mixed with barley seeds gave rise to stromata and ascospores in the soil, which served as primary inoculum (Figure 7-6). The coordination of ascospore formation and flowering was essential for an epidemic

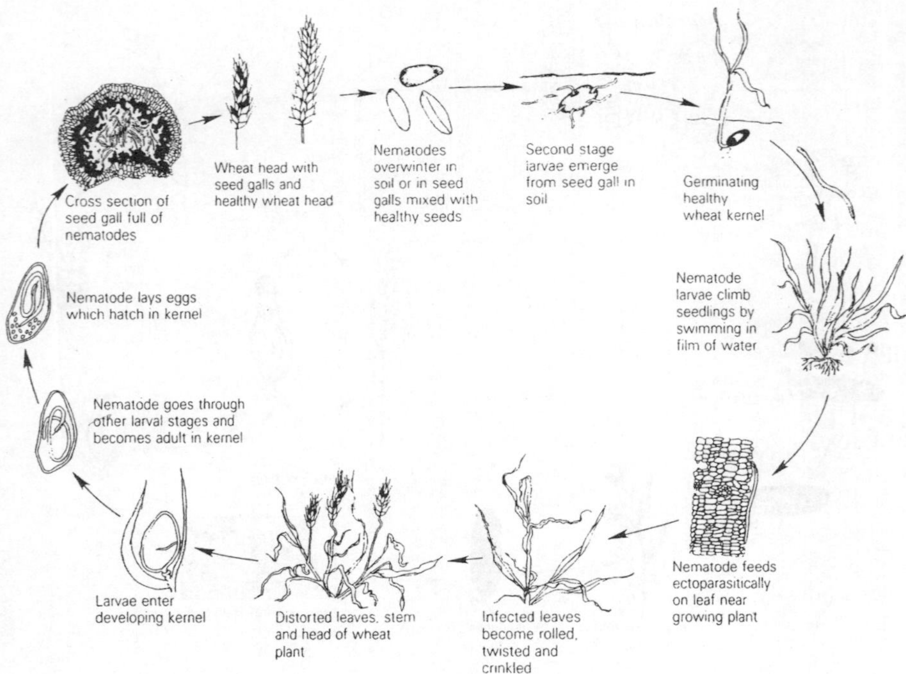

Figure 7-7 Disease cycle of *Anguina tritici* (seed gall nematode) on wheat (*Triticum aestivum*). (Adapted from Agrios, G. N., *Plant Pathology*, Academic Press, New York, 1988, 737. With permission.)

to develop. Closed flowers may have escaped infection. The hyphae penetrated the ovary near the ovule base and spread intercellularly in the ovary wall, and later intracellularly in the ovule.[109] Production of secondary inoculum was common. Conidia were disseminated chiefly by insects.[57] Sorghum lines with female sterility were highly susceptible to the ergot fungus. Sterile flowers were susceptible and fertile flowers resisted infection.[110] Wheat plants may have been infected by *C. purpurea* before onset of heading until the milky-seed stage. Infection at later stages did not affect the embryo, and the fungus remained either as sporulating hyphae in the ventral groove or as sterile hyphae between the pericarp and aleurone layers.[111]

Reports on the infection path of *Claviceps fusiformis* in pearl millet are conflicting; it may have taken place through the ovary wall[112,113] and/or the stigma and style.[114]

The nematode galls of *Anguina tritici, A. agrostis,* and *A. agropyronifloris* were introduced into the soil through seeds (Figure 7-7). Galls produced by *Anguina tritici* contained 2000 to 20,000 or more larvae per gall. Second-stage larvae emerged from galls, moved as far as 30 cm in the soil to emerging seedlings, and then were carried mechanically by the growing points to the developing earheads, and, ultimately, entered floral primordia. Larvae fed ectoparasitically on leaves and stems. Second-stage larvae underwent molts and produced third- and fourth-stage larvae and then adults, which invaded primordial tissues. Gall formation may have resulted from larval invasion of undifferentiated flower buds, staminate and carpellate tissues, and tissues lying between the stamens and carpels. Staminate tissues generally were preferred. Flower primordia usually contained both male and female nematodes. Females laid eggs, which hatched and produced first-stage larvae, which molted and by harvest produced second-stage larvae, which survived in the galls. More than one gall may have been produced in a single floret. The galls were a yellow-black.[115,116]

Similarly *A. agrostis* in *Agrostis* spp.[117] and *A. agropyronifloris* in *Agropyron smithii* produced seed galls.[118] Cool, moist weather favored disease development. A moisture film was essential for movement of nematode larvae from soil to the growing point of wheat.[119]

Cysts of *Heterodera glycines* were introduced into the soil with infested soil carried with soybean seeds. The cysts were full of eggs containing second-stage larvae, which directly penetrated young primary roots and apical meristems of secondary roots. Within a day, larvae penetrated cortical tissue with their stylets inserted in cortical or endodermal tissue or pericyclic cells. Within 2 to 6 d, pierced cells became hypertrophied, cell walls dissolved, and the clumping of nuclei from contiguous cells resulted in the development of a syncytia, a multinucleate mass of protoplasm surrounded by a common cell wall and enlarged cells.[120,121]

Dodder seeds survived in infested fields or mixed with the seeds of crop plants, such as alfalfa, clove, onion, sugarbeet, several ornamentals, etc. Dodder seeds at germination produced a slender, yellowish shoot without roots, which on contact with a susceptible host encircled the host plant, invaded with haustoria, and climbed up the plant. Later the dodder became completely dependent on the host for nutrients and water. The dodder plant developed flowers and produced seeds, which fell in soil or mixed with seeds.[28]

II. SEED INOCULATION

Seeds generally are inoculated for testing pathogenicity and cultivar resistance to seed- and soilborne pathogens, pathogens causing pre- and postemergence damping off, root rot, foot rot, systemic infections such as down mildew and smuts, and pathogens producing symptoms on seedlings. The kind and quantity of inoculum is important. Spore inoculum generally gives better results than hyphal inoculum. Inoculated seeds are planted in a sterile substrate, preferably sand or vermiculite, or on moist filter paper. The seeds are surface disinfested with a nonpersistent disinfectant and washed with sterile water before inoculation. The simplest inoculation method is to coat seeds with pathogen spores or a concentrated suspension of young hyphae. With bacterial pathogens or fungi with small spores, seeds immersed in a suspension subjected to 150 to 200 mm Hg of vacuum give good results. In certain cases, such as inoculating *Fusarium* into wheat,[122] seeds are immersed in a spore suspension with occasional aeration for about 24 h.

Pathogenicity and resistance to smut and bunt fungi are tested by inoculating the seeds with chlamydospores or teliospores. Seeds are immersed first in 1:300 formalin to water solution for about 1 h, then washed in running tap water for 30 min to eliminate any inoculum. The process loosens hulls of barley and oat seeds. After washing, the seeds are dried for 24 h at room temperature, and then 100 g of seeds are shaken with 0.5 to 0.1 g of spores.[123-126] Dehulling or loosening of the hulls of barley and oat seeds prior to inoculation is essential to obtain a high percentage of infection. Generally, dehulling reduces seedling emergence, especially if planted deeply.[127,128] Some workers remove the hull and plant the embryo 2 cm deep.[127] Dehulling seeds after soaking for 30 to 60 min in tapwater may not affect emergence, and seeds can be stored in paper or cloth bags or can be inoculated by soaking them in a spore suspension (2 g spores/l) for 15 min and then drying them at room temperature for 24 h.[128]

Mildew fungi can be established by coating seeds with oospores and planting them in sterile soil. Inoculation with conidia does not cause infection.[129] When rust fungi are seedborne, such as *Puccinia calcitrapae* var. *centaureae* in safflower, infection is obtained by seed inoculation. Safflower seeds are dipped into fresh egg albumin, coated with teliospores, and planted in cool soil.[130] The pots or trays in which the seeds are planted are covered with a polyethylene sheet or moist muslin cloth.[131]

Resistance of some crops to certain pathogens can be tested by seed inoculation. Inoculated seeds are placed on moist filter paper in culture plates on growing cultures of the pathogen, or on filter paper soaked in a spore suspension (Figure 7-8). Graham et al.[132] tested alfalfa resistance to *Colletotrichum trifolii* and *Stemphylium botryosum* using the following method. Streak the spores of the pathogen on an agar medium in culture plates for 5 d; then distribute about 75 surface-disinfested and dried seeds over the culture. Incubate in the dark at room temperature until most seeds have germinated, and then incubate under light. After 10 to 14 d, remove the seedlings and rate for resistance. Results are similar to those obtained by spraying seedlings grown in the greenhouse and maintaining them in a humid chamber for 3 d.[132]

Mesterhazy[133] used the filter paper method to test the pathogenicity of *Fusarium* to wheat. A double layer of filter paper was submerged in a spore suspension and placed in sterile culture plates. Surface-disinfested seeds were placed over the paper and incubated. After a few days, seedlings were rated for resistance.

Mohammad and Mahmood[134] distinguished between barley seedlings resistant and susceptible to the *Drechslera graminea* by placing dehulled seeds on a fungus culture in plates and then covering with an agar disk of the fungus from another plate. Disease was rated after 3 to 4 d.

Dehulling of seeds is laborious and presents the risk of low emergence. Methods for inoculating seeds with hulls, especially with bunt and covered smut fungi, were developed. To inoculate barley and wheat seeds with hulls, seeds were surface disinfected in 0.25% NaOCl solution and smeared over soil extract agar and incubated under continuous light at 5°C until germination occurred.[135,136]

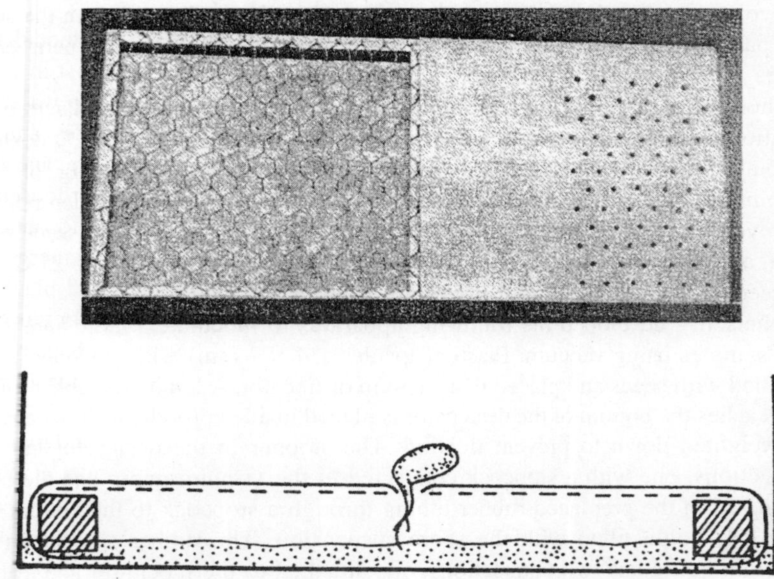

Figure 7-8 Zinc container with frame partially covered with filter paper (*above*), and a schematic diagram of the cross section of the container (*below*) for testing resistance of germinating seeds to fungal pathogens. (Adapted from Glessen, A. C. V. D. and Steenbergen, A. V., *Euphytica*, 6, 90, 1957, from Dhingra, O. D. and Sinclair, J. B., *Basic Plant Pathology Methods,* 2nd ed., CRC Press, Boca Raton, FL, 1995. With permission.)

The treated seeds and 2 ml of water were placed into a culture plate of the test fungus. The seeds were stirred with a glass rod until covered with spores, then placed in vermiculite in the culture plates and covered with a 2-cm layer of moist vermiculite. The dishes were covered and incubated at 5 to 10°C until seedlings began to push up the cover; then they were transplanted to a soil-sand mixture in pots. Cherewick[137] inoculated oats with germinating sporodia of *Ustilago segetum* var. *avenae* in a similar manner. The inoculated seeds were kept in a moist chamber for 24 h, then planted in sterile moist soil.

Another method is to place 200 g of machine-threshed seeds in vials and cover them with a spore suspension containing 1 g spores/l; shake for 2 min, and after 15 min pour off the spore suspension and invert the vials on clean blotting paper to absorb all free water. Incubate the vials with seed for 16 to 20 h at 20°C, and then transfer them to paper bags and dry for 3 to 4 d. Either plant or store for future use.[138,139] Popp and Cherewick[140,141] developed a method to inoculate manually threshed barley or oat seeds with smut fungi. Prepare a spore suspension in a high-speed homogenizer with a metal container. Replace the sharp blades with two blunt blades of the same size and shape made from 2-mm thick stainless steel. The blades are soldered to keep them in line, and all sharp corners are rounded-off. Place the seeds in the jar containing 300 to 400 ml of spore suspension for each 10,000 kernels, and agitate for 10 to 25 s. Pour the contents through a sieve to separate the seeds from the suspension. Immediately place the seeds in a paper bag, and dry for 2 d at room temperature. Slow drying permits the spores to germinate and develop hyphae beneath the hull.

Inoculating seeds in hull by applying vacuum gave results similar to using agitation. Add 250 g seeds to 1 L of a spore suspension in water, 1% dextrose or potato-dextrose broth, and subject to vacuum of 100 to 120 mm Hg for 10 to 20 min. Suddenly release the vacuum. Intermittent vacuum and sudden releases improve efficiency. After pouring off the spore suspension, dry the seeds overnight, and then keep at 90% or more relative humidity for 2 d at 20 to 25°C. Dry the seeds under forced air and store until used.[123,139,142]

Nielsen[143] developed the following apparatus to inoculate a large number of seed samples using vacuum. Plastic capsules (34 × 7 mm) with perforated ends are filled with seeds and placed in a vacuum desiccator. A length of rubber tubing that reaches the bottom of the desiccator is placed inside beforehand. The capsules are weighted down to prevent floating. The stopper in the desiccator has two connections, one with a stopcock connected to the vacuum pump and the other leading from the preplaced rubber tubing through a stopcock to the bottom of a 6- to 8-L bottle filled with the spore suspension. The rubber stopper on the inoculum bottle has two connections, one attached to the inoculum line leading to the desiccator. The seed-filled capsules are placed in the desiccator and evacuated with the stopcock of the inoculum line closed. After evacuation, the inoculum line is opened and evacuation continues. When the spore suspension covers the capsules to about 5 cm, the inoculum stopcock is closed and evacuation continues for 10 min. The vacuum is broken suddenly. The excess spore suspension is drawn back to the storage bottle by applying a gentle vacuum. The seeds in the capsule are dried by spreading the capsules over a wire mesh, and then are planted or stored in paper bags.

Germination of seeds to a certain stage prior to inoculation has been successful in the establishment of pathogens that are difficult to establish by inoculating nongerminated seeds. This method was used for screening for resistance to anthracnose fungi in bean and soybean, downy mildew fungi in maize and soybean, and bunt and covered and loose smut fungi in cereals. Kiesling[144] germinated barley seeds for 24 h at 20°C and inoculated them by placing a mixture of compatible sporidia of *U. segetum* var. *segetum* on the base of the coleoptile. Another method was to place seeds between layers of wet paper until the first signs of germination appeared. Then enough hull was removed from each seed to expose the embryo, and the seeds were placed on a culture medium on which a mixture of compatible sporidia were germinating. The young coleoptile were in direct contact with the inoculum. Seedlings were incubated until they pushed up the plate lid and then were transplanted.[145]

Inoculation of germinating maize or soybean seeds with oospores of downy mildew fungi gives a high percentage of systemically infected plants compared to inoculation of nongerminating seeds. Inoculation of maize seeds gives no infection. Germinated sunflower seeds with 4-mm long rootlets were immersed in a zoospore suspension of *Plasmopara halstedii* for 5 to 6 h at 18°C and then transplanted to vermiculite to test resistance. After 10 to 14 d, the seedlings were placed in a moist chamber to induce fungus sporulation on cotyledons.[146] Grabe and Dunleavy[147] placed a drop of an oospore suspension or oospores with talc between the cotyledons of germinating soybean seeds before planting them into soil. Maize seeds germinated for 2 to 3 d under an inoculum suspension of *Peronosclerospora sorghi* prepared to 500 μg/ml of Tween®80 and were incubated in a moist chamber for 4 h before planting in soil.[148]

Immersing surface-disinfested bean seeds previously germinated for 2 to 3 d in the dark in a spore suspension was the best method of testing for resistance to *C. lindemuthianum.* Incubated seeds were placed on moist filter paper, and after 3 to 4 d were rated for disease.[149] Spraying germinated seeds also can be used to study resistance to anthracnose in bean and to *Botrytis cinerea* in sunflower (Figure 7-8).[150]

The establishment of maize and wheat loose smut fungi by seed inoculations, as is done for barley and oat smuts, is difficult. To obtain high levels of infection, treat seeds in hot water; then surface disinfect with 1% NaOCl. Germinate seeds, and when the coleoptile is about 5- to 10-mm long, clip off about 2 mm from the tip. Immerse the clipped seedling in a spore suspension prepared in sterile water containing 0.02% Tween®20 and apply 50 to 80 mm Hg of vacuum for 2 to 5 min. Plant the seedlings in sterile soil or vermiculite with the shoot upward.[151-153] Cutting wheat coleoptiles is not essential if seedlings with 1- to 2-mm long coleoptiles are used.[140] The technique was not successful for inoculating pearl millet with an oospore suspension of the downy mildew fungus.[129] Stevens et al.[154] injected a mixture of compatible sporidia of *U. zeae* into plumules slightly above coleoptiles when they were 10- to 15-mm long and placed them at 90% relative humidity for 2 to 5 d at 30°C before planting in soil.

Inoculating shallow-planted germinating seeds by application of a concentrated spore suspension to the soil surface or as a drenched sterile soil also gave satisfactory results. The method was used to inoculate safflower with *Puccinia calcitrapae* var. *centaureae*[155] and *Trifolium* with *Aureobasidium caulivora*,[156] and cereals with bunt fungi.[136,157]

The following method is most useful for testing pathogenicity of seedborne bacterial pathogens. Surface disinfect seeds with a nonpersistent disinfectant, and soak the seeds in the bacterial inoculum for 24 h. Occasionally aerate the mixture. Place the seeds over two layers of filter paper in culture plates. For large seeds, trays can be used. Pour the bacterial suspension over the paper and seeds, and incubate at room temperature until seeds germinate. If moisture is needed, add more inoculum. When germination is evident, place the paper with seeds in pots with sterile soil and cover with the sterile soil. Cover the pots with a plastic sheet until plants emerge. Symptoms are seen on cotyledons and primary leaves. A large number of plants can be tested in a small space, and the same amount of inoculum is applied to all the plants.[158,159] Inoculation by spraying a 10^8 cfu/ml suspension of *X. c.* pv. *phaseoli* on beans over the racemes of immature pods after wounding has resulted in seed infection.[160]

REFERENCES

1. Bennett, C. W. and Esau, K., Further studies on the relation of the curly top virus to plant tissues, *J. Agric. Res.*, 53, 595, 1936.
2. Roongruangsree, U. T., Olson, L. W., and Lange, L., The seedborne inoculum of *Peronospora manshurica*, causal agent of soybean downy mildew, *J. Phytopathol.*, 123, 233, 1988.
3. Neergaard, P., *Seed Pathology*, Vol. I, 2nd ed., Macmillan Press, New York, 1979, 839.
4. Shetty, H. S., Neergaard, P., and Mathur, S. B., Demonstration of seed transmission of downy mildew or green ear disease, *Sclerospora graminicola* in pearl millet, *Pennisetum typhoides, Proc. Indian Nat. Sci. Acad.*, 43, 201, 1977.
5. Shetty, H. S., Mathur, S. B., and Neergaard, P., *Sclerospora graminicola* in pearl millet seeds and its transmission, *Trans. Br. Mycol. Soc.*, 74, 127, 1980.
6. Bain, D. C. and Alford, W. A., Evidence that downy mildew (*Sclerospora sorghi)* of sorghum is seed-borne, *Plant Dis. Rep.*, 53, 802, 1969.
7. Jones, B. L., Leeper, J. C., Frederiksen, R. A., *Sclerospora sorghi* in corn: its location in carpellate flowers and mature seed, *Phytopathology*, 62, 817, 1972.
8. Safeeulla, K. M., *Biology and Control of Downy Mildews of Pearl Millet, Sorghum and Finger Millet,* Wesley Press, Mysore, India, 1976, 304.
9. Novotelnova, N. S., Nature of parasitism of the pathogen of downy mildew of sunflowers at seed infection, *Bot. Zh.*, 48, 845, 1963.
10. Delanoe, D., Biologic et epidemiologie du mildiou du touresol (*Plasmopara helianthi*) CETIOM (Cent. Tech. Interprof. Ol. Metrop.), *Inf. Technol.*, 29, 1, 1972.
11. Melhus, E. L., The presence of mycelium and oospores of certain downy mildews in the seeds of their hosts, *Iowa St. Coll. J. Sci.*, 5, 185, 1931.
12. McKay, R., Field studies on downy mildew of sugarbeet, Irish Sugar Co. Ltd., Dublin, 1957.
13. Williams, R. J., Pawar, M. N., and Huibers-Govaert, I., Factors affecting staining of *Sclerospora graminicola* oospores with triphenyl tetrazolium chloride, *Phytopathology*, 70, 1092, 1980.
14. Semangoen, H., Studies on downy mildew of maize in Indonesia, with special reference to the perennation of the fungus, *Indian Phytopathol.*, 23, 307, 1970.
15. Bains, S. S. and Jhooty, J. S., Distribution, spread and perpetuation of *Peronosclerospora philippinensis* in Punjab, *Indian Phytopathol.*, 35, 566, 1982.
16. Ullstrup, A. J., Observations on crazy top of corn, *Phytopathology*, 42, 675, 1952.
17. Singh, R. S., Chaube, H. S., Khanna, R. N., and Joshi, M. M., Internally seedborne nature of two downy mildews on corn, *Plant Dis. Rep.*, 51, 1010, 1967.
18. Advinula, B. A. and Exconde, O. R., Seed transmission of *Sclerospora philippinensis* Weston in maize, *Philipp. Agric.*, 59, 214, 1975.
19. Rathore, R. S., Siradhana, B. S., and Mathur, K., Transmission and location of a *Peronosclerospora heteropogoni* in maize seeds, *Seed Sci. Technol.*, 15, 101, 1987.
20. Bains, S. S. and Dhaliwal, H. S., Downy mildew of maize, in *Plant Diseases of International Importance,* Vol. I, *Diseases of Cereals and Pulses*, Singh, U. S., Mukhopadhyay, A. N., Kumar, J., and Chaube, H. S., Eds., Prentice Hall, New York, 1992, 212.
21. Rao, B. W., Shetty, H. S., and Safeeulla, K. M., Production of *Peronosclerospora sorghi* oospores in maize seeds and further studies on the seedborne nature of the fungus, *Indian Phytopathol.*, 37, 278, 1984.

22. Lamka, G. L., Hill, J. H., McGee, D. C., and Braun, E. J., Development of an immunosorbent assay for seedborne *Erwinia stewartii* in corn seeds, *Phytopathology,* 81, 839, 1991.
23. Mink, G. I., Kraft, J., Knesek, J., and Jafri, A., A seed-borne virus of pea, *Phytopathology,* 59, 1342, 1969.
24. Hampton, R. O., Dynamics of symptom development of the seed-borne pea fizzletop virus, *Phytopathology,* 62, 268, 1972.
25. Alconero, R., Weeden, N. F., Gonsalves, D., and Fox, D. T., Loss of genetic diversity in pea germplasm by the elimination of individuals infected by pea seed-borne mosaic virus, *Ann. Appl. Biol.,* 106, 357, 1985.
26. Khetarpal, R. K. and Maury, Y., Seed transmission of pea seed-borne mosaic virus in peas: early and late expression of the virus in the progeny, *J. Phytopathol.*, 129, 265, 1990.
27. Pesic, Z. and Hiruki, C., Differences in the incidence of alfalfa mosaic virus in seed coat and embryo of alfalfa seed, *Can. J. Plant Pathol.,* 8, 39, 1986.
28. Agrios, G. N., *Plant Pathology,* 3rd ed., Academic Press, New York, 1988, 803.
29. Babadoost, M. and Herbert, T. T., Incidence of *Septoria nodorum* in wheat seed and its effects on plant growth and grain yield, *Plant Dis.*, 68, 125, 1984.
30. Agarwal, K., Singh, T., Singh, D., and Mathur, S. B., Studies on glume blotch disease of wheat. II. Transference of seedborne inoculum of *Septoria nodorum* from seed to seedling, *Phytomorphology,* 36, 291, 1986.
31. Raghavendra, S. and Safeeulla, K. M., Seed transmission of the downy mildew of finger millet (Ragi), *Seeds and Farms*, April 1978, 9, 1978.
32. Smedegaard-Petersen, V., Cross fertility and genetic relationship between *Pyrenophora teres* and *P. graminea*, the cause of net blotch and leaf stripe of barley, *Seed Sci. Technol.*, 11, 673, 1983.
33. Bandyopadhyay, R., Mughogho, L. K., and Satyanarayana, M. V., Systemic infection of sorghum by *Acremonium strictum* and its transmission through seed, *Plant Dis.*, 71, 647, 1987.
34. Bhatt, J. C. and Singh, R. A., Blast of rice, in *Plant Diseases of International Importance,* Vol. I, *Diseases of Cereals and Pulses*, Singh, U. S., Mukhopadhyay, A. N., Kumar, J., and Chaube, H. S., Eds., Prentice Hall, New York, 80, 1992.
35. Chung, H. S. and Lee, C. U., Detection and transmission of *Pyricularia oryzae* in germinating rice seed, *Seed Sci. Technol.*, 11, 625, 1983.
36. Ou, S. H., *Rice Diseases*, Int. Mycol. Inst., Kew, Surrey, U.K., 1985, 380.
37. Lawrence, E. B., Nelson, P. E., and Ayers, J. E., Histopathology of sweet corn seed and plants infected with *Fusarium moniliforme* and *F. oxysporum, Phytopathology,* 71, 379, 1981.
38. Tiffany, L. H., Delayed sporulation of *Colletotrichum* on soybean, *Phytopathology,* 41, 975, 1951.
39. Burkholder, W. H., The bacterial blight of the beans: a systemic disease, *Phytopathology,* 11, 61, 1921.
40. Zaumeyer, W. J., The bacterial blight of beans caused by *Bacterium phaseoli, U.S. Dep. Agric. Tech. Bull.*, 186, 86, 1930.
41. Zaumeyer, W. J., Comparative pathological histology of three bacterial diseases of bean, *J. Agric. Res.,* 44, 605, 1932.
42. Nelson, R., Investigations in the mosaic disease of bean (*Phaseolus vulgaris* L.), *Mich. Agric. Exp. Stn. Tech. Bull.,* 118, 71, 1932.
43. Agarwal, V. K., Nene, Y. L., and Beniwal, S. P. S., Detection of bean common mosaic virus in urdbean (*Phaseolus mungo*) seeds, *Seed Sci. Technol.,* 5, 619, 1977.

44. Porto, M. D. M. and Hagedorn, D. J., Seed transmission of a Brazilian isolate of soybean mosaic virus, *Phytopathology,* 65, 713, 1975.
45. Gilmer, R. M. and Wilks, J. M., Seed transmission of tobacco mosaic virus in apple and pear, *Phytopathology,* 57, 214, 1967.
46. Wallace, J. M. and Drake, R. J., A high rate of seed transmisson of avocado sun blotch virus from symptomless trees and the origin of such trees, *Phytopathology,* 52, 237, 1962.
47. George, J. A. and Davidson, T. R., Pollen transmission of necrotic ring spot and sour cherry yellows viruses from tree to tree, *Can. J. Plant Sci.*, 43, 276, 1963.
48. Callahan, K. L., Pollen transmission of elm mosaic virus, *Phytopathology,* 47, 5, 1957.
49. Lister, R. M. and Murant, A. F., Seed-transmission of nematode-borne viruses, *Ann. Appl. Biol.,* 59, 49, 1967.
50. Mink, G. I., Kraft, J., Knesek, J., and Jafri, A., A seed-borne virus of peas, *Phytopathology,* 59, 1342, 1969.
51. Hampton, R. O., Dynamics of symptom development of the seed-borne pea Fizzletop virus, *Phytopathology,* 62, 268, 1972.
52. Maude, R. B. and Presly, A. H., Neck rot (*Botrytis allii*) of bulb onions. I. Seed-borne infection and its relationship to the disease in the onion crop, *Ann. Appl. Biol.,* 86, 163, 1977.
53. Pavgi, M. S. and Mukhopadhyay, A. N., Development of coriander fruit infected by *Protomyces macrosporus* Unger, *Cytologia,* 37, 619, 1972.
54. Tabai, H., Anatomical studies of rice plants affected with bacterial leaf blight, with special reference to stomatal infection at the coleoptile and the foliage leaf sheath of rice seedling., *Ann. Pythopathol. Soc. Jpn.*, 33, 12, 1967.
55. Shakya, D. D., Chung, H. S., and Vinther, F., Transmission of *Pseudomonas avenae*, the cause of bacterial stripe of rice, *J. Phytopathol.*, 116, 92, 1986.
56. Gilbertson, R. L. and Maxwell, D. P., Common bacterial blight of bean, in *Plant Diseases of International Importance,* Vol. II, *Diseases of Vegetables and Oil Seed Crops*, Chaube, H. S., Singh, U. S., Mukhopadhyay, A. N., and Kumar, J., Eds., Prentice Hall, New York, 1992, 81.
57. Walker, J. C., *Plant Pathology,* 3rd ed., McGraw-Hill, New York, 1969, 819.
58. Zaumeyer, W. J., Seed infection by *Bacterium phaseoli, Phytopathology,* 19, 96, 1929.
59. Demski, J. W., Tobacco mosaic virus is seed-borne in pimento peppers, *Plant Dis.,* 65, 723, 1981.
60. Crowley, N. C., Studies on the seed transmission of plant virus diseases, *Aust. J. Biol. Sci.*, 10, 449, 1957.
61. Inouye, T., Inouye, N., Asatani, M., and Mitsuhata, K., Studies on cucumber green mottle mosaic virus in Japan, *Ber. Ohara Inst. Landwirtsch. Biol. Okayama Univ.,* 14, 49, 1967.
62. Phatak, H. C., Seed transmitted plant viruses — general aspects, in *Seed Pathology Problems and Progress*, Yorinori, J. T., Sinclair, J. B., Mehta, Y. R., and Mohan, S. K., Eds., Fundacao Instituto Agronomico do Parana — IAPAR, Londrina, Brazil, 1979, 141.
63. Trione, E. J., Dwarf bunt of wheat and its importance in international wheat trade, *Plant Dis.*, 66, 1083, 1982.
64. Fernandez, J. A., Duran, R., and Schafer, J. F., Histological aspects of dwarf bunt resistance in wheat, *Phytopathology,* 68, 1417, 1978.
65. Singh, R. S., *Plant Diseases*, Oxford & IBH Publ. Co., New Delhi, 1990, 608.

66. Swinburne, T. R., Infection by *Tilletia caries* (DC.) Tul., the causal organism of bunt, *Trans. Br. Mycol. Soc.*, 46, 145, 1963.
67. Fischer, G. W. and Holton, C. S., *Biology and Control of the Smut Fungi*, Ronald Press, New York, 1957, 622.
68. Mills, J. T., The development of loose smut (*Ustilago avenae*) in the oat plant with observations on spore formation, *Trans. Br. Mycol. Soc.*, 49, 651, 1966.
69. Brandwein, P. F., Seedling invasion of the covered smut of oats, *Phytopathology*, 34, 481, 1944.
70. Person, C. and Cherewick, W. J., Infection multiplicity in *Ustilago, Can. J. Genet. Cytol.*, 6, 12, 1964.
71. Sampson, K., The biology of oat smuts. II. Varietal resistance, *Ann. Appl. Biol.*, 16, 65, 1929.
72. Groth, J. V. and Person, C. O., Smutting patterns in barley and some plant growth effects caused by *Ustilago hordei, Phytopathology*, 68, 477, 1978.
73. Walter, J. M., The mode of entrance of *Ustilago zeae* into corn, *Phytopathology*, 24, 1012, 1934.
74. Ramakrishnan, T. S., Diseases of Millets, Indian Council of Agricultural Research, New Delhi, 1963, 152.
75. Ling, L., The history of infection of susceptible and resistant selfed lines of rye by the rye smut fungus, *Urocystis occulta, Phytopathology*, 30, 926, 1940.
76. Calvert, O. H. and Thomas, C. A., Some factors affecting seed transmission of safflower rust, *Phytopathology*, 44, 609, 1954.
77. Pederson, V. D., Downy mildew of soybeans, *Diss. Abstr.*, 22, 703, 1961.
78. Hildebrand, A. A. and Koch, L. W., A study of systemic infection by downy mildew soybean with special reference to symptomatology, economic significance and control, *Sci. Agric.*, 31, 505, 1951.
79. Demski, J. W., Tobacco mosaic virus is seedborne in pimento peppers, *Plant Dis.*, 65, 723, 1981.
80. Goto, K. and Fukatsu, R., Studies on the white tip of rice plants. II. Number and distribution of the nematode on the affected plants, *Ann. Phytopathol. Soc. Jpn.*, 16, 57, 1952.
81. Fukano, H., Ecological studies on white tip disease of rice plant caused by *Aphelenchoides besseyi* Christie and its control, *Fukuoka Agric. Exp. Stn. Bull.*, 18, 108, 1962.
82. Goodey, T., Further observation on *Anguillulina dipsaci* infestation of the onion scape and inflorescence, *J. Helminthol.*, 21, 60, 1945.
83. Newhall, A. G., Pathogenesis of *Ditylenchus dipsaci* in seedlings of *Allium cepa, Phytopathology*, 33, 61, 1943.
84. Dekker, J., Inwendige ontsmetting van door *Ascochyta pisi* aangetaste erwtezaden met de antibiotica rimocidine en piraricne, bevens enkele aspecten van het parasitisme van deze schimmel, *Tijdschr. Platenziekten*, 63, 65, 1957.
85. Neergaard, P., *Seed Pathology*, Vols. 1 and 2, Macmillan, London, 1977, 1187.
86. Gardner, M. W., Cladosporium leaf mold of tomato: fruit invasion and seed transmission, *J. Agric Res.*, 31, 519, 1925.
87. Kuniyasu, K., Seed transmission of Fusarium wilt of bottle-gourd *Lagenaria siceraria*, used in root stock of watermelon. III. Course of seedling infection by the seedborne pathogen, *Fusarium oxysporum* f. sp. *lagenarium* Matuo and Yamamoto, *Ann. Phytopathol. Soc. Jpn.*, 43, 270, 1977.
88. Miller, C. S., and Colhoun, J., Fusarium diseases of cereals. VI. Epidemiology of *Fusarium nivale* on wheat, *Trans. Br. Mycol. Soc.*, 52, 195, 1969.

89. Jordan, V. W. L., An etiology of barley net blotch caused by *Pyrenophora teres* and some effects on yield, *Plant Pathol.,* 30, 77, 1981.
90. Holmes, S. J. I. and Colhoun, J., Infection of wheat seedlings by *Septoria nodorum* in relation to environmental factors, *Trans. Br. Mycol. Soc.,* 57, 493, 1971.
91. Hewett, P. D., The field behavior of seed-borne *Ascochyta fabae* and disease control in field beans, *Ann. Appl. Biol.,* 74, 287, 1973.
92. Wallen, V. R. and Galway, D. A., Studies on the biology and control of *Ascochyta fabae* on faba bean, *Can. Plant Dis. Surv.,* 57, 31, 1977.
93. Chung, H. S. and Lee, C. U., Detection and transmission of *Pyricularia oryzae* in germinating rice seed, *Seed Sci. Technol.,* 11, 625, 1983.
94. Gabrielson, R. L., Epidemiology and control of blackleg disease of crucifers, *Seed Sci. Technol.,* 11, 749, 1983.
95. Taylor, J. D., Dudley, C. L., and Presly, L., Studies of halo-blight seed infection and disease transmission in dwarf beans, *Ann. Appl. Biol.,* 93, 267, 1979.
96. Singh, R. A. and Pavgi, J. S., Development of sorus in kernel bunt of rice, *Il Riso,* 22, 243, 1973.
97. Singh, R. S., *Plant Diseases,* Oxford and IBH, New Delhi, 1990, 608.
98. Goates, B. J., Histology of infection of wheat by *Tilletia indica,* the karnal bunt pathogen, *Phytopathology,* 78, 1434, 1988.
99. Warham, E. J., Karnal bunt of wheat, in *Plant Diseases of International Importance,* Vol. I, *Diseases of Cereals and Pulses*, Singh, U. S., Mukhopadhyay, A. N., Kumar, J., and Chaube, H. S., Eds., Prentice Hall, New York, 1992, 1.
100. Singh, D. V. and Dhaliwal, H. S., Establishment and spread of karnal bunt in wheat spikes, *Seed Res.,* 16, 200, 1988.
101. Royer, M. H. and Rytter, J., Artificial inoculation of wheat with *Tilletia indica* from Mexico and India, *Plant Dis.,* 69, 317, 1985.
102. Cashion, N. L. and Luttrell, E. S., Host–parasite relationships in karnal bunt of wheat, *Phytopathology,* 78, 75, 1988.
103. Thakur, R. P. and King, S. B., Smut Disease of Pearl Millet, ICRISAT Inform. Bull. No.25, ICRISAT, Patancheru, India, 1988, 17.
104. Huang, C. S., Detection of *Aphelenchoides besseyi* in rice seeds and correlation between seed infection and crop performance, *Seed Sci. Technol.,* 11, 691, 1983.
105. Huang, C. S. and Huang, S. P., Bionomics of white tip nematode, *Aphelenchoides besseyi* in rice florets and developing grains, *Bot. Bull. Acad. Sin.,* 13, 1, 1972.
106. Sanwal, K. C., Nematodes associated with seed, *Seed Res.,* 2, 1, 1974.
107. Thakur, R. P. and Chahal, S. S., Problems and strategies in the control of ergot and smut in pearl millet, Proc. Int. Pearl Millet Workshop, ICRISAT, 1986, 1987, 173.
108. Thakur, R. P. and King, S. B., Ergot of pearl millet, in *Plant Diseases of International Importance*, Vol. I, *Diseases of Cereals and Pulses*, Singh, U. S., Mukhopadhyay, A. N., Kumar, J., and Chaube, H. S., Eds., Prentice Hall, New York, 1992, 302.
109. Campbell, W. P., Infection of barley by *Claviceps purpurea, Can. J. Bot.,* 36, 615, 1958.
110. Futrell, M. C. and Webster, O. J., Ergot infection and sterility in grain sorghum, *Plant Dis. Rep.,* 49, 680, 1965.
111. Rapilly, F., Some notes on the infection of wheat by *C. purpurea, C. R. Acad. Agric. Fr.,* 51, 1265, 1965.
112. Reddy, K. D., Govindaswamy, C. V., and Vidhyasekaran, P., Studies on ergot disease of cumbu (*Pennisetum typhoides*), *Madras Agric. J.,* 56, 367, 1969.

113. Sundaram, M. V., Ergot of bajra, in Advances in Mycology and Plant Pathology, Prof. R. N. Tandon's Birthday Celebration Comm., New Delhi, 155, 1975.
114. Thakur, R. P. and Williams, R. J., Pollination effects on pearl millet ergot, *Phytopathology,* 70, 80, 1980.
115. Leukel, R. W., Investigations on the nematode disease of cereals caused by *Tylenchus tritici, J. Agric. Res.,* 27, 925, 1924.
116. Midha, S. K. and Swarup, G., Studies on the wheat seed galls caused by *Anguina tritici, Indian J. Nematol.,* 4, 53, 1974.
117. Courtney, W. D. and Howell, H. B., Investigations on the bentgrass nematode, *Anguina agrostis* (Steinbuch, 1799) Filipjev 1936, *Plant Dis. Rep.,* 36, 75, 1952.
118. Norton, D. C. and Sass, J. E., Pathological changes in *Agropyron smithii* induced by *Anguina agropyronifloris, Phytopathology,* 56, 769, 1966.
119. Thorne, G., *Principles of Nematology,* McGraw-Hill, New York, 1961, 553.
120. Endo, B. Y., Penetration and development of *Heterodera glycines* in soybean roots and related anatomical changes, *Phytopathology,* 54, 79, 1964.
121. Endo, B. Y., Histological response of resistant and susceptible varieties and backcross progeny to entry and development of *Heterodera glycines, Phytopathology,* 55, 375, 1965.
122. Mesterhazy, A., Comparative analysis of artifical inoculation methods with *Fusarium* spp. on winter wheat varieties, *Phytopathol. Z.,* 93, 12, 1978.
123. American Phytopathological Society, Greenhouse method for testing dust seed treatments to control certain cereal smuts, *Phytopathology,* 34, 401, 1944.
124. Kendrick, E. L., Race groups of *Tilletia caries* and *Tilletia foetida* for varietal-resistance testing, *Phytopathology,* 51, 537, 1961.
125. Rodenhiser, H. A. and Holton, C. S., Physiological races of *Tilletia tritici* and *T. levis, J. Agric. Res.,* 55, 483, 1937.
126. Rodenhiser, H. A. and Taylor, J. W., Effect of soil type, soil sterilization, and soil reaction on bunt infection at different incubation temperatures, *Phytopathology,* 30, 400, 1940.
127. Schafer, J. F., Dickson, J. G., and Shands, H. L., Barley seedling response to covered smut infection, *Phytopathology,* 52, 1157, 1962.
128. Srivastava, S. N. and Srivastava, D. P., A modified inoculation technique of covered smut of barley, *Indian Phytopathol.,* 23, 726, 1970.
129. Pu, M. H. and Szu, T. M., Some studies on downy mildew of millet, *Phytopathology,* 39, 512, 1949.
130. Calvert, O. H. and Thomas, C. A., Some factors affecting seed transmission of safflower rust, *Phytopathology,* 44, 609, 1954.
131. Prasada, R. and Chothia, H. P., Studies on safflower rust in India, *Phytopathology,* 40, 363, 1950.
132. Graham, J. H., Devine, T. E., McMurtrey, J. E., and Fleck, D. L., Agar plate method for selecting alfalfa for resistance to *Colletotrichum trifolii, Plant Dis. Rep.,* 59, 382, 1975.
133. Mesterhazy, A., Comparative analysis of artificial inoculation methods with *Fusarium* spp. on winter wheat varieties, *Phytopathol. Z.,* 93, 12, 1978.
134. Mohammad, A. and Mahmood, M., Inoculation techniques in Helminthosporium stripe of barley, *Plant Dis. Rep.,* 58, 32, 1974.
135. Hoffmann, J. A., Kendrik, E. L., and Meiners, J. P., Pathogenic races of *Tilletia contraversa* in the Pacific Northwest, *Phytopathology,* 52, 1153, 1962.
136. Meiners, J. P., Methods of infecting wheat with the dwarf bunt fungus, *Phytopathology,* 49, 4, 1959.

137. Cherewick, W. J., Smut, an additional cause of oat blast, *Phytopathology,* 55, 1368, 1965.
138. Groth, H. V. and Person, C. O., Estimating the efficiency of partial-vacuum inoculation of barley with *Ustilago hordei, Phytopathology,* 66, 65, 1976.
139. Leukel, R. W., Factors influencing infection of barley by loose smut, *Phytopathology,* 26, 630, 1936.
140. Popp, W. and Cherewick, W. J., An improved method of smut inoculation, *Phytopathology,* 42, 472, 1952.
141. Popp, W. and Cherewick, W. J., An improved method of inoculating seed of oats and barley with smut, *Phytopathology,* 43, 697, 1953.
142. Western, J. H., Sexual fusion in *Ustilago avenae* under natural conditions, *Phytopathology,* 27, 547, 1937.
143. Nielsen, J., A collection of cultivars of oats immune or highly resistant to smut, *Can. J. Plant Sci.,* 57, 199, 1977.
144. Kiesling, R. L., Effect of temperature and point of inoculation on the symptomatology of barley covered smut, *Phytopathology,* 52, 16, 1962.
145. Fischer, G. W., Induced hybridization in graminicolous smut fungi. I. *Ustilago hordei x U. bullata, Phytopathology,* 41, 839, 1951.
146. Zazzerini, A., A method for determining the resistance of sunflower to *Plasmopara helianthus* (Novot), *Rev. Patol. Veg.,* 15, 5, 1979.
147. Grabe, D. F. and Dunleavy, J., Physiologic specialization in *Peronospora manshurica, Phytopathology,* 49, 791, 1959.
148. Barredo, F. C. and Exconde, O. R., Inoculation of pregerminated corn seeds, conidial production during the day and the use of Tween 80 in inoculating Philippine corn downy mildew, *Philipp. Agric.,* 55, 42, 1971.
149. Champion, M. R., Brunet, D., Mauduit, M. L., and Ilami, R., Method for testing the resistance of bean varieties to anthracnose [*Colletotrichum lindemuthianum* (Sau. and Magn.) Briosi and Cav.], *Acad. Agric. Fr.,* 59, 951, 1973.
150. Giessen, A. C. V. D. and Steenbergen, A. V., A new method of testing beans for anthracnose, *Euphytica,* 6, 90, 1957.
151. Kavanagh, T., Inoculating barley seedlings with *Ustilago nuda* and wheat seedling with *U. tritici, Phytopathology,* 51, 175, 1961.
152. Rowell, J. B. and DeVay, J. E., Factors and results in the partial-vacuum inoculation of seedling corn with *Ustilago zeae, Phytopathology,* 42, 17, 1952.
153. Rowell, J. B. and DeVay, J. E., Factors affecting the partial-vacuum inoculation of seedling corn with *Ustilago zeae, Phytopathology,* 43, 654, 1953.
154. Stevens, K., Melhus, I. E., Semeniuk, G., and Tiffany, L., A new method of inoculating some Maydeae with *Ustilago zeae* (Beckm.) Unger, *Phytopathology,* 36, 411, 1946.
155. Zimmer, D. E., Hypocotyl reaction to rust infection as a measure of resistance of safflower, *Phytopathology,* 52, 1177, 1962.
156. Helms, K., Variation in susceptibility of cultivars of *Trifolium subterraneum* to *Kabatiella caulivora* and in pathogenicity of isolates of the fungus as shown in germination-inoculation tests, *Aust. J. Agric. Res.,* 26, 647, 1975.
157. Hodgges, C. F. and Britton, M. P., Infection of Merion bluegrass, *Poa pratensis,* by stripe smut, *Ustilago striiformis, Phytopathology,* 59, 301, 1969.
158. Rangarajan, M. and Chakravarti, B. P., Bacterial stalk rot of maize in Rajasthan, effect on seed germination and varietal susceptibility, *Indian Phytopathol.,* 23, 470, 1970.

159. Bain, D. C., Observations on resistance to black rot in cabbage, *Plant Dis. Rep.*, 35, 200, 1951.

160. Dos Marques, A. S., Parente, P. M., Machado, F.O.C., and Santana, C. R., Screening of inoculation methods to produce bean seeds contaminated with *Xanthomonas campestris* pv. *phaseoli* for experimental purposes, *Fitopatol. Brasileria*, 19, 178, 1994.

CHAPTER 8

Factors Affecting Seed Transmission

Approximately 3000 seedborne microorganisms and viruses, many of which are plant pathogens, have been recorded on over 800 host plants.[1] Successful establishment of seedborne inoculum in the field is an important factor in the epidemiology of many pathogens. Organisms may be seedborne but not seed transmitted. Seed transmission often is complex, depending on several interactions. A number of factors influence successful seed transmission and establishment of infection in a subsequent crop.

I. CROP SPECIES

The establishment of seedborne infection in the field depends on the host cultivar. Seeds of a resistant cultivar may not be infected, or even if infected, seed transmission may not occur.

The lack of seed transmission of the loose smut pathogen of wheat, *Ustilago segetum* var. *tritici*, in certain cultivars may be due to embryo resistance, a noncompatible reaction with the pathogen resulting in seedling death or mature plant resistance.[2-6] Embryos of the wheat cv. Keystone were susceptible to *U. segetum* var. *tritici* hyphae, and normally seeds were infected, but mature plants were resistant.[7] Similarly, embryo infection of the barley cv. Emir by *U. segetum* var. *tritici* was not transmitted to the plant.[8] Resistance to this fungus in wheat cultivars could have been expressed when seedlings outgrew smut fungal hyphae.[5] In the barley cv. Jet, embryos could have been infected, but an incompatible reaction generally prevented development of infected shoots; however, if infection resulted in stunted seedlings, they might have partially recovered and produced healthy tillers.[9] Partial earhead infection of barley or wheat by loose smut fungi could have been caused by the fungus not keeping pace with ear development.[10] In rye cultivars susceptible to *Urocystis occulta,* hyphal development was rapid and abundant, whereas in resistant cultivars, it was restricted and did not spread into meristematic regions.[11]

Resistance to *U. segetum* var. *avenae* transmission in oats is expressed as necrosis of epidermal cell walls and restriction of hyphal growth. However, depending on the resistance level, the host reacts in a variety of ways. In resistant lines, hyphae might not have penetrated the cell wall. In cultivars with less resistance, penetration could be achieved, and in 7 d necrosis developed in the surrounding host tissues and the hyphae died; or hyphae were found after 7 d in the seedling coleoptile and mesocotyl, but degenerated in less than 21 d; or hyphae were abundant, penetrated deeper into host tissues, and after 21 d, remained as hyphae in the mesocotyl but did not invade the growing point and meristem; or in susceptible cultivars the growing point and meristem were invaded.[12]

In initial stages of infection, hyphae of *U. segetum* var. *segetum* developed equally in barley coleoptile tissues, regardless of the degree of cultivar resistance. Hyphae degeneration began after 5 to 6 d, with the number of degenerated hyphae corresponding to the degree of resistance. Cell nuclei reacted to hyphal penetration by increasing in size. Vacuolization of protoplasm and lysis of the hyphal walls also were observed.[13]

The spore load of *Tilletia caries* that caused appreciable (9.5%) bunt in wheat in the susceptible cv. Jenkins Club gave a smut-free crop in the resistant cv. Marquis.[14] *T. caries* penetrated resistant and susceptible wheat seedlings, but did not develop beyond penetration and development of hyphae within epidermal cells of resistant cultivars. In cultivars with less resistance, the fungus progressed into deeper parts of coleoptile and sheath tissues of the earliest true leaves almost as readily as in a susceptible cultivar until seedlings were 9-d-old, after which it was restricted. In susceptible cultivars it developed in young leaf blades, nodes, internodes, and growing points.[15]

Pseudomonas syringae pv. *glycinea* (soybean bacterial blight) developed larger populations on germinating soybean seeds of a susceptible cultivar than of a resistant one, thus demonstrating pre-emergence host–pathogen specificity.[16] Seed transmission of a particular virus varies with the species or cultivar of a plant, as well as with the virus or virus strain. For example, seed transmission of BCMV varied among bean cultivars from 1 to 75%,[17,18] BSMV among barley cultivars from 15.6 to 64.6%,[19] alfalfa mosaic virus among alfalfa cultivars from 0.6 to 10.3%,[20] and lettuce mosaic virus among lettuce cultivars from 1 to 8%.[21] Seed transmission of alfalfa mosaic virus in alfalfa varied from 15.5 to 27.5%, depending upon cultivar.[22] A 3% seed transmission of tobacco ringspot virus (strain 98) was found in lettuce seeds of cv. Paris Island Cos, but none was found in seeds of the cv. Imperial 615.[23] Grogan et al.[24] reported 5.1% seed transmission of squash mosaic virus through seeds of zucchini squash, whereas in seeds of Early Summer Golden and Crookneck squash cultivars, no transmission occurred. Similarly, Kennedy and Cooper[25] reported 20.6% seed transmission of SbMV in seeds of the soybean cv. Harosoy but none in Merit. No seed transmission of peanut mottle virus occurred in four cultivars of large-seeded peanuts (Early Runner, Florigiant, Florunner, Virginia Bunch 67), but there was transmission in four small-seeded cultivars (Argentine, Spancross, Starr, Tifspan).[26] The rate of peanut stripe virus seed transmission was affected by peanut genotype and length of infection. Incidence of seed transmission varied from less than 1 to 50% in the 935 peanut germ plasm lines. Peanut cultivars and lines of the Spanish type showed higher seed transmission rates than those of other types.[27] According to the frequency of seed transmission, BCMV was better adapted to bean than to any other host–plant species.[28]

The degree of seed transmission can vary among individual plants within cultivars. Seeds from individual soybean plants of the same cultivar showed a range of 0 to 35% seed transmission of SbMV,[29] while plants infected with tobacco ringspot virus showed 100% seed transmission.[30] Peanut seeds from 7 of 30 plants produced virus-free seeds, and the other 23 produced seeds with peanut mottle virus ranging from 0.5 to 8.3%.[26]

Reasons for variation in seed transmission among cultivars may be the genetic makeup of the cultivars involved or differences in severity of strains on cultivars. The barley cv. Hypana has a lower level of seed transmission of BSMV than the cv. Atlas, in part due to the greater genetic variation in Atlas.[31] The interaction between BSMV and various barley cultivars fall into three categories: (1) very susceptible with little or no seed produced; (2) tolerant, in which some strains of the virus survived indefinitely; and (3) resistant, in which plants either did not become infected or, did become infected, but seed transmission was low or absent. Thus seed transmission depended upon the nature of the cultivar and survivability of the virus strain.[32]

II. ENVIRONMENT

Seed transmission of pathogens and their establishment and development in the host are influenced by environmental conditions, with moisture and temperature being the most important factors. These factors also affect seed germination, spore germination, the infection process, and subsequent inoculum spread. High relative humidity (>80%) and an average temperature of 30°C favored smut development in pearl millet.[33] The maximum temperature from 19 to 21°C and minimum of 8 to 11°C followed by intermittent rainfall during anthesis created conditions favorable for karnal bunt infection in wheat.[34] *Bipolaris sorokiniana* infection in wheat was favored by warm temperatures and high relative humidities (70 to 85%).[35] A moderate temperature of 20 to 30°C, with a relative humidity of 80% or more and air currents, favor *Claviceps fusiformis* infection in pearl millet.[36] Seedlings grown from infected maize seeds by *Penicillium oxalicum* had more rotted mesocotyls and lower shoot height at 15 to 20°C than at 25 and 30°C.[37] The rapid spread of *Ascochyta fabae* f. sp. *lentis* in lentil in western Canada in the late 1970s was due to the use of poor-quality seeds by growers and conditions favorable for seed transmission and plant-to-plant spread. Transmission from infected seeds to aerial portions of seedlings occurred at 0.03% in dry soil at 15°C. When soil moisture was maintained near field capacity, transmission frequency was highest at 8°C. In a growth chamber at 8°C and with soil moisture near field capacity, transmission frequency was 0.07, 0.21, and 0.39% from infected seeds with no discoloration, slight discoloration, and large lesions, respectively.[38]

A. Moisture

Atmosphere and soil moisture play an important role in seedling infection and subsequent disease establishment and spread. Soil moisture influences spore germination and viability as well as emergence. For example, the occurrence of loose smut of oats, was higher at 30%, lower at 60%, and lowest at 80% water-holding capacity.[39] Excessive moisture reduced oxygen and thus inhibited spore germination. Similarly, the incidence of *U. segetum* var. *avenae* in oats at 20% water-holding capacity and 20°C could be as high as 94%, while at 60% water-holding capacity, it was 48%.[40] Similarly, seedling sorghum infection by *Sporisorium sorghi* was favored by low soil moisture and temperature.[41] In contrast, covered smut of barley was favored by high soil moisture (50% water-holding capacity) rather than low (40%).[42] Dry soil favored infection and development of *Urocystis occulta* in rye.[43] Soil moisture had a variable effect on bunt of wheat caused by *Tilletia caries*; the percentage infection was 55, 22, and 11 at 40, 20, and 80% water-holding capacity, respectively.[44]

Low soil moisture (48.8% water-holding capacity) promoted spore germination of *T. caries* and *T. laevis* and delayed development of susceptible wheat seedlings, and at higher moistures (85% moisture-holding capacity) seedling infection was least.[45] A combination of 10°C and an 11, 13, 18, or 24% soil moisture capacity was favorable for wheat bunt, but infection dropped off rapidly at 10% and was absent at 9%. At all moisture levels, infection was greater at 10°C and intermediate at 15 and 5°C, with none developing at 25°C.[46] The incidence of preemergence death of wheat seedlings due to *Fusarium avenaceum, F. culmorum, Microdochium nivale,*[47-49] *Stagonospora nodorum*[50] was higher in dry than wet soils. High soil moisture inhibited seed germination and development of *Rhynchosporium secalis* hyphae, thus preventing infection of barley seeds during germination.[51] Pearl millet smut caused by *Maeziomyces bullatus* was favored by humid conditions and continuous cropping of pearl millet in the same field. The incidence was low under drier conditions.[52] *Phomopsis* reduced emergence most when soybean seeds were incubated in dry soil.[53] If the soil remained moist (15% and above) for 7 to 10 d during February and the first week of March, high sporulation of *T. indica* occurred. Secondary sporidia were produced at the ear-emergence stage during this period. Any coincidence of the above factors resulted in a high proportion of disease in wheat.[54] Soil moisture had a considerable influence on transmission of *P. syringae* pv. *pisi* in peas from seeds to seedlings. An equation was derived that described the relationship between the proportion of seedlings infected (P) and the soil water stress(s),

$$Mp_a: -In\ (-In(I - P)) = 0.64\ In(s) + 4.5$$

Although agreement between the observed and predicted transmission rates was poor, years of severe and slight disease transmission were successfully predicted.[55]

B. Temperature

Temperature affects spore germination, the infection process, and disease development. The optimum temperature for infection of wheat plants by seedborne *T. caries* and *T. laevis* ranged from 5 to 10°C.[45,46,56] In this range, germinating spores were highly infective and wheat seed germination was slow; thus only few plants escaped infection. A range of 15 to 20°C favored wheat seed germination more than spore germination; when spores and seeds germinated at 25°C, many plants escape infection. The escaped mechanism is not known, but has been attributed to rapid growth of the host. Wheat plants were susceptible to infection at 15 to 25°C, provided that the fungus spores germinated in the soil and were in an infective stage.[57] A range of 10 to 25°C and 15 to 20°C was optimum for infection of barley by *Ustilago segetum* var. *avenae* and var. *segetum,* respectively.[58,59] Infection of wheat by *Urocystis agropyri* was optimum at 10 to 20°C and low at 5°C, with no infection taking place at 25°C, regardless of soil moisture.[60] The two barley cultivars, one susceptible, Odessa, and one moderately resistant, Persicum, to the covered smut fungus (*U. segetum* var. *segetum*) showed parallel reactions at the same temperature. Susceptible barley cultivars developed over 95% covered smut at 12°C soil and 16°C atmospheric temperature or 20°C soil and 24°C atmospheric temperature and at diurnal alternations of the two temperature regimes. Resistant cultivars developed no symptoms at 12°C soil and 16°C atmospheric temperature, with a low percentage of sori, primarily of inflorescences, occurring at the other temperature regimes.[59]

With loose smut of wheat, smutted earheads, earheads with reduced teliospore formation, and slender green heads with few to no teliospores were produced at 18.3, 23.8, and 29.8°C, respectively.[61] Maximum development of *U. segetum* var. *tritici* races C_1 and C_3 occurred on the wheat cv. Kota at 23°C, whereas at 15 to 20°C, the incidence of loose smut was reduced, and at 6°C the reduction was highly significant. At the lower temperatures hyphae in ears often failed to sporulate.[62]

Infection of germinating barley seedlings by seedborne *D. graminea* was favored by soil temperatures below 15°C and was reduced sharply at near 20°C and above.[63] The rate of infection by *M. nivale* in barley was slower at 10 than 20°C, but the percentage of coleoptiles infected at 10°C was greater than at 20°C after they had grown to similar lengths. Infection occurred in the mesocotyl soon after extrusion of the coleorhiza.[64] An 8-h d at less than 20°C, and a 4.6-h d at more than 30°C were more favorable to ergot fungal infection (70 to 72% severity, 32 to 49% sclerotial infection) than at 21 to 35°C with 6.4-h d at more than 30°C (53% severity, 23% sclerotial formation) in pearl millet. The minimum temperature was more critical for ergot infection than the maximum.[65] In carrot, seed to seedling transmission of *A. dauci* was highest at 25°C.[66] The germination process of teliospores of *T. controversa* was slow (1 to 3 months) and had an obligate requirement for low temperature (1 to 10°C). Teliospores did not germinate above 12°C.[67] In *Brassica, A. brassicae* infection was favored by warm (17 to 25°C) wet weather during flowering and pod-filling stages.[68] The percentages of systemically infected soybean plants by *Peronospora manshurica* were 16% at 15°C, 1% at 20°C, and nil at 25°C.[69] Sunflower plants raised from *Macrophomina phaseolina*-infected seeds showed severe pre- and postemergence mortality at 35°C. The ratio of seed infection to seed transmission was 1:08. At 25°C, all the plants raised from infected seeds remained green, but the pathogen persisted in cotyledons and roots.[70] Planting winter barley at 12°C caused yield losses when seeds were infected with *D. graminea*.[71]

For *Sporisorium sorghi* to develop in sorghum, the mean maximum soil temperature during the infection period should be at or below 24°C.[72,73] A maximum temperature range between 19 to 20° ± 1°C and a minimum of 8 to 10° ± 1°C followed by intermittent rains during wheat anthesis was favorable for *Tilletia indica* infection.[74] Wheat seeds infected with *T. indica* (10% in 1970–71 and 1.5% in 1971–72) failed to establish field infection due to unfavorable environmental conditions.[75]

Transmission of *Drechslera graminea* in barley seeds increased between 12 to 15°C and decreased or was prevented above 15°C in field plantings of naturally infected seeds.[76] Soybean seeds encrusted with *Peronospora manshurica* oospores (Figure 3-13) showed systemic infection in 40% of the seedlings at 13°C soil temperature due to slow seed germination, whereas at 18°C and above, none of the seedlings showed systemic infection.[77] Maize kernels infected with *B. maydis* resulted in 1 and 8% wilted seedlings after 3 weeks at 13 and 22°C, respectively.[78] Safflower seedlings infected by *Puccinia calcitrapae* var. *centaureae* were 96.1, 76.2, 67.3, and 29.3% at 5, 10, 15, and 20°C, respectively.[79]

Seedling infection from seeds with *Sclerospora graminicola* hyphae occurred by planting pearl millet seeds for 12 h under artificial daylight at 23 to 25°C.[80] Disease development caused by *Microdochium nivale* in wheat occurred at high levels at low temperatures in dry soils and at low levels in warm dry soil.[49] The percentage of infected barley seedlings by *X. campestris* pv. *translucens* was 6% at 10°C and 77% at 35°C.[81]

Seed transmission of some viruses is influenced by the environment, with temperature being a major factor. Crowley[82] reported no transmission of BCMV in seeds of bean plants grown at 17 to 19°C, whereas 16 to 25% transmission was obtained from seeds produced at 20°C. Singh et al.[83] found 3% seed transmission of BSMV in barley line C.I.5020, and none in C.I.3212, C.I.3212.I, or C.I.4219 when plants were grown at 16°C, but when plants were grown at 24°C, seed transmission in the latter three lines ranged from 7 to 28%. At 20.8 to 39°C under intense light, no seed transmission of BCMV occurred in urdbean but at 20 to 30°C under diffused light, 10% seed transmission was detected.[84] Transmission of different nepoviruses through *Stellaria media* seeds was affected differently by ambient temperature during seed production. Raspberry ringspot and tomato black ring (Scottish isolates) viruses were seed transmitted at 14, 18, or 22°C; arabis mosaic virus at 14°C; and strawberry latent ringspot and tomato blackring (German isolates) viruses at 22°C.[85]

C. Wind-blown Rain

Wind blown rain was essential for the spread of seedborne inoculum and initiation of disease such as bacterial blight of soybeans.[86] Secondary spread of halo blight of bean was recorded after hail storms but not rain storms in Idaho in 1966.[87] The absence of rainfall during the growing season in California precluded spread of *P. syringae* pv. *lachrymans* in cucumber seed fields.[88]

D. Light

Increased levels of bunt occur in the wheat cvs. Hope and Marquis with increased day length. Hope plants developed 64.1, 17.5, and 0.8% bunt; and Marquis plants 32.7, 17.8, and 1.9%, exposed to light for 24, 10 to 11, and 8 h daily, respectively.[89] The wheat cv. Canus was resistant to three races of *T. laevis* and five of *T. caries* under both day length conditions, but resistance to this race was reduced under long days.[90] Systemic transmission of *Peronospora parasitica* in cabbage seedlings occurred at the cotyledon stage with <16-h d.[91] Systemic symptoms of seedborne viruses often were masked under high light intensity.

III. INOCULUM

Seed transmission can be influenced by the amount of inoculum as well as the type, virulence, and location of inoculum in seeds. Cucumber mosaic virus (CMV) contains three single-stranded positive-sense genomic RNAs — RNA-1, -2 and -3. CMV RNA-1, cogoverning viral replication and affecting virus movement, influenced seed transmission of recombined CMV-Pg and CMV-Le genomic RNAs.[92]

A. Minimum Effective Level of Inoculum for Seed Transmission and Establishment in Seedlings or Plants

A spore load of 36,000 to 150,000 per seed was required to produce the maximum infection of bunt in wheat.[14] The incidence of *T. caries* in the less susceptible cv. Austrobankut and in a susceptible one, cv. Stamm 101, was proportional to the spore load (3 to 3500 spores per seed).[93] The incidence of *Microdochium nivale* in wheat tended to increase with an increase in the spore load up to 50,000.[49] A heavier spore load of *F. avenaceum* was required for disease development than of *F. culmorum*.[94] Rice seeds with few conidia grouped at one infection site without hyphae, few conidia scattered all over the surface of the seed coat without hyphae, or light to heavy sporulation along with light to profuse hyphal growth of *Bipolaris oryzae* resulted in 46, 80, and 92 to 100% losses in seedlings, respectively (Figure 8-1, Tables 8-1 and 8-2).[95] *Phoma pinodella* and *Mycosphaerella pinodes* caused greater yield reductions than *Asco-*

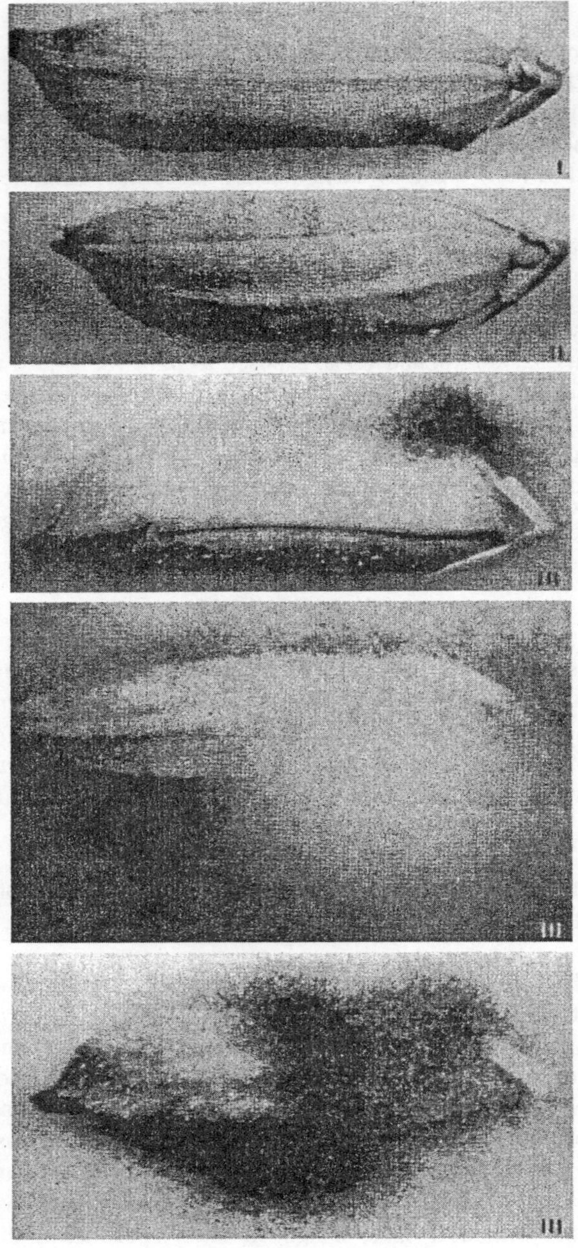

Figure 8-1 Categories (I-III) of infection by *Bipolaris oryzae* on rice (*Oryza sativa*) seeds incubated on blotters for 7 d at 20°C under 12/12 h alternating cycles of near UV light and darkness. (From Aulakh, K. S., Mathur, S. B., and Neergaard, P., *Seed Sci. Technol.*, 2, 385, 1974. With permission.)

Table 8-1 Occurrence of *Bipolaris Oryzae* in Rice Seed Lots Obtained from Different Countries (400 Seeds Tested per Sample by Blotter Method)

Country	No. of seed lots examined	No. of infected seed lots	Range of infection percentages in infected seed lots									
			1–5	6–10	11–20	21–30	31–40	41–50	51–60	61–70	71–80	81–90
Argentina	1	1	1	0	0	0	0	0	0	0	0	0
Ceylon	5	5	1	3	1	0	0	0	0	0	0	0
Egypt	21	21	7	11	1	1	1	0	0	0	0	0
Ghana	41	39	6	7	15	5	3	1	1	1	0	0
India	81	60	25	13	10	8	1	2	1	0	0	0
Indonesia	31	31	7	4	5	3	5	0	3	1	1	2
Iran	5	5	2	0	0	1	0	0	2	0	0	0
Japan	4	2	2	0	0	0	0	0	0	0	0	0
Korea (S)	27	27	5	1	7	11	3	0	0	0	0	0
Nepal	35	28	16	3	8	1	0	0	0	0	0	0
Nigeria	229	158	108	20	15	6	4	0	5	0	0	0
Pakistan	1	1	0	1	0	0	0	0	0	0	0	0
Philippines	62	35	18	3	7	5	2	0	0	0	0	0
Portugal	4	4	1	2	1	0	0	0	0	0	0	0
S. Viet Nam	4	4	1	3	0	0	0	0	0	0	0	0
Thailand	135	101	49	21	19	5	1	2	4	0	0	0
Total	686	522	249	92	89	46	20	50	16	2	1	2

From Aulakh, et al., *Seed Sci. Technol.*, 2, 385, 1974. With permission.

chyta pisi because lower seed infection by *P. pinodella* and *M. pinodes* can cause greater yield loss. Higher levels of *A. pisi* in seeds was necessary for similar losses in yield.[96] Transmission of *Burkholderia solanacearum* in pepper seeds occurred at an infestation level of 1000 but not 50 propagules per seed.[97]

Inaba[69] concluded that the percentage of infected spinach seedlings by the downy mildew fungus was positively correlated with the degree of oospores infestation of seeds. Heavy infection by *F. moniliforme* in maize caused seed rot and seedling blight, but plants from lightly infected seeds grew without symptoms. However, the fungus was detected inside the stem of healthy seedlings.[98] In wheat, bunted spikes due to *T. controversa* resulted only when heavily infested seeds (≥1 g teliospores/kg equivalent to 20,000 teliospores/seed) were planted in disease-conducive locations.[99] The growth of fava bean plants was reduced when the area of the seed coat infected by *Ascochyta fabae* exceeded 25%.[100]

P. syringae pv. *syringae* significantly reduced emergence of kidney and pinto beans when infiltrated with 3×10^4 or 10^6 cfu/ml. Emergence of beans treated with 300 or 3 cfu/ml were not significantly different from the control. Only infiltration with 3×10^6 cfu/ml of *P. s.* pv. *phaseolicola* significantly reduced emergence of pinto beans, but infiltration with 3×10^4 or 10^6 cfu/ml significantly reduced emergence of kidney beans.[101] The white tip nematode has been detected in discolored and deformed as well as healthy-looking rice seeds. The highest recorded count in one seed was 121, with an average of 22 nematodes per seed. For a susceptible cultivar, the critical inoculum was 30 viable nematodes per 100 seeds, assuming a tolerance limit of 5% yield loss. Two infested seeds were sufficient to cause disease spread.[102] There was a correlation between seed symptoms and the population of *X. c.* pv. *phaseoli* in bean seeds. The minimum population of *X. c.* pv. *phaseoli* required to initiate infection in the field was 10^2 cfu/seed, while a 0.2% seed infection resulted in serious disease incidence.[103]

Seed transmission of bean common mosaic virus in *Vigna mungo* was correlated with the amount of virus present in the embryonic axis and later in primary leaves. Virus concentration varied in different tissues; the mean amount of virus in the three cultivars was 48 to 1234 ng per embryonic axis, 15 to 24 ng per cotyledon, and 12 to 20 ng per testa. The infection of primary leaves through seeds resulted in a systemic infection if the amount of virus in primary leaves exceeded 100 ng per 100 mg of tissue. Cultivars that resisted seed transmission contained relatively small amounts of the virus in the embryonic axes.[104]

B. Inoculum Location

Popp[105] reported that infection of wheat plumule buds by *Ustilago segetum* var. *tritici* was correlated with adult plant infection, but Khanzada et al.[106] did not observe plumule bud infection and found that scutellar infection of wheat embryo produced infection in adult plants. *Alternaria brassicicola* was externally and internally seedborne in Brassicas, with seedling infection correlated with the latter.[107] Infection of oats by *U. segetum* var. *avenae* occurred more frequently in plants from infected inner grains of the second flower in a spikelet than outer grains of the first flower. Spores adhering to the glume exterior were incapable of causing infection because seedlings cannot be infected from spores in such a position until hyphae have traversed the glume length, a distance of about 1 cm. Spores on the glume generally are incapable of forming sufficiently long hyphae to reach seedlings. Spores within glumes were more favorably situated to cause infection.[108]

Cowpea banding mosaic virus was found primarily in the plumule bud, hence leading to a higher rate of seed transmission, whereas sunnhemp mosaic virus was found primarily in the testa and to a minor extent in the plumular bud in cowpea, hence leading to lower seed transmission.[109] Peanut stripe virus is not seed transmitted in soybeans, although infective virus particles were detected from seed coats of immature soybean seeds by ELISA. As seeds matured, there was still a positive reaction from the seed coats, but the particles were not infective. Neither infective virus nor serologically detectable PStV was recovered from cotyledons or the embryo axes of mature seeds.[110] Alfalfa mosaic virus was detected by ELISA in seeds, seedlings, seed coats, and embryos of alfalfa. Virus incidence in seeds (20.6%) was significantly higher than in seedlings (7.3%). The virus was detected more frequently in seed coats than in embryos. The difference in detection levels in seeds and seedlings was related to the high incidence of virus in the seed coats and the low incidence in the embryos. Virus remaining on the seed coat did not serve as a source of seedling infection.[111] PSbMV was detected in the seed coat of chickpea at rate of 1.81%, but no transmission was detected in the embryo axes.[112]

C. Type of Inoculum

Beet seeds contaminated by *Uromyces betae* teliospores may give rise to spermogonia on hypocotyls originating from basidiospore infection; thus, teliospore-contaminated seed clusters are a potential risk for introduction of *U. betae* into new areas. However, the risk of beet rust spread by seedborne uredospores was less, because beet seed germination requires a longer time than uredospore germination. Thus, it is possible that plants will not be available for infection when the uredospores germinate.[113] Seedlings from safflower seeds covered with uredospores of *Puccinia calcitrapae* var. *centaureae* showed no symptoms after 1 month, while those grown from seeds covered with teliospores were 90% rusted.[79] Mixed populations of *Claviceps purpurea* conidia may give rise to a spherical fructification in wheat consisting of the components of the mixture, even at a ratio of 500:1, to form a heterogeneous sclerotium in which the extent of differentiation of a sexual stage is dominated by a strain that rarely produced stromata. The necrosis of affected spikelets that occurred in response to the interparasitic competition was recognized as a novel phenomenon in ergot–host relationships.[114]

Different strains of the same virus can differ in their degree of seed transmission. Different seed transmissions of soybean mosaic virus isolates and strains in various soybean cultivars has been shown.[115] Grogan and Schnathorst[23] found that tobacco ringspot virus strain 98 was transmitted up to 3% in lettuce seeds of cv. Paris Island Cos, whereas a calico strain was not. Transmission of bean yellow mosaic virus in cowpea varies with the strain, ranging from 0 to 55%.[116] The bean-infecting strain of SBMV was not seed transmitted in cowpea,[117,118] while the strain infecting cowpea was transmitted. The serological group IA of squash mosaic virus was seed transmissible in cantaloupe, honeydew melon, pumpkin, and scalloped summer squash, but group IIA was transmitted only in pumpkin and squash seeds.[119] Similarly, a strain of squash mosaic virus, SMV-W, was seed transmitted in watermelon but the SMV-C strain was not.[120] Four isolates of peanut mottle virus differ in frequency of seed transmission in the peanut cv. Starr: $M_1 = 0.3\%$, $M_2 = 0\%$, $M_3 = 8.5\%$, and $N = 0\%$. Isolate M_2 was not seed transmitted in large-seeded peanuts but was at a very low frequency (0.23%) in small-seeded peanuts.[26] The tobacco streak virus isolate A-TSV was transmitted in soybean, but W-TSV isolate from tobacco was not.[121] Maximum earcockle infection of wheat was obtained with two nematode galls (approximately 2×10^4 larvae per 1000 g soil), and any increase in inoculum caused a reduction in infection due to competition for food.[122]

IV. SURVIVAL OF INOCULUM

Seedborne pathogens must survive various stages of seed development and storage for successful transmission. Seed transmission of pathogen can vary from country to country and region to region, depending upon weather and storage conditions. Seedborne pathogens, in general, survive longer in temperate than in tropical regions. Survivability of seedborne inoculum is discussed in Chapter 6.

When freshly extracted wet or dried discolored tomato seeds were planted in a steamed soil mix or in field soil in the greenhouse, some of the seedlings from wet seeds were infected with *Phytophthora infestans* (av. 26%), whereas seedlings emerging from dry seeds were healthy.[123] It was concluded that seed transmission of *P. infestans* in tomato did not occur on commercial tomato seed because the external hyphae were destroyed by fermentation and acid treatment of seeds, and both internal and external hyphae were destroyed by seed-drying operations. However, fungus transmission may occur in the field through infected or infested seeds when the fruit and seed remain wet. Infected and infested seeds germinate in 2 to 3 weeks and give rise to some blighted seedlings, and windborne sporangia may be carried to healthy plants.

V. CULTURAL PRACTICES

Cultural practices can affect establishment of seedborne infection in the field, but available data are sparse or inadequate.

A. Soil Type

Soil type affects seed germination rate and seedling growth. Seed transmission of loose smut in barley was higher in a clay–sand mixture than in heavier soil.[59]

Symptoms of *Stagonospora nodorum* in wheat was more marked on seedlings grown in loam than on those in chalky soil with a high moisture content.[124] The most severe damage by *D. teres* in barley was observed in dry soils (PF 3–4) at 12°C. On young barley seedlings, form *maculata* caused coleoptile symptoms and form *teres* induced mostly foliar symptoms. Foliar necrosis resulted from systemic invasion by the pathogen located in seeds.[125] *Tilletia cariès* and *T. laevis* infection in a susceptible wheat cultivar was affected by soil type and temperature during the infection period. There was a difference in infection level in seedlings of cvs. Marquis and Thatcher in Hempstead silt loam and Mendon loam soil at 10 and 15°C, but at 5°C cv. Marquis developed 72.5 and 30.3% bunt in Mendon loam and Hempstead silt loam soil, respectively.[126]

B. Soil Reaction

Information on the effect of soil pH on seed transmission is scanty. Neutral soils generally favored disease development. Seed transmission of *Ustilago segetum* var. *segetum* in oats varied from 4 to 12%, 64 to 92%, and 8 to 18% at pH 4.6,7.4, and 8.6, respectively.[40] Similarly, seed transmission of *T. caries* in wheat was 42% at pH 7.9, compared to 6% at pH 5.6.[46] In contrast, seed transmission of *U. segetum* var. *segetum* in barley was favored by acid soil, with the transmission rate approximately double of that found in alkaline soils.[42] In rye, the highest level of *Urocystis occulta* was recorded when seeds were planted in soil at pH 7.36.[43] The severity of *Microdochium nivale* in wheat increased with increased pH.[49]

C. Seeding Rate

Seeding rate can influence tiller infection. In barley, the number of smutted tillers decreased by one half to four times with increased seeding rate.[127] The density of sowing had no effect on the percentage of infected plants, except on smutted ears. Any condition that increased tillering lowered the percentage of diseased ears.[128]

D. Depth of Sowing

Deep sowing and cool temperatures, which tend to slow germination and plant growth, increase frequency of *T. caries*.[129,130] These conditions lengthen the period of seedling susceptibility. Tiemann[131] found that a low rate of seed transmission of barley loose smut was associated with a moderate depth of planting compared to either shallow or deep seeding. Pedersen[132] found that seed transmission of barley loose smut decreased with an increase in sowing depth: 40 and 20% at 4- and 12-cm depth of sowing, respectively. The incidence of dwarf bunt in wheat was highest when seeds were planted at or near the soil surface.[133] As the depth of sowing increased from 2.5 to 15 cm, the disease index of *M. nivale* in wheat increased.[49] Deep sowing increased infection of *T. caries* in winter wheat and was correlated with delayed seedling maturation.[130] Maximum ear cockle infection in wheat occurred when nematode galls were placed with seeds at a 2-cm depth,[122] with a decrease in infection if galls were placed deeper than the seeds because larvae failed to reach the seedlings (Table 8-3).[122,134]

When cucumber mosaic virus-infected narrow-leafed lupin (*Lupinus angustifolius*) seeds were sown at different soil depths, depths of 8 and 11 cm decreased the incidence of seed-infected plants by 15 and 50%, respectively, compared with sowing at 5 cm.[135]

E. Sowing Time

The emergence of loose smutted ears in barley extends over a period of 3 to 4 weeks. Late sowing had no effect on the number of diseased ears but increased the rate of emergence.[136] Seed transmission of BSMV was higher in spring-seeded barley than when the same cultivars were sown in the autumn.[137]

F. Fertilizers

Gassner and Kirchhoff[138] reported that seed transmission of the loose-smut pathogen in barley was less when optimum fertilizer levels were used compared to low levels. In contrast, Pedersen[132] found that the fertilizer levels had no effect on seed transmission. Increased N fertilizer resulted in increased incidence of *S. nodorum* in wheat seeds[139] but reduced *F. moniliforme* infection in maize kernels.[140] The incidence of *Tilletia barclayana* in rice was highest in fields recently dressed with N and planted with long-grain rather than short-or medium-grain cultivars.[141]

G. Planting Method

The incidence of tobacco mosaic virus in pepper was higher in transplanted seedings; few if any plants became infected when seeds were planted directly without transplanting. Observable systemic symptoms on seedlings took 10 to 50 d after transplanting to develop.[142]

VI. SEED ABNORMALITIES

Abnormal seeds may give rise to a higher level of transmission compared to normal seeds. Transmission of the loose smut pathogens of barley and wheat was higher in small compared to large seeds.[127,131,143-146] Transmission of the loose smut fungus of barley was 3.2, 8.7, and 13.5 in large, medium, and small seeds with 1000-grain weights of 45.1, 33.6, and 24.5, respectively.[147]

Large barley seeds had less of the loose smut pathogen than small ones. The average percentage of smutted plants ranged from 8.2 for large (retained on a 2.58-mm slotted screen) to 26.7 for small seeds (passed in a 2.38 mm screen, retained on a 1.98 mm screen). Field germination was lowest for the small seeds, with large seeds producing vigorous, tolerant seedlings of rapid growth.[143]

Table 8-2 Effect of *Bipolaris oryzae* Seed Infection, Referred to Different Categories of Severity on Seeds and Seedlings on Blotter and, After Transfer, in Pots. Percentages of Infection and Loss

Seed sample number	1					2					3					4				
Categories	I	II	III	IV	V	I	II	III	IV	V	I	II	III	IV	V	I	II	III	IV	V
Infection on blotter	10	10	7	8	14	14	13	3	9	5	4	8	3	6	7	7	12	2	3	3
Loss after transfer																				
Preemergence	40	90	86	100	100	29	38	100	100	100	25	40	100	100	100	43	75	50	100	100
Postemergence	10	0	0	0	0	36	38	0	0	0	25	40	0	0	0	14	0	50	0	0
Total loss	50	90	86	100	100	65	76	100	100	100	50	80	100	100	100	57	75	100	100	100

Seed sample number	5					6					7					8				
Categories	I	II	III	IV	V	I	II	III	IV	V	I	II	III	IV	V	I	II	III	IV	V
Infection on blotter	11	3	1	7	2	1	0	3	0	6	6	2	0	1	1	2	4	0	1	2
Loss after transfer																				
Preemergence	18	33	100	84	100	0	0	66	0	100	33	50	0	100	100	0	25	0	100	100
Postemergence	27	33	0	16	0	0	0	0	0	0	17	50	0	0	0	50	50	0	0	0
Total loss	45	66	100	100	100	0	0	66	0	100	50	100	0	100	100	50	75	0	100	100

Note: I = few conidia at one point, no mycelial growth; II = few conidia scattered all over the surface of the seed, no mycelial growth; III = few conidia with light mycelial growth; IV = light sporulation with profuse mycelial growth; and V = heavy sporulation with profuse mycelial growth.

From Aulakh, K. S., et al., *Seed Sci. Technol.*, 2, 385, 1994. With permission.

Table 8-3 Effect of Depth of Placement of *Anguina tritici* Galls in Soil on the Incidence of Ear Cockle of Wheat

Depth (cm)[a]	Ear cockle infection (%)	No. of tillers/pot	No. of galls/pot	Grain yield/ear (g)
2	74.4	9.3	87.6	0.4
4	35.5	8.0	48.3	0.7
6	25.3	6.7	13.3	1.0
8	16.6	4.0	2.3	1.0

[a] Galls and seeds placed at same depth.

From Midha, S. K. and Swarup, G., Indian J. Nematol., 2, 97, 1972. With permission.

Loose smut fungus infection in small seeds of dwarf wheat cvs. Sonora-64 and Sonalika was higher than in large seeds. A negative correlation between seed weight and loose smut fungus infection was recorded. In the tall cv. Agra Local, loose smut fungus infection in large seeds was significantly higher than in short ones. A high positive correlation was obtained between seed weight and infection. A lower percentage of seed transmission than seed infection occurred using the embryo count method since smaller seeds were removed during processing.[148]

Small pea seeds were found to transmit pea seedborne mosaic virus at a rate higher than medium and large pea seeds.[149] Small seeds had a higher level of seed transmission of viruses than large ones. The rate of seed transmission of peanut mottle virus in peanut was 3.7% in small seeds (<6.00 mm) compared to 0 to 0.9% in large seeds (>6.5 to 7.9 mm).[150] Seed transmission of peanut stunt virus in peanut was low in seeds that were large enough for planting, and seeds that transmitted the virus tended to have low germination and gave rise to weak seedlings. Thus seed transmission of the virus under field conditions was unlikely to occur.[151]

Seed transmission of pea seedborne mosaic virus in pea was correlated with small size, abnormal shape, or seeds with cracked seed coats, but not with normal seeds.[152] Squash mosaic virus was transmitted at a higher percentage in light-weight, poorly filled, and deformed squash seeds compared to heavy, well-filled ones.[153] The separation of lettuce mosaic virus-infected lettuce seeds into heavy and lightweight portions by a vertical airstream concentrated the virus-containing seeds in the lightweight portions.[154] This is in contrast to reports that separation of lettuce seeds into light and heavy fractions by aspiration or gravity failed to increase or decrease the percentage of lettuce mosaic virus in any of the fractions.[21]

VII. SEED GERMINATION

Seed germination and emergence can favor seed transmission of pathogens by seedling growth rate. Slow growth may favor rapid infection or may limit infection. Cotyledons may carry the seed coat along with it, or the seed coat may remain in the soil. In either case, depending upon the pathogen involved, it can effect infection of aerial parts. There are two patterns of seedling emergence.

A. Epigeal

The cotyledons and enclosed plumule (epicotyl) are carried up during hypocotyl growth, emerge from the soil, and become green and photosynthetic (Figure 8-2B). Ultimately, cotyledons wither and drop. Epigeal germination occurs, for example, in seeds of castor bean, cucumber, French bean, lettuce, onion, peanut, and soybean. Epigeal germination favors infection of aerial parts by *C. michiganensis* subsp. *michiganensis* in tomato, *P. syringae* pv. *phaseolicola* in beans, *X. campestris* pv. *campestris* in cabbage, and *X. campestris* pv. *carotae* in carrot.[155]

In bottle gourd, seed coats sometimes remain attached to "pegs" on seedlings after germination and serve as a source of primary inoculum for *F. oxysporum* f. sp. *lagenarium*. The infection rate of seedlings with pegs accompanied by a seed coat was 14 to 18%, while the infection rate of those having their seed coats carried by the cotyledons was 2 to 3%. During germination, the fungus, which is latent in the seed coat prior to germination, multiplied and penetrated the seedling from the lower side of the epidermis.[156]

B. Hypogeal

Cotyledons remain in the soil as the plumule (epicotyl) elongates and emerges (Figure 8-2A). Hypogeal germination occurs in seeds of barley, fava bean, maize, pea, and wheat. It favors root and stem infection. This type of germination limits seed transmission of bacteria that infect only aerial parts, such as *R. fascians* in *Tropaeolum majus*,[157] *P. syringae* pv. *pisi* in pea,[158] *X. campestris* pv. *translucens* in cereal,[81] and *C. michiganensis* subsp. *nebraskensis* in maize.[159]

Seedling infection, especially by smut fungi, is influenced by germination rate. The susceptibility of sorghum to *Sporisorium sorghi* was limited to the period between sowing and seedling emergence. The length of susceptibility depended on the sowing date.[160]

VIII. SEED LEACHATES

Seeds produce sugars and amino acids during early stages of imbibition.[161] The chemical composition of leachates may interfere with spore germination.[162] Seed leachates of oil crops, cumin, and pepper inhibit spore germination of many seedborne fungi.[163-165] Six sugars, twelve amino acids, and five organic acids were found in seed leachates of *Vigna radiata* and *V. mungo*. The interaction between the abiotic and biotic environment was mediated by the occurrence of seed leachates in relation to seedborne pathogens.[166]

Seed leachates directly affect seedborne fungi by contributing to their nutritional status prior to penetration, or by inhibiting their saprophytic or pathogenic activity. Leachates from *V. radiata* retarded the spore germination of *Alternaria alternata, Aspergillus flavus, Colletotrichum truncatum, Fusarium moniliforme, F. oxysporum, F. poae,* and *F. pallidoroseum*. They stimulated spore germination of *Curvularia lunata, Exserohilum rostratum, Phoma lingam,* etc.[167] Seed coat

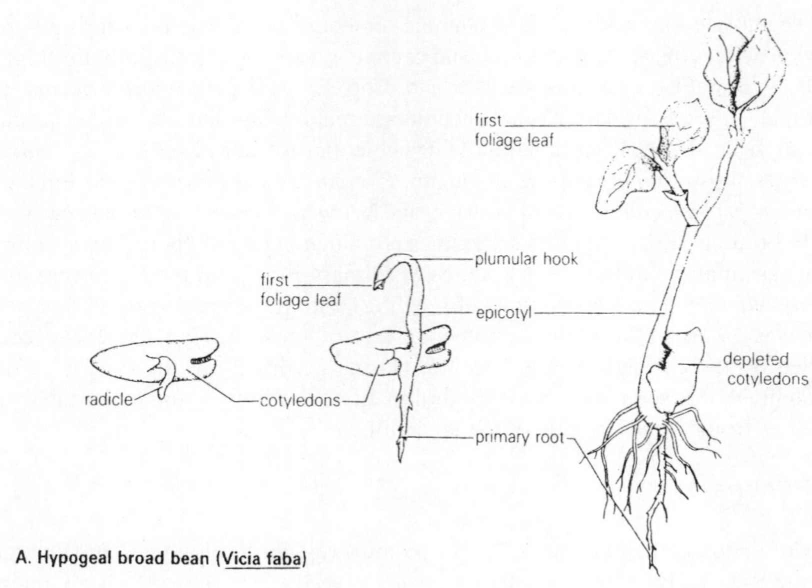

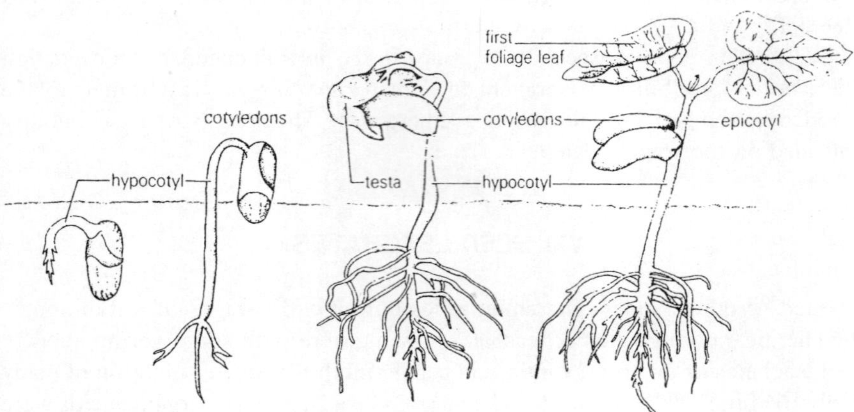

Figure 8-2 Two types of germination. (**A**) Hypogeal germination of fava bean (*Vicia faba*). (**B**) Epigeal germination of bean (*Phaseolus vulgaris*). (Adapted from Bewley, J. D., and Black, M., *Physiology and Biochemistry of Seed in Relation to Germination,* Part I, *Development, Germination and Growth*, Springer-Verlag, New York, 1978, 125.)

leachates of pigeon pea inhibited spore germination of *Curvularia geniculata, Dothiorella hawaiiensis*, and *F. oxysporum*. The inhibitory action of seed coat leachates on fungal spore germination may be due to antifungal substances, which act as defensive agents against seed infection or seed transmission.[168-170]

IX. PRESENCE OF OTHER MICROFLORA

Seeds harbor a wide range of microflora and viruses, and some of them affect seed transmission and establishment of infection in the field due to antagonistic action. The resistance of an oat cultivar from Brazil to *Bipolaris victoriae* was due to antagonistic mycoflora such as *Chaetomium cochlioides* and *C. globosum* on the seed surface.[171] When infected seeds were planted in soil, these fungi produced a metabolite, cochliodinol.[172] Indications are that microorganisms antagonistic to *Pyrenophora avenae* in oats inhibited seed transmission.[173] Bamberg et al.[174] demonstrated an interaction between *Tilletia caries* and *T. laevis*, both of which can infect alone but eliminate each other when in combination. Seeds inoculated with *T. laevis* and sown in soil infested with *T. caries* resulted in reduced seedling infection by *T. caries* compared to noninoculated seeds. In contrast, Berend[175] found no evidence of antagonism when seeds were inoculated with a mixture of the two fungi. *C. tritici* was unable to cause tundu disease in wheat in absence of the nematode *Anguina tritici*.[122] An antagonistic bacterium in sesame seeds inhibited the growth of *P. syringae* pv. *sesami* in culture.[176] *Claviceps purpurea* has infected wheat ovaries parasitized by *Tilletia caries* and has become the dominant pathogen, displacing the bunted ovary. Infection by *C. purpurea* occurred toward the base of the bunt-infected ovary. The invasion by *C. purpurea* was essentially a displacement rather than a replacement phenomenon, which characterizes *C. purpurea* parasitism of healthy ovaries. The establishment of the sphacelium and subsequent differentiation to sclerotial tissue was more rapid in bunted rather than in nonbunted ovaries. Thus bunted ovaries presented less of a barrier to infection and sclerotial development than healthy ones.[177] Seed transmission of melon necrotic spot virus in *Cucumis melo* occurred when infected seeds were sown in soil containing the virus-free fungus (*Olpidium radicale*) soil; 10 to 40% of the seedlings became infected. No infection was observed when the affected muskmelon seeds were sown in soil without the fungus.[178] Cowpea stunt of cowpea was caused by a combination of cucumber mosaic virus and blackeye cowpea mosaic virus, which are synergistic. Both were seed transmitted and carried by the same aphid alone or in combination. The viruses were transmitted from double-infected plants either separately or together, and caused single or double infections, respectively, in cowpeas.[179]

REFERENCES

1. Richardson, M. J., *An Annotated List of Seed Borne Diseases,* Commonwealth Mycology Institute, Kew, Surrey, U.K., 1990, 387.

2. Bever, W. M., Embryo test not reliable for determining percentage of loose smut infection in wheat, *Ill. Res.*, 2, 18, 1960.
3. Gaskin, T. A. and Schafer, J. F., Some histological and genetic relationships of resistance of wheat to loose smut, *Phytopathology,* 52, 602, 1962.
4. Mantle, P. G., Further observations on an abnormal reaction of wheat to loose smut, *Trans. Br. Mycol. Soc.,* 44, 529, 1960.
5. Ohms, R. E. and Bever, W. M., Effect of *Ustilago tritici* infection in third internode elongation in resistant and susceptible winter wheat, *Phytopathology,* 44, 500, 1954.
6. Oort, A. J. F., Hypersensitivity of wheat to loose smut, *Tijdschr. Plantenziekten,* 50, 73, 1944.
7. Batts, C. C. V. and Jeater, A., The reaction of wheat varieties to loose smut as determined by embryo, seedling and adult plant tests, *Ann. Appl. Biol.,* 46, 23, 1958.
8. Hewett, P. D., Resistance to barley loose smut (*Ustilago nuda*) in the variety Emir, *Trans. Br. Mycol. Soc.,* 59, 330, 1972.
9. Mumford, D. L. E. and Rasmusson, D. C., Resistance of barley to *Ustilago nuda* after embryo infection, *Phytopathology,* 53, 125, 1963.
10. Ribeiro, V. do. M. A. M., Investigations into the pathogenicity of certain races of *Ustilago nuda, Trans. Br. Mycol. Soc.,* 46, 49, 1963.
11. Ling, L., The histology of infection of susceptible and resistant selfed lines of rye by the rye smut fungus, *Urocystis occulta, Phytopathology,* 30, 926, 1940.
12. Western, J. H., The biology of oat smuts. IV. The invasion of some susceptible and resistant oat varieties, including Markton, by selected biological species of smut (*Ustilago avenae* (Pers.) Jens. and *Ustilago kolleri* Wille), *Ann. Appl. Biol.,* 23, 245, 1936.
13. Ponirovskii, V. N., On the histology of the parasitism of *U. hordei* in barley shoots, *Tr. Khark. Skh. Inst.,* 38, 157, 1962.
14. Heald, F. D., The relation of spore load to the percent of stinking smut appearing in the crop, *Phytopathology,* 11, 269, 1921.
15. Woolman, H. M., Infection phenomena and host reactions caused by *Tilletia tritici* in susceptible and nonsusceptible varieties of wheat, *Phytopathology,* 20, 637, 1930.
16. Laurence, J. A. and Kennedy, B. W., Population changes of *Pseudomonas glycinea* on germinating soybean seeds, *Phytopathology,* 64, 1470, 1974.
17. Fajardo, T. G., Progress on experimental work with the transmission of bean mosaic, *Phytopathology,* 18, 155, 1928.
18. Smith, F. L. and Hewitt, W. B., Varietal susceptibility to common bean mosaic and transmission through seed, *Calif. Agric. Exp. Stn. Bull.,* 621, 18, 1938.
19. McNeal, F. H. and Afanasiev, M. M., Transmission of barley stripe mosaic through the seed in eleven varieties of spring wheat, *Plant Dis. Rep.,* 39, 460, 1955.
20. Hemmati, K. and McLean, D. L., Gamete–seed transmission of alfalfa mosaic virus and its effect on seed germination and yield in alfalfa plants, *Phytopathology,* 67, 576, 1977.
21. Grogan, R. G. and Bardin, R., Some aspects concerning the seed transmission of lettuce mosaic virus, *Phytopathology,* 40, 965, 1950.
22. Babovic, M. V., The transmission rate of alfalfa mosaic virus by lucerne seed, *Acta Biol. Yugosl.,* 13, 83, 1976.
23. Grogan, R. G. and Schnathorst, W. C., Tobacco ring spot virus the cause of lettuce calico, *Plant Dis. Rep.,* 39, 803, 1955.

24. Grogan, R. G., Hall, D. H., and Kimble, K. A., Cucurbit mosaic viruses in California, *Phytopathology,* 49, 366, 1959.
25. Kennedy, B. W. and Cooper R. L., Association of virus infection with mottling of soybean seed coats, *Phytopathology,* 57, 35, 1967.
26. Adams, D. B. and Kuhn, C. W., Seed transmission of peanut mottle virus, *Phytopathology,* 67, 1126, 1977.
27. Xu, Z., Chen, K., Zhang, Z., and Chen. J., Seed transmission of peanut stripe virus in peanut, *Plant Dis.,* 75, 723, 1991.
28. Edwardson, J. R. and Christie, R. G., Viruses infecting forage legumes, Agric. Exp. Inst. Food Agric. Sci. Univ. Florida Mono. No. 14, 742, 1986.
29. Kendrick, J. B. and Gardner, M. W., Soybean mosaic seed transmission and effect on yield, *J. Agric. Res.,* 27, 91, 1924.
30. Athow, K. L. and Bancroft, J. B., Development and transmission of tobacco ringspot virus in soybean, *Phytopathology,* 49, 697, 1959.
31. Carroll, T. W. and Chapman, S. R., Variation in embryo infection and seed transmission of barley stripe mosaic virus within and between two cultivars of barley, *Phytopathology,* 60, 1079, 1970.
32. Timian, R. G., The range of symbiosis of barley and barley stripe mosaic virus, *Phytopathology,* 64, 342, 1974.
33. Thakur, R. P. and Chahal, S. S., Problems and strategies in the control of ergot and smut in pearl millet, ICRISAT Proc. Int. Pearl Millet Workshop, 1986, 1987, 173.
34. Joshi, L. M., Singh, D. V., and Srivastava, K. D., Meteorological conditions governing the spread of karnal bunt (*Neovossia indica*) of wheat, Fourth Int. Congr. Plant Pathol., 1983, 217.
35. Lapis, D. B., Insect pests and diseases of wheat in the Philippines, Proc. Int. Symp. Wheat for More Tropical Environments, CIMMYT, El Batan, Mexico, 1985, 152.
36. Thakur, R. P. and King, S. B., Ergot disease of pearl millet, ICRISAT Inf. Bull. No. 24, International Crops Research Institute for the Semi-arid Tropics, Patancheru, India, 1988, 24.
37. Halfon-Meiri, A. and Solel, Z., Factors affecting seedling blight of sweet corn caused by seedborne *Penicillium oxalicum, Plant Dis.,* 74, 36, 1990.
38. Gossen, B. D. and Morrall, R. A. A., Transmission of *Ascochyta lentis* from infected lentil seed and plant residue, *Can. J. Plant Pathol.,* 8, 28, 1986.
39. Jones, E. S., Influence of temperature, moisture and oxygen on spore germination of *Ustilago avenae, J. Agric. Res.,* 24, 577, 1923.
40. Reed, G. M and Faris, J. A., Influence of environmental factors on the infection of sorghums and oats by smuts, *Am. J. Bot.,* 11, 518, 1924.
41. Kulkarni, G. S., Conditions influencing the distribution of grain smut (*Sphacelotheca sorghi*) of jowar (sorghum) in India, *Agric. J. India,* 17, 159, 1922.
42. Faris, J. A., Factors influencing infection of *Hordeum sativum* by *Ustilago hordei, Am. J. Bot.,* 11, 189, 1924.
43. Ling, L., Factors affecting infection in rye smut and subsequent development of the fungus in the host, *Phytopathology,* 31, 617, 1941.
44. Rabien, H., On the germination and infection conditions of *Tilletia tritici, Arb. Biol. Reichsanst. Land Forstwirtsch, Berlin Dahlem,* 14, 297, 1924.
45. Gibs, W., Modifications in susceptibility to bunt due to external conditions, *J. Landwirtsch.,* 72, 111, 1924.

46. Kendrick, E. L. and Purdy, L. H., Influence of environmental factors on the development of wheat bunt in the Pacific Northwest. III. Effect of temperature on time and establishment of infection by races of *Tilletia caries* and *T. foetida, Phytopathology,* 52, 621, 1962.
47. Colhoun, J., Taylor, G. S., and Tomlinson, R., Fusarium diseases of cereals. II. Infection of seedlings by *F. culmorum* and *F. avenaceum* in relation to environmental factors, *Trans. Br. Mycol. Soc.,* 51, 397, 1968.
48. Malalasekera, R. A. P. and Colhoun J., Fusarium diseases of cereals. III. Water relations and infection of wheat seedlings by *Fusarium culmorum, Trans. Br. Mycol. Soc.,* 51, 711, 1968.
49. Millar, C. S. and Colhoun, J., Fusarium diseases of cereals. VI. Epidemiology of *Fusarium nivale* on wheat, *Trans. Br. Mycol. Soc.,* 52, 195, 1969.
50. Holmes, S. J. I. and Colhoun, J., Infection of wheat seedlings by *Septoria nodorum* in relation to environmental factors, *Trans. Br. Mycol. Soc.,* 57, 493, 1971.
51. Skoropad, W. P., Seed and seedling infection of barley by *Rhynchosporium secalis, Phytopathology,* 49, 623, 1959.
52. Thakur, R. P. and King, S. B., Smut disease of pearl millet. ICRISAT Inf. Bull. No. 25, ICRISAT, Patancheru, India, 1988, 17.
53. Gleason, M. L. and Ferriss, R. S., Influence of soil water potential on performance of soybean seeds infected by *Phomopsis* sp., *Phytopathology,* 75, 1236, 1985.
54. Aujla, S. S., Sharma, I., and Gill, K. S., Effect of soil moisture and temperature on teliospore germination of *Neovossia indica, Indian Phytopathol.,* 43, 223, 1990.
55. Roberts, S. J., Effect of soil moisture on the transmission of pea bacterial blight (*Pseudomonas syringae* pv. *pisi*) from seed to seedling, *Plant Pathol.,* 41, 136, 1992.
56. Leukel, R. W., Studies on bunt or stinking smut of wheat and its control, *U. S. Dep. Agric. Tech. Bull.,* 582, 47, 1937.
57. Purdy, L. H. and Kendrick, E. L., Influence of environmental factors on the development of wheat bunt in the Pacific Northwest. IV. Effect of soil temperature and soil moisture on infection by soilborne spores, *Phytopathology,* 53, 416, 1963.
58. Schafer, J. F., Dickson, J. G., and Shands, H. L., Effect of temperature on covered smut expression in two barley varieties, *Phytopathology,* 52, 1161, 1962.
59. Leukel R. W., Factors influencing infection of barley by loose smut, *Phytopathology,* 26, 630, 1936.
60. Purdy, L. H., Soil moisture and soil temperature, their influence on infection by the wheat flag smut fungus and control of the disease by three seed-treatment fungicides, *Phytopathology,* 56, 98, 1966.
61. Kavanagh, T., Temperature in relation to loose smut in barley and wheat, *Phytopathology,* 51, 189, 1961.
62. Dean, W. M., The effect of temperature on loose smut of wheat (*Ustilago nuda*), *Ann. Appl. Biol.,* 64, 75, 1969.
63. Teviotdale, B. L. and Hall, D. H., Factors affecting inoculum development and seed transmission of *Helminthosporium gramineum, Phytopathology,* 66, 295, 1976.
64. Perry, D. A., Snow rot on winter barley, Annual Report, Scottish Crop Research Institute, Invergowrie, Dundee, U. K., 1984, 98.
65. Thakur, R. P., Rao, V. P., and King, S. B., Influence of temperature and wetness duration on infection of pearl millet by *Claviceps fusiformis, Phytopathology,* 81, 835, 1991.

66. Maude, R. B., Spencer, A., Brocklehurst, P. A., Gott, K. A., and Bambridge, J. M., The biology and control of *Alternaria dauci* (leaf blight) on carrot seeds, 35th Annual Report 1984, National Vegetable Research Station, Wellesbourne, Warwick, U.K., 1985, 81.
67. Trione, E. J., Dwarf bunt of wheat and its importance in international wheat trade, *Plant Dis.*, 66, 1083, 1982.
68. Davies, J. M. L., Diseases of oil seed crops, in *Oilseed Rape*, Scarisbrick, D. H. and Daniels, R. W., Eds., Collins Professional and Technical Books, Williams Collins Sons, London, 1986, 309.
69. Inaba, T., Seed transmission of downy mildews of spinach and soybean, *JARJA*, 19, 26, 1985.
70. Raut, J. G., Transmission of seed-borne *Macrophomina phaseolina* in sunflower, *Seed Sci. Technol.*, 11, 807, 1983.
71. Cappelli, C. and Torre, G. D., *Drechslera graminea* (Rabenh. ex Schlecht) Shoemaker in barley seed samples and its transmission in the field in the years 1984–1986, *Ann. della Facolta di Agraria Universita degli studi di Perugia,* 40, 85, 1988.
72. Hsi, C. H., Environment and sorghum kernel smut, *Phytopathology,* 48, 22, 1958.
73. Melchers, L. E. and Hansing, E. D., The influence of environmental conditions at planting time on sorghum kernel smut infection, *Am. J. Bot.,* 25, 17, 1938.
74. Joshi, L. M., Singh, D., and Srivastava, K. D., Meteorological conditions in relation to incidence of Karnal bunt of wheat in India, in 3rd Int. Symp. Plant Pathology, Indian Agric. Res. Stat., New Delhi, 1981, 11.
75. Agarwal, V. K., Singh, O. V., and Singh, A., A note on certification standards for the Karnal bunt disease of wheat, *Seed Res.,* 1, 96, 1973.
76. Teviotdale, B. L. and Hall, D. H., Factors affecting inoculum development and seed transmission of *Helminthosporium gramineum, Phytopathology,* 66, 295,1976.
77. Lehman, S. G., Systemic infection of soybean by *Peronospora manshurica* as affected by temperature, *J. Elisha Mitchell Sci. Soc.,* 69, 83, 1953.
78. Kommedahl, T. and Lang, D. S., Temperature effects on seedling wilt from corn kernels infected with *Helminthosporium maydis, Phytopathology,* 62, 770, 1972.
79. Calvert, O. H. and Thomas, C. A., Some factors affecting seed transmission of safflower rust, *Phytopathology,* 44, 609, 1954.
80. Shetty, H. S., Mathur, S. B., and Neergaard, P., Occurrence of *Sclerospora graminicola* (Sacc.) Schroet. inoculum in pearl millet (*Pennisetum typhoides* (Burm.) Stapf and Hubb.) seeds and its transmission, *3rd Int. Congr. Plant Pathology,* P. Parey, Berlin, 1978, 120.
81. Wallin J. R., Seed and seedling infection of barley, bromegrass, and wheat by *Xanthomonas translucens* var. *cerealis, Phytopathology,* 36, 446, 1946.
82. Crowley, N. C., Studies on the seed transmission of plant virus diseases, *Aust. J. Biol. Sci.,* 10, 449, 1957.
83. Singh, G. P., Arny, D. C., and Pound, G. S., Studies on the stripe mosaic of barley, including effects of temperature and age of host on disease development and seed infection, *Phytopathology,* 50, 290, 1960.
84. Agarwal, V. K., Nene, Y. L., and Beniwal, S. P. S., Detection of bean common mosaic virus in urdbean (*Phaseolus mungo*) seeds, *Seed Sci. Technol.,* 5, 619, 1977.
85. Hanada, K. and Harrison, B. D., Effects of virus genotype and temperature on seed transmission of nepoviruses, *Ann. Appl. Biol.,* 85, 79, 1977.

86. Daft, G. C. and Leben, C., Bacterial blight of soybeans: epidemiology of blight outbreaks, *Phytopathology,* 62, 57, 1972.
87. Guthrie, J. W., Factors influencing halo blight transmission from externally contaminated *Phaseolus vulgaris* seed, *Phytopathology,* 60, 371, 1970.
88. Grogan, R. G., Lucas, L. T., and Kimble, K. A., Angular leaf spot of cucumber in California, *Plant Dis. Rep.,* 55, 3, 1971.
89. Rodenhiser, H. A. and Taylor, J. W., Studies on environmental factors affecting infection and the development of bunt in wheat, *Phytopathology,* 30, 40, 1940.
90. Rodenhiser, H. A. and Taylor J. W., The effect of photoperiodism on the development of bunt in two spring wheats, *Phytopathology,* 33, 240, 1943.
91. Polyakov, I. M. and Vladimirskaya, M. E., The role of light conditions in the resistance of cabbage to false powdery mildew, *Tr. Vses. Inst. Zashch. Rast.,* 21, 18, 1964.
92. Hampton, R. O. and Francki, R. I. B., RNA-1 dependent seed transmissibility of cucumber mosaic virus in *Phaseolus vulgaris, Phytopathology,* 82, 127, 1992.
93. Glaeser, G., The extent of field attack by wheat bunt (*T. caries*) in relation to seed infestation, *Pflasch Ber.,* 26, 33, 1961.
94. Malalasekera, R. A. P. and Colhourn, J., Fusarium diseases of cereals. V. A technique for the examination of wheat seed infected with *Fusarium culmorum, Trans. Br. Mycol. Soc.,* 52, 187, 1969.
95. Aulakh, K. S., Mathur, S. B., and Neergaard, P., Comparison of seed-borne infection of *Drechslera oryzae* as recorded on blotter and in soil, *Seed Sci. Technol.,* 2, 385, 1974.
96. Wallen, V. R., Field evaluation and the importance of the *Ascochyta* complex in peas, *Can. J. Plant Sci.,* 45, 27, 1965.
97. Moffett, M. L., Wood, B. A., and Hayward, A. C., Seed and soil: sources of inoculum for the colonisation of the foliage of solanaceous hosts by *Pseudomonas solanacearum, Ann. Appl. Biol.,* 98, 403, 1981.
98. Kim, W. G., Oh, I. S., Yu, S. H., and Park, J. S., *Fusarium moniliforme* detected in seeds of corn and its pathological significance, *Korean J. Mycol.,* 12, 105, 1984.
99. Grey, W. E., Mathre, D. E., Hoffmann, J. A., Powelson, R. L., and Fernandez, J. A., Importance of seedborne *Tilletia controversa* for infection of winter wheat and its relationship to international commerce, *Plant Dis.,* 70, 122, 1986.
100. Madeira, A. C., Clark, J. A., Rossall, S., and McArthur, A. J., A classification for seeds of fababean infected by *Ascochyta fabae, Fabis Newsl.,* 30, 48, 1992.
101. Venette, J. R., Reduction of bean seed emergence by vacuum infiltration with bacterial pathogens, *Phytopathology,* 75, 1385, 1985.
102. Mew, T. W., Gergon, E. B., and Merca, S. D., Impact of seedborne pathogens in rice germplasm exchange, *Seed Sci. Technol.,* 18, 441, 1990.
103. Opio, A.F., Teri, J.M., and Allen, D.J., Studies on seed transmission of *Xanthomonas campestris* pv. *phaseoli* in common beans in Uganda, *Afr. Crop Sci. J.,* 1, 59, 1993.
104. Varma, A., Krishnareddy, M., and Malathi, V. G., Influence of the amount of blackgram mottle virus in different tissues on transmission through the seeds of *Vigna mungo, Plant Pathol.,* 41, 274, 1992.
105. Popp, W., A new approach to the embryo test for predicting loose smut of wheat in adult plants, *Phytopathology,* 49, 75, 1959.
106. Khanzada, A. K., Rennie, W. J., Mathur, S. B., and Neergaard, P., Evaluation of two routine embryo test procedures for assessing the incidence of loose smut infection in seed samples of wheat (*Triticum aestivum*), *Seed Sci. Technol.,* 8, 363, 1980.

107. Maude, R. B. and Humpherson-Jones, F. M., Studies on the seed-borne phases of dark leaf spot (*Alternaria brassicicola*) and grey leaf spot (*Alternaria brassicae*) of Brassicas, *Ann. Appl. Biol.*, 95, 311, 1980.
108. Zade, A., Recent investigation on the life-history and control of loose smut of oats [*Ustilago avenae* (Pers) Jens.], *Angew. Bot.*, 6, 113, 1924.
109 Gupta, M. D. and Summanwar, A. S., The location of two mosaic viruses in cowpea seeds, *Seed Sci. Technol.*, 8, 203, 1980.
110. Warmick, D. and Demski, J. W., Susceptibility and resistance of soybean to peanut stripe virus, *Plant Dis.*, 72, 19, 1988.
111. Pesic, Z. and Hiruki, C., Differences in the incidence of alfalfa mosaic virus in seedcoat and embryo of alfalfa seed, *Can. J. Plant Pathol.*, 8, 39, 1986.
112. Makkouk, K. M., Kumari, S. G., and Bos, L., Pea seed-borne mosaic virus: occurrence in faba bean *Vicia faba* and lentil *Lens culinaris* in West Asia and North Africa, and further information on host range, transmission characteristics and purification, *Neth. J. Plant Pathol.*, 99, 115, 1993.
113. Emdal, P. S. and Foldo, N. E., Seed-borne inoculum of *Uromyces betae*, *Seed Sci. Technol.*, 7, 93, 1979.
114. Swan, D. J. and Mantle, P. G., Parasitic interactions between *Claviceps purpurea* strains in wheat and an acute necrotic host response, *Mycol. Res.*, 95, 807, 1991.
115. Schmidt, H. E., Bean mosaics, in *Plant Diseases of International Importance*, Vol. II. *Diseases of Vegetables and Oil Seed Crops*, Chaube, H. S., Singh, U. S., Mukhopadhyay, A. N., and Kumar, J., Eds., Prentice Hall, New York, 1992, 40.
116. Anderson, C. W., Seed transmission of three viruses in cowpea, *Phytopathology,* 47, 515, 1957.
117. Cheo, P. C., Effect of seed maturation on inhibition of southern bean mosaic virus in bean, *Phytopathology,* 45, 17, 1955.
118. Shepherd, R. J. and Fulton, R. W., Identity of a seed-borne virus of cowpea, *Phytopathology,* 52, 489, 1962.
119. Nelson, M. R. and Knuhtsen, H. K., Squash mosaic virus variability: epidemiological consequences of differences in seed transmission frequency between strains, *Phytopathology,* 63, 918, 1973.
120 Nelson, M. R. and Knuhtsen, H. K., Relation of seed transmission to the epidemiology of squash mosaic virus strains, *Phytopathology,* 59, 1042, 1969.
121 Ghanekar, A. M. and Schwenk, F. W., Seed transmission and distribution of tobacco streak virus in six cultivars of soybean, *Phytopathology,* 64, 112, 1974.
122. Midha, S. K. and Swarup, G., Factors affecting development of ear-cockle and tundu diseases of wheat, *Indian J. Nematol.*, 2, 97, 1972.
123. Vartanian, V. G. and Endo, R. M., Survival of *Phytophthora infestans* in seeds extracted from tomato fruits, *Phytopathology,* 75, 375, 1985.
124. Baker, C. J., Morphology of seedling infection by *Leptosphaeria nodorum*, *Trans. Br. Mycol. Soc.*, 56, 306, 1971.
125. Youcef-Benkada, M., Bendahmane, B. S., Sy, A. A., Barrault, G., and Albertini, L., Effects of inoculation of barley inflorescences with *Drechslera teres* upon the location of seedborne inoculum and its transmission to seedlings as modified by temperature and soil mixture, *Plant Pathol.*, 43, 350, 1994.
126. Rodenhiser, H. A. and Taylor J. W., Effects of soil type, soil sterilization and soil reaction on bunt infection at different incubation temperatures, *Phytopathology,* 30, 400, 1940.

127. Doling, D. A., The influence of seedling competition on the amount of loose smut (*Ustilago nuda* (Jens.) Rostr.) appearing in barley crops, *Ann. Appl. Biol.*, 54, 91, 1964.
128. Milan, A., Sul "carbone volante" del grano in rapporto all'accestimento delle piante, *Nuovo G. Bot. Ital.*, 46, 149, 1939.
129. Jones, G. H. and Seif-El-Nasr, A. El G., The influence of sowing depth and moisture on smut diseases and the prospects of a new method of control, *Ann. Appl. Biol.*, 27, 35, 1940.
130. Swinburne, T. R., Infection by *Tilletia caries* (DC.) Tul, the causal organism of bunt, *Trans. Br. Mycol. Soc.*, 46, 145, 1963.
131. Tiemann, A., Untersuchungen über die Empfanglichkeit des Sommerweizens für *Ustilago tritici* und den Einfluss der äusseren Bedingungen dieser Krankheit, *Khün-Arch.*, 9, 405, 1925.
132. Pedersen, P. N., Investigations on the influence of growth conditions on the attacks of loose smut of barley, *Acad. Scand.*, 15, 245, 1965.
133. Meiners, J. P., Kendrick, E. L., and Holton, C. S., Depth of seeding as a factor in the incidence of dwarf bunt and its possible relationship to spore germination on or near the soil surface, *Plant Dis. Rep.*, 40, 242, 1956.
134. Leukel, R. W., Investigations on the nematode disease of cereals caused by *Tylenchus tritici, J. Agric. Res.*, 27, 925, 1924.
135. Jones, R. A. C. and Proudlove, W., Further studies on cucumber mosaic virus infection of narrow-leafed lupin (*Lupinus angustifolius*): seed-borne infection, aphid transmission, spread and effects on grain yield, *Ann. Appl. Biol.*, 118, 319, 1991.
136. Hewett, P. D., Loose smut in winter barley: comparisons between embryo infection and the production of diseased ears in the field, *J. Nat. Inst. Agric. Bot.*, 15, 231, 1980.
137. Slack, S. A., Shepherd, R. J., and Hall, D. H., Spread of seed-borne barley stripe mosaic virus and effects of the virus on barley in California, *Phytopathology*, 65, 1218, 1975.
138. Gassner, G. and Kirchhoff, H., Zur Frage der Beeinflussung des Flugbrandbefalls durch Umweltfaktoren und chemische Beizmittel, *Phytopathol. Z.*, 7, 487, 1934.
139. Olsson, L., The influence of certain factors on the occurrence of seedling injuring fungi in the resulting crop of cereal seed, *Seed Sci. Technol.*, 7, 235, 1979.
140. Ooka, J. J. and Kommedahl, T., Kernel infected with *Fusarium moniliforme* in corn cultivars with opaque-2-endosperm or male sterile cytoplasm, *Plant Dis. Rep.*, 61, 162, 1977.
141. Templeton, G. E., Johnson, T. H., and Henry, S. E., Kernel smut of rice, *Ark. Farm Res.*, 9, 10,1960.
142. Demski, J. W., Tobacco mosaic virus is seedborne in pimento peppers, *Plant Dis.*, 65, 723, 1981.
143. Krull, C. F., Robayo, G., Valbuena, L. A., Luis, A., Rico, G., Castibalco, L. E., and Bravo, L. E., Influence of seed size on the incidence of loose smut in Funza barley, *Plant Dis. Rep.*, 50, 101, 1966.
144. Kuznetsova, A. F., The effect of seed size on infection of barley by loose smut, *Ref Zh. Rastenievod.*, 855, 1971.
145. Lavery, P., The relationship between seed size and the incidence of loose smut in three winter barleys, *Proc. Indian Acad. Sci.*, 74, 155, 1965.

146. Taylor, J. W., Effect of the continuous selection of large and small wheat seed on yield, bushel weight, varietal purity, and loose smut infection, *J. Am. Soc. Agron.*, 20, 856, 1928.

147. McFadden, A. D., Kaufmann, M. L., Russell, R. C., and Tyner, L. E., Association between seed size and the incidence of loose smut in barley, *Can. J. Plant Sci.*, 40, 611, 1960.

148. Agarwal, V. K., Seed-borne fungi and viruses of some important crops, Experimental Station, Research Bulletin No. 108, G. B. Pant University of Agricultural Technology, Pantnagar, India, 1981, 144.

149. Masmoudi, K., Khetarpal, R. K. and Maury, Y., Seed transmission of pea seedborne mosaic virus, Proc. 1st European Conf. Grain Legumes, France, 1992, 317.

150. Paguio, O. R. and Kuhn, C. W., Incidence and sources of inoculum of peanut mottle virus and its effect on peanut, *Phytopathology,* 64, 60, 1974.

151. Troutman, J. L., Bailey, W. K., and Thomas, C. A., Seed transmission of peanut stunt virus, *Phytopathology,* 57, 1280, 1967.

152. Stevenson, W. R. and Hagedorn, D. J., Further studies on seed transmission of pea seed-borne mosaic virus in *Pisum sativum, Plant Dis. Rep.*, 57, 248, 1973.

153. Middleton, J. T., Seed transmission of squash mosaic virus, *Phytopathology,* 34, 405, 1944.

154. Ryder, E. J. and Johnson, A. S., A method for indexing lettuce seeds for seed-borne lettuce mosaic virus by airstream separation of light from heavy seeds, *Plant Dis. Rep.,* 58, 1037, 1974.

155. Schuster, M. L. and Coyne, D. P., Survival mechanism of phytopathogenic bacteria, *Annu. Rev. Phytopathol.,* 12, 199, 1974.

156. Kuniyasu, K., Seed transmission of Fusarium wilt of bottle gourd *Lagenaria siceraria,* used as root stock of watermelon. III. Course of seedling infection by the seed-borne pathogen, *Fusarium oxysporum* f. sp. *lagenarium* Matuo and Yamamoto, *Ann. Phytopathol. Soc. Jpn.,* 43, 270, 1977.

157. Baker, K. F., Bacterial fasciation disease of ornamental plants in California, *Plant Dis. Rep.,* 34, 121, 1950.

158. Skoric, V., Bacterial blight of peas: overwintering, dissemination, and pathological histology, *Phytopathology,* 17, 611, 1927.

159. Schuster, M. L., Hoff, B., Mandel, M., and Lazar, I., Leaf freckles and wilt, a new corn disease, *Proc. Annu. Corn Sorghum Res. Conf.* (Chicago), 27, 176, 1973.

160. El-Helaly, A. F. and Ibrahim, I. A., Host parasite relationship of *Sphacelotheca sorghi* on sorghum, *Phytopathology,* 47, 620, 1957.

161. Larson, L. A. and Beenvers, H., Amino acid metabolism in young pea seedlings, *Plant Physiol.,* 40, 424, 1966.

162. Kandaswamy, D., Kesavan, R., Ramaswamy, K., and Prasad, N. N., Occurrence of microbial inhibitors in the exudates of certain leguminous seeds, *Indian J. Microbiol.,* 14, 25, 1974.

163. Chaturvedi, S. N., Muralia, R. N., and Sirdhana, B. S., The influence of cumin seed exudates on fungal spore germination, *Plant Soil,* 40, 49, 1974.

164. Dhawale, S. D. and Kodmelvar, R. V., Studies on mycoflora of chili seed, *Seed Res.,* 6, 23, 1978.

165. Mishra, R. R. and Kanaujia, R. S., Studies on certain aspects of seed-borne fungi, *Indian Phytopathol.,* 26, 284, 1965.

166. Takayanki, K. and Murakame, K., Rapid germinability test with exudates from seeds, *Proc. Int. Seed Test. Assoc.,* 34, 243, 1968.

167. Saxena, R. M. and Gupta, J. S., Effect of seed leachates on spore germination of seed-borne fungi on *Vigna radiata, Indian Phytopathol.*, 35, 236, 1982.
168. Charya, M. A. S. and Reddy, S. M., Effect of *Cajanus cajan* seed coat leachates on germination of some seedborne fungi, *Indian Phytopathol.*, 33, 112, 1980.
169. Ark, P. A. and Thompson, J. P., Antibiotic properties of the seeds of wheat and barley, *Plant Dis. Rep.*, 42, 959, 1958.
170. Srivastava, V. B. and Mishra, R. R., Fungal inhibitory agents in seed coats, *Phytopathol. Mediterr.*, 10, 127, 1971.
171. Tveit, M. and Moore, M. B., Isolates of *Chaetomium* that protect oats from *Helminthosporium victoriae, Phytopathology,* 44, 686, 1954.
172. Brewer, D., Jeram, W. A., Meiler, D., and Taylor, A., The toxicity of cochliodinol, an antibiotic metabolite of *Chaetomium* spp., *Can J. Microbiol.*, 16, 433, 1970.
173. Old, K. M., Mercury tolerant *Pyrenophora avenae* in seed oats, *Trans Br. Mycol. Soc.*, 51, 525, 1968.
174. Bamberg, R. H., Holton, C. S., Rodenhiser, H. A., and Woodward, R. W., Wheat dwarf bunt depressed by common bunt, *Phytopathology,* 37, 556, 1947.
175. Berend, I., The occurrence of bunt fungi in wheat inoculated by *Tilletia caries* and *T. foetida, Acta Phytopathol. Acad. Sci. Hung.*, 8, 365, 1973.
176. Vajavat, R. M. and Chakravarti, B. P., Survival of *Pseudomonas sesami* and effect of an antagonistic bacterium isolated from seeds on the control of the disease in seed, *Indian Phytopathol.*, 31, 286, 1978.
177. Willingale, J. and Mantle, P.G., Interaction between *Claviceps purpurea* and *Tilletia caries* in wheat, *Trans. Br. Mycol. Soc.*, 89, 145, 1987.
178. Furuki, I., Hibi, T., Honda, Y., Saito, T., and Komuro, T., Comparison of the biological characteristics between melon necrotic spot virus and cucumber necrosis virus, *Ann. Phytopathol. Soc. Jpn.*, 46, 419, 1980.
179. Pio Ribeiro, G., Wyatt, S. D., and Kuhn, C. W., Cowpea stunt: a disease caused by a synergistic interaction of two viruses, *Phytopathology,* 68, 1260, 1978.

CHAPTER 9

Epidemiology and Inoculum Thresholds of Seedborne Pathogens

I. EPIDEMIOLOGY

Epidemiology is the study of the development and spread of disease inoculum and of the factors affecting these processes,[1] and deals with the effects of the biotic (host and pathogen) and abiotic environment. Van der Plank[2] defined epidemics as the science of disease in populations involving the persistence and spread of inoculum and environmental factors affecting disease incidence on particular plant populations. Epidemiological studies are important in disease forecasting and disease control. Van der Plank[2] applied mathematical models to epidemiological studies of plant diseases. Plant disease epidemic refers to the development and rapid spread of a disease on a particular kind of crop cultivated over a large area.[3] Epidemics develop when a susceptible cultivar is planted over a large area in the presence of a virulent pathogen coupled with a favorable environment. The development and decline of epidemics is a balance between inoculum potential and disease potential:

Disease severity	=	Inoculum potential (inoculum density × capacity)	×	Disease potential (proneness × susceptibility)

Inoculum potential is the number of infective propagules (inoculum density) and their pathogenic capacity. Disease potential is host susceptibility, which may be influenced by an unfavorable environment, nutritional imbalances, and/or a susceptible growth stage. Host susceptibility is controlled genetically, and development of an epidemic depends on the number of infective propagules, their pathogenic capacity, host susceptibility, and the effect of the environment on pathogen virulence and host proneness to disease. These same factors affect the epidemiology of seedborne pathogens.

The rate of seed infection to seed transmission or plant infection and subsequent establishment in the field and further spread of inoculum depends on the host, pathogen(s), environment, transmitting agents, and their interactions. Yield losses may be high, even with a low-percentage seed infection by certain pathogens. For example, as few as two Brassica seeds per 10,000 infected with *X. campestris* pv. *campestris*,[4] less than 1% *X. campestris* pv. *vesicatoria*-infected tomato seeds,[5] or 0.02% *P. syringae* pv. *phaseolicola*-infected seeds in bean[6] can produce disease epidemics caused by these pathogens under suitable environmental conditions. The incidence of *X. campestris* pv. *vignicola* may be 62% from an initial inoculum of 1% infected cowpea seeds.[7] If lettuce seed lots carry over 0.5% lettuce mosaic virus, significant yield losses may result if aphids are prevalent.[8]

Environmental conditions greatly influence the epidemiology of seedborne pathogens. During the summer in India (relative humidity 20 to 80%, and 25 to 34°C) the incidence of cowpea blight (*X. campestris* pv. *vignicola*) reaches 62% from an initial inoculum of 1% infected seeds, but the incidence of greengram leaf spot (*X. campestris* pv. *phaseoli*) is only 0, 3, and 32% from 1, 10, and 100% initial seed infestation, respectively. However, during the rainy season (relative humidity 50 to 95%, 24 to 30°C), both diseases become severe.[7] As few as two infected seeds per 10,000 can cause an epidemic of blackrot (*X. campestris* pv. *campestris*) in cabbage. However, the environmental conditions influence the spread of primary inoculum. In 1976, seed infestations of 0.12, 0.06, and 0.02% resulted in epidemics. In 1977, epidemics resulted from infestations of 0.05% but not 0.01%.[4] A population of 10^3 to 10^4 *X. campestris* pv. *phaseoli* per bean seed was required for production of infected plants under field conditions.[9]

A highly significant correlation was found between barley seed infection with *Pyrenophora graminea* and plant infection, tiller infection, and yield reduction. The ratios of infected seeds to infected seedlings equaled 1:0.4, of infected tillers to yield loss equaled 1:0.9; and of infected seeds to yield loss equaled 1:0.3. The threshold of seed infection at which production was not significantly lower than the control was 14%. Therefore, seed treatment was advisable under northern and central Italy conditions, when the percentage of seed infection in commercial seed lots was above this level. However, a tolerance near zero was recommended in prebasic and basic seeds.[10] There was a direct correlation (r = 0.76) between the level of seed infection of *A. brassicae* and seedling infection in Brassica.[11] The ratio of seedborne infection to seed rot and seedling mortality was 1:0.88 in canola and mustard.[12] As the incidence of seed infection by *Stagonospora nodorum* at planting of wheat increased from 1 to 40%, the intensity of subsequent disease increased but the relationship of seed infection to disease was nonlinear. Seed infection of 10% supplied enough inoculum to cause a severe epidemic (Figure 9-1).[13] A significantly high correlation (r = 0.86) between seed infection to plant infection was observed in loose smut of wheat, but the ratio of seed infection to plant infection differed between cultivars.[14] The ratio of *A. raphani* seed infection of radish to infected plants approached 1:1.[15] A high correlation (r = 0.76, *P* = 0.01) was shown between safflower seed contamination by teliospores of *Puccinia calcitrapae* var. *centaureae* and incidence of disease.[16]

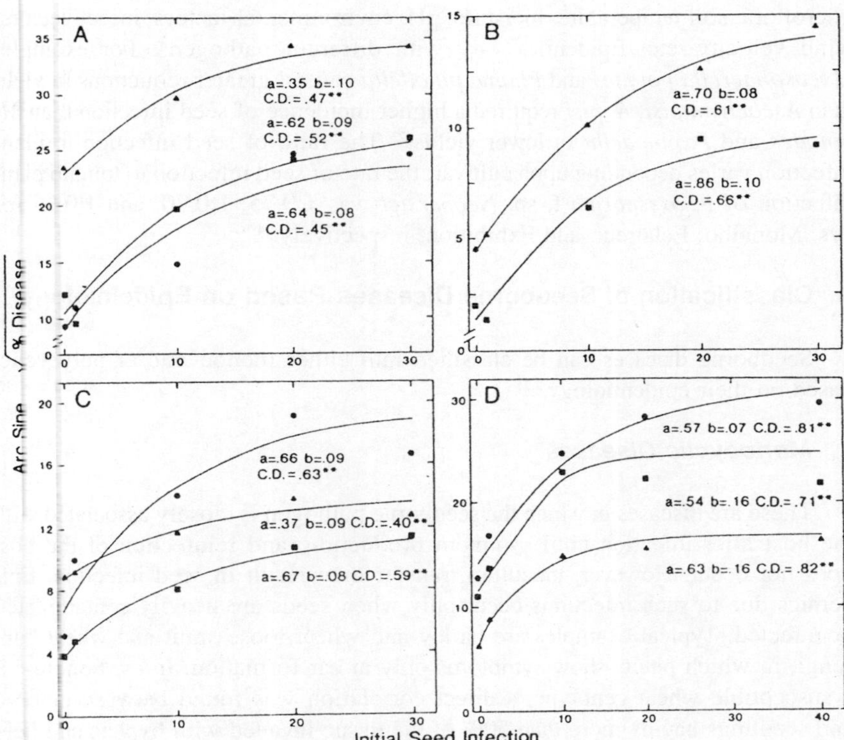

Figure 9-1 Effects of percent seed infection at planting on the development of *Stagonospora nodorum* on the upper leaves and heads of Coker 68-19 wheat (*Triticum aestivum*) at two locations in Florida. **A** = Gainesville, 1982; **B** = Gainesville, 1983; **C** = Quincy, 1983; **D** = Quincy, 1984. The model used for data analysis had the following characteristics: $y = y_{max}$ [1 − a exp(−bt)], where y = sine percent disease, y_{max} = sine maximum percent disease, a = position parameter for seed infection, b = rate parameter, and t = percent seed infection at planting, Δ = percent disease on head, = % disease on flag leaf, and •= % disease on second leaf. (From Luke, H. H., et al., *Plant Dis.*, 70, 252, 1986. With permission.)

In France, 5 infected bean seeds per 1000 resulted in an epidemic of halo blight of bean, whereas 1 per 20,000 did not.[17] Incidence of *X. c.* pv. *campestris* in crucifers was high in field plots that contained 0.03% infected plants, but less disease developed when the initial level was 0.01%.[18] Seedborne *C. m.* subsp. *michiganensis* did not affect tomato seed germination or seedling emergence, but primary cankers developed where the incidence had a ratio of 10:8 and 10:9 during 1987 and 1988, respectively. The seedborne pathogen also reduced yield.[19] *C. m.* subsp. *nebraskensis* was transmitted to maize seedlings at 0.1 to 0.4% from seeds inoculated by vacuum infiltration.[20] The possibility of transmission of maize whiteline mosaic virus in maize was less than 0.01%.[21] Coconut cadang-cadang viroid was seed transmitted at a rate of 1 in 300 coconuts.[22]

The threshold level for pathogenicity of any seedborne inoculum is not constant since it is affected by inoculum level, inoculum location, seed- and soilborne microflora, soil temperature, moisture, pH, environmental factors, insect vectors, wind velocity, etc. Epidemics vary with different pathogens. For example, *Mycosphaerella pinodes* and *Phoma pinodella* caused greater reductions in yield than *Ascochyta pisi*. *A. pisi* required a higher incidence of seed infection than *M. pinodes* and *P. pinodella* to lower yields.[23] The ratio of seed infection to plant infection varies depending upon cultivar; the rate of seed infection to tomato plant infection of *F. oxysporum* f. sp. *lycopersici* was 1:0.75, 1:0.70, and 1:0.36 for cvs. Monalbo, Eclaireur, and Exhibition, respectively.[24]

A. Classification of Seedborne Diseases Based on Epidemiology

Seedborne diseases can be classified into either monocyclic or polycyclic, based on their epidemiology.[2,25]

1. Monocyclic Diseases

These are diseases in which the seedborne pathogen is closely associated with the host after infection until symptom production, and reinfection of the host does not occur. However, inoculum transfer may result in seed infection. Epidemics due to such infections occur only when seeds are heavily contaminated or infected. Typical examples are barley and wheat loose smut and wheat bunt fungi, in which plants show symptoms only at ear formation. In cv. Sonora-64, a susceptible wheat genotype, a direct correlation was found between embryo and seedlings having more than 50% of the tissue invaded with hyphae and field expression of the disease.[26]

A system was developed in India for predicting loose smut of wheat utilizing the seed infection and transmission ratio. Seed infection of wheat loose smut was determined by an embryo count and seed transmission ratio based on tiller infection in the field. The data were fitted into the equation

$$Y = EK$$

for calculating the predicted value of loose smut incidence, where Y = predicted incidence of loose smut, E = percent loose smut infection in the seed lot, and K = the constant for seed transmission of the loose smut pathogen in the cultivar used based on the previous 3 years.[27]

In the United Kingdom, the leaf stripe of barley pathogen does not reinfect leaves and was considered as a monocyclic disease. The initial seedborne inoculum is the total inoculum available each year.[28] The diseases *Gloeotinia granigena* (blind seed disease) and *Claviceps purpurea* (ergot), which invaded only inflorescence, also were regarded as monocyclic diseases by Hewett,[28] but conidia produced by both fungi on infected plants resulted in secondary spread. These diseases should be considered polycyclic diseases.

It is easier to control monocyclic diseases because the use of clean seed can be achieved using seed certification or seed treatment with systemic or nonsystemic fungicides.

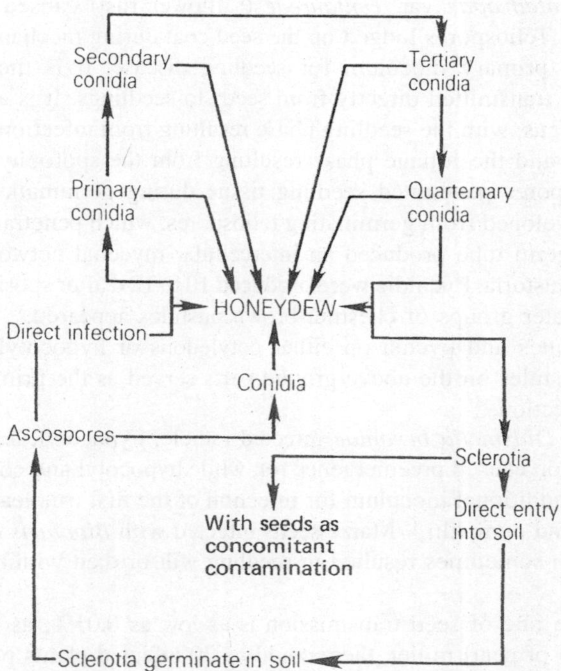

Figure 9-2 Life cycle of *Claviceps fusiformis* (ergot of pearl millet [*Pennisetum glaucum*]). (Adapted from Prakash, H. S., *Trans. Br. Mycol. Soc.*, 81, 65, 1983.)

2. *Polycyclic Diseases*

These are diseases in which seedborne pathogens after infection produce inoculum on the host capable of being carried to other host plants, which may result in disease (Figure 9-2). The seedborne inoculum gives rise to infected seedlings scattered throughout the field. The inoculum multiples repeatedly and spreads throughout the growing period of the host when conditions are favorable.[2] The inoculum may be carried by rain splash, wind, vectors, and other means. Epidemics of polycyclic diseases are influenced by the environment. Under favorable conditions even a low initial seedborne inoculum can result in an epidemic. Many seedborne pathogens belong to this group. Some examples are given.

Fava bean seeds infected with *A. fabae* produced seedlings with leaf lesions, inoculum from which spread for up to 10 m in an average season and usually infected the new seed crop. Seed lots with 1% infected seeds were suitable for crop production, but little or no *A. fabae* was tolerated in seeds intended for multiplication. Infection in British-grown commercial seeds was reduced by seed selection. Standards adapted in the seed scheme eliminated *A. fabae*. Theoretically, 1% seed infection gave one diseased seedling per 40 m^2 (18 cm between rows).[29] *A. fabae* in *Vicia faba* increased from initial seed infection levels of 8.4 to 20% at harvest.[30]

Puccinia calcitrapae var. *centaureae* (safflower rust) caused seedling and foliar diseases. Teliospores lodged on the seed coat during mechanical threshing and served as primary inoculum for seedling disease. It is the only known seedborne rust transmitted directly from seeds to seedlings. It is a macrocyclic, autoecious fungus, with the seedling phase resulting from infection by the gametophytic stage and the foliage phase resulting from the sporophytic stage.[31] In tests, basidiospores penetrated seedling tissue during germination.[32] Sporidial germ tubes developed from germinating teliospores, which penetrated host tissue directly. The germ tube produced an intercellular mycelial network and lobed, intracellular haustoria. Pycnidia were produced 10 to 12 d after sporidial infection, and 4 to 7 d later groups of chestnut-brown pustules appeared.[33] Infected parts exhibited pustules and pycnia on either cotyledons or hypocotyls or on both. Pycnia and pustules on the above-ground parts served as the primary inoculum for foliage infection.[34]

Seedborne *Didymella bryoniae* infected radicle, hypocotyl, and cotyledons. Radicle infection caused preemergence rot, while hypocotyl and cotyledon infection provided additional inoculum for infection of the first true leaves and stems of cucumber and pumpkin.[35] Maize seeds infected with *Bipolaris maydis* race T on germination sometimes resulted in seedling wilt or died within a few weeks of planting.[3]

Even if the rate of seed transmission is as low as 0.01%, as in the case of downy mildew of pearl millet, there could be 30 infected plants per acre. These plants have produced sporangial inoculum sufficient for secondary spread and caused an epidemic. In India, the seed rate for pearl millet was 3 kg/acre (1000 grains in 10 g), which gave about 300,000 plants. A plant infected by *S. graminicola* has produced 35,000 sporangia per cm^2 of the infected leaf in 1 h during the night.[36]

Bacteria have a short generation time; thus even a low inoculum level can result in an epidemic.[37] For example, *P. syringae* pv. *phaseolicola* (halo blight of beans) is seedborne, and infected seeds may give rise to infected seedlings, which serve as a source of primary inoculum. Rain splash and wind help further disease spread. In the United States, secondary infections were found up to 26 m away from primary infection sites.[6] Taylor[38] reported that only 1 of 10 infected seeds produced infected seedlings. Rapid disease spread resulted in severe epidemics. Taylor et al.[39] used a Van der Plank[2] equation as an epidemic model for halo blight in bean:

$$r = \frac{2.3}{t_2 - t_1}\left(\log 10 \frac{X_2}{1 - x_2} - \log 10 \frac{X_1}{1 - x_1}\right)$$

where x_1 = the initial inoculum (or disease) at time t_1, x_2 = the inoculum (or disease) present at time t_2, and r = infection rate expressed as per unit per d (where $t_2 - t_1$ is measured in days).

They proposed this equation as an appropriate one for a polycyclic disease. The values of r vary with climatic differences; from 0.17 to 0.22 and 0.10 to 0.16

in Wisconsin and the United Kingdom, respectively.[6,40,41] Under British conditions it was predicted that 0.025% seed infection would give rise to 0.0025% primary infection and that this would produce 4% infection in the mature crop with a *r*-value of 0.15.[39] The model has been used to determine tolerance levels for seed infection and to compare the effectiveness of foliar sprays and seed treatments. A reduction in primary inoculum derived from infected seeds either by exclusion through seed testing or seed treatment gave effective disease control.[39] Seed treatment reduced initial inoculum (x_1) and foliar sprays affected the rate of disease increase (r). Effective control of halo blight was achieved by seed treatment with the antibiotics kasugamycin or streptomycin, which reduced seed infection by 98%,[41] or with the foliar sprays with copper oxychloride or treptomycin sulfate.[40,41]

It was found that 5 crucifer seedlings per 10,000 infected with *X. c.* pv. *campestris* resulted in a high incidence of black rot, but a single diseased seedling did not. Laboratory seed assays capable of detecting 1 infected seed per 10,000 could predict field severity of black rot. Thus it was suggested that a tolerance of 1 infected seed per 10,000 be accepted for direct seeding of cabbage for head production, but a 0 tolerance be accepted for seedbed production.[42,43] Infected cabbage seeds with *X. c.* pv. *campestris* up to 1% resulted in a high disease incidence in the direction of the prevailing wind.[44] The severity of black rot infection in brussels sprouts in central New York State was related inversely to the distance from infected transplants.[45]

Twelve bean seeds per acre infected with *P. s.* pv. *phaseolicola* caused a severe epidemic of halo blight. Diseased plants could be found 20 to 25 m from the inoculum source at harvest. Rain splash dispersal disseminated the bacterium under Wisconsin conditions.[6]

Seed transmission is the primary source of inoculum and the most important factor in the epidemiology of certain plant viruses, such as BSMV in barley and wheat, BCMV in bean, broadbean stain virus in broadbean, eggplant mosaic virus in eggplant, lettuce mosaic virus in lettuce, peanut mottle virus in peanut, squash mosaic virus in cantaloupe, TMV in tomato, and nepoviruses. The spread of SbMV from soybean seedlings infected from seeds depended on the percent of infected seedlings, timing, numbers, and species composition of transient alate aphids, and wind and other environmental factors. The pattern of spread decreased as the distance from infected seedlings increased. A spread up to 50 m was noted from an initial inoculum source.[46] From 1959 to 1960, lettuce crops from seed <0.1% infected with lettuce mosaic virus contained 0 to 4.5% infected plants, whereas crops from 2.5 to 5.3% infected seeds contained 25 to 96% infected plants.[47]

A unique study of the annual occurrence of a seedborne fungus was made by Shortt et al.[48] In the 3-year study of seed decay of soybean caused by *Phomopsis* in Illinois, disease incidence was highest in 1977, lowest in 1976, and intermediate in 1975. A low positive correlation was found between disease incidence and rainfall during pod fill, indicating that moisture, rather than temperature or geographic area, is the dominant environmental factor in disease development. Maturity dates of cultivars interacted with changing weather conditions to affect disease incidence.

II. INOCULUM THRESHOLDS

The inoculum threshold of seedborne pathogens is the amount of seed infection or infestation that will cause disease in the field under a favorable environment and lead to economic losses. Inoculum thresholds must be established only when clean seeds are used as a control.[49] The need for establishing inoculum thresholds is greater than ever because of increased international seed movement and need for reasonable phytosanitary requirements. When a pathogen does not occur in a country or area, it is logical to protect the area by quarantine and insist that only pathogen- or disease-free seeds be allowed to enter. However, when a pathogen is present in a country, clean seeds may still be used for disease management. Finite levels of seed infection may be acceptable if this level does not result in economic loss. "Clean" seeds are preferred over "pathogen-free" or "disease-free" seeds in control programs. Clean seeds are seeds with a pathogen level that will not result in economic loss. Zero tolerance, that is, pathogen- or disease-free seeds, becomes an impossible and unrealistic requirement under seed conditions.[49] Although increasingly sensitive seed-testing techniques are being developed and used, seed health testing for seedborne pathogens can never ensure that there is absolutely no infection or infestation.[49] In addition, these terms create a false sense of security when the determination is based on a nil test result.[50]

Inoculum threshold is an important concept in seed pathology and disease control through the use of clean seeds. It validates seed testing, and through certification programs, it can evaluate the efficacy of seed treatment and provide realistic guidelines for the development and implementation of disease control programs in seed production areas. These programs are necessary for production and shipment of clean seeds.[49] The threshold level must be zero for a pathogen not in an area protected by established quarantine. Sampling methods routinely used for seed health testing cannot predict a zero level and are not suitable for quarantine purposes. However, for disease management purposes, some infection level above zero, a threshold level predicting no effect, should be established based on field experience with seeds known to contain different infection levels.[51]

A. Fungi

A single fungal spore can cause disease, but it requires thousands to establish a parasitic relationship resulting in disease. This relationship is similar to threshold levels of seedborne disease. The amount of inoculum necessary for disease development has been discussed in Chapter 8.

B. Bacteria

Because bacteria spread quickly under suitable environmental conditions, low levels of seedborne inoculum result in severe epidemics. This is a problem with crops grown from seedling transplants. A single infected plant in a transplant bed potentially could provide enough inoculum to infect a high percentage of the remaining plants. This can be complicated if symptoms do not develop before the transplants are shipped.[18] Epidemics of halo blight of beans resulted following primary infection levels of 0.02%.[6] Guthrie et al.[52] reported that 1 infected seed in 16,000 could result in complete crop loss. One of the most comprehensive studies on using field disease for developing tolerance level for *P. s.* pv. *phaseolicola* in bean was given by Trigalet and Bidaud.[17] They assayed commercial seed lots and established a tolerance level of 1 infected seed per 20,000 for controlling halo blight of bean in France. The tolerance was based on agar isolation and a direct immunofluorescence (IF) assay. They showed that epidemics could result from seed lots containing 5 seeds per 1000, whereas only a few diseased plants resulted from seed lots containing 1 infected seed per 20,000. Seed lots containing 5 infected seeds per 10,000 yielded variable results, depending on the season.[17] In England, Van der Plank's[2] infection rate equation $r = 2.3/t_2 - t_1$ (log10 $X_2/1 - X_2 - \log_{10} X_1/1 - X_1$) has been used to determine tolerance levels for *P. s.* pv. *phaseolicola*.[38] Based on 3 years of field studies, an infection rate of 0.15 and a transmission rate of 10:1 was common in England.[38] Therefore, a seed lot with an infection level of 0.025% would not result in a field disease. However, in later studies, a transmission ratio of 2:1 was observed. This illustrates the problems one faces with epidemiological formulas and field data. In England, a green bean crop with a 4% infection at harvest was considered tolerable.[18] Taylor's[38] results in England agreed with the observations by Guthrie et al.[52] and Wharton[53] that severe crop losses occurred from primary infections of 0.1 to 0.06%, assuming seed transmission of 10:1, a level higher than the suggested tolerance level of 0.025%.[18] The results agreed with a tolerance of 0.01% for *X. c.* pv. *campestris*.[43] The results comparing laboratory assays and field disease of two wheat cultivars in Idaho showed that a tolerance was not necessary for control of *X. c.* pv. *translucens* (Table 9-1).[18]

C. Viruses

Bennett[54] reported that seed transmitted viruses were easily sap transmissible, indicating their ability to invade parenchymatous tissue. Thus such symptoms as mottling, local chlorotic or necrotic lesions, etch, and other abnormalities have their origin in parenchymatous tissue. Viruses transmitted by leafhoppers and aphids in a persistent manner are not seedborne, whereas viruses transmitted by beetles, nematodes, and aphids in a nonpersistent manner may be seedborne. Stace-Smith and Hamilton[55] categorized seedborne viruses based on the virus groups approved by the International Committee on Taxonomy of Viruses. The number of members and possible members in each group and type of injury commonly associated with seed transmission to draw definite conclusions on seed transmission are given (Table 9-2). They noted

- that seed transmission occurred in approximately 18% of those viruses recognized as possible members of established plant virus groups;
- that although seed transmission has been recorded for one or more viruses belonging to 21 of the 28 groups, it was of economic significance in only 10;

Table 9-1 Correlation of Laboratory Assay Results to Black Chaff (*Xanthomonas campestris* pv. *translucens*) of Wheat (*Triticum aestivum*) Development in the Field[a]

Infected cultivar and seed lot	No. seeds from		Laboratory assay	Field disease	
	Foundation lot	Infected lot	Mean cfu *X. c.* pv. *translucens*/ml	No. plots with black chaff	Disease rating[b]
Bliss					
1106	3,000	0	7	0	0.00
	2,970	30	70	0	0.00
	2,700	300	266	0	0.00
	0	3,000	1,205	1	0.25
Waid					
1084	3,000	0	0	0	0.00
	2,970	30	1,170	1	0.25
	2,700	300	42,400	4	2.50
	0	3,000	363,083	4	4.00

[a] Eight replications were made for four mixtures of naturally contaminated wheat seeds of Bliss lot 1106 and Wiad 1084 with seeds of Foundation Lot 1090. Four replications of 125 g of seed were assayed by plating washings onto a semiselective agar medium (XTS agar), and four replications were sown in the field 7 d later on 7 May 1983 at Kimberly, ID. Each plot was at least 4.6 m (15 ft) from an adjacent plot.

[b] Disease rating: 0 = no black chaff; 1, 2, 3, and 4 = one site with one to two infected plants, two sites with one to two infected plants; three to four sites with three to four infected plants, and numerous sites with three to four infected plants, respectively. Plants were read on 19 July. Isolation was made from leaves of plants with black chaff symptoms collected at random from each plot. Figures are the mean of four plots.

From Schaad, N. W., Phytopathology, 78, 872, 1988. With permission.

Table 9-2 Relative Importance of Seed Transmission of Plant Viruses that are Considered to be Members or Possible Members of the Recognized Virus Groups

Virus group[a]	Number of members		Type of potential injury[b]					
	In group	Seedborne	A	B	C	D	E	F
Alfalfa mosaic	1	1	+	+	+			
Bromovirus	5	1	+	+	+			
Carlavirus	47	2						
Caulimovirus	7	0						
Closterovirus	15	1						
Comovirus	15	6	+	+	+			
Cucumovirus	4	4	+	+	+			
Dianthovirus	3	0						
Geminivirus	16	1						
Hordeivirus	3	2	+	+	+			+
Ilarvirus	13	8					+	
Luteovirus	29	0						
Maize chlorotic dwarf	2	0						
Maize rayado fino[c]	4	0						
Necrovirus[c]	3	1						
Nepovirus	30	22	+	+	+			
Pea enation mosaic	1	1						
Plant reovirus	9	0						
Potexvirus	38	4						
Potyvirus	117	16	+	+	+	+	+	+
Rhabdovirus	74	1						
Rice stripe[c]	6	0						
Sobemovirus	10	2						
Tobamovirus	18	7	+	+	+			
Tobravirus	3	3	+	+	+			
Tombusvirus	11	1						
Tomato spotted wilt	1	1						
Tymovirus	18	3						
Viroids	15	5				+	+	

[a] Based on Fourth Report of the International Committee on Taxonomy of Viruses. Viroids are included for comparative purposes.
[b] A, survival of inoculum; B, dispersal of inoculum; C, primary inoculum source; D, contamination of germ plasm lines; E, contamination of virus-free planting material; and F, direct injury to crop.
[c] Group names approved at meeting of International Committee on Taxonomy of Viruses, Sendai, Japan, Sept. 1–7, 1984, (unpublished).

From Stace-Smith, R. and Hamilton, R. I., *Phytopathology*, 78, 875, 1988. With permission.

- that the injuries attributed to virus survival, virus dispersal, and as a primary inoculum source commonly were associated with and were the dominant types of potential injury caused by seedborne viruses;
- that of the 11 virus groups that contained seedborne viruses of economic significance, only the potyvirus group contained those associated with all types of injury.[55]

The maximum amount of seedborne virus inoculum tolerated without appreciable constraint to yield has been studied for only a few host–virus combinations, with the following characteristics[55]: (1) the level of seed transmission can be extremely low and still be of importance, even if a few infected seedlings arising from contaminated seed constitute the sole source of inoculum, when the virus is acquired readily and actively vectored; (2) the crop must be annual, because infected seeds would not be the sole inoculum source in perennial crops; (3) the vector is an aphid and the virus is transmitted in a nonpersistent manner; (4) the virus is confined to a narrow natural host range; (5) when the virus is sufficiently characterized to justify inclusion in a virus group, it would belong to either the potyvirus or cucumovirus group. Viruses meeting these criteria would have an inoculum threshold of zero or close to zero.[55] Those seedborne viruses that are not as actively vectored but may constitute the sole source of inoculum are epidemiologically important, and although tolerance levels are not generally stated, the levels must be low. For those seedborne viruses that have a broad natural host range, including both annual and perennial crops, seed transmission may play an insignificant role in virus epidemiology, and consequently a high inoculum level could be tolerated.[55]

III. DEVELOPMENT OF INOCULUM THRESHOLDS

Inoculum threshold levels must be developed for the average environmental conditions in which seeds are sown. Threshold levels are influenced by all factors affecting the epidemiology of each host–pathogen combination. It is difficult to establish inoculum thresholds because they are influenced by many factors.[49] Inoculum threshold levels should be fixed by a correlation between infection levels based on seed health tests and disease incidence in the field.[50] Establishment of valid tolerances for local areas coupled with accurate seed health tests have provided a powerful disease management tool that can minimize fungicide use.[51] Tolerances for some pathogens have been published.[50] The following factors may be considered for development of inoculum thresholds.

A. Planting Area

Environmental factors such as moisture and temperature affect whether infected seeds will yield infected seedlings, a primary source of inoculum. Temperature affected the percentage of wire stem disease of cabbage seedlings grown from seeds naturally infected with *A. brassicicola*. The disease became more severe with a rise in temperature (Table 9-3).[56] This has been observed also in Karnal bunt of wheat in India. During the 1969–70 harvest, seedborne infection varied from 0 to 7.5% in different cultivars, while in 1970–72, its incidence was less than 1%, and in 1974–75, the incidence reached 50%.[57,58] In Canada, depending upon the rate of seed infection, the tolerance limit for *Ascochyta* on peas varied from 2 to 6%. Therefore, tolerance varied from year to year for certain pathogens, depending on seed infection. Lettuce mosaic virus has been controlled

Table 9-3 Seedling Emergence, Weight of Cabbage (*Brassica oleracea*) Seedlings, Radial Growth of *Alternaria brassicicola In Vitro*, Seed Infection Levels, and *A. brassicicola* Wire Stem Disease at Different Temperatures

	Temperature °C				
	10	**15**	**20**	**25**	**30**
Percentage emergence at 5 d[a]	0 A	9 B	46 C	58 D	71 E
Growth (fresh weight) of cabbage seedlings (g)[a]	0.4 A	1.1 B	1.9 C	3.4 D	2.1 C
Radial growth of *A. brassicicola* (mm)[a]	29 A	53 C	67 D	71 D	44 B
Wirestem disease index following wound inoculation[b]	0.30 A	0.95 B	1.55 C	2.6 D	1.50 C

[a] Emergence and weight data from hot water-soaked lot of plants grown in washed sand. Mean fresh weights of five plants receiving one-half Hoagland's solution measured after 3 weeks. Radial growth (diameter) recorded on V-8 agar after 9 d. Significance letters based on Duncan's New Multiple Range Test ($P = .05$).

[b] Wire stem disease on wound inoculated 14-d-old plants incubated under plastic for 48 h. Disease severity was scored after 5 d on wire stem disease index; 0 = no lesion; 1 = <25% hypocotyl (hyp) necrotic; 2 = 25 to 50% hyp necrotic; 3 = 51 to 75% hyp necrotic; 4 = 76 to 99% hyp necrotic; 5 = 100% hyp necrotic (typical wire stem). Significance letters based on the use of Chi-square values in a contingency table ($P = 0.5$).

Adapted from Bassey, E. O. and Gabrielson, R. L., *Seed Sci. Technol.*, 11, 403, 1983.

by using seed certified to have no infected seeds in 30,000 in California and 9 in 2000 in The Netherlands.[59] The Netherlands has cooler weather, hence a lower aphid population; therefore, farmers have practiced multiple cropping systems to break the disease cycle. However, California has warmer weather and high populations of aphids. Lettuce is grown during almost the whole year in California. Thus there is no interruption of lettuce production to break the disease cycle.[49] Therefore, tolerance has depended upon the planting area. Some pathogens did not infect a crop in certain areas due to unfavorable environmental conditions. For such areas, a relaxation in tolerance could be allowed. However, in areas where infected seeds served as a source of soil contamination or where the infection could occur only pathogen-free seeds should have been planted.

B. Role of Seedborne Inoculum in Disease Development

Inoculum threshold levels should be determined by a correlation between seed infection level established by seed health testing and field disease damage data established by well-designed experiments.[49] However, due to the complexity of seed transmission inoculum, thresholds for most seedborne pathogens have not been established experimentally. Most of these have been set arbitrarily or with only field observation data. There are few seedborne pathogens, namely *D. phaseolorum* var. *sojae*, *Phomopsis* sp.,[60] *P. lingam*,[61] *X. c.* pv. *campestris*, *P. s.* pv. *phaseolicola*,[62] *Ustilago segetum* var. *tritici*, lettuce mosaic virus, and barley stripe mosaic virus, for which inoculum thresholds have been established with supportive correlation data. Tolerance varies depending on the rate of establishment, infection, and spread. For the majority of seedborne pathogens, only information on their seedborne nature is available. Transmission rate and spread in the field are not well understood. Under favorable conditions, 12 bean seeds per acre infected with *P. s.* pv. *phaseolicola* caused severe epidemics of bacterial blight.[6] Only 0.1% seed transmission of LMV led to severe field infection when aphid vectors were active.[63] Therefore, a tolerance of 0.003% is allowed in California for this virus. A correlation was shown between *Ustilago segetum* var. *tritici* infection of seeds and loose smut in barley and wheat.[64-66] Thus disease development could be measured by seed infection. Seed transmission becomes important if there is potential for pathogen spread from infected seeds. The tolerance must be near zero in such cases. The pathogens are carried within and on seeds. This inoculum must survive, grow, reproduce, and attack seedlings in the soil or seedling, until it becomes a source of primary inoculum. However, some pathogens survive in soil for a limited time in the absence of host (*S. apiicola*), or seedlings may rot before reaching the soil surface. Thus much of the inoculum indicated by seed health tests may be buried in the soil.[51] Theoretically one infected seed results in one infected plant, but generally this is not the case under field conditions. Wire stem disease of *Brassica* crops caused by *A. brassicicola* may be on the surface or below the seed coat, with seedling infection closely correlated with internal seedborne inoculum.[11] Bassey and Gabrielson[56] reported that at 20°C, wire stem disease was closely associated with deep-seated inoculum. However, at 30°C wire stem was better correlated with total inoculum. Virulence may range from avirulent to virulent for a particular pathogen. Normally seed health tests do not reveal the virulence of the pathogen. In case of *P. lingam* several strains varying in virulence have been reported.[67] Only the most virulent strains were able to spread in the field and cause damage in cabbage plants in Wisconsin.[68] The virulence was determined by seedling inoculation.[51] The potential of secondary inoculum spread in the field ultimately influenced disease severity and inoculum threshold levels. In the case of monocyclic pathogens, there was no repeating cycle of infection, and the disease incidence was closely related to initial seed infection. However, in polycyclic pathogens, a low level of seedborne inoculum resulted in severe losses. Thus for a monocyclic pathogen a finite level of seed infection may be prescribed, but in polycyclic ones a very low inoculum threshold or tolerance should be allowed.[51]

C. Host Susceptibility

Threshold levels will be lower for highly susceptible cultivars than for relatively resistant ones. A classic experiment was conducted by Heald,[69] who artificially inoculated *T. caries* spores in seedlots of resistant cv. Marquis and susceptible cv. Jenkins Club wheat to determine inoculum thresholds. The inoculum threshold level for Marquis was between 542 and 5043 spores per grain, whereas for Jenkins Club it was 104 spores per grain.[69]

D. Economic Loss Due to Seedborne Pathogens

Losses caused by seedborne pathogens vary depending upon virulence, host range, and environment. If a pathogen causes heavy losses, the level of infection allowed for tolerances should be low. The role of seed infection, ratio of subsequent transmission in the field, and yield loss should determine tolerance for different pathogens.

E. Perpetuation of Inoculum by Other Means

Inoculum may survive on weed hosts or infected crop residues. It is difficult to differentiate losses caused by seedborne inoculum under such situations. The losses caused by seedborne *S. nodorum* have been quantified separately in wheat.[70]

F. Cropping Practices

Cropping practices also affect seed transmission, and hence inoculum thresholds. In 1973, low levels of seedborne *Phoma lingam* inoculum spread to cabbage transplant fields and resulted in severe losses, whereas no disease was found when this same seedlot was direct seeded.[67] Seedborne diseases could be serious when transplants were produced in semisterile media under high moisture levels.[68] It has been reported that seedborne inoculum was not important in oilseed canola, which also may be seeded directly. Inoculum threshold levels must be established for average conditions under which the crop is grown. Thus if a pathogen is solely seedborne, strict tolerances can be followed, and infected seeds should not be sent into areas where the pathogen has not been detected. But if a pathogen is soilborne, its seedborne nature becomes secondary in disease development. In such cases, tolerance levels may be relaxed.

G. Influence of Seed Treatment on Seedborne Infection

If a seedborne pathogen can be eliminated by seed treatment, then tolerance levels should be relaxed. Most chemical seed treatments are not effective against bacterial and viral pathogens, and certification standards are necessary.

H. Certification Program

Different inoculum thresholds need to be established for seeds at different levels in a certification program. The threshold for breeder seeds must be lower than that for foundation or certified seeds. It should be zero for quarantine purposes.[49]

I. Seed Processing Procedures

A number of plant pathogen propagules can be separated from healthy seeds during processing, such as weed seeds, sclerotia, galls, etc. For such propagules, seed processing must be taken into consideration while developing inoculum thresholds.

J. Seed Health Tests

To establish a threshold level, the number of seeds infected must be determined. Ideally, the seed health testing method should be accurate, cheap, quick, reproducible, and sensitive. Accurate estimates of pathogen levels in a seed lot are essential for establishing usable inoculum thresholds of seedborne pathogens. In any seed health test, one must guard against false positives and false negatives. A false positive is the identification of a dead fungal pycnidium (*Septoria apiicola*) or isolation of a strain with low virulence (*P. lingam*) or a morphologically similar nonpathogenic fungus (*P. herbarum* instead of *P. lingam*). False positives can be avoided by including a pathogenicity test.[67] A false negative occurs when the test fails to reveal a virulent pathogen in the seed. This can be avoided by including a known infected control in each test.[67] Also, sample size and replication must be adequate to measure significant amounts of infection or infestation.[67] An effective ELISA threshold cannot be established for BCMV in bean because of uneven virus distribution in individual seed parts. False positive seed tests are attributable to virus or viral antigen present only in the cotyledon or other seed parts, which was not transmitted to the seedling, whereas false negative tests are attributable to BCMV infection of the primary axis without infection or contamination of the testa or cotyledons. The latter also can be due to the virus being present below concentrations that the assay can detect.[71] It is hoped that the seeds will be pathogen free if a zero tolerance for a pathogen is established. But it is difficult or impossible to predict that a seed lot is pathogen free based on seed health tests because of errors inherent in sampling procedures. As the samples tested become larger, the predicted infection percentage becomes lower but never zero.

REFERENCES

1. Zakods, J. C. and Schein, R. D., *Epidemiology and Plant Disease Management*, Oxford University Press, London, 1979.
2. Van der Plank, J. E., *Plant Disease: Epidemics and Control,* Academic Press, New York, 1963, 349.
3. Agrios, G. N., *Plant Pathology,* 3rd ed., Academic Press, New York, 1988, 804.
4. Schaad, N. W., Sitterly, W. R., and Humaydan, H., Relationship of levels of seedborne *Xanthomonas campestris* as determined by laboratory assays to the development of black rot of cabbage, in 3rd Int. Congr. Plant Pathol., Munich, 66, 1978.
5. Cox, R. S., The role of bacterial spot in tomato production in South Florida, *Plant Dis. Rep.,* 50, 699, 1966.
6. Walker, J. C. and Patel, P. N., Splash dispersal and wind as factors in epidemiology of halo blight of bean, *Phytopathology,* 54, 140, 1964.
7. Shekhawat, G. S. and Patel, P. N., Seed transmission and spread of bacterial blight of cowpea and leaf spot of greengram in summer and monsoon seasons, *Plant Dis. Rep.,* 61, 390, 1977.
8. Broadbent, L., Tinsley, T. W., Buddin, W., and Roberts, E. T., The spread of lettuce mosaic in the field, *Ann. Appl. Biol.,* 38, 689, 1951.

9. Weller, D. M. and Saettler, A. W., Evaluation of seed-borne *Xanthomonas phaseoli* and *X. phaseoli* var. *fuscans* as primary inocula in bean blight, *Phytopathology,* 70, 148, 1980.
10. Porta-Puglia, A., Delogu, G., and Vannacci, G., *Pyrenophora graminea* on winter barley seed: effect on disease incidence and yield losses, *J. Phytopathol.*, 117, 26, 1986.
11. Maude, R. B. and Humpherson-Jones, F. M., Studies on the seedborne phases of dark leaf spot (*Alternaria brassicicola*) and grey leaf spot (*Alternaria brassicae*) of Brassicas, *Ann. Appl. Biol.*, 95, 311, 1980.
12. Chahal, A. S. and King, M. S., Some aspects of seedborne infection of *Alternaria brassicae* in rape and mustard cultivars in the Punjab, *Indian J. Mycol. Plant Pathol.,* 9, 51, 1979.
13. Luke, H. H., Barnett, R. O., and Pfahler, P. L., Development of *Septoria nodorum* blotch on wheat from infected and treated seed, *Plant Dis.*, 70, 252, 1986.
14. Agarwal, V. K., Verma, H. S., Agarwal, M., and Gupta, R. K., Studies on loose smut of wheat. II. Factors influencing seed infection and seed transmission, *Seed Res.*, 12, 72, 1984.
15. Vannacci, G. and Pecchia, S., Location and transmission of seedborne *Alternaria raphani* Groves and Skolko in *Raphanus sativus* L.: a case study, *Arch. für Phytopathol. Pflanzensch.*, 24, 305, 1988.
16. Cappelli, C. and Zazzerini, A., Safflower rust (*Puccinia carthami* Cda.) in Italy: seed contamination, seed-plant transmission and seed dressing for disease control, *Phytopathol. Medit.*, 27, 145, 1988.
17. Trigalet, A. and Bidaud, P., Some aspects of epidemiology of bean halo blight, in Station de Pathologie Vegetale et Phytopathol., Proc. Fourth Int. Conf. Plant Pathogenic Bacteria, Anger, France, 1978, 895.
18. Schaad, N. W., Inoculum thresholds of seedborne pathogens: bacteria, *Phytopathology,* 78, 872, 1988.
19. Dhanvantari, B. N., Effect of seed extraction methods and seed treatment on control of tomato bacterial canker, *Can. J. Plant. Pathol.,* 11, 400, 1989.
20. Biddle, J. A., McGee, D. C., and Braun, E. J., Seed transmission of *Clavibacter michiganense* subsp. *nebraskense* in corn, *Plant Dis.*, 74, 908, 1990.
21. Lourie, R., Gordon, D. T., Knoke, J. K., Gingery, R. E., Bradfute, O. E., and Lipps, P. E., Maize white line mosaic virus in Ohio, *Plant Dis.*, 66, 167, 1982.
22. Randles, J. W. and Imperial, J. S., Coconut cadang-cadang viroid, in Descriptions of Plant Viruses, Commonwealth Mycological Institute, Assoc. Appl. Biol., Kew, Surrey, U.K., 1984.
23. Wallen, V. R., Field evaluation and the importance of the *Ascochyta* complex on peas, *Can. J. Plant Sci.*, 45, 27, 1965.
24. Besri, M., Phases of the transmission of *Fusarium oxysporum* f. sp. *lycopersici* and *Verticillium dahliae* by seeds of some tomato varieties, *Phytopathol. Z.*, 93, 148, 1978.
25. Hewett, P. D., Epidemiology of seed-borne disease, in *Plant Disease Epidemiology,* Scott, P. R. and Bainbridge, A., Eds., Blackwell Scientific, Oxford, 1978, 167.
26. Rewal, H. S. and Jhooty, J. S., Correlation between embryo, seedling and field infection of loose smut of wheat, *Indian Phytopathol.*, 35, 571, 1982.
27. Verma, H. S., Singh, A., and Agarwal, V. K., A system for prediction of loose smut incidence in wheat, in 3rd Int. Symp. Plant Pathology, New Delhi, 1981, 5.

28. Hewett, P. D., Disease testing in a seed improvement programme, in *Seed Pathology — Problems and Progress,* Yorinori, J. T., Sinclair, J. B., Mehta, Y. R., and Mohan, S. K., Eds., Fundacao Instituto Agronomico do Paraná — IAPAR, Londrina, Brazil, 1979, 72.
29. Hewett, P. D., The field behavior of seed-borne *Ascochyta fabae* and disease control in field beans, *Ann. Appl. Biol.,* 74, 287, 1973.
30. Gaunt, R. E. and Liew, R. S., Control strategies of *Ascochyta fabae* in New Zealand field and broadbean crops, *Seed Sci. Technol.,* 9, 707, 1981.
31. Zimmer, D. E., Rust infection and histological response of susceptible and resistant safflower, *Phytopathology,* 55, 296, 1965.
32. Calvert, O. H. and Thomas, C. A., Some factors affecting seed transmission of safflower rust, *Phytopathology,* 44, 609, 1954.
33. Zimmer, D. E., Spore stages and life cycle of *Puccinia carthami, Phytopathology,* 53, 316, 1963.
34. Halfon-Meiri, A., Seed transmission of safflower rust (*Puccinia carthami*) in Israel, *Seed Sci. Technol.,* 11, 835, 1983.
35. Lee, D.-H., Mathur, S. B., and Neergaard, P., Detection and location of seed-borne inoculum of *Didymella bryoniae* and its transmission in seedlings of cucumber and pumpkin, *Phytopathol. Z.,* 109, 301, 1984.
36. Safeeulla, K. M., Biology and Control of the Downy Mildews of Pearlmillet, Sorghum and Finger Millet, Downy Mildew Res. Lab., Mysore University, Mysore, India, 1976, 304.
37. Schuster, M. L. and Coyne, D. P., Survival mechanism of phytopathogenic bacteria, *Annu. Rev. Phytopathol.,* 12, 199, 1974.
38. Taylor, J. D., The quantitative estimation of the infection of bean seed with *Pseudomonas phaseolicola* (Burkh.) Dowson, *Ann. Appl. Biol.,* 66, 29, 1970.
39. Taylor, J. D., Dudley, C. L., and Presly, L., Studies of halo-blight seed infection and disease transmission in dwarf beans, *Ann. Appl. Biol.,* 93, 267, 1979.
40. Taylor J. D., Field studies on halo-blight of beans (*Pseudomonas phaseolicola*) and its control by foliar sprays, *Ann. Appl. Biol.,* 70, 191, 1972.
41. Taylor, J. D. and Dudley, C. L., Seed treatment for the control of halo blight of beans (*Pseudomonas phaseolicola*), *Ann. Appl. Biol.,* 85, 223, 1977.
42. Schaad, N. W. and Kendrick, R., A qualitative method of detecting *Xanthomonas campestris* in crucifer seed, *Phytopathology,* 65, 1034, 1975.
43. Schaad, N. W., Sitterly, W. R., and Hymaydan, H., Relationship of incidence of seed-borne *Xanthomonas campestris* to black rot of crucifers, *Plant Dis.,* 64, 91, 1980.
44. Cytrynomicz, L. E. and Fieldhouse, D. J., Epidemiology of *Xanthomonas campestris* on field grown cabbage, *Phytopathology,* 70, 668, 1980.
45. Hunter, J. E., Abawi, G. S., and Becker, R. F., Observations on the source and spread of *Xanthomonas campestris* in an epidemic of black rot in New York, *Plant Dis. Rep.,* 59, 384, 1975.
46. Irwin, M. E. and Goodman, R. M., Ecology and control of soybean mosaic virus, in *Plant Disease and Vectors,* Maramorosch, K. and Harris, K. F., Eds., Academic Press, New York, 1981, 181.
47. Tomlinson, J. A., Control of lettuce mosaic by the use of healthy seed, *Plant Pathol.,* 11, 61, 1962.
48. Shortt, B. J., Grybauskas, A. P., Tenne, F. D., and Sinclair, J. B., Epidemiology of Phomopsis seed decay of soybean in Illinois, *Plant Dis.,* 65, 62, 1981.

49. Kuan, T. L., Inoculum thresholds of seedborne pathogens, an overview, *Phytopathology,* 78, 867, 1988.
50. Hewett, P. D., Seed standards for disease in certification, *J. Nat. Agric. Bot.,* 15, 373, 1981.
51. Gabrielson, R. L., Inoculum thresholds of seedborne pathogens: fungi, *Phytopathology,* 78, 868, 1988.
52. Guthrie, J. W., Huber, D. M., and Fenwick, H. S., Serological detection of halo blight, *Plant Dis. Rep.,* 49, 297, 1965.
53. Wharton, A. L., Detection of infection by *Pseudomonas phaseolicola* (Burkh.) Dowson in white-seeded dwarf bean seed stocks, *Ann. Appl. Biol.,* 60, 305, 1967.
54. Bennett, C. W., Seed transmission of plant viruses, *Adv. Virus Res.,* 14, 221, 1969.
55. Stace-Smith, R. and Hamilton, R. I., Inoculum thresholds of seedborne pathogens: viruses, *Phytopathology,* 78, 875, 1988.
56. Bassey, E. O. and Gabrielson, R. L., The effects of humidity, seed infection levels, temperature and nutrient stress on cabbage seedling disease caused by *Alternaria brassicicola, Seed Sci. Technol.,* 11, 403, 1983.
57. Agarwal, V. K., Singh, O. V., and Singh, A., A note on certification standards for the karnal bunt disease of wheat, *Seed Res.,* 1, 97, 1973.
58. Agarwal, V. K., Singh, A., and Verma, H. S., Outbreak of karnal bunt of wheat, *FAO Plant Prot. Bull.,* 24, 99, 1976.
59. Grogan, R. G., Control of lettuce mosaic with virus free seed, *Plant Dis.,* 64, 446, 1980.
60. Garzonio, D. M. and McGee, D. C., Comparison of seeds and crop residue as source of inoculum for pod and stem blight of soybean, *Plant Dis.,* 67, 1374, 1982.
61. Gabrielson, R. L., Mulanax, M. W., Matsuoka, K., Williams, P. H., Whiteaker, G. P., and Maguire, J. D., Fungicidal eradication of seedborne *Phoma lingam* of crucifers, *Plant Dis. Rep.,* 61, 118, 1977.
62. Taylor, J. D., Phelps, K., and Dudley, C. L., Epidemiology and strategy for the control of halo-blight of beans, *Ann. Appl. Biol.,* 93, 167, 1979.
63. Zink, F. W., Grogan, R. G., and Wetch, J. E., The effect of the percentage of seed transmission upon subsequent spread of lettuce mosaic virus, *Phytopathology,* 46, 662, 1956.
64. Agarwal, V. K., Agarwal, M., Verma, H. S., and Gupta, R. K., Studies on loose smut on wheat. III. Influence of infection on plant morphology and control through seed treatment with carboxin, *Seed Res.,* 10, 79, 1982.
65. Morton, D. J., A quick method of preparing barley embryos for loose smut examination, *Phytopathology,* 50, 270, 1960.
66. Rennie, W. J. and Seaton, R. D., Loose smut of barley. The embryo test as a means of assessing loose smut infection in seed stocks, *Seed Sci. Technol.,* 3, 697, 1975.
67. Gabrielson, R. L., Blackleg diseases of crucifers caused by *Leptosphaeria maculans (Phoma lingam)* and its control, *Seed Sci. Technol.,* 11, 749, 1983.
68. Bonman, J. M., Gabrielson, R. L., Williams, P. H., and Delwiche, P. A., Virulence of *Phoma lingam* to cabbage, *Plant Dis.,* 65, 865, 1981.
69. Heald, F. D., The relation of spore load to the per cent of stinking smut appearing in the crop, *Phytopathology,* 11, 269, 1921.
70. Babadoost, M. and Hebert, T. T., Incidence of *Septoria nodorum* in wheat seed and its effects on plant growth and grain yield, *Plant Dis.,* 68, 125, 1984.
71. Klein, R. E., Wyatt, S. D., Kaiser, W. J., and Mink. G. I., Comparative immunoassays of bean common mosaic virus in individual bean (*Phaseolus vulgaris*) seed and bulked bean seed samples, *Plant Dis.,* 76, 57, 1992.

CHAPTER 10

Nonparasitic Seed Disorders

Nontransmissible seed disorders are nonpathogenic, noninfectious, and may or may not lead to deleterious effects on seeds or seedlings. Nontransmissible disorders are due to several factors.

I. GENETIC EFFECTS

The genetic background of a host genotype can result in seed abnormalities. For example, in flax, seed coat splitting in yellow-seeded cultivars is common.[1] Eggplant, pepper, and tomato seed breaks commonly occur in the seed coat, resulting in a chamber between the seed coat and endosperm, which provides an entry point for pathogens.[2,3] Defective fertilization in barley and wheat can result in production of embryoless seeds.[4,5] The fruit pox and gold fleck diseases of tomato are inherited and are transmitted through seeds for four generations. Fruit pox appears as incipient lesions on immature fruits, which later rupture and become necrotic before the fruits turn red. Gold fleck appears as small lesions on immature fruit, which turn golden yellow on ripe fruit. Fruit pox is controlled by a recessive gene and gold fleck by a dominant one. It is possible to eliminate fruit pox and gold fleck within two generations by roguing affected plants in segregating populations. The cv. Florida MH-1, free from fruit pox and gold fleck, was released using this approach.[6]

Stevenson et al.[7] reported a seed-transmitted genetic disease in cv. Chico III tomato known as corky stunt, which is expressed as shortened internodes, proliferated axillary buds, malformed petioles, roughened lower petiole surface, and malformed and corked fruit. Corky lesions blemish the fruit surface. The optimum temperature for symptom expression is 22°C. The disease is controlled by a single recessive gene.

Figure 10-1 Severe mechanical damage on soybean (*Glycine max*) seeds. (From J. B. Sinclair, Ed., *Compendium of Soybean Diseases*, APS Press, St. Paul, MN, 1982, 49. With permission.)

II. MECHANICAL INJURIES

Operations such as harvesting, threshing, processing, and postharvest handling may result in mechanical injury to seeds (Figure 10-1). Injuries range from cracking or splitting of the seed coat to fracturing of embryo parts. The degree of mechanical injury depends upon cultivar susceptibility, seed moisture content, and machinery used. Seeds that are too dry become brittle. Wet seeds suffer from impact damage of embryos. Baldhead or snakehead in lima bean seedlings is caused by mechanical impact damage on the embryo during threshing. The seedlings may lack a growing point, radicle, or cotyledon(s).[8] Machines with high cylinder speeds result in damaged seed coats. Seed coat injuries provide entry points for pathogenic and saprophytic microorganisms. Injured seeds are more susceptible to the phytotoxic effects of fungicide seed treatments, especially mercurials. Severely injured seeds may fail to germinate or may produce weak seedlings.

III. ENVIRONMENTAL EFFECTS

Humidity and temperature are factors that have the greatest effect on nonparasitic disorders.

A. Temperature

In general, moist seeds are more sensitive to extremes of low rather than high temperatures, with the degree of sensitivity depending upon seed moisture content. Dry seeds usually can tolerate low temperatures (–185 to –192°C).[9] Frost damage to seeds in the field has been reported in temperate countries. Cold-damaged canola seeds become brown and shrunken.[10] Frost can injure developing seeds of cauliflower,[11] which may fail to germinate or produce weak, damaged seedlings.[9] Heating of seeds during storage because of metabolic activities may result in seed damage, resulting in the production of abnormal seedlings.[9]

B. Humidity

Harvesting moist seeds or stacking a harvested crop in the field can maintain or raise seed moisture content, which results in softening or cracking of seed coats. These conditions favor development of storage fungi. Under humid conditions, radish seeds develop a gray discoloration due to swelling of the subepidermal parenchyma, which results in distortion or cracking of the epidermis. This facilitates invasion by *Alternaria*.[12,13] Desiccation of soybean seeds during maturation results in cracking and wrinkling of the seed coat. Such seeds have a lower emergence than nondamaged seeds.[14] Hollow heart, which is one of the most important physiological disorders of pea seeds, is due to sudden drying of immature seeds at high temperatures.[15,16] Hollow heart symptoms appear as cavities or depressions on the adaxial face of cotyledons. It is a common disorder of wrinkled pea seeds and peas having compound starch grains (Figure 10-2).[17,18] The cavities contain cells with reduced contents. The hollow heart symptoms are determined by soaking pea seeds in distilled water for 16 to 18 h at 20 to 25°C, followed by splitting the cotyledons apart and storing them for 24 h at 20 to

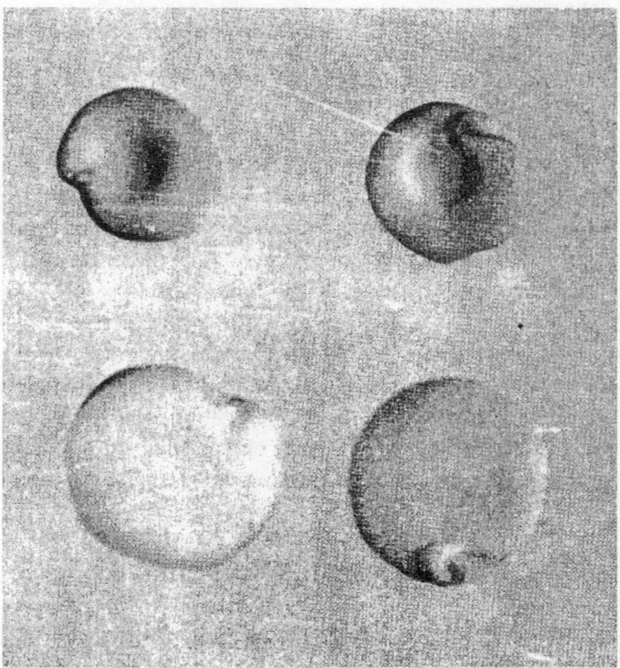

Figure 10-2 Split pea (*Pisum sativum*) seeds of cultivar CG 141 (*above*) showing cotyledons affected by hollow heart disease compared to nonaffected seeds (*below*).

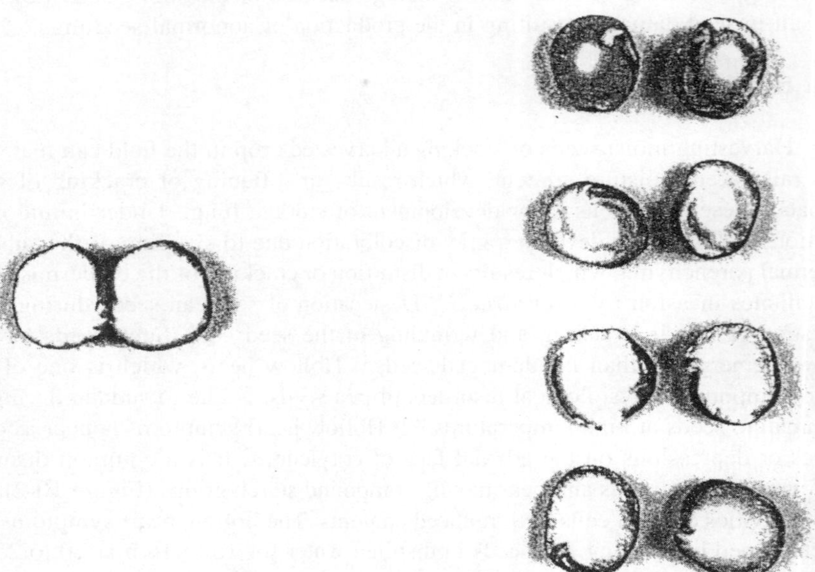

Figure 10-3 Tetrazolium staining pattern of the adaxial cotyledon surfaces of pea (*Pisum sativum*) seeds with symptoms of hollow heart; (*right*) seeds with deteriorated tissues; (*left*) unaffected seeds. (From Don, R., et al., *Seed Sci. Technol.*, 12, 707, 1984. With permission.)

25°C. The cavities become conspicuous because they lose water during drying and take more time to recover during imbibition.[17]

Although hollow heart becomes apparent during germination, it develops during seed maturation. Water stress, high ambient temperature during maturation, and premature drying of harvested seeds induce the disorder: seeds with hollow heart germinate normally under ideal conditions, but the disorder predisposes seedlings to attack by pathogenic fungi or may cause poor emergence. Hollow heart in pea seedlings is related to environmental conditions during germination. Controlled deterioration and water stress tests increase the amount of hollow heart. However, irrespective of germination conditions, hollow heart is related to the quantity of deteriorated tissues near the center of the adaxial surface of cotyledons of ungerminated seeds (Figure 10-3).[19]

IV. MINERAL DEFICIENCIES

Mineral deficiencies in soil may result in poor plant growth and thus poor quality seed in the form of reduced seed size, abnormal seed shape, and low viability.

In nitrogen (N) deficiency, soils having low N and high potash and phosphorus result in the N deficiency symptoms in cereal seeds. Nitrogen deficiency in wheat is known as yellow berry. Symptoms appear on seeds as light yellowish spots covering all or part of the seed. Such seeds germinate normally but may result in poor market value. Affected seeds have a high starch content but are deficient in protein. Nitrogen deficiency is corrected by N application.

In manganese (Mn) deficiency, symptoms have been reported in fava bean, common bean, pea, and many other crops. However, symptoms of Mn deficiency are more apparent in pea. The disorder, known as marsh spot, is common in England (Figure 10-4).[20] Symptoms appear as discolored lesions in the center of the adaxial face of the cotyledons.[17,21] Marsh spot is common in cultivars that possess simple and large starch granules in the cotyledons. Affected cells secrete a pigment into enlarged air spaces due to the depletion of protoplasmic contents.[17] Seed coats of severely affected peas are brownish, with sunken spot. Affected seeds germinate, resulting in weak and poorly developed seedlings. Two sprays of 1% Mn sulfate — first, just after the close of flowering and second, about 3 weeks later — reduce incidence of this disorder.[22] An application of Mn sulfate to soil at 100 to 200 kg/ha at flowering also controls marsh spot.[22,23] The incidence of marsh spot in a seed lot is determined by soaking pea seeds in distilled water for 16 to 18 h at 20 to 25°C.[24] The cotyledons are split apart and examined immediately.[17]

Potash (K) deficiency, in cucumber results in tapered seeds.[25] Marsh spot of peas, primarily due to Mn deficiency, also has been reported due to K deficiency.[26]

Boron deficiency, which causes hollow heart of peanut in Florida, resulted in a discoloration and rotting of seeds.[27]

V. INSECT DAMAGE

Insects cause distinctive damage to seeds; the damage is easily detected either by the presence of the insects or characteristic damage to the seeds. Insects such a lygus bugs (*Lygus campestris, L. elisus, L. hesperus, L. oblineatus,* and *L. sallei*) cause injuries to carrot, celery, coriander, and other crop seeds. The seed embryo is replaced by a cavity, and the seed coat and endosperm remain undamaged. Such seeds appear normal but fail to germinate.[3] *L. hesperus* and *L. elisus* cause pits on lima bean seeds where the initial injury appears as water-soaked areas around a small hole, which shrivels to give an irregular pit that enlarges as the seeds mature. Such seeds may not attain normal size and appear shriveled. Pitted areas vary from tiny, sunken pinpoints to large, irregular, crater-like yellow or brown spots in cotyledons. The cavities are filled with a brown, granular mass of necrotic cells and starch granules, the latter of which are particularly evident before the seeds dry.[28]

Insect injury to seeds often is followed by invasion of microorganisms or is accompanied by the introduction of a pathogen. Careful observation is required to determine the damage caused by the insect and by an associated pathogen.

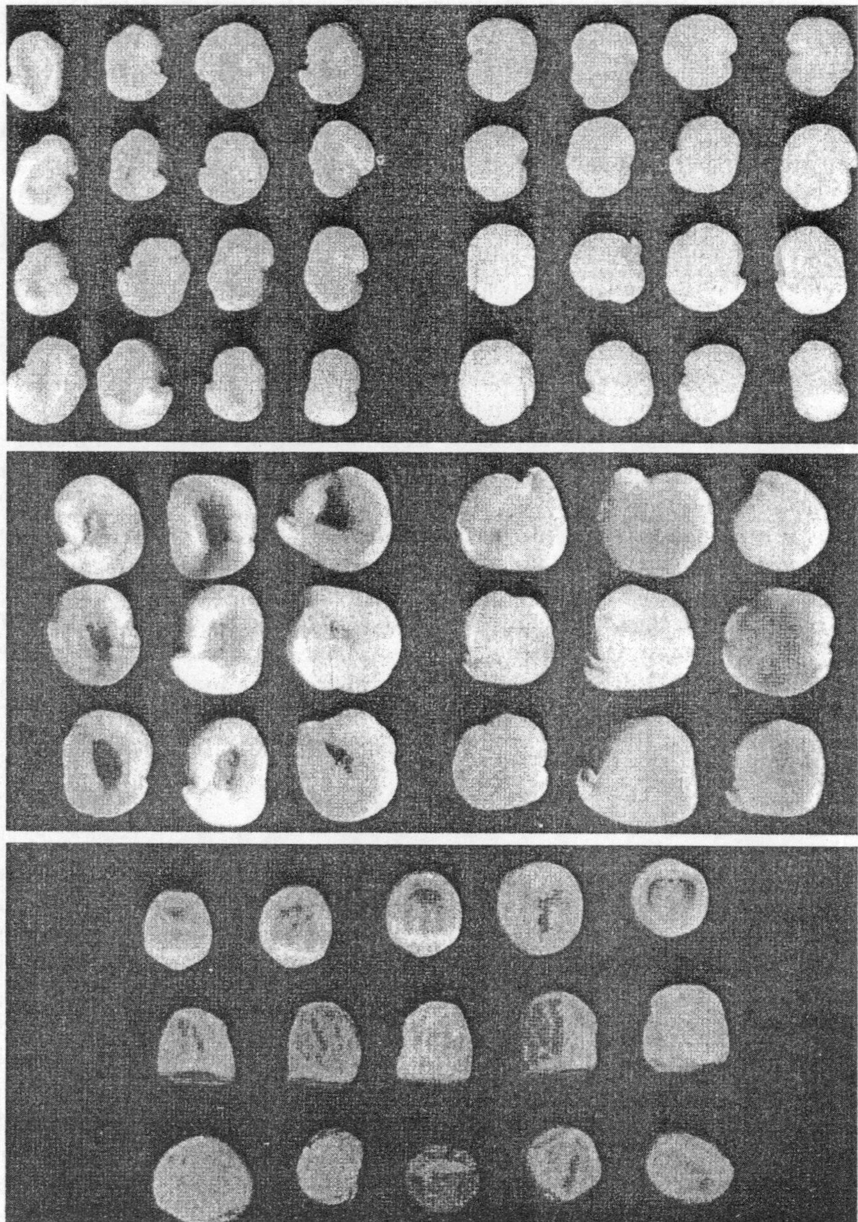

Figure 10-4 Hollow heart and marsh spot symptoms on pea (*Pisum sativum,*) seeds. *Top*: Split seeds of cv. Dark Skin Perfection showing hollow heart affected cotyledons (*left*) and nonaffected cotyledons (*right*); *middle*: split seeds of cv. Koroza showing marsh spot affected cotyledons (*left*) and nonaffected cotyledons (*right*); *bottom*: seeds of cv. Mingomark showing squarish (*top*), conical (*middle*), and irregular (*bottom*) shapes. (From Singh, D., *Seed Sci. Technol.*, 2, 443, 1974. With permission.)

REFERENCES

1. Kommedahl, T., Christensen, J. J., Culbertson, J. O., and Moore, M. B., The prevalence and importance of damaged seed flax, *Minn. Agric. Exp. Stn. Tech. Bull.*, 215, 1, 1955.
2. Baker, K. F., Seed transmission of *Rhizoctonia solani* in relation to control of seedling damping-off, *Phytopathology*, 37, 912, 1947.
3. Baker, K. F., Seed pathology, in *Seed Biology*, Vol. 2, Kozlowski, T. T., Ed., Academic Press, New York, 1972, 317.
4. Harlan, H. V. and Pope, M. N., Some cases of apparent single fertilization in barley, *Am. J. Bot.*, 12, 50, 1925.
5. Lyone, M. E., The occurrence and behavior of embryoless wheat seeds, *J. Agric. Res.*, 36, 631, 1928.
6. Crill, P., Burgis, D. S., Jones, J. P., and Strobel, J. W., The fruit pox and gold fleck syndromes of tomato, *Phytopathology*, 63, 1285, 1973.
7. Stevenson, W. R., Tigchelaar, E. C., and Jackson, A. O., Corky stunt: a genetic disease of tomato, *Phytopathology*, 66, 132, 1976.
8. Borthwick, H. A., Thresher injury in baby limabeans, *J. Agric. Res.*, 44, 503, 1932.
9. Neergaard, P., *Seed Pathology*, Vols. 1 and 2, Macmillan, London, 1977, 1187.
10. Pape, H. and Harle, A., Rapsschaden dürch Maifröste, *Mitt. Schweiz. Landwirtsch.*, 31, 3, 1943.
11. Stapel, C. and Bovien, P., Mark-froafgrodernes Sygdomme og Skadedyr, *Det Kgl. Danske-Landhusholdningasselskab*, Copenhagen, 1943, 234.
12. Jorgensen, J., Nogel undersogelser over ärsagerne til gräfarvningen af frø af gul sennep (*Sinapis alba*), *Statsføkontrollen København*, Bertning for det 96. Arbejdä fra 1 Juli 1966 til 30 Juni, 78, 1967.
13. Neergaard, P., *Danish species of Alternaria and Stemphylium*, Einar Munksgaard, Copenhagen, 1945, 560.
14. Moore, R. P., Wet, then dry, means poor soybean germination, *N.C. Agric. Exp. Res. Farming*, 18, 12, 1960.
15. Allen, J. D., Hollow heart of pea seed, *N.Z. J. Agric. Res.*, 4, 286, 1961.
16. Perry, D. A. and Harrison, J. G., Causes and development of hollow heart in pea seed, *Ann. Appl. Biol.*, 73, 95, 1973.
17. Singh, D., Occurrence and histology of hollow heart and marsh spot in peas, *Seed Sci. Technol.*, 2, 443, 1974.
18. Agarwal, V. K. and Gupta, R. K., Further reports of hollow heart in pea seeds, *Seed Res.*, 8, 75, 1980.
19. Don, R., Bustamente, L., Rennie, W. J., and Seddon, M. G., Hollow heart of pea (*Pisum sativum*), *Seed Sci. Technol.*, 12, 707, 1984.
20. Furneaux, B. S. and Glasscock, H. H., Soils in relation to marsh spot of pea seed, *J. Agric. Sci.*, 26, 59, 1936.
21. Perry, D. A. and Howell, P. J., Symptoms and nature of hollow heart in pea seed, *Plant Pathol.*, 14, 111, 1965.
22. Koopman, C., The influence of manganese sulfate spraying on marsh spot of Schokker peas, *Tijdschr. Plantenziekten*, 43, 64, 1937.
23. Lewis, A. H., Manganese deficiency in crops. I. Spraying pea crops with solutions of manganese salts to eliminate marsh spot, *Emp. J. Exp. Agric.*, 7, 150, 1939.
24. Ovinge, A., Marsh spot in Schokker peas. *Tijdschr. Plantenziekten*, 43, 67, 1937.
25. Hoffman, I. C., Potash starvation in the greenhouse, *Better Crops Plant Food*, 18, 10, 1933.

26. De Bruijn, H. L. G., Kwade harten van de erwten, *Tijdschr. Plantenziekten,* 39, 281, 1933.
27. Harris, H. D. and Gilman, R. L., Effect of boron on peanuts, *Soil Sci.*, 84, 233, 1957.
28. Baker, K. F., Snyder, W. C., and Holland, A. H., Lygus bug injury of limabean in California, *Phytopathology,* 36, 493, 1946.

CHAPTER 11

Deterioration of Seeds by Storage Fungi

The storage of seeds as grain and other staple food and feed free of fungal contamination is of importance to consumers of all nations. It has greater importance to nations that earn income from the export of agricultural commodities. An international association, the Group for Assistance on Systems Relating to Grain After-Harvest (GASGA), is a voluntary association linked with donor operations that promote the preservation of grain in storage. GASGA has the following objectives:[1]

1. To stimulate technical cooperation in all aspects of grain postharvest research and development in both developing and developed countries;
2. To identify and seek ways of meeting research and development training, and information needs in the postharvest subsector;
3. To serve an advisory role and provide a forum for exchange of technical information, experience and ideas for a global collaborative approach to fungal and mycotoxin problems.

The grain industry involves the transfer of seed and seed products from the farmer or producer to the processor and then on to the consumer. During this process, seeds are graded, priced, purchased, cleaned, stored, transported, marketed and shipped, preserving their quality. World losses during processing have been estimated to be as high as 30% annually in some developing countries, with an average in most countries between 8 to 15%.[2] Handling grain involves many operations that can cause damage to and loss of either the seeds or their usefulness. Damage caused by breakage of the protective outer seed layers can provide sites for fungal infection or insect infestation.[3] During storage interactions occur among insects, mites, fungi, and the stored seeds. All of these factors must be considered in maintaining an efficient operation.

I. FIELD AND STORAGE FUNGI

Seed deterioration may be initiated by fungi, bacteria, or yeasts. The role of bacteria and yeasts in seed deterioration has not been investigated extensively.

Fungi that attack seeds are classified into either field or storage fungi on the basis of their ecological requirements.[4-7] Field fungi invade seeds either during development or after maturity but before harvest. Generally, damage caused by field fungi occurs in the field, with little or no damage occurring during storage. For field fungi to cause damage to occur during storage, moisture content must be in equilibrium with relative humidities above 95%, which gives a seed moisture content of 24 to 25% on a wet weight basis in seeds of barley, maize, oat, soybean, and wheat. This moisture content is far above that at which these seeds are stored; hence field fungi rarely cause damage during storage.

As a group, storage fungi do not colonize actively metabolizing plant tissue but are confined to dying, dead, or naturally dried organic material. However, Mycock et al.[8] showed that the storage fungi were opportunistic invaders of coleoptiles and other seedling tissues via stomata and surface lesions. *Aspergillus flavus* var. *columnaris* invaded the internal tissues of emerging shoots via surface wounds caused by injury during germination of maize seeds in sand or vermiculite, through stomata, and by penetrating intact cuticle. Although the mycelium was concentrated in the space between the coleoptile and primary leaves, both tissues were infected. Thus the distinction between field fungi and storage fungi may not be clear-cut, since the latter may gain access to seeds prior to harvest.

Storage fungi are adapted to live and grow without free water, and some not only endure but require an environment without free water. Storage fungi are those that grow in seeds when moisture content is in equilibrium, with 65 to 90% relative humidities. Storage fungi normally do not play a role in disease development in the field but play a major role in seed deterioration in storage. Storage fungi mainly are species of *Aspergillus*: *A. amstelodami, A. candidus, A. chevalieri, A. flavus, A. repens, A. restrictus, A. ruber,* and *A. ochraceus*, and *Penicillium*,[4,6] which usually invades seeds with a moisture content above 16% and at low temperatures.[5] *P. viridicatum* is a common storage fungus that invades dent corn kernels stored at 19 to 24% moisture content between 8 to 24°C.[9] *Penicillium* is more common in temperate than in tropical regions. Monographs on *Aspergillus*[10] and *Penicillium*[11] deal with species identification. In addition, species of *Mucor* and *Rhizopus*,[12] *Absidia*,[13] *Byssochlamys*,[14] and *Candida* and *Hansenula*[15] have been reported on stored seeds.

II. INVASION BY STORAGE FUNGI

The storage fungi of seeds comprise mainly species of *Aspergillus* and *Penicillium.* Members of these groups normally do not infect seeds prior to storage but invade under conditions prevailing in storage (Figure 11-1). A storage fungus gains access to seed tissues through injuries to the pericarp or testa.[16] In addition,

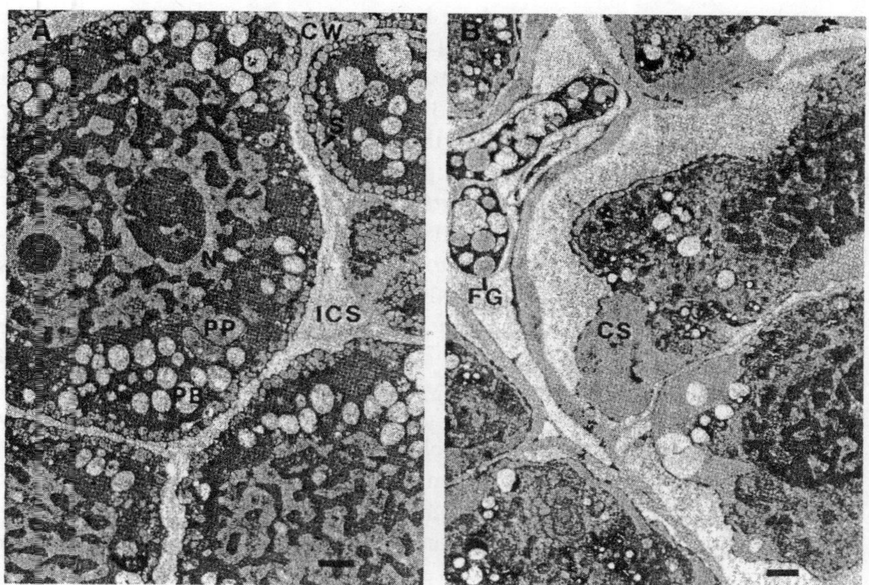

Figure 11-1 Electron micrographs of wheat (*Triticum aestivum*) embryo tissue of (**A**) control and (**B**) storage fungus infected seed showing cell wall (CW), nucleus (N), intercellular space (ICS), proplastids (P.P), protein bodies (PB), spherosomes (S), coalesced spherosomes (CS), and fungal hyphae (FG). Bars = 1μm. (From Anderson, J. D. and Baker, J. E., *Phytopathology*, 73, 321, 1983. With permission.)

maturing grain can be damaged on the parent plant by birds or insects, and fungi can readily infect these damaged tissues. It has become increasingly clear that species of the seed storage fungi can infect maturing seeds on the parent plant.[17-19]

In maize, the pericarp covers the entire seed and is discontinuous at the micropyle. This pore in the testa is surrounded by the peducle. Storage fungi are encountered in or on the caryopses, which are not externally damaged. *A. flavus* var. *columnaris* may gain access to seed tissues via the micropyle and peduncle or penetrate through microscopic cracks in the seed surface. The peduncle scar tissue is loose parenchyma, which allows access for advancing fungal mycelium.[8,20] Smart et al.[21] showed that *A. flavus* grew from a wound inoculation site to infest nonwounded spikelets in two ways in maize: (1) superficially over surfaces of the rachis and spikelets invading at the junction of the tracts with their rachilla, or (2) through the internal airspace of the rachis and spikelets. They described *A. flavus* development in maize ears and seeds. The fungus entered the rachilla of adjacent spikelets from the rachis and tracts, then grew through the rachilla parenchyma to the floral axis and innermost layers of the pericarp (the endocarp). The hyphae were inter- and intracellular in the floral axis and inside the seed. Entry into the seed from the endocarp was not across the black layer; random tears in the testa over the embryo were the immediate pathway. Invasion of maize seeds by *A. flavus* occurred via the silks. The fungus grew from the ear tip to the base by colonizing the silks, then the glumes. By the milk stage, *A. flavus* colonized seed surfaces but had rarely penetrated the cob pith.[22] *A. flavus* colonized external silk and the exposed tip of maize cobs and rapidly moved down into the ears, where it infected physiologically mature seeds.[23] *A. flavus* penetrated maize seeds through the pericarp[24] or silk.[25] Maize seeds were used as a food source by the storage fungi, probably utilizing the embryo and cotyledons as a nutrient supply. *A. flavus* var. *columnaris* produced highly active proteases and lipases in its extracellular exudate, and these facilitated breakdown and utilization of cell components of the embryo and scutellum.[26] Also, direct invasion by *A. flavus* of developing peanut pods in soil has been assumed the path for the eventual contamination of seeds with aflatoxin. However, *A. flavus* may have been present in the developing peanut ovary at the tip of the peg before it was pushed into the soil.[22] *A. flavus* may have entered cottonseeds via the vascular tissues. Hyphae were located throughout the cotyledon and in the nonlignified layers of the seed coat.[27] Cotton bolls infected by *A. flavus* displayed a bright greenish yellow fluorescence (BGYF) in the cotton fibers. This compound was formed when plant peroxidases reacted with the kojic acid synthesized by *A. flavus* hyphae and have been used to screen cottonseeds for field infection by *A. flavus*.[22] In cotton, *A. flavus* invaded bolls via the vascular system after inoculation of the nectaries and other natural openings.[28] In oil seeds, seed coats are made up of cellulose and pectin; hence the ability of storage fungi to secrete cellulolytic and pectinolytic enzymes becomes important for invasion and colonization of seed tissues. *A. candidus, A. fumigatus,* and *R. solani* produced cellulolytic enzymes in safflower and sesame seeds.[29] White-seeded peanuts were highly susceptible to *in vitro* colonization by *A. flavus*, but no correlation was found between seed coat color and seed rots and aflaroot diseases caused by *A. flavus*.[30]

A. flavus was found in endosperm, scutellum, and embryo of maize seeds.[31] The source of *A. flavus* inoculum for infection was not known, but increased levels of airborne conidia were associated with increased seed infection. Because *A. flavus* is predominantly soilborne, the soil was assumed to be the primary source of airborne inoculum. *A. flavus* also produces sporogenic sclerotia.[32] Drought conditions contribute to increased airborne conidia because the fungus survives longer in drier soils and is disseminated easily by windblown dry soil containing the conidia. Inoculations of maize at 17, 20, and 23 d after milk stage with *A. flavus* resulted in the severe ear rot.[33] The rate of infection and ultimate mycelial location of *A. oryzae, A. sydowii, A. chevalieri,* and *P. pinophilum* in maize seeds related to extracellular enzyme produced by the individual species.[34]

The systemic transmission of storage fungi can be one mode of seed infection. *A. flavus* var. *columnaris* from infected seeds was isolated from roots, stems, and leaves of plants produced from infected seeds at all stages during growth, as well as from male and female inflorescence, indicating that storage fungus can be systemically transmitted in maize.[35] Infection by *A. flavus* was greater on the basal than apical seeds of peanut.[36]

Table 11-1 Effects and Consequences of Fungal Activities on Stored Seeds

Gross effects	Type of damage	Consequence
Adverse quality changes	Dull appearance	
	Musty odors	Possible degrading
	Visible fungi	
	Reduced germination	Rejection for seed purposes
	Germ damage, discoloration	Downgrading
	Increased free fatty acids	Rejection for processing
	Promote mite growth	Possible degrading
Aggregation of seeds	Clogging of pipes, augers	Interruption of operations
	Sticking to bin walls	Uneven pressure effects, partial bin collapse
	Bridging of bin contents	Dangerous air spaces
	Aggregation and/or fusion of bin contents	Chipping out costs, unusable facilities
Extreme heating of seeds	Binburning	Damage to product and premises
		Possible degrading, rejection, extra costs
		Could lead to fireburning, explosions
Contamination of seeds by harmful substances	Mycotoxins	Livestock poisoning, feed refusal
		Rejection of shipments
		Loss of markets
		Chronic human health problems
	Respiratory/allergenic effects	Breathing problems in animals and man
		New employment of seed handlers may be required

Adapted from Mills, J. T., in Crop Science Society of America, 33rd Annu. Meeting, Ottawa, 1985, 38. With permission.

III. LOSSES

The role of storage fungi in grain deterioration has been underestimated. Storage fungi that grow in seeds during storage and transport cause a germination decrease, an increase in visible mold, discoloration, musty or sour odors, caking, and chemical and nutritional changes, reduced processing quality, and formation of mycotoxins. These changes reduce grade and price, contributing to customer dissatisfaction when the grain is marketed.[37] The major effects and consequences of fungal activities on stored seeds are presented in Table 11-1.[38]

A. Decrease in Germinability

A high, uniform germination is desired if seeds are to be used for planting or malting, or as edible sprouts. Decreased germinability can be caused by mechanical or physiological reasons or by storage fungi. If caused by storage fungi, the effect on germination is influenced by seed moisture content, storage period, storage temperature, and other factors.

Seed aging involves degradation that reduces seed germination and decreases seedling vigor, or results in death. These changes can be accelerated by microorganisms. Seed deterioration increases when moisture and temperature conditions are favorable for microorganisms. Seed aging involves a sequence of physiological and ultrastructural changes that occur independently of microorganism; however, pathogens produce metabolites that increase the deterioration rate.[39] The loss of stored seed viability results from aging or from infection by storage fungi. *A. ruber* was highly pathogenic to stored pea seeds. It produced a toxin that killed the embryonic root–shoot axes when its mycelium was primarily confined to the testa.[40] Mitochondria from pea tissues infected with *A. ruber* have been damaged.[41] The mitochondria from the axes of aged infected peas had a lower respiration rate than those from axes of noninfected peas; and the mitochondria respiration rate from infected peas was not stimulated by ADP (adenosine 5-diphosphate). *A. flavus* var. *columnaris* invaded internal tissues of emerging shoots via surface wounds caused by physical injury during germination in the sand or vermiculite, through stomata, or by penetrating the intact cuticle of maize.[20]

Storage fungi that invade seeds are classified generally as saprophytes, but experimental evidence has shown that most storage fungi invade seed embryos preferentially and reduce germination. Seed samples of barley, maize, peas, sorghum, soybean, and wheat stored at moisture contents and temperatures favorable for growth of storage fungi, but free of fungi, retained germinability of 95 to 100% for some months, whereas in similar samples inoculated with storage fungi, germinability was reduced to zero or near zero (Table 11-2).[42-47]

Rice seed samples stored at 20 and 30°C with a 14% moisture content decreased in germinability and increase in storage fungi (*A. candidus, A. glaucus,* and *A. restrictus*) in proportion to increasing storage time.[48] Sunflower seeds stored at moisture contents of 10, 12, and 14% and 3 to 5, 8 to 10, and 27 to 28°C and infested with *Alternaria, Aspergillus glaucus,* and *Penicillium* decreased in germinability in proportion to increasing moisture content, temperature, and storage time.[49] Maize seeds invaded with *A. flavus* at 19 to 20% moisture content and 20 to 25°C had 13% germination, whereas noninoculated seeds had 97% germination after 74 d.[47] Maize seeds free of storage fungi stored at 17% moisture content and 25°C retained a 98% germination after 12 weeks, whereas those inoculated with storage fungi germinated at 6%.[50] *A. candidus* and *Penicillium* reduced barley seed germination.[51,52] Sorghum seeds with a moisture content of 14.5% on a wet weight basis and invaded by storage fungi had a decreased germinability in proportion to the increased moisture content and storage period.[53] A higher recovery of *A. flavus* and *A. melleus* was correlated with decreased germination in soybean (Figure 11-2).[54,55] *P. oxalicum* reduced germination in maize seeds infected at the full-silk stage.[56] *A. glaucus* and *A. niger* reduced germination in soybean, with *A. niger* reducing it more than *A. glaucus*. An application of eight to ten spores per seed of *A. niger* was as effective in reducing germination as an application of 50 to 60 spores per seed of *A. glaucus*.[57]

Table 11-2 Loss of Seed Germinability Caused by Storage Fungi

Crop	Moisture content (% wet wt)	Temp. (°C)	Storage period	Treatment	Germination (%)	Ref.
Pisum sativum (pea)	Not given	30	6 months	Noninoculated	97	43
				Inoculated with storage fungi	0	
Triticum aestivum (wheat)	17.3–17.8	22–25	60–68 d	Noninoculated	100	44
				Inoculated with *Aspergillus ochraceus*	2	
	16–16.4	25	2 months	Noninoculated	90	45
				Inoculated with *A. amstelodami*, *A. candidus*, and *A. restrictus*	27	
Zea mays (maize)	17–18	15	2 years	Noninoculated	96	46
				Inoculated with storage fungi	0	
	19.1–19.9	27–32	74 d	Noninoculated	97	47
				Inoculated with *A. flavus*	13	

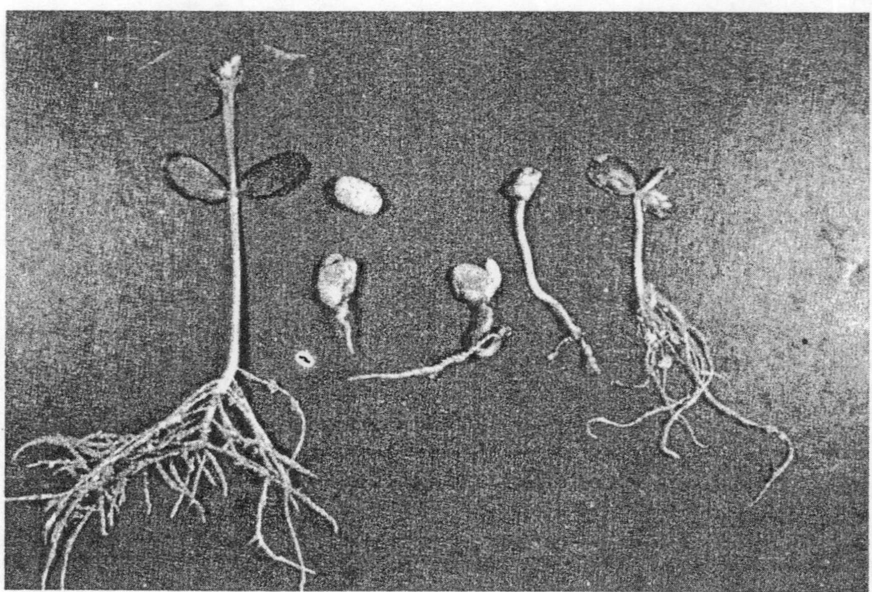

Figure 11-2 Reduced germination and vigor of soybean (*Glycine max*) seedlings from seeds infested with *Aspergillus melleus* (*right*), compared to a noninfected seedling (*left*). (From J. B. Sinclair, Ed., *Compendium of Soybean Diseases*, 2nd ed., APS Press, St. Paul, MN, 1982, 75. With permission.)

Ghosh and Nandi[58] recorded a reduction in germination of wheat seeds by storage fungi. The decrease in germination and seedling growth was due to diffusible toxic substances produced by these fungi in cultural filtrates. Damage to maize seedlings from *P. oxalicum*-infected seeds was expressed more in sterile sandy loan soil than in sand.[59] Inhibition of maize seedling growth and the occurrence of wilt increased as inoculum concentration of *P. oxalicium* increased.[59]

Reduced germination can vary depending upon the species of *Aspergillus* involved. Pea seeds stored at 85% relative humidity and 30°C were killed in 3 months when inoculated with *A. flavus*, in 6 months with *A. candidus* and *A. ruber,* and in 8 months with *A. restrictus,* whereas noninoculated seeds maintained a 97% germination for 6 months.[43]

Loss in germination due to storage fungi may be attributed to several factors. A toxin produced by *A. ruber* killed tissues in embryonic axes of pea seeds in advance of infection.[60] Wheat seeds infected with *Aspergillus*, when inbibed with water, became a jelly-like mass, suggesting that cell wall-degrading enzymes were involved. In contrast, pea and squash embryos were killed without physical invasion by fungi, indicating the involvement of diffusible toxins.[61] Mitochondria isolated from the embryonic axes of *A. ruber*-infected pea seeds were less active than those from noninfected seeds, suggesting that mitochrondia damage by *A. ruber* played a role in seed deterioration.[62]

B. Discoloration and Shrinkage of the Grains

Aspergillus candidus, especially *A. restrictus,* and *A. glaucus,* to a lesser extent, can cause jet black discoloration in wheat seeds. Extensive invasion may result in discoloration of the whole seed. Such seeds produce musty odors.[6] The official U.S. standards for malting barley permit 4% damaged kernels. Damaged kernels are defined as "mold-damaged kernels (major): Kernels and pieces of kernels of barley which are weathered and contain considerable evidence of molds," and "mold-damaged kernels (minor): Kernels and pieces of kernels of barley containing slight evidence of mold."[63]

Aspergillus and *Penicillium* cause discoloration and shrinkage of barley seeds.[52] Maize artificially inoculated with *A. candidus* and stored at 18% moisture content and at 25°C developed a dark brown discoloration at the germ ends after 4.5 months.[64] Wheat seeds inoculated with *A. candidus* with 16.0 to 16.4% moisture content and stored for 3 months at 25°C showed 79% seeds with dark germ ends and 6% germination compared to 95% germination of noninoculated seeds.[45]

C. Heating

A seed is a living organism that respires. During respiration, oxygen and starch are converted to carbon dioxide, water, and energy. An increase in storage temperature increases the respiration rate. Nutrients being respirated lead to loss in weight and quality. Seeds also contain moisture. A high moisture content is conducive to infestation with fungi. Cereal and legume seeds have a low heat conductivity, so that the effects of temperature fluctuations are noticeable over short distances or long periods. This results in heat accumulation in pockets.[65] Heat damage also may be due to fungal activity and not heat, although both factors are present when stored grain undergoes heating due to fungal growth. Heat damage is determined by the visual evaluation of seed discoloration and is a component of the grading standards used by state and federal grain inspection services. For example, sunflower seeds have maximum limits of 0.5% for grade No. 1 and over 1% for sample grade. According to the Minnesota Department of Agriculture, Grain Inspection Division, "heat damaged sunflower seed shall be seed and pieces of seed, which, when sliced open, show evidence for meats that have been discolored by heat."[66] However, Robertson et al.[67] reported that heat damage scores do not always accurately reflect sunflower seed and oil quality. A more quantitative evaluation, such as percent free-fatty acids (FFA), should be used in conjunction with the more traditional criteria for grading sunflower seeds. Several samples of No. 1 seeds with little or no heat damage were unacceptable because of high FFA values from overheating during drying.[67] *Aspergillus candidus* and *A. flavus* colonization can result in heating up to 55°C and may be maintained for several weeks. The resulting hot spots may then "burn themselves out," and thermophilic bacteria and fungi may take over and raise the temperature to 75°C. At this temperature, short-chain hydrocarbons may be produced and, when exposed to air, oxidize rapidly and result in fire. This heating to ignition occurs more commonly in maize and soybean, resulting in destruction of seeds and storage structures.[6]

The temperature of stored seeds as grain always increases due to seed respiration. In addition, respiration by insects and storage fungi also can increase storage temperature. The rate of respiration and thus the rise in temperature are higher in seeds invaded by storage fungi. This has been shown for soybeans[68] and wheat.[69,70] A temperature increase from 50 to 55°C and a parallel increase in respiration were associated with proliferation of *A. flavus* and *A. glaucus* in soybeans.[71] Temperature increases accompanied by increased water content in barley seeds from 15 to 24% favored development of storage fungi.[52]

It is difficult to separate increases in temperature due to storage fungi and insects. In temperate zones or tropics, temperature increases are due to fungi, which develop in moist seeds, but in the dry tropics, insect-induced dry-grain heating is possible.[71] Temperatures may rise to above 60°C when induced by fungi, whereas insect-induced hot spots never exceed 46°C. Hot zones generally rise vertically through the bulk when induced by fungi and expand laterally when induced by insects because of insect mobility.[72]

D. Spoilage in Nutritive Value

Storage fungi use stored seeds as a substrate, resulting in the chemical breakdown of nutrients.

1. Increase in Fatty Acid Value (FAV)

Seed contain two major lipids, functional and storage. Storage lipids primarily are neutral triglycerides. Functional lipids can be grouped in several classes, glycolipids, phospholipids, sterols, sterol esters, sterol ester glucosides, etc., and are present in membranes, subcellular organelles, and other compartmentalized structures. When seeds are damaged by improper storage conditions or are exposed to certain microorganisms, lipid degradation occurs. Lipase and lipoxygenase are the principal enzymes involved in lipid degradation in seeds.[73] Lipase may arise through the activation of preexisting materials, de novo synthesis, or through production by microorganisms. In each case, the mechanism is similar and the end result is the same — an increase in FFA. Lipoxygenase oxidizes polyunsaturated fatty acids and esters to hydroperoxides, which degrade to aldehydes, acids, ketones, and other low molecular weight compounds. Several of these compounds cause off-flavor and off-odor in stored seeds.[73] FFA is an important parameter of oilseed quality. According to the National Cottonseed Products Association Trading rules, grade No.1 sunflower seeds shall contain not more than 1.8% FFA in the oil and seeds, and seeds containing over 3% FFA shall be sample grade and subject to rejection.[67]

Seed deterioration is accompanied by an increase in the fatty acid value (FAV), and the FAV level in any given sample provides an indication of the stage of deterioration and seed storability. When the FAV increases, the flavor and odor of fatty acids make the seeds rancid. The level of FFA produced varies depending upon species and isolates within a species of storage fungi.

An increase in FFA was correlated with the presence of storage fungi in maize[74] and wheat.[75] The FAV of sound maize seeds was low but increased with increasing levels of damaged kernels[76] and invasion by fungi.[77,78] The increase in FAV may be due to the action of seed lipases or microflora on the lipids in seeds.[78] FAV increased more rapidly in broken maize seeds inoculated with *A. glaucus* and *A. restrictus* stored at 15.5% moisture content and 30°C than in broken seeds free of fungi and similarly stored.[78,79]

2. Biochemical Changes in Nutritional Value

Seeds contain a high concentration of nutrients and are easily storable due to low moisture content. Seeds contain a large amount of oil, proteins, and vitamins. In some seeds, the endosperm, which constitutes the nutritional reserves for the embryo, largely consists of starch. Legumes do not have endosperm. Instead, the cotyledons are developed to a thick and fleshy nutritive tissue.[65] Most seed proteins are stored within protein bodies. During imbibition and germination, proteins are hydrolyzed to polypeptides and free amino acids by peptidases and proteases and translocated to the developing embryo.[80] Microorganisms grow luxuriantly on seeds and decompose proteins to low molecular weight components, resulting in the formation of free amino acids. In addition, there is a depletion of some enzymes and an intensification of others and/or production of new multiple molecular forms of enzymes. Many of these enzymes in seeds remain active during the infection period. Thus the biochemical mechanisms operating in the saprophyte–seed interaction vary efficiently and systemically enhance fungal growth at the expense of the seed reserves.[81] Maximum endopolygalucturonase (endo-PG) was produced by *A. flavus*, followed by *R. solani*. *A. flavus* is the best lipase producer, followed by *A. fumigatus* and *R. solani*. Reduction in germinability and oil content and increase in FAV were more pronounced in seeds inoculated with *A. flavus, A. fumigatus*, or *R. solani* than in those with *A. candidus* and *A. sydowii*.[82] Endo-PG, which randomly cleaves the α-1,4-glucoside bond between the galacturonosyl moieties in the polymer of galacturonic acid, was considered one of the primary enzymes for maceration.[29]

Peanut cotyledonary tissue contains 25 to 30% protein (arachin, conarachin, and albumins). Arachin and conarachin account for 78% of peanut seed proteins.[83] Cultivar differences in peanut seed polypeptide composition are known, and electrophoretic patterns of peanut proteins are complex. Seed-invading microorganisms change these patterns, because protein molecules are altered by microbial enzymatic action and new proteins of microbial origin are synthesized. Such a situation occurs when the aflatoxin producing or nontoxigenic strains of *Aspergillus* colonize shells and seeds. These changes have been detected electrophoretically at 2 to 3 d after inoculation of seeds with *Aspergillus*.[81] Novel polypeptides were detected in viable peanut cotyledons during early stages of infection by *Aspergillus*.[84] *A. niger* and *A. flavus* caused spoilage of stored peanuts and reduced quality. There was loss in oil content, which was acrid and colored. There was loss in unsaturated fatty acid content, namely linoleic, linolenic, and oleic acids.[85] Healthy Bengal gram seeds contained 14 amino acids, whereas 12 were in *A. niger*-infected seeds. Proline and dihydroxyphenyl alanine were absent. The remaining amino acids were in lower content, indicating deterioration in seed quality.[86] *A. flavus* and *A. alternata* caused a large increase in oil content and iodine at high relative humidity but saponification and FFA contents decreased in sunflower.[87]

Oxygen uptake of radish seeds increased due to seedborne *A. flavus* with an extended storage period to 20 d and increased relative humidity. The activity of starch phosphorylase, fructose diphosphate aldolase, pyruvic, α-ketoglutaric, succinic acid dehydrogenase, peroxidase, ascorbic acid oxidase, catalase, and ATPase was enhanced, in addition to an increase in pyruvic acid and total keto acid but a decrease in α-ketoglutaric acid. The enhanced activity of all the enzymes indicated an increase in respiratory metabolism of the seeds due to *A. flavus*. This respiration enhancement was at the cost of seed nutrient, causing proportional loss in dry weight.[88]

There was a continuous loss in protein content of cowpea seeds infested by *A. flavus*.[89] This may have been due to hydrolysis of seed protein by hydrolytic enzymes, as was reported for peanuts infected by *A. parasiticus*.[81]

E. Production of Toxins

Storage fungi may produce mycotoxins that are injurious to man and animals on consumption.

IV. CONDITIONS FAVORING STORAGE FUNGI DEVELOPMENT

High temperature, high relative humidity, and moisture content of stored products are favorable for the development of pest microorganisms. Respiration of pests and the stored produce releases heat and moisture, which further improves conditions, leading to an increase in pest populations. Rainfall, ground moisture, and a drop in temperature increase relative humidity. Rain water and ground moisture may be absorbed directly by the seeds. High relative humidity leads to a rise in the moisture content of stored product and under certain conditions to condensation. If no measures are taken to counteract this, considerable losses are likely to occur. Good storage hygiene, controlled ventilation, pest control, and drying of the produce can maintain the quality of the stored product.[65]

The growth rate at different active water (a_w) and water/temperature relationships for storage fungi has been summarized.[65] When temperature and relative humidity were high, *A. flavus* colonized silks and invaded developing maize seeds. Exposed silks were susceptible to colonization by the fungus and provided a suitable infection court for entry into intact seeds.[25] The incidence of *A. niger* was higher on onion seeds produced in temperate than in hot climates.[90] Ramakrishna et al.[91] showed the effect of water activity and temperature on spore germination and the initial growth and interaction of fungi (*A. flavus, F. poae, Hyphopichia burtonii*, and *P. verrucosum*) on seed surfaces.

The invasion and successful establishment of and seed deterioration by storage fungi are influenced by a number of factors, which may act alone or in combination.

A. Moisture Content

Water activity (a_w) quantifies the relationship between moisture in foods and the ability of microorganisms to grow on them.[65] The water activity is defined as a ratio:

$$a_w = P/P_o$$

where P = the partial pressure of water vapor in the test material, and P_0 = the saturation vapor pressure for pure water under the same conditions. Water activity is numerically equal to equilibrium relative humidity (ERH) expressed as a decimal. If a food sample is held at a constant temperature in a sealed enclosure until the water in the sample reaches equilibrium with the water vapor in the surrounding air, then

$$a_w \text{ (food)} = \text{ERH (air)}/100$$

Fungi require water for growth, and the tolerance of low a_w by different classes of fungi often is sharply defined. Certain storage fungi that occur in dried foods have the ability to grow at lower a_w levels than other organisms. In many practical situations, a_w is the dominant environmental factor controlling growth of fungi and hence determine the stability of stored products. A knowledge of fungal water relations enables the prediction of storage life of commodities, and a knowledge of the water relations of mycotoxin production will assist in understanding the potential for mycotoxin formation. The degree of tolerance to low a_w is simply expressed in terms of the minimum a_w at which fungal spore germination and hyphal growth can occur. Fungi able to grow at a low a_w are termed *xerophiles* (dryness loving). Commodities stored at humidities between 75 and 85% ERH are susceptible to attack by xerophilic fungi.[92] Seed infection levels are affected significantly by water potentials on the day of anthesis and inoculation of cottonseed with the highest infection level in seed from plants with water potentials between –1.6 and 1.9 MPa.[93] At 75% relative humidity *Aspergillus* and *Penicillium* growth was extensive, and FAV increased rapidly in sunflower seeds stored at 10, 20, 30, and 40°C. Seed germination decreased rapidly at 40°C, and insect infestation increased during storage.[94]

Moisture content of seeds prior to storage is the most important factor for establishment, development, and growth of storage fungi during storage. Various storage fungi have different moisture content requirements below which they fail to develop or may remain dormant with seeds without causing any damage (Table 11-3).[95,96] The role of moisture content in grain invasion has been reviewed.[4,5]

Table 11-3 Moisture Content Necessary for Growth of Storage Fungi[96]

Fungus	Crop		
	Glycine max (soybean)	*Sorghum bicolor* (sorghum)	*Triticum aestivum* (wheat); *Zea mays* (maize)
Aspergillus restrictus	12.0–12.5	14.0–14.5	13.5–14.5
A. glaucus	12.5–13.0	14.0–15.0	14.0–14.5
A. candidus	14.5–15.0	16.0–16.5	15.0–15.5
A. ochraceus	14.5–15.0	16.0–16.5	15.0–15.5
A. flavus	17.0–17.5	19.0–19.5	18.0–18.5
Penicillium	16.0–18.5	17.0–19.5	16.5–19.0

Generally, storage fungi grow at moisture contents in equilibrium with relative humidities of 65 to 90%. The moisture content of some common seeds in equilibrium with relative humidities within this range are presented in Table 11-4.[4,6,42] Each species within a genus has a lower limit of relative humidity below which it will not grow (Table 11-5).[42] The upper relative humidity limit is determined in part by the nature of the fungus and in part by competition. The quantity of physically bound "free water" within cereals generally determines the storability of seeds.

A moisture content lower than 13% retards development of all types of microorganisms in seeds, and below 10% insects fail to develop. The moisture content within a seed lot may be different in different pockets and also may change from season to season.[97] The lower limit of moisture content that permits invasion of sunflower seeds by *Aspergillus glaucus* and *A. restrictus* is about 6%, and at moisture contents below 6.5%, invasion is very low.[98] *A. flavus* does not invade maize seeds stored with moisture contents below 17.5% wet weight basis but does invade at 18.5% and above. At 18.5% moisture content, 25 and 35°C favor invasion by *A. ochraceus* and *A. candidus,* respectively.[47]

Table 11-4 Moisture Content, Wet Weight Basis of Common Grains and Seeds at Equilibrium with Relative Humidities of 65 to 85% at 20 to 25°C

Relative humidity (%)	Crop				
	Glycine max (soybean) (%)	*Helianthus annuus* (sunflower) anchenes (%)	Oryza sativa Rough (%)	Oryza sativa Polished (%)	*Triticum aestivum* (wheat); *Zea mays* (maize) (%)
65	12.5	8.0	12.5	14.0	12.5–13.5
70	13.0	9.0	13.5	15.0	13.5–14.0
75	14.0	10.0	14.0	15.5	14.5–15.0
80	16.0	11.0	15.0	16.5	16.0–16.5
85	18.0	13.0	16.5	17.5	18.0–18.5

Note: The figures are approximate; the equilibrium moisture content of a given kind of seed at a given relative humidity will vary with several factors.

Note: Oil type; the confectionary types have equilibrium moisture contents about 1% higher.

From Christensen, C. M. and Kaufmann, H. H., *Annu. Rev. Phytopathol.*, 3, 69, 1965, and Christensen, C. M., *Seed Sci. Technol.*, 1, 547, 1973. With permission.

Table 11-5 Minimum Relative Humidity that Permits Growth of Various Storage Fungi

Fungus	Min. relative humidity (%)
Aspergillus halophilicus	65
A. restrictus	70
A. repens	73
A. candidus, A. ochraceus	80
A. flavus	85
Pencillium spp.	85–95

From Christensen, C. M., *Seed Sci. Technol.*, 1, 547, 1973. With permission.

An increase in moisture content from 15 to 24% and temperature from 20 to 30°C in barley results in an increase of *Aspergillus* and *Penicillium* and decreased germination.[52]

B. Temperature

Storability and development of storage fungi within seeds are influenced by atmospheric, seed, and intergranular air temperature. Atmospheric heat slowly enters the seed bulk. In addition, heat produced by fungi, insects, and other organisms is considerably higher than the heat produced by seeds. Most storage fungi do not develop below 0°C, mites below 5°C, and insects below 15°C. The minimun, optimum, and maximum temperature requirement for the growth of most storage fungi is 0 to 5°C, 30 to 33°C, and 50 to 55°C, respectively. Low temperature can be used as a substitute for control of high moisture content because at temperatures below 10°C storage fungi that invade seeds at moisture contents in equilibrium with relative humidities up to 85% will grow very slowly.[42] Some *Penicillia* grow slowly at –5°C when the moisture content is in equilibrium with a relative humidity of 90% or more.[42] Papavizas and Christensen[99] found that wheat seed with a moisture content of up to 10% can be stored without deterioration for 1 year at 10°C or below, and with a moisture content of up to 18% for 19 months at –5°C. Rice seeds with an initial moisture content of 12 to 14% were stored without deterioration for 465 d at 5 and 15°C.[48] Soybean seeds stored at 15°C germinated above 95% after 24 weeks at 12.1 to 16.5% moisture content.[100] The optimum temperature for growth of *Aspergillus flavus* and *A. niger* on agar was 30 and 35°C and the optimum for *A. fumigatus* was 40°C.[101]

Although temperature is an important factor in fungal growth, commodities usually are stored under conditions suitable for fungal growth. Temperatures below 20°C tend to favor cold-tolerant fungi, such as *Cladosporium* and *Penicillium*, while higher storage temperatures favor *Aspergillus*. Under tropical conditions, stored products are more susceptible to *Aspergillus* than other fungi, since *Aspergilli* are favored by a combination of low a_w and relatively high storage temperatures.[102] Parasitic ability of *A. flavus* is enhanced at high temperature (30 to 40°C) in maize.[103]

C. Gas Tension

Both reduction in oxygen tension and increase in carbon dioxide concentrations can affect growth of fungi. These factors are important in the storage of commodities, where such conditions are used primarily for insect control.[102]

D. Physical Damage of the Seed

Poor harvest and postharvest conditions, such as mechanical damage during lifting, slow drying, improper storage facilities, and rewetting of pods, are ideal for *A. flavus* invasion.[22]

Damaged or injured seeds are more susceptible than whole seeds to invasion by storage fungi and to deterioration by these fungi after invasion.[104] Care in threshing procedures to eliminate or reduce seed damage helps preserve germinability in wheat seeds stored under moist conditions. In studies, germinability of machine-threshed seeds decreased faster and resulted in fewer living seeds at 80, 85, and 90% relative humidity than hand-threshed seeds.[105] A high level of injury on barley seeds reduced germination faster than a low level of injury.[52] Loss of germinability in damaged maize seeds was greater than in nongerminated seeds.[46] Loss of germination in damaged wheat seeds inoculated with *Penicillium* spores was greater than in undamaged seeds.[105]

E. Degree of Seed Infestation/Invasion Prior to Storage

Seeds infected in the field are likely to deteriorate faster because storage fungi continue to develop at lower moisture and temperature conditions.[42] Maize seeds invaded by storage fungi prior to storage deteriorate more rapidly under conditions favorable to the fungi than seeds free of storage fungi.[46]

Methods developed for in-storage drying in temperate regions have been successful. However, the microbiological status of seeds in the system needs assessment, as commodities that carry high microbial loads can deteriorate rapidly, particularly in humid tropical climates. However, if loading is delayed, the microbiological quality of rice is affected even in the dry season. A 100 times increase in *A. flavus* was detected in the dry season, when it took 7 d to fill a 750 ton storage bin in Malaysia.[106] Seeds treated with 1% Luprosil (99.9% propionic acid) 3 months after treatment yielded no infection by *A. flavus* and *P. cyclopium.*[107]

F. Admixtures with the Seed

Any admixture with seeds, such as plant parts, broken seeds, weed seeds, soil, or field insects such as grasshoppers and crickets, can serve as a substrate for growth of storage fungi or insects. Admixtures usually are more moisture absorbent and retentive, and thus more susceptible to storage fungi colonization. Weed seeds may have a higher moisture content than cultivated crop seeds.[42]

G. Length of Storage

Length of storage influences storage fungi. The chances of damage are less for seeds stored for a few weeks or months compared to seeds stored for years at a given moisture or temperature safe for storage. Moisture content and temperature are likely to change due to metabolic activities of storage fungi and insects; hence it becomes essential to periodically check the condition of stored seeds.

H. Insect and Mite Infestation

Infestation of grain by insects and mites can accelerate deterioration by storage fungi. Insects and mites can carry fungal spore, thus infesting previously noninfested seeds. In addition, they increase the moisture content of seeds through the release of water from their digestive processes.[4,42,108] Respiration releases carbon dioxide and water. Cereal seeds infested by the granary weevil, *Sitophilus granarius*, and stored for 3 months at 75% relative humidity showed an increase in moisture content from 12.1% to between 17.6 and 23.0%, compared to 14.6 and 14.8% in weevil-free seeds.[109] The increase in storage fungi was correlated with insect infestation.[110,111] The mites, *Acarus siro* and *Tyrophagus castellanii,* carried *Aspergillus glaucus* spores on their bodies and in their digestive tract and feces.[112] An increase in moisture content by the gram moth, *Sitotroga cerealella*, was responsible for increased population of *A. amstelodami, A. repens,* and *A. ruber.*[111] Bottles of wheat seeds infested with *Aeroglyphus robustus* mites had higher levels of *A. glaucus* infection than did those containing *Lepidoglyphus destructor.*[113] The incidence of the *A. flavus* group and aflatoxins was higher in maize seeds damaged by insects (*Sitophilus oryzae* and *Tribolium castaneum*) than in nondamaged samples.[114] Infection and contamination by *A. flavus* of noninjured peanut seeds was enhanced by injury to peanut pods by the lesser cornstalk borer, *Elasmopalus lignosellus*.[115]

Some insect pests such as the rice weevil *(Sitophilus oryzae)* develop and feed inside seeds consuming the entire seed, while some species such as the red flour beetle feed on the germ, thus reducing germination.[116] *S. oryzae* may transmit *A. flavus* from infected to healthy rice seeds.[117] Maize seed infection by *A. flavus* and *A. niger* is promoted by insect wounding of seeds.[118] *Sitophilus oryzae* in maize and *Rhyzopertha dominica* in peanut carry seedborne fungal spores, and such infestation escalates the process of seed deterioration in storage.[119]

Mills[120] developed classifications of diverse insect interactions influencing seed deterioration based on preharvest, harvest, postharvest, and postplanting. Preharvest interactions were divided into anthesis and seed enlargement phases. The interactions were classified further into several categories based on type of insect and on whether the insect that transports the fungal pathogen to the host damages the host by feeding with mandibulate or sucking mouth parts or participates in a symbiotic relationship (Figure 11-3). Insects transport and provide entry for fungi into seeds through oviposition and feeding wounds. *A. flavus* is

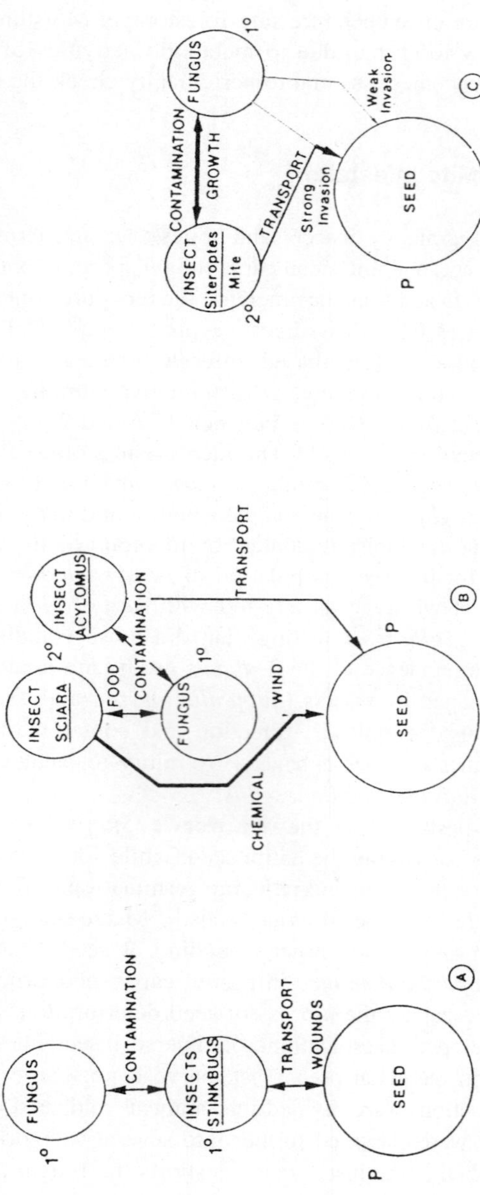

Figure 11-3 Schemas showing examples of ecological relationships among insects, deteriorative fungi, and seeds: **(A)** *Nematospora* on soybean; **(B)** *Claviceps* on rye; **(C)** *Nigrospora* on cotton; **(D)** *Aspergillus flavus* on cotton; **(E)** *Pythium ultimum* on squash; **(F)** *Phytophthora* on cacao; **(G)** *Aspergillus* spp. on wheat. Arrows → = strong, → = direct, and → = indirect insect–fungus influences on seed deterioration. The symbols P, 1°, and 2° indicate producer and primary and secondary consumers, respectively. (From Mills, J. T., *Phytopathology*, 73, 330, 1983. With permission.)

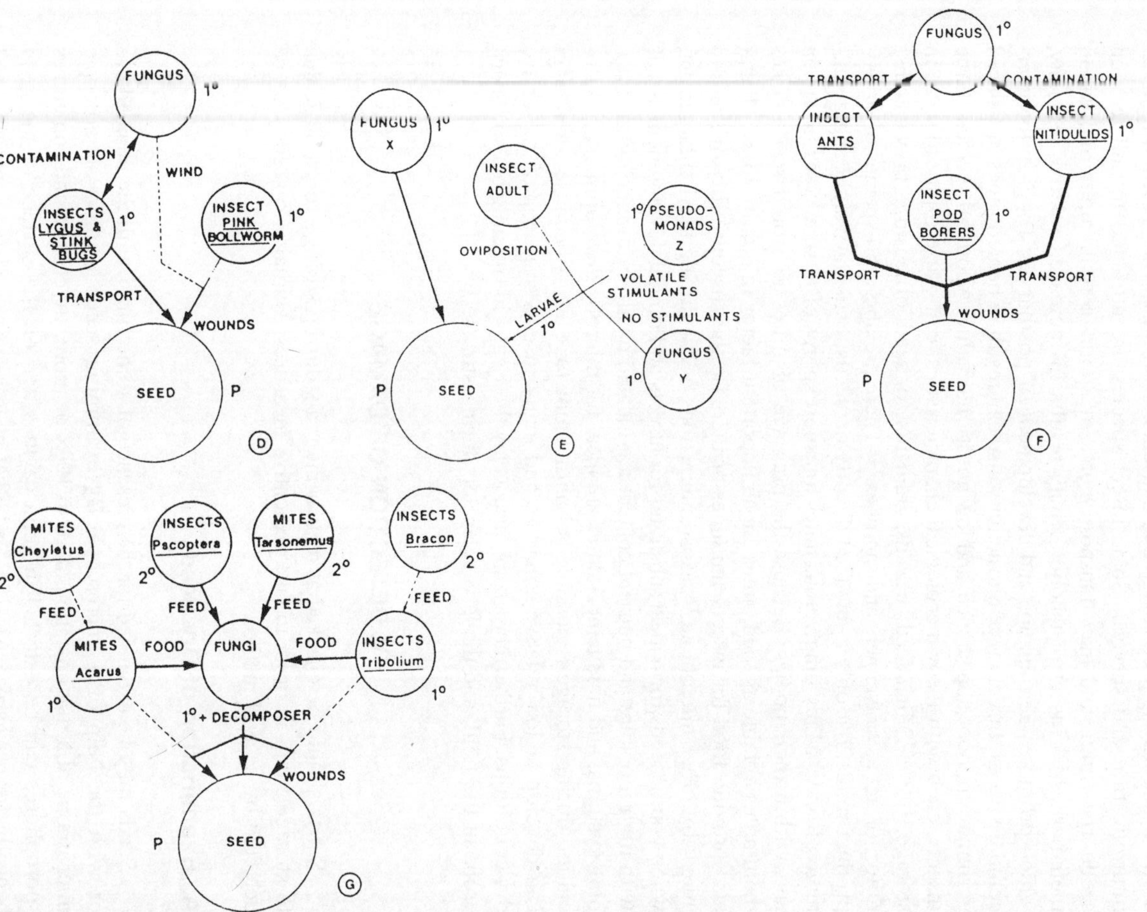

Figure 11-3 *Continued.*

soilborne and wind disseminated and is transported externally and internally by lygus and stinkbugs, which frequently visit cotton bolls.[121] Access to bolls is via exit tunnels made by mature larvae of the pinkboll worm and possibly via oviposition wounds made by other insects. Insects feed on intact or damaged seeds, transforming seed energy and nutrients to biomass, thus permitting insect multiplication and survival. Some insect species are attracted to and feed on seedborne fungi, and may transport spores of spoilage fungi upon and within their bodies.[122] Insect metabolic activity results in increased relative humidity, which promotes germination and development of postharvest fungi in the intercellular spaces. Insect metabolic activity also can result in heated seeds of reduced germination, poor appearance, and low nutritional value. About 50 species of true insects, each with its own ecological characteristics, can cause injury to seeds in storage.[116,120,123] Mites feed on seeds, grain dust, or seedborne fungi, prey on other mites, or act as scavengers and saprobes.[124] There are about 50 species of stored-product mites, but only a few feed directly on seeds, causing weight and germination loss.[125] Many fungivorous mites transport spores of postharvest fungi upon and within their bodies. Because the presence of mites in stored seeds makes them unacceptable as food, mites are an important factor in quality loss.[120] Fungi also provide food for insects and mites or increase the susceptibility of seeds to attack by other microbiota.[120] Most interactions occur during seed enlargement, when young seeds form ideal substrates for insects and fungal development. Seeds with high moisture and nutrient content are located easily in monocultural cropping systems, and field temperatures are ideal for microbe development.[112] Association during storage is possible at limiting moisture levels because some fungi can develop at low relative humidity, and many insects are adapted to survive dry conditions. Furthermore, some insects lay their eggs in ripening but still moist crops in the field, and the larvae develop during storage.[116]

V. DETECTION OF DAMAGE

Seeds should be checked at regular intervals during storage for damage due to storage fungi in order to take preventive measures. Damage to seeds may be detected by several methods.

A. Examination of Dry Seeds

Seeds may be examined for the presence of some fungi with the unaided eye or with the help of a microscope. By examining seeds using a stereoscopic microscope at a magnification of 10 to 100× or more, mycelium and sometimes sporophores can be detected in germ cavities and on the surface of the germ beneath the pericarps.[6] Also, in seeds that have been stored for some time and are beginning to be invaded by storage fungi, mycelium and sometimes microsclerotia of species of *Aspergillus* and other fungi are common in the inner layer of cells of the pericarp.[6]

Seed damage by the invasion of storage fungi shows one or more of the following symptoms: weakening of the embryo, embryo death, discoloration of the embryo or seed, mustiness, and/or complete decay.[39] Discoloration of the germ may be detected by removing the pericarp. Discoloration may be complete or restricted to the tip of the seed. Such germs are likely to be moldy and later turn dark brown.

B. Isolation of Fungi

The common method used for routine examination of seeds is plating surface-disinfected seeds on either potato-dextrose agar (PDA) or acidified potato-dextrose agar (A-PDA) for growth and identification of fungi.[6] Seeds usually are surface disinfected with a 2% NaOCl. This mild disinfectant may kill mycelium within cells of the outer pericarp layers, but it does not eliminate contaminating spores on the outside of the seeds.[6] When mercury was used as a surface disinfectant, it was adsorbed on outer tissues of seeds, so it disinfected and inhibited growth of some fungi present in the seeds. However, if the mercury was precipitated to an insoluble form by a 5-min dip in M/5 solution of sodium thiosulfate, other fungi in the seeds, even if sensitive to mercury, could grow.[126]

For particular foods, including stored products, direct rather than dilution plating of samples was considered effective. In direct plating, seeds are placed on suitable agar after surface sterilization in 10% chlorine bleach for 2 min. Such media are universally of high water content and thus are suited for detection of fungi associated with growing plants, but not of xerophilic fungi associated with stored products. A major problem in the detection of fungi from stored products is the presence of rapidly growing fungi that overgrow culture plates, rendering counting and isolation of important fungi difficult or impossible. The incorporation of the fungicide dichloran in the medium inhibits such growth.[102,127] Media commonly used for detection of storage fungi are presented.

1. *Dichloran Rose Bengal Chloramphenicol Agar (DRBC)*

This medium has sufficient nutrients for growth of many storage fungi, including species of *Alternaria, Aspergillus, Curvularia, Fusarium,* and *Penicillium,* and all major mycotoxin-producing fungi. It consists of 10 g glucose, 5 g peptone, 1 g KH_2PO_4, 0.5 g $MgSO_4 \cdot 7H_2O$, 25 mg Rose Bengal (5% in water, 0.5 ml), 2 mg dichloran (0.2% in ethanol, 1 ml), 0.1 g chloramphenicol, 15 g agar in 1 L distilled water. After the addition of all ingredients, the medium is sterilized by autoclaving for 15 min at 121°C. It is stored away from light, which causes slow decomposition of Rose Bengal. Chloramphenicol is not affected by autoclaving and has long-term stability.[128]

2. *Dichloran 18% Glycerol Agar (DG18)*

DG18 has lower water activity than DRBC and was developed for the isolation and enumeration of xerophiles.[102,127] It consists of 10 g glucose, 5 g peptone,

1 g KH_2PO_4, 0.5 g $MgSO_4 \cdot 7H_2O$, 220 g glycerol, 2 mg dichloran (0.2% in ethanol, 1.0 ml), 0.1 g chloramphenicol, 15 g agar in 1 L distilled water. Minor ingredients and agar are added to 800 ml distilled water and steamed to dissolve the agar. Then distilled water is added until the mixture measures 1 L. Glycerol (18% W/W) is added, and the medium is autoclaved for 15 min at 121°C. The final a_w is 0.955.[102] This medium was developed by Hocking and Pitt[127,129] for isolation of *Aspergillus, Fusarium,* and *Penicillium.* It has been used routinely at the Department of Biotechnology, Lyngby, Denmark.[130] Seeds are surface disinfected with a mild chemical such as NaOCl with 0.4% available chlorine for 2 min. Surface disinfected seeds are plated on the medium at 5 to 10 seeds per culture dish, depending on the seed size. The plates are incubated for 7 d at 25°C under alternating cycles of 12 h near UV light and darkness. Fungal colonies are visible at different magnifications using a stereomicroscope. *Aspergillus* colonies consist of heads in different colors. All colonies showing different colors are subcultured separately on oatmeal agar. Colonies of *Penicillium* consist of brush-like structures showing different colors of grayish to blue-green. Cultures are made from colonies showing different colors and types of *Penicillium* on Czapek yeast autolysate agar medium (CYA).[130] Separate cultures in culture plates containing oatmeal agar (60 g oatmeal, 12.5 g agar, and 1000 ml distilled water) are prepared from *Aspergillus* colonies growing on DG18 medium showing different colors. Inoculated dishes are incubated in complete darkness for 7 d at 25°C. *Aspergillus* has been identified by plating at three points on dishes of Czapek agar (CZ) (30 g sucrose, 3 g $NaNO_3$, 1 g K_2HPO_4, 0.5 g KCl, 0.5 g $MgSO_4$ $7H_2O$, 0.01 g $FeSO_4 \cdot 7H_2O$, 15 g agar in 1000 ml distilled water, and 1 ml trace metal solution). For the three-point inoculation, approximately 0.5 ml of molten agar (0.2%) and detergent (0.05%) is placed in a small vial with screw cap and sterilized. A needle carrying conidia or other propagules is dipped in it and mixed. The medium is inoculated with a flamed needle dipped into the spore suspension holding the culture plate upside down. This results in better inoculation, and any fine droplets formed on the needle drop off. The dishes are incubated upright in the dark for 7 d at 25°C. Colony characters are recorded and slides are prepared in lactic acid for determination of morphological characters using a bright-field microscope.[130] Likewise, isolates of different *Penicillium* have been grown on Czapek yeast autolysate agar (3 g $NaNO_3$, 1 g K_2HPO_4, 0.5 g KCl, 0.5 g $MgSO_4$ $7H_2O$, 0.01 g $FeSO_4 \cdot 7H_2O$, 5 g yeast extracts, 30 g sucrose, 15 g agar in 1000 ml distilled water and 0.5 g trace metal solution · [trace metal solution (1 g $ZnSO_4$ $7H_2O$, 0.5 g $CuSO_4 \cdot 5H_2O$, 100 ml water). *Penicillia* have been identified by inoculation onto fresh CYA plates at three points, as described previously and then incubated in the dark for 7 d at 25°C. Colony characters have been recorded and slides prepared in lactic acid for morphological characters.[130]

3. *Aspergillus flavus* and *A. parasiticus* Agar (AFPA)

This medium is recommended for detection and enumeration of *A. flavus* and other aflatoxigenic fungi in maize, nuts, spices, and other commodities.[131] The medium consists of 10 g peptone, 20 g yeast extract, 0.5 g ferric ammonium citrate, 0.1 g chloramphenicol, 2 mg dichloran (0.2% in ethanol, 1 ml), 15 g agar, and 1 L distilled water. After addition of all ingredients, the medium is sterilized by autoclaving for 15 min at 121°C. The final pH is near 6.2. Colonies of *A. flavus* and *A. parasiticus* are distinguished by bright orange-yellow reverse colors on grains incubated on this medium for 48 to 60 h at 30°C.

4. Dichloran Peptone Chloramphenicol Agar (DCPA)

This medium has been used to identify *Fusarium* because it induces formation of macroconidia. It is also effective for isolation of field fungi, such as *Alternaria, Bipolaris,* and *Curvularia*. It consists of 15 g peptone, 1 g KH_2PO_4, 0.5 g $MgSO_4$ $7H_2O$, 0.1 g chloramphenicol, 2 mg dichloran (0.2% in ethanol, 1 ml), and 15 g agar in 1 L distilled water.[132] After addition of all ingredients, the medium is sterilized by autoclaving for 15 min at 121°C.

5. Dilution Cultures

A weighed amount or counted number of seeds are shaken for 1 min in a suspension of 0.12% sterile agar in water and later 1 ml aliquots are placed in each of several culture plates in the same suspension. The dishes are swirled to disperse the suspension; then they are incubated and the colonies counted. This method simply provides an estimation of fungi in the seeds.[6]

6. Blotter Method

Moist blotters are used for detection of storage fungi. *Penicillium* grows well on barley seeds after incubation on moist filter paper without surface disinfection.[133] Filter paper soaked with water and 7.5% common salt has been found to be superior for detection of *A. glaucus* and *Penicillium*.[134] The deep freeze blotter procedure with soaking blotters in a dichloran suspension (0.04%) prevents development of *Rhizopus* and assists in detection of *Penicillium oxalicum* from maize seeds.[59]

C. Bright-field Microscopic Examination

Microscopic examination of barley seeds has been used to show an increase in mycelial growth with increasing moisture content (15 to 24%) and rising temperature (20 to 30°C).[52]

D. Observation Under UV Light

A characteristic bright greenish-yellow fluorescence under UV (≥365 nm) was seen on dead maize seeds with *A. flavus* compared with other fungi.[135]

E. Measurements of Gases

Fungal activity and growth in stored seeds can be monitored by measuring respiration rates and adenosine triphosphate (ATP). The ATP assay is convenient, fast, and relatively inexpensive, and it correlates well with other methods, such as estimation of chitin and ergosterol, agar plate method, etc. It is a valuable alternative to more complicated and time-consuming methods for measuring ongoing fungal activity.[136]

The measurement of CO_2 production is used to determine losses in seed quality and to predict storability of maize.[137] The production of 7.4 g CO_2/1 kg of dry matter of maize seeds results in about 0.5% loss in dry matter.[138] At moisture contents that permit fungal growth, three major gases are involved — butanol, ethanol, and methanol — in maize and soybean seeds. Their production is correlated with fungal growth as measured by chitin content of seeds and plating on agar media.[139]

F. Determination of Fatty Acid Value (FAV)

Increases in the FAV are due to the activities of storage fungi. Normally, seeds with low FAV are preferred over those with high FAV. However, deterioration may occur without an increase in FAV.

G. Moldy Smell

A heavy mycelial growth in barley usually is accompanied by moldy odor, but such an odor may occur without visible mycelial growth.[52]

H. Collection of Seed Exudates

Exudates from germinating seeds are influenced by temperature, soil moisture, injuries, dormancy, and oxygen tension. The simplest way of collecting the exudates from seeds is to place 10 to 12 g of surface-disinfested seeds in 15 to 25 ml sterile distilled water in a sterile container. After the desired interval, the liquid is removed and the seeds are washed in a small amount of distilled water; then the wash water is pooled with liquid collected earlier. The maximum collection period should not exceed the time it takes for the radicle to reach the length of the seed. At the time of collection the sterility of the solution is checked by streaking a loopful on nutrient agar.[140-143]

Loria and Lacy[144] placed individual seeds in moist sand in 25-mm diameter test tubes (cotton plugged and autoclaved) and covered them with sterile sand. After germination, seeds were removed aseptically, rinsed in the same tube with sterile distilled water, and placed on PDA to check for sterility. The sand from each tube was washed with distilled water to obtain the exudate.

Hayman[141,142] developed a simple system that permits repeated collection of exudates from a single seed at the desired time intervals without disturbing or damaging the germinating seeds. A glass tube (5 cm long and 2.2 cm diameter),

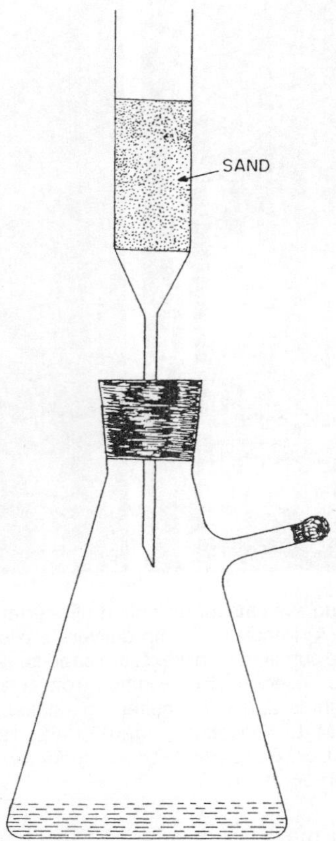

Figure 11-4 Schematic diagram of the system for collecting exudates from a single seed. (Adapted from Hayman, D. S., *Can. J. Bot.*, 47, 1521, 1969. From Dhingra, O. D. and Sinclair, J. B., *Basic Plant Pathology Methods*, CRC Press, Boca Raton, FL, 1985. With permission.)

sealed at one end with a 3.5 cm piece of capillary tubing, was half filled with sand (Figure 11-4). The free end of the capillary was fitted into a rubber stopper in a 25-ml Erlenmeyer flask with side arm and sealed with silicon. The sand tube and the side arm were plugged with cotton. Several assemblies thus were prepared and autoclaved. A single surface-disinfested seed was placed at uniform depth in each tube. At predetermined intervals, the exudates were collected by rinsing the sand with 10 ml of sterile distilled water in two portions of 5 ml each. The exudates were collected in the flask below by applying a mild vacuum.

The collection of exudates by continuous leaching has been used by Schlub and Schmitthenner.[145] A Buchner funnel (40-mm diameter) was filled with 5 ml of 1-mm glass beads supported on a fine mesh nylon screen. The mouth of the funnel was covered with an aluminum cap containing four holes and a glass tube sealed with tygon tubing to deliver 3 ml/h H_2O for leaching. The assembly was autoclaved, and a number of seeds were buried halfway into the beads and then

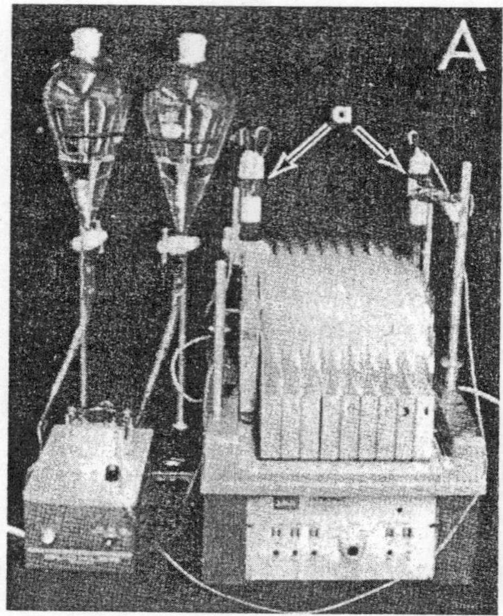

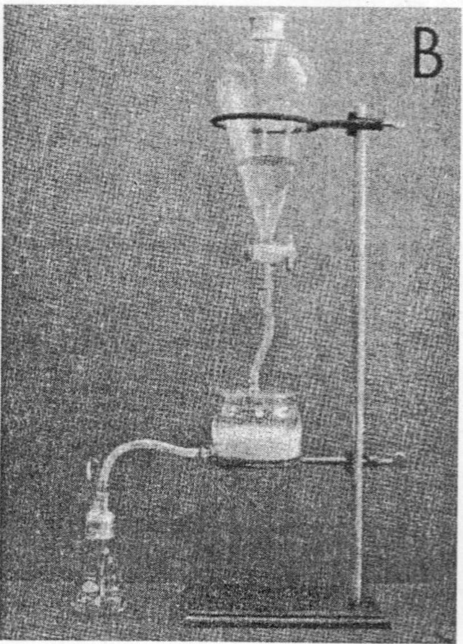

Figure 11-5 Sterile leaching systems for collection of exudates from a single seed during germination. (**A**) Periplastic pump delivers a predetermined amount of water per hour to the substratum in which the seed is germinating, with the leachings being collected every h. (**B**) Exudates from a number of seeds germinating in a substratum in a modified culture are collected every h. (From Short, G. E. and Lacy, M. L., *Phytopathology*, 66, 182, 1976; and from Dhingra, O. D. and Sinclair, J. B., *Basic Plant Pathology Methods*, CRC Press, Boca Raton, FL, 1985. With permission.)

covered with 9 ml of 3-1 mm glass beads. The leachates were collected in test tubes containing a 1-ml solution of 500 μg/ml each of chloramphenicol and streptomycin.

Short and Lacy[146] collected exudates from seeds for 1 h (Figure 11-5). A glass tube, 90 mm long and 25 mm diameter with a rubber stopper on each end, was filled with 20 g of 1-mm glass beads. A separatory funnel was connected to the tube with tygon tubing. The bottom rubber stopper was fitted with a 7-mm diameter drain hole. Following sterilization of the apparatus, sterile distilled water was poured aseptically into the funnel and a surface-disinfested seed was placed within the glass bead matrix. The glass cylinder could have been covered with aluminum foil to exclude light. The water was percolated through the glass bead matrix at 10 ml/h. A fraction collector was used to collect the leachates at 1-h intervals in test tubes containing 10 ml 50% ethanol to control microbial growth.

I. Immunoassays

Some attempts have been made to detect and quantify fungi in stored cereals by immunological techniques using polyclonal antisera.[147] Immunodiagnostic assays (ELISA and dip-stick assay) were developed using monoclonal antibodies (P 101 and EC 6) for detection of *Humicola lanuginosa* and *Penicillium island-*

icum in seeds.[148,149] Mice were immunized by injecting, directly into the peritoneum; cell-free surface washings of agar-slant cultures of *H. lanuginosa* and *P. islandicum*. Hybridoma supernatants were screened by ELISA. Most McAbs raised cross-reacted with other storage fungi and/or noninfected rice seeds but three of the *P. islandicum* McAbs were species specific. One IgM antibody to *H. lanuginosa* did not recognize antigens from rice seeds and cross-reacted significantly with only *P. variabile*. The results of immunofluorescence showed that McAb P101, which is a Ig G1 antibody raised against *P. islandicum*, bound to fungal antigens present in the hyphal wall. The Ig *M* McAb EC 6 raised against *H. lanuginosa* also bound to fungal antigens present in hyphal walls and immature aleuriospores.[148] *A. repens* in maize seed extracts could be measured over a range of 10 to 4000 μg/g maize by an inhibition radio immunoassay (IRIA).[150]

J. Ergosterol Estimation

Ergosterol, a sterol found in fungi but not in higher plants, has been used to measure fungal biomass. Ergosterol content can quantify fungal biomass of storage fungi.[136]

K. Scanning Electron Microscopy

Scanning electron microscopic studies on *A. flavus* var. *columnaris* have shown hyphae in the xylem of the peduncle of maize seeds.[8] Klich et al.[28] reported the presence of *A. flavus* in the xylem of the outer pigment layer of cottonseeds.

VI. CONTROL

The losses due to deterioration of seeds by storage fungi can be reduced by the following methods.

A. Avoiding Damage to Seeds During Harvesting, Processing, and Threshing

Storage fungi gain entry to seeds at any stage after maturity to harvest and during processing and threshing. Any damage to seeds results in the entry of storage fungi and poses a problem during storage. Such seeds are likely to carry a heavy invasion of fungi. Proper precautions at harvest and during postharvest operations help reduce occurrence of storage fungi. Seeds that are clean, sound, undamaged, and dried are ideal for storage.

B. Storage Conditions

The most important factors in seed storage are seed type, storage period, container type or storage method, seed temperature at delivery time, moisture content at delivery, foreign material content at delivery, protection from physical damage and spoilage, and predominant relative humidity of the atmosphere.

Efficient storage means delivering seeds to the consumer in the same condition they were received. The single most important factor is moisture. If this is higher than the equilibrium moisture content (EMC) at 70% relative humidity, deterioration is certain to take place, even after 2 or 3 months and at 5°C.[3] For stored products, the combination of a low a_w and temperature, reduced oxygen, and/or increased carbon dioxide levels will influence storage life. Under natural storage conditions, a_w is the dominant factor determining the storage life of commodities. Drying a product quickly and keeping it dry remain the most effective methods for ensuring that fungi do not invade stored products. The use of fumigants for insect control in bulk storage of seeds can have the benefit of controlling fungal growth and mycotoxin production. Fumigants in common use for insect control include carbon dioxide, methyl bromide, and phosphine. Phosphine fumigation is used in the tropics because it is cheap and effective, and application is easy. In normal fumigation practice, concentrations of phosphine up to about 3 g/m^3 are used. Phosphine is an inhibitor of respiratory enzymes.[151] Phosphine, even at low levels (0.1 g/m^3), can slow the development of storage fungi in seeds where the moisture content is above the levels normally accepted for safe storage. Fumigation with phosphine also can reduce mycotoxin production. However, phosphine is effective only against growing fungi and has little effect on dormant conidia.[129]

The most effective means of avoiding damage to seeds by storage fungi is to maintain storage conditions that prevent fungal development.[42] Seed samples should be examined periodically for moisture content, germination, and fungi. Any damage, if detected early, may be checked by taking suitable precautions. This may be achieved by drying seeds to a safe moisture level before storage and then frequently aerating during storage to maintain moisture and temperature below which the storage fungi do not develop. Aeration is the most important method to uniformly reduce seed temperature to 5 to 10°C throughout the storage bin. At such temperatures, storage fungi grow slowly, and insects and mites are dormant. Aeration also lowers moisture content of seeds, but seeds, being hygroscopic, gain moisture produced by insects, mites, and fungi and absorb atmospheric moisture. Navarro et al.[152] have found that wheat seeds maintain 90% germination after 22 months in a metal bin equipped with an aeration system. Seed drying can be achieved using high-speed dryers, solar energy, unheated air, or partially heated air. The latter methods result in physically better seeds, but there are more chances of infection by storage fungi.

C. Reducing Seed Moisture to Safe Limits

Fungi begin growth at a relative humidity above 65 to 70%. The safe moisture content for food stuffs for long-term storage, therefore, are those that provide an equilibrium at a relative humidity of 65 to 70%.[61] The values vary with differences in the chemical composition of various types of stored produce. Seeds with high lipid content, for example, will have a higher equilibrium moisture content than cereals, which are composed largely of starch.[65]

The most effective means of avoiding seed damage is to dry seeds to a moisture content at which no storage fungi will develop. Damage by storage fungi will not occur in storage bins at moisture contents below 13% for starchy cereal seeds, such as barley, maize, rice, sorghum, and wheat, below 12% for soybeans, and below 10% for flax seeds.[4] The ability of the container to minimize moisture gain during storage is as important as low initial moisture content in preserving seed viability in the tropics. Unprotected seeds absorb moisture from the air until equilibrium is reached. This is especially critical in the tropics, where wide, daily fluctuations of relative humidity occur. The key to preserving seed viability under the humid conditions in the tropics is to dry seeds to a level of moisture content below which no storage fungi will develop. For soybeans, to avoid deterioration, seeds are dried to 8.6% moisture content on a dry weight basis and stored in a container that minimizes moisture gain from the atmosphere while in storage.[97] To avoid damage to sunflower seeds, the moisture content should be less than 6.5%.[49] For barley, a moisture content of 15% at 20°C retains germination up to 41 months.[52] Rapid moisture equilibration between wet and dry maize indicates that seeds can be blended with minimal risk of fungal damage or aflatoxin contamination if the average moisture content of the blend is low enough to prevent fungal growth.[153]

D. Seed Treatment

Soybean oil at 200 μg/ml sprayed onto maize and soybean seeds as it goes into storage has proved an effective means of reducing development of storage fungi. It is technically feasible to combine soybean oil (200 μg/ml) and thiabendazole (20 μg/ml) and then apply them to maize and soybean seeds at rates recommended for storage fungi control and dust suppression without any problem of rancidity.[154] The best control of *Aspergillus* and *Penicillium* in soybean seeds was achieved by iprodione or thiabendazole in soybean or mineral oil rather than in water, allowing for maintenance of quality seeds.[155]

A number of chemicals reduce development of storage fungi. An ideal chemical must have low mammalian toxicity and long lasting microbial-inhibiting properties.[137] If seeds are to be used for planting, the chemical should not have an adverse effect on germination. Matz and Milner[156] found that an addition of 0.2% or more of propylene oxide to wheat seeds checked deterioration of damp wheat but resulted in a heavy reduction in germination. Herting and Drury[157] reported that isobutyric acid was effective in controlling fungi, but treated seeds failed to germinate.

Propionic acid salts are used by bakers to prevent fungal development by preventing development of bacteria, fungi, and yeasts. Also, they block the activity of enzymes that decompose carbohydrates, and they have no adverse toxicological effects. They also check the formation of aflatoxin and overheating of wet cereals. In maize at 30% moisture content, propionic acid at 9.5 g/kg eliminated *Aspergil-*

Table 11-6 Application Rates of Propionic Acid in Accordance with Different Moisture Contents

Moisture content (%)	Application rate (%)	Crop: *Zea mays* (maize) (g/100 kg)
20	0.5	500
25	0.7–0.8	700–800
30	1.1–1.2	1100–1200
35	1.4–1.5	1400–1500
40	1.7–1.8	1700–1800
45	1.0–2.1	2000–2100

Note: The figures apply to undiluted acid and fresh cereal or maize.

From Sauer, D. B., Hodges, T. O., Burroughs, R., and Converse, H. H., *Trans. Am. Soc. Agric. Eng.*, 18, 1162, 1975. With permission.

lus and *Penicillium* up to 120 d.[158] The rate of infection by *A. candidus* was reduced significantly with sodium propionate (5000 μg/ml) in rice seeds (12.77 to 13% moisture) stored from 6 months at 75% and for 4 months at 85% relative humidity. It suppressed *A. candidus* but not *A. glaucus.*[159] Application of propionic acid at 2 ml/kg to maize seeds stored in cloth bags inhibited fungal growth for 12 months without any loss of either germination or nutritional components.[160]

Propionic acid (1000 μg/ml) is used to control fungi in undried stored cereals and pulses destined for animal feed in temperate countries.[161] Acetic and formic acids are less effective, but these are used in combination with propionic acid.[162] Methylene bispropionate is equal or superior to propionic acid. It breaks down into formaldehyde and propionic acid.[163,164] There are fungi and bacteria that have varying degrees of tolerance to propionic acid and other organic acids; therefore, a combination of two or more chemicals may be useful.[156] The rate of application of propionic acid depends upon the moisture content and storage period (Table 11-6).[165] Application of benomyl or thiourea decreased maize seed infection by *A. flavus* during storage and maintained high seed quality.[166] Some lipids applied as seed treatments were effective in preventing adult development of the Mexican bean weevil, *Zabrotes subfasciatus*, while others decreased seed infection by *A. ruber.*[167]

E. Resistance

Disease control by using plant resistance is an ideal, simple, and practical method, which is less costly than other approaches. The seed coat is the most important component in resistance of seeds to deterioration. The seeds of most wild cottons are impermeable to water through the chalaza, whereas cultivated cottons are permeable. The impermeable seed character imparts immunity to deterioration as well as to infection by *A. flavus* and elaboration of aflatoxin.[168] Infection of ripened permeable cottonseeds occurs through the chalaza; it is the only portion of the thick tannin-impregnated seed coat that is permeable. No differences have been observed in the degree of cottonseed deterioration by *A.*

flavus that could be attributed to seed coat permeability, since impermeable or low permeability phenotypes are expressed only upon seeds.[169] Other mechanisms of resistance are prophylaxis by physically impenetrable barriers, such as inhibition of fungal growth by phenolic compounds, and restriction of nutrient availability to microorganisms outside the seeds. The localized tannin in the nucellus may serve to protect the embryo from attack by microorganisms.[170] Kirsi[171] reported a protease from barley embryos inhibitory to *Aspergillus* proteases. Halim et al.[172] observed that a trypsin inhibitor from maize inhibited fungal growth and proposed that it may serve to protect seeds against fungal invasion. There was a negative relationship between chitinase activity and fungal infection in maize seeds by *A. flavus*. Endogenous chitinase in fungal-resistant seeds may act on chitin in cell walls of fungi and inhibit their invasion. Chitinase catalyses inhibit hydrolysis of chitin into N-acetylglucosamine.[173] Calcium, potash, and total phosphate content of both the testa and germs of cowpea seeds susceptible to *A. flavus* were low compared with those of partially and highly resistant seeds, while the opposite was true for sodium and zinc content.[174] Treatment of soybean seeds with benomyl increased volatile aldehyde compounds (VAC), indicating control of *A. glaucus* and *A. niger* and increased germination.[57]

Seed germination, visible fungi, number of fungal propagules of *Aspergillus glaucus* and *Penicillium*, and ergosterol contents were similar in undamaged and damaged maize seeds inoculated with three species of *A. glaucus* group, but the genotypes were affected differently; resistant genotypes were resistant compared to susceptible ones.[175] Although peanut cultivars were resistant to *A. flavus* colonization, the susceptibility of all genotypes was higher under drought stress at fruit ripening.[176]

To identify resistant cultivars, reliable and practical screening techniques are needed. Pupipat et al.[177] developed a practical field and laboratory screening technique for resistance to *A. flavus* infection and aflatoxin contamination in maize. For large scale field inoculation of maize, plants were injected with an *A. flavus* suspension inoculum of 10^7 conidia/ml concentration onto the silk channel at 3 weeks after 100% silking. For laboratory evaluation, seeds were inoculated by dipping into a suspension of 10^7 conidia/ml and then incubated in a sterile culture plate with two layers of filter paper on top of three layers of straw paper. Only the second layer was moistened by soaking in water. White-seeded lines of peanuts were susceptible to *in vitro* seed colonization by *A. flavus*, but no correlation was found between seed coat color and seed rots or aflaroot disease caused by *A. flavus*.[178] Some breeding lines of peanut possessed some level of resistance to *A. flavus*, thus reflecting the presence of genes for resistance.[179] Selection for resistance to seed infection by *A. flavus* should be done at 60 d after the midsilk stage. Maize should be harvested as early as possible after physiological maturity when drying facilities are available to reduce seed moisture to a safe storage level to avoid aflatoxin production. However, aflatoxin contamination can increase rapidly in early-harvested seeds that are not dried properly.[180]

REFERENCES

1. Weber, E. J., Objectives of the conference, in *Fungi and Mycotoxins in Stored Products*, Champ, B. R., Highley, E., Hocking, A. D., and Pitts, J. I., Eds., *Proc. Int. Conf.*, Bangkok, April 1991, 9.
2. Williams, P. C., Storage of grains and seeds, in *Mycotoxins and Animal Foods*, Smith, J. E. and Henderson, R. S., Eds., CRC Press, Boca Raton, FL, 1991, 721.
3. Williams, P. C., The principles of grain handling and transportation, in *Mycotoxins and Animal Foods*, Smith, J. E. and Henderson, R. S., Eds., CRC Press, Boca Raton, FL, 1991, 247.
4. Christensen, C.M. and Sauer, D. B., Microflora, in *Storage of Cereal Grains and Their Products,* Christensen, C. M., Ed., American Association of Cereal Chemists, St. Paul, MN, 1982, 219.
5. Christensen, C. M. and Kaufmann, H. H., Deterioration of stored grains by fungi, *Annu. Rev. Phytopathol.,* 3, 69, 1965.
6. Christensen, C. M., Fungi and seed quality, in *Handbook of Applied Mycology,* Vol. III, *Foods and Feeds,* Arora, D. K., Mukerji, K. G., and Marth, E. H., Eds., Marcel Dekker, New York, 1991, 99.
7. Raper, K. B. and Fennell, D. I., *The Genus Aspergillus*, R. E. Krieger, Malabar, FL, 1977.
8. Mycock, D. J., Rukenberg, F. H. J., and Berjak, P., Infection of maize seedlings by *Aspergillus flavus* var. *columnaris var. nov., Seed Sci. Technol.*, 18, 693, 1990.
9. Caldwell, R. W., William, W. C., and John, T., The occurrence and toxicity of Indian isolates of *Penicillium viridicatum*, Proc. 2nd Int. Congr. Plant Pathol., St. Paul, MN, 1973, 410.
10. Raper, K. B. and Fennell, D. I., *The Genus Aspergillus*, Williams & Wilkins, Baltimore, 1965, 686.
11. Pitt, J. I., *The Genus Penicillium*, Academic Press, New York, 1979, 634.
12. Clarke, J. H., Fungi in stored products, *PANS*, 15, 473, 1969.
13. Ellis, J. J. and Hesseltine, C. W., Species of *Absidia* with ovoid sporangiospores. II., *Sabourandia*, 5, 59, 1966.
14. Brown, A. H. S. and Smith, G., The genus *Paecilomyces* Bainier and its perfect stage *Byssochlamys* Westling, *Trans. Br. Mycol. Soc.,* 40, 17, 1957.
15. Wickerham, L. J., Taxomony of Yeasts, U. S. Dep. Agric. Tech. Bull., No. 1029, Washington, D. C., 56, 1951.
16. Christensen, C. M. and Kaufmann, H. H., Microflora, in Storage of Cereal Grains and Their Products, Christensen, C. M., Ed., American Association of Cereal Chemists, St. Paul, MN, 1974, 59.
17. Hesseltine, C. W. and Bothast, R. J., Mold development in ears of corn from tasseling to harvest, *Mycologia*, 69, 328, 1977.
18. Marsh, S. F. and Payne, G. A., Preharvest infection of corn silks and kernels by *Aspergillus flavus, Phytopathology,* 74, 1284, 1984.
19. McLean, M. and Berjak, P., Maize grains and their associated mycoflora — a micro-ecological consideration, *Seed Sci. Technol.*, 15, 831, 1987.
20. Mycock, D. J., Lloyd, H. L., and Berjak, P., Micropylar infection of post-harvest caryopses of *Zea mays* by *Aspergillus flavus* var. *columnaris var. nov., Seed Sci. Technol.*, 16, 647, 1988.
21. Smart, M. G., Wicklow, D. T., and Caldwell, R. W., Pathogenesis in Aspergillus ear rot of maize: light microscopy of fungal spread from wounds, *Phytopathology*, 80, 1287, 1990.

22. Diener, U. L., Cole, R. J., Sanders, T. H., Payne, G. A., Lee, L. S., and Klich, M. A., Epidemiology of aflatoxin formation by *Aspergillus flavus, Annu. Rev. Phytopathol.*, 25, 249, 1987.
23. Payne, G. A., *Aspergillus flavus* infection in maize: silks and kernels, in *Aflotoxins in Maize, Proc. Workshop,* Zuber, M. S., Lillehoj, E. B., and Renfro, B. J., Eds., CIMMYT, Mexico, 1987, 119.
24. Zummo, N. and Scott, G. E., Cob and kernel infection by *Aspergillus flavus* and *Fusarium moniliforme* in inoculated field grown maize ears, *Plant Dis.*, 74, 627, 1990.
25. Jones, R. K., Duncan, H. E., and Hamilton, P. B., Planting date, harvest date and irrigation effects on infection and aflatoxin production by *Aspergillus flavus* in field corn, *Phytopathology,* 71, 810, 1981.
26. McLean, M., Mycock, D. J., and Berjak, P., A preliminary investigation of extracellular enzyme production by some species of *Aspergillus*, *South African J. Bot.*, 51, 425, 1983.
27. Huizar, H. E., Bertke, C. C., Klich, M. A., and Aroson, J. M., Cytochemical localization and ultrastructure of *Aspergillus* in cottonseed, *Mycopathology,* 110, 43, 1990.
28. Klich, M. A., Thomas, S. H., and Mellon, J. E., Field studies on the mode of entry of *Aspergillus flavus* into cotton seeds, *Mycologia*, 76, 665, 1984.
29. Barmore, C. R. and Brown, G. E., Role of pectolytic enzymes and galacturonic acid in citrus fruit decay caused by *Penicillium digitatum, Phytopathology,* 69, 675, 1979.
30. Singh, A. K., Mehan, V. K., Mengesha, M. H., and Jambunathan, R., Inhibition sites, leachates and fungal colonization of seeds of selected groundnut germplasm lines with different seed test colours, *Oleagineax*, 47, 579, 1992.
31. Zummo, N., Localization of infection by *Fusarium moniliforme, Cephalosporium acremonium,* and *Aspergillus flavus* in corn kernels in Mississippi, *Phytopathology,* 75, 1330, 1985.
32. Wicklow, D. T. and Donahue, J. E., Sporogenic germination of sclerotia in *Aspergillus flavus* and *A. parasiticus*, *Trans. Br. Mycol. Soc.*, 82, 621, 1984.
33. Campbell, K. W. and White, D. G., An inoculation device to evaluate maize for resistance to ear rot and aflatoxin production by *Aspergillus flavus, Plant Dis.*, 78, 778, 1994.
34. Mycock, D. J. and Berjak, P., In defence of aldehyde osmium fixation and critical point drying for characterization of seed-storage fungi by scanning electron microscopy, *J. Microsc.*, 163, 321, 1991.
35. Mycock, D. J., Rijkenberg, F. H. J., and Berjak, P., Systemic transmission of *Aspergillus flavus* var. *columnaris* from one maize seed generation to the next, *Seed Sci. Technol.*, 20, 1, 1992.
36. Mehan, V. K., Ba, A., Ramakrishna, N., and McDonald, D., Preharvest *Aspergillus flavus* infection of apical and basal seed in pods of groundnut genotypes, *Int. Arachis Newsl.*, 10, 16, 1991.
37. Sauer, D. B., Effects of fungal deterioration on grain: nutritional value, toxicity, germination, *Int. J. Food Microbiol.*, 7, 267, 1988.
38. Mills, J. T., Post harvest insect–fungus associations affecting seed deterioration, in Physiological–Pathological Interactions Affecting Seed Deterioration, West, S. H., Ed., Crop Science Society of America Special Publ. No. 12, Madison, WI., 39, 1986.

39. Cherry, J. P., Protein degradation during seed deterioration, *Phytopathology*, 73, 317, 1983.
40. Harman, G. E., Deterioration of stored pea seed by *Aspergillus ruber*: extraction and properties of a toxin, *Phytopathology*, 62, 206, 1972.
41. Harman, G. E. and Drury, R. E., Respiration of pea seeds (*Pisum sativum*) infected with *Aspergillus ruber*, *Phytopathology*, 63, 1040, 1973.
42. Christensen, C. M., Loss of viability in storage: microflora, *Seed Sci. Technol.*, 1, 547, 1973.
43. Fields, R. W. and King, T. H., Influence of storage fungi on the deterioration of stored pea seed, *Phytopathology*, 52, 336, 1962.
44. Christensen, C. M., Invasion of stored wheat by *Aspergillus ochraceus*, *Cereal Chem.*, 39, 100, 1962.
45. Papavizas, G. C. and Christensen, C. M., Grain storage studies. XXIX. Effect of invasion by individual species of *Aspergillus* upon germination and development of discolored germs in wheat, *Cereal Chem.*, 37, 197, 1960.
46. Qasem, S. A. and Christensen, C.M., Effect of moisture content and temperature on invasion of stored corn by *Aspergillus flavus*, *Phytopathology*, 50, 703, 1960.
47. Lopez, L. C. and Christensen, C. M., Effect of moisture content and temperature on invasion of stored corn by *Aspergillus flavus*, *Phytopathology*, 57, 588, 1967.
48. Fanse, H. A. and Christensen, C. M., Invasion by storage fungi of rough rice in commercial storage and in the laboratory, *Phytopathology*, 60, 228, 1970.
49. Christensen, C.M., Factors affecting invasion of sunflower seeds by storage fungi, *Phytopathology*, 59, 1699, 1969.
50. Moreno, E., Lopez, L. C., and Christensen, C. M., Loss of germination of stored corn from invasion by storage fungi, *Phytopathology*, 55, 125, 1965.
51. Tuite, J. F. and Christensen, C. M., Grain storage studies. XVI. Influence of storage conditions upon the fungus flora of barley seed, *Cereal Chem.*, 32, 1, 1955.
52. Welling, B., Fungus flora and germination of barley, *Tidsskr. Planteavl.*, 73, 291, 1969.
53. Christensen, C. M., Moisture content, moisture transfer and invasion of stored sorghum seeds by fungi, *Phytopathology*, 60, 280, 1970.
54. Dhingra, O. D., Nicholson, J. F., and Sinclair, J. B., Influence of temperature on recovery of *Aspergillus flavus* from soybean seed, *Plant Dis. Rep.*, 57, 185, 1973.
55. Ellis, M. A., Ilyas, M. B., and Sinclair, J. B., Effect of cultivar and growing region on internally seedborne fungi and *Aspergillus melleus* pathogenicity in soybean, *Plant Dis. Rep.*, 58, 332, 1974.
56. Caldwell, R. W., Tuite, J., and Carlton, W. W., Pathogenicity of *Penicillia* to ear rots, *Phytopathology*, 71, 175, 1981.
57. Gupta, I. J., Schmitthenner, A. F., and McDonald M. B., Effect of storage fungi on seed vigour of soybean, *Seed Sci. Technol.*, 21, 581, 1993.
58. Ghosh, J. and Nandi, B., Deteriorative abilities of some common storage fungi of wheat, *Seed Sci. Technol.*, 14, 141, 1986.
59. Halfon-Meiri, A. and Solel, Z., Factors affecting seedling blight of sweet corn caused by seed borne *Penicillium oxalicum*, *Plant Dis.*, 74, 36, 1990.
60. Harman, G. E. and Nash, G., Deterioration of stored pea seed by *Aspergillus ruber*: evidence for involvement of a toxin, *Phytopathology*, 62, 209, 1972.
61. Harman, G. E. and Pfleger, F. L., Pathogenicity and infection sites of *Aspergillus* species in stored seeds, *Phytopathology*, 64, 139, 1974.
62. Harman, G. E. and Drury, R. E., Respiration of pea seeds (*Pisum sativum*) infected with *Aspergillus ruber*, *Phytopathology*, 63, 1040, 1973.

63. U. S. Department of Agriculture Federal Grain Inspection Service, *The Official United States Standards for Grains,* U. S. Government Printing Office, Washington, D.C., 1978.
64. Qasem, S. A. and Christensen, C. M., Influence of moisture content, temperature, and time on the deterioration of stored corn by fungi, *Phytopathology,* 48, 544, 1958.
65. Gwinner, J., Harnisch, R., and Muck, O., Manual on the Preservation of Post-Harvest Grain Losses, Post-Harvest Project, Hamburg, 1990, 294.
66. Minnesota Department of Agriculture, Grain Inspection Division, 1979.
67. Robertson, J. A., Roberts, R. G., and Chapman, G. W., Jr., An evaluation of "heat damage" and fungi in relation to sunflower seed quality, *Phytopathology,* 75, 142, 1985.
68. Milner, M. and Geddes, W. F., Grain storage studies. II. The effect of aeration, temperature, and time on the respiration of soybeans containing excessive moisture, *Cereal Chem.,* 22, 484, 1945.
69. Carter, E. P., Role of fungi in the heating of moist wheat, U. S. Dept. of Agriculture Circular No. 838, Washington, D.C., 1950.
70. Hummel, B. C. W., Cuendet, L. S., Christensen, C. M., and Geddes, W. F., Grain storage studies. XIII. Comparative changes in respiration, viability, and chemical composition of mold-free and mold-contaminated wheat upon storage, *Cereal Chem.,* 31, 143, 1954.
71. Milner, M. and Geddes, W. F., Grain storage studies. III. The relation between moisture content, mold growth, and respiration of soybeans, *Cereal Chem.,* 23, 225, 1946.
72. Howe, R. W., Loss of viability of seed in storage attributable to infestations of insects and mites, *Seed Sci. Technol.,* 1, 563, 1973.
73. St. Angelo, A. J. and Ory, R. L., Lipid degradation during seed deterioration, *Phytopathology,* 73, 315, 1983.
74. Nagel, C. M. and Semeniuk, G., Some mold-induced changes in shelled corn, *Plant Physiol.,* 22, 20, 1947.
75. Milner, M., Christensen, C. M., and Geddes, W. F., Grain storage studies. VII. Influence of certain mold inhibitors on respiration of moist wheat, *Cereal Chem.,* 24, 507, 1947.
76. Baker, D., Neustadt, M. H., and Zeleny, L., Application of the fat acidity test as an index of grain deterioration, *Cereal Chem.,* 34, 226, 1957.
77. Bottomley, R. A., Christensen, C. M., and Geddes, W. F., Grain storage studies. X. The influence of aeration, time and moisture content on fat acidity, non-reducing sugars and mold flora of stored yellow corn, *Cereal Chem.,* 29, 53, 1952.
78. Christensen, C. M., Some changes in No. 2 corn stored two years at moisture contents of 14.5 and 15.2 percent and temperatures of 12, 20 and 25C, *Cereal Chem.,* 44, 95, 1967.
79. Sauer, D. B. and Christensen, C. M., Some factors affecting increase in fat acidity values in corn, *Phytopathology,* 59, 108, 1969.
80. Pernollet, J. C., Protein bodies of seeds: ultra-structure, biochemistry, biosynthesis and degradation, *Phytochemistry,* 17, 1473, 1978.
81. Cherry, J. P., Young, C. T., and Beuchat, L. R., Changes in proteins and total amino acids of peanuts (*Arachis hypogaea*) infested with *Aspergillus parasiticus, Can. J. Bot.,* 53, 2639, 1975.
82. Bose, A. and Nandi, B., Role of enzymes of storage fungi in deterioration of stored safflower and sesame seeds, *Seed Res.,* 13, 19, 1985.

83. Irving, G. W., Fontaine, T. D., and Warner, R., Electrophoretic investigation of peanut proteins, *Arch. Biochem.,* 79, 475, 1946.
84. Szerszen, J. B. and Pettit, R. E., Detection and partial characterization of new polypeptides in peanut cotyledons associated with early stages of infection by *Aspergillus flavus, Phytopathology,* 80, 1432, 1990.
85. Vaidya, A. and Dharam Vir, Changes in the oil in stored groundnut due to *Aspergillus niger* and *A. flavus, Indian Phytopathol.,* 42, 525, 1989.
86. Pande, A., Changes in amino acid composition of bengalgram seeds due to *Aspergillus niger* infection, *Biovigyanam,* 11, 115, 1985.
87. Prasad, T. and Singh, B. K., Effect of relative humidity on oil properties of fungal infested sunflower seeds, *Biol. Bull. India,* 5, 85, 1983.
88. Sao, R. N., Singh, R. N., Narayan, N., Singh, S. P., Kumar, S., and Prasad, B. K., Respiration and viability of stored radish seeds due to seedborne *Aspergillus flavus, Indian Phytopathol.,* 43, 197, 1990.
89. Vijaya Kumari, P. and Karan, D., Deterioration of cowpea seeds in storage by *Aspergillus flavus, Indian Phytopathol.,* 34, 222, 1981.
90. Hayden, N. J., and Maude, R. B., The role of seedborne *Aspergillus niger* in transmission of black mould of onion, *Plant Pathol.*, 41, 573, 1992.
91. Ramakrishna, N., Lacey, J., and Smith, J. E., Effects of water activity and temperature on the growth of fungi interacting on barley grains, *Mycol. Res.,* 97, 1393, 1993.
92. Pitt, J. I. and Hocking, A. D., Significance of fungi in stored products, in Fungi and Mycotoxins in Stored Products, Champ, B. R., Highley, E., Hocking, A. D., and Pitt, J. I., Eds., ACIAR Proc. Int. Conf., Bangkok, 1991.
93. Klich, M. A., Relation of plant water potential at flowering to subsequent cottonseed infection by *Aspergillus flavus, Phytopathology,* 77, 739, 1987.
94. White, N. D. G. and Jayas, D. S., Microbial infection and quality deterioration of sunflower seeds as affected by temperature and moisture content during storage and the suitability of the seeds for insect or mite infestation, *Can. J. Plant Sci.,* 73, 303, 1993.
95. Kennedy, B. W., Moisture content, mold invasion, and seed viability in stored soybeans, *Phytopathology,* 54, 771, 1964.
96. Kennedy, B. W., The occurrence of *Aspergillus* spp. on stored seeds, in *Seed Pathology — Problems and Progress,* Yorinori, J. T., Sinclair, J. B., Mehta, Y. R., and Mohan, S. K., Eds., IAPAR, Londrina, Brazil, 1979, 257.
97. Tenne, F. D., Ravalo, E. J., Sinclair, J. B., and Rodda, E. D., Changes in viability and microflora of soybean seeds stored under various conditions in Puerto Rico. *J. Agric. Univ. P. R.,* 62, 255, 1978.
98. Christensen, C. M., Moisture content of sunflower seeds in relation to invasion by storage fungi, *Plant Dis. Rep.,* 56, 173, 1972.
99. Papavizas, G. C. and Christensen, C. M., Grain storage studies. XXVI. Fungus invasion and deterioration of wheats stored at low temperatures and moisture contents of 15 to 18 percent, *Cereal Chem.,* 35, 27, 1958.
100. Dorworth, C. E. and Christensen, C. M., Influence of moisture content, temperature, and storage time upon changes in fungus flora, germinability and fat acidity values of soybeans, *Phytopathology,* 58, 1457, 1968.
101. Hayden, N. J. and Maude, R. B., The effect of heat on the growth and recovery of *Aspergillus* spp. from the mycoflora of onion seeds, *Plant Pathol.,* 43, 627, 1994.

102. Pitt, J. I. and Hocking, A. D., Significance of fungi in stored products, in Fungi and Mycotoxins in Stored Products, Champ, B. R., Highley, E., Hocking, A. D., and Pitt, J. I., Eds., Proc. Intern. Conf. Bangkok, ICIAR No. 36, 1991, 16.
103. Payne, G. A., Thompson, D. L., Lillehoj, E. B., Zuber, M. S., and Adkins, C. R., Effect of temperature on the preharvest infection of maize kernels by *Aspergillus flavus, Phytopathology,* 78, 1376, 1988.
104. Kulik, M. M., Retention of germinability and invasion by storage fungi of hand-threshed and machine-threshed wheat seeds in storage, *Seed Sci. Technol.,* 1, 805, 1973.
105. Hurd, A. M., Seedcoat injury and viability of seeds of wheat and barley as factors in susceptibility to molds and fungicides, *J. Agric. Res.,* 21, 99, 1921.
106. Rashidan, A. and Ibni Hajar, R., Microbiological quality of paddy during in-store drying, in Fungi and Mycotoxin in Stored Products, Champ, B. R., Highley, E., Hocking, A. D., and Pitt, J. I., Eds. Proc. Int. Conf. Bangkok, ACIAR Proc. No. 36, 1991, 220.
107. Sulaiman, E. D. and Hussain, S. S., Chemical control of *Aspergillus flavus* and *Penicillium cyclopium* in storage, *Pakistan J. Sci. Ind. Res.,* 27, 363, 1984.
108. Christensen, J. J., Variability in the microflora in barley kernels, *Plant Dis. Rep.,* 47, 635, 1964.
109. Agarwal, N. S., Christensen, C. M., and Hodson, A. C., Grain storage fungi associated with the grain weevil, *J. Econ. Entomol.,* 50, 659, 1957.
110. Christensen, C. M. and Hodson, A. C., Development of granary weevils and storage fungi in columns of wheat, II., *J. Econ. Entomol.,* 53, 375, 1960.
111. Misra, C. P., Christensen, C. M., and Hodson, A. C., The angoimois grain moth, *Sitotroga cerealella,* and storage fungi, *J. Econ. Entomol.,* 54, 1032, 1960.
112. Griffiths, D. A., Hodson, A. C., and Christensen, C. M., Grain storage fungi associated with mites, *J. Econ. Entomol.,* 52, 514, 1959.
113. White, N. D. G., Henderson, L. P., and Sinha, R. N., Effects of infestations by three-stored product mites on fat acidity, seed germination, and microflora of stored wheat, *J. Econ. Entomol.,* 72, 763, 1979.
114. Sinha, K. K. and Sinha, A. K., Impact of stored grain pests on seed deterioration and aflatoxin contamination in maize, *J. Stored Prod. Res.,* 28, 211, 1992.
115. Lynch, R. E. and Wilson, D. M., Enhanced infection of peanut, *Arachis hypogaea* L., seeds with *Aspergillus flavus* group fungi due to external scarification of peanut pods by the lesser cornstalk borer, *Elasmopalus lignosellus* (Zeller), *Peanut Sci.,* 18, 110, 1991.
116. Howe, R. W., Insects attacking seeds during storage, in *Seed Biology,* Vol. III, *Insects and Seed Collection, Storage, Testing and Certification,* Kozlowski, T. T., Ed. Academic Press, New York, 1972, 247.
117. Pande, N. and Mehrotra, B. S., Rice weevil (*Sitophilus oryzae* Linn.) vector for oxigenic fungi, *Nat. Acad. Sci. Lett.,* India, 11, 3, 1988.
118. Wicklow, D. T., Horn, B. W., Shotwell, O. L., Hesseltine, C. W., and Caldwell, R. W., Fungal interference with *Aspergillus flavus* infection and aflatoxin contamination of maize grown in a controlled environment, *Phytophathology,* 78, 68, 1988.
119. Mukherjee, P. S. and Nandi, B., Insect–fungus associations influencing seed deterioration in storage, *J. Mycopathol. Res.,* 31, 87, 1993.
120. Mills, J. T., Insect–fungus associations influencing seed deterioration, *Phytopatholcgy,* 73, 330, 1983.

henson, L. W., and Russell, T. E., The association of *Aspergillus flavus* with 'erous and other insects infecting cotton bracts and foliage, *Phytopathology,* 2, 1974.

122. Griffiths, D. A., Hodson, A. C., and Christensen, C. M., Grain storage fungi associated with mites, *J. Econ. Entomol.*, 52, 514, 1959.
123. Cotton, R. T., and Wibur, D. A., Insects, in Storage of Cereal Grains and Their Products, Christensen, C. M., Ed., American Association of Cereal Chemists, St. Paul, MN, 549, 1974.
124. Sinha, R. N., Role of Acarina in the stored grain eco-system, in *Recent Advances in Acarology,* Vol. I, Rodriguez, J. G., Ed., Academic Press, New York, 1979, 263.
125. Jeffrey, I. G., A survey of the mite fauna of Scottish farms, *J. Stored Prod. Res.*, 12, 149, 1976.
126. Moore, M. B. and Olien, C. R., Mercury bichloride as a surface disinfectant for cereal seeds, *Phytopathology,* 42, 471, 1952.
127. Hocking, A. D. and Pitt, J. I., Dichloran-glycerol medium for enumeration of xerophilic fungi from low-moisture foods, *Appl. Environ. Microbiol.*, 39, 488, 1980.
128. King, A. D., Pitt, J. I., Beuchat, L. R., and Corry, J. E. L., *Methods for the Mycological Examination of Food,* Plenum Press, New York, 1986.
129. Hocking A. D., Isolation and identification of xerophilic fungi in stored commodities, in *Fungi and Mycotoxins in Stored Products,* Champ, B. R., Highley, E., Hocking, A. D., and Pitt, J. I., Eds., Proc. Int. Conf., Bangkok, ACIAR Proc. No. 36, 1991, 270.
130. Singh, K., Frisvad, J. C., Thrane, U., and Mathur, S. B., *An Illustrated Manual on Identification of Some Seedborne Aspergilli, Fusaria, Penicillia and Other Mycotoxins,* Danish Govt. Inst. Seed Pathol. Developing Countries, Copenhagen, 1991, 133.
131. Hocking, A. D., Aflatoxigenic fungi and their detection, *Food Technol.*, Australia, 34, 236, 1982.
132. Andrews, S. and Pitt, J. I., Selective medium for isolation of *Fusarium* species and Dematiaceous Hypomycetes from cereals, *Appl. Environ. Microbiol.*, 51, 1235, 1986.
133. Jorgensen, J., Changes in the microflora of barley seed stored with a high moisture content, *Tidsskr. Planteavl.*, 74, 425, 1970.
134. Mills, J. T., Sinha, R. N., and Wallace, H. A. H., Multivariate evaluation of isolation techniques for fungi associated with stored rape seed, *Phytopathology*, 68, 1520, 1978.
135. Wicklow, D. T. and Hesseltine, C. W., Fluorescence produced by *Aspergillus flavus* in association with other fungi in autoclaved corn kernels, *Phytopathology,* 69, 589, 1979.
136. Kaspersson, A., Detection of fungi in stored grain, British Crop Protein Council Monograph No. 37, 1987, 71.
137. Tuite, J. and Foster, G. H., Control of storage diseases of grain, *Annu. Rev. Phytopathol.*, 17, 343, 1979.
138. Steele, J. L., Saul, R. A., and Hukill, W. V., Deterioration of shelled corn as measured by carbon dioxide production, *Trans. ASAE,* 12, 685, 1969.
139. Donald, W. W. and Mirocha, C. J., Volatiles of grain storage fungi, *Proc. Am. Phytopathol. Soc.*, 1, 104, 1974.

140. Brookhouser, L. W. and Weinhold, A. R., Induction of polygalacturonase from *Rhizoctonia solani* by cotton seed and hypocotyl exudates, *Phytopathology,* 69, 599, 1979.
141. Hayman, D. S., The influence of temperature on the exudation of nutrients from cotton seeds and on preemergence damping-off by *Rhizoctonia solani, Can. J. Bot.,* 47, 1663, 1969.
142. Hayman, D. S., A note on the quantitative determination of carbohydrate in exudate from single cotton seeds and its significance, *Can. J. Bot.,* 47, 1521, 1969.
143. Kraft, J. M. and Erwin, D. C., Stimulation of *Pythium aphanidermatum* by exudates from mung bean seeds, *Phytopathology,* 57, 866, 1967.
144. Loria, R. and Lacy, M. L., Mechanism of increased susceptibility of bleached pea seeds to seed and seedling rot, *Phytopathology,* 69, 573, 1979.
145. Schlub, R. L. and Schmitthenner, A. F., Effects of soybean seed coat cracks on seed exudation and seedling quality in soil infested with *Pythium ultimum, Phytopathology,* 68, 1186, 1978.
146. Short, G. E. and Lacy, M. L., Carbohydrate exudation from pea seeds: effect of cultivar, seed age, seed color, and temperature, *Phytopathology,* 66, 182, 1976.
147. Clarke, J. H., MacNicoll, A. D., and Norman, J. A., Immunological detection of fungi in plants including stored cereal, *CAB Int. Biodeterioration Suppl.,* 22, 123, 1986.
148. Dewey, F. M., MacDonald, M. M., and Phillips, S. I., Development of monoclonal antibody-ELISA-DOT Blot and Dip-Stick immunoassays for *Humicola lanuginosa* in rice, *J. Gen. Microbiol.,* 135, 361, 1989.
149. Phillips, S. I., Twiddy, D. R., Wareing, P. W., and Dewey, F. M., Monoclonal antibodies for the detection of fungi in stored products, in Fungi and Mycotoxins in Stored Products, Champ, B. R., Highley, E., Hocking, A. D., and Pitt, J. I., Eds., Proc. Int. Conf. Bangkok, ACIAR Proc. No. 36, 1991.
150. Martin, S. L., Tuite, J., and Diekman, M. A., Inhibition radio-immunoassay for *Aspergillus repens* compared with other indices of fungal growth in stored corn, *Cereal Chem.,* 66, 139, 1990.
151. Nakakita, H., Katsumata, Y., and Ozawa, T., The effect of phosphine on respiration of rat liver mitochondria, *J. Biochem.,* 69, 589, 1971.
152. Navarro, S., Donahaye, E., and Calderson, M., Observations on prolonged grain storage with forced aeration in Israel, *J. Stored Prod. Res.,* 5, 73, 1969.
153. Sauer, D. B. and Burrough, R., Fungal growth, aflatoxin production and moisture equilibrium in mixtures of wet and dry corn, *Phytopathology,* 70, 516, 1980.
154. McGee, D. C. and Misra, M. K., Soybean oil application for dust suppression and control of storage molds in corn and soybeans, *Iowa Seed Sci.,* 10, 6, 1988.
155. White D. G. and Toman, J., Jr., Effects of postharvest oil and fungicide application on storage fungi in corn following high temperature drying, *Plant Dis.,* 78, 38, 1994.
156. Matz, S. A. and Milner, M., Inhibition of respiration and preservation of damp wheat by means of organic chemicals, *Cereal Chem.,* 28, 196, 1951.
157. Herting, D. C. and Drury, E. E., Antifungal activity of volatile fatty acids on grains, *Cereal Chem.,* 51, 74, 1974.
158. Dhanraj, K. S., Chohan, J. S., Sunar, M. S., and Sone Lal, Preliminary studies in the use of luprosil in the prevention of microbial spoilage of wet maize during storage, *Indian Phytopathol.,* 26, 63, 1973.
159. Schroeder, H. W., Sodium propionate and infra-red drying for control of fungi infecting rough rice (*Oryza sativa*), *Phytopathology,* 54, 858, 1964.

160. Kumar, S., Sinha, R. K., and Prasad, T., Propionic acid as a preservative of maize grains in traditional storage in India, *J. Stored Prod. Res.*, 29, 89, 1993.
161. Kozakiewicz, Z. and Clarke, J. H., Toxicity of propionic acid to some pre-harvest and post-harvest fungi of stored grain, in 2nd Int. Congr. Plant Pathol., St. Paul, MN, 1973, 60.
162. Hall, G. E., Hill, L. D., Hatfield, E. E., and Jensen, A. H., Propionic-acetic acid for high moisture corn preservation, *Trans. Am. Soc. Agric. Eng.,* 17, 379, 1974.
163. Bothast, R. J., Black L. T., Wilson, L. L., and Hatfield, E. E., Methylene-bis propionate preservation of high moisture corn, *J. Anim. Sci.,* 46, 484, 1978.
164. Sauer, D. B., Hodges, T. O., Burroughs, R., and Converse, H. H., Comparison of propionic acid and methylene bis propionate as grain preservatives, *Trans. Am. Soc. Agric. Eng.,* 18, 1162, 1975.
165. Zwick, W., A new conservation method for tapioca and maize, in Proc. Indian Sci. Congr., Kharagpur, 1972, 8.
166. Aly, H. Y., Control of storage molds of moist corn grains during storage, *Fac. Agri. Univ. Cairo Bull.*, 43 (Suppl. 1), 353, 1992.
167. Hall, J. S. and Harman, G. E., Protection of stored legume seeds against attack by storage fungi and weevils: mechanism of action of lipoidal and oil seed treatments, *Crop Prot.*, 10, 375, 1991.
168. Mayna, R. Y., Harper, G. A., Franz, A. O., Lee, L. S., and Goldblatt, L. A., Retardation of the elaboration of aflatoxin in cotton seed by impermeability of the seed coats, *Crop Sci.,* 9, 147, 1969.
169. Halloin, J. M., Lee, L. S., and Cotty, P. J., Pre-ripening damage to cotton seed by *Aspergillus flavus* is not influenced by seed coat permeability, *J. Am. Oil Chem. Soc.,* 68, 522, 1991.
170. Halloin, J. M., Deterioration resistance mechanisms in seeds, *Phytopathology,* 73, 335, 1983.
171. Kirsi, M., Proteinase inhibitors in germinating barley embryos, *Physiol. Plant,* 32, 89, 1974.
172. Halim, A. H., Wassom, C. E., Mitchell, H. L., and Edmunds, L. K., Suppression of fungal growth of isolated trypsin inhibitors of corn grain, *J. Agric. Food Chem.*, 21, 118, 1973.
173. Chokethaworn, N., Suttajit, M., Vanitanakome, N., Limtraku, P., Iamsupasit, C., and Chutkaew, C., Chitinase activity in maize seeds and their fungal resistance, in *Fungi and Mycotoxin in Stored Products,* Champ, B. R., Highley, E., Hocking, A. D., and Pitt, J. I., Eds., Proc. Int. Conf. Bangkok, ACIAR Proc. No. 36, 1991, 254.
174. Zohri, A. A., Suitability of some cowpea cultivars for *Aspergillus flavus* growth and aflatoxin production, *African J. Mycol. Biotechnol.,* 1, 87, 1993.
175. Tuite, J., Koh-Knox, C., Stroshine, R., Cantone, F. A., and Bauman, L. F., Effect of physical damage to corn kernels on the development of *Penicillium* species and *Aspergillus glaucus* in storage, *Phytopathology,* 75, 1137, 1985.
176. Mazzani, C. and Layrisse, A., Field resistance of genotypes of peanut (*Arachis hypogaea L.*) to seed infection by *Aspergillus* spp., *Phytopathol. Mediterr.*, 31, 95, 1992.
177. Pupipat, U., Juthawantana, P., Kittithamkul, C., Sukjaimit, S., and Wongyala, P., Field and laboratory techniques for evaluating resistance of maize to *Aspergillus flavus,* in *Fungi and Mycotoxin in Stored Products,* Champ, B. R., Highley, E., Hocking, A. D., and Pitt, J. I., Eds., Proc. Int. Conf. Bangkok, ACIAR Proc. No. 36, 1991, 255.

178. Singh, A. K., Mehan, V. K., Mengesha, M. H., and Jambunathan, R., Inhibition sites, leachates, and fungal colonization of seeds of selected groundnut germplasm lines with different seed test colours, *Oleagineax*, 47, 579, 1992.
179. Waliyar, F., Ba, A., Hassan, H., Bonkoungou, S., and Bose, J. P., Sources of resistance to *Aspergillus flavus* and aflatoxin contamination in groundnut genotypes in West Africa, *Plant Dis.*, 78, 704, 1994.
180. Scott, G. E. and Zummo, N., Kernel infection and aflatoxin production in maize by *Aspergillus flavus* relative to inoculation and harvest dates, *Plant Dis.*, 78, 123, 1994.

CHAPTER 12

Mycotoxins and Mycotoxicoses

Mycotoxin is derived from the Greek word "mykes," meaning fungus and the Latin word "toxicum" meaning poison.[1] Mycotoxins are secondary fungal metabolites that cause pathological or undesirable physiological responses in humans and other animals (Table 12-1). Mycotoxicoses are diseases caused by the ingestion of foods or feeds contaminated by mycotoxins.[2] References to mycotoxin poisoning appear in the Bible,[3] and ergotism has been blamed for traumatic incidents of Western history, from the French Revolution to the Salem witch trials in the United States.[4] *F. moniliforme* associated with maize seeds caused human toxicosis in 1881. This is almost 60 years earlier than 1940, when the first record of toxicosis in animals caused by aflatoxin-producing Aspergilli was thought to have occurred; the first record does not appear until 30 years later in humans.[5]

The study of mycotoxicology actually began in 1891, when Sakaki, in Japan, demonstrated that an ethanol extract from moldy, unpolished yellow rice was fatal to dogs, guinea pigs, and rabbits, causing paralysis of the central nervous system. The sale of yellow rice subsequently was banned in Japan in 1910.[6,7] Penicillic acid, isolated from *P. puberulum* on moldy maize in Nebraska, was toxic to animals when injected at 200 to 300 mg/kg body weight.[8] This was the first reliable account of toxin production by a fungus in pure culture.[7] Miyake et al.[9] published a study on *P. toxicarium*, which produced a highly toxic metabolite named citreoviridin from yellow rice.

Alimentary toxic aleukia (ATA) was described prior to 1900 and was associated with the ingestion of overwintered grain. During World War II, Russians were forced to eat grain left in the field. Thousands of people were affected, resulting in the elimination of an entire village. This syndrome was caused by T-2/neosolaniol/T-2 tetraol toxicosis produced by *F. sporotrichioides*, which grows on wet grain left in the field.[10] Red mold poisoning was reported in many parts of rural Japan in the 1950s[11] and was due to deoxynivalenol (DON) contamination. Thousands of people were affected by DON toxicosis in the Kashmir Valley in India in 1987. Flour that had approximately 10 mg/kg trichothecenes was used

Table 12-1 Possible Involvement of Mycotoxins in Some Human Diseases

Implicated mycotoxins	Mycotoxicosis
Aflatoxin	Aflatoxicosis (acute) Hepatocarcinogensis Indian childhood cirrhosis Reye's syndrome Encephalopathy and fatty degeneration of the viscera
Amatoxins, agaritine amanitins, phallotoxins, etc.	Kwashiorkor, mushroom poisoning
Citreoviridin	Cardiac beriberi
Cyclopiazonic acid	Kodua poisoning
Ergot alkaloids	Ergotism
Gliotoxin	Lung disease
Moniliformin	Onyalai disease
Ochratoxin	Balkan nephropathy, renal tumors
Trichothecenes	
DAS, HT-2	Alimentary toxic aleukia
FUS-X	Akakabi-byo disease
F. oxysporum metabolites	Kaschin-Bech disease
S. alternans metabolites	Stachybotryotoxicosis
Fusarium metabolites	Esophageal cancer
Fusarium metabolites	Red Mold disease
DON, DAS, T-2	Yellow Rain
Tenuazonic acid	Onyalai disease
Zearalenone	Cervical cancer
Premature thelarche	

Adapted from Pohland, A. E. and Wood, G. E., in *Mycotoxins in Food*, Krogh, P., Ed., Academic Press, New York, 1987, 35. With permission.

as food.[12] Fusarins were described in 1984. Fusarin C is a mutagen of similar potency as sterigmatocystin and aflatoxin B_1. Fusarins occurred in maize seeds in high esophageal cancer areas of China and South America.[13]

Turkey X disease, caused by an aflatoxin produced by *Aspergillus flavus*, resulted in the death of 100,000 turkey poults in England in 1960. Moldy peanut meal from Africa and South America was fed to the poults, which led to the first experimental evidence that a fungus could produce toxins. The evidence was believed to date from 1960. However, various food contaminations that were correlated with aflatoxins resulted in acute or chronic diseases in humans in various parts of the world before 1960. During World War II, Korean prisoners of war who consumed fungal-damaged maize suffered from toxic hepatitis. Sporadic cases of acute liver disease due to consumption of aflatoxin-contaminated cassava in Uganda and rice in Taiwan were reported in 1967.[14]

Outbreaks of aflatoxin hepatitis in humans, with supporting epidemiological, mycological, mycotoxic, and pathological evidence, were reported in 1975 from India and in 1982 from Kenya.[15,16] In developing countries, the poorest quality seeds have been fed to animals. Seeds containing mycotoxins resulted in low animal productivity, thus contributing indirectly to poor human nutrition. In extreme cases, there could have been mycotoxic residues in eggs, meat, or milk.[17]

FAO of the United Nations estimated that 25% of the world's food crop is affected by mycotoxins each year.[18] The significance of mycotoxins in international trade is increasingly being recognized by both developed and developing countries. During the past few years, the export of agricultural commodities, such as copra, cottonseeds, peanuts, and pistachio nuts or their derivatives have been affected. Often, and especially in developing countries, the best quality of these commodities, which are free from mycotoxins, are exported, while the substandard products are distributed and sold within the country. This practice has the potential of unfavorable consequences in either the health of the local population or productivity of the animals fed with contaminated or substandard feed.[19] Thai maize has been noted for its bright yellow color and high protein content. However, samples have been found to contain unacceptably high aflatoxin levels and are therefore discounted or rejected by foreign buyers.[20] Hence, aflatoxin contamination has posed serious problems in commerce and international trade because of stringent quality standards on aflatoxin contamination by importing countries.[21]

I. MYCOTOXINS

Currently, over 300 mycotoxins have been described; most research has been concentrated on aflatoxins and trichothecenes and on secondary groups, such as citrinin, cyclopiazonic acid, ochratoxins, patulin, sterigmatocystin, and zearalenone. The chemical structures of some important mycotoxins are shown in Figure 12-1.[2,22]

A comparison of the major mode of action of aflatoxin, ochratoxin, and trichothecene on the immune response in animals has been published.[23]

A. Aflatoxins

Aflatoxins are mycotoxins produced by *Aspergillus flavus* and *A. parasiticus*. Other species may produce other toxins. *A. flavus* produces aflatoxin B_1 and B_2, and *A. parasiticus* aflatoxin G_1, G_2, M_1, as well as B_1 and B_2. Aflatoxin B_1 is the most potent and carcinogenic naturally occurring substance known, causing liver damage to most domestic and experimental animals and humans.[24,25] Aflatoxins, the most common mycotoxins, have been investigated all over the world (Table 12-2). They are basically difuranocoumarin compounds and include aflatoxin B_1, B_2, B_{2a}, B_3, G_1, GM_1, G_2, G_{2a}, M_1, M_2, M_{2a}, GM_2, P_1, Q_1, R_0, RB_1, RB_2, AFL, AFLH, AFLM, and methoxy, ethoxy,and acetoxy derivatives. The most common aflatoxins are B_1, found mainly in agricultural commodities, and M_1, secreted in milk by dairy cattle consuming aflatoxin-contaminated commodities.[26] M_1 is a toxic metabolite derived from B_1, which occurs in milk when aflatoxin-contaminated feed is ingested by dairy cattle.[14]

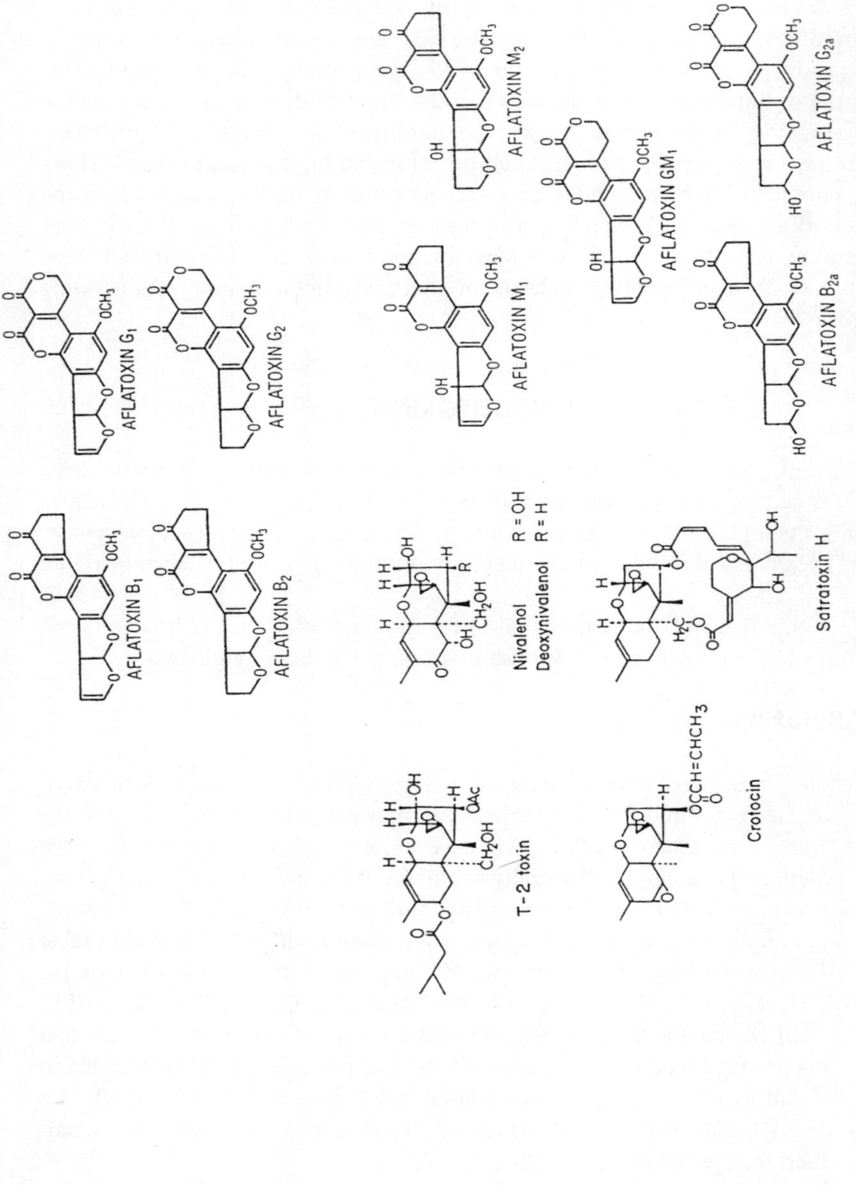

Figure 12-1 Chemical structures of some important aflatoxins and trichothecenes. (Adapted from Palmgren, M. S., Hayes, A. W., and Ueno, Y., in *Mycotoxins in Food*, Krogh, P., Ed., Academic Press, New York, 1987, 68 and 123. With permission.)

Table 12-2 Occurrence of Aflatoxins in Food and Feeds

General sources	Crop/product
Animal feeds	Extractions from coconut, cottonseeds, and peanut
Animal products	Cheese, fish, milk, shrimp
Cereal grains	Maize, rice, sorghum, wheat
Fermented products	Alcohol, beer, sauces, wines
Oil seeds	Coconut, cottonseed, linseed, peanut, soybean, sunflower
Pulses	Various beans (Africa), peas (Africa/Asia)
Root crops	Cassava, sweet potato
Tree nuts	Almonds, arecanuts, peanuts, pistachio, walnuts
Vegetable oils	Coconut oil, cottonseed oil, olive oil, peanut oil
Vegetable products	Cocoa, coffee, figs, peaches

From Bhat, R. V., Aflatoxins: successes and failures of three decades of research, in *Fungi and Mycotoxins in Stored Products*, Champ, B. R., Highley, E., Hocking, A. D., and Pitt, J. I., Eds., Proc. Int. Conf. Bangkok, ACIAR Proc. No. 36, 1991, 170. With permission.

In most cases, aflatoxins are formed after harvest, particularly when harvest takes place during cyclones, floods, or unseasonal rains, or because of the improper storage of insufficiently dried commodities. The occurrence of aflatoxins in maize and peanuts in the field has received attention, particularly in the United States. Factors that affect aflatoxin production include the fungal isolate, inoculum potential, crop cultivar, relative humidity, temperature, insect–microbial interaction, and storage conditions.[14] A number of animals, such as cattle, chickens, dogs, ducklings, ferrets, guinea pigs, hamsters, horses, mice, monkeys, pigs, quail, rabbits, rainbow trout, rats, sheep, and swine, are affected by aflatoxins.[14]

A. flavus is a saprophyte or weak parasite that has increased parasitic ability above 30°C. Exposed maize silk is susceptible and provides an infection court to kernels, where it produces aflatoxin.[27] Aflatoxin concentration (AC) in maize can be determined using the following formula:

$$\text{AC} = \text{inoculum potential} \times \text{disease potential}$$

where inoculum potential = inoculum density (number of propagules having a variable amount of toxin-producing ability) and disease potential = the extent of predisposition of the host.[28]

A flavus and its aflatoxin can cause serious problems in maize prior to harvest.[29] Invasion of *A. flavus* and aflatoxin formation occurs in the field prior to harvest and during curing.[30] *A. flavus* infection in peanut has been significantly correlated with aflatoxin production.[31]

Aflatoxins cause human fatalities in the Third World when heavily contaminated maize and rice are consumed under famine conditions. Diseases such as hepatitis, hepatocarcinoma, and Reye's Syndrome are caused by consumption of aflatoxin-contaminated food. Aflatoxins interfere with the immune system in certain animals[26] and are carcinogenic and mutagenic.

The liver is the major target organ of aflatoxins. Malnourished animals are more susceptible than well-nourished ones.[14] Aflatoxin carcinogenicity in laboratory animals including primates has been proven unequivocally. In addition to liver tumors, AFB_1 induced tumors in the colon, kidneys, nervous system, and stomach. Immunocytochemical studies have quantified the DNA adduct level in various tissues.[32] The International Agency for Research on Cancer placed aflatoxin on its list of human carcinogens.[33] A number of methods, namely BGYG, HPLC, immunological, minocolum, and TLC, have been used for aflatoxin analysis. Immunological assay kits are available for analyzing for aflatoxin B_1, total aflatoxins (B_1, B_2, G_1, and G_2) and M_1.[2]

Lipids are important in determining growth and AFB_1 production by *A. parasiticus*. Peanut cotyledons with high lipid content support a high production of AFB_1.[34] Aflatoxin concentrations were significantly ($P = 0.0001$) correlated with *A. flavus*-group populations in shells and seeds ($r = 0.70$ and 0.57, respectively).[35] Different *Aspergilli* (*flavus, parasiticus*, and *nidulans*) produce different intermediates and end products of the aflatoxin pathway and are useful in studying the maize–*Aspergillus*–mycotoxin interaction. NoR mutants may be useful tools to identify likely infection sites in maize seeds, and the genetically characterized *A. nidulans* may be useful in helping identify global regulatory mechanisms in the maize–*Aspergillus*–mycotoxin interaction.[36]

B. Ochratoxin A (OA)

Ochratoxin A is produced by *Aspergillus alliaceus, A. melleus, A. ochraceus, A. ostianus, A. petrakii, A. sclerotiorum*, and *A. sulphureus*; and *Penicillium aurantiogriseum, P. chrysogenum, P. puberulum, P. purpurescens, P. variabile, P. verrucosum*, and *P. viridicatum*.[37] The major source of *P. verrucosum* and its toxin is barley grown in temperate zones. Because it is fat soluble and not readily extracted, it accumulates in fatty tissues. In northern Europe, where barley, which may be contaminated with *P. verrucosum*, forms a major diet of pigs, bacon and pork may contain high levels of the toxin. Thus ochratoxin may pose a serious health risk to humans, especially in rural areas where pigs are not subjected to vigorous inspection.[7] In humans, it can be responsible for kidney degeneration, which can lead to death in extreme cases.[7] It is an acute nephrotoxin, with an oral LD_{50} of 20 mg/kg in young rats and 3.6 mg/kg in day-old chicks. It is lethal to dogs, mice, and pigs.[38] It can act as a carcinogen, hepatotoxin, nephrotoxin, mutagen, teratogen, and immunosuppressive agent. It affects enzyme activity, including those involved with carbohydrate metabolism and protein synthesis, and specifically phosphoenolpyruvate carboxykinase and phenylalanyl-tRNA synthetase. It occurs in cereals, grains, and other plant products, animal feeds, meats, and human tissues in countries throughout the world.[37]

C. Trichothecenes

Trichothecenes are produced by several fungal genera: *Cephalosporium, Cylindrocarpon, Fusarium, Myrothecium, Phomopsis, Stachybotrys, Trichothecium*, and *Verticimonosporium*.[22] However, most trichothecenes that have been isolated and characterized are from *Fusarium* (Table 12-3). There are about 100

Table 12-3 *Fusaria* Producing Trichothecenes

Anamorph	Teleomorph	Trichothecenes produced
F. avenaceum	*Gibberella avenacea*	Neosolaniol, T-2 toxin, diacetoxyscripenol, deoxynivalenol (= vomitoxin)
F. culmorum		Neosolaniol, T-2 toxin, HT-2 toxin, diacetoxyscirpenol
F. equiseti	*Gibberella intricans*	Diacetoxyscirpenol, neosolaniol, T-2 toxin
F. graminearum	*Gibberella zeae*	Fusarenon X, nivalenol
F. baccata	*Gibberella lateritium*	Diacetoxyscirpenol, T-2 toxin
F. sporotrichioides		T-2 toxin

known trichothecenes, but information on their occurrence in foodstuffs is limited mainly to dexoxynivalenol (vomitoxin) (DON), nivalenol (NIV), T-2 toxin (T-2), HT-2 toxin (4-deacetyl T-2, HT-2), and diacetoxyscirpenol (DAS), all produced by *Fusarium*. *F. graminearum* is the primary species responsible for natural contamination of cereals by DON and NIV. *F. sporotrichioides* is a major producer of T-2 and HT-2 toxins, and *F. poae* is a minor T-2 producer. *F. equiseti, F. poae, F. roseum, F. sulphureum*, and other species produce DAS.[22,39] Fungi differ in their ability to produce individual tricothecenes. For example, *F. graminearum* isolates from cereals in northern Japan produce DON, while those that produce NIV are found in central Japan.[40]

Trichothecenes are divided into three groups: Group A is composed of basic Δ^9-trichothecenes, such as the T-2 toxin, diacetoxyscripenol and neosolaniol; Group B is composed of 8-ketotrichothecenes, such as nivalenol and deoxynivalenol (vomitoxin); and Group C is composed of macrocyclic trichothecenes, such as roridin A and verrucarin A.[2] These are known as the agents of ATA (*F. graminearum*), yellow rain (*F. roseum*), and acute toxicoses (*F. poae*). Little was known about trichothecenes until the late 1970s. Acute trichothecene toxicity depends on the chemical structure of the toxin. Group A is more toxic than Group B. In addition to ATA, other known symptoms, toxicities, and toxicoses caused by trichothecenes include vomiting, feed refusal, skin irritation, hemorrhagic lesions in the gut, necrosis of the esophageal mucosa, teratogenicity, and carcinogenicity.[2] There are reports of contamination of agricultural products by deoxynivalenol, nivalenol, and T-2 toxin. Cereal seeds and feed worldwide often are contaminated with trichothecenes.[22]

For the chemical analysis of trichothecenes, more reliance should be placed on data from gas chromatography-mass spectrometry (GC-MS) with selected ion monitoring rather than from thin-layer chromatography (TLC). Capillary gas chromatography (Cap GC) is more specific than packed column GC or liquid chromatography. In some cases, trichothecenes have been isolated and characterized from foodstuffs by procedures such as infrared spectroscopy (IR), molecular magnetic resonance spectroscopy (NMR), and mass spectrometry[22] and by HPLC and immunological assays such as ELISA.[2] Chemical analytical methods are difficult and time-consuming, and immunological techniques are highly useful for the analysis of number of samples. Assay kits using ELISA are commercially available.[2]

D. Ergot Alkaloids

Ergot alkaloids produced by *Claviceps* are associated with rye and triticale in northern latitudes.[18] Alkaloid consumption may result in gangrene, leading to necrosis of extremities; central nervous system effects include ataxia, convulsions, and paralysis, and gastrointestinal disorders. These afflictions result in reduced weight gains, reproductive efficiency, and milk production.[18] *Acremonium,* an endophyte that infects many varieties of tall fescue in North America, produces ergot alkaloids.[41] Ruminants grazing on infested plants during cold winters develop lameness (fescue foot), appetite loss, arched back, sloughing of hooves, foot deformities, swelling of hind limbs, and weight loss.[42]

E. Patulin

Patulin is produced by many species of *Aspergillus* (*A. clavatus*) and *Penicillium (P. patulum* and *P. expansum)* spp.[26] Patulin initially was considered as a cure of common colds, but later it was found too toxic for use.[7] Patulin was also at first considered carcinogenic, but now this is considered unlikely. The toxicity by oral intake is low, so it poses little threat to humans.[7] It is produced by *P. expansum* in rotten apples and has been reported in apple juice. However, fermentation during the production of cider destroys the toxin.[43]

F. Zearalenone

Zearalenone, which is produced by *F. roseum* and *F. tricinctum,* occurs along with other trichothecenes. It results in estrogenic and anabolic activity in animals and is associated with reddening and swelling of the vulva, uterine enlargement, and vaginal and rectal prolapse.[18] It occurs in cereals, primarily maize, and in animal feeds.[26] It may be present in cereal crops in cooler, moist regions of the northeastern United States and in eastern Canada.[18]

G. Penicillic Acid

Penicillic acid is produced by *P. aurantiogriseum* and *P. puberulum* and occurs in maize and poultry diets. It produces carcinogenic, cardiotoxic, and cytotoxic effects. However, it has not been shown to cause animal or human mycotoxicosis.[26]

H. Sterigmatocystin

Produced by species of *Aspergillus, Bipolaris, Drechslera,* and *Penicillium*, sterigmatocystin is one of the precursors in the aflatoxin biosynthetic pathway.[44] This toxin has been shown to be mutagenic[45] and carcinogenic.[46]

I. Citreoviridin

Citreoviridin is produced by *P. ochrosalmoneum* and *P. citreonigrum.*[7] It is a neurotoxin, acutely toxic to mice, with an intraperitoneal and oral LD_{50} of 7.5

mg/kg and 20 mg/kg, respectively.[6] The disease, beriberi, known in Japan as acute cardiac beriberi, has been associated with consumption of moldy "yellow rice," which resulted in a ban on the sale of yellow rice in Japan in 1910.[6] The disease subsequently disappeared.[7] In several animal species it causes vomiting, convulsions, ascending paralysis, and respiratory arrest.

J. Citrinin

Citrinin was discovered as an antibiotic during the 1940s but proved too toxic for therapeutic use. It is considered a hazardous mycotoxin.[7] It is produced by *P. citrinum*, *P. verrucosum*, and *Aspergillus terreus*. Citrinin is a nephrotoxin that occurs in apple juice, barley, oats, peanuts, rye, and wheat.[43] Human kidney damage occurs with prolonged ingestion. It is a significant renal toxin to monogastric domestic animals, including domestic birds, dogs, and pigs, and it results in watery diarrhea, increased food consumption, and reduced weight due to kidney degeneration in chickens, ducklings, and turkeys.[7] The oral LD_{50} in mice is 110 mg/kg.[38]

II. FUNGI KNOWN TO PRODUCE MYCOTOXINS

Important mycotoxins are produced by different species of fungi.[47] Mycotoxins may be produced by field fungi, such as *Cladosporium, Claviceps purpurea, Diplodia maydis, Fusarium, Phomopsis*, and *Sclerotinia sclerotiorum*, and by storage fungi, such as *Aspergillus* and *Penicillium. Aspergilli* are common saprophytes on dried agricultural products. A review on mycotoxins in seeds and their implication for stored product research has been published.[48] Some of the important toxigenic *Aspergilli* and their mycotoxins are summarized in Table 12-4.[49] Sources of primary inoculum of *A. flavus* are spores in the soil, mycelia that overwinter in crop debris and litter on the soil, insects, and sclerotia in the soil.[50] *A. flavus* spores carried by insects such as the stink bug *(Chorochroa sayi)* and lygus bug *(Lygus hesperus)* may serve as primary inoculum sources. Conidia production on floral parts and leaves can be important sources of secondary inoculum. Conidia are airborne and readily dispersed by air movements.[50]

Penicillia produce a range of toxic metabolites with varied structure that are toxic to animals or humans. Most are produced under wet conditions, are only acutely toxic, or are formed by species commonly found in soils or other nonfood sources. The most toxic species grow at low temperatures, and hence are commonly found in temperate rather than in tropical zones. The most important toxin is ochratoxin A, produced by *P. verrucosum,* which is common in European barley.[7] The mycotoxins produced by *Penicillium*, with their relative toxicity to human and animals, and less toxic mycotoxins are presented in Table 12-5.[51] A number of *Fusaria* produce mycotoxins (Table 12-3), but the important toxigenic species are *F. equiseti, F. graminearum, F. moniliforme, F. poae*, and *F. sporotrichioides,* known or strongly suspected to cause mycotoxicoses in animals and man. The mycotoxins produced by these species are presented.[52] The most impor-

Table 12-4 The Toxigenic Potential of some *Aspergillus*

Group	Potentially toxic metabolites
Aspergillus candidus	Chlorflavonin, terphenylin
A. clavatus	Patulin, ascladiol, cytochalasin E, tryptoquivaline
A. flavus	Aflatoxins, kojic acid, maltoryzine, aflatrem
A. fumigatus	Viriditoxin, gliotoxin, fumitremorgin A, B, fumagillin, verruculogen, TR-2
A. glaucus	Echinulin, xanthocillin X, gliotoxin
A. nidulans	Emerin, sterigmatocystin, nidurufin
A. niger	Malformins, oxalic acid
A. ochraceus	Ochratoxins, penicillic acid, destruxin B
A. terreus	Aranotin, gliotoxin
A. ustus	Austocytins, austamide, austdiol, deoxybrevianamide
A. versicolor	Sterigmatocystin, cyclopiazonic acid, versimide, nidurufin, sterigmatin

Adapted from Moss, M. O., in *Mycotoxins in Food*, Krogh, P., Ed., Academic Press, New York, 1987, 3. With permission.

tant toxigenic *Fusaria* are: *F. graminearum* (Grp 2), producing estrogenic, emetic, and feed refusal syndromes, and akakabi-byo mainly due to zearalenone and type B trichothecenes; *F. equiseti*, producing degnala disease and fescue foot, probably due to butenolide and type A trichothecenes, and tibial dyschondroplasia, probably due to type A trichothecenes, and hemorrhagic syndrome; *F. moniliforme*, producing human esophageal cancer and equine leucoencephalomalacia, probably due to fumonisins; *F. sporotrichioides,* producing alimentary toxic aleukia, hemorrhagic syndrome, fescue foot, bean hull poisoning, and akakabi-byo, possibly due to type A trichothecenes.[53,54]

Alternaria, a common contaminant of cereal seeds, produces an array of toxic secondary metabolites, including alternariol and its monomethyl ether, tenuazonic acid, altenuene, and altertoxins.[54,55] Feed grain may from time to time contain *Alternaria* toxins, which, if ingested, give rise to a toxicosis and reduced productivity.[56] *Alternaria* is a common contaminant of cereal seeds in Australia, and most isolates are toxigenic. Animals ingesting *Alternaria*-infested seeds may suffer production drops.[57] *A. alternata* toxin causes human esophageal cancer. It has been isolated from maize in areas of high incidence of esophageal cancer, and contamination by *A. alternata* in the area of high incidence of the disease is greater than that in an area of low incidence.[58]

III. FACTORS AFFECTING MYCOTOXIN PRODUCTION

Mycotoxin production is influenced by the same factors that affect seed infection. For example, aflatoxin contamination in maize is attributed to at least two mechanisms, high spore loads and increased drought stress during seed-fill,[28] coupled with temperatures favorable for growth of *A. flavus*.[28,59] The processes of seed deterioration and mycotoxin formation by a storage fungi are influenced

Table 12-5 A Partial List of *Penicillium* spp. That Produce Toxins

Anamorph	Toxin produced
Penicillium aurantiogrisum	Citrinin, cyclopiazonic acid, ochratoxin A, penicillic acid, penitrem A
P. brunneum	Rugulosin
P. chrysogenum	Ochratoxin A, patulin, penicillic acid
P. chrzaszczi	Citrinin
P. citreonigrum	Citrinin, citreoviridin
P. citrinum	Citrinin
P. claviforme	Patulin
P. crustosum	Penitrem A
P. expansum	Patulin, citrinin
P. fernelliae	Penicillic acid
P. granulatum	Patulin
P. griseofulvum	Patulin
P. implicatum	Citrinin
P. islandicum	Cyclochlorotin, islanditoxin, luteoskyrin
P. lividum	Citrinin
P. melinii	Patulin
P. notatum	Xanthocillin X
P. novae-zeelandiae	Patulin
P. paraherquei	Verruculogen
P. paxilli	Paxilline
P. piscarium	Fumitremorgin B, penicillic acid, verruculogen
P. purpurescens	Citrinin, ochratoxin A
P. purpurogenum	Glaucanic acid, rubratoxins
P. roquefortii	Citrinin, patulin, P.R. toxin, roquefortine
P. rugulosum	Patulin
P. spinulosum	Citrinin, penitrem A
P. terrestre	Patulin
P. variabile	Ochratoxin A, patulin, rugulosin
P. verrucosum	Verruculogen
P. viridicatum	Citrinin, ochratoxins, rubrosulphin, viomellein, viopurpurin, xanthomegnin
P. wortmannii	Rugulosin

Adapted from Moss, M. O., in *Mycotoxins in Food*, Krogh, P., Ed., Academic Press, New York, 1987, 3. With permission.

by many factors.[60] There was a significant correlation between infection of peanuts by *A. flavus* and colonization (r = .950) and colonization and aflatoxin content (r = 0.387). Genotypes differed significantly for percent infection and colonization. However, there was no relation between sugar content and infection, colonization, or aflatoxin content.[61] Seeds are contaminated by fungal spores at the beginning of storage, and fungal growth depends on available heat, oxygen, time, and moisture; from these spores subsequent toxin development depends on the presence and predominance of the toxigenic strains, sufficient warmth for fungal development, a substrate of suitable composition for that strain, and time for fungal metabolism to produce toxin.[60] Factors that effect mycotoxin production follow.

A. Host Species

Substrate factors are involved in the contamination process, since the natural occurrence of aflatoxins is restricted to certain agricultural commodities, although *A. flavus* will grow on and form aflatoxins in many sterilized inoculated substrates. Invasion of cotton bolls and subsequent aflatoxin formation by *A. flavus* in cottonseeds depends on fungal growth on lint or fibers, unless penetration occurs through the vascular system.[62]

Aflatoxin production varies among cultivars. Fifteen maize cultivars were grouped on the basis of the quantity of aflatoxin B_1 produced by *A. parasiticus*: (1) resistant Him-123, Ganga-2, Ganga-5 and EH 2420; (2) moderately resistant EH 2310, EH 2380, Hi Starch and Hunis; and (3) susceptible Vikram, Decann-101, Vijay, Early Composite, A51-54, NLD, and Diara. This variation was due to seed inhibitors.[63] Nagarajan and Bhat[64] reported opaque-2 maize resistant to *A. parasiticus* aflatoxin under artificial conditions and identified a low molecular weight protein as the inhibitory factor.

Seed invasion of peanuts occurs under favorable conditions by direct or indirect pod penetration. However, seeds of cultivars whose pods are penetrated by *A. flavus* may escape with difficulty. Resistant peanut pod tissues contain compact sclerotized cells with a high lignin content, and resistant seed coats contain tightly packed cells with an outer layer of wax platelets.[65] The hilum of resistant cultivars is small and closed; in susceptible ones it is large and open, thus allowing for a higher level of infection.[66] Since resistance is due to anatomical features, any damage to the seed coat can render a resistant type susceptible. All cultivars, both resistant and susceptible, should be stored under humidity and temperature conditions unfavorable for fungal deterioration, because under conditions favorable for fungal development, resistant genotypes may become contaminated with aflatoxin.[67]

Seed surface lipids (SSL) play a key role in supporting aflatoxin formation in oily and starchy seeds following infection by *A. parasiticus*. Fungal growth and aflatoxin production occur on oily seeds when SSL levels are greater than 0.15% of the total oil content of seeds, with a ratio between triglycerides and free-fatty acids of SSL >1. If free-fatty acids predominate over triglycerides, growth can be inhibited because of the toxicity of the unsaturated free-fatty acids of SSL to *Aspergillus*. *A. parasiticus* proliferates on starchy seeds in the germ region, the area richest in lipids.[68]

A. flavus infection and aflatoxin contamination are lower in peanut seeds of all genotypes harvested from vertisols (silty clay loam) than in those from alfisols (light sandy and red sandy loam). Vertisols also have significantly lower populations of *A. flavus* than alfisols. Irrespective of soil types, *A. flavus*-resistant genotypes show lower levels of seed infection than do *A. flavus*-susceptible genotypes.[69] Sugar content of maize seeds is an overriding factor contributing to differences among genotypes for aflatoxin contamination by *A. flavus*. Mycelial growth is enhanced by the presence of a sugary endosperm.[70] The maize cv. Suwan Composite, which has a tight, incomplete husk and an intermediate maturity period, showed the maximum natural contamination of aflatoxin from *A.*

flavus. The cv. M_9, with a tight, complete husk and long (105 d) maturity period, was most susceptible. The cv. Diara Composite, with a loose, complete husk and 80-d maturity period, was less prone to aflatoxin contamination.[71] Maize seeds are borne singly on paired spikelets. The vascular system of any spikelet pair converges at their insertion into one of the vascular bundles serving the entire ear. The data of Mertz et al.[72] implied that toxin moved into the ear. If toxin moves independently of hyphae in infected ears, it must do so through the vascular system. Such movement would lead to a distinct pattern of toxin accumulation. Smart et al.[73] concluded that long-distance aflatoxin transport did not occur in infected ears independent of the hyphae. Cotton cultivars in which the intercarpellary membrane matures rapidly may have a lower risk of aflatoxin contamination by checking the spread of *A. flavus*.[74]

B. Fungal Invasion

A high infection level results in high mycotoxin production. Aflatoxin concentration in maize was correlated linearly with the number of *A. flavus*-infected seeds.[27] Seed infection and aflatoxin contamination always were greater in silk-inoculated plots than in naturally infected plots.[75] Simultaneous inoculation of wounded 28 to 32 d old cotton bolls with toxigenic and atoxigenic *A. flavus* isolates led to lower levels of aflatoxin B_1 (AFB_1) in the cottonseeds at maturity than in bolls inoculated with a toxigenic isolate alone. Less AFB_1 was detected when the atoxigenic isolates was introduced into the wound 1 d before inoculation with a toxigenic isolate than when both isolates were inoculated at the same time. In contrast, toxin levels at maturity were not reduced when the atoxigenic isolate was introduced 1 d after the toxigenic isolate.[76]

C. Agronomic Practices

1. Planting Time

Planting date directly affected aflatoxin production by influencing maturity time. Late-planted maize had a greater chance for preharvest aflatoxin production due to *A. flavus* than an early-planted crop.[59] Late harvesting of peanuts resulted in an increase in aflatoxin.[30]

2. Soil Type

There was a lower risk of preharvest *A. flavus* infection and aflatoxin contamination in peanuts grown in vertisols than in alfisols.[69]

3. Fertilizer

Plant stress associated with reduced fertilization increased the incidence of aflatoxin.[77,78] Maize produced under low N stress was a better substrate for preharvest aflatoxin production than that produced with adequate N; less aflatoxin

B_1 was detected in maize seeds in treatments receiving a high rate (145.7 kg/ha) than in those receiving a low rate (11.2 kg/ha) of N.[59] Aflatoxin concentration in maize seeds due to *A. flavus* was correlated negatively with yield and with silk, leaf, and grain N. Nitrogen stress significantly increased aflatoxin contamination. Nitrogen stress also may result from drought stress. Reduced uptake of water results in decreased translocation and mobilization of N.[79] The presence of dense populations of plants and a weed canopy in maize, which results in low fertility, contributes to aflatoxin contamination. The competition of weeds for available nutrients and water is a form of stress that has been attributed to the aflatoxin in maize seeds as well as leading to lower yields.[80]

4. Drought Stress

Stress conditions due to low rainfall or late planting can reduce yield, predispose plants to fungal infection, and increase aflatoxin production. Drought stress predisposed maize to aflatoxin contamination by increasing the exposure time of susceptible silks to airborne *A. flavus* spores.[81] Drought stress altered the uptake and translocation of N in maize.[82] The amino acid proline was essential for aflatoxin synthesis by *A. flavus* in peanuts.[83] High proline levels were found in maize under drought conditions[84] and N stress.[85] Low soil moisture also favored invasion of peanut pods and seeds by *A. flavus*, with subsequent aflatoxin production.[30] Although several factors contribute to levels of aflatoxin in a crop, water stress is a major factor affecting aflatoxin contamination, because subsoiling as well as irrigation reduces aflatoxin contamination.[75] Drought and temperature stresses cause increased accumulation and/or synthesis of carbohydrates and certain polypeptides and thus may enhance *A. flavus* invasion and aflatoxin production in peanut.[86]

5. Pod and Kernel Damage

Seeds from peanut pods with growth cracks or other such damage are prone to infection and aflatoxin production. The level of aflatoxin in sound peanut seeds and in those from broken pods was <0.005 and >2.000 μg/ml, respectively.[87] Invasion of pods by *A. flavus* prior to digging was associated with biological and physical damage to pod shells. Seeds inside broken pods had extensive fungal growth compared to those from nondamaged pods.[30]

Whole maize seeds have lower levels of external and internal aflatoxin B_1 than wounded kernels, indicating that the pericarp and aleurone layers contribute to the defense of the seed against the fungus.[88] No AFB_1 aflatoxin was produced on sound, viable cottonseeds inoculated with *A. flavus,* whereas cracked viable seeds gave levels similar to those produced on autoclaved seeds.[89] *A. flavus* spores inoculated onto viable and nonviable soybean seeds showed no difference in growth or sporulation. However, the rate of aflatoxin accumulation in viable seeds was lower than that in nonviable seeds. Studies with liquid cultures of *A. flavus* showed the presence of glyceollin at concentrations lower than that detected in viable seeds, which could reduce the accumulation rate of AFB_1. Under normal conditions the relatively low level of aflatoxin in soybean seeds could be due to elicitation of glyceollin production when attacked.[90]

6. Harvesting

Aflatoxin was 3 to 12 times higher in cottonseeds harvested near the end than at the beginning of harvest.[91]

D. Moisture Content

Moisture content and relative humidity surrounding a substrate are the most important factors for aflatoxin production. Invasion of peanut pods and seeds by *A. flavus* in the field occurred rapidly when kernel moisture content was 12 to 20%. After harvest, *A. flavus* invasion was most rapid when seed moisture content was 14 to 30%. Aflatoxin production in mature, sound seeds was limited by 83% relative humidity.[30] Aflatoxin can be formed in peanut seeds at moisture contents in equilibrium with relative humidity as low as 85%, but significant quantities are produced at 88, 90, and 99%.[50] When *A. flavus, F. graminearum*, and *P. viridicatum* grew on irradiated maize and rice seeds, only aflatoxin and zearalenone were formed in maize and only ochratoxin A in rice. Aflatoxin production was highest at 0.95 and 0.98 A_w (water activity) at 25°C, zearalenone at 0.98 A_w at 25°C and 0.95 A_w at 16°C, while large quantities of ochratoxin A were produced at 0.90, 0.95, and 0.98 A_w at 25°C.[92] Ochratoxin A reached maximum levels of 11.8 and 0.11 mg/kg at 19 and 15% moisture content after 44 to 48 weeks, while citrinin reached levels of 80.0 and 0.65 mg/kg, respectively, in the same period in wheat.[93]

E. Temperature

The highest level of aflatoxin B_1 and G_1, 3.4 and 2.6 µg/ml, respectively, was detected at 19.7% moisture and 34°C in maize but not at 3 or 9°C.[94] Infection of maize seeds by *A. flavus* was favored by warm (32 to 38°C) rather than by cool (21 to 26°C) temperatures.[27] Jones[28] suggested that high temperatures (>30°C) and high relative humidity (>85%) favored maize infection by *A. flavus* and accounted for the higher incidence of aflatoxin in southern rather than northern United States regions. The optimum temperature for aflatoxin B_1 and G_1 production in rice is 28°C.[95] Ochratoxin A produced by *P. verrucosum* was found in wheat at 20.5% moisture and 15 to 20°C.[96] Peanut seeds from nondamaged pods grown under drought conditions at a mean soil temperature of 24.8°C or lower were not contaminated. Under similar conditions, seeds grown at 25.7, 26.3, or 27.8°C had lower concentrations of aflatoxin, but at 29.0, 29.6, or 30.5°C, seeds were heavily contaminated. Seeds from pods grown at 31.3°C were aflatoxin-free, as were seeds from pods grown under irrigation, including an irrigated plot at 34.5°C.[97] *A. parasiticus* grew best and produced the highest amount of aflatoxins at 25°C and grew least with least aflatoxin at 12°C. No growth nor aflatoxin production by *A. parasiticus* was detected at 5°C.[98]

F. Oxygen and Carbon Dioxide Concentration

Aflatoxin production in peanut seeds decreased with increasing CO_2 from 0.03 to 100% and reduced O_2. Decreased aflatoxin occurred when O_2 was reduced from 5 to 1% in combination with 0, 20, or 80% CO_2 during storage for 2 weeks at 30°C and 6 weeks at 15°C.[99]

G. pH

A. parasiticus produced less growth and aflatoxin at pH 3.5 than at pH 5.0 or 6.5 at 12°C.[98]

H. Mycoflora Competition

Individual grains can be infected by a number of fungi. Microbial competition influences mycotoxin production. *A. flavus* has been associated with several other microorganisms in stored seeds. Microbial competition or microbial break down products lower aflatoxin levels in peanut seeds. These microorganisms either breakdown aflatoxin or restrict the development of *A. flavus*. *A. niger* prevented penetration of peanut pods by *A. flavus* and several other fungi.[30] The successful establishment of *A. flavus* in maize and subsequent aflatoxin contamination depended upon the sequence in which *A. flavus* reached the seeds in relation to colonization by other fungi and the biological properties of the coinvading fungi. No aflatoxins were detected when *A. niger* or *Trichoderma viride* were paired with *A. flavus* in maize at harvest.[100] This may have resulted in variation in aflatoxin levels among field samples. *A. alternata*, *Cladosporium cladosporioides*, and *Curvularia lunata* are less competitive than *A. flavus*. *Acremonium strictum*, *Fusarium moniliforme*, *Nigrospora oryzae*, *Penicillium funiculosum*, *P. oxalicum*, and *P. variabile* are more competitive than *A. flavus*. Aflatoxin B_1 was recovered in higher amounts (1.41 μg/ml) in the former than in the latter group pairings (0.85 μg/ml).[100] *Aspergillus parasiticus* was limited from full development by *A. flavus*. Hence, *A. flavus* was found routinely in naturally aflatoxin-contaminated maize, and *A. parasiticus* was found only rarely.[101] Aflatoxin B_1 production by an *A. flavus* isolate from maize was inhibited when *A. flavus* was in competition with other fungi.[102] Atoxigenic strains of *A. flavus* may have interfered with the contamination process by physically excluding the toxigenic strain during infection and/or competing for nutrients required for aflatoxin biosynthesis in cotton bolls.[103] Toxin production by *A. parasiticus* may have been reduced as a result of competition with *A. candidus*. When the number of *A. parasiticus* conidia in an inoculum was equal to or less than the number of *A. candidus* conidia, *A. candidus* became the dominant species in rice seeds after 7 d at 25 to 35°C. Only small amounts or no aflatoxin was produced under these conditions.[104]

A. flavus and *A. niger* are widespread colonists in developing maize ears and invade both wounded and noninjured seeds. Both species frequently infect the same seed. However, *A. niger* does not prevent aflatoxin formation by *A. flavus* when the two are wound-inoculated simultaneously into developing ears. This may be because the substrate pH of the wound-inoculated seeds is not lowered enough by *A. niger* to inhibit aflatoxin production.[105] *F. moniliforme* was recovered frequently from symptomless maize seeds from ears inoculated in the field with *A. flavus* in Mississippi. *F. moniliforme* was found to inhibit seed infection by *A. flavus* in inoculated ears and resulted in reduced aflatoxin contamination in these seeds.[106] Metabolites from *Neurospora* and *Rhizopus* inhibited growth and aflatoxin B_1 production of *A. flavus* and *A. parasiticus* in peanut. In addition, *Neurospora* and *Rhizopus* degraded aflatoxin B_1.[107] Seeds contaminated with trichothecenes produced by *Fusarium* may cause increased aflatoxin formation with the growth of *A. flavus* and *A. parasiticus*.[5]

I. Insect Damage

Insects can spread infection from seed to seed within a maize ear or provide sites for growth and sporulation on seed surfaces injured by feeding.[81] Damage to maize by the corn earworm, *Helicoverpa zea,* or by the European cornborer, *Ostrinia nubilalis,* resulted in 10 and 11% aflatoxin B_1 concentration, respectively, at harvest.[81] Damage by the fall army worm, *Spodoptera frugiperda,* along with prolonged drought resulted in high levels of aflatoxin B_1 contamination in the Alabama maize crop in 1977.[108] Although damage is not a prerequisite for aflatoxin formation, the incidence of *A. flavus* and levels of aflatoxin are higher in damaged seeds and other inedible components of peanuts than in edible, sound mature seeds.[97] Insect wounds provide infection courts and hasten seed dry down to levels more favorable for *A. flavus* growth and aflatoxin production than for other fungi. If the fungus is present on the surface of seeds, it will invade injured seeds.[109] In cotton, exit holes made by mature pink boll worm larvae resulted in increased infection by *A. flavus* and aflatoxin contamination.[110] Maize seeds damaged by *Carpophilus* spp., *Dolycoris indicus, Graptostephus servus,* and *Nezara viridula* had higher levels of BGY-fluorescence and aflatoxin contamination than insect-free samples.[111] Relatively little toxin was formed in bolls produced near the end of the season, and most toxin occurred in bolls near the soil line due to pink boll worm infestation.[112]

IV. EFFECTS OF MYCOTOXINS

Losses due to mycotoxins can be significant, depending on the amount of mycotoxin produced.[28] Losses are expressed in any one or a combination of the following:

A. Yield Reduction

Yield losses caused by aflatoxins have not been estimated accurately. In 1979 about 90% of the maize crop in the southeastern United States was contaminated with aflatoxins, resulting in an estimated loss of $32 million in North Carolina alone.[28,29] A significant ($P = 0.05$) correlation was reported between aflatoxin B_1

production and yield reduction.[59,81] Losses directly attributed to aflatoxin production exceeded $2 million in Alabama in 1977.[108] Yield reductions in maize were 523 and 610 kg/ha where aflatoxin B_1 levels were highest, 37 and 38 µg/ml, respectively.[113]

B. Inhibition of Seed Germination

Mycotoxins can inhibit seed germination. Lettuce hypocotyl elongation was inhibited by aflatoxin.[114] Aflatoxin B_1 inhibited chlorophyll formation and seed germination in cowpea.[115] Aflatoxin in soybean reduced germination and radicle length by 50% in malts.[116] Aflatoxin B_1 inhibited germination and respiration in maize seeds by inhibiting synthesis of α-amylase.[117] Aflatoxin B_1 inhibited chlorophyll, nucleic acids, and protein syntheses of germinating maize seeds, which resulted in albinism in affected seedlings.[118]

C. Units for Trade

Mycotoxin concentration may exceed government standards for trade and be unfit for human consumption.

D. Extra Cost for Seed Analysis

The mycotoxin problem in seeds requires additional costs for seed analyses to determine mycotoxin levels prior to marketing.

E. Extra Cost to Dry Infected Grain

If seeds are infested by fungi and contaminated by mycotoxins, they can be dried so that mycotoxin production is inhibited.

F. Additional Cost of Detoxifying

Mycotoxin-contaminated seeds require detoxification prior to release for marketing.

G. Mycotoxicoses

Intake of moldy seeds is one of the common causes of mycotoxicosis. Ergotism is the first known human mycotoxicosis. During the Middle Ages in central Europe, a disease that occurred irregularly was associated with the consumption of rye bread made with flour contaminated by toxigenic strains of *Claviceps paspali* and *C. purpurea*. The disease was known as St. Anthony's Fire because affected individuals got relief from burning sensations in the hands and feet by making a pilgrimage to St. Anthony's shrine. The relief resulted because the patient left the contaminated area. In the 17th century, it was found that alkaloids produced by ergot fungi were responsible for the disease.[119] Certain ergot alkaloids also are produced in culture by *Aspergilli* and *Penicillia*.[120] Ergot alkaloids result in peripheral vasoconstriction and gangrene and affect the nervous system.[119,121] These alkaloids in controlled amounts are used for the treatment of certain human diseases and to induce abortions in animals and humans.

Mycotoxins result in both acute and chronic toxicities. Acute effects result in rapid, readily noticeable fatal diseases. Some mycotoxins are acutely toxic to the liver, while others attack the kidneys, central nervous system, or circulatory system. For example, aflatoxin B_1 is a potent liver toxin, and less than 20 μg has a lethal effect on ducklings. The chronic effect results from the carcinogenicity of aflatoxin B_1. A diet containing only 0.1 μg/kg aflatoxin B_1 resulted in liver tumors in rainbow trout.[43] The immunosuppressing effects of mycotoxins could be important in human health in developing countries. In addition to their hepatocarcinogenicity, aflatoxins cause reduced T-cell function, diminished antibody response, and suppressed phagocyte activity in animals. Similarly, ochratoxin modulates various aspects of the immune system at the cellular and humoral levels. Ochratoxin exposure induces killer-cell activity and suppresses antibody exposure. *Fusarium* inhibits the production and functioning of macrophages.[5,122]

Mycotoxins produce toxic reactions in animals by contact or inhalation. Intake of low mycotoxin concentrations reduced mental alertness, physical abilities, and feed intake. Intake of moderate concentrations reduced the activity of the immune system and increased susceptibility to other diseases. Intake of high concentrations resulted in endemic nephropathy, primary liver carcinoma, bile duct proliferation, necrosis of various organs, leukopenia, endema, hyperplasia of the parathyroid, hematopoietic activity in the bone marrow, kidney damage, respiratory paralysis, and eventually death in extreme cases.[123] Physiological damage by mycotoxins varies with individual species, their age, state of health, degree of exposure, and other factors. Occurrence of mycotoxicosis is governed by the quantity of toxic food ingested, kind of cereal ingested, time lag in symptom development, toxic concentration in the food, sensitivity and age of the individual, sex, nutritional status, season of harvesting, weather, and altitude of the crop production area.[124] Mycotoxicosis has four clinical symptoms: it is noncommunicable; drugs and antibiotics are ineffective; outbreaks are associated with a specific food or feed; and foodstuffs show active fungal growth upon examination.[125] Since 1961, when turkey X disease was described and attention was focused on the capability of fungi to produce toxigenic compounds, scientists have screened thousands of fungal isolates and have shown that hundreds of compounds were produced. These can be divided into three categories.

1. *Mycotoxins and mycotoxicoses*

These are compounds causing economic problems or real illness in humans or domestic animals.[7]

Aflatoxins produced by *A. flavus, A. nominus,* and *A. parasiticus* cause or are a contributing factor to liver cancer in humans in Africa, Southeast Asia, and other countries. Aflatoxins initiate aflatoxicoses, which are carcinogenic, mutagenic, and teratogenic. Cotton, dried maize, millet, peanut, and rice seeds are the principal sources of aflatoxin. Left-over cooked rice is a significant source of dietary aflatoxins in Thailand. Acute aflatoxin poisoning in humans was reported from Taiwan, Thailand, and Uganda, where the incidence of liver cancer is high. It was suggested that aflatoxin might have a role in Reye's syndrome.[125] Liver damage is the primary toxicological consequence due to aflatoxin, but it can cause chromosomal aberrations and DNA breakage in animal and plant cells. The chronic intake of aflatoxin results in tumor formation. AFM_1 was found in the liver and kidneys of sows fed aflatoxin-contaminated maize (500 to 750 μg/kg aflatoxin). Aflatoxin fed to gestating sows reduced growth and altered plasma enzymes in piglets.[126] Aflatoxin B_1 affected liver, including swollen hepatocytes, fatty degeneration, bile duct hyperplasia, hepatic necrosis, hepatocellular necrosis, biliary cell proliferation of the periportal areas, liver cirrhosis, vacuolation in hepatocytes, interlobular fibrosis, periportal lipidosis, and periportal lymphocytic infiltration in barrows, chicks, ducklings, goslings, guinea pigs, poults, and turkeys.[127]

The presence of ochratoxin A produced by *P. verrucosum* in barley, which is a staple pig feed, was found in Danish pigs in the 1970s. There were excessive levels in the blood and fat of humans and pigs in central and northern Europe. The result was the development of nephritis sometimes followed by renal failure and premature death. Leucoencephalomalacia in horses was a result of feeding maize seeds infected with *F. moniliforme*, which produced fumonisins.[7]

Human esophageal cancer, a problem in southern Africa and China, is a disease without a defined cause. It has been linked with maize seed consumption. Trichothecenes are commonly produced *Fusarium* toxins. The trichothecenes of most concern are deoxynivalenol (DON) and nivalenol (NIV), which are less toxic than T-2 but more prevalent. Produced by *F. graminearum* in wheat in Australia, Canada, and Japan, NIV and DON, like all trichothecenes, are immunosuppressive and cause vomiting and feed refusal in pigs.[7]

The alimentary toxic aleukia (ATA) observed in Russia during the 19th century, which was also widespread during World War II and the postwar years until 1947, is caused by consumption of moldy seeds. *Fusarium poae* and *F. sporotrichioides* developed on seeds overwintered in the field. They produced trichothecenes, such as butenolide, neosolaniol, T-2 tetraol, zearalenone, but primarily the T-2 toxin. ATA can be fatal.[121] Typical symptoms include hemorrhagic rash on the skin; leukopenia agranulocytosis; bleeding from the nose, throat, gums and genitals; necrotic angina; hemorrhagic diathesis; sepsis; and exhaustion of the bone marrow. Joffe[121] reported that ATA developed after consumption of 2 kg food prepared from toxic overwintered food, and death occurred after ingesting 6 kg or more. Prosomillet and wheat contained the highest levels of toxin, which also occurred in barley, buckwheat, oats, and rye seeds. Symptoms first appeared after 2 to 3 weeks, and death occurred 6 to 8 weeks after ingesting toxic seeds. The toxic material does not pass into the milk of a sick mother. The disease causes the highest mortality between the ages of 8 and 50 years. There is variation in the amount of toxin in harvested seeds, depending upon the season. Overwintered seeds within the same field differ in their toxicity. Grains harvested during spring thaws were highly toxic, whereas those harvested in the autumn were almost free of toxin. A number of mycoflora have been isolated from the over-wintered cereal crops in toxic seeds of barley, oats, rye and wheat.[121,124]

Trichothecenes in food may have been associated with esophagitis and esophageal cancer in Linxian, China.[128] The fungi may cause three types of mycotoxicotic diseases in animals.[129] Toxigenic *Fusarium* such as *F. acuminatum, F. avenaceum*, and *F. heterosporum* were a potential hazard to grazing animals because of moniliformin production.[130]

2. *Mycotoxins of Lesser Importance*

These are compounds with demonstrated toxicity, known to occur naturally in toxic concentration, which from time to time cause sickness or debility of a less serious nature or are confined to a minor crop or restricted area.[131]

Aspergilli produce a variety of more or less toxic compounds. Cyclopiazonic acid (CPA) often is produced by *A. flavus*, with or without simultaneous aflatoxin production. Because of the prevalence of *A. flavus,* CPA is widespread, and involvement in animal and human disease should not be overlooked. Lupinosis, caused by *Phomopsis* produced during growth of *Phomopsis leptostromiformis* in lupins, also infects lupin seeds, and is of concern since lupin seeds are sold for human consumption. Lupinosis is characterized by jaundice and the development of yellow, fatty livers. Death may occur within a few days of disease onset.[131]

Diplodiosis occurred in cattle grazing on maize plants infected with *Diplodia maydis* after harvest. Diplodiosis is a neurotoxicosis, characterized in cattle by ataxia, a peculiar walk, lack of coordination, and often paralysis and death. *Penicillia* produce a range of toxins, but few qualify as causing significant mycotoxicoses. However, toxins produced by *Alternaria* in grains, especially sorghum, are of significance. Tenuazonic acid causes unthriftiness in chickens and probably in other animals.[131]

3. *Nontoxic Fungal Metabolites*

These are compounds that do not produce a significant disease syndrome or that occur in significant concentrations but lack demonstrated toxicity under natural conditions. These are citreoviridin, janthitrems, rubratoxins A and B, sterigmatocystin, viomellein, xanthomegnin, and *Penicillium islandicum* toxins. The fungi involved are *Penicillium verrucosum* (rubratoxin), *P. janthinellum* (janthitrems), *P. paxilli, P. simplicissimum* (verrucologen), *P. viridicatum* and *P. aurantiogriseum* (viomellein and xanthomegnin), *A. versicolor* (sterigmatocystin), *P. citreonigrum* (citreoviridin), and *P. islandicum (P. islandicum* toxins).[132]

V. DETECTION

It is difficult to differentiate different groups of mycotoxins simultaneously by a single method. A number of methods have been used for analyzing different groups. The analytical method used depends on the material and purpose of the analysis. The main steps for the analysis are sampling and sample preparation, extraction, cleanup and/or preparation for analysis, analysis (quantitative and qualitative), and finally confirmation. Chances for error occur at sampling, at subsampling, and during analysis.[2]

A. Sampling

The level of mycotoxin contamination varies depending on the sample(s) used, in part because mycotoxins are found near the area where the toxigenic fungi develop. Mycotoxins are not distributed uniformly within a peanut kernel. Therefore, a large sample is required for the analysis of peanuts — three 22-kg samples [133] or three 10.5-kg samples[26] from one 20-ton batch. The sampling should be at random from a recently blended or processed sample. Because fungi may develop in pockets, mycotoxin production may occur in pockets.

B. Extraction

Chloroform and methanol commonly are used for extraction of mycotoxins. Acetone, acetonitrile, and ethyl acetate also have been used. Extracted samples are cleaned using liquid–liquid extraction, solid–liquid extraction, or column chromatography.[2]

C. Analysis

Chromatographic techniques such as thin-layer chromatography (TLC), gas–liquid chromatography (GLC), high performance liquid chromatography (HPLC), high performance thin-layer chromatography (HPTLC), colorimetry, fluorometry, or other optical or electrical methods are used.

Two-dimensional TLC, mass spectrometry (MS), or chemical derivatization are used for qualitative analysis. Immunological techniques, such as enzyme-linked immunosorbent assay (ELISA), or radioimmunoassay (RIA), are useful for analyzing large numbers of samples. TLC followed by HPLC has been widely used. Several ready-to-use kits for aflatoxin detection based on ELISA are marketed.[14]

1. Thin-Layer Chromatography (TLC)

TLC is an effective, rapid screening technique, used either to detect the presence of a mycotoxin or a predetermined toxin level. The latter is useful especially as a quality control method where only samples that exceed a maximum tolerance level require rejection. The volume of a sample applied to the plate is adjusted so that the mycotoxin is observed only if the specified level is exceeded. Approximately 20 samples can be screened simultaneously on a 20 × 20 cm TLC plate. This technique has been used widely since its development for aflatoxin detection in the early 1960s.[134] The Association of Official Analytical Chemists recommended TLC methods for the detection of aflatoxin B_1, B_2, G_1, G_2, and M in cocoa beans, coconut, copra, cottonseed products, dairy products (eggs, milk, cheese), maize, peanuts and peanut products, and soybeans; ochratoxin A and B have been detected by these methods in barley and green coffee, patulin has been found in apple juice, sterigmatocystin has appeared in barley and wheat, and zearalenone has been detected in maize.[43, 135]

Reverse-phase TLC (RPTLC) has been used for detection of ochratoxin A in moldy grain. The method results in high levels of recovery and requires less solvent and preparation time.[136] The limit of detection is 1 to 2 μg/kg aflatoxin B_1 and 25 μg/kg zearlenone in maize.[137] The advantages include low cost and the ability to segregate components from other interfering compounds on the plates. However, the method lacks precision and is time consuming. Aflatoxins B_1, B_2, G_1, G_2, and M fluoresce under UV light, and subnanograms are detected on a plate. Aflatoxins can be quantified by comparing, either visually or densitometrically, the fluorescent intensities of the toxin within a sample with known quantities of standard aflatoxin solutions.[43]

2. Gas Liquid Chromatography

Gas liquid chromatography (GLC) has been used for mycotoxins quantification, particularly with patulin, zearalenone, and the trichothecenes (deoxynivalenol and T-2 toxin). The techniques used include mass spectrometry (MS), electron capture (ECD), and flame ionization detection (FID). The mass spectrometric method usually is sensitive and selective.[43] Zearalenol and a zearalenone were detected in maize by liquid chromatography using a C_{18} column and a mobile phase consisting of ratio of 35:25:40 CH_3CN:MeOH:H_2O, and 0.02 mol/l sodium acetate at pH 6.5. The toxins also were detected with an electrochemical detector at +0.95V vs. Ag/AgCl.[138] A reliable liquid chromatography method for the quantitative determination of citreoviridin in maize and rice was developed. Minimum detection limits were 10 μg for a citreoviridin standard and 2 μg/g for citreoviridin added to maize.[139] Capillary gas chromatography in combination with a selective mass detector (ion trap) detected 1 to 5 μg/kg trichothecene, deoxynivalenol, 3-acetyl-deoxynivalenol, nivalenol, T-2 toxin, HT-2 toxin, and diacetoxyscirpenol in cereals, with a recovery from spiked cereals of 78 to 89%.[140]

3. High Performance Liquid Chromatography (HPLC)

HPLC methods are precise, selective, and sensitive, with a provision for automation. HPLC quantification methods have been developed for aflatoxin, citrinin, deoxynivalenol, ochratoxin A, patulin, penicillic acid, T-2 toxin, and zearalenone.[43,134] The various classes have little in common except for an oxygenated structure with moderate or high polarity and low volatility. HPLC is suited for separation of such compounds and has become widely used in analytical laboratories.[134] Improved sensitivity and selectivity have been developed for detection of DON, fusarenon-X, and nivalenol using HPLC with online, postcolumn photolysis and oxidative amperometric detection (HPLC-hv-EC).[141] These methods have been used for normal and reverse phase systems in conjunction with UV absorption fluorescence, differential refraction, mass spectrometric, and amparometric detection techniques. A fully automated system comprising microcomputer-controlled sample injection, solvent recovery, detection, and interpretation methods was installed at the Tropical Development and Research Institute, London.[43] Precise HPLC techniques have been described for aflatoxin analysis in maize and peanuts.[43]

4. High Performance Thin-Layer Chromatography (HPTLC)

This method overcomes some of the disadvantages of TLC such as lack of precision because of sampling errors and plate development and interpretation. HPTLC helps reduce these errors by automating sample application, improving the uniformity of the adsorbent layer on TLC plates, and developing plates under controlled conditions.[43] For HPTLC, a sample of less than 1 μl is required, compared to 20 μl for TLC. Thus, due to reduced spot size to 1 mm or less, about 60 spots can be applied on a 10 × 20 cm HPTLC plate. A small number of standard spots is required because of the high degree of uniformity associated with the adsorbent layer. As little as 5 pg of pure aflatoxin (B_1, G_1, and M_1) can be detected by HPTLC, whereas, using TLC, only about 500 pg can be detected.[142] Lee et al.[141] separated pure aflatoxins (B_1, B_2, G_2, G_2, M_1, M_2), citrin, luteoskyrin, ochratoxin A, patulin, penicillic acid, sterigmatocystin, and zearalenone by the HPTLC multimycotoxin method using continuous multiple development. A bidirectional HPTLC method for determining low aflatoxin levels in maize extracts was evaluated. The method had a limited detection of 0.8 for B_1, 0.4 for B_2, 1.7 for G_1, and 0.4 for G_2 μg/kg. Compared with two-dimensional HPTLC methods, a 15- to 30-fold improvement in sample capacity per plate was achieved, and positioning errors were reduced during densitometry.[143]

5. The Fluorotoxinmeter (FTM)

Aflatoxins B and G are adsorbed onto a minicolumn as a discrete band. The fluorescence of the band under UV light was measured by a fluorotoxinmeter, which was calibrated to give a read-out of the total aflatoxin content in μg/kg.[43,144] The detection limit of deoxynivalenol by fluorometric analysis in seeds was 0.1 μg/kg.[145] Aflatoxin B_1 and B_2 fluoresce blue, and G_1 and G_2 fluoresce green under UV light. Nonfluorescent samples may contain aflatoxin, but the amount will be extremely low. In cottonseed lots, about 25% of nonfluorescing, weathered samples contained about 0.03 μg/ml compared with those that fluoresced greenish-yellow (9 to 30 μg/ml).[146] Aflatoxin concentration was 400 to 2300× higher in fluorescent than in nonfluorescent cottonseeds.[147] In addition to aflatoxin, *A. flavus* produced kojic acid on the lint. Host peroxidases converted the kojic acid to a substance that exhibited green-yellow fluorescence (BGYF).[148] The presence of BGYF in lint indicated that *A. flavus* infected the lint before boll maturity. Bright greenish-yellow fluorescence (BGYF) of maize has been used as a presumptive technique for aflatoxin contamination. Dickens and Whitaker[149] demonstrated that kernels with BGYF had higher aflatoxin levels than the non-BGYF kernels.

However, Tucker et al.[150] showed the presence of aflatoxin in nonfluorescent seeds. In rice inoculated with *P. citrinum*, fluorescence due to citrinin was observed only on the seed surface after 14 d at 28°C. The fungus digested the cell walls and starch in the surface layer. The rice was inoculated with *A. flavus*, and fluorescence due to aflatoxin was found in the whole seed after 7 to 8 d at 28°C. Most of walls of starchy cells disappeared because of fungal digestion.[151]

6. Differential Pulse Polarography

This method has been used for estimating aflatoxin B_1 and vomitoxin in maize. The detection limit for aflatoxin B_1 in maize and rice is 1000 μg/kg, whereas for vomitoxin it is 50 μg/kg in maize.[43,152,153]

7. Immunoassay Methods

Immunoassays for mycotoxin detection in ecosystems have progressed rapidly. Monoclonal and polyclonal antibodies with diverse specificity against mycotoxins have been developed. The availability of such antibodies has resulted in the establishment of simple, specific, sensitive, and reliable immunoassay protocols for mycotoxins in feeds, foods, and biological samples.[154] Kits are available commercially in formats ranging from simple yes-or-no tests through semiquantitative to fully quantitative procedures.[155, 156] ELISA, radioimmunoassay, and immunoaffinity assays have been used for the mycotoxin detection.

The ELISA and radioimmunoassay are based on the binding between unlabeled mycotoxin in the sample and labeled mycotoxins present in the assay system at specific binding sites of the antibody molecules. A radioactive mycotoxin is used as a labeled ligand in the RIA, whereas either toxin–enzyme conjugate or antibody–enzyme conjugate is used for ELISA. The immunoaffinity assay involves the use of an antibody column, which traps a specific mycotoxin in the column. The toxin is eluted for subsequent analysis.[155] Since aflatoxins are low molecular weight compounds, they do not possess antigenicity. However, if an aflatoxin molecule is conjugated with a protein, such as bovine serum albumin, it can be used to produce a specific antiserum. The oxime derivative of aflatoxin B_1 (aflatoxin B_1-carboxymetyloxime) conjugated to bovine serum albumin was suitable for production of aflatoxin B_1-specific antibodies of high titer.[157] A modification[158] was suitable for the detection of aflatoxin B in peanut seeds.[159] Commercially obtained hapten for aflatoxin B_1 (oxime bovine serum albumin) was used to prepare a polyclonal antiserum in rabbits. The oxime was adsorbed to wells of an ELISA plate. Bovine serum albumin was added to saturate the wells. Antiserum produced for oxime-BSA was diluted to 1:40,000 and mixed either with various concentrations of a pure aflatoxin B_1 standard or with peanut seeds with antisera containing various concentrations of aflatoxin, and then incubated for 1 h at 37°C. Prior to mixing, filtered methanol and 10-fold dilutions in the saline of the extract were used. Test samples and pure toxin, following incubation with antiserum, were added to ELISA plates, precoated with the oxime-BSA. Ig present in the antiserum, but not neutralized by toxin, adsorbed to oxime-BSA. In the final step, Ig attached to oxime-BSA was detected by alkaline phosphatage conjugated with Fc-specific Ig. The intensity of color produced by the substrate, P-nitrophenyl phosphate, was inversely proportional to the concentration of toxin present. By employing a standard curve prepared from pure toxin, it was possible to estimate the aflatoxin B_1 concentrations in test sample. Aflatoxin compounds were extracted using a combination of procedures.[160]

ELISA methods have been developed for detection of a number of mycotoxins, including the aflatoxins, deoxynivalenol, T-2 toxin, and zearalenone, in feeds and feed ingredients.[161] A rapid ELISA method for estimating aflatoxin in cottonseed, maize, peanuts, rice, and mixed feeds has been recommended by the Association of Official Analytical Chemists (AOAC).[162] Solid-phase ELISA kits have been developed for estimating aflatoxins, ochratoxin A, T-2 toxin, and zearalenone in a variety of commodities. An "immunodot" cup test, where the antibody is immobilized on a disk in the center of a small plastic cup, has been approved by the AOAC as an official first action screen for aflatoxin in cottonseeds, maize, and peanuts.[163] Card tests also have been developed, where the antibody is immobilized within a small indentation on a card the size of a credit card. Such tests were developed for estimating aflatoxin, ochratoxin A, T-2 toxin, and zearalenone in maize.[162]

The time for extraction, filtration, and estimation analysis for solid phase ELISA kits is about 5 to 10 min. Dorner and Cole[164] have compared rapid ELISA and solid-phase ELISA with HPLC for detection of aflatoxins in peanuts. Good agreement was found between the methods when the aflatoxin concentration of the sample was greater than 10 μg/kg.

Different mycotoxins, ELISA-assayed extraction procedures, cleaning prior to ELISA use, detection limits of ELISA, and samples analyzed by different workers have been summarized.[165] The sensitivity of ELISA and working ranges of different mycotoxins have been published.[165] Immunoassay protocols have been established for aflatoxin analysis and for mycotoxins other than aflatoxins in various agricultural commodities and dairy products. The sensitivities of these immunoassays and screening methods have been published.[154] Ez-screen quick card and Afla-10 cup tests reliably (>95%) identified samples of peanuts containing >10 μg/L aflatoxin. Both were reliable (95%) in identifying samples that were negative for aflatoxin.[162] An ELISA method is more accurate, precise, and cost effective for estimation of aflatoxin B_1 in maize, peanut butter (5 μg/kg), and wheat.[166] ELISA can determine ochratoxin A up to 10 pg per well in barley.[167] ELISA is recommended for aflatoxin B_1 detection in peanuts as an alternative to TLC.[168] Trucksess et al.[169] recommended ELISA be adopted for screening aflatoxin B_1, B_2, and G_1 contamination at ≥20 μg/kg in cottonseeds and peanut butter, and at ≥30 μg/kg in maize and raw peanuts. T-2 toxin was detected by indirect competitive ELISA in barley, oats, and wheat.[161] A competitive inhibition (CI) ELISA for ergotamine was developed using polyclonal antibodies produced in rabbits by immunization with an ergotamine–albumin conjugate. The assay was specific for ergot alkaloids having a phenylalanine moiety in the peptide portion of the molecule. These include ergocristine, ergostine, and ergotamine, which are produced by *Claviceps*. The assay detected ergotamine in spike-grain samples at 10 μg/kg. Target alkaloids were detected in ergot sclerotia from wheat and tall fescue and in fescue seeds infected by *Acremonium coenophialum*.[170] ELISA showed potential for aflatoxin B_1 detection in agricultural commodities in Thailand at 0.1 μg/L per assay.[171] Likewise, RIA was developed for detection of aflatoxin B_1 and T-2.[166,172] Results from RIA analysis for nivalenol in barley agreed with gas chromatography analyses.[173] The detection limit for aflatoxin in malting barley by RIA was 0.2 to 1.0 μg/kg.[174] The test strip format (dip stick) immunological assays detected mycotoxins up to 0.6 ng/ml for aflatoxin B_1, 0.2 ng/ml for aflatoxin M_1, 2.0 ng/ml for ochratoxin A, 1 ng/ml for T-2 toxin, 0.2 ng/ml for diacetoxyscirpenol, 10 ng/ml for 3-acetyl-deoxynivalenol, 15 ng/ml for roridin A, and 5 ng/L for zearalenone. A simple extraction procedure detected AFB_1, OA, T-2 toxin, and ZEA in spiked maize samples by test strip EIA at levels of 15, 100, 20, and 80 ng/g, respectively.[175] ELISA tests indicated more aflatoxins than were actually present in 17% of maize samples by two commercially available ELISA kits (92 Ez-screen and 36 Agri-Screen) differing in the form of the antibody surface.[176]

Integration of an ELISA with conventional chromatographic methods proved the versatility of ELISA as a research tool and allowed for rapid assessment of aflatoxin in individual cottonseeds, parts of cottonseeds, and composite samples of seeds from individual cotton bolls. An aliquot of extract was used to screen samples using ELISA. Negative samples were identified, and toxin levels between 1 and 70 μg/kg were quantitated. Samples with toxin levels beyond the upper limits of ELISA were subjected to liquid chromatography or TLC. Toxin-negative portions of individual seeds with high-positive toxin in another portion also were identified.[177]

VI. CONTROL

The ideal control of mycotoxin production is to provide seed storage conditions that inhibit fungal invasion and subsequent mycotoxin production. Seeds can be contaminated with mycotoxin at preharvest. Preventive measures to avoid fungal penetration and subsequent mycotoxin formation can be achieved by (1) planting resistant host species, (2) proper agronomic practices, (3) harvesting at the optimum stage of maturity, (4) drying after harvesting, and (5) chemical spraying.[178] The methods used currently in industrialized countries can be divided into two categories, physical methods designed to reduce fungal growth (seed drying, cold storage) and chemical methods (the use of fungicides and pesticides), especially in the form of fumigants.[179] However, in all cases, the following criteria must be met: (1) high effectiveness, (2) absence of an effect on the gustatory properties of the food, (3) lack of toxicity to the consumer, and (4) safety of the personnel involved.[179] Once seeds are invaded and toxins produced, it is difficult to detoxify them. Procedures for control of storage fungi are discussed in Chapter 11. Mycotoxin production in seeds can be reduced or eliminated by one or a combination of the following methods.

A. Storage Conditions

Seeds should be stored under conditions that retard or prevent growth of storage fungi. Invasion and growth of storage fungi largely depend on moisture, temperature, and oxygen supply. Most toxicogenic *Aspergilli* do not grow at 4°C or below.[180] With small seeds, relative humidity of less than 75% (13 to 15% moisture, wet weight basis) inhibited growth of storage fungi.[181] Storage under an O_2-free atmosphere also checks invasion by storage fungi.[180] Storage of commodities in a modified atmosphere containing high (>60%) levels of CO_2 to prevent insect infestation also inhibits fungal growth and mycotoxin production, while atmospheres of N need to contain >1% O_2 to retard fungal growth. Mycotoxin production is more sensitive to fumigation and modified atmospheric conditions than is fungal growth but may still occur if other conditions (temperature and a_w) are favorable.[182] Concentrations of CO_2 between 20 and 60% prevent or significantly reduce mycotoxin production by some *Aspergillus, Fusarium,* and *Penicillium.* Reduction of O_2 is less effective in preventing mycotoxin production.[182] Production of zearalenone by *F. equiseti* was inhibited almost completely in high moisture maize seeds (27%) kept under high CO_2 levels (60, 40, or 20%) with either 5 or 20% O_2.[183] Treatment with the fumigant phosphine is widespread for storage insect control. It has been used 2 g/ton for 5 d and has reduced maize seed infection by *Eurotium chevalieri* and *E. repens* and has reduced the population of *Aspergillus wentii.*[184] Growth of *A. flavus* and aflatoxin production decreased with increased phosphine concentration from 241.2 (air control) to 65.3 μg/L (3.5 mg phosphine/L).[185]

B. Sorting of Grains

Damaged and broken seeds are likely to carry more fungal-infected seeds than nondamaged seeds. Physically damaged seeds may be separated from nondamaged ones. Sclerotia of *C. purpurea* can be separated from healthy seeds by immersing contaminated seed lots in water. The sclerotia float to the top and are removed. In most warehouses, mills, and markets, depending upon the quantity of grain handled, either manual labor or air-screen cleaners generally are used to remove broken and infected seeds. The process indirectly reduces insect infestation, fungal growth, and mycotoxin production. Use of gravity separators ensures efficient separation of the good from the inferior seeds.[186]

Techniques for sorting based on color or other visual characteristics have been used most for aflatoxin control in peanuts. The most effective method is to remove off-color suspect seeds by electronic color-sorting. Such equipment views individual seeds and removes those that differ in intensity from a given standard.[187] Removal of loose-shelled seeds (LSK) and small peanut pods by belt-screening reduced aflatoxin by an average of 35%. Belt-screening removed 97% of the LSK, and 4% of sound mature seeds, and sound splits (SMK + SS). Further removal of other edibles (OE), oil stock (OS), LSK, and damaged kernels (DK) from the peanuts riding over (OVERS) the belt-screen reduced aflatoxin levels from an average of 110.7 μg/L in the unscreened load to 3.8 μg/L in SMK + SS. The OE, OS, LSK, and DK were removed from the OVERS through the use of slotted screens and by sorting.[188] A device capable of measuring fluorescence intensities from peanut surfaces and of physically rejecting peanuts having undesired fluorescence is being developed. It operates at a feed rate of 22,000 peanut halves/h. The entire surface of each peanut is examined at 10 to 20 discrete spatial regions. Fluorescence intensities from each region are used to make accept or reject decisions.[189] Wheat seeds severely damaged by Fusarium head blight in Canada were thin and shriveled, and highly concentrated into the least-dense fraction. Deoxynivalenol was highly concentrated in these fractions, while the most dense fractions contained low levels compared with unfractioned seeds.[190]

The efficacy was tested of two physical methods, sieving and dehulling, in reducing DON and zearalenone (ZEA) concentrations in contaminated barley, maize, and wheat containing 5 to 23 and 0.5 to 1.2 mg/kg DON and ZEA, respectively. The seeds were segregated into fractions of differing particle size by sieving through a series of screens. The retained fractions containing the longer particles (+9 mesh for barley; +9 mesh for wheat; +16 mesh for maize) contained 67 to 83% less toxin than that present in whole seeds. Removing the hull from barley prior to sieving resulted in a further 16% reduction in DON from the +9 mesh fraction. The amount of material lost during the sieving was 34, 69, and 55% of the total material for intact ground barley, maize, and wheat, respectively.[18] In maize, separated into three particle sizes with U. S. standards sieves #20 and #60, size 1 would not pass through a #20 screen, size 2 passed through a #20 screen but not a #60 size, and size 3 passed through a #60 screen. Size 2 contained the highest levels of aflatoxin, whereas size 1 had the lowest.[191]

C. Cultural Operations

Generally, seed invasion occurs at any time from initial seed development to harvest and storage. Cultural operations that prevent fungal invasion are useful in controlling mycotoxin production.

1. Irrigation

Aflatoxin contamination of maize is a complex problem, influenced by several factors. High temperature, water stress, and insects all play an important role in aflatoxin contamination in preharvest maize. Under drought stress, all three factors favor aflatoxin accumulation. One effective cultural practice for the control of aflatoxin in maize is to avoid water stress; irrigation, where possible, can be very effective.[192] Where tillage causes hard pans, subsoiling may be an effective alternative to irrigation.[75] Jones et al.[81] showed a reduction in aflatoxin contamination by irrigation. Associated with this reduction were fewer airborne conidia and fewer infected maize seeds. Aflatoxin levels in naturally infected maize exceeded 80 μg/kg in nonirrigated, nonsubsoiled plots and exceeded 1200 μg/kg in a year with low rainfall.

2. *Harvest stage*

In maize, reduction in aflatoxin B_1 is associated with early planting and low harvest moisture (28% vs. 18%). Planting dates and cultivar combinations that promote silking during periods of high airborne spore loads often result in a large number of infected maize seeds.[59,81] The collection of maize seeds dried to 14% moisture content within 48 h of harvest showed the lowest level of aflatoxin. This combined system of grain management and drying, referred to as the UTP system (U.K.–Thai Project system), reduced aflatoxin levels to a mean value of 2.5 μg/L, whereas maize at local merchants and regional silos had more than 50 and 100 μg/L, respectively.[20]

Delaying cotton harvest until dew has evaporated, avoiding storage of damp cotton, and ginning of harvested cotton by the third day after harvest minimized aflatoxin production by *A. flavus*.[193] Aflatoxin levels were higher in cottonseeds from ground-cleaned harvesters than from first-picked spindle harvests.[193] Aflatoxin levels may increase in fully mature cottonseeds before or after harvest. Toxin increases occurred over a range of 16 to 37°C if the relative humidity was 93% or greater. Early harvest was recommended to improve management of aflatoxin contamination of Arizona cottonseeds. Prolonged exposure of mature cotton to field conditions resulted in increased aflatoxin levels.[194] Harvesting maize as early as possible limited aflatoxin contamination.[109]

D. Chemical Treatment

Prevention of mold growth by chemicals, especially propionic acid, limits mycotoxin production. The insecticide dichlorvos inhibited aflatoxin production.[195] Aflatoxin production was avoided in peanuts by harvesting at maturity, inverting plants in wind rows for drying, and spraying the pods immediately with 5% propionic acid, 0.1% sorbic acid, or 0.15% chlorothalonil.[196] The use of ammonium and sodium bicarbonate on maize seeds reduced the growth of *A. parasiticus*, but only sodium bicarbonate reduced aflatoxin.[197] Some success has been reported in the control of *A. flavus* in peanut pods by foliar sprays or soil application. Mixon et al.[198] reported a reduction in peanut seed colonization after a gypsum treatment and observed that gypsum enhanced the control of *A. flavus* in plots treated with *Trichoderma harzianum* and PCNB + fensulfothion. Draughon[199] reported that the insecticides bux, carbaryl, and diphonate reduced the aflatoxin levels in naturally infected maize by more than 85%. Ferulic acid inhibited aflatoxin production by 70%, and Foltaf and Bavistin inhibited it by 95%. Treatment with *Annona squamosa* inhibited productivity 90%, whereas cardamom and turmeric inhibited aflatoxin synthesis almost completely.[200] Ammonium benzoate at 0.4 and 0.8 g per 100 g in potato-dextrose broth showed complete inhibition of *A. flavus* and *A. parasiticus*, respectively. When used as an antifungal agent, 10 spores/g of both *A. flavus* and *A. parasiticus* were inhibited by ammonium benzoate at 0.64 and 0.32 g per 100 g, respectively, in preheated peanuts and maize.[201] Deoxynivalenol was nearly completely destroyed after treatment with sodium bisulfite of maize, which could then be fed to swine.[202]

Hydrogen peroxide solution treatment has been used for separating aflatoxin-contaminated peanut seeds from sound ones. For peanuts containing 50 ng/L aflatoxin, a 0.08% hydrogen peroxide treatment for 0.7 min was required. This procedure resulted in a maximum sinker fraction yield of 85.5% of the original lot with a residual aflatoxin content of ≥5 ng/L.[203] It was found that spraying rice crops that are physiologically mature with 5% salt solution advanced the harvest stage. This treatment reduced the moisture content of seeds and straw, and the possibility of infection by spoilage fungi was reduced.[186] Mineral oil and highly saturated soybean oil reduced the levels of pea seed infection by *A. flavus* by 50% when applied to seeds at 5 ml/kg. All oils reduced weevil oviposition and decreased embryo and/or larval survival.[204] The aflatoxin producing potential of *A. flavus* populations associated with crop production was reduced in cottonseeds by applications of an atoxigenic strain of *A. flavus* in soil.[205]

E. Biological Control

There is variation among *A. flavus* strains in the quantity of aflatoxin produced in cottonseeds. Atoxigenic strains may have a potential as biological control agents for reducing aflatoxin contamination.[76] The use of an atoxigenic strain at a tenfold higher spore concentration than a toxigenic strain resulted in a significant reduction in aflatoxin B_1 if the former strain was introduced within 16 h after the toxigenic strain.[76] Treatment of maize with mycocurb, a mold inhibitor, inhibited *A. flavus* growth and aflatoxin levels.[206]

Controls based on genetic engineering for aflatoxin prevention have been suggested. One suggestion was to alter the fungus to a nontoxic form, either by producing genetically engineered *A. parasiticus* or by selecting aggressive non-toxigenic biocompetitve strains. In this approach, the gene that controls production of the toxin was altered or removed. Methyl transferase responsible for conversion of aflatoxin precursors found in the late stages of the pathway to aflatoxin B_1 has been isolated. It may be possible to genetically engineer a fungus that does not produce aflatoxin and competes with the potential toxin-producing strains for a niche in the environment of a cottonseed, maize, or peanut field. Another approach is to alter the plants to make them more resistant to the fungus by locating genes responsible for production of phytoalexins or introducing genes in cotton plants that code for resistant strains.[14]

F. Detoxification

While prevention of fungal growth and aflatoxin production in foods is the best method of controlling contamination, it is sometimes not possible. Detoxification becomes a second line of defense. Among many chemicals that have been examined for their ability to destroy aflatoxins, chlorine has shown promise.[207] Chlorine gas is more acceptable than other chemicals because it is used in the food industry to bleach flour, purify water, and clean food processing equipment. Chlorine gas treatment has the potential for detoxifying alfatoxins in copra meal and peanuts. More than 75% degradation of aflatoxin B_1 occurred in copra meal and peanuts after treatment with 16 and 35 mg chlorine gas per gram of substrate, respectively.[208] The most effective and suitable chemical for aflatoxin detoxification is ammonium bicarbonate (3%), which reduces aflatoxin B_1 by 80% in maize and peanuts. The mechanism of detoxification by ammonium released from the ammonium salt may involve the destruction of the lactone ring of aflatoxin B_1 into the nontoxic aflatoxin.[209,210] Aflatoxins are insoluble in water but soluble in organic solvents. It was found that the aflatoxin content of peanut meal was reduced from 110 to 8 μg/L by extraction with 10% aqueous acetone at 120°C.[211] Most modern oil extraction units now use solvent extraction, which reduced mycotoxin levels in the oil.[186] Cottonseed contamination with *A. flavus* contained aflatoxins B_1, B_2, G_1, and G_2. Cooking for 2 to 3 h at 90 to 100°C slightly reduced aflatoxin levels. Extraction with hexane reduced levels by 75 to 95%.[212] Heat treatment of oil cake for 2 h at 135°C reduced aflatoxin in peanut, but it affected the nutritional quality of the cakes. Treatment with HCl or H_2O_2 has been adopted.[186]

Aflatoxins are moderately stable to the roasting process.[213] However, dry roasting of naturally contaminated peanuts resulted in the destruction of aflatoxins B_1 and G_1; and B_2 and G_2 to 40 to 50% and 20 to 40%, respectively.[214] Microwave roasting of peanuts completely eliminated aflatoxins.[215] Aflatoxins were affected by exposure to UV light. However, the practical use of such treatment for destroying aflatoxins is questionable.[187]

Nonsterile barley inoculated with *A. ochraceus* conidia and then irradiated with ^{60}CO gamma rays and high energy electrons and incubated at 25% moisture content and 28°C showed reduced ochratoxin A with increased doses of irradiation.[216]

G. Regulatory Measures

A number of countries have prescribed regulations for some mycotoxins because of their harmful effects on animals and humans. Some 60 countries have currently enacted or proposed regulations for mycotoxins in food and feedstuffs, 17 of them in Africa and Asia.[217] Various factors influence the establishment of mycotoxin tolerances, such as availability of toxicological and dietary exposure data, the distribution of mycotoxins throughout commodities, legislation of other trading countries, and availability of analytical methods.[217]

Most European countries have limits for total aflatoxin of 10 μg/kg or less, and the trend is to 5 μg/kg or even 1 μg/kg in import commodities like copra, cottonseeds, figs, maize, peanuts, tree nuts, etc.[218]

In the European Community (EC) regulatory control for presence of mycotoxins applies primarily to aflatoxins, although some countries in addition control ochratoxin A and patulin. In the EC, there are limits for aflatoxin levels in animal feeds, but no agreement exists on the limits in human food. Although limits have not been finalized, they are likely to be 50 μg/kg for aflatoxin B_1 in simple feedstuffs that will not be used for dairy cattle, calves, or lambs; 20 μg/kg for pigs and poultry, and 10 μg/kg for other animals including dairy cattle. The only mycotoxins controlled by individual EC member states are ochratoxin A and patulin. Patulin must not exceed 50 mg/kg in fruit juices in Belgium, France, or Greece. In France the limit is 20 μg/kg for ochratoxin A in cereals, and in Greece 20 μg/kg in coffee beans.[219]

In the United States, aflatoxin levels cannot exceed 20 μg/kg for all feedstuffs and feedstuff ingredients, with the exception of cottonseeds intended for beef cattle, pigs, and poultry feed, where the limit is 300 μg/kg. All human foods are controlled similarly at the same limit. For whole milk, low fat milk, and skim milk there is a limit of 0.5 μg/kg for aflatoxin M_1. For deoxynivalenol there is a advisory limit range from 1 mg/kg for finished wheat and wheat products for human consumption to 4 mg/kg for wheat and wheat products for feed ingredients.[219] According to the legislation adopted in the former USSR, the maximum content of deoxygnivalenol in wheat cannot exceed 0.5 to 1 μg/g in grain for human consumption and 1 μg/g in grain for animal feed.[145] In Asia all mycotoxin regulation involves aflatoxins, and most regulations are concerned with aflatoxin B_1. Only five African countries are known to have regulations for aflatoxins.[217] Regulations for mycotoxins other than aflatoxins hardly exist in Africa and Asia. In India a limit of 30 μg/kg has been proposed for deoxynivalenol in wheat, whereas a limit of 0.01% for ergot in pearl millet was applied.[217]

H. Use of Resistant Cultivars

Cultivars may be developed with resistance to mycotoxin-producing strains of the fungus. Such cultivars would provide a poor substrate for toxin production. These may be either resistant to fungal invasion or to mycotoxin formation.[220] Cultivars may be artificially inoculated to determine their resistance or susceptibility to infection and mycotoxin production. Several peanut genotypes have the inherent ability to resist *A. flavus* invasion and aflatoxin contamination. *A. flavus* and *A. parasiticus* are considered low level pathogens in healthy peanut tissue, although the fungi can invade the fruits near maturity.[47] Peanut seed resistance is due to cuticular wax accumulation, seed coat structure, presence of cracks or detachment of the epidermal foundation, concentration of low molecular weight peptide-like compounds, and tanin concentrations.[221] Peanut cultivars with seeds resistant to insect attack, a rapid drying rate, and resistance to harvest and handling damage, fungal attack and penetration, or mycotoxin production may not have mycotoxin production problems.[29,222] Some of the methanol-extracted and water-soluble tanins from peanut testa and cotyledons significantly inhibited *A. parasiticus* growth and reduced aflatoxin.[223]

Genetic resistance to maize seed infection by toxin-producing fungi and production of toxin before harvest have been studied by (1) indirect protection of developing seeds provided by husk cover and antinutritional substances in silks, (2) direct protection provided by intrinsic presence of elicitation of phytoalexin-type substances that block insect and fungal development, (3) direct protection provided by external integuments of developing seeds, and (4) indirect protection provided by shifts in monomeric sugars and amino acids in conjunction with moisture that selects for nontoxigenic microbial populations.[178]

The percentage of maize seeds containing *A. flavus* was significantly correlated with total aflatoxin concentration. Thus measurement of seed infection appeared as a valid predictive estimate for aflatoxin concentration in ears. For pinbar inoculation, a plastic resin bar with 35 sewing needles in a single row, with 6-mm pointed ends protruding from the bar were dipped into a conidial suspension of 2×10^7 conidia per milliliter. These needles were pushed through the husk and into the underlying seeds.[150] Only the pinbar technique permitted complete separation of maize hybrids into groups based on relative susceptibility to seed infection by *A. flavus*.[150] Resistant seeds of pulses to aflatoxin production after infection with *A. flavus* showed that total phenol and protein were greater in resistant cultivars, while total sugar was greater in susceptible ones.[224]

REFERENCES

1. Forgacs, J. and Carll, W. T., Mycotoxicoses, *Adv. Vet. Sci.*, 7, 273, 1962.
2. Goto, T. Mycotoxins: current situation, *Food Rev. Int.*, 6, 265, 1990.
3. Hesseltine, C. W., Global significance of mycotoxins, in *Mycotoxins and Phycotoxins,* Steyn, P. S. and Vleggaar, R., Eds., Elsevier, Amsterdam, 1986, 16.
4. Matossian, M. K., *Poisons of the Past*, Yale University Press, 1989, 190.
5. Miller, J. D., Significance of grain mycotoxins for health and nutrition, in Fungi and Mycotoxins in Stored Products, Champ, B. R., Highley, E., Hocking, A. D., and Pitt, J. I., Eds., Proc. Int. Conf. Bangkok, ACIAR Proc. No. 36, 1991, 126.
6. Ueno, Y. and Ueno, I., Isolation and acute toxicity of citreoviridin, a neurotoxic mycotoxin of *Penicillium citreoviride* Biourge, *Jpn. J. Exp. Med.*, 42, 91, 1972.
7. Pitt, J. I., After 30 years: real mycotoxins, real mycotoxicoses, *Aust. Mycotoxin Newsl.*, 2, 1, 1991.
8. Alsberg, C. L. and Black, O. F., Contributions to the study of maize deterioration, biochemical and toxicological investigations of *Penicillium puberulum* and *Penicillium stoloniferum*, Bull. Bur. Anim. Ind., USDA No. 270, 1913, 1.
9. Miyake, I., Naito, H., and Sumeda, H., *Rep. Res. Inst. Rice Improvement*, 1, 1, 1940.
10. Mirocha, C. J., Pawlosky, R. A., Chatterjee, K., Watson, S., and Hayes, W., Analysis of *Fusarium* toxins in various samples implicated in biological warfare in Southeast Asia, *J. Assoc. Off. Anal. Chem.*, 66, 1485, 1983.
11. Udagawa, S., Mycotoxicoses — the present problems and prevention of mycotoxins, *Asian Med. J.*, 31, 599, 1988.
12. Bhat, R. V., Beedu, S. R., Ramakrishna, Y., and Munshi, K. L., Outbreak of trichothecene toxicosis associated with the consumption of mould-damaged wheat products in Kashmir Valley, *Lancet*, 33, 1989.
13. Gelderblom, W. C. A., Thiel, P. G., Marasas, W. F. O., and Vander Merwe, J., Natural occurrence of fusaric, a mutagan produced by *Fusarium moniliforme* in corn, *J. Agric. Food Chem.*, 37, 1064, 1989.
14. Bhat, R. V., Aflatoxins: successes and failures of three decades of research, in *Fungi and Mycotoxins in Stored Products,* Champ, B. R., Highley, E., Hocking, A. D., and Pitt, J. I., Eds, Proc. Int. Conf. Bangkok, ACIAR Proc. No. 36, 1991, 170.
15. Krishnamachari, K. A. V. R., Bhat, R. V., Nagarajan, V., and Tilak, T. B. G., Hepatitis due to aflatoxicosis: an outbreak in Western India, *Lancet*, 1, 1061, 1975.

16. Ngindu, A., Johnson, B. K., Kenya, P. R., Ngira, J. A., Oohing, D. M., Nandwa, H., Omundi, T. N., Jansen, A. J., Ngare, W., Kaviti, J. N., Gatei, D., and Siongok, T. A., Outbreak of acute hepatitis caused by aflatoxin poisoning in Kenya, *Lancet*, 1, 1346, 1982.
17. Kuiper-Goodman, T., Risk assessment to humans of mycotoxins in animal-derived products, *Vet. Hum. Toxicol.*, 32 (Suppl.), 6, 1990.
18. Trenholm, H. L., Charmley, L. L., Prelusky, D. B., and Bryden, W. L., Safety of mycotoxins in animal feeds and approaches to detoxification, in *Fungi and Mycotoxins in Stored Products*, Champ, B. R., Highley, E., Hocking, A. D., and Pitt, J. I., Eds., ACIAR Proc. No. 36, Proc. Int. Conf. Bangkok, 1991, 136.
19. Dawson, R. J., A global view of the mycotoxin problem, in *Fungi and Mycotoxin in Stored Products*, Champ, B. R., Highley, E., Hocking, A. D., and Pitt, J. I., Eds., ACIAR Proc. No. 36, Proc. Int. Conf. Bangkok, 1991, 22.
20. Cutler, M., Strategies for managing spoilage fungi and mycotoxins: a case study in Thailand, in *Fungi and Mycotoxins in Stored Products*, Champ, B. R., Highley, E., Hocking, A. D., and Pitt, J. I., Eds., ACIAR Proc. No. 36, Proc. Int. Conf. Bangkok, 1991, 168.
21. Bhat, R. V., Mould deterioration of agricultural commodity during transit: problems faced by developing countries, *Int. J. Food Microbiol.*, 7, 219, 1988.
22. Scott, P. M., The natural occurrence of trichothecenes, in *Trichothecene Mycotoxicoses, Pathophysiological Effects,* Vol. I, Beasely, V. R., Ed., CRC Press, Boca Raton, FL, 1989, 1.
23. Pier, A. C., The influence of mycotoxin on the immune system, in *Mycotoxin and Animal Foods*, Smith, J. E. and Henderson, R. S., Eds., CRC Press, Boca Raton, FL, 1991, 489.
24. Diener, U. L., Pettit, R. E., and Cole, J. R., Aflatoxins and other mycotoxins in peanuts, in Peanut Science and Technology, Pattee, H. E. and Young, L. T., Eds., American Peanut Research Educational Society, Yokum, TX, 1982, 486.
25. CAST, Aflatoxins and Other Mycotoxins: an Agricultural Perspective, Council of Agricultural Science Technology Report No. 80, University of Iowa, Ames, 1979, 56.
26. Coker, R. D., Jones, B. D., and Nagler, M. J., Mycotoxin training manual, Tropical Development Research Institute, London, Section A-10, 1984.
27. Jones, R. K., Duncan, H. E., Payne, G. A., and Leonard, K. J., Factors influencing infection by *Aspergillus flavus* in silk-inoculated corn, *Plant Dis.*, 64, 859, 1980.
28. Jones, R. K., Epidemiology and management of aflatoxins and other mycotoxins, in *Plant Disease: An Advanced Treatise*, Vol. IV, Horsfall, J. G. and Cowling, E. B., Eds., Academic Press, New York, 1979, 381.
29. Zuber, M. S. and Lillehoj, E. B., Status of the aflatoxin problem in corn, *J. Environ. Qual.*, 8, 1, 1979.
30. Diener, U. L., Deterioration of peanut quality caused by fungi, in Peanuts — Culture and Uses, American Peanut Research Educational Association, Stillwater, OK, 1973, 523.
31. Waliyar, F., Ba, A., Hassan, H., Bonkoungou, S., and Bose, J. P., Sources of resistance to *Aspergillus flavus* and aflatoxin contamination in groundnut genotypes in West Africa, *Plant Dis.*, 78, 704, 1994.
32. Wild, C. P., Jiang, Y., Sabbioni, G., Chapot, B., and Montesano, R., Evaluation of methods for quantitation of aflatoxin-albumin adducts and their application to human exposure assessment, *Cancer Res.*, 50, 245, 1990.

33. Hansen, T. J., Quantitative testing for mycotoxins, *Cereal Foods World*, 38, 346, 1993.
34. Reddy, M. J., Shetty, H. S., Fanelli, C., and Lacey, J., Role of seed lipids in *Aspergillus parasiticus* growth and aflatoxin production, *J. Sci. Food. Agric.*, 59, 177, 1992.
35. Brenneman, T. B., Wilson, D. M., and Beaver, R. W., Effects of diniconazole on *Aspergillus* populations and aflatoxin formation in peanut under irrigated and nonirrigated conditions, *Plant Dis.*, 77, 608, 1993.
36. Keller, N. P., Butchko, R. A. E., Starr, B., and Phillips, T. D., A visual pattern of mycotoxin production in maize kernels by *Aspergillus* spp., *Phytopathology*, 84, 483, 1994.
37. Marquardt, R. R. and Frohlich, A., Ochratoxin A: an important western Canadian storage mycotoxin, *Can. J. Physiol. Pharmacol.*, 68, 991, 1990.
38. Scott, P. M., *Penicillium* mycotoxins, in *Mycotoxic Fungi, Mycotoxins, Mycotoxicoses, An Encyclopedic Handbook*, Vol. I, *Mycotoxigenic Fungi*, Wyllie, T. D. and Morehouse, L. G., Eds., Marcel Dekker, New York, 1977, 283.
39. Ichinoe, M. and Kurata, H., Trichothecene-producing fungi, in *Developments in Food Science*, Vol. IV, *Trichothecenes: Chemical, Biological and Toxicological Aspects*, Ueno, Y., Ed., Kodansha, Tokyo and Elsevier, Amsterdam, 1983, 73.
40. Ichinoe, M., Hagiwara, H., and Kurata, H., Distribution of trichothecene-producing fungi in barley and wheat fields in Japan, in *Developments in Food Science*, Vol. 7, *Toxigenic Fungi — Their Toxins and Health Hazard*; Kurata, H. and Ueno, Y., Eds., Kodansha, Tokyo, and Elsevier, Amsterdam, 1984, 190.
41. Bacon, C. W., Lyons, P. C., Porter, P. K., and Robbins, J. D., Ergot toxicity from endophyte-infected grasses: a review, *Agron. J.*, 78, 106, 1986.
42. Hemken, R. W., Jackson, J. A., Jr., and Boling, J. A., Toxic factors in tall fescue, *J. Anim. Sci.*, 58, 1011, 1984.
43. Coker, R. D., High performance liquid chromatography and other chemical quantification methods used in the analysis of mycotoxins in foods, in *Analysis of Food Contaminants*, Gilbert, J., Ed., Elsevier, London, 1984, 207.
44. Singh, R. and Hsich, D. P. H., Aflatoxin biosynthetic pathway: elucidation by using blocked mutants of *Aspergillus parasiticus*, *Arch. Biochem. Biophys.*, 178, 285, 1977.
45. McCann, J., Choi, E., Yamasaki, E., and Ames, B. N., Detection of carcinogens as mutagens in the salmonella microsome test: assay of 300 chemicals, *Proc. Natl. Acad. Sci. U.S.A.*, 72, 5135, 1975.
46. Imaida, K., Hirose, M., Ogiso, T., Kurata, Y., and Ito, N., *Cancer Lett.*, 16, 137, 1982.
47. Moss, M. O., The environmental factors controlling mycotoxin formation, in *Mycotoxins and Animal Foods*, Smith, J. E. and Henderson, R. S., Eds., CRC Press, Boca Raton, FL, 1991, 37.
48. Miller, J. D., Fungi and mycotoxins in grain: implications for stored product research, *J. Stored Prod. Res.*, 31, 1, 1995.
49. Smith, J. E. and Ross, K., The toxigenic Aspergilli, in *Mycotoxins and Animal Foods*, Smith, J. E. and Henderson, R. S., Eds., CRC Press, Boca Raton, FL, 1991, 101.
50. Diener, U. L., Cole, R. J., Sanders, T. H., Payne, G. A., Lee, L. S., and Klich, M. A., Epidemiology of aflatoxin formation by *Aspergillus flavus*, *Annu. Rev. Phytopathol.*, 25, 249, 1987.

51 Pitt, J. I. and Leistner, L., Toxigenic *Penicillium* species, in *Mycotoxins and Animal Foods,* Smith, J. E. and Henderson, R. S., Eds., CRC Press, Boca Raton, FL, 1991, 81.

52. Marasas, W. F. O., Toxigenic Fusaria, in *Mycotoxins and Animal Foods*, Smith, J. E. and Henderson, R. S., Eds., CRC Press, Boca Raton, FL, 1991, 119.

53. Marasas, W. F. O., Nelson, P. E., and Toussoun, T. A., Taxonomy of toxigenic *Fusaria*, in *Trichothecenes and Other Mycotoxins*, Lacey, J., Ed., John Wiley and Sons, New York, 1985, 571.

54. Blaney, B. J., *Fusarium* and *Alternaria* toxins, in *Fungi and Mycotoxins in Stored Products*, Champ, B. R., Highley, E., Hocking, A. D., and Pitt, J. I., Eds., Proc. Int. Conf. Bangkok, ACIAR Proc. No. 36, 1991, 86.

55. Watson, D. H., An assessment of food contamination by toxic products of *Alternaria, J. Food Prot.*, 47, 485, 1984.

56. Andrews, S. and Lukas, S., Speciation and toxin profile of *Alternaria* isolates from Australian cereal grains used in the production of poultry rations, Proc. Mycotoxin Symposium, University of Sydney, Australia, 1988, 8.

57. Bakau, W. J. K., Bryden, W. L., and Burges, L. W., Toxicity of feed grain moulded with *Alternaria,* in *Fungi and Mycotoxins in Stored Products*, Champ, B. R., Highey, E., Hocking, A. D., and Pitt, J. I., Eds., Proc. Int. Conf. Bangkok, ACIAR Proc. No. 36, 1991, 225.

58. Liu, G. T., Qian, Y. Z., Zhang, P., Dong, Z. M., Shi, Z. Y., Zhen, Y. Z., Miao, J., and Xu, Y. M., Relationship between *Alternaria alternata* and oesophageal cancer, in IARC Sci. Publ. No. 105, O'Neill, I. K., Chen, J., and Bartsch, H., Eds., Lyon, France, 1991, 258.

59. Jones, R. K. and Duncan, H. E., Effect of nitrogen fertilizer, planting date, and harvest date on aflatoxin production in corn inoculated with *Aspergillus flavus, Plant Dis.*, 65, 741, 1981.

60. Mills, J. T., Postharvest insect–fungus associations affecting seed deterioration, in *Physiological–Pathological Interactions Affecting Seed Deterioration,* West, S. H., Ed., Crop Sci. Soc. Am., Special Publ. No. 12, 39, 1986.

61. Ghewande, M. P., Nagaraj, G., Desai, S., and Narayan, P., Screening of groundnut bold seeded genotypes for resistance to *Aspergillus flavus* seed colonization and less aflatoxin production, *Seed Sci. Technol.*, 21, 45, 1993.

62. Klich, M. A., Lee, L. S., and Huizar, H. E., Occurrence of *Aspergillus flavus* in vegetative tissue of cotton plants and its relation to seed infection, *Mycopathology*, 95, 171, 1986.

63. Bilgrami, K. S., Misra, R. S., Prasad, T., and Sinha, K. K., Screening of different varieties of maize for aflatoxin production by *Aspergillus parasiticus*, *Indian Phytopathol.,* 35, 376, 1982.

64. Nagarajan, V. and Bhat, R. V., Factor responsible for varietal differences in aflatoxin production in maize, *J. Agric. Food Chem.*, 20, 911, 1972.

65. Pettit, R. E., Taber, R. A., and Schroeder, H. W., Barriers to mold and mycotoxin damage in peanuts and pecans, in Third Int. Congr. Plant Pathol., 1978, 269.

66. Rodricks, J. V., *Mycotoxins and Other Fungal Related Food Problems,* in Advances in Chemistry Series 149, American Chemical Society, Washington, D.C., 1976.

67. Wilson, D. M., Mixon, A. C., and Troeger, J. M., Aflatoxin contamination of peanuts resistant to seed invasion by *Aspergillus flavus*, *Phytopathology*, 67, 922, 1977.

68. Luca, C. de, et al., Surface lipids of seeds support both *Aspergillus parasiticus* growth and aflatoxin production, *J. Toxicol. Toxins Rev.*, 8, 339, 1989.
69. Mehan, V. K., Mayee, C. D., Jayanthi, S., and McDonald, D., Preharvest seed infection by *Aspergillus flavus* group fungi and subsequent aflatoxin contamination in groundnuts in relation to soil types, *Plant Soil*, 136, 239, 1991.
70. Widstrom, N. W., McMillian, W. W., Wilson, D. M., Gaswood, D. L., and Glover, D. V., Growth characteristics of *Aspergillus flavus* on agar infused with maize kernel homogenates and aflatoxin contamination of whole kernel samples, *Phytopathology*, 74, 887, 1984.
71. Bilgrami, K. S., Masood, A., Ranjan, K. S., and Sinha, A. K., Impact of cob husks and maturity period on aflatoxin contamination in preharvest Kharif maize, *Indian Phytopathol.*, 43, 508, 1990.
72. Mertz, D., Lee, D., Zuber, M., and Lillehoj, E. B., Uptake and metabolism of aflatoxin by *Zea mays*, *J. Agric. Food Chem.*, 28, 963, 1980.
73. Smart, M. G., Wicklow, D. T., and Caldwell, R. W., Pathogenesis in *Aspergillus* ear rot of maize: light microscopy of fungal spread from wounds, *Phytopathology*, 80, 1287, 1990.
74. Cotty, P. J., Effects of cultivar and boll age on aflatoxin in cotton seed after inoculation with *Aspergillus flavus* at simulated exit holes of the pink boll worm, *Plant Dis.*, 73, 489, 1989.
75. Payne, G. A., Cassel, D. K., and Adkins, C. R., Reduction of aflatoxin contamination in corn by irrigation and tillage, *Phytopathology*, 76, 679, 1986.
76. Cotty, P. J., Effect of atoxigenic strains of *Aspergillus flavus* on aflatoxin contamination of developing cotton seed, *Plant Dis.*, 74, 233, 1990.
77. Anderson, H. W., Nehring, E. W., and Wichser, W. R., Aflatoxin contamination of corn in the field, *J. Agric. Food Chem.*, 23, 775, 1975.
78. Lillehoj, E. B. and Zuber, M. S., Aflatoxin problem in corn and possible solutions, Proc. 30th Annu. Corn Sorghum Res. Conf., Am. Seed Trade Assoc., Pub. No. 30, 1975, 230.
79. Payne, G. A., Kamprath, E. J., and Adkins, C. R., Increased aflatoxin contamination in nitrogen stressed corn, *Plant Dis.*, 73, 556, 1989.
80. Cobb, W. Y., Aflatoxin in the south eastern United States. Was 1977 exceptional? *Q. Bull. Assoc. Food Drug Off.*, 43, 99, 1979.
81. Jones, R. K., Duncan, H. E., and Hamilton, P. E., Planting data, harvest date, and irrigation effects on infection and aflatoxin production by *Aspergillus flavus* in field corn, *Phytopathology*, 71, 810, 1981.
82. Younis, M. A., Pauli, A. W., Mitchell, H. L., and Strickler, S. C., Temperature and its interaction with light and moisture in nitrogen metabolism of corn (*Zea mays* L.) seedlings, *Crop Sci.*, 5, 321, 1965.
83. Naik, M. V., Modi, V., and Patel, N. C., Studies on aflatoxin synthesis in *Aspergillus flavus*, *Indian J. Exp. Biol.*, 8, 345, 1970.
84. Goring, H. and Thien, B. H., Influence of nutrient deficiency on proline accumulation in the cytoplasm of *Zea mays* L. seedlings, *Biochem. Physiol. Pflanz.*, 174, 9, 1979.
85. Hsiao, T. C., Plant responses to water stress, *Annu. Rev. Plant Physiol.*, 24, 519, 1973.
86. Musingo, M. N. and Basha, S. M., Effect of drought and temperature stress on peanut (*Arachis hypogaea* L.) seed composition, *J. Plant Physiol.*, 134, 710, 1989.
87. Schroeder, H. W. and Ashworth, L. J., Aflatoxins in Spanish peanuts in relation to pod and kernel condition, *Phytopathology*, 55, 464, 1965.

88. Wallin, J. R., Production of aflatoxin in wounded and whole maize kernels by *Aspergillus flavus*, *Plant Dis.*, 70, 429, 1986.
89. El-Naghy, M. A., Mazen, M. B., and Fadl-Allah, E. M., Production of aflatoxin B_1 by *Aspergillus flavus* isolated from stored cotton seeds with different substrates, *World J. Microbiol. Biotechnol.*, 7, 67, 1991.
90. Song, D., A biochemical mechanism that suppresses aflatoxin accumulation in soybean seeds, *Acta Microbiol. Sin.*, 31, 169, 1991.
91. Cotty, P. J., Effect of harvest date on aflatoxin contamination of cotton seed, *Plant Dis.*, 75, 312, 1991.
92. Cuero, R. G., Smith, J. E., and Lacey, J., Interaction of water activity, temperature, and substrate on mycotoxin production by *Aspergillus flavus, Penicillium viridicatum* and *Fusarium graminearum* in irradiated grains, *Trans. Br. Mycol. Soc.*, 89, 221, 1987.
93. Abramson, D., Mills, J. T., and Sinha, R. N., Mycotoxin production in amber durum wheat stored at 15 and 19% moisture content, *Food Addit. Contam.*, 7, 617, 1990.
94. Hunter, J. H. and Tuite, J. F., The growth of storage fungi and aflatoxin production in corn as related to moisture and temperature, *Phytopathology*, 57, 816, 1967.
95. Schroeder, H. W. and Hein, H., Aflatoxins: production of the toxins *in vitro* in relation to temperature, *Appl. Microbiol.*, 15, 441, 1967.
96. Abramson, D., Sinha, R. N., and Mills, J. T., Mycotoxin formation in moist wheat under controlled temperatures, *Mycopathology*, 79, 87, 1982.
97. Blankenship, P. D., Cole, R. J., Sanders, T. H., and Hill, R. A., Effect of geocarposphere temperature on preharvest colonization of drought stressed peanuts by *Aspergillus flavus* and subsequent aflatoxin contamination, *Mycopathology*, 85, 69, 1984.
98. Bullerman, L. B., Interactive effects of temperature and pH on mycotoxin production, *Lebensm. Wiss. U. Technol.*, 18, 197, 1985.
99. Landers, K. E., Davis, N. D., and Diener, U. L., Influence of atmospheric gases on aflatoxin production by *Aspergillus flavus* in peanuts, *Phytopathology*, 57, 1086, 1967.
100. Wicklow, D. T., Hesseltine, C. W., Shotwell, O. L., and Adams, G. L., Interference, competition and aflatoxin levels in corn, *Phytopathology*, 70, 761, 1980.
101. Calvert, O. H., Lillehoj, E. B., Kwolek, W. F., and Zuber, M. S., Aflatoxin B_1 and G_1 production in developing *Zea mays* kernels from mixed inocula of *Aspergillus flavus* and *A. parasiticus, Phytopathology*, 68, 501, 1978.
102. Choudhary, A. K. and Sinha, K. K., Competition between a toxigenic *Aspergillus flavus* strain and other fungi on stored maize kernels, *J. Stored Prod. Res.*, 29, 75, 1993.
103. Cotty, P. J. and Bayman, P., Competitive exclusion of a toxigenic strain of *Aspergillus flavus* by an atoxigenic strain, *Phytopathology*, 83, 1283, 1993.
104. Boller, R. A. and Schroeder, H. W., Influence of temperature on production of aflatoxin in rice by *Aspergillus parsiticus, Phytopathology*, 64, 283, 1974.
105. Wicklow, D. T., Horn, B. W., and Shotwell, O. L., Aflatoxin formation in preharvest maize ears coinoculated with *Aspergillus flavus* and *Aspergillus niger*, *Mycologia*, 79, 679, 1987.
106. Zummo, N. and Scott, G. E., Interaction of *Fusarium moniliforme* and *Aspergillus flavus* on kernel infection and aflatoxin contamination in maize ears, *Plant Dis.*, 76, 771, 1992.

107. Nout, M. J. R., Effect of *Rhizopus* and *Neurospora* spp. on growth of *Aspergillus flavus* and *A. parasiticus* and accumulation of aflatoxin B_1 in groundnut, *Mycol. Res.*, 93, 518, 1989.
108. Gray, F. A., Faw, W. F., and Boutwell, J. L., The 1977 corn–aflatoxin epiphyotic in Alabama, *Plant Dis.*, 66, 221, 1982.
109. Payne, G. A., Hagler, W. M., Jr., and Adkins, C. R., Aflatoxin accumulation in inoculated ears of field grown maize, *Plant Dis.*, 72, 422, 1988.
110. Russell, T. E., Watson, T. F., and Ryan, G. F., Field accumulation of aflatoxin in cotton seed as influenced by irrigation termination dates and pink bollworm infestation, *Appl. Environ. Microbiol.*, 31, 711, 1976.
111. Sinha, A. K. and Rajan, K. S., BGY-fluorescence and aflatoxin contamination in insect-damaged maize crop, *Indian Phytopathol.*, 42, 514, 1989.
112. Russell, T. E., Lee, L. S., and Buco, S., Seasonal formation of aflatoxins in cottonseed produced in Arizona and California, *Plant Dis.*, 71, 174, 1987.
113. Wallin, J. P., Minor, H., and Rottinghaus, G. E., Maize yield and the incidence of aflatoxin, *Phytopathology*, 75, 1282, 1985.
114. Crisan, E. V., Effects of aflatoxin on germination and growth of lettuce, *Appl. Microbiol.*, 25, 342, 1973.
115. Adekunle, A. A. and Bassir, O., Effects of aflatoxin B_1 and palmitoxins Bo and Go on the germination and leaf color of cowpea, *Mycopathol. Mycol. Appl.*, 51, 299, 1973.
116. Hamada, A. S. and Megalla, S. E., Effect of malting and roasting on reduction of aflatoxin in contaminated soybeans, *Mycopathology*, 79, 3, 1982.
117. Misra, R. S. and Tripathi, R. K., Effect of aflatoxin B_1 on germination, respiration and α-amylase in maize, *J. Plant Dis. Prot.*, 87, 155, 1980.
118. Sinha, K. K. and Kumari, P., Effect of aflatoxin B_1 on some biochemical changes in germinating maize seeds, *Indian Phytopathol.*, 42, 519, 1989.
119. Shank, R. C., Mycotoxicoses of man: dietary and epidemiological conditions, in *Mycotoxicoses of Man and Plants, Mycotoxin Control and Regulatory Aspects*, Vol. III, Wyllie, T. D. and Morehouse, L. G., Eds., Marcel Dekker, New York, 1978, 1.
120. El-Refai, A. M. H., Sallam, L. A. R., and Naim, N., The alkaloids of fungi, I. The formation of ergotine alkaloids by representative mold fungi, *Jpn. J. Microbiol.*, 14, 91, 1970.
121. Joffe, A. Z., Toxin production by cereal fungi causing toxic alimentary aleukia in man, in *Mycotoxins in Foodstuffs*, Wogan, C. N., Ed., MIT Press, Cambridge, 1965, 77.
122. Pestka, J. J. and Bondy, G. S., Alteration of immune function following dietary mycotoxin exposure, *Can. J. Physiol. Phamacol.*, 68, 1009, 1990.
123. Pettit, R. E. and Taber, R. A., Introduction and histological perspectives in mycotoxicology research, in *Symposium on Mycotoxicology of Food and Feed Contamination,* Proc. Am. Phytopathol. Soc., 3, 1976, 99.
124. Joffe, A. Z., *Fusarium poae* and *F. sporotrichioides* as principal causal agents of alimentary toxic aleukia, in *Mycotoxic Fungi, Mycotoxins and Mycotoxicoses,* Vol. III, Wyllie, T. D. and Morehouse, L. G., Eds., Marcel Dekker, New York, 1978, 21.
125. Detroy, R. W., Lillehoj, E. B., and Cieglar, A., Aflatoxins and related compounds, in *Microbial Toxins*, Vol. 6, Cieglar, A., Kadis, S., and Ajl, S. J., Eds., Academic Press, New York, 1971, 1.

126. McKnight, C. R., Hagler, W. M., Jr., Jones, E. E., and Armstrong, W. D., Mycotoxins, biotoxins, wood decay, air quality, cultural properties, general biodeterioration and degradation, in *Proc. 3rd Meeting Pan American Biodeterioration Soc.*, 1989, Washington, D.C., Llewellyn, G. C. and O'Rear, C. E., Eds., Plenum Press, New York, 1990, 129.
127. Champ, B. R., Highley, E., Hocking, A. D., and Pitt, J. I., Fungi and Mycotoxins in Stored Products, Proc. Int. Conf. Bangkok, ACIAR Proc. No. 36, 1991, 270.
128. Hsia, C. C., Wu, J. L., Lu, X. Q., and Li, Y. S., Natural occurrence and clastogenic effects of nivalenol, deoxynivalenol, 3-acetyl-deoxynivalenol, 15-acetyl-deoxynivalenol and zearalenone in corn from a high-risk area of esophageal cancer, *Cancer Det. Prev.*, 13, 79, 1988.
129. Austwick, P. K. C., Mycotoxins, *Br. Med. Bull.*, 31, 222, 1975.
130. Lamprecht, S. C., Marasas, W. F. O., Theil, P. G., Schnieder, D. J., and Knox-Davies, P. S., Incidence and toxigenicity of seedborne-*Fusarium* species from annual *Medicago* species in South Africa, *Phytopathology*, 76, 1040, 1986.
131. Pitt, J. I., After 30 years, real mycotoxins of lesser importance, *Aust. Mycotoxin Newsl.*, 3 (2), 1992.
132. Pitt, J. I., Aflatoxigenic fungi, *Aust. Mycotoxin Newsl.*, 3 (1), 1992.
133. Dickens, J. W. and Whitaker, T. B., Some mycotoxins, in *Environmental Carcinogens Selected Methods of Analysis,* Vol. 5, Egan, H., Ed., I.A.R.C., Lyon, 1982, 17.
134. Shepherd, M. J., High performance liquid chromatography and its application to the analysis of mycotoxins, in *Modern Methods in the Analysis and Structural Elucidation of Mycotoxin,* Cole, R. S., Ed., Academic Press, New York, 1986, 293.
135. Stoloff, L., *Official Methods of Analysis*, 13th ed., Horwitz, W., Ed. Assoc. Official Analyt. Chem., Washington, D.C., 1980.
136. Frohlich, A. A., Marquardt, R. R., and Bernatsky, A., Quantitation of ochratoxin A: use of reverse phase thin-layer chromatography for sample clean-up followed by liquid chromatography or direct fluorescence measurement, *J. Assoc. Off. Anal. Chem.,* 71, 949, 1988.
137. Fulgueira, C. L. and Bracalenti, B. J. C. de, Rapid and economical method to detect aflatoxins and zearalenone in animal feed, *Rev. Microbiol.,* 20, 210, 1989.
138. Ware, G. M., Francis, O. J., Kuan, S. S., and Carman, A. S., Determination of zearalenol and zearalenone using electro-chemical detection, *Anal. Lett.,* 22, 2335, 1989.
139. Stubblefield, R. D., Greer, J. I., and Shotwell, O. L., Liquid chromatographic method for determination of citreoviridin in corn and rice, *J. Assoc. Off. Anal. Chem*, 71, 721, 1988.
140. Schwadorf, K. and Muller, H. M., Determination of trichothecenes in cereals by gas chromatography with ion trap detection, *Chromatographic,* 32, 137, 1991.
141. Childress, W. L., Krull, I. S., and Selavka, C. M., Determination of deoxynivalenol (DON, vomitoxin) in wheat by high-performance liquid chromatography with photolysis and electro-chemical detection (HPLC-hv-EC), *J. Chromatographic Sci.,* 28, 76, 1990.
142. Lee, K. Y., Poole, C. F., and Zlatkis, A., Simultaneous multi-mycotoxin determination by high performance thin layer chromatography, *Ann. Chem.*, 52, 837, 1980.
143. Tomlins, K. I., Jewers, K., Coker, R. D., and Nagler, M. J., A bi-directional HPTLC development method for detection of low levels of aflatoxin in maize extracts, *Chromatographia*, 27, 49, 1989.

144. Velasco, J., Fluorometric measurement of aflatoxin absorbed in florisil in mini-columns, *J. Assoc. Off. Anal. Chem.*, 58, 757, 1975.
145. Leonov, A. N., Soboleva, N. A., and Kononenko, G. P., Fluorometric analysis of deoxynivalenol in contaminated grain, *Sov. Agric. Sci.*, 7, 17, 1989.
146. Ashworth, L. J., McMeans, J. L., Pyle, J. L., Brown, C. M., Osgood, J. W., and Ponton, R. E., Aflatoxins in cotton seeds: influence of weathering on toxin content of seeds and on a method for mechanically sorting seed lots, *Phytopathology,* 58, 102, 1968.
147. McMeans, J. L. and Ashworth, L. J., Preharvest occurrence of *Aspergillus flavus* and aflatoxins in California cotton seed, *Phytopathology,* 56, 889, 1966.
148. Marsh, P. B., Simpson, M. E., Ferretti, R. J., Mesola, G. V., Donosa, J., Craig, G. O., Trucksess, M. W., and Work, P. S., Mechanism of formation of a fluorescence in cotton fiber associated with aflatoxins in the seeds at harvest, *J. Agric. Food Chem.*, 17, 468, 1969.
149. Dickens, J. W. and Whitaker, T. B., Bright greenish-yellow fluorescence and aflatoxin in naturally harvested yellow corn marketed in North Carolina, *J. Am. Oil Chem. Soc.*, 58, 973A, 1981.
150. Tucker, D. H., Jr., Trevathan, L. E., King, S. B., and Scott, G. E., Effect of four inoculation techniques on infection and aflatoxin concentration of resistant and susceptible corn hybrids inoculated with *Aspergillus flavus, Phytopathology,* 76, 290, 1986.
151. Takahashi, H., et al., Distribution of citrinin and aflatoxins in steamed milled rice kernels inoculated with *Penicillium citrinum* and *Aspergillus flavus, Maikotokishin,* 31, 49, 1990.
152. Palmisano F., Visconti, A., Bottalico, A., Lerario, P., and Zambonin, P. G., Differential pulse polarography of trichothecene toxins: detection of deoxynivalenol in corn, *Analyst,* 106, 992, 1981.
153. Smyth, M. R., Lawellin, D. W., and Osteryoung, J. G., Polarographic study of aflatoxins B_1, B_2, G_1 and G_2: application of differential pulse polarography to the determination of aflatoxin B_1 in various foodstuffs, *Analyst,* 104, 73, 1979.
154. Chu, F. S., Development and use of immunoassays in the detection of ecologically important mycotoxins, in *Handbook of Applied Mycology* Vol. V, *Mycotoxins in Ecological Systems,* Bhatnagar, D., Lillehoj, E. B., and Arora, D. K., Eds. Marcel Dekker, New York, 1992, 87.
155. Chu, F. S., Immunoassays for mycotoxins: current state of the art, commercial and epidemiological applications, *Vet. Hum. Toxicol.,* 32 (Suppl.), 42, 1990.
156. Morgan, M. R. A., Mycotoxin immunoassay with special reference to ELISA, *Tetrahedron,* 45, 2237, 1989.
157. Gaur, P. K., El-Nakib, O., and Chu, F. S., Comparison of antibody production against aflatoxin B_1 in goats and rabbits, *Appl. Environ. Microbiol.*, 40, 678, 1980.
158. Morgan, M. R. A., Kang, A. S., and Chan, H. W. S., Aflatoxin determination in peanut butter by enzyme-linked immunosorbent assay, *J. Sci. Food Agric.,* 37, 908, 1986.
159. Reddy, D. V. R., Nambiar, P. T. C., Rajeswari, R., Mehan, V. K., Anjaiah, V., and McDonald, D., Potential of enzyme linked immuno-sorbent assay for detecting viruses, fungi, bacteria, mycoplasma-like organisms, mycotoxin and harmones, in Bio-technology, in Tropical Crop Improvement: Proc. Int. Biotechnol. Workshop, ICRISAT, India, 1988, 43.

160. Wieman, D. M., White, G. M., Taraba, J. L., Ross, I. J., Hicks, C. L., and Langlois, B. E., Production of aflatoxin on damaged corn under controlled environmental conditions, *Trans. Am. Soc. Agric. Eng.*, 29, 1150, 1986.
161. Kawamura, O., Nagayama, S., Sato, S., Ohtani, K., Suguira, Y., Tanaka, T., and Ueno, Y., Survey of T-2 toxin in cereals by an indirect enzyme-linked immunosorbent assay, *Food Agric. Immunol.*, 2, 173, 1990.
162. Coker, R. D., Recent developments in methods for sampling and analysis of mycotoxin, in *Fungi and Mycotoxins in Stored Products,* Champ, B. R., Highley, E., Hocking, A. D., and Pitt, J. I., Eds., Proc. Int. Conf. Bangkok, ACIAR Proc. No. 36, 1991, 115.
163. Trucksess, M. W., Young, K., Donahue, K. F., Morris, K., and Lewis, E., Comparison of two immunochemical methods with thin layer chromatographic method for determination of aflatoxins, *J. Assoc. Off. Anal. Chem.,* 73, 425, 1990.
164. Dorner, J. W. and Cole, R. J., Comparison of two ELISA screening tests with liquid chromatography for determination of aflatoxins in raw peanuts, *J. Assoc. Off. Anal. Chem.,* 72, 962, 1989.
165. Candlish, A. A. G., The determination of mycotoxins in animal feeds by biological methods, in *Mycotoxins and Animal Foods,* Smith, J. E. and Henderson, R. S., Eds., CRC Press, Boca Raton, FL, 1991, 223.
166. El-Nakib, O., Pestka, J. J., and Chu, S. S., *J. Assoc. Off. Anal. Chem.*, 64, 1077, 1981.
167. Morgan, M. R. A., Matthew, J. A., McNurney, R., and Chan, H. W. S., Proc. Fifth Int. Sym. Mycotoxins and Phycotoxins, UPAC/WHO, Vienna, 1982, 32.
168. Figueira, A. C., Taylor, K. D. A., and Barlow, P. J., ELISA determination of aflatoxin levels in whole nuts, *Food Agric. Immunol.*, 2, 125, 1990.
169. Trucksess, M. W., Stack, M.E., Nesheim, S., Park, D. L., and Pohland, A. E., Enzyme-linked immunosorbent assay of aflatoxins B_1, B_2 and G_1 in corn, cottonseed, peanuts, peanut butter, and poultry feed: collaborative study, *J. Assoc. Off. Anal. Chem.,* 72, 957, 1989.
170. Shelby, R. A. and Kelley, V. C., An immunoassay for ergotamine and related alkaloids, *J. Agric. Food Chem.,* 38, 1130, 1990.
171. Sanimtong, A. and Tanboon-Ek, P., Detection of aflatoxin B_1 by enzyme-linked immunosorbent assay in Thailand, in *Fungi and Mycotoxins in Stored Products,* Champ, B. R., Highley, E., Hocking, A. D., and Pitt, J. I., Eds., Proc. Int. Conf. Bangkok, ACIAR Proc. No. 36, 1991, 227.
172. Lee, S. and Chu, F. S., *J. Assoc. Off. Anal. Chem.*, 64, 150, 1981.
173. Teshima, R., Hirai, K., Sato, M., Ikebuchi, H., Ichinoe, M., and Terao, T., Radioimmunoassay for nivalenol in barley, *Appl. Environ. Microbiol.*, 56, 764, 1990.
174. Fukal, L., Prosek, J., and Rakosova, A., Radiochemical aflatoxin determination in barley, malt and beer, *Monatsschrift Brauwissenschaft,* 43, 212, 1990.
175. Schneider, E., Dietrich, R., Martlbauer, E., Usleber, E., and Terplan, G., Detection of aflatoxins, trichothecenes, ochratoxin A, and zearalenone by test strip enzyme immunoassays: a rapid method for screening cereals for mycotoxins, *Food Agric. Immunol.*, 3, 185, 1991.
176. Wilcke, W. F., Ehrich, M. F., and Ko, K. W., Using ELISA kits to test corn for aflatoxin, *Appl. Agric. Res.,* 5, 32, 1990.
177. Lee, L. S., Wall, J. H., Cotty, P. J., and Bayman, P., Integration of enzyme-linked immunosorbent assay with conventional chromatographic procedures for quantitation of aflatoxin in individual cotton bolls, seeds, and seed sections, *J. Assoc. Off. Anal. Chem.,* 73, 581, 1990.

178. Lisker, N. and Lillehoj, E. B., Prevention of mycotoxin contamination (principally aflatoxins and *Fusarium* toxins) at the preharvest stage, in *Mycotoxins and Animal Foods,* Smith, J. E. and Henderson, K. S., Eds., CRC Press, Boca Raton, FL, 1991, 689.
179. Leitao, J., de Saint Blanquat, G., Bailly, J. R., and Derache, R., Preventive measures for microflora and mycotoxin production in foodstuffs, *Arch. Environ. Contam. Toxicol.,* 19, 437, 1990.
180. Mislivec, P. B., Mycotoxin production by conidial fungi, in *Biology of Conidial Fungi,* Vol. 2, Cole, G. T. and Kendrick, J., Eds., Academic Press, New York, 1981, 38.
181. Christensen, C. M. and Kaufmann, H. H., Deterioration of stored grains by fungi, *Annu. Rev. Phytopathol.,* 3, 69, 1965.
182. Hocking, A. D., Effects of fumigation and modified atmosphere storage on growth of fungi and production of mycotoxins in stored grains, in *Fungi and Mycotoxins in Stored Products,* Champ, B. R., Highley, E., Hocking, A. D., and Pitt, J. I., Eds., Proc. Int. Conf. Bangkok, ACIAR Proc. No. 36, 1991, 145.
183. Paster, N., Blumenthal-Yonassi, J., Barkai-Golas, R., and Menasherov, M., Production of zearalenone *in vitro* and in corn grains stored under modified atmospheres, *Int. J. Food Microbiol.,* 12, 157, 1991.
184. Dharamputra, O. S., Tjitrosomo, H. S. S., Sidik, M., and Umaly, R. C., The effects of phosphine on some biological aspects of *Aspergillus flavus,* in *Fungi and Mycotoxins in Stored Products,* Champ, B. R., Highley, E., Hocking, A. D., and Pitt, J. I., Eds., Proc. Int. Conf. Bangkok, ACIAR Proc. No. 36, 1991, 244.
185. Dharamputra, O. S., Tjitrosomo, H. S. S., Sidik, M., and Halid, H., The effects of phosphine on storage fungi of maize, Symp. Pests of Stored Products, Bogor, Indonesia, 1991.
186. Dakasinamurthy, A. and Shukla, B. D., Problems and perspectives of spoilage fungi and mycotoxins in India, in *Fungi and Mycotoxins in Stored Products,* Champ, B. R., Highley, E., Hocking, A. D., and Pitt, J. I., Eds., Proc. Int. Conf. Bangkok, ACIAR Proc. No. 36, 217, 1991.
187. West, D. T. and Bullerman, L. B., Physical and chemical separation of mycotoxins from agricultural products, in *Mycotoxins and Animal Foods,* Smith, J. E. and Henderson, R. S., Eds. CRC Press, Boca Raton, FL, 1991, 777.
188. Dowell, F. E., Dorner, J. W., Cole, R. J., and Davidson, J. I., Jr., Aflatoxin reduction by screening farmers stock peanuts, *Peanut Sci.,* 17, 6, 1990.
189. Pelletier, M. J., Spetz, W. L., and Aultz, T. R., Fluorescence sorting instrument for the removal of aflatoxin from large numbers of peanuts, *Rev. Sci. Instrum.,* 62, 1926, 1991.
190. Tkachuk, R., Dexter, J. E., Tipples, K. H., and Nowicki, T. W., Removal by specific gravity table of tombstone kernels and associated trichothecenes from wheat infected with Fusarium head blight, *Chem. Chem.,* 68, 428, 1991.
191. Stauffer, C. S., Wallin, J. R., and Zuber, M. S., The influence of *Aspergillus flavus* on particle sizes of maize kernels and aflatoxin levels, *Phytopathology,* 75, 1282, 1985.
192. Payne, G. A., Cassel, D. K., and Adkins, C. R., Reduction of aflatoxin levels in maize due to irrigation and tillage, *Phytopathology,* 75, 1283, 1985.
193. Griffin, A. C. and Schroeder, H. W., Aflatoxin in cotton after harvesting, *Phytopathology,* 68, 119, 1978.
194. Russell, T. E., Von Bretzel, P., and Easely, J., Harvesting method effects on aflatoxin levels in Arizona cotton seed, *Phytopathology,* 71, 359, 1981.

195. Schroeder, H. W., Cole, R. J., Grigsby, R. D., and Hein, H., Jr., Inhibition of aflatoxin production and tentative identification of an aflatoxin intermediate versiconal acetate from treatment with dichlorvos, *Appl. Microbiol.*, 27, 394, 1974.
196. Madan, S. L. and Chohan, J. S., Efficacy of antimicrobial chemicals to control post-harvest occurrence of *Aspergillus flavus* in groundnut kernels, *Indian Phytopathol.*, 31, 57, 1978.
197. Montville, T. J. and Goldstein, P. K., Sodium bicarbonate inhibition of aflatoxigenesis in corn, *J. Food Prot.*, 52, 45, 1989.
198. Mixon, A. C., Bell, D. K., and Wilson, D. M., Effect of chemical and biological agents on the incidence of *Aspergillus flavus* and aflatoxin contamination of peanut seed, *Phytopathology,* 74, 1440, 1984.
199. Draughon, F. A., Control or suppression of aflatoxin production with pesticides, in Aflatoxin and *Aspergillus flavus* in Corn, Diener, U. L., Asquith, R. L., and Dickens, J. W., Eds., Southern Coop. Series Bull. 279, Auburn, AL, 1983.
200. Sahay, S. S. and Ranjan, K. S., Efficacy of certain chemicals, fungicides, species and plant extracts in the control of aflatoxin production by *Aspergillus flavus* in mustard, *Proc. Natl. Acad. Sci. India Sect. B.*, 60, 169, 1990.
201. Niyomca, P., Chinanonwate, N., and Suttajit, M., Potential of ammonium benzoate for aflatoxin control, in *Fungi and Mycotoxins in Stored Products,* Champ, B. R., Highley, E., Hocking, A. D., and Pitt, J. I., Eds., Proc. Int. Conf. Bangkok, ACIAR Proc. No. 36, 1991, 251.
202. Young, J. C., Trenholm, H. L., Friend, D. W., and Prelusky, D. B., Detoxification with sodium bisulfite and evaluation of the effects when pure mycotoxin or contaminated corn was treated and given to pigs, *J. Agric. Food Chem.,* 35, 259, 1987.
203. Clavero, M. R. S., Hung, Y. C., Beuchat, L. R., and Nakayama, T., Separation of aflatoxin-contaminated kernels from sound kernels by hydrogen peroxide treatment, *J. Food Prot.*, 56, 130, 1993.
204. Hall, J. S. and Harman, G. E., Efficacy of oil treatments of legume seeds for control of *Aspergillus* and *Zabrotes, Crop Prot.,* 10, 315, 1991.
205. Cotty, P. J., Influence of field application of an atoxigenic strain of *Aspergillus flavus* on the populations of *A. flavus* infecting cotton bolls and on the aflatoxin content of cotton seed, *Phytopathology,* 84, 1270, 1994.
206. Pupipat, U., Control of aflatoxin in corn by a mold-inhibitor — Mycocurb, Proc. Jpn. Assoc. Microbiol., Suppl. No. 1, 1988, 75.
207. Sen, A. C., Wei, C. I., Fernando, S. Y., Toth, J., Ahmed, E. M., and Dunaif, G. E., Reduction of mutagenicity and toxicity of aflatoxin B_1 by chlorine gas treatment, *Food Chem. Toxicol.*, 26, 745, 1988.
208. Samarajeewa, U., Fernando, S. Y., Ahmed, E. M., and Wei, C. I., Degradation of aflatoxin B_1 and loss of mutagenicity in peanuts and copra meal on chlorine gas treatment, in *Fungi and Mycotoxins in Stored Products,* Champ, B. R., Highley, E., Hocking, A. D., and Pitt, J. I., Eds., Proc. Int. Conf. Bangkok, ACIAR Proc. No. 36, 1991, 249.
209. O-ari-Yakul, N., Wangjaisuk, S., Banjerdanpongchai, R., Vinjtkotkumneun, U., Suttajit, M., and Chokethaworn, N., Chemical detoxification of aflatoxin B_1, in *Fungi and Mycotoxins in Stored Products,* Champ, B. R., Highley, E., Hocking, A. D., and Pitt, J. I., Eds., Int. Conf. Bangkok, ACIAR Proc. No. 36, 1991, 252.
210. Park, D. I., Lee, L. S., Price, R. I., and Pohland, A. E., Review of the contamination of aflatoxin by ammoniation: current status and regulation, *J. Assoc. Off. Anal. Chem.,* 71, 685, 1988.

211. Pruthi, J. S., Mycotoxins in foods and feeds — their detection, estimation, preventive and curative measures, *Bull. Grain Technol.*, 21, 63, 1978.

212. Khalafallah, A. E. M., Elraheim, A. A., Badawy, A., and Guergues, S. N., Biochemical studies on aflatoxins during processing of cottonseed oil, *Egyptian J. Food Sci.*, 17, 185, 1989.

213. Scott, P. M., Effects of food processing on mycotoxins, *J. Food Prot.*, 47, 489, 1984.

214. Waltking, A. C., Fate of aflatoxin during roasting and storage of contaminated peanut products, *J. Assoc. Off. Anal. Chem.*, 54, 533, 1971.

215. Luter, L., Wyslouzil, W., and Kashyap, S. C., The destruction of aflatoxins in peanuts by over- and microwave-roasting, *Can. Inst. Food Sci. Technol. J.*, 15, 236, 1982.

216. Chelack, W. S., Borsa, J., Marguardt, R. R., and Frohlich, A. A., Role of the competitive microbial flora in the radiation-induced enhancement of ochratoxin production by *Aspergillus alutaceus* var. *alutaceus* NRRL 3174, *Appl. Environ. Microbiol.*, 57, 2492, 1991.

217. Van Egmond, H. P., Regulatory aspects of mycotoxin in Asia and Africa, in *Fungi and Mycotoxins in Stored Products,* Champ, B. R., Highley, E., Hocking, A. D., and Pitt, J. I., Eds., Proc. Int. Conf. Bangkok, ACIAR Proc. No. 36, 1991, 198.

218. Pitt, J. I. The ongoing problem of aflatoxin in international trade, *Aust. Mycotoxin Newsl.*, 49, 1, 1993.

219. Gilbert, J., Regulatory aspects of mycotoxins in the European Community and USA, in *Fungi and Mycotoxins in Stored Products,* Champ, B. R., Highley, E., Hocking, A. D., and Pitt, J. I., Eds., Proc. Int. Conf. Bangkok, ACIAR Proc. No. 36, 1991, 194.

220. Shotwell, O. L., Hesseltine, C. W., Goulden, M. L., and Vanderbraft, E. E., Survey of corn for aflatoxin, zearalenone and ochratoxin, *Cereal Chem.*, 47, 700, 1970.

221. Sanders, T. H. and Mixon, A. C., Effect of peanut tannins on per cent seed colonization and *in vitro* growth by *Aspergillus parasiticus, Mycopathology,* 66, 169, 1978.

222. Tuite, J. and Foster, G. H., Control of storage diseases of grain, *Annu. Rev. Phytopathol.*, 17, 343, 1979.

223. Azaizeh, H. A., Pettit, R. E., Sarr, B. A., and Phillips, T. D., Effect of peanut tanin extracts on growth of *Aspergillus parasiticus* and aflatoxin production, *Mycopathology,* 110, 125, 1990.

224. Singh, P., Bhagat, S., and Ahmed, S. K., Aflatoxin elaboration and nutritional deterioration in some pulse cultivars during infestation with *A. flavus, J. Food Sci. Technol.*, 27, 60, 1990.

Chapter 13

Control of Seedborne Pathogens

The principles of plant disease control have been defined and discussed in many texts available in personal or public libraries and will not be elaborated on here. A flow chart summarizing the general classification of plant disease control methods is presented (Figure 13-1). Generally, no single method, except for immunity, will provide complete control for any single disease. The control of most plant diseases involves the use of more than one measure in a disease control strategy similar to programs developed by entomologists in integrated pest management systems.

Control of seedborne pathogens and diseases is attained through integrated disease management systems. For example, the production of wheat seeds free of loose smut fungal infection in India has been attempted through an integrated approach using the following guidelines:[1] (1) production of seeds of relatively resistant cultivars, (2) production of seeds in areas isolated (about 150 m) from commercial plots, (3) field inspections to meet the requirements for certification, i.e., maximum permissible infection of 0.1 and 0.5% for foundation and certified seed production plots, respectively, (4) roguing infected plants, (5) testing of seeds using the embryo-count method to identify heavily infected seed lots, and (6) seed treatment with systemic fungicides.

The seedborne inoculum of *Phaeosphaeria maculans* in *Brassica oleracea* has been reduced to nondetectable levels in the United States using the following: (1) seed production in the summer in dry areas of the West Coast, (2) use of resistant cultivars, (3) crop rotation, (4) reduction of residues, (5) seed health testing and (6) seed treatment with benomyl.[2]

In New Zealand the control strategies for *Ascochyta fabae* on fava bean have included selection of seed lines with a low level of infection based on laboratory tests, seed treatment, and foliar sprays to control secondary inoculum. To attain this integrated approach, the Ministry of Agriculture and Fisheries charges for a standard health test, and some commercial firms have their own testing facilities. Seeds used for processing or as dried beans have less than 1% seed infection. To

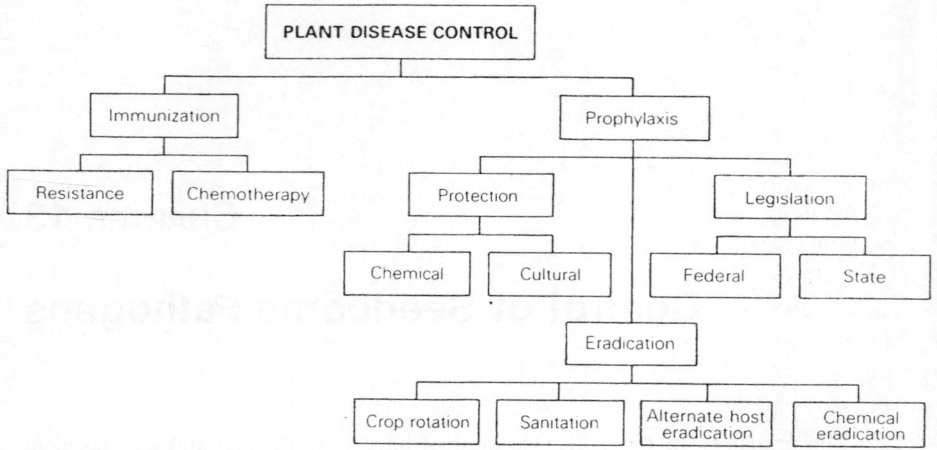

Figure 13-1 General classification of various methods for the control of plant diseases. (Adapted from Sharvelle, E. G., *Plant Disease Control*, AVI Publishing, Westport, CT, 1979, 311; and Tarr, S. A. J., *Principles of Plant Pathology*, Macmillan, London, 1972, 632.)

avoid inoculum buildup in seeds, they are treated with a methylbenzimidazole carbamate (MBC) compound (50 g/kg) plus captan (100 g/kg seed). If disease is present in field, foliar sprays containing chlorothalonil (2.5 kg/ha) are applied at early- and late-pod stages to prevent lesion formation on pods. Foliar sprays have reduced seed infection from 22.3 to 7.4%.[3]

Control of a seedborne pathogen is not likely control of a disease if the pathogen survives by alternative means or if attempts at control are directed only toward control of seedborne inoculum.

In this chapter, methods that directly assist in the prevention of seed infection in the field or control of seedborne infection and infestation are discussed.

I. SELECTION OF SEED PRODUCTION AREAS

Seeds should be produced in areas where the pathogens of major concern are unable to establish or maintain themselves at critical levels during periods of seed development. Areas with low rainfall and relative humidity generally are favorable for production of high-quality seeds with low inoculum levels. Walker[4] identified western Washington state as an area for producing cabbage seeds. The Skagit Valley produces approximately 80% of the cabbage seeds required for the United States and 30% for the world.[5] This moderately cool maritime climate, usually with dry summers, is ideal for production of high-quality cabbage seed relatively free of *X. c.* pv. *campestris* and *Phoma lingam.*[6] However, two blackleg epidemics in the eastern and mideastern growing areas occurred in 1947 and 1973 because of infected seeds produced in western Washington in 1946 and 1972, respectively.

The epidemic was due in 1946 to a virulent strain of *P. lingam* and wet weather at pod set, and in 1972 to the introduction of a new hybrid. An integrated disease-control approach to produce relatively pathogen-free seeds is now in effect and includes seed health testing, benomyl seed treatment of infected stock seeds, and rotation of seed beds and production fields.[2]

Most of the United States bean seed industry is located in the irrigated desert of southern Idaho because of the dry climate. Due to epidemics of bacterial bean blight in the early 1960s, Idaho adopted strict regulations, including laboratory tests and field trials.[7,8] All seeds grown in Idaho pass laboratory assays. Seeds grown for export are field inspected several times each season, and a zero tolerance is enforced. Infected fields are destroyed by plowing within 5 d. The system improved seed health quality. After the blight epidemics in 1984, state investigators developed an improved, rapid, sensitive method of assaying bean seeds for *P. s.* pv. *phaseolicola* and *P. s.* pv. *syringae*.[9]

A safe source of alfalfa seeds comes from areas in southern California where the crop is grown as a seed and not as a forage crop. Seeds are taken from crops in their first or second cycle of vegetative growth. This is because *C. m.* subsp. *insidiosus* spreads by cutter bars, and in older crops there is a risk of disease increase. Also, the dense growth of older crops makes detection of an occasional wilted plant impractical. Although alfalfa bacterial wilt occurred in the United Kingdom, almost all seeds are imported with the aim to keep levels of infection low.[10] Weather and soil conditions in northern Illinois are more conducive to plant growth and high-quality soybean seeds production than in southern Illinois. Seed weight and percent radicle emergence are greatest from northern plots, whereas the number of fungus-infected and noninfected nonviable seeds is greatest from southern plots. The number of seeds infected by *Aspergillus*, *Chaetomium*, and *Phomopsis* is highest in northern plots, whereas the number of seeds infected by *Alternaria, C. kikuchii, Cladosporium, Colletotrichum, Fusarium, M. phaseolina, Nematospora coryli, Penicillium,* and *Phoma* is highest in southern plots.[11] Conditions in northern Illinois include adequate rainfall. In southern Illinois there is limited soil moisture and inadequate rainfall. The extent of *Ustilago segetum* var. *tritici* infection in barley depends on wind direction at flowering, being maximum at 40 m and minimum at 80 m from the source. It is recommended that barley seeds for sowing be separated by at least 100 m from crops grown for consumption in Romania.[12] Control measures for *Plasmopara halstedii* in sunflower in France include seed production at least 50 km from the source of infection in the previous year and treatment of all seeds.[13]

The Piedmont area is favored over the coastal plains of Georgia for seed production of cottonseeds because low rainfall during boll opening results in reduced infection by *Alternaria alternata, Fusarium oxysporum, F. roseum, Glomerella gossypii, Lasiodiplodia theobromae*, and *Nigrospora sphaerica*.[7] Seed infection by *Stagonospora nodorum* is less when wheat is produced in the mountain region than in the southern part of Georgia.[14] *Ascochyta pisi, Mycosphaerella pinodes, Phoma*, and *Phoma pinodella* occur frequently in pea seed samples produced in the eastern United States and Canada, but seed stocks produced in the Palouse districts of Idaho and Washington generally are free of these fungi and commonly used throughout the United States.[15] Because of low rainfall, pea seeds devoid of these fungi are produced from infected seeds grown in the Imperial and Temecula valleys of southern California.[15]

Sugarbeet seed lots produced along the Pacific Coast of North America or in the Mediterranean region of Europe are unlikely to carry *Phoma betae.*[16] A high level of *Alternaria zinniae* has occurred in *Zinnia elegans* seeds produced in coastal valleys of California, while those from the dry interior valleys have been free of the pathogen.[17] Bean seeds free of *P. s.* pv. *phaseolicola* have been produced under semiarid conditions in Idaho where there is zero tolerance of the bacterium in seed production fields.[18,19]

Lettuce seeds free of LMV have been obtained annually from the Swan Hill area of Victoria and New South Wales, Australia since 1954. Vectors of the virus are not found during most of the growing season because of unfavorable climate and ecological conditions. The seeds produced in the Swan Hill area are used in the southern Victoria districts, where the spread of mosaic from infected seeds has been a problem.[20] Peanut seeds free of PMV are produced for Georgia farmers by growing breeder seeds under insect-free conditions in screenhouses where individual plants are indexed and rogued. Seeds then are grown in isolation from other peanuts or in a location free of the aphid vectors.[21] Peanut stripe virus can be eliminated from peanut seeds, and fields in Florida can remain virus-free for several growing seasons provided they are not close to inoculum sources.[22]

II. CROP MANAGEMENT

Proper crop management can help in the production of relatively pathogen-free seeds.

A. High-Quality Seeds

Planting seeds should be as free of the pathogen as possible. Seeds should come from a carefully maintained, genetically pure block and should be grown in selected production fields. They should be cleaned commercially and treated either chemically or nonchemically.

B. Seeding Rate

The seeding rate should not be excessive. A high seeding rate can increase the number of foci of primary infection that may develop in the field, which ultimately can result in higher disease incidence and seed infection.

C. Planting Time

Planting should occur at a time when requirements of the host do not correspond with the pathogen, and thus the plants may escape infection. Winter wheat

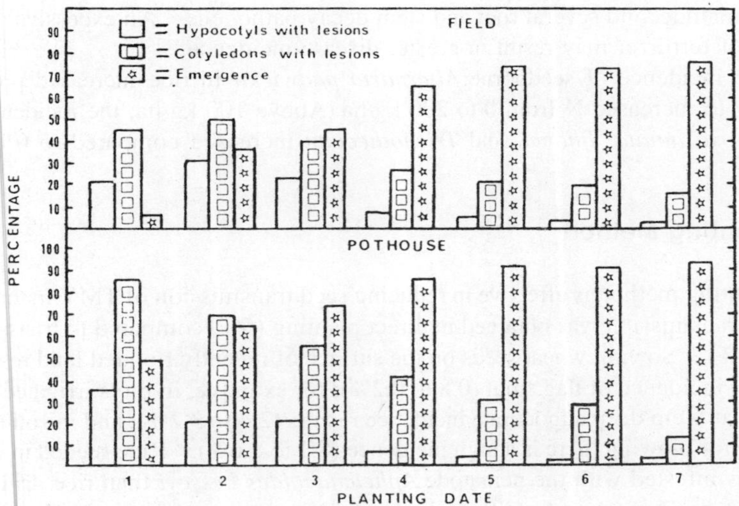

Figure 13-2 Percentage emergence of soybean (*Glycine max*) seeds infected with *Colletotrichum truncatum* (anthracnose) planted on seven planting dates in pothouse and field tests during the monsoon season in India, 1972, and percentage of hypocotyls and cotyledons of resulting plants that showed lesions. (From Nicholson, J. F. and Sinclair, J. B., *Plant Dis. Rep.*, 57, 770, 1973. With permission.)

sown in early autumn may escape infection from *Tilletia caries* and *T. laevis* because plants pass the susceptible stage before bunt spores germinate. Crops sown later may become infected.[23] Planting oat seeds in early spring reduces the incidence of *Ustilago segetum* var. *avenae*. Germination of infected oat seeds at 7°C reduces smut incidence, and at 0 to 2.5°C, reduction in smut is equivalent to using disinfected seeds. In this temperature range, the fungus does not develop.[24] In India and Puerto Rico, adjusting planting schedules so that soybeans mature at the end of (or out of) the rainy season reduces the amount of seedborne *Colletotrichum truncatum* (Figure 13-2). Although seed yields are lower than those produced during the rainy season, seed quality is superior.[25]

D. Burning

Burning grass seed production fields in Oregon destroys inoculum of *Gloeotinia granigena* (blind seed disease) and *Claviceps purpurea* (ergot).[23]

E. Balanced Fertility

Adequate, balanced soil fertility coupled with near neutral pH is important in reducing seed infection. Plants under stress from deficient or toxic levels of nutrients are more susceptible to disease than those grown in soil with well-balanced fertility. Insufficient phosphorus or potash in soybean can increase losses from bacterial blight, bacterial pustule, charcoal rot, pod and stem blight, soybean cyst nematode, and several root and stem decay pathogens.[25] An excessive application of fertilizer may result in greater disease incidence.

The incidence of seedborne *Alternaria padwickii* in rice increased proportionally to increased N from 0 to 200 kg/ha. Above 100 kg/ha, the incidence of *Curvularia lunata, Phoma,* and *Trichothecium* increased compared to 0 or 50 kg/ha.[26]

F. Planting Method

Planting method is effective in reducing seed transmission of TMV in tomato. A low transmission was obtained in direct planting (5%) compared to transplanting (71%).[27] Sowing wheat seeds on the surface of recently flooded land resulted in a low incidence of flag smut (0.8 to 0.2%); for example, in plots irrigated after sowing at 4-cm deep a moderate incidence results (2.4 to 3.2%); and in soil moist enough for plowing there is a high incidence (8.1 to 8.6%).[28] Rice seeded in water was less infested with the nematode *Aphelenchoides besseyi* than rice drilled in and then flooded when 6- to 9-cm high. Quiescent nematodes probably revived in water, moved about, and died before seed emergence.[29]

G. Spacing

Reduced spacing between plants favors seedborne infections. Close spacing results in high humidity among plants, which can be conducive to heavy seedborne infections. At 15 cm between rice plants, there was a higher percentage of seedborne *Alternaria alternata, A. longissima, A. padwickii, Bipolaris oryzae, C. lunata, Fusarium pallidoroseum*, and *Sclerotium* than at wider distances.[26] Soybean seeds from narrow-row (25 cm) compared to wide-row (76 cm) spacing resulted in a higher recovery of total fungi and bacteria, which adversely affected the quality of clean seeds.[30]

H. Depth of Planting

Depth of planting greatly influences seed transmission of smuts. Shallow planting in wet soils protects wheat plants from *Urocystis agropyi.*[28]

I. Water Management

Water management practices influence disease development.[25] Irrigation can be timed to reduce water stress. Water management strategies vary, depending upon the growing area, most common pathogens, soil types, etc. Irrigation, especially at the seed-development stage, may favor seed infection. Irrigation time and the amount of water should be controlled so that the relative humidity is not raised to such an extent that it becomes conducive for seed infection. Overhead irrigation should be avoided where possible.

J. Crop Rotation

Crop rotation and clean tillage play an important role in controlling seedborne pathogens because many important bacterial and fungal pathogens survive between crops on or in crop debris. Rotating soybeans with a nonhost crop every second year is effective for reducing most foliage and stem pathogens, and rotation every third year sharply reduces soybean cyst nematodes.[25] Recommended tillage practices and rotation lengths vary with region and soil type. Soybean seed infection by *Phomopsis longicolla* can be reduced by rotating soybean fields with maize.[23]

K. Isolation Distances

The distance between seed production and commercial plots has been worked out for reducing seedborne loose smut of barley and wheat. The distance between plots may vary from region to region depending on weather conditions. In Canada, the problems of proper isolation of stock seed production areas for barley is governed by legislation.[31] Similar legislation exists in Germany for stock seed production of barley and wheat; seed crops are rejected if loose smut is encountered in a field within a distance of less than 50 m of the seed field and if upwind of the prevailing wind direction.[32] In Holland, Oort[33] demonstrated that a minimum distance of 100 m from infected fields is necessary for barley grown under seed certification programs. Barley and wheat crops should be isolated by at least 50 m from any source of loose smut infection for production of certified seeds in the United Kingdom.

L. Roguing

Roguing should be practiced where possible. In seed production fields, infected plants should be rogued and destroyed. Roguing has been followed successfully in the control of loose smut of barley and wheat.

M. Foliar Fungicide Sprays

Fungicide sprays are used to control bacteria and fungi that cause disease on foliage and seeds of field crops. Many of the microorganisms that cause disease of roots, stems, leaves, and reproductive structures are seedborne. The control of plant diseases in the field may indirectly or directly affect seed infection.

The use of preharvest fungicide sprays to control internally seedborne fungi has been reported in a number of crops with results that are promising in the tropics, the subtropics, and temperate regions.[34] However, the use of preharvest foliar fungicides for the control of internally seedborne fungi in seed is complicated. When disease pressure is high and environmental conditions are favorable for disease development and when fungicides are applied correctly, favorable results often are recorded. Benefits from fungicide application include less fungal infection in the seeds, larger seeds, higher yields, higher seed germination, and seeds of high quality. The close association between cultivar, cropping history, tillage practices, and other environmental conditions such as weather and number of sprays at harvest, time of harvest, and epidemiology of the disease influences the incidences of seedborne fungi and their control with foliar fungicides. The use of fungicide sprays should be a part of a disease control strategy to reduce losses from fungi that are seedborne.[34]

1. Bean

No seed infection by *P. s.* pv. *phaseolicola* and *X. c.* pv. *phaseoli* was recorded from plants sprayed with copper oxide every 7 d, and 0.28% infection by *P. s.* pv. *phaseolicola* and none by *X. c.* pv. *phaseoli* were recorded from plants sprayed at 14-d intervals.[35] Chlorothalonil (0.1% a.i.) sprays two to three times prevented *Ascochyta fabae* infection in fava bean.[36]

2. Brassica

Three sprays of iprodione (50% a.i.) at 0.5 to 1 kg a.i. per hectare on cabbage seed crops at 3-week intervals from the young green-pod stage until cutting controlled pod and seed infection by *A. brassicicola.* Only a few seeds were infected, and seed yield and germination were improved.[37] Applications of iprodione (1.12 kg a.i.)to cabbage seed fields with 117 ml Biofilm spreader-sticker in 90 l/ha made at early-, mid-, and late-pod stages at 3-week intervals effectively controlled *A. brassicae* and *A. brassicicola* on plants and reduced the incidence of the pathogens in seeds.[38]

3. Maize

Foliar sprays of triphenyltin acetate and copper oxychloride reduced seedborne *F. moniliforme* and *Curvularia pallescens* and white streak on maize kernels. The percentage of streaking in seeds was 39.2 from nonsprayed plots, 36 in those sprayed with Bordeaux mixture, 31.2 with fentin hydroxide, 27.6 with copper oxychloride plus zineb, 26.8 with triphenyltin chloride, 17.6 with mancozeb, 11.8 with triphenyltin acetate, and 9 with copper oxychloride.[39]

4. Okra

Injury by spotted bollworm (*Earias fabia* and *E. insulana*) increased the incidence of *Aspergillus flavus, F. moniliforme, F. oxysporum, F. pallidoroseum,* and *Macrophomina phaseolina* in okra seeds. A preharvest spray with the insecticide monocrotophos (0.1%) reduced the incidence of these fungi. A carbendazim (0.1%) + monocrotophos spray almost completely controlled these fungi.[40]

5. Peanut

Carbendazim, chlorothalonil, and mancozeb sprays controlled *Cercospora arachidicola* and *C. personata* as well as increased oil content of seeds and reduced the free fatty-acid content and saponification value of the oil. Carbendazim sprays increased the iodine value of the oil.[41]

6. Pigeon Pea

Pigeon pea seeds from plants sprayed two to four times with benomyl (2.2 kg/ha) produced fewer seedborne fungal infections.[42]

7. Rice

Spraying rice with the following fungicides reduced seed infection by *Alternaria longissima, A. padwickii, Bipolaris oryzae, C. lunata,* and *Sclerotium*: fentin hydroxide, IBP (Kitazin), triphenyltin chloride, mancozeb, and zineb.[43] The incidence of *A. padwickii* and *B. oryzae* was lowest in seeds when chlorothalonil (1.5 kg/ha), mancozeb (5 kg/ha), carboxin (2.51 l/ha), and sisthane (0.64 l/ha) were applied before the dough stage. Two applications of chlorothalonil controlled *B. oryzae*. All the fungicides controlled *A. padwickii.*[44]

8. Sorghum

Weekly applications of benomyl + captan (0.5 kg + 0.5 kg/ha) from boot stage to physiological maturity significantly reduced *C. lunata* and completely controlled *F. moniliforme* in sorghum seeds. Fungicide application increased seed yield, 1000-seed weight, and germination.[45] A spray of mancozeb (0.2%) was more effective in reducing sorghum ergot than one of captafol (0.2%) and ziram (0.2%).[46]

Three sprays of propionic acid (1.0%), thiram (0.2%), and triadimefon (0.1%) in the field starting at anthesis until physiological maturity at 10-d intervals resulted in a reduction in *C. lunata* in sorghum seeds. Carbendazim (0.1%) and propionic acid significantly reduced infection by *F. moniliforme* and *Phoma sorghina* in seeds. The use of captafol (0.2%) and triadimefon was better for control of *F. moniliforme.* Seed vigor, viability, and germination were improved due to fungicidal sprays. Thus the use of fungicides should be based on the relative cultivar susceptibility to seed fungal pathogens.[47]

9. Soybean

The strongest evidence that fungicide sprays improve seed quality by controlling seedborne pathogens comes from soybeans.[48] Prasartsee et al.[49] found that soybean seeds from plots sprayed with benomyl 50 WP + zinc + maneb 80

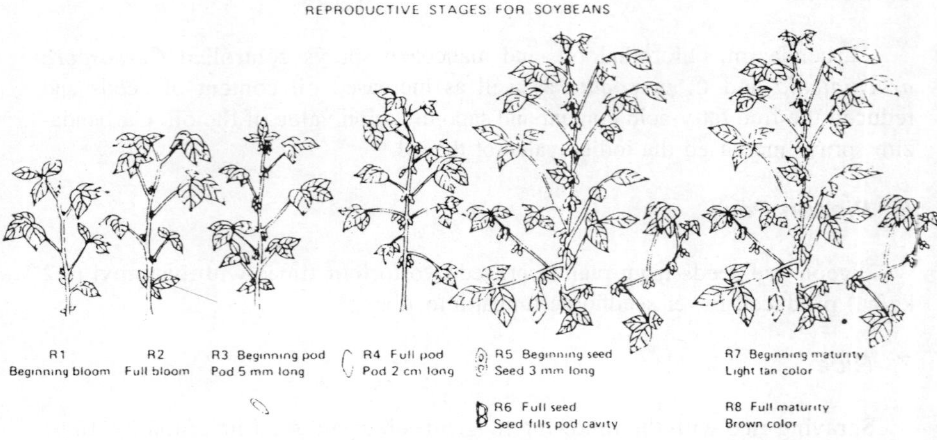

Figure 13-3 Reproductive stages of soybean (*Glycine max*). (Adapted from Walla, W. J., Ed., Soybean Diseases Atlas, Texas A & M University, College Station, TX, 1979; from Sinclair, J. B., Ed., *Compendium of Soybean Diseases,* APS Press, St. Paul, MN, 1982, 2. With permission.)

WP, thiophanate methyl 70 WP, benomyl 50 WP, chlorothalonil, zinc + maneb 80 WP, or thiabendazole 98.5 WP had significantly (P = 0.05) less seedborne *Phomopsis* than those from nonsprayed plants. Similarly, Ellis et al.[50-53] showed that soybean plants grown in Illinois and sprayed with benomyl produced seeds with significantly (P = 0.05) less Phomopsis seed decay than nonsprayed plants and that the occurrence of *Phomopsis* was significantly correlated (r = –0.71) with germination in culture. Benomyl sprays reduced losses due to this disease even more so after delayed harvest.[48] In Brazil, fungicide sprays with benomyl, captafol, cercobin, chlorothalonil, and thiabendazole significantly (P = 0.05) reduced seed coat and embryo infection by *Phomopsis,* and germination was significantly higher for seeds from sprayed plants than nonsprayed plants.[54] Benomyl or a zinc ion–maneb complex sprayed on soybeans in Puerto Rico controlled seedborne *Cercospora kikuchii.*[55] The incidence of most fungi in soybean seeds, *Diaporthe phaseolorum* var. *sojae, C. kikuchii, Colletotrichum truncatum,* and *Glomerella glycines* were reduced in seeds after two sprays of benomyl (1.1 kg/ha).[56]

Foliar fungicides can be used to improve seed quality through control of seedborne anthracnose (*C. truncatum*), Cercospora seed decay (*C. sojina*), purple seed strain (*C. kikuchii*), Phomopsis seed decay (*P. longicolla*), and others. Application of fungicides on seed production fields is necessary when wet conditions above 25°C prevail during growth stages R1 (beginning bloom) and R5 (beginning seed formation) (Figure 13-3; Table 13-1). Under these conditions, these diseases can cause serious seed quality problems. In general, fungicides applied between growth stages R2 and R5 provide improved seed quality by reducing fungal seed infection. With increasingly more restrictive grading standards, reduced fungal

Table 13-1 Growth Stage Key for Soybeans

Stage	Description[a]
V^1	Completely unrolled leaf at the unifoliolate node
V^2	Completely unrolled leaf at the first node above the unifoliolate node
V^3	Three nodes on main stem beginning with the unifoliolate node
V_N	*N* nodes on the main stem beginning with unifoliolate node
R^1	One flower at any node
R^2	Flower at node immediately below the uppermost node with a completely unrolled leaf
R^3	Pod 0.5-cm (1/4 in) long at one of the four uppermost nodes with a completely unrolled leaf
R^4	Pod 2-cm (3/4 in) long at one of the four uppermost nodes with a completely unrolled leaf
R^5	Beans beginning to develop (can be felt when the pod is squeezed) at one of the four uppermost nodes with a completely unrolled leaf
R^6	Pod green containing full-size beans at one of the four uppermost nodes with a completely unrolled leaf
R^7	Pods yellowing, 50% of leaves yellow, physiological maturity
R^8	95% of pods brown, harvest maturity

[a] Use a key to indicate the stage of soybean growth. Determine the vegetative stages by counting the number of nodes on the main stem, beginning with the unifoliolate node, which has a completely unrolled leaf.

Adapted from Fehr, W. R., Caviness, C. E., Burmood, D. T., and Pennington, J. S., *Crop Sci.*, 11, 929, 1971. With permission.

infections may increase the value of grain lots.[57] Thiophanate-methyl is the most effective foliar fungicide for reducing *C. kikuchii* in soybean seeds. A single application made at 15 to 20 d after the R3 growth stage was effective, and seeds were protected from infection for 30 d. The fungicide also had a curative effect and reduced seed invasion by the pathogen after infection occurred.[58] Benomyl, thiabendazole, and thiophanate methyl are commonly used foliar fungicides for soybeans. Chlorothalonil is not widely used because of its high cost. Some synoptic prediction systems (checklists) have been developed in the United States to aid both grain and seed producers in making decisions about foliar fungicide applications (Tables 13-2 and 13-3).[59] A single spray of benomyl (1.21 kg/ha) at growth stage R6 gave lowest infection by *Phomopsis*.[58]

The plant growth stage at which fungicides are applied is important in the control of seedborne infection (Figure 13-1; Tables 13-2 and 13-3). Infection by *Phomopsis* in soybean usually occurs after the R6 growth stage, although pods could be infected earlier.[60] The most effective time to apply a fungicide is at the R6 to R7 growth stages, before seed infection takes place.[61] Benomyl applied at 1.2 g/l, either 3 d before or on the day of inoculation reduced *Phomopsis* pod and seed infection; when applied 3 d after inoculation, it was less effective.[61] Fungicide application just before seed infection took place ensured an optimum concentration of methylbenzimidazole carbamate, a fungitoxic product of benomyl, in the seeds at the time when protection was needed most. Seed germination was higher and occurrence of *Phomopsis* and total fungi lower when fungicides were applied at R4 or R3 plus R4 stages.[62] One benomyl (0.05%)

Table 13-2 Illinois Checklist Used to Deteremine Whether Foliar Fungicide Should be Applied to Soybeans

Risk factor	Point value if answer is yes[a]	
Rainfall, dew, and humidity up to early bloom and pod set are:		
Below normal	0	
Normal	2	
Above normal	4	
Soybeans were grown in the field last year	2–3	—
Chisel plow, disk, or no-tillage was used	1	—
Pynidia (black specks) are visible on fallen petioles[b] and Septoria brown spot is obvious on the lower leaves	2	—
Early-maturing cultivar (not full-season)	1–2	—
Soybeans are to be used or sold for seed	6	—
Yield potential is better than 35 bushels per acre	2	—
Seed quality at planting time is less than 85% germination in a warm test	1	—
Other conditions that favor disease development (weather forecast with a 30-d period of greater than normal rainfall and a field history of disease)	1–3	—
	Total	—

[a] If the total value is 15 or more, application will probably mean increased yields and higher seed quality.

[b] Only brown, fallen petioles should be assayed, and more than two-thirds to three-fourths of these petioles should show pycnidia.

From Kirby, H. W., Shurtleff, M. C., Eastburn, D. M., and Edwards, D. I., in *1992 Illinois Pest Control Handbook*, University of Illinois at Urbana-Champaign, 1992. With permission.

spray before flowering and another during flowering were as effective as five sprays of benomyl after flowering to control *C. kikuchii* of soybean.[63] Fungi may gain entry to soybean seeds in unopened pods through cracks in pod walls, insect injuries, and/or by systemic infection.[64-66] Benomyl may persist in plant parts for up to 30 d after spraying.

Foliar fungicides are used to control anthracnose, Cercospora leaf blight, frogeye leaf spot, Phomopsis seed decay, purple seed stain, Septoria brown spot, soybean rust, stem blight, and stem canker on soybeans.[25] All except the soybean rust fungus are seedborne. Growers in tropical and subtropical areas and seed producers in all regions may benefit from the proper application of foliar fungicides. Benefits from fungicide application include higher yields, better seed germination, less fungal seed infection, larger seeds, and seeds that appear cleaner. If seedborne fungi are controlled effectively by foliar fungicides, seed dressing treatment may not be necessary. Checklists have been developed at the University of Illinois at Champaign-Urbana (Table 13-2) and at the University of Kentucky, Lexington (Table 13-3) to help determine conditions when these diseases are likely to be severe enough to apply a fungicide(s). At present, two fungicides, benomyl and thiabendazole, are approved for use as foliar fungicides on soybean in the United States. To control both foliar and late season diseases and to achieve the highest yield and seed quality, fungicide protection should be provided from R2 growth stage to R8 growth stage. Growers who wish to increase yield should

Table 13-3 A Point System for Determining Whether to Apply Foliar Fungicides to Soybeans (*Glycine max*) for Kentucky

Risk factor	Point value[a]
Cropping history	
Soybeans grown previous 2 or more years	3
Soybeans grown previous year	2
Soybeans not grown previous year	0
Cultivar selection	
Early-season cultivar	3
Mid-season cultivar	2
Late-season cultivar[b]	0
Planting date	
Before May 20	3
Between May 20 & June 20	2
After June 20	0
Rainfall[c]	
Below normal	0
Near normal (±2 cm [0.5 in])	2
Above normal	4

[a] If the point total is 11 or more, a fungicide should be applied; if it is 9 or 10, a fungicide may be beneficial; if it is 8 or less, fungicide should not be applied.

[b] Foliar fungicides should not be used on late-season cultivars.

[c] Based on recorded precipitation and predicted rainfall during seed development and maturation from growth stage R_2 to R_7.

From Stuckey, R. E., Jacques, R. M., TeKrony, D. M., and Egil, D. B., Foliar Fungicides Can Improve Soybean Seed Quality, Kentucky Seed Improvement Association, Lexington, KY, 1981. With permission.

begin fungicide application during R6 growth stage, while seedsmen interested in seed quality should time applications to ensure that pods are protected until harvest.[25] In 1978, approximately 3.5 million acres of soybeans were sprayed with fungicides in the United States to control soybean diseases and to reduce seedborne inoculum.[67] Spraying soybeans with fungicides is profitable particularly when Phomopsis seed decay is a problem.[68] Latent infection of many soybean fungal pathogens can influence the use of foliar fungicides.[68]

10. Wheat

Field studies in southeastern England with winter wheat during the 1973–75 seasons showed that a single benomyl or benomyl + maneb spray applied between flag leaf and ear emergence protected the leaf and ear and resulted in less seedborne infection of *Septoria tritici* and *Stagonospora nodorum.*[69] Wheat seed infection by *F. graminearum* was reduced by spraying plants with benomyl + mancozeb, or benomyl or methylbenzimidazole carbamate alone.[70]

11. Kentucky Bluegrass

A single application of flusilazole, propiconazole, or tebuconazole significantly reduced ergot (*C. purpurea*) in Kentucky Bluegrass. Sclerotia and panicle exudate were reduced to zero, with a single application of flusilazole alone or flusilazole, propiconazole, or tebuconazole combined with a wetting agent. Fungicide application at preanthesis controlled disease more than at mid- or late anthesis. Most fungicides applied at mid- or late anthesis reduced germination. Seed germination was not reduced by preanthesis application except at the high rates of flusilazole.[71]

N. Insect Control

Insect control reduces stress and injuries where pathogens may enter and may reduce virus spread. An insecticide spray schedule to control insect infestation can be useful in maintaining low levels of seed infection.

O. Weed Control

Weed control is important because weeds compete for space, nutrients, and water and can be hosts to pathogens. Plants under stress from weed competition are more susceptible to most pathogens. Hand hoeing and mechanical and chemical weed control methods vary from one region to another. Herbicides must be used with care to avoid plant damage. Evidence is increasing that many weeds in and around soybean fields are susceptible to soybean pathogens, including seedborne pathogens such as *Colletotrichum* and *Phomopsis* spp., and the SbMV and TRSV.[25]

There was a significant correlation between development of weeds (*Alternanthaera ficoidea, Braccharis*, and *Commelina diffusa*) and the occurrence of *Colletotrichum truncatum, F. pallidoroseum,* and *P. sojae* in soybean seeds.[72] Weeds, such as *Abutilon theophrasti, Amaranthus spinosus, Leonotis nepetaefolia*, and *Leonurus sibiricus,* can act as alternative hosts or provide a microclimate of prolonged humidity favoring seed infection.[73,74]

P. Harvesting

Seeds should be harvested immediately when mature. The longer seeds remain in the field after maturity, the greater the chance for invasion by pathogenic bacteria and fungi, especially under warm, moist conditions. Harvesting and threshing plots with a low seed infection rate should be worked prior to those having a high infection rate to avoid mixing healthy and diseased seeds or contaminating healthy seed lots. Clean cottonseeds have been contaminated by *X. c.* pv. *malvacearum*-infected dry refuse in or on harvesting equipment, trailers, or trucks used to take picked cotton to gins.[75]

III. SEED TREATMENT

Seed treatment is a biological, chemical, mechanical, or physical process designed to mitigate externally or internally seed- or soilborne microorganisms, resulting in the emergence of a healthy seedling and, subsequently, a healthy plant. Seeds may be treated to promote good seedling establishment, to minimize yield loss, or to maintain and improve quality, and to avoid further spread of pathogens.

A. Biological Control

Biological control is the reduction of inoculum intensity or disease-producing activities of a pathogen or parasite in its active or dormant state, by one or more organisms, accomplished naturally or through manipulation of the environment, host, or antagonist, or by mass introduction of one or more antagonists.[76] *Trichoderma viride* was the first fungus demonstrated as an antagonist for control of soilborne pathogens, such as *Rhizoctonia solani.*[77] Biological control of soilborne fungi has been studied extensively but demonstrated in few cases.[76] Results from studies on the control of seedborne pathogens and better plant stands through the application of antagonistic fungi and bacteria to seeds have been inconsistent.

Antagonists applied commonly to seeds are *Bacillus subtilis, Chaetomium, Penicillium oxalicum,* and *Trichoderma.* Their application to seeds reduces seedborne fungi and results in vigorous seedlings.

Pea seeds treated with *P. oxalicum* produced plant stands, vine, and pod weights equal to those produced from captan-treated seeds and were significantly better than those produced from nontreated seeds.[78] The number of spores per seed necessary to protect a plant varies with inoculum concentration of the pathogen in the soil. For treating pea seeds, approximately 6×10^6 of *P. oxalicum* per seed provided protection for pre- and postemergence damping-off similar to that of captan.[78] Spores of *P. oxalicum* germinated and produced hyphae that grew between root hairs and on the root surface.[79]

Trichoderma hamatum applied to pea or radish seeds controlled seed rot in soil infested with *Pythium* or *R. solani* in soil at between 17 and 34°C. Seeds were treated with a conidial suspension at a concentration equal to or greater than 10^6/ml. Seed treatment with *Bradyrhizobium* and *T. hamatum* had no adverse effect on the nodulating activity of the former or the protective ability of the latter.[80] *B. subtilis* and *Chaetomium globosum* applied to maize seeds controlled seedling blight caused by *F. moniliforme* and *F. graminearum.* The treatment increased emergence, root vigor, plant fresh weight, root dry weight, and seedling stand. Treating kernels with either microorganism gave control equal to treatment with captan or thiram when the soil was under 20°C.[81,82] A *Chaetomium* isolate from an oat seed protected oat seedlings from seedborne *Bipolaris victoriae.*[83] Certain isolates of *C. globosum* and *C. cochlioides* controlled Fusarium blight (*Microdochium nivale*) in oats as effectively as an organic mercury seed treatment.[84] Seed treatment of sweet corn and wheat with *T. harzianum* was an effective component of crop management[85] and when used on snap beans reduced *R. solani* damping-off in acid soils.[86] As a treatment of seeds and 2-week-old seedlings, *Pseudomonas fluorescens* strain Pf2-79r could be an effective biological control agent of *Tilletia laevis*.[87]

Reduction in seedborne inoculum of *A. brassicae* was reported using sporulating cultures of *C. globosum* and/or *Epicoccum nigrum*.[88] A formulation of Mycostop® was prepared from spores and mycelium of *Streptomyces* isolated from peat. Treatment of cabbage seeds gave control of *A. brassicicola* and partial control of *R. solani*. Seed treatment of cereals with Mycostop® reduced seedborne root rots caused by several fungi.[89] The conidiophores and mycelia of *Botrytis cinerea* from fava bean seeds were parasitized and sclerotia killed after treatment with *Gliocladium catenulatum*.[90] *T. viride* was antagonistic to *Microdochium nivale* in barley.[91]

Bacteria associated with certain seeds can be antagonistic to plant pathogenic bacteria and may be useful in plant disease management.[92] *Erwinia herbicola* was found to be the main resident bacterium of cottonseeds. Its strains were effective in reducing seedborne infection of *X. c.* pv. *malvacearum*.[93] Leben[94] reported that soybean seedling infection by *P. s.* pv. *glycinea* was reduced when seeds were treated with an antagonistic bacterium isolated from the seed coat. Similarly, bacteria isolated from seeds of sesamum, peppers, and crucifers were antagonistic to *P. s.* pv. *sesami, X. c.* pv. *vesicatoria*, and *X. c.* pv. *campestris,* respectively.[95-97] *Erwinia herbicola* and *Penicillium oxalicum* were as effective as hot water (30 min at 50°C), solar heat, streptocycline (100 μg/ml) + captan (0.2%) and streptocycline (100 μg/ml) + Agallol (0.2%) for the eradication of *X. c.* pv. *vignicola* from cowpea seeds.[98]

Treating flax seeds with *B. mesentericus* reduced seedling disease caused by *Colletotrichum* and *Fusarium*.[99] *B. subtilis* and *Streptomyces* applied to wheat seeds reduced the effects of *R. solani* and stimulated seedling growth.[100] Five of nine *B. subtilis* isolates applied to maize seeds and sown in *F. graminearum*-infested soil in the greenhouse at 18°C resulted in significantly higher stands. The use of antagonistic microorganisms may be as effective as captan in reducing the incidence of seedborne infection.[101] Transmission of *Aureobasidium lini* and *Colletotrichum linicola* in flax seeds was lower in natural than in sterilized soil due to microorganism activity. A bacterium was found that caused conidial disintegration of *C. linicola* in soil.[102]

Treating cottonseeds with spore suspension of *Epicoccum nigrum* + *Chaetomium globosum* improved germination and reduced the seedborne *A. alternata, Aspergillus flavus, F. moniliforme*, and *Trichothecium roseum*.[103] *Gliocladium roseum* and *Trichoderma harzianum* were effective in reducing *A. brassicicola* seed infection and increasing seedling emergence in broccoli.[104]

Soaking carrot seeds in a suspension of *B. subtilis* strain T99 improved germination and seedling health in the presence of seedborne *A. radicina*. Application of small amounts of metiram (as Polyram-combi) and iprodione (as Rovral) enhanced the effectiveness of the antagonist.[105] *Bacillus* and an isolate of *Trichoderma harzianum* increased the percentage of healthy seedlings grown from

Alternaria linicola-infected linseed seeds. With a few exceptions, antagonists reduced root and shoot lengths.[106]

Bacteriophages have been used to control seedborne bacteria. Maize seeds treated with a phage and then inoculated with *Erwinia stewartii* developed 1.4%, compared to 18% infected plants from nontreated seeds.[107,108] The spermosphere environment can affect root colonization of *B. subtilis* strain GB03 applied as seed treatment to cottonseeds, thereby possibly affecting the biological control or plant growth promoting potential of the inoculants.[109]

TMV in tomato seeds was partially inactivated or inhibited during germination by normal physiological processes or through soil microflora. Five days after sowing, levels of infected seeds dropped from 94 to 44%.[110]

The use of biological control agents has not been found completely practical under all field conditions. Results have been inconsistent because of variables such as types of coating materials used as carriers of antagonists, moisture and temperature effects, and duration of storage after treatment.

1. Testing Antagonists for Seed Treatment

Biological seed treatments differ from chemical ones in that microorganisms are alive and must grow well if they are to be useful. Therefore, a biological seed treatment must allow for prolific growth of the biological agent. The major factors that restrict growth of desirable organisms on seeds are competitive microflora and unfavorable edaphic factors, such as nutrient and moisture status and pH.[111] An important criterion for a successful biological seed treatment is the preparation of a biomass of high quality. Also, biological seed treatment preparations must have a long shelf life.[111] For commercial production the biomass must survive under low moisture conditions. Microbial preparations must be stored in bulk prior to use and then stored after treatment until planted.

Seed treatment with antibiotic-producing antagonists has been used in various crops for control of pre- and postemergence damping-off and seedling root rot diseases.[101,112-116] Antagonists are grown in a liquid or on an agar medium. When bacteria are grown on a liquid medium, they are agitated during incubation, while *Ascomycotina* and other fungi are cultured without agitation. Bacterial cells are collected by centrifugation and resuspended in physiological saline or used directly. *Ascomycotina* and other fungal cultures are filtered and the mat blended in sterile water and used as a seed treatment. When the antagonists are suspended in water, the seeds are immersed in the suspension for 15 min to about 1 h, then air dried at moderate temperatures and planted either immediately or stored in a refrigerator in paper bags. The dry treatment generally is used for fungal spores. A weighed quantity of spores is added to seeds slightly moistened by water or an adhesive agent such as 4% carboxyl-methylcellulose or gum arabic. The seeds and spores are mixed so that all seeds are covered by the spores. Rolling moist seeds over a heavily sporulating colony of the antagonist on agar media also is useful. Rolling is used for inoculation with bacterial antagonists. The bacterium is grown on an agar medium, and after 48 h the colonies are scraped off and mixed with a small amount of distilled water in a culture plate. The seeds are rolled over the viscous fluid until most of the liquid has been absorbed; then they are air dried at room temperature.

To determine the efficacy of the antagonists as seed treatment, one or two standard seed treatment fungicides should be included in the tests for comparison. Initial testing is done on blotter paper or sterile sand in pathogen-infested soil of different types and finally in the field.

B. Chemical Method

Application of chemicals to seeds is the cheapest and most effective means of controlling most seedborne pathogens. Fungicidal seed treatment may kill or inhibit seedborne pathogens and may form a protective zone around seeds that can reduce seed decay and seedling blight caused by soilborne pathogens, resulting in healthy and vigorous seedlings. The use of fungicides as seed treatments is the most widely followed disease control practice used in all crops.[117,118] The first mention of seed treatment for control of plant diseases was by Caius Plinius Secundua, Pliny the Elder, in a Roman agriculture encyclopedia, *Historia Naturalis (circa* 23 to 79 A. D.). It referred to wheat smut control by "...steeping of the seed for planting in wine or mixing bruised cypress leaves with it is to be recommended."[118] The importance of good seeds was realized in India in Surpala's *Vrksayurveda* (800 A. D.): "A seed which has been steeped in milk and rubbed well in cowdung and after drying, again rubbed repeatedly in honey and the powder of vidanga (*Embelia ribes*) grows without fail."[117]

It was an accident that led to the general use of seed treatment for control of plant diseases. In 1670, a sailing vessel loaded with wheat encountered a storm in the Bristol Channel. The wheat, saturated with salt water, was grounded. Farmers along the coast salvaged some of the seeds and used it as planting seeds. The seeds from such plants were free of stinking smut, whereas seeds produced from native seeds were infested. This led to the soaking of wheat seeds in salt brine in England during the 17th and 18th centuries for control of stinking smut of wheat. In 1637, Remnant reported on the use of salt as a wheat seed treatment for control of stinking smut of wheat. In 1733, Jethro Tull published the same method. Later, Schulthess, in 1760, suggested the use of a seed treatment for control of stinking smut of wheat. The use of copper compounds was supported by Tillet in 1755 and Prevost in 1807.[117,118]

In 1913, Reihm suggested the use of organic mercurial fungicides as a seed treatment for control of stinking smut of wheat. Following World War II, organic mercurials were introduced commercially for treatment of vegetables and small grain seeds. However, mercury poisoning of a family in New Mexico as a result of consuming a pig fed with mercury-treated wheat seeds in 1969-70 resulted in a ban on the use of mercury in the United States.

Two quinone compounds were introduced as seed protectants in the 1940s. The first organic seed protectant, chloranil, was introduced by Cunningham and Sharvelle in 1940. Felix and Terhorst introduced dichlone in 1943, a dithiocarbamate seed protectant; thiram, introduced in 1941 by Harrington for control of turf diseases, was recommended as a seed protectant in 1951. In 1952, a heterocyclic compound, captan, was introduced as the "miracle" seed protectant fungicide. The discovery of the systemic fungicide carboxin by UniRoyal Chemical Co. in 1966 revolutionized the control of plant diseases by seed treatment. Carboxin replaced all nonchemical methods for the control of loose smut of wheat. This led subsequently to the introduction of large numbers of systemic fungicides. Benomyl was introduced by E. I. du Pont in 1968. Other compounds, such as fenfuram, metalaxyl, and others, were introduced later by other companies. Antibiotics, such as streptomycin and tetracyclines, also are used for the control of seedborne bacteria.[117,118] In the early 1980s loose smut reached damaging levels in winter barley as carboxin-resistant strains of the fungus were introduced on seeds from Europe. A change to triadimenol or flutriafol fungicide during seed multiplication reduced the disease to manageable levels.[119]

The chemical natures and trade names of fungicides marketed in different countries for seed treatment are given in Table 13-4.[117,118]

1. *Formulations*

Seed treatment chemicals are available in different formulations.[117,118]

Wettable powders (WP) — This type of formulation usually contains a wetting agent. It aids in dispersal of the chemical in water. Water-dispersible formulations are preferred for slurry or mist seed treatment. Most fungicides are available as water-dispersible powders.

Dusts — Dust formulations are dry powders used for slurry or mist treatments.

Slurries or suspensions — The chemical is mixed with a liquid in high concentrations and diluted at the time of seed treatment.

2. *Categories of Chemical Seed Treatment*

Control through chemical seed treatment may be classified in the following categories.[120]

a. Seed Disinfection — This refers to the control of inoculum established within the seed or seed coat tissues. Such pathogens are controlled by thermotherapy (hot air, oil, or water treatment), or by systemic or mercurial fungicides, which are absorbed or penetrate or diffuse inside the seed. Seed disinfection chemicals are specific and are applied after an assessment of the seedborne inoculum.

b. Seed Disinfestation — This refers to the control of the pathogens that are externally or passively present on the seed surface. It is easier to control such pathogens by fungicide seed treatment than those microorganisms that infect seeds.

c. Seed Protection — This method involves treatment with a fungicide that protects the seed and seedling from seed- and soilborne microflora. Many soilborne fungi are facultative parasites which, under suitable environmental conditions, cause seed rot and seedling blight. Ideally, all seeds planted under favorable

Table 13-4 Fungicides Used as Seed Treatments to Control Seedborne Pathogens

Common name	Trade name	Active ingredient
Nonsystemic fungicides		
Sulfur fungicides		
Inorganic sulfur		
Elemental sulfur		Sulfur
Organic sulfur (carbamates)	Hexaferb, Coronet, Ferbek, Fermate, Fermocide, Ferradow, Karbam black	
Ferbam		Ferric dimethyl dithiocarbamate
Maneb	Dithane M-22,Manzate, MEB, MnEBD	Manganese ethylenebisdithiocarbamate
Thiram	Arasan, Hexathir, Nomersan, Fermide, Fernacol, Thirid, TMTD, Thylate, Tersan, Fernasan, Spottrete	Tetramethyl thiuram disulfide
Zineb	Dithane Z-78, Hexathane, Lonacol, Parzate C, duPont Fungicide A	Zinc ethylene bisdithiocarbamate
Quinone fungicides		
Chloranil	Spergon	2,3,5,6,-Tetrachloro-1,4-benzoquinone
Dichlone	Phygon, Phygon XL	2,3-Dichloro-1,4-napthoquinone
Heterocyclic nitrogenous compounds		
Captan	Captan, Esso fungicide 406, Orthocide, Vancide 89, Captan, Vondcaptan	*N*-(Trichloromethylthio)-4-cyclohexene 1,2 dicarboximide
Captafol	Difolatan, Difosan, Sanspor, Sulfonimide	*cis-N*-(1,1,2,2-tetrachloroethylthio)-4-cy clohexene-1,2 dicarboximide
Systemic fungicides		
Triadimenol	Baytan Universal	(Combination product — with fuberida zole & imazalil)
Triadimenol	Baytan 15SD	1-(4-Chlorophenoxy)-3,3-dimethyl 1,(1,2,4-triazole-1-yl)butanone
Triadimefon	Bayleton 25 WP	1-(4-Chlorophenoxy)-3,3-dimethyl 1-1-(1-H-1,2 4-triazol-1-yl)- 2-butanone
Benomyl	Benlate, Grex, Tersan 1991, Ultra sofril	Methyl-(1-butylcarbamoyl)-2-benzimidazole carbamate

Carbendazim	BAS 3460, Bavistin, Derosal, MBC	Methyl-2 benzimidazole carbamate
Carboxin	Vitavax, DCMO	5,6-Dihydro-2methyl-1, 4-oxathiin,3-3 carboxanilide
Fenfuram	Panoram	*N*-phenyl-2-methylfuram-3-carboxamide 2,5-dimethyl-3-furanilide
Furcarbanil	BAS3191F	2,5-dimethyl-3-furanilide
Metalaxyl	Apron 35SD	Methyl *DL*-*N*-(2,6-dimethyl phenyl)-*N*-2- methoxyacetyl)-alaninate (35% metalaxyl)
Oxycarboxin	DCMOD, Plantvax	2,3-Dihydro-5-carboxanilido-6-methyl-1,4-oxathiin-4,4-dioxide
Pyracarbolid	Sicarol	2-Methyl-5,6-dihydro-4H pyran-3-car boxalic acid anilide
Tridemorph	Calixin M	11% tridemorph + 36% maneb (11% 2,6-dimethyl-4-tridecyl morpholine + 36% maneb)
Quintozene + ethazol	Terracoat L-205	Pentachloronitrobenzene (23.2% + Terrazole 5.8%) PCNB + 5-ethoxy-3 (trichloromethyl)-1,2,4-thiadiazole
Carboxin + thiram	Vitavax 200	37.5% carboxin + 37.5% thiram
Antibiotics		
Cycloheximide	Actidione	β-(2-(3,5-Dimethyl-2-oxocyclohexyl)-2-hydroxyethyl) glutarimide
Streptomycin	Agrimycin-17, Agri-step, Rimocidin, Chemform	2,4-Diguanidino-3,56-trihydroxycyclohexyl-5-deoxy-2-O(2-deoxy-23 methylamino α-glucopyranosyl)-3-formyl pentantofuranoside
Miscellaneous fungicides		
PCNB	Quintozene	Pentachloronitrobenzene
Lesan	Dexon	Sodium-*p*-(dimethyl amino benzene) diazosulfonate

From Sharvelle, E. G., *Plant Disease Control*, AVI Publishing, Westport, CT, 1979, 331. With permission.

environmental conditions should have enough vigor to germinate and emerge. However, poor emergence often is a problem under field conditions. The lack of emergence or seed rot may be attributed to genetic, mechanical, physiological, or pathological factors. There are situations in which good quality seeds germinate and emerge poorly because of the association with seed rot caused by certain pathogens. This type of seed rot is common with seeds of most crop plants, especially legumes and vegetables. There may be both pre- and postemergence damping-off.

Both seed- and soilborne microorganisms are responsible for causing seed decay. Many of the pathogenic seedborne fungi are species of *Alternaria, Aspergillus, Bipolaris, Botrytis, Cephalosporium, Cercospora, Colletotrichum, Curvularia, Cladosporium, Drechslera, Fusarium, Phoma, Phomopsis,* and *Verticillium.* Common soilborne fungi associated with seed rots are species of *Botrytis, Fusarium, Phytophthora, Pythium,* and *Rhizoctonia.* Environmental conditions, such as an excess of moisture, heavy soils, careless handling, and presence of weeds, also favor seed rots. The application of a general seed protectant to seeds helps in producing better emergence and vigorous seedlings. Seed-protectant chemicals differ from crop to crop and from region to region.[121]

3. Method of Treatment

Method of treatment includes any process used for the addition of materials to seeds; in its simplest form, it is the direct application of a material to seeds. Seed coating generally is used to denote the application of a useful material to the seed without changing its general size or shape.[111]

Dry seed treatment — Seeds can be treated in small amounts in seed treaters (Figures 13-4 and 13-5). However, some fungicide is lost to the sides and base of a treater, and the chemical may not stick to the seed surface. Generally, this method is used by farmers who treat their own seed. Dry seed treatment generally is recommended for legume seeds, especially soybean, where the seeds are apt to be damaged during seed processing, and in cases when the slurry or dip methods are not applicable due to water absorption by the seeds, or for other reasons where the dry method is more reliable.

Dry powders applied to seeds are used widely as planter box treatments. These materials do not adhere well to the seed surface and result in a poor mixture, lack of uniform coverage, and dust problems.[111,122]

Seed dip — This method involves dipping seeds into a fungicide solution for a time depending upon the fungicide and type of seed. It is used when the seed coat is thick. It assists in the absorption of the chemical. The seed dip method generally is used just prior to planting. Streptomycin was used as a soak treatment for eradication of deep-seated bacterial infections in beet[123] and bean seeds.[124] Immersion of crucifer seeds into a tetracycline solution (<3000 μg/ml for 30 min) reduced *X. c.* pv. *campestris* on naturally infected seeds to 1% and on inoculated seed to <8%.[125]

Soaking tomato seeds in 0.6 M HCl or 0.05% O-hydroxydiphenyl for 15 min reduced seedborne infection of *P. infestans.*[126] In Korea, rice seeds soaked in a

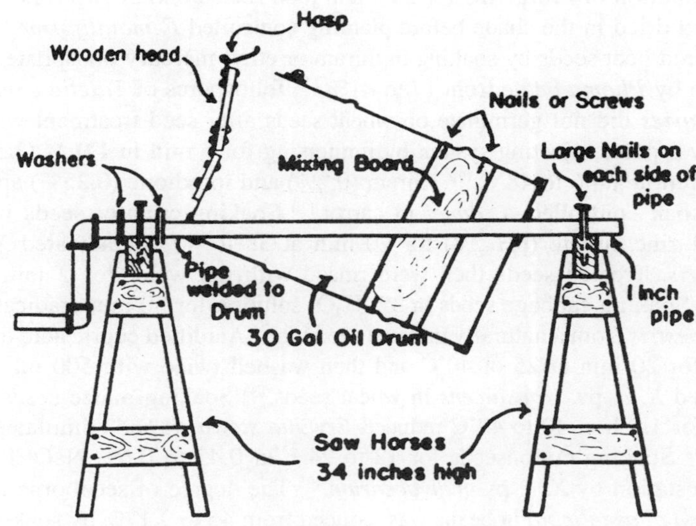

Figure 13-4 An oil drum or rotary seed treater. (From Sharvelle, E. G., *Plant Disease Control*, AVI Publishing, Westport, CT, 1979, 203. With permission.)

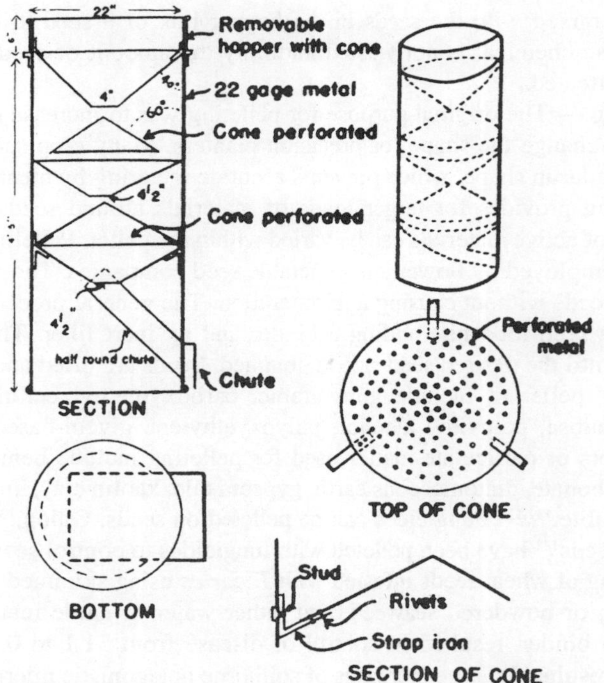

Figure 13-5 A gravity-type seed treater. (From Sharvelle, E. G., *Plant Disease Control*, AVI Publishing, Westport, CT, 1979, 210. With permission.)

1:2000 solution of Proraz-EC for 24 h and then rinsed one or two times in fresh water and dried in the shade before planting controlled *F. moniliforme.*[127] Treatment of red beet seeds by soaking in thiram or ethyl mercury phosphate reduced infection by *Phoma betae* from 17 to <1%.[128] Teliospores of *Tilletia controversa* and *T. caries* did not germinate on wheat seeds after seed treatment with H_2O_2 vapor by a pulse injecting system by immersing for 5 min in 1.0 *M* (3.5% w/v) H_2O_2 solution at 46 to 48°C.[129] Thiram (0.2%) and iprodione (0.25%) applied as a 24-h soak controlled *A. dauci* in carrot.[130] Soaking crucifer seeds in 0.1 *M* acidified zinc sulfate (pH 2.8) for 20 min at 38 to 40°C eradicated *X. c.* pv. *campestris.* Treated seeds then were rinsed with tap water for 2 min and air dried.[131] Soaking cabbage seeds in 3% H_2O_2 solution for 30 min eradicated *X. c.* pv. *campestris* from a naturally infested seed lot.[132] Acidified cupric acid treatment (0.5%) for 20 min at 25 or 40°C and then washed twice with 500 ml of water eradicated *X. c.* pv. *translucens* in wheat seeds.[133] Soaking maize seeds in antibiotics for 1.5 h at 40 to 47°C reduced *Erwinia stewartii* and stimulated germination.[134] Soaking cottonseeds for 12 to 24 h in 0.45 to 0.6% NaOCl reduced seed infestation by *X. c.* pv. *malvacearum.*[135] The degree of seedborne infection of *P. s.* pv. *phaseolicola* in beans was reduced from 45 to 9.17% by soaking seeds in (1) 1% copper sulfate for 30 min, (2) 2% copper acetate in 0.005 N acetic acid for 30 min, or (3) 0.05 to 0.08% streptomycin sulfate for 60 min with a water rinse and a second soak in 0.5% NaOCl for 30 min.[136]

Planter or hopper box treatment — In this method the amount of fungicide required is mixed with the seeds in the hopper box of a seed drill just before planting. The chemical is not wasted and only the amount of seed required for planting is treated.

Pelleting — The original purpose for pelleting was to increase seed size and weight and change the shape for precision planters. Many crop seeds are small and/or irregular in shape, which prevents accurate metering by mechanical planters. Pelleting provides for larger loads of materials around seeds, and spatial orientation of active materials can be varied within the pellet. Pelleting techniques have been employed by flower and vegetable seed companies. The aim is to coat individual seeds without causing agglomeration. The general procedure is to roll or tumble a seed lot while adding a binder and an inert filler. This process is continued until the desired seed size is obtained. Seeds are dried and stored.[111,137] Binders for pelleting include gum arabic, carboxymethyl cellulose, gelatin, methyl cellulose, polyvinyl alcohol, polyoxyethylene glycol-based waxes, and starch. Filters or particulate matter used for pelleting include bentonite zeolite, calcium carbonate, diatomaceous earth, gypsum talc, kaolin clay, limestone, peat, and vermiculite.[111,138] Fungicides can be pelleted on seeds. Onion,[139] pine,[140] and sugarbeet seeds[141] have been pelleted with fungicides to control seedborne fungi. The pelleting of wheat seeds infested with *T. caries* using skimmed milk powder, wheat flour, or powdered seaweed, and either water or cattle manure compost extract as a binder, resulted in control of disease from 31.1 to 0.2–1.3%. The treatments resulted in large increases of soilborne antagonistic microorganism on the seed surface, especially aerobic spore-forming bacteria, and also inhibited *T.*

caries teliospore germination.[142] A limitation of this method is the slow absorption of fungicide by seeds. The method is used only if other methods are unsuitable.

Fumigation — Eradication of seedborne pathogens with fumigants requires additional seed treatment with a seed protectant against soilborne microflora. Fumigation of high quality soybean seeds with formaldehyde for 2 h, propylene oxide for 9 h, or hydrazoic acid for 53 h results in 82, 83, and 75% bacteria-free seeds, respectively, and eliminates most surface fungi. Formaldehyde or propylene oxide fumigation does not decrease seed vigor or increase seed leachate conductivity.[143] Ethylene oxide gas controls *F. avenaceum* in clover, *P. s.* pv. *phaseolicola* in bean, and *P. s.* pv. *glycinea* and *X. c.* pv. *glycines* in soybean seeds with vacuum fumigation. Seeds are fumigated under a sustained vacuum in a heated 17-m^3 chamber. The gas is introduced as a 90% ethylene oxide + 10% CO_2 mixture. Seeds are kept in polyethylene bags, which are opened prior to fumigation and resealed immediately after treatment.[144] Application of methyl bromide (240 g/m^3 for 24 h) at atmospheric pressure repeated four times prevented teliospore germination of *T. controversa* in low moisture content (10.2% moisture) in wheat seeds. Teliospore-infested wheat seeds (14.7% moisture) are six times more sensitive to methyl bromide than low moisture content (10.2 to 12.4% moisture) wheat seeds. *Tilletia caries* was more sensitive to the fumigant than *T. controversa*. Fumigation doses that reduced teliospore germination were high and caused a reduction in seed germination.[145] *Ditylenchus dipsaci* was destroyed in alfalfa seeds by fumigation with 80 g/m^3 methyl bromide for 16 h at atmospheric pressure or 50 g/m^3 for 20 h under vacuum. No harmful effects on seed germination were found 2 h after treatment.[146] However, when nematodes occurred inside cotyledons (i.e., *D. dipsaci* in beans), the level of fumigation required to eradicate the infestation was phytotoxic and did not eradicate the nematode.[147] For fumigation, seed moisture content should not exceed 12% and is more suitable for small seeds such as alfalfa and onion.[148] Methyl bromide dosages at 24.5 mg/l killed *Pratylenchus brachyurus* larvae in peanut shells. Dosages of 44.6 and 50.9 mg/l killed all but one or two nematodes in whole pods. However, 15% reduction in seed germination occurred at the 50.9 mg/l dosage.[149] Onion seeds can be disinfected of the nematode *Ditylenchus dipsaci* by fumigating with 40 oz of methyl bromide per 1000 ft^3 for 24 h. This also kills nematodes in stem pieces. There are no deleterious effects on seed viability.[150] Satisfactory control of *Anguina agrostis* in bent grass (*Agrostis tenuis*) seeds was obtained by fumigating infested seeds of about 12% moisture with methyl bromide.[151]

Soaking rice seeds in a 0.2% solution of mancozeb + monocrotophos followed by vacuum fumigation with methyl bromide (32 g/m^3) for 2 h at 30°C eliminated *Aphelenchoides besseyi*. Vacuum fumigation substituted by atmospheric fumigation using aluminum phosphide (fosetyl) (9.3 g/m^3) was equally effective.[152]

Slurry treatment — Active ingredients may be dispersed or suspended in water to form a slurry. Slurry application improves uniformity and helps overcome problems associated with dry powder application (Figure 13-6). Slurry treatments may include the use of adhesives, such as binders, glues, or stickers, to improve retention of materials. Adhesives used for this application include dextran, gum arabic, methyl cellulose, paraffin, or vegetable oils.[111,138]

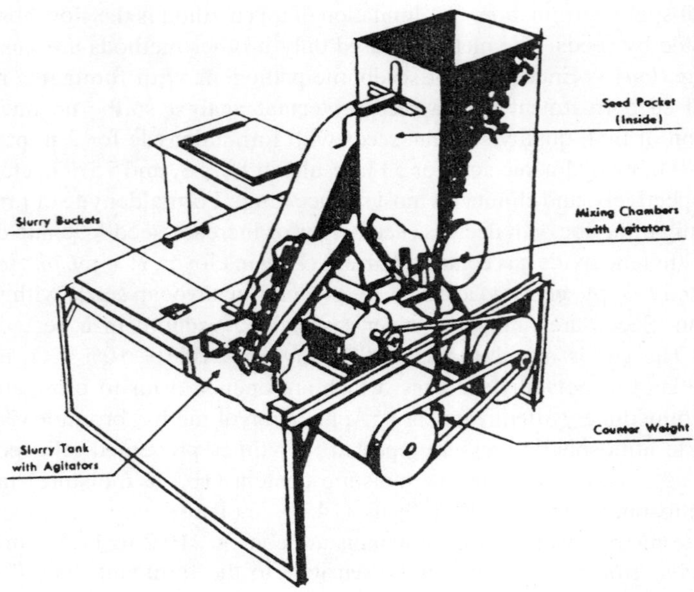

Figure 13-6 A slurry seed treater. (From Sharvelle, E. G., *Plant Disease Control*, AVI Publishing, Westport, CT, 1979, 212. With permission.)

Using spectrophotometric methods, it was shown that slurry formulations of thiram adhered to seeds (80% of target dose) more than dust formulations (38% of target dose).[153] In slurry treatments, water-dispersible fungicide formulations are mixed in water so that a slurry formation results, which is applied to seeds. The slurry treatment is used primarily to treat seeds after harvest during seed processing. The seeds are coated with the fungicide. It is easy to differentiate between treated and nontreated seed lots. Most fungicides used for seed treatment can be treated using a slurry method.[117,118]

Mist treatment — In this technique, a fungicide suspension is broken into fine droplets during the treatment process. This is considered to be the most effective means of treating seeds. The seeds are coated uniformly with the fungicide. Mistomatic treaters currently are used by many seed processing agencies throughout the world.[117,118]

Film coating — Active materials are dispersed or dissolved in a liquid adhesive and applied to seeds either with a fluidized-bed treater or pharmaceutical coating film. It permits the application of multiple coatings and can increase seed weight from 1 to 10%. Recovery rates have been reported as high as 90%, and seed-to-seed variability has been low.[111,137] Compared with dust and slurry applications, seed film coating with Baytan Flowable and Baytan DS (fuberidazole + triadiamenol), Baytan 1M (fuberidazole + triadimenol + imazalil) and Ferrax (ethirimol + flutriafol + thiabendazole) gave improvements in operator safety, seed-to-seed chemical distribution, and handling at drilling.[154] Control of the following barley pathogens was reported: *Phaeosphaeria nodorum, P. graminea, P. teres,* and *U. segetum* var. *tritici.*[154]

Seed Priming — Seed priming or osmoconditioning is a presowing hydration treatment developed to improve seedling establishment.[111] It controls imbibition by osmotic means, allowing uptake of water by the seed to initiate germination but to delay radicle emergence. Because most seeds reach the same stage during priming, germination is rapid and synchronous after sowing. Priming can involve up to 14 d in polyethylene glycol (PEG) solution at 15°C. Seeds can be primed in small quantities by placing them on filter paper moistened with PEG, or in large quantities in PEG bubble columns or stirred bioreactors.[155] Priming large seed quantities followed by drying and then coating with a polymer film containing pesticides or other agents is an integrated process, termed "process engineering of seeds."[154]

Iprodione (0.1% a.i.) or thiram (0.1% a.i.) in polyethylene glycol using the filter paper or bubble column priming system partially reduced *A. dauci* in carrot. Either dusting of primed and dried seeds with iprodione or application of that fungicide in a polymer filmcoat as the final stage in the processing of carrot seeds was necessary for control. The addition of thiram in the priming fluid followed by an application of iprodione to primed and dried seeds improved emergence and yield.[156,157]

4. Application

Seed treatment with chemicals has been used for the control of seedborne bacteria, fungi, nematodes, and viruses. Important seedborne bacteria and fungi controlled through seed treatment are given (Table 13-5).

Control of seedborne viruses through chemical seed treatment has been attempted. Treatment of pepper seeds with calcium or NaOCl, hydrochloric acid, or trisodium phosphate controlled TMV in seedlings.[260] Treating tomato seeds with 10% trisodium phosphate or extracting with concentrated hydrochloric acid (10 ml per 25 pounds of fruit pulp) reduced TMV.[261,262] Soaking tomato seeds in 1% trisodium orthophosphate for 15 min and then in 0.525% NaOCl for 30 min reduced seed transmission of TMV.[263] Soaking cowpea seeds in malic hydrazide at 40, 100, and 400 μg/ml for 90 min, 2-thiouracil at 500 or 700 μg/ml for 60 min naphathalene acetic acid (NAA) at 40 μg/ml for 240 min, and teepol (5 or 10%) for 240 min eliminated cowpea banding mosaic virus without affecting viability.[264] LMV was inactivated in polyethyleneglycol-imbibed lettuce seeds after 6 to 10 d at 40°C, or 16 d at 38°C. Dry seed treatment for 21 d at either 22, 38, or 40°C did not inactivate the virus. Treated seeds were stored following treatment.[265] Dipping urd bean seeds in 0.05% 2-thiouracil and 1% 8-azaguanine for 30 min inactivated the leaf crinkle virus in seeds without affecting germination.[266]

The nematode *A. besseyi* was controlled in rice seeds using 3-methyl-5-ethyl rhodanine, 3-P chlorophenyl-5-methylrhodanine, or 3-P-chlorophenyl-5-ethyl rhodanine.[267] Rice seeds soaked in 20% emulsifiable concentrate of rhodanine at 1:100–300 for 24 h or 1:400–500 for 48 h at 15°C killed up to 95% of the nematodes.[268]

Table 13-5 Control of Important Seedborne Bacterial and Fungal Pathogens Through Chemical Seed Treatment

Crop	Disease	Pathogen	Chemical	Dosage	Ref.
Allium cepa (onion)	Gray mold	*Botrytis acalda*	Benomyl	0.1%	158
	Smut	*Urocystis cepulae*	Ferbam	0.15–0.2%	159
			Pelleting with captan or thiram		118
	Seed rot	Several fungi	Captan 50 WP	156 g/45 kg	118
			Pelleting with captan 50 WP	303 g/4.5 kg	118
			Thiram 75 WP	156 g/45 kg	118
Apium graveolens (celery)	Early blight	*Cercospora apii*	Thiram	—	118
Arachis hypogaea (peanut)	Seed rot	Several fungi	Captan + quintozene (3:1)	0.3%	160
Avena sativa (oats)	Loose smut	*Ustilago segetum* var. *avenae*	Furcarbanil	1%	161
	Covered smut	*U. segetum* var. *segetum*	Carboxin	0.2%	162
			Pyracabolid	0.1%	163
	Seed rot	Several fungi	Captan 50 WP	57–85 g/bu	118
			Mancozeb 80 WP	43 g/bu	118
Beta vulgaris (beet)	Blackleg	*Phoma betae*	Thiram	0.25%	164
	Seed rot	Several fungi	Benomyl	0.25% soak	164
			Captan 83 WP	0.25% soak	164
			Ceresan dry (1% Hg)	0.25%	165
			Dexon 50 WP	85 g/45 kg	118
			Quintozene	0.25%	165
			Thiram 50 WP	99 g/45 kg	118
			Thiram	0.25%	166
Brassica (crucifers)	Black spot	*Alternaria brassicicola*	Iprodione	2.5 g/kg	167
	Gray leaf spot	*A. brassicae*	Iprodione	2.5 g/kg	168
			Thiram	0.15%–0.25%	169
	Black leg	*P. lingam*	Benomyl	0.25%	6
	Black rot	*Xanthomonas c.* pv. *campestris*	Agrimycin 100	0.01%	170
			+ Bangtan	0.2%	
			Streptocycline	0.01%	170
			+ Bangtan	0.20%	

	Seed rot	Several fungi	Captan 50 WP	51 g/45 k	118
			Thiram 75 WP	85 g/45 kg	118
Capsicum frutescens (pepper)	Damping-off, anthracnose	Colletotrichum capsici	Thiram	0.2%	171
	Wilt	*Fusarium oxysporum*	Bavistin	0.2%	172
	Seed rot	Several fungi	Captan 83 WP	0.2%	173
			Captan 50 WP	128 g/45 kg	118
			Thiram 75 WP	156 g/45 kg	118
Cicer arietinum (chick pea)	Blight	*Phoma rabiei*	Bavistin + Thiram (1:1)	3 g/kg	174
			Benomyl	0.3%	175
			Calixin M	0.3%	176
			Granosan	—	177
			Pimaracin-12 h immersion	150 µg/ml	178
			Rovral	3 g/kg	179
			Thiabendazole	0.3%, 5g/kg	174, 175
			Thiram	0.3%	179
Coriandrum sativum (coriander)	Stem gall	*Protomyces macrosporus*	Thiram	0.25%	180
Corchorus capsularis (jute)	Stem rot	*Macrophomina phaseolina*	Captan	0.6%	181
	Seed rot	Several fungi	Thiram 75 WP	0.25%	182
Cucurbita (cucurbits)	Wilt	*F. oxysporum*	Benomyl	—	183
	Seed rot	Several fungi	Captan 50 WP	2.5 tsp/1.9 kg	118
			Dexon 70 WP	1.25 tsp/1.9 kg	118
			Thiram 75 WP	3.75 tsp/1.9 kg	118
Daucus carota (carrot)	Leaf blight	*A. dauci*	Captan 50 WP	112 g/45 kg	118
			Iprodine	—	157
	Seed rot	Several fungi	Thiram 50 WP	85–113 g/45 kg	118
Glycine max (soybean)	Purple stain	*Cercospora kikuchii*	Benomyl + thiram (1:1)	0.4%	52
	Downy mildew	*Peronospora manshurica*	Spergon	0.25%	184
			Thiram	0.25%	184
	Seed rot	Several fungi	Captan 83 WP	0.5%	185
			Thiram 75 WP	0.5%	186

Table 13-5 Control of Important Seedborne Bacterial and Fungal Pathogens Through Chemical Seed Treatment (continued)

Crop	Disease	Pathogen	Chemical	Dosage	Ref.
Gossypium (cotton)	Seedling blight	*Rhizoctonia solani*	Mancozeb	0.25%	187
	Blackarm	*X. campestris* pv. *malvacearum*	Thiram	0.25%	188
	Seed rot	Several fungi	Captafol	0.25%	189
Helianthus annuus (sunflower)	Downy mildew	*Plasmopara halstedii*	Apron 35 SD	0.6%	190
	Soft rot, blight	*Sclerotinia sclerotiorum*	7.7% ethyl Hg-*p*-toluene sulfanilide	—	191
	Seed rot	Several fungi	Captan 75 WP	0.25%	192
			Mancozeb 80 WP	0.25%	192
Hordeum vulgare (barley)	Leaf stripe	*Drechslera graminea*	Benlate	0.2%	193
			Carboxin	0.15%	194
			Sisthane	0.2%	195
	Covered smut	*U. segetum* var. *segetum*	Carboxin	0.25%	195, 196
			Bavistin	2 g/kg	197
			Bayleton	2 g/kg	197
			Panoram	2 g/kg	197
			PMA/Ceresan M	0.2%	198
			Vitavax	2 g/kg	197
			Vitavax + Thiram	2 g/kg	198
	Loose smut	*U. segetum* var. *tritici*	Baytan U, Benlate T	0.2 g/kg	193
			Furcarbanil	0.2 g/kg	199
			Pyracarbolid	25 g a.i./1000 kg	163
			Sisthane	0.2 g/kg	193
	Net blotch	*D. teres*	Benlate 50	1.66 g/kg	200
	Seed rot	Several fungi	Captan 80 WP	57–85 g/bu	118
Lactuca sativa (lettuce)	Seed rot	Several fungi	Thiram 75 WP	85 g/45 kg	118
Lens culinaris (lentil)	Leaf blight	*A. fabae* f. sp. *lentis*	Thiabendazole	0.35 g a.i./kg	201
	Seed rot	Several fungi	Captan	0.2–0.3%	202

Linum usitatissimum (flax)	Grey mold	*B. cinerea*	Iprodione	5–10 g/kg	203
	Leaf blight	*A. linicola*	Iprodione	5–10 g/kg	203
	Seed rot	Several fungi	Thiram 75 WP	0.25%	204
Lycopersicon esculentum (tomato)	Blight	*Pseudomonas syringae* pv. *tomato*	NaOCl dip	—	205
Nicotiana tabacum (tobacco)	Anthracnose	*Colletotrichum tabacum*	Thiram	0.12%–0.24%	206
Oryza sativa (rice)	Bakanae	*F. moniliforme*	Benomyl	2.5 g/kg	207, 208
			Benomyl + Thiram	2.5 g/kg	208
			Benomyl T	2.5 g/kg	208
			Triforine (Saprol)	—	209
	Brown spot	*B. oryzae*	Fenapanii	—	210
			Panoctine (quazatine)	—	210
			Triforine (Saprol)	—	209, 210
			Bavistin	—	211
			Captan	0.25%	212
			Mancozeb	0.3%	213
	Stackburn	*Alternaria padwickii*	Carboxin	3 g/kg	214
			Homai	3 g/kg	214
			Kasugamycin	3 g/kg	214
			Mancozeb	3 g/kg	214
			Mancozeb	0.25%	207
			Triforine		209
	Blight	*Pseudomonas fuscovaginae*	Kasugamycin	0.2 g a.i./kg	215
	White tip	*Aphelenchoides besseyi*	Benomyl	—	117
			Furadan 75 PM	0.5%	216
	Seed rot	Several fungi	Captan 75 WP	0.25%	118
	Seed rot	*F. pallidoroseum*	Triforine (Saprol)	—	209
Pennisetum glaucum (pearl millet)	Downy mildew	*Sclerospora graminicola*	Ridomil	0.2% a.i.	217
Phaseolus vulgaris (bean)	Anthracnose	*C. truncatum*	Dichlone	0.25%	218
	Halo blight	*P. syringae* pv. *phaseolicola*	Kasugamycin	0.025%	219
			Streptomycin	1 h at 5 μ/ml	219

Table 13-5 Control of Important Seedborne Bacterial and Fungal Pathogens Through Chemical Seed Treatment (continued)

Crop	Disease	Pathogen	Chemical	Dosage	Ref.
	Common blight	*X. campestris* pv. *phaseoli*	Streptomycin	1 h at 5 µg/ml	220
	Charcoal rot	*Macrophomina phaseolina*	Benomyl	2.5 g/kg	221
			Vitavax	2.5 g/kg	221
	Seed rot	Several fungi	Captan 50 WP	43 g/45 kg	118
			Chloranil	0.2 to 0.25%	222
			Dexon 70 WP	28 g/45 kg	118
			Thiram 75 WP	43 g/kg	118
Pisum sativum (pea)	Leaf and pod spot	*A. pinodes*	Benomyl	0.4%	223
			Granosan	0.3%	224
	Downy mildew	*Peronospora pisi*	Apron 35 SD	—	225
	Blight	*Pseudomonas syringae* pv. *pisi*	Streptomycin sulfate	2.5 g a.i./kg	226
	Seed rot	Several fungi	Captan 83 WP	0.25%	227
			Chloranil	0.2%	228
			Dexon 70 WP	20 g/45 kg	118
			Thiram 75 WP	57 g/45 kg	118
Raphanus sativus (radish)	Leaf spot	*Alternaria raphani*	Thiram	0.15%–0.25%	229
Secale cereale (rye)	Seed rot	Several fungi	Mancozeb 80 WP	43 g/bu	118
Sesamum indicum (sesame)	Root rot	*Macrophomina phaseolina*	Captan	0.3%	230
Solanum melongena (eggplant)	Seed rot	Several fungi	Thiram 75 WP	112 g/45 kg	118
Sorghum bicolor (sorghum)	Downy mildew	*Peronosclerospora sorghi*	Apron 35 SD	0.2%	231
	Grain smut	*Sphacelotheca sorghi*	Chloranil	0.2%	232
			Dichlone	0.12	233
			Elemental sulfur	0.5%	234
			Thiram		235

	Zonate leaf spot	*G. sorghi*	Bavistin	1 g/kg	236
			Tecto 50	2 g/kg	236
	Seed rot	Several fungi	Thiram 75 WP	0.25%	237
Spinacia oleracea (spinach)	Seed rot	Several fungi	Thiram 50 WP	85–113 g/45 kg	118
Triticum aestivum (wheat)	Leaf spot	*B. sorokiniana*	Imazalil	—	238
			Iprodione	2 g/kg	238
	Snow mold	*Microdochium nivale*	Benomyl	0.1%	239
			Benomyl + Radosan	(1 g + 2 g/kg)	239
	Hill bunt	*Tilletia caries*	Carboxin	0.25%	240
		T. laevis	Chloranil	0.1%	241
			Mancozeb	0.25%	242
	Flag smut	*Urocystis agropyri*	Oxycarboxin	0.3%	243
			Pyracarbolid	0.5%	117
	Loose smut	*Ustilago segetum* var. *tritici*	Bavistin	0.25%	244
			Baytan U	0.2%	245
			Benomyl	0.25%	244
			Carboxin	0.25%	246–248
			Fenfuram	0.25%	245
			Furcarbanil	0.2%	249
			Sicarol	0.2%	244
			Vitavax 200 (37.5% carboxin + 37.5% thiram)	0.4%	244
	Seed rot	Several fungi	Benomyl 50 WP	57–85 g/bu	250
			Captan 80 WP		118
			Mancozeb 80 WP	43 b/bu	118
			Thiram 75 WP	0.25%	250
Vigna unquiculata (cowpea)	Blight	*X. campestris* pv. *vignicola*	Streptocycline + agallol (0.2%)	100 µg/ml	251
			Streptocycline + captan (0.2%)	100 µg/ml	251

Table 13-5 Control of Important Seedborne Bacterial and Fungal Pathogens Through Chemical Seed Treatment (continued)

Crop	Disease	Pathogen	Chemical	Dosage	Ref.
Zea mays (maize)	Anthracnose	*C. graminicola*	Terracoat L-205	0.8%	252
	Leaf blight	*B. maydis*	Carboxin + thiram	0.25%	253
	Head smut	*Sphacelotheca reiliana*	Carboxin	0.25–4%	254
	Seed rot	*Several fungi*	Captafol	0.12%	255
			Captan 50 WP	14–57 g/45 kg	118
			Chloranil	0.2%	256
			Dexon 70 WP	71–113 g/45 kg	118
			Thiram 50 WP	71–113 g/45 kg	118
Zinnia elegans (zinnia)	Seedling blight	*A. zinniae*	Dithane M-45	3 g/kg	257
			Thiram	0.25%	258
	Blight	*X. campestris* pv. *zinniae*	30 min soak in 10,500 µg/ml NaOCl		259

[a] a.i. = active ingredient.

5. General Guidelines for Seed Treatment

a. Use of a Good Seed-Treatment Chemical — A number of chemicals of different formulations are marketed throughout the world. A good seed treatment chemical should (1) be effective under different agro-climatic conditions, (2) not be phytotoxic, (3) be safe to operators during handling and sowing and to wildlife, (4) not leave harmful residues in plants or in the soil, (5) be compatible with other seed treatment chemicals, (6) be compatible with the *Bradyrhizobium* inoculum, if necessary, and (7) be economically beneficial.[117,118]

b. Objectives of Seed Treatment — Seeds are treated for control of seedborne inoculum. However, a 100% control may not be achieved. Seeds may be treated during export or import at quarantine stations. For quarantine, a 100% eradication of seedborne inoculum is desirable. However, there are few examples where complete control of seedborne inoculum has been achieved through fungicide seed treatment. Careful attention is required in determining the type of inoculum, inoculum location, and amount and host cultivar before seeds are fungicide treated.[117,118] For eradication of inoculum, highly selective fungicides, mostly systemics, are useful.[117,118] Seed treatment with 0.3% thiabendazole and tridemorph was effective in eradicating *Ascochyta rabiei* from chickpea seeds,[175,269] whereas 0.3% of a mixture of 30% benomyl + 30% thiram eradicated *F. oxysporum* f. sp. *ciceri*.[270]

c. Selective Seed Treatment — With systemic fungicide development and improved methods for seed health testing, it is now a practice to have selective seed treatments. Healthy, good quality seeds need not be treated if there are no pathological or germination problems. Guidelines were developed by the Illinois Crop Improvement Association, Urbana, and may be used as a rough guide to relative seed health (Table 13-6).[25] Seeds that meet the criteria for healthy, vigorous seeds do not require fungicide seed treatment unless they are planted under adverse conditions at a reduced seeding rate in a seed production field. With below average to low quality seeds, a fungicide seed treatment would increase stands and possible yields.[25]

Table 13-6 Soybean Seed Quality Ratings

Quality text	Healthy, vigorous seeds	Below average to low-quality seeds
Warm germination (23–25˚C)	85% or more	75–84%
Cold soil emergence (15˚C)	70% or more	60–69%
Vigor test (23–25˚C)	74% or more	55–73%
Diseased seeds	10% or less	11–20%

From Sinclair, J. B. and Backman, P. A., Eds., *Compendium of Soybean Diseases*, 3rd ed., APS Press, St. Paul, MN, 1989, 106. With permission.

Seed treatments for soybeans are useful when (1) quality is fair to low because of fungal infection, (2) seeds must be planted in seedbeds where germination will be delayed, or (3) low seeding rates must be used (Table 13-6). Seed treatments are not required when high-quality seeds are planted at an appropriate rate under conditions favorable for rapid germination and emergence.[271] Seed treatments are not useful when seed quality is low because of mechanical damage or physiological factors. However, fungicide seed treatments should be used routinely for (1) seeds in germ plasm collections, (2) breeder's seeds, and (3) seed used for certified seed production.[57] The embryo test may indicate the need for seed treatment of barley against the loose smut fungus.[272] Treatment of first-generation wheat seed lots may depend on laboratory results for loose smut.[273] The fluorescent, freezing blotter, and sand methods are used to assess *S. nodorum* incidence in wheat. The freezing blotter and fluorescent methods are applicable equally, but the latter is quicker. The sand method may be best to forecast seedling disease and determine the need for seed treatment.[274] Control of *T. controversa* can be achieved only by seed treatment with a systemic fungicide such as thiabendazole. Treatment with hexachlorobenzene (HCB) was not effective.[275] However, in winter wheat when infection occurred 3 to 6 months after HCB application and when the HCB concentration in the seedling zones was below an effective threshold, it was not effective. Thiabendazole is the only fungicide approved for use against dwarf bunt in the United States.[275] Seed treatment with carbendazim at 2.5 g/kg, carbendazim + captan (1:1) at 2.5 g/kg, and carbendazim + thiram (1:1) at 2.5 g/kg reduced the seedborne infection of *F. moniliforme* and *Phoma sorghina* associated with sorghum seeds. *C. lunata* infection was reduced when seeds were treated with thiram alone or carbendazim + thiram (1:1) at 2.5 g/kg. Such treatment resulted in high germination and seedling vigor.[47,276] Treatment of wheat seeds with carboxin for control of loose smut depended upon internal infection.[277]

Seed treatment is needed in Europe and Scandinavia to ensure emergence of wheat seed infected by *Phaeosphaeria nodorum,* but is unnecessary in New Zealand where seed infection is unknown.[278] Klitgard and Jorgensen[279] stated that wheat seed lots were treated only when laboratory tests confirm *P. nodorum* infection. Seedborne fungi of cereals in Denmark, such as *Drechslera graminea, Tilletia caries, Urocystis occulta*, and *Ustilago segetum* var. *avenae* were controlled by applying organomercury fungicides to prebasic, basic, and first generation certified seeds, and nonmercurial fungicides to second generation certified seeds.[280] *D. graminea, B. sorokiniana, D. teres, F. avenaceum, F. culmorum, F. graminearum,* and *Microdochium nivale* are important pathogens occurring in barley seeds. The effect on yield due to seed treatment was negative if the number of seedlings with discolored roots, as determined by the Doyer filter-paper method, was below 15% and positive if above 15%. Therefore, if infection is above 15%, seed treatment is recommended.[281]

In Norway it is a goal to reduce pesticide use, including seed treatments. Seedlots of spring cereals are tested routinely for pathogens and assessed for treatment. Seeds are treated when infections, as a percentage of infected seeds, reach the following levels: ≥5% *P. graminea* and *D. teres* in barley, ≥10% *D. teres* in barley, ≥25% *Fusarium* in barley, ≥10% *Bipolaris sorokiniana* in barley,

≥25% *D. avenae* in oats, ≥15% *Fusarium* in wheat, and ≥5% *Stagnospora nodorum* in wheat.[282] Barley with more than 4% infection of *P. graminea* is certified only when treated with a fungicide in Scotland.[283] Guazatine Plus is effective against *X. c.* pv. *undulosa* above 200 µg/ml. It reduced the disease severity and increased yields up to 39.7%, depending on the wheat cultivar and level of contamination in the seeds.[284]

d. Dosage — Seed-treatment dosage depends upon the fungicide, inoculum type, location, level, geographic area, physical condition of seeds, etc. The average recommended dose of a fungicide for seed treatment is 2.5 g/kg seed. For soybeans, a higher (4.5 g/kg) dosage of thiram or thiram + captan was required for maintaining a plant stand under Tarai condition of Uttar Pradesh, India.[186] *Alternaria brassicicola* in cabbage seeds with up to 61.5% infection was eradicated using 2.5 g a.i./kg iprodione 50% WP, but higher doses were required for higher infection levels to eradicate the fungus.[167] Wheat seed treatment with carboxin at 50 to 250 g/qt significantly decreased loose smut incidence below the control. Disease control increased with an increase in treatment dose. The average disease control is 45 and 90% at 50 and 250 g/qt, respectively.[248]

e. Method of Application — To obtain maximum benefits of a seed treatment, a suitable method must be used. The selection of a method depends on the fungicide, type of infection, and crop species. *P. syringae* pv. *phaseolicola* in bean seeds was controlled by soaking seeds in a streptomycin solution. Pelleting seeds with streptomycin sulfate failed to control infection because of slow absorption of streptomycin sulfate from a pellet compared with rapid uptake from a solution.[285] Mistomatic® seed treaters are preferred over dry or slurry treaters because of a uniform distribution of chemicals on seeds. Leguminous seeds often are subjected to a dry seed treatment.

f. Combination of a Systemic and a Nonsystemic Fungicide — Systemic fungicides are highly selective. Thus a nonsystemic fungicide should be mixed with a systemic fungicide to ensure a broad spectrum of control. Such a combination inhibits or controls the important seedborne pathogens and gives protection against soilborne seed decay and seedling blight pathogens. A combination of benomyl + thiram (1:1) at 0.4% was effective for control of seedborne *C. kikuchii* and seed decay and seedling blights of soybeans.[63]

g. Location and Amount of Inoculum — The location of inoculum also governs the type of treatment. For example, control of *Ascochyta pinodes* in pea seeds was obtained by treatment with granosan at 3 kg/ton when mycelium was found on the seed coat, but not in the endosperm up to 2 mm or more.[224]

Alternaria tenuissima, Aspergillus, Cladosporium, F. pallidoroseum, Lasiodiplodia theobromae, Penicillium, and *Phomopsis* were isolated from pigeon pea seed coats and occasionally from embryo tissues. Captan and thiram moved into seed coat tissues but not into the embryo. Benomyl penetrated the seed coat and embryo; thus captan and thiram were effective against fungi in seed coat.[286]

X. c. pv. *campestris* is deep seated in crucifer seed tissues, resists chemical treatment, and is controlled by hot-water treatment. Surface contaminants yield to both chemical and hot-water treatment.[287] Hot-water treatment (30 min at 50 to 52°C) may not eradicate deep-seated infection by the bacterium in crucifers.[288]

h. Resistance to Chemicals — Continuous use of a fungicide may result in the development of resistance to that chemical by the pathogen. Lack of control of *Phaeosphaeria nodorum* may be due to resistance to phenyl mercury acetate.[289] Oat leaf spot, caused by *Pyrenophora avenae,* was controlled by seed treatment with ethyl mercury chloride dust in Scotland in the 1920s, but not in the 1960s because of fungal resistance.[290] Kuiper[291] described *T. caries* resistance to hexachlorobenzene in Australia and associated it with its fungicidal specificity. Mercury-tolerant strains of *P. avenae* were reported in oat seeds.[292]

i. Phytotoxicity — Seed treatment chemicals should not be phytotoxic. Streptomycin was phytotoxic to soybean seedlings in the field but was useful for control of *Bacillus subtilis* seed decay in culture plates and in vermiculite but not in soil.[293] *X. c.* pv. *campestris* was eradicated from Brassica seeds by soaking for 1 h in a 500 μg/ml solution of aureomycin, streptomycin, or terramycin. These chemicals were phytotoxic, but if the soak was followed by a water rinse and a soak in 0.5% (w/v) NaOCl for 30 min, phytotoxicity was eliminated. NaOCl oxidizes residual antibiotic on seeds or under the seed coats.[294]

j. Combination of Seed Treatment and Foliar Spray — Seed treatment alone may not be effective in controlling seedborne pathogens. Seed treatment followed by a fungicidal spray may be more beneficial in disease control than seed treatment alone. Metalaxyl 25 WP at 1 g a.i. per kilogram seed and one spray 40 d after planting at 1 g a.i. per liter (750 l/ha) controlled both systemic infection and local lesions of *Peronosclerospora sorghi* in sorghum. Seed treatment alone did not protect against either late systemic infection of main shoots or nodal tillers, or against local lesions on leaves.[295] However, seed treatment with Apron 35 SD (5.8 g/kg) + Ridomil spray MZ 72 WP at 4 g/l controlled *S. graminicola* in pearl millet.[296] Seedborne pathogens of cereals were effectively controlled for more than 50 years by the routine and extensive use of organomercury fungicides in Scotland, but mercury-based treatments were withdrawn in 1992.[297]

k. National Laws — The use of seed-treatment chemicals is regulated by national laws. In the United States, federal and state regulations ban the use of mercury fungicides. In the Netherlands organomercury compounds are restricted to cereal seed for multiplication purposes.[5] In Sweden the use of mercurial compounds is limited to spring barley, wheat, and oats, with the seeds to be treated assessed by laboratory testing.[298] In Denmark organomercurials are permitted for cereals up to and including first-generation certified seeds.[298] Under EEC rules it has been illegal to treat seeds with mercury since March 1992.[299] All mercury compounds were removed from registration in Japan in 1970, and manufacture was prohibited after 1971.[300]

l. Seed Coloration — Seeds may be dyed by seedsmen to identify seed lots or improve seed appearance. The dyes are used at safe rates.[301]

m. Use of Nonaqueous and Aqueous Solvents for Fungicide Infusion into Seeds — With surface seed treatment, significant amounts of chemical are lost to the air and soil, thus reducing its efficiency and polluting the environment. Also, fungicides are not always uniformly distributed on seed surfaces and may not be absorbed by seeds. These problems can be overcome by incorporating chemicals into seeds using volatile solvents. The solvents carry antimicrobial materials into the seed coat and then evaporate when air dried, thus leaving the fungicide or antibiotic inside the seeds. No water is used; thus the seed coat remains firm and unswollen. Dichloromethane (DCM) is used to introduce germination inhibitors,[302] amino acids,[303] erythromycin,[304] penicillin,[305] or systemic fungicides in dry seeds.[306] In addition, acetone and polyethyleneglycol (PEG) have been used to incorporate antibiotics and systemic fungicides into soybean[305,307,308] and vegetable seeds.[309] Heydecker et al.[310] reported that seeds of several vegetables or ornamentals soaked in aerated solutions of PEG partially imbibed water and freshly soaked seeds failed to germinate. PEG was a better solvent than DCM because of its low cost and nontoxicity to humans.[311]

Systemic fungicides are infused easily into soybean seeds and are effective for a relatively long time. In addition, less fungicide is used, and because infusion reduces the amount of material on the seed coat, fungicides are less likely to interact with *Bradyrhizobium* inoculum. Infusion of systemic fungicides with acetone into soybean seeds controlled damping-off caused by *Phytophthora sojae.* Infusion of benomyl with DCM reduced seedborne *Phomopsis.* Soaking soybean seeds in a PEG 6000 solution containing either penicillin G or streptomycin sulfate inhibited *B. subtilis.*[25,308,312] Incorporation of quintozene with acetone or DCM controlled storage decay on soybean seeds by *A. ruber.*[309] Fungicide solubility was not related to seed uptake. Benomyl, sisthane, and thiabendazole were detected in soybean seeds treated with acetone, DCM, or PEG. However, carboxin was detected only in seeds treated with PEG. A greater reduction of seedborne *Phomopsis* occurred when benomyl and sisthane were infused with PEG than with either acetone or DCM. No treatment was 100% effective. Acetone and, to a greater extent, DCM killed *Phomopsis* within the seed coat; however, they damaged exposed cotyledonary tissues. The efficacy of any treatment depends on fungicide and solvent carrier combination.[313] Benomyl, fenapronil, or thiabendazole but not chlorothalonil or quintozene were carried into soybean seed coats on soaking seeds in a mixture of each fungicide with either benzene or ethanol for 30 min or longer. None of the fungicides were carried into the endosperm.[314] Polyethylene glycol solutions effectively introduced streptomycin into bean seeds and reduced internal populations of *X. c.* pv. *phaseoli* with few phytotoxic effects.[315]

Using an organic solvent infusion, fungicides such as carbendazim or ethylene thiosulphonate penetrated dormant cottonseeds and controlled Fusarium and Verticillium wilt.[316] Carrot seed treatment with iprodione (5 g a.i. per kilogram) gave 97% control of *A. dauci* when the seeds were dried after priming with PEG.[317]

Sunflower seed treatment with benomyl, iprodione, procymidone, or vinclozolin eliminated seedborne *S. sclerotiorum*. With the exception of benomyl, fungicides were more effective applied in acetone solutions than applied as a dry seed dressing at 100 g a.i. per 100 kilograms. Acetone infusion had no deleterious effect on germination or vigor. Less fungicide was needed using infusion than a dust treatment.[318] The low surface tension of acetone aided penetration of the fungicide solution into crevices and surface irregularities on the seed coat as well as into the testa. It is also possible that acetone may have removed germination inhibitors from the seed coat.[318]

Spinach latent ringspot virus was eradicated from *Nicotiana xanthi* seeds imbibed in a PEG solution for 20 to 40 d at 40°C.[319] The BSMV was reduced in barley seeds when seeds imbibed 10 μg a.i. per gram seed dimethylsulfoxide solution of Tilorone R 11567 DA (23.2 μg/ml).[320] Lettuce seeds imbibed PEG solutions on absorbent paper for 6 to 10 d at 40°C and then at 20°C inactivated lettuce mosaic virus.[321]

Solvent incorporation of fungicides has been used in other crops. Soaking asparagus seeds in 1.5% benomyl in acetone for 24 h (on a shaker), followed by surface sterilization in a 10% Clorox® solution controlled seedling infection by *F. moniliforme* and *F. oxysporum*.[322] More carboxin or ethazol was found within or on embryos in cottonseeds treated with acetone infusion than when applied directly.[323] Infusion of fungicides into pea seeds using acetone required less fungicide than when applied as a slurry to control seed rot or damping-off.[324,325] Acetone, benzene, or ethanol was effective equally in transporting either benomyl or thiabendazole into bean seeds. It overcame the disadvantage of treating seeds in an aqueous fungicide suspension or with powdered fungicides.[326] DCM was superior to trichloromethane or carbontetrachloride in facilitating benomyl, TBZ, and thiophanate movement into bean seeds. There was an interaction between seed type, solvent, and fungicide in fungicide infusion into the seed coat.[326]

n. Seed Treatment and Bradyrhizobium Inoculum — Proper plant stand and nodulation are two important factors in successful leguminous crop production. Seeds often are treated with a fungicide and then *Bradyrhizobium*. Studies made on the compatibility of seed treatment fungicides and inoculum in soybeans were reviewed (Table 13-7).[327,328] Copper fungicides, oxycarboxin, and quintozene adversely affected nodulation while aureofungin (a heptaine antibiotic, *N*-methyl para-amino acetophenone and mycosamine), benzimidazoles (benomyl, carbendazim, thiabendazole), and zineb had no adverse effect. Reports on the efficacy of seed treatment with captan, carboxin, chloranil, dichlone, mercurials, and thiram have been inconsistent.

o. Fungicide–Cultivar Interaction — Most workers use a single cultivar for evaluating fungicide efficacy against seedborne infection. However, cultivars may react differently to the same seed treatment. Maude and Shuring[341] using the wheat cv. Capelle Desprez reported that a higher dosage of carboxin (8 oz per 100 pounds) was required for control of loose smut.

Table 13-7 Effect of Various Seed Treatment Chemical on Soybean Nodulation

Decrease	No adverse effect	Increase
Captan[328]	Aureofungin[332]	Benomyl[329]
Carboxin[329]	Benomyl[331]	Dichlone[337]
Chloranil[330]	Captan[331,334,335]	Molybdenum[337]
Copper fungicides[331]	Carbendazim[334,336]	
	Carboxin[327,337]	
	Chloranil[338]	
Oxycarboxin[329]		
Quintozene[328]	Thiabendazole[339]	
Thiram[332,333]	Thiram[185,328,331,334,340]	
	Zineb[332,337]	

p. Persistence of Chemicals on Seeds — Most seed-producing agencies and a majority of progressive farmers treat seeds before sowing. Although considerable information is available about the efficacy of seed treatment fungicides against seedborne pathogens, little is known about their effectiveness when treated seeds are stored for some time. Aureofungin and piomy (antibiotics), mancozeb and thiram, and thiabendazole lose their fungicidal properties after 1 year. Some systemic fungicides, such as benomyl, carbendazim, carboxin, and thiophanate methyl, retain their activity after 1 year and do not affect germination of rice seeds.[342] Persistence of thiram on soybean seeds gradually decreases with increased storage. Maximum degradation occurs when seeds are stored in cloth bags and to a lesser degree in paper and alkathene-lined jute bags.[343]

q. Time of Seed Treatment — Soybean seeds rarely are treated with a fungicide because of the risk of financial loss to seedsmen. Once seeds are treated, they can be used only for planting but not for food, feed, or processing.[344] Unlike maize seeds, soybean seeds cannot be stored for over 1 year and maintain high viability. For this reason, soybean seeds are treated with a fungicide frequently at planting time, and only the quantity of seeds required is treated. Captan and thiram alone or in combination, are used for broad spectrum control and may be used in combination with narrow spectrum compounds such as carboxin or metalaxyl (Table 13-7).

r. Eradicative Seed Treatment — In practice, the elimination of a pathogen is impossible, and seed treatment only reduces the number of infected seeds in a seed lot at or below that number at which they can give rise to disease outbreaks of economic importance. The effectiveness of a seed treatment can be estimated by laboratory and greenhouse tests on a limited number of seeds and seedlings and must be confirmed in the field to ensure that the treatment reduces primary inoculum to less than the tolerance limit set for the pathogen. Eradicative seed treatments are most useful where the seed is the primary or only source of a pathogen.[345] Calixin-M, Benlate, Tec-to-60 and Topsin-M proved effective in the eradication of seedborne inoculum of *A. fabae* f. sp. *lentis* in lentil.[346] Fungicides effective in controlling seedborne pathogens in the laboratory may not perform well under field conditions, where the interactions of environment, soil, and seed treatment may affect their efficacy. This has been demonstrated in seed treatment trials for control of *Phoma rabiei* in chickpea and *A. fabae* in fava bean by Kaiser and Hannan[347] and Kharbanda and Bernier.[348] Several compounds controlled a pathogen in the laboratory but were either phytotoxic or ineffective in field tests. Seed treatment of soybeans by metalaxyl at 30 mg/100 kg gave partial control (40%), whereas 200 mg/100 kg gave complete control of *Phytophthora sojae.*[349] Difenoconazole at 0.12 g a.i./kg controlled *T. controversa* in wheat.[350]

In laboratory tests, *Phoma rabiei* was eradicated from chickpea seeds treated with benomyl and thiabendazole at 3 g a.i./kg or in combination with captan. However, in greenhouse trials an infected seedling was detected occasionally.[347] Treatment with carboxin + thiram (0.75 + 0.75 g a.i./kg) and fenfuram (1 g a.i./kg) eradicated *Puccinia calcitrapae* var. *centaureae* in safflower.[351] *Diaporthe phaseolorum* and *Neocosmospora vasinfecta* were eliminated from *Aspalanthus linearis* seeds treated with H_2SO_4 for 40, 45, or 50 min.[352]

Clavibacter m. subsp. *michiganensis* was eradicated from tomato seeds soaked in 0.6 *M* HCl for 5 h, 0.25 or 0.50% acidified cupric acetate for 20 min, or water for 20 min at 52°C. The treatments also eradicated saprophytic bacteria.[353] *P. pseudoalcoligenes* subsp. *citrulli* was eliminated from watermelon seeds by soaking in a 1 mg/ml streptomycin solution for 16 h.[354] Its seedborne infection in muskmelon was controlled by soaking seeds for 5 min in NaOCl (5 mg/ml), followed by a 30-min rinse in running tap water.[355]

X. c. pv. *campestris* in crucifer seeds was reduced to undetectable levels after seeds were slurry-treated with CaOCl at 10 to 20 g a.i. per kilogram seeds and sealed in containers for 16 h; however, the treatment reduced seed germination of some lots after 6 months.[356]

s. Testing Treated Seeds — Testing treated seeds for commercial seed treatment quality is part of the Official Seed Certification Scheme for cereals in Denmark. Two methods are used, colorimetric and bioassay. The first is based on the provision that fungicides used for seed treatment must contain a red stain, which can be washed off by a 50% water–alcohol mixture. The color intensity is estimated using a standard scale. In the bioassay, different test organisms such as *Arthrobacter globiformis, Bacillus cereus,* or *Bipolaris sorokiniana* are used. The test organism is spread on meat–peptone agar and the seeds sown on the plates. After incubation, an inhibition zone will develop around each seed treated with a fungicide to which the organism is susceptible, and the diameter of the zone is a quantitative expression of the fungicide on the seed.[298]

t. Disposal of Treated Seed — The disposal of pesticide-treated maize seeds has been a problem in the United States. Captan has been used widely as a seed protectant but has caused difficulties in disposing of leftover seeds. As a result of overproduction, obsolete hybrids, or decreased viability, many bushels of maize seeds are earmarked for disposal annually. Although captan is not toxic, it is mutagenic, carcinogenic, or teratogenic to certain organisms. Captan can be removed from maize seeds and degraded rapidly in an 0.5 N NaOH plus 5% detergent solution. A strong base is more efficient than sodium bicarbonate. Addition of a detergent enhances detreatment, and agitation improves efficiency. Under optimal conditions, 0.1 to 0.2 μg/ml captan remained on seeds. The primary breakdown product of captan is 4-cyclohexene-1,2-dicarboximide, which is degraded further in an alkaline solution after 24 h.[357]

6. How to Avoid Damage to Bradyrhizobium Inoculum

The application of *Bradyrhizobium* inoculum to seeds does not ensure nodulation, and the presence of nodules does not always mean a high level of N fixation. Factors such as soil moisture, pH, texture, *Bradyrhizobium* strain, and method of application can affect nodulation at time of inoculation.[358] Captan, carboxin, and thiram may be compatible with *Bradyrhizobium* in plate-count tests, but in field studies on nodule counts at 2 weeks, only thiram was found compatible on soybeans.[328] The chemicals may reduce the number and/or size of nodules. Damage to nodulation by seed treatments can be reduced by:

1. Use of a peat-based medium: Apply the inoculum in a peat base immediately prior to planting using the wet method with sucrose in the inoculating fluid. Quintozene, quintozene plus ethazol, and thiram are least toxic when applied to cowpea seeds before inoculation with a peat inoculant.[359]
2. Maintenance of high soil moisture at planting time.
3. Use of higher doses of inoculum: High doses of inoculum may offset fungicide injury. When captan and zineb were used, increased *Bradyrhizobium* levels did not increase nodulation. However, when using carboxin, increased nodulation was observed with increased inoculum.[327]
4. Application of *Bradyrhizobium* at planting time: Treated seed should be sown immediately after inoculation. Curley and Burton[328] found that 40% of *Bradyrhizobium* applied to seeds survived as long as 1 h. Captan and carboxin reduced *Bradyrhizobium* viability by less than 20% in 1 h, whereas quintozene killed 78% of the *Bradyrhizobium.* Thiram had no adverse effect. Carboxin was found compatible when seeds were planted within 4 h of inoculation but not up to 24 h.
5. Use of fungicide-resistant mutant strains of *Bradyrhizobium*: *Bradyrhizobium* strains differ in compatibility with fungicides. A combination of a fungicide with a tolerant strain may give nodulation higher than the control. Odeyemi[360] reported that mutant strains of *B. meliloti* or *B. phaseoli* plus thiram on alfalfa, or plus chloranil on beans, or a cowpea strain plus dichlone on cowpeas produced 100% more nodulation than wild strains. *Bradyrhizobium* L × 717 plus captan, L × 717 plus carboxin, L × 718 plus thiram, or L × 718 plus quintozene seed treatment resulted in higher nodulation than the control in soybeans.[327]
6. Use of granular *Bradyrhizobium* inoculum: Granular inoculants may be used to prevent fungicide damage to inoculum since the inoculum does not come in direct contact with the fungicide. Granular applicators are used to drop inoculant near peanut seeds during planting.[360] Major drawbacks are that inoculum is expensive and planters must be fitted with a granular applicator.

7. Use of an organic solvent as a carrier for incorporating fungicide into seeds: Sinclair[312] summarized data showing that using anhydrous DCM as a carrier to introduce chemicals into seeds facilitated the movement of methylbenzimidazole carbamate (MBC) and thiabendazole into dormant soybean seeds. The solvent carried the antimicrobial material into the seed coat and evaporated when allowed to air dry, thus leaving the fungicides in the seed coat. The chances of interference of *Bradyrhizobium* inoculum by the fungicide inside the seed is less than with dry or slurry seed treatment.

C. Mechanical Methods

Seed lots may harbor inert materials, colonized plant debris, pathogenic fungal propagules, etc. In addition, the seeds may be discolored, distorted, small, or enlarged due to infections. Such seed lots should be cleaned mechanically. The use of clean seeds can reduce seedborne inoculum but does not eliminate it. The use of clean seeds for control of cucumber mosaic virus and TMV was stressed by Bewley and Corbett.[361] The use of ergot sclerotia-free seeds in pearl millet reduced the primary inoculum level. This can be done by hand-picking, either with a gravity separator or by immersing contaminated seeds in a 10% NaCl solution to remove floating sclerotia. There is no efficient method available to separate sclerotia from seeds on a large scale. Fungal sclerotia associated with *Poa pratensis, P. trivalis,* and *Raphanus sativus* and ergot with *Cucumis sativus* can be removed with a gravity separator. Sclerotia in sunflower seeds are difficult to remove; however, larger ones can be removed by an air screen cleaner, lighter ones by an air separator, and small ones by an indented cylinder of no. 7 or 8. However, these machines do not ensure complete extraction of sclerotia.[362] Fermentation of fruit pulp from field-grown tomatoes for 24 or 48 h at 20 or 25°C successfully disinfected seeds from *Clavibacter m.* subsp. *michiganensis*.[363]

Processing, screening, and sieving can remove inert material, plant debris, and fungal fruiting bodies. *F. oxysporum* f. sp. *betae* is an external contaminant in sugarbeet seeds. Commercial processing of seeds reduced the amount of infested seeds and seed transmission.[364] LMV occurred in seeds of light weight, and such seeds were separated from heavier, noninfected seeds.[365] Separation of LMV-infected seeds was done using a vertical air stream.[366] The removal of small barley seeds reduced BSMV.[367] The removal of pea seeds with growth cracks reduced pea seedborne mosaic virus from 33 to 4%.[368]

Removal of the infested pulp from vegetable seeds at extraction helps reduce seedborne inoculum. *C. m.* subsp. *michiganenis* was eradicated from tomato seeds by extracting pulp with HCl followed by drying in a tumbler for 3 h at 66°C.[369] When extracted with HCl (25 ml commercial HCl added to pulp from 2.27 kg fruit), seeds of the tomato cvs. Potentate and V 548 did not show seed transmission of TMV.[370] Cleaning seeds removed fruit pulp remnants that carry TMV in tomato and cucumber green mottle virus in cucumber seeds.

Seeds with the wheat gall nematode have a lower specific gravity and can be separated from noninfected seeds in a 20% NaCl solution. The seeds are dried and treated with a suitable seed protectant. Sunflower seeds were stirred in warm water (2 l/kg seed) for 15 min; then floating seeds were removed and dried, and no sclerotia of *S. sclerotiorum* were found.[371] Soybean seeds readily absorb water, which makes the seed coat expand, become slippery, and pull away from the embryo. A nonaqueous mixture of glycerine and polyethyleneglycol 400 (60:40) can be used to separate lightweight, infected seeds from heavier, noninfected soybean seeds.[372] The separation of light, poor quality, discolored cottonseeds by flotation in water after delinting in sulfuric acid helped in removing seeds infected with *X. c.* pv. *malvacearum*.[373]

D. Physical Methods

Thermotherapy of seeds is commonly used to control certain seedborne pathogens while leaving the host tissue viable. Many media have been used for heat treatment, including water, steam, air, carbon tetrachloride, petroleum oils, and microwave radiation. However, seeds of large-seeded legumes, such as soybeans, rarely are heated in hot water because they quickly imbibe water, swell, and slough off their seedcoats. Physical methods or thermotherapy are some of the oldest methods used for control of seedborne pathogens. Thermotherapy is used for control of seedborne pathogens when seed treatment chemicals are not available. Jensen demonstrated in 1882 that the hyphae and spores of *P. infestans* in infected potato tubers were killed after a 4-h hot air treatment at 40°C. In 1888, he found that hot-water treatment of dry seeds controlled *Ustilago segetum* var. *tritici*.[280] This method was followed in Denmark and later adopted in the United States on the recommendation of Swingle in 1892.[118]

The principle of thermotherapy is that microorganisms are killed or viruses destroyed at temperatures not injurious to seeds.[374] Seeds with a low moisture content are ideal for heat therapy. Seeds with a high moisture content are killed at lower temperatures than those with a low moisture content. Various theories have been proposed on the cause of seed death from exposure to high temperature,[374] including denaturation of proteins, lipid liberation, hormone destruction, tissue asphyxiation, depletion of food reserves, and metabolic injury with or without accumulation of toxic intermediates. The methods used for the application of heat to seeds are the following.

1. Hot-Water Treatment

Hot water is used widely to control seedborne pathogens, especially bacteria and viruses, and includes the following steps:[374]

Selection of seeds — Seeds that can withstand hot-water treatment are preferred. Legumes and soybean seeds cannot be subjected to hot water because their seed coat swells and sloughs off. Pathogens easily controlled by fungicide seed treatment usually are not subjected to hot-water treatment. Hot-water treatment is recommended for seeds with deep-seated infection that cannot be eradicated by other means.

Presoaking the seed — Seeds may or may not be soaked in water before treatment, depending on the crop and pathogen. Presoaking is done to replace air between embryos and seed coats with water, which is a better heat conductor. The water soak may stimulate pathogen growth, by which the pathogen becomes more heat susceptible. The duration of soak (4 to 12 h) may not be sufficient for water penetration, hydration of tissue,. and initiation of pathogen growth in all seeds.

Chalara elegans was eliminated from peanut seeds by hot water treatment for 15 min at 50°C, 10 min at 55°C, or 2, 5, or 10 min at 60°C.[375] Hot water (50°C for 30 min) or NaOCl (0.1 or 1.0% at 50°C for 30 min) eradicated *Alternaria radicina* from infested carrot seeds with a minimal reduction in germination.[376] A hot water soak (50°C for 20 min) proved better than fungicidal seed treatment for reducing seedborne *Alternaria porri* and *Stemphylium vescicarium* from onion seeds, although germination and emergence were reduced.[377]

X. c. pv. *malvacearum* was eliminated from cottonseeds placed in hot water for 20 min at 65°C. Acid delinting was suitable for external control of the pathogen only. Thermotherapy should always be used in quarantine applications.[378]

Preheating — After soaking in cool water, seeds are heated for 1 to 2 min at 9 to 10°C below the temperature of the final treatment. This is done to counter the cooling effect of soaking in cool water.

Hot-water soak — The temperature and time required varies with the pathogen and crop. Timing must be precise; otherwise seed viability is lost. A large volume of water helps maintain a constant temperature. Seeds should be packed loosely in porous bags, screen boxes, or frames, with sufficient provision for an ample water flow. Water in tanks may be heated by thermostatically controlled, electric immersion heaters or by steam pipes.

Cooling — Treated seeds are spread out for cooling and drying immediately after treatment.

Drying — Seeds are dried quickly to prevent sprouting.

Post-treatment — The application of a seed protectant fungicide maintains seed viability and avoids seed decay due to soilborne microflora.

Hot-water treatment is not applicable for seeds whose seed coats rupture when they absorb water, such as legume seeds, or that exude mucilaginous materials, which stick seeds together on drying, such as flax seeds. Carbon tetrachloride and oils were suggested as alternatives to hot water treatment. Onion seed treatment with either benomyl plus thiram at rate of 2.5 + 2.5 g a.i./kg or in hot water (15 min at 60°C), eliminated *A. niger* in naturally infected seeds.[379] The use of NaOCl (0.13 *M*) at ≤55°C kills teliospores of *T. controversa* on wheat seeds.[380] The amount of *P. s.* pv. *pisi* in pea was eliminated using a hot water soak for 15 min at 55 or 60°C, with a reduction in germination of 5 to 20% depending on cultivars.[381] Watson et al.[382] tested various nonaqueous fluids to find a medium for heating seeds of large-seeded legumes. They found green and lima bean seeds survived longer when treated in motor oil than in water at 90°C, and seeds heated for 60 min in boiling CCl_4 (76.8°C) also survived. Soybean oil was used for heating soybean seeds to control seedborne fungi, especially *Phomopsis*.[383] Treatments that decreased *Phomopsis* and increased the germination potential of pathogen-free seeds ranged from 5 min at 70°C to 10 sec at 140°C.[383] Seeds placed in soybean oil at 21°C did not imbibe oil, swell, or slough off their seed coats.[384,385]

Cabbage seed thermotherapy often results in injury expressed as reduced or delayed germination. PEG treatment of heat-damaged cabbage seeds increased germination and emergence and restored germination rate and emergence. Cabbage seeds were dipped in PEG 6000 (305 g/l H_2O) for 14 d at 15°C. The treatment stimulated a physiological repair mechanism, which may operate in fully imbibed but dormant seeds. There is a possibility that the pathogen also may recover, but pathogens do not have the same metabolic reserves.[304] Heat treatment of maize seeds in a water bath for 5 min at 60°C eliminated *F. moniliforme*.[386] Pretreatment of maize seeds in water for 4 h at 18 to 22°C and then 5 min at 60°C eliminated *F. moniliforme*.[387] Treatment of soybean seeds with a moisture content 7.5% or lower with palm, soybean, or sunflower oil from 2 to 15 min at 80 to 90°C significantly reduced *Alternaria*, *C. kikuchii*, and *Phomopsis*.[385] *C. kikuchii* was eradicated after 5 min at 90°C. The germination potential of treated seeds was increased with a decrease in recovery of *Phomopsis*.[383,384]

Hot water was used to eliminate pathogens from seeds and planting material, especially when the pathogens were deep seated and out of reach of nonsystemic fungicides. It has been used to control seedborne infection of *X. o.* pv. *oryzae* in rice,[388] *X. c.* pv. *phaseoli* in bean,[389] *X. c.* pv. *cucurbitae* in pumpkin,[390] *X. c.* pv. *manihotis* in cassava,[391] *X. c.* pv. *campestris* in crucifer,[392] and *X. c.* pv. *vignicola* in cowpea (30 min at 50°C).[393] For cowpea seeds, care was needed during treatment because the difference between the thermal death point of the bacterium (49°C) and temperature injurious to the seeds was narrow, with a complete loss at 53°C.[393] Rice seeds were presoaked in water to activate the dormant larvae of *A. besseyi*, and subsequent soaking for 7 min at 51°C killed the nematodes. This treatment resulted in slight germination loss.[148] Hot water treatment is inexpensive, but because seed temperature must be raised quickly, only small quantities can be treated at one time, and germination, particularly of older seed samples, may be reduced.

An advantage of using heated vegetable oils rather than heated water to treat large-seeded legumes and other crop seeds is that the seeds are not damaged by inbibition of water. The advantage of using heated vegetable oils rather than fungicides is that the seeds can be used for feed. However, vegetable oil thermotherapy is not practical for large-scale commercial use because of the difficulty of heating seeds uniformly. However, it can be used to treat small seed lots in plant quarantine.[344]

2. Hot Air Treatment

Hot air is less effective than hot water in the control of seedborne pathogens. The advantage of hot air treatment is that it is easy and seeds are less damaged. In Indonesia maize seeds are dried to 13 to 14% moisture to control *S. maydis* and *S. sacchari* infection.[394] Control of *F. solani* f. sp. *cucurbitae* was obtained after 3 d at 75°C or 2 d at 80°C without loss of germination of cucumber seeds. *Verticillium tricorpus* infection was eliminated after 6 d at 75°C or 5 d at 80°C, with a slight effect on germination of tomato seeds. Treatment for 5 d at 85°C gave control of *Colletotrichum capsici* on pepper seeds with some reduction in germination. In these tests all seeds were heated uniformly in a revolving cage installed in a dry heat chamber.[395] Drying tomato seeds in an oven for 6 h at 29.5 to 37.5°C eliminated *P. infestans* from discolored seeds from infected fruits.[396] Dry heat treatment of rice seeds for 6 d at 65°C eradicated *Pseudomonas avenae, P. fuscovaginae*, and *P. glumae*.[397] Dry treatment of tomato seeds for 2 d at 78°C reduced tomato mosaic virus without an effect on germination. This method was more effective than a 30-min soak in trisodium phosphate.[398] *Anguina tritici* galls were killed after 10 min at 60°C.[399] Exposure to dry heat (50°C for 72 h or 60°C for 24 h) gave 99% disinfection of bean seeds artificially contaminated with *P. s.* pv. *phaseolicola*.[400] *B. solanacearum* was not detected in peanut seeds with a water content below 8.9% after drying at room temperature or <10% after drying in sunlight. Thus, transmission may be prevented by dry preservation.[401] Dry heat treatment of *Capsicum* seeds for 7 d at 70°C gave control of capsicum mosaic virus.[402] *Acremonium coenophialum* was eradicated from germ plasm and genotypes of tall fescue either by heat treatment for 7 d at 47° in a mixture of 75% glycerol plus 25% water mixture that produced a relative humidity of 45% or by long-term storage (>3 years) of seeds.[403]

3. Solar Heat Treatment

Solar heat treatment is effective for controlling loose smut of wheat for small seed quantities during hot summer months. During the last week of May or first week of June in north India, where summers are hot (above 35°C), seeds are soaked in water for 4 h in the morning and then dried in the sun before storing.[404] For solar heat treatment of cowpea seeds, seeds were soaked for 4 h in tap water (38°C) from 0800 to 1200 on a calm, bright, and sunny day and then dried for 5 h under the sun's rays for the control of *X. c.* pv. *vignicola*. This method was used in areas where day temperatures reach 45°C or above.[393]

4. Aerated Steam Treatment

The use of aerated steam is safer than hot water and more effective than hot air. The heat capacity of water vapor is about half that of water and 2.5 times that of air; hence the temperature and time required may be higher than that of hot water and lower than that of hot air.[374] Drying of seeds is easier, loss in germination is low, temperature control is easy, seed coats of legumes remain intact, and flax seeds do not become sticky. *Itersonilia pastinacea* in parsnip plant debris associated with seeds was inactivated with a steam/air mixture for 30 min at 45.5°C.[405]

Seed moisture content affects the control of seedborne fungal pathogens by aerated steam because both pathogens and seeds are more sensitive to it in moist than in dry seeds. A pathogen is difficult to eradicate in seeds with a low moisture content. Seeds are treated first at a low temperature for 60 min followed by exposure to a higher temperature for 30 min. Using this method, *A. carthami* in sunflower and *F. moniliforme* in sweet corn seeds were eradicated at 54 and 62°C for 30 min, respectively, while seed viability was slightly affected.[406] In treating seeds to control or eradicate seedborne fungi or bacteria, aerated steam has advantages over hot water, because there is no water soak. This avoids damage to some seeds such as bean and soybean, and rapid drying and cooling are facilitated by the blower that forms part of the equipment. Complete eradication of *Septoria apiicola* was obtained in celery after treatment for 30 min at 56°C, of *F. moniliforme* in sweet corn after treatment for 30 min at 64°C, of *C. michiganensis* subsp. *michiganensis* in tomato seeds after treatment for 30 min at 56°C, and of *X. campestris* pv. *campestris* in cabbage seeds after treatment for 30 min at 54°C.[407] Dry heat treatment for 11 d at 71, 75, or 84°C eliminated *X. c.* pv. *translucens* from barley seeds without germination reduction. In contrast, the pathogen was eliminated from moderately infested seeds after exposure for 4 d to 72°C. The treatment decreased bacterial populations more rapidly the first day than on subsequent days. Reduction in germination was negligible for seeds treated for 7 d or less at 71 or 72°C.[408] Aerated-steam equipment with a temperature control box equipped with a manually operated choke located on the fan intake, a modification of the equipment developed by Setchell,[409] was used for treating seed in an aerated system. The lack of suitable equipment has prevented the wider use of aerated steam for treatment of seedborne pathogens. Apart from a steam generator, which requires substantial capital outlay, the steaming box and blower are relatively inexpensive.

5. Radiation

Electromagnetic radiation has been studied for control of seedborne pathogens. Soybean seeds were used to test the efficiency of disinfection by radiation, high voltage electric currents, ultrasonic radiation, and very high frequency (VHF) radio waves. On passing electricity at 4 kW/g for 30 sec through seeds, bacterial infection decreased from 5.9 to 2%, the degree of disinfection depending on voltage and exposure. Bacterial infection in plants grown from seeds exposed to ultrasonic radiation decreased with exposure for 15 min of 21.3 kc/sec. Cotyledonary bacteriosis (*P. s.* pv. *glycinea,* and *R. solanacearum*) and angular leaf spot (*P. s.* pv. *glycinea*) were 16.5 and 40.4%, respectively, compared with 30.8 and 73.2% in control on soybean plants. No specific results were obtained with VHF waves, although angular leaf spot bacteria were inhibited. Germination of seeds was not affected.[410] Treating tobacco seeds with 625-W microwave radiation for 20 min eliminated *E. c.* subsp. *carotovora* without affecting germination. The number of infected seeds declined by 68 and 99% by a 10- and 15-min treatment, respectively. Infected seeds had a 5.3% moisture content.[411] Microwave radiation (1420 W) of single-celled fungal spores of *A. flavus, A. niger, C. truncatum*, and *C. lindemuthianum* resulted in lower viability than radiation of multicelled spores of *F. oxysporum* and *B. sorokiniana*, and dark spores of *A. niger* and *B. sorokiniana* were less affected than hyaline spores in peanut, popcorn, soybean, wheat, or bean.[412]

Irradiation of soybean seeds (moisture content 6.5 to 7.8%) with 1000, 2000, to 3000 Gy ^{60}Co eliminated most seedborne fungi, but highest doses reduced seed germination; 60 and 29.7% germination occurred after 4 months of storage following irradiation with 1000 and 2000 Gy, respectively, compared with 100% germination after 250 or 500 Gy. Oil content after 4 months of storage was not affected by irradiation treatment, but the composition and quality varied with a decrease in unsaturated fatty acids, an increase in acidity and saponification value, and a decrease in the iodine value with increased irradiation.[413]

Irradiation with microwaves (1420 W, 2450 MHz) has a potential of eliminating fungal propagules in seeds. Irradiation of single-celled spores resulted in lower viabilities, but multicelled spores were less affected; dark spores were less affected than hyaline spores.[412]

Barley seeds treated with 5, 10, 20, or 30 passages of laser treatment prior to sowing resulted in a sevenfold decrease in infection by *U. segetum* var. *segetum* and no root rot caused by *Fusarium, D. graminea*, or *B. sorokiniana*. With an increase in the number of passages from 10 to 30, seed disinfection increased. Seed infection by *Fusarium* decreased by 2.6 to 4 times, and by *D. graminea* and *B. sorokiniana* by 1 to 3 times.[414] Microwave oven treatment of casava seeds (1400 W heating power, 2450 MHz) for 120 sec eradicated *Cladosporium, Colletotrichum, Diplodia, Fusarium*, and *X. c.* pv. *manihotis*. Effectiveness of this treatment depended on reaching an optimum of 77°C. An arasan dust treatment after microwave exposure reduced seed reinfestation.[415] Gamma radiation (^{60}Co source), neutron particle (^{252}Cf source) reduced *Acremonium coenophialum* from tall fescue seeds.[416]

Gamma irradiation of barley seeds from an 1800-^{60}Co source from O roentgens to a lethal dose increased the ratio of healthy plants over plants infected with barley stripe mosaic virus. This may be due to a lethal effect on infected seeds, on the virus in some seeds, or on symptom expression in seedlings.[417] Radiation of *Prunus* seeds at higher than 20,000 rads caused reduction in seed transmission of *Prunus* necrotic ringspot and prune dwarf viruses; seed viability also was reduced.[418]

The exposure of rice seeds to ultrasound had no nematicide effect alone but increased the efficacy of several fungicide solutions. A 94% mortality was achieved using a 0.5 to 1% solution of oxamyl 25 EC and 21.5 KHz ultrasonic generator against *A. besseyi*.[419]

6. Factors Governing Heat Therapy

Factors influencing results of heat therapy are seed moisture content, dormancy, age, vigor, and physical condition, as well as cultivar susceptibility and amount and location of inoculum. Reduction of seedborne inoculum without affecting seed viability is not always attained.[374]

7. Application

Hot-water treatment used correctly can eliminate most seedborne bacteria and fungi without affecting germination. Hot-water treatment can be used on seeds of cabbage, carrots, cucumber, eggplant, lettuce, pepper, radish, spinach, tomato, and turnip. The procedure involves placing seeds in a loosely woven cotton bag to half full and warming them in water for 10 min at 38°C, before placing the bags in a waterbath at the recommended time and temperature required. Seeds are dipped in cold water immediately after treatment and dried. The seeds are treated with a seed protectant before planting. Pathogens controlled with hot water are listed in Table 13-8.

Some seedborne viruses can be inactivated by thermotherapy, but most viruses in dry seeds are resistant to heat. Reddick and Stewart[434] reported that bean seeds treated with dry heat for 10 min at 80°C or up to 40 min at 75°C had reduced survival of BCMV but still had 50% transmission. Similarly, hot-water treatment for 15 min at 70°C still showed transmission of BCMV. In every case the virus survived where seeds survived. TRSV in soybean seeds is not inactivated by heat treatment without destroying seed viability.[435] By treating dry lettuce seeds above 100°C, LMV could be reduced, but germination was so reduced that the effort was not worthwhile.[436] There is no loss of BSMV in barley seeds treated for 30 min at 130°C.[437]

Exposure of seeds to high temperature for short time periods is not effective in eliminating viruses from seeds. Long periods at low temperature are effective in reducing certain viruses. TMV was eliminated from tomato seeds after 2 d at 50 to 52°C followed by 1 d at 78 to 80°C.[438] Treatment of tomato seeds for 22 d at 72°C reduced seed transmission of TMV without loss in germination.[439] Heating dry tomato seeds for 3 d at 76°C also reduced TMV.[440] In Germany, dry-heat treatment of tomato seeds for 24 h at 80°C is officially advocated for control of TMV.[441,442] Cucumber green mottle virus can be eradicated from cucumber seeds by hot-water treatment for 3 d at 76°C.[443] In a commercial trial, 4500 symptomless cucumber plants were raised from previously infected seeds treated for 3 d at 70°C.[444] Cowpea banding mosaic virus was reduced in cowpea seeds by hot-water treatment for 40 min at 45°C, or for 20 min at 50°C; with hot air for 50 min at 55°C, or for 20 min at 65°C; or a dry treatment for 15 min at 65°C followed by 2, 4, or 8 d at 30°C.[264] Cowpea mosaic virus was inactivated in cowpea seeds when freshly infected seeds were exposed for 4 d to 30°C or 15 min at 55°C, followed by 4 d at 25°C.[445] Hot air-treated vegetable marrow seeds for 2 d at 70°C or for 4 weeks at 40°C, or hot water treated for 6 min at 55°C eliminated cucumber mosaic virus.[446]

Soaking *A. tritici* galls for 10 min at 55°C killed 97% of the larvae, whereas 83% of the larvae were killed in dry sand after 10 min at 55°± 1°C. After 15 min, all larvae were dead in both treatments.[447]

8. *Limitations of Thermotherapy*

Thermotherapy for the control of seedborne pathogens has limitations,[448,449] including that (1) reduced germination is possible, (2) seed coats may swell and split, (3) seeds may stick together, (4) maintaining temperature and time requirements is difficult, (5) the treatment may not be as effective as a fungicide treatment, (6) it is difficult to process, (7) the mode of action is not known, (8) deep-seated infections may not be eliminated.

Table 13-8 Control of Important Seedborne Pathogens Through Hot Water Treatment of Seeds

Crop	Disease	Pathogen	Treatment[a]	Ref.
Arachis hypogaea (peanut)	Testa nematode	*Aphelenchoides arachidis*	15 min cool water 60°C for 5 min	420
Aspalanthus linearis (roosibus tea)		*Diaporthe phaseolorum*	20 min at 50°C	352
		Neocosmospora vasinfecta	10 min at 55°C	352
Brassica (broccoli, cauliflower, Chinese cabbage, collard, kale, kohlrabi, rape, rutabaga, turnip)	Black rot	*Xanthomonas campestris* pv. *campestris*	30 min at 50°C	170, 421
Brassica (brussels sprouts, cabbage)	Seed rot		25 min at 50°C	421
Brassica (mustard, cress, radish)	Seed rot		15 min at 50°C	421
Brassica napus	Black spot	*Alternaria brassicicola*	30 min at 50°C	422
C. sativus (cucumber)	Seedling blight	*Pseudomonas syringae* pv. *lachrymans*	75% R. H. for 3 d at 50°C	423
Cyamposis tetragonoloba (Cluster bean, guar)	Blight	*X. campestris* pv. *cyamopsidis*	10 min at 56°C	424
Daucus carota (carrot)	Seed rot		25 min at 50°C	421
Dipsacus spp. (teasel)	Stem eelworm	*Ditylenchus dipsaci*	1 h at 50°C or 2 h at 48.8°C	425
Lactuca sativa (lettuce)	Leaf spot	*X. campestris* pv. *vitians*	1 to 4 d hot air at 70°C	426
Lycospericon esculentum (tomato)	Black speck	*P. syringae* pv. *tomato*	60 min at 52°C	205, 421, 427
Nicotiana tabacum (tobacco)	Hollow stalk	*Erwinia carotovora* pv. *carotovora*	12 min at 50°C	428
Oryza sativa (rice)	Udbatta	*Ephelis oryzae*	10 min at 54°C	429
	White tip	*A. besseyi*	24 h cool water	430
			15 min at 51 to 53°C	431
			15 min at 54 to 57°C	
Pennisetum glaucum (pearl millet)	Downy mildew	*Sclerospora graminicola*	10 min at 55°C	432
Solanum melongena (eggplant)	Seed rot		25 min at 50°C	421
Spinacia oleracea (spinach)	Seed rot		25 min at 50°C	421
Tropaeolum majus (nasturtium)	Fascians disease	Rhodococcus fascians	1 h cool water, 30 min at 51.7°C	433

[a] R. H. = relative humidity.

IV. CERTIFICATION

Certification ensures that seedlots meet certain quality standards and that the history of each lot is traceable. It involves testing seeds before sowing and after harvest and crop inspection for compliance with standards, including isolation and freedom from weed seeds and diseases.[5] Certification programs usually follow breeders' seeds through successive multiplication. Through certification, certain seedborne pathogens have been controlled, and spread to new areas has been checked. Presently, there is more emphasis on seed testing for seedborne pathogens in seed certification programs than on disease incidence in the field. Field inspection is useful in rejecting seedlots with a high incidence of seed-transmitted pathogens, but apparent absence of disease does not guarantee pathogen-free seeds. The causal agents of anthracnose, bacterial blight, frogeye leaf spot, Phomopsis seed decay, pod and stem blight, stem canker, and soybean mosaic may be present in soybean seeds and plants without producing symptoms (asymptomatic) until plants begin to mature.[25,68] Cafati and Saettler[450] emphasized that tests to detect seedborne *X. c.* pv. *phaseoli* should be included in any production program for certified blight-free *Phaseolus acutifolius* and *P. vulgaris* seeds, because seed transmission of the bacteria occurs in symptomless seeds from symptomless pods of both resistant and susceptible genotypes.

Loose smut of barley and wheat was controlled through seed certification programs in India, Scotland, and Sweden.[1,451] Certification was based on seed production in isolation, field inspection, laboratory evaluation of seedborne infection by the embryo count method, and seed treatment with a systemic fungicide. In England, the standard for field approval for barley and wheat is one smutted ear per 10,000 ears. Lower standards are applied to susceptible cultivars of barley and wheat, with no more than 1 in 2000 in commercial crops and 1 in 5000 in crops intended for multiplication.[452] Field inspection for loose smut was discontinued in many places in favor of a laboratory embryo count test, which permits 0.2% infection.[453] The embryo test for detecting loose smut hyphae in barley seeds is used at the Official Seed Testing Station for Scotland. The test is carried out on samples of commercial and farm-saved seeds submitted for certification. A standard of 0.2% infection, determined by a laboratory test on 1000 embryos, is set in conjunction with the Scottish Central Seed Certification Scheme.[451]

The embryo test for detection of loose smut in wheat seeds is used by the Uttar Pradesh Seeds and Tarai Development Corporation Ltd., India. Seed lots with up to 0.5% infection are certified without seed treatment and those between 0.51 to 2% are treated. Seed lots with more than 2% infection are not used for multiplication. Use of the embryo test and the necessity of treating seed with more than 0.5% infection has restricted loose smut infection by the loose smut fungus.[1] A standard of no *Phoma lingam* in 1000 prebasic and basic seeds for multiplication has been proposed for *Brassica* seed certification in the United Kingdom. Similar standards were set for *Phoma betae* on red beet, *Phoma apiicola* and *Septoria apiicola* on celery, *Colletotrichum lindemuthianum* and *P. s.* pv. *phaseolicola* on bean, and *Aschochyta fabae* on fava bean.[5] Infection of *A. fabae* in British-grown commercial bean seeds was reduced by seed selection.[454]

In Australia bean certification began in the 1930s in New South Wales and Victoria as a means to control anthracnose and bacterial blight. The scheme was based on strict field inspections and was a success.[455] The sale of uncertified cereal seeds for sowing is not permitted within the European Economic Community (EEC).[273] All seeds sold in the United Kingdom meet minimum standards for quality, including seedborne pathogens, mainly ergot and loose smut fungi for cereals, *Phoma lingam* on Brassicas, *P. betae* on red beet, *C. lindemuthianum, A. fabae,* and *P. s.* pv. *phaseolicola* on beans, lettuce mosaic virus on lettuce, *Phoma apiicola* and *Septoria apiicola* on celery, and *Ascochyta* and *Phoma* on peas.[456-458] The EEC marketing directives are for ergot of cereals, *S. sclerotiorum* for oil and fiber crops, and *A. fabae* for beans.[457] In Australia, a seed testing service for cucumber mosaic virus in *Lupinus angustifolius* is available to growers who cannot sow seeds with >0.5% infection. This tolerance limit was allowed for plants with CMV symptoms during inspection of crops grown for certified seeds.[459] Field assessment of loose smut is unreliable unless the crop is inspected at the correct time and care is taken in counting diseased and healthy ears. The expression of loose smut in a field occurs over a long period, and early visits by inspectors may fail to record diseased ears that still are to appear. A late visit means that many of the spores may have dispersed, and it is difficult to attain an estimate of smutted ears when bare rachises are all that remain. In New South Wales certification of soybean seeds allows trace levels of *S. sclerotiorum* in crops nominated for certified seed production.[460] There was a correlation between crop infection and seed infection in New South Wales, and the rules for production of certified seed ensures little chance of seedborne infection by *S. sclerotiorum.*[461] In the United States combined losses due to black leg and black rot of crucifers, *Phoma lingam,* and *X. c.* pv. *campestris,* amounted to 10% of the national cabbage production in 1973. Local losses were catastrophic: 70% of production was lost in Minnesota and 100% in some New York fields. An official seed health certification program was established based on laboratory testing, and a zero tolerance level out of 30,000 seeds was set. As a consequence, American seed stocks of cabbage were cleaned up and considerable losses prevented.[455]

A rigid seed certification program contributed to elimination of seedborne bacteria from bean seed stocks in Idaho. Using serological or greenhouse evaluations or both, seed samples can be shown to be free of seedborne bacteria.[462] The Michigan bean seed program involves production of early multiplications in California and Idaho, with a standard of permitting fields 0.005% blighted plants (*P. s.* pv. *phaseolicola*) and zero tolerance in seeds using a seedling infection test with leachates from 2.27 kg seeds.[463] In the Canadian Scheme, disease standards for fuscous blight in beans include zero tolerance in a test on a 5 kg sample. If blight is detected, resistant foundation seeds must show negative results in a test on a 3.4 kg sample.[5,464]

Resistant bean cultivars may not show symptoms, yet *X. c.* pv. *phaseoli* can be seed transmitted.[465] *X. c.* pv. *campestris* grows best at 30°C, whereas crucifers are cool season plants. Thus symptoms can be masked at 15 to 20°C, but become obvious at 25 to 30°C. Since temperatures generally are low during the vegetative growth stage of crucifer seed plants, one would not expect diseased plants to be detected easily.[466] Hence seed assays are the most reliable method for determining seed infection.

The use of LMV-free seed stocks of lettuce controlled LMV in the Imperial Valley, CA since 1969. The program reduced yield losses by 95 to 100%. Lettuce seeds are grown in insect-free greenhouses, infected plants are rogued, and seeds are produced from virus-free plants. Seeds are certified by growing-on or infectivity tests. Seed lots with no infection in 30,000 seeds are used for planting. In Europe, tolerance for this virus is 0.1%.[467-470]

The incidence and severity of BSMV in barley have been reduced through seed certification and through planting virus-free seeds in the United States since 1970. Seeds are tested for BSMV in barley embryos using serological methods.[471] Two certification programs exist. A complete generation certification program initiated in Montana in 1966 involves the production of foundation, registered, and certified seeds. Seeds are harvested from an inspected field with no diseased plants, and samples are tested for the virus using the sodium dodecyl sulfate (SDS)-disk test. In North Dakota, a limited generation certification program is followed. Foundation seed fields are examined, and seed lots are tested for BSMV using the latex flocculation method. Both programs have prevented millions of dollars of crop loss due to BSMV.[472] In Canada, pea seedborne mosaic virus-infected pea lines are screened and eradicated on the basis of routine examination of seed lots.[473] Seed production plot inspection also is used in certifying for absence of pea early browning virus in peas in the Netherlands. Over one diseased plant per 100 m^2 renders the seed unfit.[474] Soybean plants are inspected five times for SbMV and screened using ELISA. Infected plants are rogued. Up to 20 leaflets per line, one from each of 20 plants, are tested. Lines giving a positive ELISA reaction are reinspected, rogued, and assayed again. Seeds with negative ELISA are used for further multiplication.[475] A satisfactory control of LMV under Netherlands conditions was obtained by using seeds indexed to 0 in 2200 seeds sown in the Rohloff grown-on seedling test.[476] In 1980–81, the Rohloff test was applied with ELISA, and because of a good correlation, ELISA is the standard test used for detection of LMV.[477] The use of ELISA to detect BSMV in barley was adopted by the Montana Seed Certification Program. The assay replaced a double immunodiffusion technique facilitated by the SDS-disk test, which detected a single infected embryo in 200. It detected 0.5 µg/ml BSMV or 1 infected seedling among 250 in a composite sample using P-nitrophenyl phosphate as substrate.[478]

Canada and many states in the United States have established fruit tree virus certification programs to provide fruit tree growers with virus-tested trees.[479] Prunus necrotic ringspot virus (PNRV) and prune dwarf virus (PDV) occur worldwide and are seed transmitted in many *Prunus* spp. Consequently, most stone fruit certification programs specify that *Prunus* seedlings must be tested for these viruses and meet certification standards of fewer than 5% virus-infected plants. ELISA procedures to detect PNRV and PDV have been used since 1981 to test all *Prunus* seed lots planted in Montana and Washington.[480]

A. Setting Certification Standards

Most certification standards are based on field inspections and/or seed health testing. Problems with standards based only on field inspections are that (1) evaluation of field infection depends upon the skill of certification personnel, (2) the crop may be infected but the seeds may not, (3) symptoms may be masked, making it difficult to detect the disease in the field, and (4) the correlation between disease incidence and seed infection is not known for a majority of seedborne pathogens. Another problem is in certification standards based on percent of seedborne infection. Data often are lacking about the role of seedborne inoculum in disease development and subsequent yield losses (i.e., economic thresholds); this is a major reason for lack of seed certification standards for seedborne pathogens.

Setting certification standards on seed health testing is the most appropriate approach for a majority of plant pathogens because it is the seed infection level that governs the pathogen transmission through seeds.[481] Where seed health tests are designed to provide data based on individual seed infection, the data are qualitative and best suited for analysis by binomial distribution. One cannot predict a zero infection count with results taken from one sample. One can use binomial tables to predict an appropriate sample size when tests yield qualitative data. However, it is possible to predict infection levels with massed data when a zero pathogen count is required.[482]

Setting certification standards on seed health testing are the most appropriate approach for a majority of plant pathogens because it is the level of seed infection which governs the transmission of pathogens through seeds. The permissible limit may vary depending on the following factors.[481]

1. Relative Role of Seedborne Infection in Disease Development

Certification standards vary depending on their potential rate of establishment, infection, and spread. For the majority of seedborne pathogens, information only on their seedborne nature is available. Transmission rate and spread in the field are not well understood. Under favorable conditions, 12 bean seeds per acre infected with *P. syringae* pv. *phaseolicola* can cause severe epidemics of bacterial blight.[483] Only 0.1% seed transmission of LMV leads to severe field infection when aphid vectors are active.[484] Therefore, a tolerance of 0.003% is allowed in California for this virus.[485] A close correlation exists between *Ustilago segetum* var. *tritici* infection of seeds and loose smut development in barley and wheat.[1,451,486] Thus disease development can be measured by seed infection. Seed tansmission becomes important if there is potential for spread of a pathogen from infected seeds to plants. The tolerance must be near zero in such cases.

2. Perpetuation of the Seedborne Inoculum by Other Means

Many seedborne fungal and bacterial pathogens are soilborne. If a pathogen is only seedborne, strict certification standards can be followed, and infected seeds should not be sent into areas where the pathogen has not been detected. But if a pathogen is soilborne, its seedborne nature becomes secondary in disease development. In such cases, certification standards may be relaxed.

3. *Factors Affecting Seed Transmission*

The environment plays an important role in seed transmission. This has been observed in Karnal bunt of wheat (*Tilletia indica*) in the Nainital Tarai of India. During the 1969–70 harvest, seedborne infection varied from 0 to 7.5% in different cultivars, while in 1970–72 its incidence was less than 1%, and in 1974–75 the incidence reached 50%.[487,488] In Canada; depending upon the rate of seed infection, the tolerance limit for *Ascochyta* on peas varied from 2 to 6%.[485] Therefore, seed certification standards may vary from year to year for certain pathogens depending upon seed infection.

4. *Economic Loss Due to Seedborne Pathogens*

Losses caused by seedborne pathogens vary depending upon virulence, host range, and environment. If a pathogen causes heavy losses, the level of infection allowed for certification should be low. The role of seed infection, ratio of subsequent transmission in the field, and yield loss should determine seed certification standards for different pathogens.

5. *Planting Area*

Certification standards also depend upon the planting area. Some pathogens may not infect a crop in certain areas due to unfavorable environmental conditions. For such areas, a relaxation in seed certification standards can be allowed. However, in areas where infected seeds may serve as a source of soil contamination, or where the infection can occur, only pathogen-free seeds should be planted.

6. *Influence of Seed Treatment on Seedborne Infection*

If a seedborne pathogen can be eliminated by seed treatment, then certification standards should be relaxed. Most chemical seed treatments are not effective against bacterial and viral pathogens and certification standards are necessary for them

7. *Seed Processing Procedures*

A number of plant pathogen propagules can be separated from healthy seeds during processing, such as seeds (*Cuscuta, Orobanche*), sclerotia (ergot, *Sclerotium*, *Sclerotinia,* etc.), and galls (ear cockle), etc. For such pathogens, seed processing must be taken into consideration while developing certification standards.

Table 13-9 Import of Some Major Crop Seeds Prohibited in the United States Due to Infections

Crop	Country	Disease
Gossypium (cotton)	All	Viruses, various diseases
Sorghum bicolor (sorghum)	Africa, Asia, Brazil	Smuts
Lens culinaris (lentil)	South America	Rust
Oryza sativa (rice)	All	Smuts, viruses, various diseases
Triticum aestivum (wheat)	Asia, Australia, Eastern Europe	Flag smut
Zea mays (maize)	Africa, Asia	Downy mildews

From Mathys, G. and Baker, E. A., An appraisal of the effectiveness of quarantines, *Annu. Rev. Phytopathol.*, 18, 85, 1980.

V. PLANT QUARANTINE

The terms frequently used in plant quarantine are *hazard*, *risk*, and *safeguards*.[489] *Hazard* is the danger that a specified pathogen is known or perceived to present to the agriculture of the importing country should the pathogen gain entry on imported items and subsequently become established. *Risk* is the chance that a hazardous organism will enter and become established. Since quarantine risks often have not been quantified, quarantine officers generally refer to the risk level as low, medium, or high. Risks associated with germ plasm vary with the collection site (Table 13-9).[489] For example, rice seeds collected from the wild are of greater risk than those collected from a research station. Whole plants may be a greater risk than seeds, since not all pathogens are seedborne. *Safeguards* are actions taken to reduce the risk of introducing hazardous organisms. Safeguards include rules and regulations, permits, phytosanitary certificates, inspection treatment, isolation, passage through a quarantine greenhouse, seed health testing, and other measures. The principal features of a safeguard program are the option of the importing country either to deny a permit for higher-risk material for commercial purpose or to issue a permit for plant material imported for scientific purposes. For scientific purposes, a special permit that specifies conditions or safeguards is issued by the importing country. For the international exchange of seeds as germ plasm, safeguards may be used at origin, upon entry, or after entry to reduce to an acceptable level the chances that seeds will serve as an effective pathway for the establishment of pathogens of quarantine importance.[489]

Quarantine is derived from the Latin word *quarantum,* meaning 40. It refers to the 40-d period of detention of ships arriving from countries with bubonic plague and cholera in the Middle Ages. The first such quarantine was imposed in Venice in 1374.[490] Present quarantine laws now include plants. Plant quarantines, promulgated by a government or group of governments, restrict entry of plants, plant products, soil, cultures of living organisms, packing materials, and commodities, as well as their containers and means of conveyance, to protect agriculture and the environment from avoidable damage by hazardous organisms. They exclude dangerous organisms while permitting plants and plant products to enter.[489] The term *exclusion* conveys this objective more clearly than *plant quarantine.* Exclusion relates to keeping organisms out; plant quarantine relates to keeping plants out. This concept has been recognized by the California Department of Agriculture, which employs a "Detection and Exclusion Officer" rather than a "Plant Quarantine Officer".[491]

The importance of plant quarantine has increased because of the increase in exchange of seeds or grains for consumption along with better means of transportation. The international exchange of plants or their parts is practiced widely to improve crops of a country and their genetic base.[492] In addition, shiploads of grains for consumption or large quantities of seeds for direct sowing are imported in many countries.[493] Even minute quantities of soil and plant debris contaminating true seeds can disseminate pathogens.

A large number of plant pathogens have spread over the world through seeds.[453] Wheat bunt appeared for the first time in the Sacramento Valley, CA in 1854 on plants grown from seeds imported from Australia.[493] Peanut rust was introduced into Brunei on peanuts imported for consumption but used for planting.[494] *Marasmius perniciosus* (witches broom of cacao) was introduced in 1974 and 1975 into South America with seeds from Trinidad. The seed produced up to 70% infected seedlings.[495] Peanut rust was introduced from Brazil to the United States on peanut seeds.[496] A sequence of grapevine introductions from the United States, commencing about 1845, resulted in three successive catastrophes to the European grape industry owing to the introduction of powdery *(Uncinula necator)* and downy *(Plasmopara viticola)* mildews. The potato famine in Ireland in 1845 was caused by *Phytophthora infestans* introduced with seed potatoes from Peru to Belgium or France about 1842 to 1844. Coffee rust appeared in Sri Lanka about 1869 and virtually eliminated Arabian-type coffee trees. Before tea cultivation could be established, the island was impoverished, and concern about the spread of this disease led the Dutch to proclaim, in 1877, plant quarantine legislation. The prohibition of importation of coffee plants and seeds from Sri Lanka to Indonesia was the first enactment of plant quarantine legislation. The American chestnut, which made up a quarter of the native trees of the eastern United States, became infected with *Cryphonectria parasitica*, believed to have entered the United States on nursery stock from the Orient. The disease was detected first in 1904, and within 25 years the species had been eliminated. *Peronospora tabacina* (tobacco blue mold), until 1958, was found only in Australia and the United States, but that year it was reported in England. In a few years it spread to all European tobacco-growing areas, parts of North Africa, and the Near East. Europe was especially hard hit in 1960, when the losses were estimated at $25 million.[497,498] *X. o.* pv. *oryzae* was introduced in the Caribbean region and South America through the importation of either infected or infested seeds.[499] Broadbean true mosaic virus was detected for the first time in fava bean in South Australia in plants from seeds imported from the United Kingdom.[500] Introduction of rice seeds infected with foreign races of *P. oryzae* led to catastrophic blast epidemics in Nigeria, South Korea, and Upper Volta, where cultivars resistant to endemic races were attacked by races introduced with seeds.[501] In

1942, the bacterial canker organism of tomato was introduced into England with seeds from the United States; *Gloeotinia granigena* on rye grass seeds from New Zealand to Oregon, in 1940; and *X. c.* pv. *campestris* on cabbage seeds from Europe to India.[501] Two rice pathogens, *X. oryzae* pv. *oryzae* and *X. oryzae* pv. *oryzicola,* once confined to Asia, were established in West Africa and Brazil where none of the local cultivars were resistant.[502] Hop germ plasm from the United States may represent a potential source of world wide dissemination of hop latent viroid.[503]

The race of the nematode *Ditylenchus dipsaci*, present in Sweden, does not attack alfalfa, but a new race that attacked alfalfa was introduced on imported seeds.[504]

A. Components of Plant Quarantine

Plant protection and quarantine (PPQ) programs, or the plant health or quarantine services in most countries, usually have three components.[505]

1. Exclusion of pathogens and pests of quarantine and economic significance that might inadvertently be moved along manmade pathways when articles are imported, or induction of the risk of introducing such hazardous organisms to an acceptable level.
2. Containment, suppression, and eradication of exotic pathogens and pests recently introduced along natural and manmade pathways.
3. Assistance to exporters of plant products, such as fruits, vegetables, plants, cut flowers, commodities, etc., in meeting the quarantine or exclusion requirements of importing countries and, therefore, based on plant health, biologically facilitating the acceptance of imports.

1. Pest Risk Analysis

Pest risk defines the chances that a pathogen or pest of quarantine significance will enter along a manmade pathway. Risk often is expressed as low, medium, or intermediate, or high. Low risk means that there is little chance that the pathogen or pest will enter; high risk means that chances are high that the pathogen or pest could enter. An acceptable risk level means that the benefits derived after taking a risk are high enough to justify taking the risk, with safeguards, in the first place.[506] The rules and regulations governing seed entry should be based on a matching of risk with entry decisions. If the risk is low, the entry status should be liberal; if high, the entry status should be conservative; and when the risk is unknown, quarantine officers need to be conservative. The policy of an importing country might prohibit all plants of high risk and by doing so take no risk, but at the cost of receiving no benefits from crop improvement. Cost/benefit and risk/benefit factors come into play in many countries. Plant quarantine officials and plant breeders may arrive at different conclusions when given the same body of information for assessing risk for setting safeguards, standards, and criteria to grant exemptions to the prohibition.[507]

2. Movement of Pathogens

Plant pathogens and pests of quarantine importance can move along natural and human-made pathways, depending upon the life cycle of the organism, the environment through which it moves, and human activities. At the end of a pathway, the establishment of a pathogen or pest in a new area depends on the level of inoculum or pest carried by the seeds, the susceptibility of host crop(s), and the environment. From the viewpoint of plant quarantine, exchange of germ plasm creates the risk of introducing pathogens or pests of quarantine importance. The considerations are twofold; that exotic organisms or more virulent strains of existing ones might be introduced.[490] Quarantine regulations to prevent movement of feces-derived products as a potential source of introduction should be considered. Viable teliospores of *T. caries, T. controversa*, and *T. indica* are present in feces of chickens, grasshoppers *(Melanoplus sanguinipes)*, and cows.[508] Hence quarantine established to prevent movement of spore-contaminated seeds may miss an important avenue of potential introduction by animal and animal-product movements.

3. Legal Basis for Plant Quarantine

Quarantines are regulations promulgated by governments to reduce the risk of introducing hazardous pathogens and pests on articles, including seeds, from foreign areas. The legal basis of quarantine comprises (1) legislation enacted by national and sometimes state or provincial governments, (2) enabling legislation that directs the Minister of Agriculture to issue necessary rules, orders or directives or (3) legislation by a regional parliament representing groups of countries such as the EEC or Andean Pact Nations. In addition, biological standards or guidelines for promulgating rules often are suggested by regional plant protection organizations not binding on member countries.[490]

The legal umbrella that covers international plant quarantine matters is the International Plant Protection Convention of 1951 (Rome Certificate). Most countries either are signatories of the Rome Convention or follow its mandates. For signatory nations, the convention has the full force of a treaty, under which disputes may be arbitrated in an international court. The treaty is administered by the Food and Agriculture Organization of the United Nations.[495] Phytosanitary certificates such as those accompanying international shipments of germ plasm seeds are instruments of this treaty.

4. Requirements for a Quarantine Program[509]

1. A phytosanitary certificate that attests to the inspection, origin and identification of the seeds;
2. A permit requirement for added declaration on the phytosanitary certificate that the mother plants were inspected during the growing season;
3. A requirement for treatment at origin;

4. The acceptance of a special certification or safeguarding program at origin, such as may be practiced at International Agricultural Research Centers (IARC):
5. Inspection, and treatment, if necessary, upon arrival at port of entry; and
6. Isolation, special testing, or additional quarantine after entry.

B. National and International Regulations

The first plant quarantine law was passed in 1873 in Germany to prohibit importation of plants and plant products from the United States to prevent the introduction of the Colorado potato beetle; in 1875 France also imposed measures against the American pest. In 1877 the United Kingdom Destructive Insects Act prevented the introduction and spread of this beetle. In 1891 the first plant quarantine measure was initiated in the United States by setting up a seaport inspection station at San Pedro, CA, and the first U.S. quarantine law was passed in 1912. The Federal Plant Quarantine Service was established in Australia in 1909.[509] In India a Destructive Insects and Pests Act was passed in 1914. Since then, most countries have formulated quarantine regulations.

It was not until the latter half of the 19th century that the nations of the world became plant quarantine conscious. In 1877 Indonesia enacted a law prohibiting the import of coffee plants from Sri Lanka to prevent the entry of the coffee rust pathogen. In the same year four states in the United States enacted legislation to protect against certain plant pests, and in 1905 the Federal Insect Pest Act was passed, enabling the federal government to regulate, for the first time, importation and interstate movement of articles that might spread pests of quarantine importance. In 1912 the comprehensive plant quarantine act was approved in the United States. In Australia, quarantine regulations were introduced in 1909.[510] In 1971 the government of Greece prohibited introduction of rice seeds for sowing infected with *P. oryzae*.[510] A similar restriction was imposed in Chile.[510] The Plant Quarantine Station at Muguga in Kenya used postentry cultivation of rice in greenhouses for seeds introduced into Kenya, Tanzania, and Uganda.[510] Only seed from healthy plants were released. In 1980 the Philippine government introduced seed health testing of rice for *B. oryzae*, in addition to field inspections for breeder and foundation seeds. In 1982 it was made essential to carry out postentry control tests on all seeds intended for certification.[511] *P. s.* pv. *glycinea* is widespread in soybeans throughout the United States, but does not cause economic loss. However, phytosanitary regulations require that seeds exported to the EEC should be assayed for the pathogen. With a major increase in soybean seed export from the United States to the EEC since 1988, costs of testing have become an economic concern for U.S. seed exporters. A soaking test developed in France was agreed upon in 1990 as the accepted method for assaying U.S. soybean seeds to be exported to the EEC.[512]

On a global basis, the first International Plant Protection Convention (the Phylloxera Convention) was signed in 1881 with the objective of preventing the spread of severe pests. This convention was amended in 1889, 1929, and 1951. The International Plant Protection Convention (IPPC or Rome Convention) under the Food and Agriculture Organization was established to prevent the introduction and spread of diseases and pests through legislation and organizations across international boundaries.[490] This convention provided a model phytosanitary certificate (Rome certificate) to be adopted by member countries. Within this convention, ten regional plant protection organizations have been established on the basis of biogeographical areas[490,493,513,514]: European and Mediterranean Plant Protection Organization (EPPO), Inter-African Phytosanitary Council (IAPSC), Organismo International Regional de-Sanidad Agropecuria (OIRSA), Plant Protection Committee for the South-East Asia and Pacific Region (SEAPPC), Near East Plant Protection Commission (NEPPC), Comite Interamericano de Protection Agricola (CIPA), Caribbean Plant Protection Commission (CPPC), North American Plant Protection Organization (NAPPO), Organismo Bolivariano de Sanidad Agropecuria (OBSA), and Association of South East Asian Countries (ASEAN), regional grouping of Indonesia, Malaysia, the Philippines, Thailand, and Singapore. The regional organizations are concerned with the coordination of legislation and regulations within their area, agreement on the quarantine objects, inspection procedures, etc. The EPPO, IAPSC, and ASEAN have taken up in detail the question of seed quarantines.[493,513]

Many countries are aided in their efforts to develop national germ plasm systems by the International Board for Plant Genetic Resources (IBPGR), which was established in 1974 with headquarters in FAO/UN Rome. The IBPGR strives to build a world network of institutions to collect, conserve, evaluate, document, and distribute germ plasm of economic plants and their wild relatives. The IBPGR is one of 13 International Agricultural Research Centers that is funded by the Consultative Group on International Agricultural Research Centers, an informal consortium of governments, international and regional organizations, and private foundations established to promote agricultural research for the benefit of developing countries.[514,516] The principles set forth in the General Agreement on Tariffs and Trade (GATT) Uruguay Round of the Multilateral Trade Negotiations are discussed in relation to the International Plant Protection Convention (IPPC) by Hedley.[517]

1. Plant Quarantine in the United States

The control of plant introductions into the United States began in 1829, when the Congress allotted $1000 for the importation of rare plants and seeds. The office of the U.S. Patent Commissioner was authorized for the introduction of germ plasm between 1836 and 1862. With the establishment of the U. S. Department of Agriculture (USDA) in 1862, a Commissioner of Agriculture was made responsible for collection, testing, and distribution of potentially valuable plant germ plasm. A section, Seed and Plant Introduction, was established in 1898. This system has continued with minor changes.[492] Over 465,000 plants or seeds have been introduced since 1898. At present, about 7500 new introductions are made each year.[492] The Plant Protection and Quarantine program (PPQ) is planned and executed by the Animal and Plant Health Inspection Service (APHIS) of the USDA.[5,8] At present, three quarantine acts are in operation in the United States.[492]

a. Plant Quarantine Act of 1912 — The first United States federal plant quarantine law, known as the Plant Quarantine Act of 1912, was passed after the establishment of white pine blister rust and chestnut blight fungi, and the citrus canker bacterium. The Act controls the introduction of exotic pests and the spread of plant parts new to the United States and within the United States as a domestic quarantine.[519] In spite of the act, pathogens that cause the potato wart, wheat flag smut, and Dutch elm disease have been introduced into the United States.[492]

b. Organic Act of 1944 — This act is primarily for pest management strategies, but gives an authority for issuance of phytosanitary certificates in accordance with the requirements of importing states and foreign countries.

c. Federal Plant Pest Act of 1957 — This act authorizes emergency actions to prevent the introduction or interstate movement of plant pests not covered under the Act of 1912.

Nearly all imported germ plasm falls into one of three categories: restricted, postentry, or prohibited. Restricted germ plasm is inspected and chemically treated and can be imported easily. In the postentry category seeds or other material, after inspection and treatment, are grown-out under close observation. If no pest is found, then germ plasm is released. Prohibited materials must meet certain specific requirements before being imported, since they may pose a serious threat to agriculture.[492]

Postentry surveillance for the detection and interception of seedborne pathogens on introduced plants and the production of pathogen-free seeds is accomplished at regional plant introduction stations. These stations are operated by cooperative agreement between the USDA and land-grant colleges and universities. Here the plants are subjected to inspection, detection, postentry surveillance, and release of seeds.[520]

Plant introduction has played an important role in the development of a strong and diversified agriculture in the United States. Most plant introduction activities are part of the National Plant Germplasm System, which is a coordinated network of federal, state, and private institutions involved in exploration, introduction, increase, maintenance, evaluation, cataloguing, and distribution of foreign and domestic plant germ plasm. Four Regional Plant Introduction Stations are located at Ames, IA; Geneva, NY; Experiment, GA; and Pullman, WA.[519]

Waterworth[521] described the United States federal regulations for importation of plant material for propagation and procedures used for processing germ plasm of those genera in the prohibited but not the postentry, quarantine category.

2. Plant Quarantine in the United Kingdom

The plant health legislation was passed as the Destructive Insects Act of 1877, primarily to prevent the entry and establishment of the Colorado potato beetle. It was extended by the Destructive Insects and Pests Act of 1907 to check the entry of American gooseberry mildew (*Sphaerotheca mors-uvae)* and all fungi, insects, or other destructive pests of plants. To cover bacteria and viruses as well as invertebrate pests, the act was extended as the Destructive Insects and Pests Act of 1927. These three acts were consolidated and formulated in a Plant Health Act of 1967. This act was amended by the European Economic Community Act of 1972.[522] The most familiar activity is the inspection of plants or produce either before export or after import.[523]

3. *Plant Quarantine in India*

The first plant quarantine measure dates to 1906, when the danger of introducing the Mexican boll weevil had the government of India direct that all cotton imported from the New World should be admitted only after fumigation with carbon disulphide. Two regulatory measures are in operation for controlling pests, diseases, and weeds: (1) the Destructive Insects and Pests (DIP) Act of 1914 of the Central Government, which regulates the introduction of exotic diseases and pests into the country or their spread from one state or Union Territory to another and (2) the Agricultural Pests and Diseases Acts of various states, which suppress or prevent the spread of diseases and pests in areas within a State or Union Territory. Seeds were not included originally in the Act, but because of the changing situation, the government of India passed the Plants, Fruits, Seeds (Regulation of Import into India) Order of 1984, which came into effect in June 1985. The authority to implement quarantine rules and regulations framed under the DIP Act rests with the Directorate of Plant Protection, Quarantine and Storage, under the Ministry of Agriculture. Under this organization, nine seaports, ten airports, and seven borders have functioning offices for the import of plants and plant material. In addition, the government of India has approved three national institutes; National Bureau of Plant Genetic Resources for agricultural and horticultural crops, Forest Research Institute for forest plants, and Botanical Survey of India for all other plants of economic interest. All imported plant material is subjected to examination, and only healthy material is released to importers.[524,525] In addition, efforts are being made to salvage diseased material samples through fumigation, X-ray radiography, seed washing, hot water treatment, chemical seed treatment, sprays/dip, etc.[524,525]

4. *Plant Quarantine in Kenya*

At Muguga, the introduction of unwanted pathogens is prevented by growing the incoming material in isolated greenhouses, by seed health testing, by importing from selected countries that provide minimum risk, by prohibiting import of certain crop species from certain countries, by recovering healthy planting material by seed treatment, tissue culture, heat treatment, or tissue culture and the treatment combined, and by releasing only second generation seed.[526]

5. *Plant Quarantine in Australia*

The Australian quarantine service operates as a national service under a Federal Quarantine Act based on cooperation between the commonwealth and states. Policy is formulated in consultation with state Departments of Agriculture, where each state has a Chief Quarantine Officer as well as established entomological and plant pathological services. At the federal level a small group of entomologists and pathologists coordinate the national policy for plant quarantine. Plant quarantine has a research laboratory at Weston, A.C.T. In addition, it has a postentry quarantine facility for official introduction of plants and seeds under proper quarantine control. In Sydney, port inspection of cargoes began in 1889 when the Export and Import Branch of the Department of Agriculture was established. When the Commonwealth was established in 1901, the federal constitution included quarantine as the only specific health power of the new parliament. The Commonwealth Quarantine Act was passed in 1908 and came into force in 1909. The act provides measures for the exclusion, detection, observation, segregation, isolation, protection, and disinfection of vessels, persons, goods, animals, or plants, for the purpose of preventing the introduction or spread of diseases or pests affecting man, animals, or plants.[497] The importation of seeds that may carry disease is restricted by legislation. Usually only a small quantity of seeds is permitted entry and is treated prior to sowing in a postentry quarantine glasshouse. Seedlings are inspected for disease symptoms, and only seed harvested from healthy plants are released to the importer. Most seedborne pathogens intercepted have been viruses in legumes. Australian plant quarantine is testing rapid virus-detection techniques because of the shortcomings with visual detection methods and sap-indexing tests.[527] Increasing mobility through rapid air transport is a major challenge facing plant quarantine. In the Australian service, officers have power of search and are not entirely dependent upon customs. However, customs and quarantine function in collaboration at international airports. The Bureau of Customs devised, developed, and implemented an interrogation system at primary customs checkpoints in customs halls for arriving passengers. Plant quarantine officers found that a portion of travelers with agricultural or horticultural interests engaged in smuggling. Hence it became necessary to implement prosecution procedures with significant penalties as the only deterrent. Crews are handled in the same way as arriving passengers.[497]

Rice imported into Australia can contain intact or whole seeds that have been milled. Such rice with the outer protective husk still present is known as "paddy". Imported paddy rice is of concern to Australian Quarantine and Inspection Services (AQIS) and the rice industry, because such seeds can carry inoculum of various pathogens within and on the surface. The risk of an infection occurring in the Murrumbridge Irrigation Area (MIA) is considered of a low order if there are five seeds per kilogram, given that rice is used for human consumption and that imported rice and rice products are banned from movement into the MIA under New South Wales legislation. Alternate sources of disease entry, such as airborne conidia adhering to clothing of international travelers, was found a greater risk than that posed by paddy.[528]

6. ASEAN (Association of South East Asian Countries)

ASEAN was formed in 1967 by the governments of Indonesia, Malaysia, the Philippines, Singapore, and Thailand. ASEAN Regional Plant Quarantine Centre and Training Institute (PLANTI) was established in Malaysia in 1981 to upgrade plant quarantine technology through training, to develop methods, to conduct research, to organize frequent technical meetings, and to act as a repository for information pertaining to plant quarantine. Different exotic and dangerous pests have been categorized into two broad groups — noxious organisms that have not been introduced into the ASEAN region and organisms that are present in one or more ASEAN country.[529]

C. Basic Principles of Plant Quarantine

The basic principle of plant quarantine is to check the entry and spread of potentially dangerous plant pathogens and insects imported along with the germ plasm. In spite of quarantine regulations, plant pathogens have been introduced in different countries. Plant quarantine regulations have certain prerequisites. They must be[493,513,530]

1. Based on sound biological grounds. Only pests that pose a threat to major crops or forest should be taken into consideration.
2. Formulated to control or prevent the entry of pests and not to hinder trade or attainment of other objectives. Quarantine measures are for crop and not trade protection.
3. Derived from adequate legislation and operated solely under the law.
4. Modified as conditions change or further facts become available.

Those responsible for quarantine measures should be properly trained and experienced; professional workers and the public must cooperate on an international scale for effective operation of quarantine regulations.

Quarantines are only one facet of domestic pest management programs, and careful integration of measures is needed to achieve maximum efficiency.

D. Criteria for Determining Organisms for Quarantine Significance[531]

1. General

The following are general criteria considered for an organism of plant quarantine significance:

1. The organism does not occur in the country but is known to cause economic damage elsewhere and to be capable of causing damage on a susceptible host, when favorable environment conditions exist in the country.
2. The organism occurs in the country but is not widely distributed in the ecological range of its host in that country; is under a national domestic suppression, containment, or eradication program; has exotic strains of quarantine significance that do not occur in the country; and/or causes economic damage or has a potential to cause such damage on economically important crops.

3. The organism is a common pathogen established in the importing country, but government regulations require that commercial growers use pathogen-tested seed stocks so that imported stocks meet domestic standards.

2. *Criteria Based Primarily on Pathogen Characteristics*

These criteria are based on the general characteristics of a pathogen.

1. It is capable of developing a high population in a short time.
2. It is able to survive and move easily in international trade.
3. It is difficult to detect by general inspection or field survey.
4. It is capable of damaging and/or reproducing on many hosts.
5. It has a potential for rapid dispersal, especially along manmade pathways.
6. It cannot be controlled by general methods.
7. It has a potential for significant reduction in the quality and quantity of crop yields.
8. It is capable of adversely affecting the environment.

Neergaard[532] proposed that quarantine objects could be placed into three categories based on the following criteria.

Category A: Dangerous pathogens that are not present in a region of introduction and have a high or considerable epidemic potential. Pathogens belonging to this category generally occur in only trace amounts in seeds. Quarantine measures should prohibit introduction of such seeds from infested areas, or valuable seed material from infested areas must be filtered through postentry control measures by growing plants in special glasshouses under closed quarantine. Viruses and viroids belonging to this category include Arabis mosaic, barley stripe mosaic, broadbean true mosaic, peanut stunt, pea seedborne mosaic, stone fruit viruses, avocado sunblotch.

Category B: Dangerous plant pathogens that are not present in the region of introduction or present in restricted areas under effective control. Such pathogens have a moderate epidemic value. Examples are bean common mosaic virus, lettuce mosaic virus, and pea early browning virus.

Category C: These pathogens are not strict quarantine objects but may be important to the field-planting value of seeds. However, there may be a potential risk of introducing new virulent strains into the region. The seed material is tested with an emphasis on an attempt to prevent introduction of new strains. Examples are the alfalfa mosaic virus, soybean mosaic virus, and cucumber mosaic virus.

E. Problems in Plant Quarantines

Quarantines serve as a filter against the introduction of dangerous pathogens, but pathogens are still introduced. Possible reasons are that (1) it is difficult to detect all types of infectious pathogens by conventional methods, (2) methods may not be sensitive enough to detect traces of infection, (3) latent infections may pass undetected under postentry quarantine, (4) destruction of all infected or suspected material, and (5) sensitive methods for testing fungicide-treated seeds may be lacking.[521]

Plant pathogens may be introduced on inert material such as packing material, dried root bits, plant debris, soil clods. Cysts of *Heterodera schachtii* and *H. goettingiana* nematodes were intercepted on such materials.[533]

F. Organisms of Quarantine Significance

Organisms of quarantine significance may include any pathogen or pest that a government (or intergovernment organization) considers to pose a threat to the agriculture and environment of the country or region. Such organisms usually are exotic to that country or region but may include exotic strains or races of domestic strains.[491]

A pathogen that does not occur in a given country or an exotic strain of a domestic species is of quarantine significance to that country if the pathogen causes economic damage elsewhere or has a life cycle or host–pathogen interaction with a potential to cause economic damage. The importation of a pathogen into a given country also is of quarantine significance if an ongoing regional or national containment, suppression, or eradication program is directed against that pathogen.[505] The European and Mediterranean Plant Protection Organization (EPPO) has defined two types of quarantine organisms: A-1 seedborne pathogens, which include those not present in the EPPO region and A-2, those present but subject to international phytosanitary measures to prevent further spread. Two of four, and seven of 18 organisms in A-1 and A-2, respectively, are bacteria.[534] Some fungal genera and species of quarantine importance have been published.[531] Plant bacteria and viruses of quarantine significance have been published.[531]

G. Plant Quarantine Measures

The goals of regulatory actions are (1) to prevent or delay entry of the pathogen along manmade pathways; (2) if entry succeeds, to prevent infection; (3) if infection succeeds, to prevent establishment of pathogens entering on manmade and natural pathways; and (4) if established, to minimize or retard spread.

Goal 3 is implemented by eradication and Goal 4 by containment or suppression.[505]

The aim of plant quarantine is to prevent the introduction of dangerous diseases and pests or new races of a pathogen and their spread within the country. Measures suggested for effective plant quarantine are as follows:[491,493]

1. Import Control — Regulations of the Importing Country

a. Embargoes — This is the most effective measure to exclude infected plant material. However, in practice it is difficult to achieve because of more and more exchange of diverse genetic material among countries.

b. Inspection of Seed Lots — The examination of seed samples must be by the most sensitive and reliable methods for detection of dangerous pathogens listed under quarantine regulations. A sample may be subjected to more than one method. Detection of seedborne pathogens may be difficult if seeds are treated. It is difficult to eradicate infections by conventional seed treatment fungicides. Systemic fungicides help eradicate certain seedborne pathogens, if the seeds have been treated accurately. Broad spectrum fungicides that can eradicate diverse groups of pathogens are not available.

c. Postentry Quarantine — Because it is difficult to detect all types of seedborne pathogens by simple tests, it may be necessary to subject seeds to postentry quarantine. Seeds are subjected to a period of growth at a quarantine station under strict supervision in the importing country.[535,536] The plants are kept under close observations in semiisolation, so any disease that appears can be detected immediately. Plants are grown under optimum conditions so that symptoms are not masked.[536,537] Pathogen-free seeds then are produced from the imported seed for distribution. Thus, valuable germ plasm of introduced plants can be saved for breeding and crop improvement without danger of introducing prohibited pathogens.[538-540]

d. Seed Treatment — Seeds may be treated with a suitable chemical before release for further multiplications or utilization in a breeding program as an additional safety measure against chance introduction of a pathogen.

2. Export Control — Regulations of the Exporting Country

An exporting country assists the importing country in the latter's practice of exclusion, a treaty obligation for 94 signatory nations under the IPPC of 1951, but also adhered to by other countries.

a. Field Inspection of the Field Crop — The seed crop is inspected regularly for diseases. Infected plants are rogued. The crop should meet requirements of the importing country.

b. Inspection of Seed Lot — The seed lot is examined thoroughly for the presence of the microorganisms before export. The sample should meet the standards of the importing country.

c. Seed Treatment — The seed lot should be treated as per the requirements of the exporting country. However, treatment should conform to regulations of the importing country.

d. Phytosanitary Certificate — Phytosanitary certificates are issued by the exporting country along with the seeds as per the International Plant Protection Convention of 1951 (Rome Certificate or Phytosanitary Certificate). The validity of the certificate depends upon the test and testing methods. It has not been found as a safeguard as viewed by Neergaard, who stated[493]:

> Most seed importing countries have some quarantine provisions on quarantine for seed. Many of these countries require a general plant health certificate for all or nearly all kinds of seed but do not specify any pathogen at all. Consequently, it is left entirely to the discretion of the agency of the seed exporting country to decide which seedborne pathogens should be considered and which inspection procedures should be used. As a consequence, the seed importing country has no guarantee that a seed lot, accompanied by a formally duly issued phytosanitary certificate, has been correctly inspected for the presence of dangerous parasitic bacteria, fungi, nematodes and viruses; indeed most often no microbiological test at all has been carried out, before the certificate is signed and the seed may have been checked by visual inspection only, if at all. Needless to say that a certificate issued under such conditions is worse than useless, it is positively misleading.

The text of a phytosanitary certificate applies to seed lots and consignments of all plants or plant products. For most of these and for many seedborne pests, no test method is available, and zero tolerance has to be satisfied by other, sometimes indirect means. EPPO recommends specific quarantine requirements be incorporated into the phytosanitary regulations of its member countries. Exporting countries have to satisfy these, in so far as they do not involve the direct inspection of the consignment that is covered by the clause in the phytosanitary certificate, that states the consignment conforms with phytosanitary regulations of the importing country. Examples of such requirements are that the consignment: (1) should come from an area where the pest does not occur, (2) should come from a seed crop inspected during the growing season, (3) should have been treated, (4) was tested by a method involving isolation of the pathogen, (5) should have been tested by a method involving bioassay on a susceptible host, and (6) should simply have been inspected.[541] Such methods are not rigorous or sensitive but are considered adequate safeguards and satisfy the zero tolerance requirement, for quarantine pests, whether seedborne or not prescribed in the new text of the phytosanitary certificate specified in the International Plant Protection Convention.[542]

Seed lots carrying quarantined pests have to be certified before export by the phytosanitary certificate of the International Plant Protection Convention (IPPC),[542] which provides for zero tolerance. In this case, it is necessary to certify seed lots with the new text of the phytosanitary certificate specified in the IPPC and thus to require consignments be inspected according to appropriate procedures and considered free from quarantine pests and practically free from other injurious pests. The old certificate was satisfied with "thoroughly examined ... and found to be substantially free," which presented no problems for seed testing.[541]

3. *Intermediate Quarantine*

This is an international cooperative effort to lower the risk of introducing a pathogen to one country with the germ plasm from another by passing this germ plasm through isolation or quarantine in a third country. The pathogen in question should not pose a threat to the third country because either the crop is not grown there or the pathogen, even if it escapes, will not become established because of the environment.[491] Third country quarantine locations are Plant Quarantine Facility, Glenn Dale, MD; the U. S. Subtropical Horticulture Research Unit, Miami, FL; Kew Botanical Gardens, United Kingdom; Royal Imperial Institute, Wageningen, The Netherlands; and IRAT at Nogent sur Marne, France. The United States serves as a third country for the international exchange of coffee, tea, rubber, and cacao.[491]

4. Prohibition

Prohibition of the host is the most drastic action that can be taken by the quarantine service of the importing country. A regulation can specifically prohibit the host, but in actual practice several other regulatory actions can be taken that effectively result in prohibition, even though the item may not be specifically prohibited, on the basis of reviewing quarantine regulations of 125 countries found in the following categories of seed and other plant material denied for entry.

1. Plant genera or species are prohibited if they are known to be (a) hosts of pests or pathogens of quarantine significance, (b) alternate hosts of rust fungi, (c) parasitic plants or weeds, or (d) a cultivar that is genetically inferior to local ones.
2. If plant genera or species are otherwise enterable but arrive at a port of entry and are found to be (a) contaminated, infected, or with pathogens of quarantine significance against which there is no known practical and effective eradicant treatment, (b) mixed with or contaminated with prohibited species, or (c) contaminated with prohibited articles such as straw, soil, or bark, they are prohibited.
3. If plant genera or species are otherwise enterable, but the shipment arrives, (a) without a permit or phytosanitary certificate or both, (b) without the required declarations on the phytosanitary certificate, or (c) with a phytosanitary certificate that has been erased or altered, they are prohibited.
4. Plant genera or species are prohibited if it is determined that the seeds are not enterable due to pathogen risk factor or extenuating circumstances.

H. Guidelines for Import of Germ Plasm[537,538]

1. Import from a country where the pathogen(s) is absent.
2. Import from a country with an efficient plant quarantine service, so that inspection and treatment is done.
3. Obtain planting material from the safest known source within the selected country.
4. Obtain nontreated seeds so that detection of seedborne pathogens is facilitated.
5. Obtain clean, healthy-looking seeds free of any type of impurities.
6. Obtain an official certificate of freedom from pests and diseases from the exporting country.
7. Import the smallest possible amount of planting material; the smaller the amount, the less the chance of its carrying infection. It will also simplify postentry inspection.
8. Inspect material carefully on arrival and treat.

9. If other precautions are not adequate, subject the material to intermediate or postentry quarantine.
10. Salvage infected seeds (*P. calcitrapae* var. *centaureae*-infested safflower seeds were salvaged by stirring the seeds in ethyl alcohol with river sand on a test tube shaker for 30 sec).[543]

VI. DISEASE RESISTANCE

The use of disease-resistant or tolerant cultivars is the most economical and efficient way of controlling diseases. However, resistance to a pathogen may not be available in all crops. Furthermore, some pathogens, such as *Cercospora sojina*, *Heterodera glycines,* and *Peronospora manshurica* of soybean, exist in several races and others, such as SbMV, have mild and severe strains that are expressed or repressed by temperature differences. Many are latent.[67] If resistance is not available, cultivars that escape infection should be considered. The sources of resistance to soybean diseases have been summarized for selected fungal, bacterial, viral, and nematode diseases, and this information may be used as guide for selecting adapted plant material.[544]

The cultivation of a cultivar may be replaced by depending on the degree of resistance in each. Two races, T and O, of *Bipolaris maydis* cause southern blight of maize. Race T spread widely in the United States in 1970 and to a lesser extent in 1971. It produces a pathotoxin specific to cms-T cytoplasm of maize plants and infects the leaf, leaf sheath, husk, ear parts, and kernels. Race O, normally confined to the specific pathotoxin, primarily infects leaves. Because of the potential dangers of *B. maydis* race T, several countries have imposed legal restrictions on the importation or the planting of seeds having cms-T-cytoplasm (cytoplasm for male sterility, Texas or T). The obvious and most practical control of *B. maydis* race T is to produce the high-yielding hybrids without cms-T cytoplasm. The American seed industry and its counterpart in several other areas of the world have shifted to normal cytoplasm and to detasseling.[545]

In France, resistance is used to control Cercospora leaf spot in sugar beet seed production, Verticillium wilt in alfalfa, and downy mildew in sunflower. In vegetables, resistance is used to control diseases in bean, cabbage, cucumber, lettuce, melon, pea, pepper, spinach, and tomato caused by bacteria, fungi, nematode, or virus.[546] Black-seeded cultivars of *Phaseolus vulgaris* are resistant and white-seeded cultivars are susceptible to *Rhizoctonia solani* seed infection. Extracts of black-seeded cultivars contain phenolic compounds that inhibit growth of *R. solani*.[547]

Tissue culture technique can be used for production of pathogen-free seeds. It is possible to culture soybean seedlings from the embryonic axis of a seed and grow them to maturity. These plants will produce pathogen-free seeds.[548]

Most sorghum cultivars are susceptible to seedborne fungal pathogens (*C. lunata*, *F. moniliforme*, and *P. sorghina*). However, there is a difference in susceptibility of different cultivars. Therefore, it is desirable to use a cultivar having resistance or tolerance to one or two pathogens. Such cultivars will be useful in different sorghum-growing countries or regions having problems with only one or two of the pathogens.[549] Evaluation of sorghum genotypes for resistance based only on field scores may not always be reliable. Resistance should be confirmed by laboratory scoring and ergosterol estimation.[550-552] Ergosterol concentration is a highly sensitive indicator of total fungal biomass, providing an accurate method for confirming resistance to the seedborne fungi in sorghum. There is an increase in ergosterol concentration during the developmental stage in fungal-susceptible seed accessions compared with the resistant ones. A concentration of 30 $\mu g\ g^{-1}$ or below in mature sorghum seeds is an indicator of the fungal-resistant seeds.[551] Association of both flavan-4-ols (polyphenolic compounds found in grain sorghum) and seed hardness imparts resistance to seed molds. The concentration of flavan-4-ols in resistant seeds was over twice as high as that in fungal-susceptible ones 30 d or more after flowering. The presence of flavan-4-ols was associated with grain sorghum with colored seed coats. Ergosterol concentration in fungal-susceptible, mature grains was ten times higher than that in fungal-resistant ones. Ergosterol concentration can be used to assess the magnitude of fungal damage in sorghum seeds using a high performance liquid chromatography.[552]

A significant negative correlation was found between the amount of *F. graminearum* ear rot in the field and the quantity of phenolic acid, especially ferulic acid, detected in maize seeds. Therefore, breeding programs aimed at attaining resistance to *F. graminearum* should incorporate genotypes containing high concentrations of ferulic acid in the seeds.[553]

The shrunken-2 (Sh 2) endosperm mutation for high levels of sugar in maize seeds is used widely in the sweet corn industry. Cultivation of Sh2 maize has been plagued by poor emergence and seedling vigor. The sucrose content accumulated at the expense of starch is one of several factors associated with the poor vigor of Sh 2 hybrids. Another factor is infection by fungal pathogens, especially *F. moniliforme*.[554] Quantity and concentration of carbohydrates were not related to infection by *F. moniliforme*. A symptomatic infection by *F. moniliforme* was less for inbreds with silks that are green and actively growing at inoculation than for inbreds with green-brown or brown silks.[555] Resistance of cv. Hanayoma soybean seeds was due to pod tissue resistance to *C. kikuchii*.

Resistance prevented hyphal penetration into seeds because of hard seed coats in wild soybeans.[556] Resistance in wild soybeans was controlled by two dominant genes having different degrees of resistance. A cultivar highly resistant to *C. kikuchii* could be developed from interspecific crosses between soybean and *Glycine soja*.[557] The resistance of certain soybean cultivars to *Phomopsis phaseoli* was due to impermeable seeds coats.[558] There was a direct relationship between resistance of bean to the halo blight organism and seed contamination by *P. s.* pv. *phaseolicola*.[559]

Hard sorghum seeds showed less mold *(F. moniliforme)* than soft seeds during development. More intense deposition of protein bodies were found in hard rather than in soft seeds, and pitted starch granules were clearly visible in soft seeds. The endosperm of hard seeds contained more protein and prolamine than that of soft seeds.[560] Relatively thin pericarp layer of the susceptible hybrid maize allow access of *F. moniliforme* into the seeds, especially through insect wounds.[561]

The cultivar susceptibility of peas to pea seedborne mosaic virus (PSbMV) is more important than self-limitation in affecting transmission and amplification of seed infection. The PSbMV is not likely to have a significant effect on field pea production in Manitoba if cultivars such as Bellevue, Century, Titan, Topper, or Trapper are grown, but if cultivars such as Express and Fortune are grown continously, significant yield reduction could result.[562] The resistance genes in soybeans can be transferred using a backcross procedure against Phomopsis seed decay.[563] A pigmented testa of sorghum is the most important trait conferring seed fungal resistance (*F. moniliforme* and *Curvularia lunata*). A red pericarp also confers resistance, although not as much. The effect of red pericarp is enhanced by the intensifier gene. The effect of a pigmented testa and red pericarp are additive.[564] Seed discoloration of lupine *(Lupinus albus)* may be used to screen white lupin germ plasm for resistance to *Phoma* and *Pleiochaeta setosa* in environments where conditions do not favor colonization of pods and seeds by saprophytes that may discolor seeds. These pathogens appear to induce more seed discoloration.[565]

Fungal-resistant sorghum cultivars exhibit significantly more hardness than fungal-susceptible ones. Seed hardness values are significantly and negatively related to ergosterol concentration and therefore, can be used as an indicator of seed fungal resistance.[566]

REFERENCES

1. Agarwal, V. K., Quality seed production at Pantnagar, *Seed Sci. Technol.*, 11, 1071, 1983.
2. Gabrielson, R. L., Black leg disease of Crucifers caused by *Leptosphaeria maculans (Phoma lingam)* and its control, *Seed Sci. Technol.*, 11, 749, 1983.
3. Liew, R. S. S. and Gaunt, R. E., Disease problems in the production of broadbean seed, 34th N. Z. Weed Pest Control Conf., 1981, 55.
4. Walker, J. C., Seed treatment and rainfall in relation to the control of cabbage black-leg, *U.S. Dep. Agric. Bull.*, 1029, 26, 1922.
5. Hewett, P. D., Regulating seed-borne disease by certification, in *Plant Health, the Scientific Basis for Administrative Control of Plant Diseases and Pests,* Ebbels, D. L. and King, J. E., Eds., Blackwell Scientific, Oxford, 1979, 163.
6. Gabrielson, R. L., Disease problems in cabbage seed crops, *Iowa Seed Sci.*, 2, 12, 1980.
7. Baker, R. F., Seed pathology, in *Seed Biology*, Vol. 2, Kozlowski, T. T., Ed., Academic Press, New York, 1972, 317.
8. Webster, D. M., Atkin, J. D., and Cross, J. E., Bacterial blights of snap beans and their control, *Plant Dis.*, 67, 935, 1983.
9. Schaad, N. W., Initial identification, in *Laboratory Guide for Identification of Plant Pathogenic Bacteria,* Schaad, N. W., Ed., APS Press, St. Paul, MN, 1988, 164.
10. Dickens, J. S. W. and Pemberton, A. W., Quarantine measures for seed in the United Kingdom, *Seed Sci Technol.*, 11, 1175, 1983.
11. Jordan, E. G., Manandhar, J. B., Thapliyal, P. N., and Sinclair, J. B., Factors affecting soybean seed quality in Illinois, *Plant Dis.*, 70, 246, 1986.

12. Tusa, C., Miclaus, D., Damian, V., Constantin, L., and Paulian, F., Capacity for spread of *Ustilago nuda* (Jens.) Rostr. on some winter barley cultivars, *Prob. Prot. Plant.*, 16, 175, 1988.
13. Guenin, M. C. de., Mildew on sunflower: a newly reappeared disease, *Phytoma*, 419, 26, 1990.
14. Roncadori, R. W., Brooks, O. L., and Perry, C. E., Effect of field exposure on fungal invasion and deterioration of cotton seed, *Phytopathology*, 62, 1137, 1972.
15. Middleton, J. T. and Snyder, W. C., The production of *Ascochyta*-free pea seed in southern California, *Phytopathology*, 37, 363, 1947.
16. Leach, L. D. and MacDonald, J. D., Seed-borne *Phoma betae* as influenced by area of sugarbeet production, seed processing and fungicidal seed treatments, *J. Am. Soc. Sugar Beet Technol.*, 19, 4, 1976.
17. Baker, K. F. and Davis, L. H., Some diseases of ornamental plants in California caused by species of *Alternaria* and *Stemphylium*, *Plant Dis. Rep.*, 34, 403, 1950.
18. Butcher, C. L., Dean, L. L., and Laferriere, L., Control of halo blight of beans in Idaho, *Plant Dis. Rep.*, 52, 295, 1968.
19. Grogan, R. G. and Kimble, K. A., The role of seed contamination in the transmission of *Pseudomonas phaseolicola* in *Phaseolus vulgaris*, *Phytopathology*, 57, 28, 1967.
20. Stubbs, L. L. and O'Loughlin, G. T., Climatic elimination of mosaic spread in lettuce seed crops in the Swan Hill region of the Murray Valley, *Aust. J. Exp. Agric. Anim. Husb.*, 2, 16, 1962.
21. Kuhn, C. W. and Demski, J. W., The relationship of peanut mottle virus to peanut production, Research Report No. 213, Department of Plant Pathology, University of Georgia, Athens, 1975.
22. Zettler, F. W., Elliott, M. S., Purciful, D. E., Mink, G. I., Gorbet, D. W., and Knauft, D. A., Production of peanut seed free of peanut stripe and peanut mottle viruses in Florida, *Plant Dis.*, 77, 747, 1993.
23. McGee, D. C., Seed pathology: its place in modern seed production, *Plant Dis.*, 65, 638, 1981.
24. Rusch, R., Investigations into the overwintering of loose smut of oats (*Ustilago avenae* (Pers.) Jens.) and the smut reducing influence of low seed-bed temperature, *Angew. Bot.*, 31, 221, 1957.
25. Sinclair, J. B. and Backman, P. A., Eds., *Compendium of Soybean Diseases*, 3rd ed., APS Press, St. Paul, MN, 1989, 106.
26. Agarwal, V. K., Singh, O. V., and Modgal, S. C., Influence of different doses of nitrogen and spacing on the seedborne infections of rice, *Indian Phytopathol.*, 28, 38, 1975.
27. Goldin, M. I. and Yurchenko, M. A., Method for the control of mosaic and streak in tomatoes, *Zashch. Rast.(Moscow)*, 6, 36, 1958.
28. Jones, G. H. and Seif-el-nasr, A. E. G., The influence of sowing depth and moisture on smut diseases, and prospects of a new method of control, *Ann. Appl. Biol.*, 27, 35, 1940.
29. Cralley, E. M., The effect of seeding methods on the severity of white tip of rice, *Phytopathology*, 47, 7, 1957.
30. Bisht, V. S., Sinclair, J. B., Hummel, J. W., and McClary, R. D., Effect of tillage systems on yield components and diseases of soybeans, *Phytopathology*, 72, 1134, 1982.
31. Clark, F. S., The development of an isolated area for the production of smut-free barley seed, *Agric. Inst. Rev. (Canada)*, 7, 37, 1952.

32. Kublan, A., Barley and wheat loose smut and its control, *Dtsch. Landwirtsch.*, 3, 353, 1952.
33. Oort, A. J. P., De verspreiding van de sporen vom taewestuifbrand *(Ustilago tritici)* door de lucht, *Tijdschr. Plantenziekten,* 46, 1, 1940.
34. Sinclair, J. B., Fungicide sprays for control of internally seedborne fungi, *Seed Sci. Technol.*, 11, 959, 1983.
35. Wimalajeewa, D. L. S. and Young, K. J., Studies on the levels of common and halo blight seed infection occurring in the field, *Aust. Plant Pathol.*, 8, 29, 1979.
36. Kharbanda, P. D. and Bernier, C. C., Effectiveness of seed and foliar application of fungicides to control Ascochyta blight of faba beans, *Can. J. Plant. Sci.*, 59, 661, 1979.
37. Humpherson-Jones, F. M. and Maude, R. B., Control of dark leaf spot *(Alternaria brassicicola)* of *Brassica oleracea* seed production crops with foliar sprays of iprodione, *Ann. Appl. Biol.*, 100, 99, 1982.
38. Babadoost, M., Gabrielson, R. L., Olson, S. A., and Mulanax, M. W., Control of *Alternaria* diseases of Brassica seed crops caused by *Alternaria brassicae* and *Alternaria brassicicola* with ground and aerial fungicide application, *Seed Sci. Technol.*, 21, 1, 1993.
39. Tripathi, H. S., Sangam, Lal, and Agarwal, V. K., Influence of fungicidal sprays on per cent seedborne incidence of *Fusarium moniliforme* and *Curvularia pallescens* in maize, *Pantnagar J. Res.*, 2, 104, 1977.
40. Vidhyasekaran, P. and Kandaswamy, T. K., Control of seed-borne pathogens in okra by preharvest sprays, *Indian Phytopathol.*, 33, 239, 1980.
41. Rajagopal, R. and Vidhyasekaran, P., Effect of fungicidal sprays on the quality of groundnut kernel and its oil content, *Indian Phytopathol.*, 36, 52, 1983.
42. Ellis, M. A. and Paschal, E. H., Effect of fungicide seed treatment on internally seed-borne fungi, germination and field emergence of pigeon pea *(Cajanus cajan)*, *Seed Sci. Technol.*, 7, 75, 1979.
43. Singh, O. V., Agarwal, V. K., and Singh, R. A., Effect of fungicidal sprays on the quantum of seedborne infection of rice, *Oryza*, 9, 103, 1972.
44. Ferrer, A., Peart, W., and Rivera, M., Control of pathogenic fungi transmitted by rice seed, *Cienc. Agropecuaria*, 3, 113, 1980.
45. Hepperly, P. R., Feliciano, C., and Sotomayor-Rios, A., Chemical control of seedborne fungi of sorghum and their association with seed quality and germination in Puerto Rico, *Plant Dis.*, 66, 902, 1982.
46. Anahosur, K. H., Chemical control of ergot of sorghum, *Indian Phytopathol.*, 32, 487, 1979.
47. Singh, D. P. and Agarwal, V. K., Effect of fungicidal sprays on grain mould incidence and seed quality in sorghum, *Bangladesh J. Bot.*, 18, 45, 1989.
48. Sinclair, J. B., Fungicide sprays for the control of seed-borne pathogens of rice, soybeans and wheat, *Seed Sci. Technol.*, 9, 697, 1981.
49. Prasartsee, C., Tenne, F. D., Ilyas, M. B., Ellis, M. A., and Sinclair, J. B., Reduction of internally seed-borne *Diaporthe phaseolorum* var. *sojae* by fungicide sprays, *Plant Dis. Rep.*, 59, 20, 1974.
50. Ellis, M. A., Foor, S. R., and Sinclair, J. B., Effect of benomyl sprays on internally-borne fungi, germination and emergence of delay harvested soybean seeds, *Phytopathol. Z.*, 85, 159, 1976.
51. Ellis, M. A., Ilyas, M. B., and Sinclair, J. B., Effect of three fungicides on internally seed-borne fungi and germination of soybean seeds, *Phytopathology,* 65, 553, 1975.

52. Ellis, M. A., Ilyas, M. B., Tenne, F. D., Sinclair, J. B., and Palm, H. L., Effect of foliar applications of benomyl on internally seed-borne fungi and pod and stem blight in soybean, *Plant Dis. Rep.*, 58, 760, 1974.
53. Ellis, M. A. and Sinclair, J. B., Effect of benomyl field sprays on internally-borne fungi, germination, and emergence of late-harvested soybean seeds, *Phytopathology*, 66, 680, 1976.
54. Bolkan, H. A. and Cupertino, F. P., Control of seedborne *Phomopsis sojae* with foliar applications of fungicides, 1976, *Fungicide Nematicide Tests*, 32, 121, 1977.
55. Tenne, F. D. and Sinclair, J. B., Control of internally seed-borne microorganisms of soybean with foliar fungicides in Puerto Rico, *Plant Dis. Rep.*, 62, 459, 1978.
56. Miller, W. A., and Roy, K. W., Effects of benomyl on the colonization of soybean leaves, pods and seeds by fungi, *Plant Dis.*, 66, 918, 1982.
57. Sinclair, J. B., Control of seedborne pathogens and diseases of soybean seeds and seedlings, *Pest. Sci.*, 37, 15, 1993.
58. Sakai, Y. and Ogawa, M., Chemical control of soybean seed disease caused by *Cercospora kikuchii* Matsumoto et Tomoyasu, *Bull. Hiroshima Prefect. Agric. Exp. Stn.*, 46, 33, 1983.
59. Tekrony, D. M., Egli, D. B., Stuckey, R. E., and Loeffler, T. M., Effect of benomyl applications on soybean seedborne fungi, seed germination and yield, *Plant Dis.*, 69, 763, 1985.
60. Kmetz, K. T., Schmitthenner, A. F., and Ellett, C. W., Soybean seed decay: prevalence of infection and symptom expression caused by *Phomopsis* sp., *Diaporthe phaseolorum* var. *sojae* and *D. phaseolorum* var. *caulivora*, *Phytopathology*, 68, 838, 1978.
61. McGee, D. C. and Brandt, C. L., Effect of foliar application of benomyl on infection of soybean seeds by *Phomopsis* in relation to time of inoculation, *Plant Dis. Rep.*, 63, 675, 1979.
62. Foor, S. R. and Sinclair, J. B., Effects of fungicide sprays on soybean maturity, yield and seed quality, *Fungicide Nematicide Tests*, 32, 122, 1977.
63. Agarwal, V. K., Singh, O. V., Thapliyal, P. N., and Malhotra, R. K., Control of purple stain disease of soybean, *Indian J. Mycol. Plant Pathol.*, 4, 1, 1974.
64. Crittenden, H. W. and Bloss, H. W., Control of *Cercospora kikuchii* and *Diaporthe phaseolorum* var. *sojae* on soybean seed, *Phytopathology*, 50, 570, 1960.
65. Kilpatrick, R. A., Fungi associated with the flowers, pods, and seeds of soybeans, *Phytopathology*, 47, 131, 1957.
66. Grahame, R. E., Personal communication, UniRoyal Chemical, Nautauck, NJ, 1981.
67. Jacobsen, B. J., Personal communication, Department of Plant Pathology, University of Illinois at Urbana-Champaign, 1984.
68. Sinclair, J. B., Latent infection of soybean plants and seeds by fungi, *Plant Dis.*, 75, 220, 1991.
69. Cook, R. J., The effect of timed fungicide sprays on yields of winter wheat in relation to *Septoria* infection period, *Plant Pathol.*, 26, 30, 1977.
70. Jacobsen, B. J., Effect of fungicides on Septoria leaf and glume blotch, *Fusarium* scab, grain yield and test weight of winter wheat, *Phytopathology*, 67, 1412, 1977.
71. Schultz, T. R., Johnston, W. J., Golob, C. T., and Maguire, J. D., Control of ergot in Kentucky bluegrass seed production using fungicides, *Plant Dis.*, 77, 685, 1993.
72. Dhingra, O. D. and da Silva, J. F., Effect of weed control on the internally seedborne fungi in soybean seeds, *Plant Dis. Rep.*, 62, 513, 1978.

73. Hepperly, P. R., Kirkpatrick, B. L., and Sinclair, J. B., *Abutilon theophrasti*: wild host for three fungal parasites of soybean, *Phytopathology,* 70, 307, 1980.
74. Cerkauskas, R. F., Dhingra, O. D., Sinclair, J. B., and Asmus, G., *Amaranthus spinosus, Leonotis nepetaefolia* and *Leonurus sibiricus* new host of *Phomopsis* spp. in Brazil, *Plant Dis.,* 67, 821, 1983.
75. Schnathorst, W. C., Eradication of *Xanthomonas malvacearum* from California through sanitation, *Plant Dis. Rep.,* 50, 168, 1966.
76. Baker, K. F. and Cook, R. J., *Biological Control of Plant Pathogens,* W. H. Freeman, San Francisco, 1974, 433.
77. Weindling, K., Studies on a lethal principle effective in the parasitic action of *Trichoderma lignorum* on *Rhizoctonia solani* and other soil fungi, *Phytopathology,* 24, 1153, 1934.
78. Windels, C. E. and Kommedahl, T., Pea cultivar effect on seed treatment with *Penicillium oxalicum* in the field, *Phytopathology,* 72, 541, 1982.
79. Windels, C. E., Growth of *Penicillium oxalicum,* a biological seed treatment on pea seeds and roots in soil, *Phytopathology,* 71, 265, 1981.
80. Harman, G. E., Chet, I., and Baker, R., Factors affecting *Trichoderma hamatum* applied to seed as biological control, *Phytopathology,* 71, 569, 1981.
81. Mew, I. C. and Kommedahl, T., Biological control of seedling blight of corn by coating kernels with antagonistic microorganisms, *Phytopathology,* 58, 1395, 1968.
82. Mew, I. C. and Kommedahl, T., Interaction among microorganisms occurring naturally and applied to pericarps of corn kernels, *Plant Dis. Rep.,* 56, 861, 1972.
83. Tveit, M. and Moore, M. B., Isolates of *Chaetomium* that protect oats from *Helminthosporium victoriae, Phytopathology,* 44, 686, 1954.
84. Tveit, M. and Wood, R. K. S., The control of Fusarium blight in oat seedlings with antagonistic species of *Chaetomium, Ann. Appl. Biol.,* 43, 538, 1955.
85. Wiley, H. B. and Kommedahl, T., Biological seed treatment in sweet corn and wheat as a component of crop management, *Phytopathology,* 71, 265, 1981.
86. Marshall, D. S., Effect of *Trichoderma harzianum* seed treatment and *Rhizoctonia solani* inoculum concentration on damping-off in snapbean in acidic soils, *Plant Dis.,* 66, 788, 1982.
87. McManus, P. S., Ravenscroft, A. V., and Fulbright, D. W., Inhibition of *Tilletia laevis* teliospore germination and suppression of common bunt of wheat by *Pseudomonas fluorescens, Plant Dis.,* 77, 1012, 1993
88. Randhawa, H. S. and Aulakh, K. S., Reduction of seedborne fungi of raya (*Brassica juncea*) by *Chaetomium globosum* and *Epicoccum purpurescens, Indian Phytopathol.,* 37, 140, 1984.
89. Tahvonen, R., Mycostop, biological formulation for control of fungal diseases, *Vaxtskyddsnotiser,* 49, 86, 1985.
90. Simay, E. I., *In vivo* occurrence of hyperparasitism of *Botrytis cinerea* Pers. by *Gliocladium catenulatum* Gilman et Abbott, *Acta Phytopathol. Entomol. Hungarica,* 23, 133, 1988.
91. Al-Hashimi, M. H. and Perry, D. A., Fungal antagonists of *Gerlachia nivalis, J. Phytopathol.,* 116, 106, 1986.
92. Schaad, N. W. and Donaldson, R. C., Comparison of two methods for detection of *Xanthomonas campestris* in infected crucifer seeds, *Seed Sci. Technol.,* 8, 383, 1980.
93. Randhawa, P. S., Singh, N. J., and Schaad, N. W., Bacterial flora of cotton seeds and biocontrol of seedling blight caused by *Xanthomonas campestris* pv. *malvacearum, Seed Sci. Technol.,* 15, 65, 1987.

94. Leben, C., Bacterial blight of soybean: seedling disease control, *Phytopathology*, 65, 844, 1975.
95. Schaad, N. W. and Donaldson, R. C., Bacteria of crucifer seeds antagonistic to *Xanthomonas campestris, Phytopathology,* 71, 902, 1981.
96. Shekhawat, P. S. and Chakravarti, B. P., Comparison of agar plate and cotyledon methods for the detection of *Xanthomonas vesicatoria* in chilli seeds, *Phytopathol. Z.*, 94, 80, 1979.
97. Vajavat, R. M. and Chakravarti, B. P., Survival of *Pseudomonas sesami* and effect of an antagonistic bacterium isolated from seeds on the control of the disease in the field, *Indian Phytopathol.*, 31, 286, 1978.
98. Jindal, K. K. and Thind, B. S., Microflora of cowpea seeds and its significance in the biological control of seedborne infection of *Xanthomonas campestris* pv. *vignicola, Seed Sci. Technol.*, 18, 393, 1990.
99. Novogrudskii, D., Beresova, E., Nachimovskaya, M., and Perviakova, M., The influence of bacterialization of flax seed on the susceptibility of seedlings to infection with parasitic fungi, *C. R. Acad. Sci. U.S.S.R., N.S.*, 14, 385, 1937.
100. Price, R. D., Merriman, P. R., and Kollmorgan, J. F., The effect of seed applications of selected soil organisms on the growth and yield of cereals and carrots, 2nd Int. Congr. Plant Pathol., St. Paul, MN, 1973, 666.
101. Kommedahl, T. and Mew, I. C., Biocontrol of corn root infection in the field by seed treatment with antagonists, *Phytopathology,* 65, 296, 1975.
102. Henry, A. W. and Campbell, J. A., Inactivation of seed-borne plant pathogens in the soil, *Can. J. Res. Sect. C.*, 16, 331, 1938.
103. Sohi, H. S., Aulakh, K. S., and Randhawa, H. S., Control of seedborne fungi of cotton by *Chaetomium globosum* and *Epicoccum purpurescens*, *Indian J. Econ.*, 15, 111, 1988.
104. Sivapalan, A., Fungi associated with broccoli seed and evaluation of fungal antagonists and fungicides for the control of seed-borne *Alternaria brassicicola*, *Seed Sci. Technol.*, 21, 237, 1993.
105. Hentschel, K. D., Biocontrol of seed-borne *Alternaria radicina* on carrots by antagonistic *Bacillus subtilis*, *Bull. SROP*, 14, 73, 1991.
106. Mercer, P.C. and Papadopolous, S., Biological control of seedborne diseases of linseed, in Proc. Meeting on Biological Control of Pests and Diseases, McCracken, A. R. and Mercer, P. C., Eds., Agriculture and Food Science Center, Belfast, 1990.
107. Thomas, R. C., A bacteriophage in relation to Stewart's disease of corn, *Phytopathology,* 25, 371, 1935.
108. Thomas, R. C., Additional facts regarding bacteriophage lytic to *Aplanobacter stewarti, Phytopathology,* 30, 602, 1940.
109. Mahaffee, W. F. and Backman, P. A., Effects of seed factors on spermosphere and rhizosphere colonization of cotton by *Bacillus subtilis* GB03, *Phytopathology,* 83, 1120, 1993.
110. Van Winckel, A., Epidemiology of tobacco mosaic virus in tomato seed, *Agricultura (Louvain),* 13, 721, 1965.
111. Taylor, A. G. and Harman, G. E., Concepts and technologies of selected seed treatments, *Annu. Rev. Phytopathol.*, 28, 321, 1990.
112. Kommedahl, T. and Windels, C. E., Evaluation of biological seed treatment for controlling rot diseases of pea, *Phytopathology,* 68, 1087, 1978.
113. Leben, C., Bacterial blight of soybean: seedling disease control, *Phytopathology,* 65, 844, 1975.

114. Merriman, P. R., Price, R. D., and Baker, K. F., The effect of inoculation of seed with antagonists of *Rhizoctonia solani* in the growth of wheat, *Aust. J. Agric. Res.*, 25, 213, 1974.
115. Merriman, P. R., Price, R. D., Kollmorgen, J. F., Piggott, T., and Ridge, E. H., Effect of seed inoculation with *Bacillus subtilis* and *Streptomyces griseus* on the growth of cereals and carrots, *Aust. J. Agric. Res.*, 25, 219, 1974.
116. Windels, C. E. and Kommedahl, T., Factors affecting *Penicillium oxalicum* as a seed protectant against seeding blight of pea, *Phytopathology*, 68, 1656, 1978.
117. Nene, Y. L. and Thapliyal, P. N., *Fungicides in Plant Disease Control*, Oxford and IBH, New Delhi, 1993, 621.
118. Sharvelle, E. G., *Plant Disease Control*, AVI Publishing, Westport, CT, 1979, 331.
119. Rennie, W. J., The need for cereal seed treatment in the U.K. in the post-mercury era, *Pesticide Outlook*, 4, 19, 1993.
120. Walker, J. C., *Plant Pathology*, McGraw-Hill, New York, 1969, 819.
121. Agarwal, V. K. and Nene, Y. L., Seedborne Diseases of Field Crops and Their Control, Indian Council of Agricultural Research, New Delhi, 1987, 69.
122. Jeffs, K. A. and Tuppen, R. J., Application of pesticides to seeds. Requirements for efficient treatment of seeds, in *Seed Treatment*, Jeffs, K. A., Ed., Thornton Health, Brit. Crop. Prot. Counc., Surrey, U.K., 1986, 17.
123. Keyworth, W. G. and Howell, J. S., Studies on silvering disease of redbeet, *Ann. Appl. Biol.*, 49, 173, 1961.
124. Taylor, J. D., Streptomycin seed treatment for peas and beans, Report of the National Vegetable Research Station, Warwick, N.Z., 1972.
125. Klisiewicz, J. M. and Pound, G. S., Studies on control of black rot of crucifers with antibiotics, *Phytopathology*, 50, 642, 1960.
126. Dhanvantari, B. N., Effect of seed extraction methods and seed treatment on control of tomato bacterial canker, *Can. J. Plant Pathol.*, 11, 400, 1989.
127. Kang, C. S., Heo, N. Y., and Heo, C., Trial on the control of seedborne rice diseases using seed disinfectants, *Ann. Agric. Chem. Res. Inst.*, 111, 113, 1986.
128. Maude, R. B. and Bambridge, J. M., Effects of seed treatments and storage on the incidence of *Phoma betae* and the viability of infected redbeet seeds, *Plant Pathol.*, 34, 435, 1985.
129. Smilanick, J. L., and Goates, B. J., Germinability of *Tilletia* spp. teliospores after hydrogen peroxide treatment, *Plant Dis.*, 78, 861, 1994.
130. Strandberg, J. O., Efficacy of fungicides against persistence of *Alternaria dauci* on carrot seed, *Plant Dis.*, 68, 39, 1984.
131. Huang, T. C. and Lee, H. L., Hot acidified zinc sulfate as seed soaking agent for the control of crucifer black rot, *Plant Prot. Bull.*, Taiwan, 30, 245, 1988.
132. Kim, B. S., Testing for detection of *Xanthomonas campestris* pv. *campestris* in crucifer seeds and seed disinfection, *Korean J. Plant Pathol.*, 2, 96, 1986.
133. Forster, R. L. and Schaad, N. W., Control of black chaff of wheat with seed treatment and a foundation seed health program, *Plant Dis.*, 72, 935, 1988.
134. Guo, Y. F., Liang, Z. O., and Huang, H., Elimination of bacterial wilt from imported corn seeds with agricultural antibiotics, *Chinese J. Biol. Control.*, 7, 30, 1991.
135 Zachowski, M. A. and Rudolph, K., Reduction of bacterial blight infestation of cotton seeds by treatment with sodium hypochlorite, *J. Phytopathol.*, 131, 53, 1991.
136. Tamietti, G., Evaluation of different seed treatments in controlling the halo blight of bean, *Informatore Fitopatol.*, 6, 47, 1982.

137. Halmer, P., Technical and commercial aspects of seed pelleting and film coating, in *Application to Seeds and Soil,* Martin, T. J., Ed., Thornton Health, Brit. Crop Prot. Counc., Surrey., U. K., 191, 1988.
138. Kitamura, S., Wantanabe, M., and Nakazama, M., Process for producing coated seed. U.S. Patent 4,250,1991.
139. Walker, J. C., Onion disease and their control, *U.S. Dep. Agric. Farmers Bull.*, 1060, 1947.
140. Hocking, D. and Jaffar, A. A., Damping-off in pine nurseries: fungicidal control by seed pelleting, *Emp. For. Rev.*, 48, 355, 1969.
141. Byford, W. J., The incidence of sugarbeet seedling diseases and effects of seed treatment in England, *Plant Pathol.*, 21, 16, 1972.
142. Becker, J. and Weltzien, H. C., Control of common bunt of wheat (*Tilletia caries* (D.C.). Tul & C. Tul.) with organic nutrients, *Zeitsch Pflanzenkrank. Pflansensch.*, 100, 49, 1993.
143. Schlub, R. L. and Schmitthenner, A. F., Disinfecting soybean seeds by fumigation, *Plant Dis. Rep.*, 61, 470, 1977.
144. Ralph, W., The potential of ethylene oxide in the production of pathogen-free seed, *Seed Sci. Technol.*, 5, 567, 1977.
145. Smilanick, J. L., Hartsell, P. L., Denis-Arrue, R., Hensen, D. J., McKinney, J. D., Tebbets, J. C., and Goates, B. J., Survival of common and dwarf bunt teliospores and intact sori after fumigation of high and low moisture content winter wheat, *Plant Dis.*, 76, 293, 1992.
146. Caubel, G., Ducom, P., and Marre, R., Methyl bromide fumigation against *Ditylenchus dipsaci* in seed or bulb lots, *EPPO Bull.*, 15, 17, 1985.
147. Baccidel Bene, G. and Cancellara, I., Prove preliminari di fumigazion can bromuro dimetile di semi di fava infestatite da *Ditylenchus dipsaci*, *Atti Giornate Fitopatol.*, 111, 1973.
148. Caubel, G., Epidemiology and control of seedborne nematodes, *Seed Sci. Technol.*, 11, 989, 1983.
149. Minton, N. A. and Gillenwater, H. B., Methyl bromide fumigation of *Pratylenchus brachyurus* in peanut shells, *J. Nematol.*, 5, 147, 1973.
150. Goodey, T., *Anguillulina dipsaci* on onion seed and its control by fumigation with methyl bromide, *J. Helminthol.*, 21, 45, 1945.
151. Hague, N. G. M., Fumigation of agricultural products. XVIII. Effect of methyl bromide on the bentgrass nematode *Anguina agrostis* (Steinbuch, 1799) Filipjev 1936, and on the germination of bent grass *Agrostis tenuis, J. Sci. Food Agric.*, 14, 577, 1963.
152. Prasad, J. and Varaprasad, K. S., Elimination of white-tip nematode, *Aphelenchoides besseyi*, from rice seed, *Fund. Appl. Nematol.*, 15, 305, 1992.
153. Maude, R. B., Presly, A. H., and Lovett, J. F., Demonstration of the adherence of thiram to pea seeds using a rapid method of spectrophotometric analysis, *Seed Sci. Technol.*, 14, 361, 1986.
154. Clayton, P. B., Presly, A. H., and Rutherford, S. R., *Some Aspects of Film Coating Agrochemical onto Seeds,* Monograph, British Crop Prot. Council, Surrey, U. K., 39, 229, 1988.
155. Bujalski, W., Nienow, A. W., Petch, G. M., and Gray, D., Scale up studies for the osmotic priming and drying of carrot seeds, *J. Agric. Eng. Res.*, 48, 287, 1992.
156. Maude, R. B., Drew, R. L. K., Gray, D., Petch, G. M., Bujalski, W., and Nienow, A. W., Strategies for control of seed-borne *Alternaria dauci* (leaf blight) of carrots in priming and process engineering systems, *Plant Pathol.*, 41, 204, 1992.

157. Maude, R. B. and Bambridge, J. M., Seed treatment control of *Alternaria dauci* (leaf blight) of naturally infected carrot seeds, *Tests Agrochemicals Cultivars*, 12, 30, 1991.
158. Maude, R. B. and Presly, A. H., Neck rot (*Botrytis allii*) of bulb onions. II. Seed-borne infection in relationship to the disease in store and the effect of seed treatment, *Ann. Appl. Biol.*, 86, 181, 1977.
159. Croxall, H. E. and Hickman, C. J., The control of onion smut, *Ann. Appl. Biol.*, 40, 176, 1954.
160. Frank, Z. R., Localisation of seed-borne inocula and combined control of Aspergillus and Rhizopus rot of groundnut seedlings by seed treatment, *Is. J. Agric. Res.*, 19, 109, 1969.
161. Pommer, E. H., The systemic activity of a new fungicide of the furan carbonic acid anilide group (BAS 3191 F), *Proc. 2nd Int. Cong. Pest. Chem.*, 5, 397, 1971.
162. Pathak, K. D., Joshi, L. M., and Renfro, B. L., Control of covered smut of oats by systemic fungicides, *Indian Phytopathol.*, 23, 693, 1970.
163. Jank, B. and Grossman, F., 2-Methyl-5-6 dihydro-4-H pyran-3-carboxylic acid, anilide: a new systemic fungicide against smut diseases, *Pest. Sci.*, 2, 43, 1971.
164. Upadhyay, J. P., Agarwal, V. K., and Mukhopadhyay, A. N., Relative efficacy of fungicidal seed treatment on emergence and seedling blight of sugarbeet, *Seed Res.*, 4, 179, 1976.
165. Sen, C., Srivastava, S. N., and Agnihotri, V. P., Seedling diseases of sugarbeet and their control, *Indian Phytopathol.*, 27, 596, 1974.
166. Gates, L. F. and Hull, R., Experiments on blackleg disease of sugarbeet seedlings, *Ann. Appl. Biol.*, 41, 541, 1954.
167. Maude, R. B. and Humpherson-Jones, F. M., Studies on the seedborne phases of dark leaf spot (*Alternaria brassicicola*) and grey leaf spot (*Alternaria brassicae*) of brassicas, *Ann. Appl. Biol.*, 95, 311, 1980.
168. Davies, J. M. L., Diseases of oilseed crop, in *Oilseed Rape*, Scarisbrick, D. H. and Daniels, R. W., Eds., Williams Collins Sons, London, 1986.
169. Huber, G. A. and Gould, C. J., Cabbage seed treatment, *Phytopathology*, 39, 869, 1949.
170. Shekhawat, P. S., Jain, M. L., and Chakravarti, B. P., Detection and seed transmission of *Xanthomonas campestris* pv. *campestris* causing black rot of cabbage and cauliflower and its control by seed treatment, *Indian Phytopathol.*, 35, 442, 1982.
171. Grover, R. K. and Bansal, R. D., Seed-borne nature of *Colletotrichum capsici* in chilli seeds and its control by seed dressing fungicides, *Indian Phytopathol.*, 23, 664, 1970.
172. Vidhyasekaran, P. and Thiagarajan, C. P., Seed-borne transmission of *Fusarium oxysporum* in chilli, *Indian Phytopathol.*, 34, 211, 1981.
173. Dharam, V. and Grewal, J. S., Efficacy of different fungicides. III. Seed disinfection in relation to damping-off of chillies (*Capsicum annuum* L.), *Indian Phytopathol.*, 14, 10, 1961.
174. Singh, G. and Singh, M., Chemical control of Ascochyta blight of chickpea, *Indian Phytopathol.*, 43, 59, 1990.
175. Kaiser, W. J., Okhovat, M., and Mossahebi, G. H., Effect of seed treatment fungicides on control of *Ascochyta rabiei* in chickpea seed infected with the pathogen, *Plant Dis. Rep.*, 57, 742, 1973.
176. Reddy, M. V., Singh, K. B., and Nene, Y. L., Further studies on Calixin M in the control of seedborne infection of Ascochyta blight in chickpea, *Int. Chickpea Newsl.*, 6, 18, 1982.

177. Lukashevich, A. I., Control measures against ascochytosis of chickpea, *J. Agric. Sci. (Moscow)*, 5, 131, 1958.
178. Zachos, D. G., Panagopulos, C. G., and MaKris, S. A., Researches on the biology, epidemiology and the control of anthracnose of chickpea, *Ann. Inst. Phytopathol. Benaki*, 5, 167, 1963.
179. Khachatryan, M. S., Seed transmission of ascochytosis infection in chickpea and the effectiveness of treatment, *Sb. Nauchn. Tr. Nauchno Issled. Zemledel. Armyarskoi*, 2, 147, 1961.
180. Nene, Y. L., Siddiqui, I. A., and Kharbanda, P. D., Control of stemgall of coriander by fungicides, *Mycopathol. Mycol. Appl.*, 29, 142, 1966.
181. Grewal, J. S. and Dharam, V., Efficacy of different fungicides. VI. Field trials for the control of stem rot of jute, *Indian Phytopathol.*, 16, 99, 1963.
182. Agarwal, V. K. and Singh, O. V., Seed-borne fungi of jute and their control, *Indian Phytopathol.*, 27, 651, 1974.
183. Ferreira, J. F. and Knox-Davies, P. S., Occurrence and control of *Fusarium oxysporum* on sweet melon seed, *Phytophylactica*, 16, 67, 1984.
184. Hildebrand, A. A., Seedborne diseases of soybean and their control, *Proc. Can. Phytopathol. Soc.*, 12, 18, 1944.
185. Nene, Y. L., Agarwal, V. K., and Srivastava, S. S. L., Influence of fungicidal seed treatment on the emergence and nodulation of soybean, *Pesticides*, 3, 26, 1969.
186. Singh, O. V., Agarwal, V. K., and Nene, Y. L., Influence of fungicidal seed treatment on the mycoflora of stored soybean seed and seedling emergence, *Indian J. Agric. Sci.*, 43, 820, 1973.
187. Al-Beldawi, A. S. and Welleed, B. C., Chemical control of *Rhizoctonia solani* Kühn on cotton seedlings, *Phytopathol. Mediterr.*, 12, 87, 1973.
188. Kotasthane, S. R. and Agarwal, S. C., Efficacy of four seed dressing fungicides in controlling black arm of cotton, *PANS*, 16, 334, 1970.
189. Paulus, A. O., Nelson, J., Dewolfe, T., House, J., and Shibuya, F., Non-mercury fungicides for control of seedling disease of cotton, *Calif. Agric.*, 27, 9, 1973.
190. Nikolov, G., Apron 35SD an effective preparation in the control of downy mildew of sunflower, *Rastit. Zash.*, 29, 40, 1981.
191. Tollenaar, H. and Bleiholder, H., Distribution of the mycelium of *Sclerotinia sclerotiorum* in sunflower seed, *Agric. Tec. Mex.*, 31, 44, 1971.
192. Singh, O. V. and Agarwal, V. K., Influence of fungicidal seed treatment on emergence and seedborne mycoflora of sunflower, *Labdev Part B*, 11, 56, 1973.
193. Lee, D. H., Control of seedborne infection of *Ustilago nuda* and *Pyrenophora graminea* on barley, *Korean J. Mycol.*, 8, 89, 1980.
194. Kingsland, G. C., Barley leaf stripe control by Vitavax, *Phytopathology*, 60, 584, 1970.
195. Moseman, J. G., Fungicidal Control of Smut Diseases of Cereals, U.S. Department of Agriculture, Circular No. 42, Washington, D.C., 1968, 42.
196. von Schmeling, B. and Kulka, M., Systemic fungicidal activity of 1,4-oxathiin derivatives, *Science*, 152, 659, 1966.
197. Mathur, A. K. and Bhatnagar, G. C., Field evaluation of different seed dressers in controlling covered smut of barley, *Indian J. Mycol. Plant Pathol.*, 17, 245, 1987.
198. Grewal, J. S. and Dharam, V., Efficacy of different fungicides. VIII. Field trials for the control of covered smut of barley (*Ustilago hordei* (Pers.) Lager.), *Indian Phytopathol.*, 17, 162, 1964.

199. Pommer, E. H. and Kradel, J., 2,5-dimethyl-Furane-3-carboxylic acid anilide (BAS 3191 F) a new active ingredient for the control of seed-borne fungus disease in cereal, in 7th Int. Congr. Plant Prot., Paris, 1970, 409.
200. Darrag, I. E. and Arafa, M. A., Effect of some systemic fungicides used as seed dressing on net blotch disease, grain yield and some agronomic characters of Giza 121 barley cultivar, *Agric. Res. Rev.*, 59, 65, 1981.
201. Morrall, R. A. A., Significance of seedborne inoculum of lentil pathogens in Western Canada, Proc. 1st European Conf. Grain Legumes, France, 1992, 313.
202. Kovacikova, E., Seed treatment of lentil and pea against some fungal diseases, *Ochr. Rost.*, 6, 117, 1970.
203. Mercer, P. C., Linseed diseases, in *Annual Report of Research Technical Work Department of Agriculture of Northern Ireland*, 1984, Belfast, 1985, 201.
204. Agarwal, V. K. and Singh, O. V., Seed-borne fungi of linseed and their response to seed treatment, *Seed Res.*, 3, 26, 1975.
205. Devash, Y., Okon, Y., and Henis, Y., Survival of *Pseudomonas tomato* in soil and seeds, *Phytopathol. Z.*, 99, 175, 1980.
206. Cole, J. S., Some control measures for anthracnose disease of tobacco, *Ann. Appl. Biol.*, 45, 542, 1957.
207. Misra, A. K. and Dharam, V., Efficacy of fungicides. VI. Effect of fungicidal seed treatment against heavy inoculum pressure of certain fungi causing discoloration of paddy seeds, *Indian Phytopathol.*, 43, 175, 1990.
208. Ou, S. H., *Rice Diseases*, CAB *Int. Mycol. Inst.*, Kew, Surrey, U.K., 1985.
209. Reddy, O. R., Evaluation of triforine (Saporal) against seed-borne pathogens of rice (*Oryza sativa*), *Int. J. Trop. Plant Dis.*, 2, 133, 1984.
210. Ranganathiah, K. G. and Gowda, N. N., Seedborne infection of rice by *Drechslera oryzae* and its control in Karnataka, *Pesticides*, 19, 44, 1985.
211. Lakshmanan, P. and Mohan, S., Effect of seed treatment on brown spot disease and their influence on rice seedlings, *Madras Agric. J.*, 75, 57, 1988.
212. Misra, A. P. and Singh, T. B., Effect of some copper and organic fungicide on the viability of paddy seeds, *Indian Phytopathol.*, 22, 264, 1969.
213. Dharam, V., Mathur, S. B., and Neergaard, P., Control of seed-borne infection of *Drechslera* spp. on barley, rice and oats with Dithane M-45, *Indian Phytopathol.*, 23, 570, 1970.
214. Shetty, S. A., Khair, A., Safeeulla, K. M., and Shetty, H. S., Effect of fungitoxicants on seedborne *Trichoconiella padwickii* in paddy, *Indian Phytopathol*, 42, 405, 1989.
215. Zeigler, R. S. and Alvarez, E., Bacterial sheath brown rot of rice caused by *Pseudomonas fuscovaginae* in Latin America, *Plant Dis.*, 71, 592, 1987.
216. Ribeiro, A. S., Seed treatment of irrigated rice for control of the spread of *Aphelenchoides besseyi,* Rio Grande do Sul, Empressa Bras. Pesqui. Agropecu., (Pelotas, Brazil), 1977, 138.
217. Muthusamy, M. and Narayanasamy, P., Seed transmission of pearlmillet downy mildew and its control, *Indian Phytopathol.*, 34, 418, 1981.
218. Cox, R. S., Stem anthracnose of lima beans and its control, *Phytopathology,* 38, 7, 1948.
219. Taylor, J. D. and Dudley, C. L., Seed treatment for the control of halo-blight of beans *(Pseudomonas phaseolicola), Ann. Appl. Biol.,* 85, 223, 1977.
220. Vlakhov, S., Kutova, I., and Koleva, P., Action of antibiotics against some bacterioses, *Rostenievdni Nauki,* 11, 123, 1974.

221. Abawi, G. S. and Pastor-Corrales, M. A., Seed transmission and effect of fungicide seed treatments against *Macrophomina phaseolina* in dry edible beans, *Turrialba*, 40, 334, 1990.
222. Anderson, A. L. and Dezcevew, D. J., Seed treatment studies for damping-off control in garden and canning beans, *Rep. Prog. Q. Bull. Mich. Agric. Exp. Stn.*, 34, 357, 1952.
223. Yoshii, K., Seed treatment of pea with benomyl to control *Mycosphaerella pinodes, Fitopatologia,* 10, 41, 1975.
224. Kirik, N. N., Influence of the depth of mycelial penetration to the causal agent of ascochytosis into pea seeds on the effectiveness of treatment, *Mikol. Fitopatol.*, 4, 419, 1970.
225. Miller, M. W. and de Whalley, C. V., The use of metalaxayl seed treatments to control pea downy mildew, Proc. Br. Crop Protection Conf., Vol. 1, Brighton, England, 1981, 341.
226. Taylor, J. D. and Dye, D. W., Evaluation of streptomycin seed treatments for the control of bacterial blight of peas (*Pseudomonas pisi* Sackett 1916), *N.Z. J. Agric. Res.,* 19, 91, 1976.
227. Nene, Y. L. and Agarwal, V. K., Influence of fungicidal seed treatment on emergence and yield of pea var. Bridger, *Pesticides,* 3, 15, 1969.
228. Crosier, W., Chemical control of seed-borne fungi during germination on testing of peas and sweetcorn, *Phytopathology,* 36, 92, 1946.
229. Tisdale, W. B., Brooks, A. N., and Townsend, G. R., Dust treatments for vegetable seed, *Bull. Fla. Agric. Exp. Stn.,* 413, 32, 1945.
230. Shukla, B. N. and Singh, B. P., Effect of fungicidal seed treatment on Macrophomina root rot of sesame (*Sesamum indicum*), *Indian J. Mycol. Plant Pathol.*, 2, 208, 1973.
231. Venugopal, M. N. and Safeeulla, K. M., Chemical control of the downy mildew of pearlmillet, sorghum and maize, *Indian J. Agric. Sci.*, 48, 537, 1978.
232. Leukel, R. W., Spergon as a seed disinfectant, *Plant Dis. Rep.*, 26, 93, 1942.
233. Webster, O. J. and Leukel, R. W., Sorghum seed treatment tests in 1958, *Plant Dis. Rep.*, 43, 348, 1959.
234. Grewal, J. S. and Dharam, V., Efficacy of different fungicides. IV. Field trials for the control of grain smut of jowar *Sphacelotheca sorghi* (Link) Clinton, *Indian Phytopathol.*, 14, 213, 1961.
235. Hansing, E. D. and Melchers, L. E., Standard and new fungicides for the control of covered smut of sorghum and their effect on stand, *Phytopathology,* 34, 1034, 1944.
236. Mathur, K., Siradhana, B. S., and Lodha, B. C., Studies on seedling blight of sorghum caused by *Gloeocercospora sorghi*, *Seed Sci. Technol.*, 15, 851, 1987.
237. Agarwal, V. K., Verma, H. S., and Singh, O. V., Treatment of sorghum seeds to control seed-borne fungi and improve emergence, *Bull. Grain Technol.*, 15, 118, 1977.
238. Lasca, C. C., Barros, B. C., Valcarini, P. J., Fregonezi, L. E., and Chiba, S., Effectiveness of fungicides for seed dressing of wheat (*Triticum aestivum* L.) to control *Helminthosporium sativum* Pammel, King and Bakke, *Biologico*, 50, 125, 1984.
239. Siljes, I. and Halbauer, V., Possibilities of treating wheat seed with systemic fungicide, *Agron. Glas.*, 33, 447, 1974.
240. Sharma, R. C., Joshi, L. M., and Pathak, K. D., Systemic fungicides for the control of hill bunt of wheat, *Indian Phytopathol.*, 24, 604, 1971.

241. Leukel, R. W., Cooperative seed treatment on small grains in 1951, *Plant Dis. Rep.*, 35, 445, 1951.
242. More, K. J. and Kuiper, J., New treatments for bunt of wheat, *Agric. Gaz. N.S.W.*, 85, 16, 1974.
243. Line, R. F., Chemical control of flag smut of wheat, *Plant Dis. Rep.*, 56, 636, 1972.
244. Tyagi, P. D., Singh, M., and Chauhan, M. S., Comparative efficacy of some systemic fungicides for controlling loose smut of wheat, *Pesticides,* 10, 26, 1976.
245. Khanzada, A. K. and Mathur, S. B., Control of loose smut of wheat by carboxin, fenfuram and triadimenol, *Seed Sci. Technol.*, 11, 947, 1984.
246. Sharma, J. K., Aujla, S. S., Sharma, Y. R., and Chauhan, J. S., Systemic fungicides for the control of loose smut of wheat, *Pesticides,* 12, 30, 1978.
247. Chatrath, M. S., Renfro, B. L., Nene, Y. L., Grover, R. K., Roy, M. R., Singh, D. V., and Gandhi, S. M., Control of loose smut of wheat with systemic fungicides, *Indian Phytopathol.*, 22, 184, 1969.
248. Agarwal, V. K., Agarwal, M., Verma, H. S., and Gupta, R. K., Studies on loose smut of wheat. III. Influence of infection on plant morphology and control through seed treatment with carboxin, *Seed Res.*, 10, 79, 1982.
249. Pommer, E. H. and Kradel, J., 2, 5-dimethyl-furane-3-carboxylic acid anilide (BAS 3191F) a new active ingredient for the control of seed-borne fungus disease in cereal, VII, Int. Congr. Plant Prot., Paris, 1970, 409.
250. Agarwal, V. K., Singh, A., and Verma, H. S., A note on emergence of wheat seed treated with different seed dressing fungicides prior to storage, *Seed Res.*, 4, 194, 1976.
251. Jindal, K. K., Thind, B. S., and Soni, P. S., Physical and chemical agents for the control of *Xanthomonas campestris* pv. *vignicola* from cowpea seeds, *Seed Sci. Technol.*, 17, 371, 1989.
252. Warren, H. L. and Nicholson, R. L., Kernel infection, seedling blight and wilt of maize caused by *Colletotrichum graminicola, Phytopathology,* 65, 620, 1975.
253. Lim, S. M. and Kinsey, J. G., Seed treatment of corn infected with *Helminthosporium maydis* race T., *Plant Dis. Rep.*, 57, 344, 1973.
254. Simpson, W. R. and Fenwick, H. S., Suppression of corn head smut with carboxin seed treatments, *Plant Dis. Rep.*, 55, 501, 1971.
255. Berger, R. D. and Wolf, E. A., Control of seed-borne and soil-borne mycoses of "Florida Sweet" corn by seed treatment, *Plant Dis. Rep.*, 58, 922, 1974.
256. Hoppe, P. E., Comparison of certain mercury and non-metallic dusts for corn seed treatment, *Phytopathology,* 33, 602, 1943.
257. Srivastava, R. N. and Gupta, J. S., Seed transmission of *Alternaria zinniae*, its location in the seed and control, *Indian Phytopathol.*, 37, 83, 1984.
258 Beaumont, A., Cleary, J. P., and Bant, J. H., Control of damping-off of zinnias caused by *Alternaria zinniae, Plant Pathol.,* 7, 52, 1958.
259. Strider, D. L., Eradication of *Xanthomonas nigromaculans* f. sp. *zinniae* in zinnia seed with sodium hypochlorite, *Plant Dis. Rep.*, 63, 873, 1979.
260. Demski, J. W., Tobacco mosaic virus is seed-borne in pimiento peppers, *Plant Dis.*, 65, 723, 1981.
261. Alexander, L. J., Inactivation of tobacco mosaic virus from tomato seed, *Phytopathology,* 50, 627, 1960.
262. Taylor, R. H., Grogan, R. G., and Kimble, K. A., Transmission of tobacco mosaic virus in tomato seed, *Phytopathology,* 51, 837, 1961.
263. Gooding, G. V., Jr., Inactivation of tobacco mosaic virus on tomato seed with trisodium orthophosphate and sodium hypochlorite, *Plant Dis. Rep.*, 59, 770, 1975.

264. Sharma, S. R. and Varma, A., Cure of seed transmitted cowpea banding mosaic disease, *Phytopathol. Z.*, 83, 144, 1975.
265. Walkey, D. G. A. and Dance, M. C., High temperature inactivation of seed borne lettuce mosaic virus, *Plant Dis. Rep.*, 63, 125, 1979.
266. Kadian, O. P., Effects of some chemicals and heat on seed transmission of urdbean leaf crinkle virus, 3rd Int. Symp. Plant Pathol., New Delhi, 1981, 166.
267. Todd, E. H. and Atkins, J. C., White tip disease of rice. II. Seed treatment studies, *Phytopathology*, 49, 184, 1959.
268. Fukano, H., Ecological studies on white tip disease of rice plant caused by *Aphelenchoides besseyi* Christie and its control, *Fukuoka Agric. Exp. Stn. Bull.*, 18, 108, 1962.
269. Reddy, M. V., Calixin M — an effective fungicide for eradication of *Ascochyta rabiei* in chickpea seed, *Int. Chickpea Newsl.*, 3, 12, 1980.
270. Haware, M. P., Nene, Y. L., and Rajeshwari, R., Eradication of *Fusarium oxysporum* f. sp. *ciceri* transmitted in chickpea seed, *Phytopathology*, 68, 1364, 1978.
271. Henning, A. A., Seed treatments and the fungal pathogens they are designed to control, in *Soybean Diseases of the North Central Region*, Wyllie, T. D. and Scott, D. H., Eds., APS Press, St. Paul, MN, 1988, 14.
272. Cappelli, C. and Torre, G. D., Forecasting attacks of barley loose smut *(Ustilago nuda* (Jens.) Rostr.) by applying the embryo test to seed batches, *Sementi Elette*, 34, 21, 1988.
273. Rennie, W. J., Richardson, M. J., and Noble, M., Seedborne pathogens and the production of quality cereal seed in Scotland, *Seed Sci. Technol.*, 11, 1115, 1983.
274. Jørgensen, J., Om en vurdering af hyggens afsvampningsbehov foretaget tidlight i saesonen, *Statsfrøkontroll. Beret.*, 109, 102, 1980.
275. Trione, E. J., Dwarf bunt of wheat and its importance in international wheat trade, *Plant Dis.*, 66, 1083, 1982.
276. Singh, D. P. and Agarwal, V. K., Control of seedborne infection of grain mold pathogens of sorghum by fungicidal seed treatment, *Indian J. Plant Pathol.*, 6, 128, 1988.
277. Agarwal, V. K., Assessment of seed-borne infection and treatment of wheat seeds for the control of loose smut, *Seed Sci. Technol.*, 9, 725, 1981.
278. Sanderson, F. R. and Hampton, J. G., Role of perfect states in the epidemiology of the common Septoria diseases of wheat, *N.Z. J. Agric. Res.*, 21, 277, 1978.
279. Klitgard, K. and Jørgensen, J., The correlation between the germination percentage determined in the laboratory and the field with samples of seeds of winter wheat infected with *Septoria nodorum*, *Statsfrøkont. Beret.*, 102, 85, 1973.
280. Hermansen, J. F. and Jørgensen, J., Historical aspects of the control of seed-borne cereal diseases in Denmark, *Seed Sci. Technol.*, 11, 1005, 1983.
281. Jørgensen, J., Disease testing of barley seed and application of test results in Denmark, *Seed Sci. Technol.*, 11, 615, 1983.
282. Brodal, G., Fungicide treatment of cereal seeds in Norway according to need, ISTA Plant Dis. Comm. Symp. Seed Health Testing, Ottawa, 1993, 15.
283. Rennie, W. J. and Cokerell, V., Sampling and testing barley seed for *Pyrenophora graminea* infection, ISTA Plant Dis. Comm. Symp. Seed Health Testing, Ottawa, 20, 1993.
284. Mehta, Y. R. and Bassoi, M. C., Gauzatine plus as a seed treatment bactericide to eradicate *Xanthomonas campestris* pv. *undulosa* from wheat seeds, *Seed Sci. Technol.*, 21, 9, 1993.

285. Ralph, W., Pelleting seed with bactericides — the effect of streptomycin on seed-borne halo blight of French-bean, *Seed Sci. Technol.*, 4, 325, 1976.
286. Ellis, M. A. and Paschal, E. H., Transfer of technology in seed pathology of tropical legumes, in *Seed Pathology — Problems and Progress,* Yorinori, J. T., Sinclair, J. B., Mehta, Y. R., and Mohan, S. K., Eds., Fundacão Inst. Agron. Paraná, IAPAR, Londrina, Brazil, 1979, 190.
287. Cook, A. A., Larson, R. H., and Walker, J. C., Relation of the black rot pathogen to cabbage seed, *Phytopathology,* 42, 316, 1952.
288. Patel, P. N., Trivedi, B. M., Rekhi, S. S., Town, P. A., and Rao, Y. P., Black rot and stump rot in cauliflower seed crops in India, *FAO Plant Prot. Bull.*, 18, 136, 1970.
289. Harrower, K. M., Tolerance of *Leptosphaeria nodorum* to an organomercurial compound, *Trans. Br. Mycol. Soc.*, 66, 523, 1976.
290. Noble, M. and Macgarvie, Q. D., Resistance to mercury of *Pyrenophora avenae* in Scottish seed oats, *Plant Pathol.*, 15, 23, 1966.
291. Kuiper, J., Failure of hexachlorobenzene to control common bunt of wheat, *Nature (London),* 206, 1219, 1965.
292. Old, K. M., Mercury tolerant *Pyrenophora avenae* in seed oats, *Trans. Br. Mycol. Soc.*, 51, 525, 1968.
293. Ellis, M. A. and Sinclair, J. B., Uptake and translocation of streptomycin by soybean seedlings, *Plant Dis. Rep.*, 58, 534, 1974.
294. Humaydan, H. S., Harman, G. E., Nedrow, B. L., and DiNitto, L. V., Eradication of *Xanthomonas campestris,* the causal agent of blackrot from Brassica seeds with antibiotics and sodium hypochlorite, *Phytopathology,* 70, 127, 1980.
295. Anahosur, K. H. and Patil, S. H., Chemical control of sorghum downy mildew in India, *Plant Dis.*, 64, 1004, 1980.
296. Pawar, N. B. and Dandnaik, B. P., Efficacy of metalaxyl fungicides against downy mildew *(Sclerospora graminicola)* of pearl millet, *Indian Phytopathol.*, 42, 290, 1989.
297. Rennie, W. J. and Cockerell, V., A review of seedborne pathogens of cereals in the post-mercury period, in *Proc. Crop Prot. Northern Br.*, UK, 17, 24, 1993.
298. Neergaard, P., Seedborne diseases in European, especially Mediterranean, crops. Problems and importance. *Ann. deli Instit. Speriment. la Patol. Veg.*, Rome, 7, 5, 1983.
299. Anonymous, Worried about mercury seed dressings being banned? Cerevax will make you happy, *Agric. Supply Ind.*, 22, 5, 1992.
300. Fujii, H., Pre-sowing treatment of rice seeds in Japan, *Seed Sci. Technol.*, 11, 951, 1983.
301. Ryker, T. C., Seed coloration, in Proc. 1959 Short Course for Seedsmen, university of Florida, Gainesville, 1959, 123.
302. Meyer, H. and Mayer, A. M., Permeation of dry seeds with chemicals: use of dichloromethane, *Science,* 171, 683, 1971.
303. Elden, M., Mayer, A. M., and Poljakoff-Mayer, A., Permeation of dry lettuce seeds with acetic anhydride and with amino acids, using dichloromethane, *Seed Sci. Technol.*, 2, 317, 1974.
304. Ralph, W., A note on antibiotic permeation of seed with dichloromethane, *Seed Sci. Technol.*, 5, 575, 1977.
305. Royce, D. J., Ellis, M. A., and Sinclair, J. B., Movement of penicillin into soybean seeds using dichloromethane, *Phytopathology,* 65, 1319, 1975.

306. Ellis, M. A., Foor, S. R., and Sinclair, J. B., Dichloromethane: nonaqueous vehicle for systemic fungicides in soybean seeds, *Phytopathology,* 66, 1249, 1976.
307. Papavizas, G. C. and Lewis, J. A., Acetone infusion of pyroxchlor into soybean seed for the control of *Phytophthora megasperma* var. *sojae, Plant Dis. Rep.*, 60, 484, 1976.
308. Hepperly, P. R. and Sinclair, J. B., Aqueous polyethylene glycol solutions for treating soybean seeds with antibiotics, *Seed Sci. Technol.,* 5, 727, 1977.
309. Tao, K. L., Khan, A. A., Harman, G. E., and Eckenrode, C. J., Practical significance of the application of chemical in organic solvents to dry seeds, *J. Am. Soc. Hortic. Sci.,* 99, 217, 1974.
310. Heydecker, W., Higgins, J., and Turner, Y. J., Invirogration of seeds, *Seed Sci. Technol.,* 3, 881, 1975.
311. Browing, E., *Toxicology and Metabolism of Industrial Solvents,* Elsevier, Amsterdam, 1965.
312. Sinclair, J. B., Soybean seed pathology, in *Seed Pathology — Problems and Progress,* Yorinori, J. T., Sinclair, J. B., Mehta, Y. R., and Mohan, S. K., Eds., Fundacão Instit. Agron. Paraná, IAPAR, Londrina, Brazil, 1979, 161.
313. Shortt, B. J. and Sinclair, J. B., Efficacy of polyethylene glycol and organic solvents for infusing fungicides into soybean seeds, *Phytopathology,* 70, 971, 1980.
314. Muchovej, J. J. and Dhingra, O. D., Benzene and ethanol for treatment of soybean seeds with systemic fungicides, *Seed Sci. Technol.,* 7, 449, 1979.
315. Liang, L. Z., Halloin, J. M., and Saettler, A. W., Use of polyethylene glycol and glycerol as carriers of antibiotics for reduction of *Xanthomonas campestris* pv. *phaseoli* in navy bean seeds, *Plant Dis*., 76, 875, 1992.
316. Shen, C. Y., Integrated management of *Fusarium* and *Verticillium* wilts of cotton in China, *Crop Prot.,* 4, 337, 1985.
317. Maude, R. B., Spencer, A., Brocklehurst, P. A., Gott, K. A., and Bambridge, J. M., The biology and control of *Alternaria dauci* (leaf blight) on carrot seeds, in 35th Annu. Rep. for 1984, National Vegetable Research Station, Wellesbourne, 1985, 81.
318. Herd, G. W. and Phillips, A. J. L., Control of seedborne *Sclerotinia sclerotiorum* by fungicidal treatment of sunflower seed, *Plant Pathol.,* 37, 202, 1988.
319. Walkey, D. G. A., Brocklehurst, P. A., and Parker, J. E., Seed transmission of viruses, in 33rd Annu. Rep. National Vegetable Research Station, Warwick, 1983, 82.
320. Miller, R. V., Carroll, T. W., and Sands, D. C., Effect of chemical seed treatments on symptoms caused by seedborne barley stripe mosaic virus in Vantage barley, *Can. J. Microbiol.,* 32, 189, 1986.
321. Drew, R. L. K. and Brocklehurst, P. A., The effect of anti-viral thermal treatments on germination of lettuce (*Lactuca sativa)* seeds and subsequent seedling development, *Ann. Appl. Biol.,* 107, 137, 1985.
322. Damicone, J. P., Cooley, D. R., and Manning, W. J., Elimination of *Fusaria* from asparagus seed, *Phytopathology,* 70, 461, 1980.
323. Oneill, N. R., Papavizas, G. C., and Lewis, J. R., Infusion and translocation of systemic fungicides applied to seeds in acetone, *Phytopathology,* 69, 690, 1979.
324. Kraft, J. M., Comparison of acetone infusion to slurry application of fungicides to pea seed, *Phytopathology,* 71, 232, 1981.
325. Kraft, J. M., Field and greenhouse studies on pea seed treatments, *Plant Dis.,* 66, 798, 1982.

326. Muchovej, J. J. and Dhingra, O. D., Acetone, benzene and ethanol for treating *Phaseolus* bean seeds in the dry state with systemic fungicides, *Seed Sci. Technol.*, 8, 351, 1980.
327. Agarwal, V. K. and Sinclair, J. B., Seed dressings and *Rhizobium* inoculum, in Soybean Seed Quality and Stand Establishment (INTSOY Ser. No. 22), Sinclair, J. B. and Jackobs, J. A., Eds, College of Agriculture, University of Illinois, Urbana-Champaign, 1981, 127.
328. Curley, R. L. and Burton, J. C., Compatibility of *Rhizobium japonicum* with chemical seed protectants, *Agron. J.*, 67, 807, 1975.
329. Furtode, A., Effect of Systemic Fungicides on Soil Microbial Population and Nodulation by *Rhizobium* spp., Thesis Abstr., College of Agriculture, Dharwar, India, 3, 286, 1977.
330. Brinkerhoff, L. A., Fink, G., Kortsen, R. A., and Swift, D., Further studies on the effect of chemical seed treatments on nodulation of legumes, *Plant Dis. Rep.*, 38, 393, 1954.
331. Kis, G., Papp, I., Bakondizamori, E., and Garner Banfalvi, A., Study of soybean seed dressing with fungicides combined with *Rhizobium* inoculation, *Novenytermeles*, 26, 147, 1977.
332. Wankhede, V. K. and Bhide, V. P., Compatibility of different fungicides with *Rhizobium japonicum, Hind. Antibiot. Bull.*, 14, 131, 1972.
333. Tu, C. M., Effects of pesticide seed treatments on *Rhizobium japonicum* and its symbiotic relationship with soybean, *Bull. Environ. Contam. Toxicol.*, 18, 190, 1977.
334. Ganacharya, N. M., Effect of fungicidal seed treatment on emergence, nodulation and grain yield of soybean in Marathwada, *J. Maharashtra Agric. Univ.*, 4, 112, 1979.
335. Hamdi, Y. A., Moharram, A. A., and Lofti, M., Effect of certain fungicides on some *Rhizobia* legume symbiotic systems, *Zentralbl. Bakeriol. Parasitenkd. Infektionskr. Hyg. Zweite Naturwiss. Abt. Allg. Landwirtsch. Tech NMkro Ea,* 3–4, 363, 1974.
336. Batalova, T. S., Zinovev, L. S., Kiselev, I. I., Kikhanina, K. A., and Masiutina, V. A., Compatibility of the bacterial fertilizer nitrogen treatment and dressing of legume seed fungicides, *Khim. Sel'sk. Khoz.*, 15, 37, 1977.
337. Nery, M. and Dobereiner, J., Effect of pre-emergence fungicides on nodulation and N_2 fixation in soybean, in Anais do Decimo Quinto Congr. Brasileiro de Ciencia do Solo, Sao Paolo, Brazil, 1976, 177.
338. Koehler, B., Results of uniform seed treatment tests on soybeans, *Plant Dis. Rep.*, Suppl. 145, 76, 1943.
339. Maggione, C. S. and Lamsanchez, A., Effect of seed treatment with thiabendazol, alone and in combination with captan, on germination and nodulation of soybean (*Glycine max* (L.) Merrill), *Cientifica*, 4, 107, 1976.
340. Backman, P. A., Effects of seed treatment fungicides on *Rhizobium* inoculants, *Highlights Agric. Res.*, 25, 14, 1978.
341. Maude, R. B. and Shuring, C. G., Seed treatment with Vitavax for the control of loose smut of wheat and barley, *Ann. Appl. Biol.*, 64, 259, 1969.
342. Dharam, V., Efficacy of fungicides. XXI. Studies on the residual efficacy of systemic fungicides on the treated seed, *Pesticides*, 14, 43, 1980.
343. Gupta, J. P. and Chatrath, M. S., Persistence of thiram on soybean seed in storage, *Indian Phytopathol.*, 36,263, 1983.

344. Sinclair, J. B., Control of seedborne pathogens and diseases of soybean seeds and seedlings, *Pest. Sci.*, 37, 15, 1993.
345. Maude, R. B., Eradicative seed treatment, *Seed Sci. Technol.*, 11, 907, 1983.
346. Iqbal, S. M., Hussain, S., Tahir, M., and Malik, B. A., Effect of fungicidal seed treatment on lentil germination and recovery of seedborne *Ascochyta fabae* f. sp. *lentis*, *LENS Newsl.*, 19, 53, 1992
347. Kaiser, W. J. and Hannan, R. M., Seed transmission of *Ascochyta rabiei* in chickpea and its control by seed-treatment fungicides, *Seed Sci. Technol.*, 16, 625, 1988.
348. Kharbanda, P. D. and Bernier, C. C., Effectiveness of seed and foliar application of fungicides to control Ascochyta blight of faba beans, *Can. J. Plant. Sci.*, 59, 661, 1979.
349. Gupta, J. P., Erwin, D. C., Eckert, J. W., and Waki, A. I., Translocation of metalaxyl in soybean plants and control of stem rot caused by *Phytophthora megasperma* f. sp. *glycinea*, *Phytopathology*, 75, 865, 1985.
350. Sitton, J. W., Line, R. F., Waldher, J. T., and Goates, B. J. Difenoconazole seed treatment for control of dwarf bunt of winter wheat, *Plant Dis.*, 77, 1148, 1993.
351. Cappelli, C. and Zazzerini, A., Safflower rust (*Puccinia carthami* Cda) in Italy: seed contamination, seed-plant transmission and seed dressing for disease control, *Phytopathol. Mediterr.*, 27, 145, 1988.
352. Smit, W. A. and Knox-Davies, P. S., Elimination of *Diaporthe phaseolorum* and *Neocosmospora vasinfecta* from rooibas tea seeds by hot-water treatment and acid scarification, *Phytophylactica*, 21, 297, 1989.
353. Fatmi, M., Schaad, N. W., and Bolkan, H. A., Seed treatments for eradicating *Clavibacter michiganensis* subsp. *michiganensis* from naturally infected tomato seed, *Plant Dis.*, 75, 383, 1991.
354. Sowell, G., Jr. and Schaad, N. W., *Pseudomonas pseudoalcoligenes* subsp. *citrulli* on watermelon: seed transmission and resistance of plant introductions, *Plant. Dis. Rep.*, 63, 437, 1979.
355. Kaiser, W. J., Plant introduction and related seed pathology research in the United States, *Seed Sci. Technol.*, 11, 1197, 1983.
356. Schultz, T., Gabrielson, R. L., and Olson, S., Control of *Xanthomonas campestris* pv. *campestris* in crucifer seed with slurry treatments of calcium hypochlorite, *Plant. Dis.*, 70, 1027, 1986.
357. Coats, J. R. and Dahm, P. A., Detoxification of captan-treated seed corn, Proc. 6th Res. Symp. Treatment Hazardous Waste, Schultz, D., Ed., U. S. Environmental Protection Agency, 1980, 94.
358. Burton, J. C., Problems in obtaining adequate inoculation of soybeans, in *World Soybean Research*, Hill, L. D., Ed., Interstate Printers, Danville, IL, 1976, 170.
359. Staphorst, J. L. and Strijdom, B. W., Effects on *Rhizobia* of fungicides applied to legume seed, *Phytophylactica*, 8, 47, 1976.
360. Odeyemi, O., Resistance of *Rhizobium* to Thiram, Spergon, and Phygon, *Diss. Abstr. Int. B.*, 38, 993, 1977.
361. Bewley, W. F. and Corbett, W., The control of cucumber and tomato mosaic disease in glasshouse by the use of clean seed, *Ann. Appl. Biol.*, 17, 260, 1930.
362. Madsen, E. and Langkilde, N. E., Eds., *ISTA Handbook for Cleaning of Agricultural and Horticultural Seeds on Small-Scale Machines*, Part II, ISTA, Zurich, 1988.
363. Dhanvantari, B. N., Effect of seed extraction methods and seed treatment on control of tomato bacterial canker, *Can. J. Plant Pathol.*, 11, 400, 1989.

364. MacDonald, J. D. and Leach, L. D., The association of *Fusarium oxysporum* f. sp. *betae* with nonprocessed and processed sugarbeet seeds, *Phytopathology*, 66, 868, 1976.
365. Tomlinson, J. A. and Faithfull, E. M., 23rd Annual Report for 1972, Nattional Vegetable Research Institute, Wellesbourne, U.K., 1973, 139.
366. Ryder, E. J. and Johnson, A. S., A method for indexing lettuce seeds for seed-borne lettuce mosaic virus by air-stream separation of light from heavy seeds, *Plant Dis. Rep.*, 58, 1037, 1974.
367. Phatak, H. C. and Summanwar, A. S., Detection of plant viruses in seeds and seed stocks, *Proc. Int. Seed. Test. Assoc.*, 32, 625, 1967.
368. Stevenson, W. R. and Hagedorn, D. J., Effect of seed size and condition on transmission of pea seed-borne mosaic virus, *Phytopathology*, 60, 1148, 1970.
369. Thyr, B. D., Webb, R. E., Jaworski, C. A., and Ratcliffe, T. J., Tomato bacterial canker: control by seed treatment, *Plant Dis. Rep.*, 57, 974, 1973.
370. Proctor, C. H. and Fry, P. R., Seed transmission of tobacco mosaic virus in tomato, *N.Z. J. Agric. Res.*, 8, 367, 1965.
371. Milenko, Y. F., Cleaning of white rot sclerotia from sunflower seeds, *Sel. Seed-Gr.* (Moscow), 29, 73, 1964.
372. Hepperly, P. R. and Sinclair, J. B., A glycerin and polyethyleneglycol solution for separating healthy and diseased soybean seeds, *Seed Sci. Technol.*, 11, 125, 1982.
373. Brinkerhoff, L. A. and Hunter, R. E., Internally infected seed as a source of inoculum for the primary cycle of bacterial blight of cotton, *Phytopathology*, 53, 1397, 1963.
374. Baker, K. F., Thermotherapy of planting material, *Phytopathology*, 52, 1244, 1962.
375. Labuschagne, N., Kotze, J. M., and Grimbeek, R. J., Control of *Chalara elegans* seed-borne infection of groundnuts, *Phytophylactica*, 23, 309, 1991.
376. Pryor, B. M., Davis, R. M., and Gilbertson, R. L., Detection and eradication of *Alternaria radicina* on carrot seed, *Plant Dis.*, 78, 452, 1994
377. Aveling, T. A. S., Snyman, H. G., and Maude, S. P., Evaluation of seed treatments for reducing *Alternaria porri* and *Stemphylium vesicarium* on onion seed, *Plant Dis.*, 77, 1009, 1993.
378. Honervogt, B. and Lehmann-Danzinger, H., Comparison of thermal and chemical treatment of cotton seed to control bacterial blight (*Xanthomonas campestris* pv. *malvacearum*), *J. Phytopathol.*, 134, 103, 1992.
379. Hayden, N. J. and Maude, R. B., The role of seedborne *Aspergillus niger* in transmission of black mould of onion, *Plant Pathol.*, 41, 573, 1992.
380. Chastain, T. G., High temperature sodium hypochlorite effects on viability of *Tilletia controversa* teliospores and wheat seed, *Crop Sci.*, 31, 1327, 1991.
381. Grondeau, C., Landonne, F., Fourmond, A., Poutier, F., and Samson, R., Attempt to eradicate *Pseudomonas syringae* pv. *pisi* from pea seeds with heat treatments, *Seed Sci. Technol.*, 20, 515, 1992.
382. Watson, R. D., Coltrin, L., and Robinson, R., The evaluation of materials for heat treatment of peas and beans, *Plant Dis. Rep.*, 35, 542, 1951.
383. Sinclair, J. B., Phomopsis seed decay of soybeans — a prototype for studying seed disease, *Plant Dis.*, 77, 329, 1993.
384. Pyndji, M. M., Sinclair, J. B., and Singh, T., Soybean seed thermotherapy with heated vegetable oils, *Plant Dis.*, 71, 213, 1987.
385. Zinnen, T. M. and Sinclair, J. B., Thermotherapy of soybean seeds to control seedborne fungi, *Phytopathology*, 72, 831, 1982.

386. Van Wyk, P. S., Scholtz, D. J., and Marasas, W. F. O., Protection of maize seedlings by *Fusarium moniliforme* against infection by *Fusarium graminearum* in the soil, *Plant Soil*, 107, 251, 1988.
387. Daniels, B. A., Elimination of *Fusarium moniliforme* from corn seed, *Plant Dis.*, 67, 609, 1983.
388. Sinha, S. K. and Nene, Y. L., Eradication of seedborne inoculum of *Xanthomonas oryzae* by hot water treatment of paddy seeds, *Plant Dis. Rep.*, 51, 882, 1967.
389. Severin, V., Investigations on prevention of common blight of bean (*Xanthomonas phaseoli*), *Anal. Inst. Ceretari Pentru Prot. Plant.*, 7, 125, 1971.
390. Moffett, M. L. and Wood, B. A., Seed treatment for bacterial spot of pumpkin, *Plant Dis. Rep.*, 63, 537, 1979.
391. Persley, G. S., Studies on the survival and transmission of *Xanthomonas manihotis* on cassava seed, *Ann. Appl. Biol.*, 93, 159, 1979.
392. Shekhawat, P. S., Jain, M. L., and Charkravarti, B. P., Detection and seed transmission of *X. campestris* pv. *campestris* and its control by seed treatment, *Indian Phytopathol.*, 35, 442, 1982.
393. Jindal, K. K., Thind, B. S., and Soni, P. S., Physical and chemical agents for the control of *Xanthomonas campestris* pv. *vignicola* from cowpea seeds, *Seed Sci. Technol.*, 17, 371, 1989.
394. Mikoshiba, H., Studies on the control of downy mildew disease of maize in tropical countries of Asia, *Bull. Trop. Agric. Res. Cent.*, 16, 62, 1983.
395. Kobayashi, T., Impact of seedborne pathogen on quarantine in Japan, *Seed Sci. Technol.*, 18, 427, 1990.
396. Vartanian, V. G. and Endo, R. M., Survival of *Phytophthora infestans* in seeds extracted from tomato fruits, *Phytopathology*, 75, 375, 1985.
397. Zeigler, R. S. and Alvarez, E., *Pseudomonas* spp. causing grain and sheath rot of rice in Latin America, Proc. 5th Int. Congr. Plant Path., Kyoto, 1988, 411.
398. Green, S. K., Hwang, L. L., and Kuo, Y. J., Epidemiology of tomato mosaic virus in Taiwan and identification of strains, *J. Plant Dis. Prot.*, 94, 386, 1987.
399. Singh, D. and Agarwal, K., Earcockle disease (*Anguina tritici* (Steinbuch) Filipjev) of wheat in Rajasthan, India, *Seed Sci. Technol.*, 15, 777, 1987.
400. Naumann, K. and Karl, H., Possibilites of disinfecting bean seeds infected with *Pseudomonas syringae* pv. *phaseolicola*, *Nachrichten. Pflanzens.* (DDR.), 42, 204, 1988.
401. Zhang, Y. X., Hua, J. Y., and He, L. Y., Effect of infected groundnut seeds on transmission of *Pseudomonas solanacearum*, *Bacterial Wilt Newsl.*, 9, 9, 1993.
402. Stijger, C. C. M. M. and Rast, A. T. B., Prospects of dry heat treatment at 70°C for disinfection of pepper seed infected with capsicum mosaic virus, *Meded. Facul. Landouwwetenschappen* (Rijksuniversiteit, Gent), 53, 473, 1988.
403. Burpee, L. L. and Bouton, J. H., Effect of eradication of endophyte *Acremonium coenophialum* on epidemics of Rhizoctonia blight in tall fescue, *Plant Dis.*, 77, 157, 1993.
404. Luthra, J. C., Solar energy treatment of wheat loose smut *Ustilago tritici* (Pers.) Rostr., *Indian Phytopathol.*, 6, 49, 1953.
405. Smith, P. R., Seed transmission of *Intersonilia pastinacae* in parsnip and its elimnation by a steam-air treatment, *Aust. J. Exp. Agric. Anim. Husb.*, 6, 441, 1966.
406. Navaratnam, S. J. and Shuttleworth, D., A simple technique to improve efficiency of aerated steam treatment for control of seed-borne fungal pathogens, Proc. Fourth Int. Cong. Plant Pathol., 1983, 230.

407. Navaratnam, S. J., Shuttleworth, D., and Wallace, D., The effect of aerated steam on six seedborne pathogens, *Aust. J. Exp. Agric. Anim. Husb.*, 20, 97, 1980.
408. Fourest, E., Rehms, L. D., Sands, D. C., Bjarko, M., and Lund, R. E., Eradication of *Xanthomonas campestris* pv. *translucens* from barley seed with dry heat treatments, *Plant Dis.*, 74, 816, 1990.
409. Setchell, P. J., Smee, L., and Heaton, J. B., A New Approach to Thermal Treatment of Seed: Control and Application, Department of Health, Canberra, 1975.
410. Krasnova, M. V., The effect of some physical factors on the causal agents of bacterioses in soybean seeds, *J. Microbiol.* (Kiev), 25, 50, 1963.
411. Hankin, I. and Shands, D. C., Microwave treatment of tobacco seed to eliminate bacteria on the seed surface, *Phytopathology*, 67, 794, 1977.
412. Cavalcante, M. J. B. and Muchovej, J. J., Microwave irradiation of seeds and selected fungal spores, *Seed Sci. Technol.*, 21, 247,1993.
413. Basyony, A. E., El-Refaei, M. I., Galal, M. S., and Barakat, M. I. E., Effect of gamma irradiation on seed-borne fungi and soybean seed components during storage, *Agric. Res. Rev.*, 67, 619, 1989.
414. Bel'skil, A. I. and Manzulenko, N. N., Effects of presowing laser treatment of barley seeds on the incidence of fungal disease of the plants, *Mikolog. Fitopathol.*, 18, 312, 1984.
415. Lozano, J. C., Laberry, R., and Bermudez, A., Microwave treatment to eradicate seed-borne pathogens in Cassava true seed, *J. Phythopathol.*, 117, 1, 1986.
416. Bagegni, A. M., Sleper, D. A., Kerr, H. D., and Morris, J. S., Viability of *Acremonium coenophialum* in tall fescue seed after ionizing radiation treatments, *Crop Sci.*, 30, 1272, 1990.
417. Halliwell, R. S. and Langston, R., Effects of gamma irradiation on symptom expression of barley stripe mosaic virus disease on two viruses *in vivo*, *Phytopathology*, 55, 1039, 1965.
418. Megahed, E. S. and Moore, A., Inactivation of necrotic ringspot and prune dwarf viruses in seeds of some *Prunus* spp., *Phytopathology*, 59, 1758, 1969.
419. Nagy, J., Increased efficacy of seed-dressing agents by ultrasound exposure of rice nematodes, *Int. Agrophysics*, 3, 291, 1987.
420. Bridge, J., Bos, W. S., Page, L. T., and McDonald, D., The biology and possible importance of *Aphelenchoides arachidis*, a seed-borne ectoparasitic nematode of ground-nut from northern Nigeria, *Nematologica*, 23, 253, 1977.
421. Sharvelle, E. G., *Plant Disease Control*, AVI Publishing, Westport, CT, 1979, 331.
422. Prasanna, K. P. R., Seed treatment for the control of *Alternaria brassicae* and *Alternaria brassicicola* infections of oilseed rape (*Brassica napus)* and its effect on seed germination, Proc. 21st ISTA Congress, Brisbane, 1986, 22.
423. Leben, C., Control of *Pseudomonas lachrymans* in cucumber seed by a temperature-relative humidity method, *Phytopathology*, 71, 235, 1981.
424. Srivastava, D. N. and Rao, Y. P., Epidemiology and control of bacterial blight of guar (*Cyamopsis tetra-gonoloba* (L.) Taub.) *Bull. Indian Phytopathol. Soc.*, 6, 1, 1970
425. Thorne, G., *Principles of Nematology*, McGraw-Hill, New York, 1961, 553.
426. Obata, K., Serizawa, S., Azegami, K., and Shirata, A., Possibility of seed transmission of *Xanthomonas campestris* pv. *vitians*, the pathogen of bacterial spot of lettuce, *Bull. Nat. Inst. Agric. Sci.*, 36, 81, 1982.
427. Devash, Y., Okon, Y., and Henis, Y., Survival of *Pseudomonas tomato* in soil and seeds, *Phytopathol. Z.*, 99, 175, 1980.

428. McIntyre, J. L., Sands, D. C., and Taylor, G. S., Overwintering, seed disinfestation, and pathogenicity studies of the tobacco hollow stalk pathogen, *Erwinia carotovora* var. *carotovora*, *Phytopathology*, 68, 435, 1978.
429. Mohanty, N. N., Control of Udbatta disease of rice, *Proc. Indian Nat. Sci. Acad.*, B37, 432, 1971.
430. Todd, E. H. and Atkins, J. G., White tip disease of rice. II. Seed treatment studies, *Phytopathology*, 49, 184, 1959.
431. Rahim, M. A. A., Nematodes associated with paddy seeds in Malaysia, *PLANTI Proc.*, 3, 187, 1988.
432. Thakur, D. P. and Kanwar, Z. S., Internal seed-borne infection and heat therapy in relation to downy mildew of *Pennisetum typhoides* Stapf. and Hubb., *Sci. Cult.*, 43, 433, 1977.
433. Baker, K. F., Bacterial fasciation disease of ornamental plants in California, *Plant Dis. Rep.*, 34, 121, 1950.
434. Reddick, D. and Stewart, V. B., Transmission of the virus of bean mosaic in seed and observations on thermal death point of seed and virus, *Phytopathology*, 9, 445, 1919.
435. Owusu, G. K., Crowley, N. C., and Francki, R. I. B., Studies of the seed transmission of tobacco ringspot virus, *Ann. Appl. Biol.*, 61, 195, 1968.
436. Rohloff, I., Trials for inactivation of lettuce mosaic virus in seeds, *Gartenbauwissenschaft*, 28, 19, 1963.
437. Timian, R. G., Heat treatments fail to inactivate barley stripe mosaic virus in seed, *Plant Dis. Rep.*, 49, 696, 1965.
438. Vovk, A. M., Inactivation of tobacco mosaic virus in tomato seed at different storage times, *Tr. Inst. Genet. Akad. Nauk. U.S.S.R.*, 28, 269, 1961.
439. Howles, R., Inactivation of tomato mosaic virus in tomato seeds, *Plant Pathol.*, 10, 160, 1961.
440. Broadbent, L. H., The epidemiology of tomato mosaic. XI. Seed transmission of TMV, *Ann. Appl. Biol.*, 56, 177, 1965.
441. Laterrot, H. and Pecaut, P., Tomato seed production. I. Rapid cleansing using pectolytic enzymes. II. Decreasing the content of tobacco mosaic virus by dry heat treatment, *Ann. Epiphyt.*, 16, 163, 1965.
442. Laterrot, H. and Pecaut, P., Incidence du traitement thermique de tomate sur la transmission du virus de la mosaique du tabac, *Ann. Epiphyt.*, 19, 159, 1968.
443. Van Dorst, H. J. M., Investigation into cucumber virus 2, *Groenten Fruit*, 22, 1519, 1967.
444. Fletcher, J. T., George, A. J., and Green, D. E., Cucumber green mottle virus, its effect on yield and its control in the Lea Valley, England, *Plant Pathol.*, 18, 16, 1969.
445. Verma, V. S., Effect of heat on seed transmission of mosaic disease of cowpea (*Vigna sinensis* Savi), *Acta Microbiol. Pol. Ser. B.*, 3, 163, 1971.
446. Sharma, Y. R. and Chohan, J. S., Control by thermotherapy of seed-borne vegetable marrow mosaic virus, *FAO Plant Prot. Bull.*, 19, 86, 1971.
447. Bloom, J. R., Lethal effect of temperature extremes on *Anguina tritici*, *Phytopathology*, 53, 347, 1963.
448. Srinivasan, M. C., Neergaard, P., and Mathur, S. B., A technique for detection of *Xanthomonas campestris* in routine seed health testing of crucifers, *Seed Sci. Technol.*, 1, 853, 1973.
449. Ralph, W., Problems in testing and control of seed-borne bacterial pathogens: a critical review, *Seed Sci. Technol.*, 5, 735, 1977.

450. Cafati, C. R. and Saettler, A. W., Transmission of *Xanthomonas phaseoli* in seed of resistant and susceptible *Phaseolus* genotypes, *Phytopathology*, 70, 638, 1980.
451. Rennie, W. J. and Seaton, R. D., Loose smut of barley. The embryo test as a means of assessing loose smut infection in seed stocks, *Seed Sci. Technol.*, 3, 697, 1975.
452. Doling, D. A., Loose smut in wheat and barley, *Agriculture* (London), 73, 523, 1966.
453. Neergaard, P., *Seed Pathology*, Vols. 1 & 2, Macmillan, London, 1977, 1187.
454. Hewett, P. D., The field behaviour of some seed-borne *Ascochyta fabae* and disease control in field beans, *Ann. Appl. Biol.*, 74, 287, 1973.
455. Neergaard, P., Screening for plant health, *Annu. Rev. Phytopathol.*, 24, 1, 1986.
456. *The Cereal Seeds Regulations 1980,* Her Majesty's Stationery Office, London, 1980.
457. Hewett, P. D., Seed standards for disease in certification, *J. Nat. Inst. Agric. Bot.*, 15, 373, 1981.
458. Jellis, G. J. and Punithalingam, E., Discovery of *Didymella fabae sp. nov.*, the teleomorph of *Ascochyta fabae*, on faba bean straw, *Plant Pathol.*, 40, 150, 1991.
459. Jones, R. A. C. and Proudlove, W., Further studies on cucumber mosaic virus infection of narrow-leafed lupin (*Lupinus angustifolius*): seedborne infection, aphid transmission, spread and effects on grain yield, *Ann. Appl. Biol.*, 118, 319, 1991.
460 Stovold, G. E. and Priest, M. J., A note on the incidence of internally-borne *Sclerotinia sclerotiorum* in soybean seed harvested in New South Wales, *Australasian Plant Pathol.*, 15, 83, 1986.
461. Steadman, J. R., Nature and epidemiological significance of infection of bean seed by *Whetzelinia sclerotiorum*, *Phytopathology*, 65, 1323, 1975.
462. Butcher, C. L., Dean, L. L., and Guthrie, J. W., Effectiveness of halo blight control in Idaho bean seed crops, *Plant Dis. Rep.*, 53, 894, 1969.
463. Copeland, L. O. and Adams, M. W., An improved seed program for maintaining disease-free seed of field beans (*Phaseolus vulgaris*), *Seed Sci. Technol.*, 3, 719, 1975.
464. Ednie, A. B. and Needham, S. M., Laboratory test for internally-borne *Xanthomonas phaseoli* and *Xanthomonas phasaeoli* var. *fuscans* in field bean *(Phaseolus vulgaris* L.) seed, *Proc. Assoc. Off. Seed Anal.*, 63, 76, 1973.
465. Cafti, C. R. and Saettler, A. W., Role of nonhost species as alternate inoculum sources of *Xanthomonas phaseoli, Plant Dis.,* 64, 194, 1980.
466. Schaad, N. W., Detection of seedborne bacterial plant pathogens, *Plant Dis.*, 66, 885, 1982.
467. Grogan, R. G., Control of lettuce mosaic with virus free seed, *Plant Dis.*, 64, 446, 1980.
468. Grogan, R. G., Welch, J. E., and Bardin, R., Common lettuce mosaic and its control by the use of mosaic free seed, *Phytopathology,* 42, 573, 1952.
469. Kimble, K. A., Grogan, R. G., Greathead, A. S., Paulus, A. O., and House, J. K., Development, application and comparison of methods for indexing lettuce seed for mosaic virus in California, *Plant Dis. Rep.*, 59, 461, 1975.
470. Marrou, J., Messiaen, C. M., and Migliori, A., Méthode de controle de l'etat sanitaire des graines de laitue, *Ann. Epiphyt.*, 18, 227, 1967.
471. Carroll, T. W., Economic importance and control of barley stripe mosaic virus in Montana, in *3rd Int. Congr. Plant Pathol.*, P. Parey, Berlin, 1978, 30.
472. Carroll, T. W., Certification schemes against barley stripe mosaic, *Seed Sci. Technol.,* 11, 1033, 1983.

473. Hampton, R. O., Mink, G. I., Hamilton, R. I., Kraft, J. M., and Meuhlbauer, F. J., Occurrence of pea seedborne mosaic virus in North American pea breeding lines, and procedures for its elimination, *Plant Dis. Rep.*, 60, 455, 1976.
474. Bos, L. and Van der Want, J. P. H., Early browning of pea, a disease caused by a soil and seedborne virus, *Tijdschr. Plantenziekten,* 68, 368, 1962.
475. Polston, J. E. and Goodman, R. M., Enzyme linked immunosorbent assay (ELISA) to produce virus free plants from soybean germplasm, *Phytopathology,* 71, 250, 1981.
476. Rohloff, I., The controlled environment room test of lettuce seed for identification of lettuce mosaic virus, *Proc. Int. Seed Test. Assoc.*, 32, 59, 1967.
477. Grogan, R. G., Lettuce mosaic virus çontrol by use of virus–indexed seed, *Seed Sci. Technol.*, 11, 1043, 1983.
478. Zaske, S. K., Carroll, T. W., and Sipes, S. K., An enzyme linked immunosorbent assay to detect barley stripe mosaic virus in barley for use in the Montana seed certification program, *Phytopathology,* 75, 1159, 1985.
479. Mink, G. I., Control of plant diseases using disease-free stocks, in *Handbook of Pest Management in Agriculture,* Pimentel, D., Ed., CRC Press, Boca Raton, FL, 1981, 316.
480. Mink, G. I. and Aichele, M. D., Detection of prunus necrotic ringspot and prune dwarf viruses in *Prunus* seed and seedlings by enzyme-linked immunosorbent assay, *Plant Dis.*, 68, 378, 1984.
481. Neergaard, P., The infection percentage as a relative value in assessing disease tolerance for seed health testing, *Proc. Int. Seed Test. Assoc.*, 27, 400, 1962.
482. Russell, T. S., Some aspects of sampling and statistics in seed health testing and the establishment of threshold levels, *Phytopathology,* 78, 880, 1988.
483. Walker, J. C. and Patel, P. N., Splash dispersal and wind as factors in epdemiology of halo blight of bean, *Phytopathology,* 54, 140, 1964.
484. Zink, F. W., Grogan, R. G., and Wetch, J. E., The effect of the percentage of seed transmission upon subsequent spread of lettuce mosaic virus, *Phytopathology,* 46, 662, 1956.
485. Baker, K. F., Seed Pathology, in *Seed Biology,* Vol. 2, Kozlowski, T. T., Ed., Academic Press, New York, 1972, 317.
486. Morton, D. J. A., Quick method of preparing barley embyros for loose smut examination, *Phytopathology,* 50, 270, 1960.
487. Agarwal, V. K., Singh, O. V., and Singh, A., A note on certification standard for the karnal bunt disease of wheat, *Seed. Res.*, 1, 97, 1973.
488. Agarwal, V. K., Singh, A., and Verma, H. S., Outbreak of karnal bunt of wheat, *FAO Plant Prot. Bull.*, 24, 99, 1976.
489. Kahn, R. P., The importance of seed health in international seed exchange, in *Rice Seed Health,* Int. Rice Res. Inst., Manila, 1988, 7.
490. Mathys, G. and Baker, E. A., An appraisal of the effectiveness of quarantines, *Annu. Rev. Phytopathol.*, 18, 85, 1980.
491. Kahn, R. P., Plant quarantine: principles, methodology and suggested approaches, in *Plant Health and Quarantine in International Transfer of Genetic Resources,* Hewitt, W. B. and Chiarappa, L., Eds., CRC Press, Boca Raton, FL, 1977, 289.
492. Waterworth, H. E. and White, G. A., Plant introductions and quarantine: the need for both, *Plant Dis.*, 66, 87, 1982.
493. Neergaard, P., A review on quarantine for seed, in *Golden Jubilee Commemoration Volume,* National Academy of Sciences, New Delhi, India, 1980.

494. Peregrine, W. T. H., Groundnut rust *(Puccinia arachidis)* in Brunei, *PANS*, 17, 318, 1971.
495. Mendez, M., Quarantine initiatives on seed pathology in the Americas, *Seed Pathol. News*, 9, 1, 1976.
496. West, E., Peanut rust, *Plant Dis. Rep.*, 15, 5, 1931.
497. *The Australian Plant Quarantine Service,* Commonwealth Department of Agriculture, Australian Government Publishing Service, Canberra, 1983, 150.
498. Karpati, J. F., Plant quarantine on a global basis, *Seed Sci. Technol.*, 11, 1145, 1983.
499. Lozano, J. C., Identification of bacterial leaf blight of rice, caused by *Xanthomonas oryzae,* in America, *Plant Dis. Rep.*, 61, 644, 1977.
500. Randles, J. W. and Duke, A. T., Three seedborne pathogens isolated from *Vicia faba* seed imported from United Kingdom, *Aust. Plant Pathol. Soc. Newsl.*, 6, 37, 1977.
501. Neergaard, P., Seed health — policy of certification and disease control, *Seed Pathol. News,* 6, 7, 1974.
502. Chiarappa, L., Man-made epidemiological hazards in major crops of developing countries, in *Plant Diseases and Vectors Ecology and Epidemiology,* Maramorosch, K. and Harris, K. F., Eds., Academic Press, New York, 1981, 319.
503. Fonseca, M. E. N., Marinho, V. L. A., Nagata, T., and Kitajima, E. W., Hop latent viroid in hop germplasm introduced into Brazil from the United States, *Plant Dis.*, 77, 952, 1993.
504. Bingefers, S., International dispersal of nematodes, *Neth. J. Plant Pathol.*, 73 (Suppl. 1), 44, 1967.
505. Kahn, R. P., Exclusion as a plant disease control strategy, *Annu. Rev. Phytopathol.*, 29, 219, 1991.
506. Neergaard, P., Risks for the EPPO region from seedborne pathogens, *EPPO Bull.*, 11, 207, 1981.
507. Food and Agricultural Organization of the United Nations, Plant quarantine announcements, *FAO Plant Prot. Bull.*, 20, 93, 1972.
508. Smilanick, J. L., Dupler, M., Goates, B. J., and Hoffmann, J. A., Germination of teliospores of karnal, dwarf and common bunt fungi after ingestion by animals, *Plant Dis.*, 70, 242, 1986.
509. Kahn, R. P., A model plant quarantine station. Principles, concepts, and requirements, in *Exotic Plant Quarantine Pests and Procedures for the Introduction of Plant Materials,* Singh, K. G. Ed., ASEAN Plant Quarantine Centre Training Institute, Serdang, Malaysia, 1983, 302–333.
510. Neergaard, P., Seed Pathology of rice, Proc. 1st Int. Symp. Plant Pathol., New Delhi, 1970, 57.
511. Sevilla, E. P. and Guerrero, F. C., Production of quality seed in the Philippines, *Seed Sci. Technol.*, 11, 1139, 1983.
512. Alvarez, E., McGee, D. C., and Braun, E., Development of an improved seed assay for *Pseudomonas syringae* pv. *glycinea* in soybean seed lots, *Iowa Seed Sci.*, 13, 1, 1991.
513. Chock, A. K., The international plant protection convention, in *Plant Health, the Scientific Basis for Administrative Control of Plant Diseases and Pests,* Ebbels, D. L. and King, J. E., Eds., Blackwell Scientific, Oxford, 1979, 1.
514. Singh, K. G., Regional ASEAN collaborations in plant quarantine, *Seed Sci. Technol.*, 11, 1189, 1983.
515. Bunting, A. H., CGIAR: the first ten years, *Span*, 21, 15, 1986.

516. Kaiser, W. J., Testing and production of healthy plant germplasm, Danish Govt. Inst. Seed Pathol. Developing Countries, Tech. Bull. No. 2., 1987, 30.
517. Hedley, J., The implications of plant quarantine principles, *FAO Plant Prot. Bull.*, 40, 131, 1992.
518. Hodge, W. H. and Erlanson, C. O., Federal plant introduction — a review, *Econ. Bot.*, 10, 299, 1956.
519. Rohwer, G. G., Plant quarantine philosophy of the United States, in *Plant Health, the Scientific Basis for Administrative Control of Plant Diseases and Pests,* Ebbels, D. L. and King, J. E., Eds., Blackwell Scientific, Oxford, 1980, 23.
520. Leppik, E. E., Introduced seed-borne pathogens endanger crop breeding and plant introduction, *FAO Plant Prot. Bull.,* 16, 57, 1968.
521. Waterworth, H., Processing foreign plant germplasm at the National Plant Germplasm Quarantine Center, *Plant Dis.*, 77, 854, 1993.
522. Ebbels, D. L. and King, J. E., Eds., *Plant Health, the Scientific Basis for Administrative Control of Plant Diseases and Pests,* Blackwell Scientific, Oxford, 1979, 322.
523. Southey, J. F., Preventing the entry of alien diseases and pests into Great Britain, in *Plant Health, the Scientific Basis for Administrative Control of Plant Diseases and Pests,* Ebbels, D. L. and King, J. E., Eds., Blackwell Scientific, Oxford, 1979, 63.
524. Agarwal, V. K., Development of seed pathology in India, Proc. Conf. Seed Sci. Technol., Yadav, T. P. and Ram, C., Eds., Haryana Agriculture University, Haryana, India, 1988, 215.
525. Joshi, N. C., Plant Quarantine in India, *Rev. Trop. Plant Pathol.,* 6, 181, 1989.
526. Olembo, S., Seed health testing at the plant quarantine station at Muguga, Kenya, *Seed Sci. Technol.,* 11, 1217, 1983.
527. Jones, D. R., Seedborne diseases and the international transfer of plant genetic resources: an Australian perspective, *Seed Sci. Technol.,* 15, 765, 1987.
528. Phillips, D., Chandrashekar, M., and McLean, G., Evaluation of potential disease and pest risks associated with paddy as a contaminant of milled rice imported into Australia, *FAO Plant Prot. Bull.*, 40, 4, 1992.
529. Singh, K. G., Regional ASEAN collaboration in plant quarantine, *Seed Sci. Technol.,* 11, 1189, 1983.
530. Morrison, L. G., Quarantine Principles and Policy, SPC Workshop and Training Course in Plant Quarantine, Suva, Fiji, SPC, Suva, 1977, 2.
531. Kahn, R. P., *Plant Protection and Quarantine,* Vol. I, *Biological Concepts,* CRC Press, Boca Raton, FL, 1989, 226.
532. Neergaard, P., Seed health in relation to the exchange of germplasm, International Board of Plant Genetic Research, 1984, 1.
533. Sethi, C. L., Nath, R. P., Mathur, V. K., and Ahuja, S., Interceptions of plant parasitic nematodes from imported seed/plant material, *Indian J. Nematol.,* 2, 89, 1972.
534. Smith, I. M., Activities of the European and Mediterranean Plant Protection Organization in relation to seedborne pathogens, *Seed Sci. Technol.,* 12, 57, 1984.
535. Sheffield, F. M. L., Requirements of a post-entry quarantine station, *FAO Plant Prot. Bull.,* 6, 149, 1958.
536. Sheffield, F. M. L., Closed quarantine procedures, *Rev. Appl. Mycol.,* 47, 1, 1968.
537. *Plant Pathologist's Pocketbook,* Commonwealth Mycological Institute, Kew, Surrey, U.K., 1974, 267.

538. Berg, G. H., Post-entry and intermediate quarantine stations, in *Plant Health and Quarantine in International Transfer of Genetic Resources,* Hewitt, W. B. and Chiarappa, L., Eds., CRC Press, Boca Raton, FL, 1977, 315.
539. Leppik, E. E., Seed-borne diseases on introduced plants, in Report No. 7, North Central Regional Plant Introduction Station, Ames, IA, 1962, 11.
540. Leppik, E. E., List of foreign pests, pathogens and weeds detected on introduced plants, in Plant Introduction Investigation Paper No. 15, U.S. Department of Agriculture, Beltsville, MD, 1969.
541. Smith, I. M., New techniques for detection and identification of seed-borne pathogens in relation to international seed exchange and the zero tolerance concept, *Seed Sci. Technol.,* 18, 461, 1990.
542. Food and Agriculture Organization (FAO), Revised text of the International Plant Protection Conventions, FAO Document No. N 6101, FAO, Rome, 1986.
543. Agarwal, P. C., Ram Nath, Usha Dev, and Majumdar, A., Quarantine salvaging of safflower seed infected with safflower rust *(Puccinia carthami), FAO Plant Prot. Bull.,* 38, 141, 1990.
544. Tisselli, O., Sinclair, J. B., and Hymowitz, T., Sources of Resistance to Selected Fungal, Bacterial, Viral and Nematode Diseases of Soybeans, INTSOY Ser. No. 10, College of Agriculture, University of Illinois at Urbana-Champaign, 1980, 134.
545. Hooker, A. L., Southern leaf blight of corn — present status and future prospects, *J. Environ. Qual.,* 1, 244, 1972.
546. Champion, R., Testing cultivars for resistance to disease in France, *Seed Sci. Technol.,* 11, 681, 1983.
547. Prasad, K. and Weigle, J. L., Association of seed coat factors with resistance to *Rhizoctonia solani* in *Phaseolus vulgaris, Phytopathology,* 66, 342, 1976.
548. Braverman, S. W., Aseptic culture of soybean and peanut embryonic axes to improve phytosanitation of plant introductions, *Seed Sci. Technol.,* 3, 725, 1975.
549. Singh, D. P. and Agarwal, V. K., Relative susceptibility of different sorghum cultivars to grain mould, *Bangladesh J. Bot.,* 16, 111, 1987.
550. Shrotria, P. K., Singh, R., and Agarwal, V. K., Evaluation techniques of sorghum (*Sorghum vulgare* Pers.) genotypes for grain mold resistance, *Bangladesh J. Bot.,* 18, 227, 1989.
551. Jambunathan, R. K., Milind, S., and Vaidya, P., Ergosterol concentration in mold-susceptible and mold resistant sorghum at different stages of grain development and its relationship to flavan-4-ols, *J. Agric. Food Chem.,* 39, 1866, 1991.
552. ICRISAT (International Crops Research Institute for the Semi-Arid Tropics), ICRISAT Report 1991, Patancheru, India, ICRISAT, 1992, 116.
553. Assabgui, R. A., Reid, L. M., Hamilton, R. I., and Arnason, J. T., Correlation of kernel (E) ferulic acid content of maize with resistance to *Fusarium graminearum, Phytopathology,* 83, 949, 1993.
554. Headrick, J. M. and Pataky, J. K., Resistance to kernel infection by *Fusarium moniliforme* in inbred lines of sweet corn and the effect of infections on emergence, *Plant Dis.,* 73, 887, 1989.
555. Headrick, J. M., Pataky, J. K., and Juvik, J. A., Relationships among carbohydrate content of kernels, conditions of silks after pollination, and the response of sweet corn inbred lines to infection of kernels by *Fusarium moniliforme, Phytopathology,* 80, 487, 1990.
556. Fujita, Y. and Suzuki, H., Histological study on the resistance in soybean *(Glycine max)* and a wild soybean *(G. soja)* to purple seed stain caused by *Cercospora kikuchii, Ann. Phytopathol. Soc. Jpn.,* 54, 151, 1988.

557. Fujita, Y., Ikeda, R., and Suzuki, H., Gene analysis of resistance in wild soybean *(Glycine soja)* to purple seed stain caused by *Cercospora kikuchii*, *Ann. Phytopathol. Soc. Jpn.*, 54, 9, 1988.
558. Kulik, M. M. and Yaklich, R. W., Soybean seed coat structures: relationship to weathering resistance and infection by the fungus *Phomopsis phaseoli*, *Crop Sci.*, 31, 108, 1991.
559. Van Den Bovenkamp, G. W., Idema, E. D., and Franken, A. A. J. M., Seed infection and field resistance to *Pseudomonas syringae* pv. *phaseolicola* in seven bean *(Phaseolus vulgaris* L.) varieties, *Plant Varieties Seeds*, 4, 107, 1991.
560. Kumari, S. R. and Chandrashekar, A., Proteins in developing sorghum endosperm that may be involved in resistance to grain moulds, *J. Soil Food Agric.*, 60, 275, 1992.
561. Hoenisch, R. W. and Davis, R. M., Relationship between kernel pericarp thickness and susceptibility to *Fusarium* ear rot in field corn, *Plant Dis.*, 78, 517, 1994.
562. Zimmer, R. C. and Lamb, R. J., Amplification and spread of pea seedborne mosaic virus in field-grown peas, *Can. J. Plant Pathol.*, 15, 17, 1993.
563. Zimmerman, M. S. and Minor, H. C., Inheritance of Phomopsis seed decay resistance in soybean P1 417479, *Crop Sci.*, 33, 96, 1993.
564. Esele, J. P., Frederiksen, R. A., and Miller, F. R., The association of genes controlling caryopsis traits with grain mold resistance in sorghum, *Phytopathology*, 83, 490, 1993.
565. Faluyi, M. A., Mather, D. E., Atlin, G. N., Merrick, L. C., and Paulitz, T. C., Field evaluation of seed, pod and stem rot in white lupine germplasm, *Plant Dis.*, 77, 926, 1993
566. Jambunathan, R., Kherdekar, M. S., and Stenhouse, J. W., Sorghum grain hardness and its relationship to mold susceptibility and mold resistance, *Agric. Food Chem.*, 40, 1403, 1992.

Appendix A Revised Scientific Names of Fungal Plant Pathogens

Previous name	New classification
Acrocylindrium oryzae	*Sarocladium oryzae* (Sawada) W. Gams & P. Hawksworth
Alternaria tenuis	*Alternaria alternata* (Fr.:Fr.) Keissl
Ascochyta imperfecta	*Phoma medicaginis* Malbr. & Roum. in Roum.
A. pinodella	*P. pinodella* (L. K. Jones) Morgan-Jones & K. B. Burch
A. pinodes	*Mycosphaerella pinodes* (Berk. & Bioxam) Vestergr.
A. rabiei	*Phoma rabiei* (Pass.) Khune & J. N. Kapoor
Botryodiplodia theobromae	*Lasiodiplodia theobromae* (Pat.) Griffon & Maubl.
Botrytis allii	*Botrytis acalda* Fresn.
Cephalosporium acremonium	*Acremonium strictum* W. Gams
C. cerealis	*Hymenella cerealis* Ellis & Everh.
C. gramineum	*H. cerealis*
C. gregatum	*Phialophora gregata* (Allington & D. W. Chamberlain) W. Gams
Ceratocystis montia	*Ophiostoma ips* (Rumbold) Nannf.
C. ulmi	*O. ulmi* (Buisman) Nannf.
C. wageneri	*O. wageneri* (D. J. Goheen & F. W. Cobb) T. C. Harrington
Cercosporidium personatum	*Phaeoisariopsis personata* (Berk. & M. A. Curtis) Arx
Colletotrichum hibisci	*Colletotrichum gloeosporioides* (Penz.) Penz. & Sacc. in Penz.
C. malvacearum	*C. malvarum* (A. Braun & Casp.) Southworth
Coryneum carpophilum	*Stigmina carpophila* (Lév.) M. B. Ellis
Diaporthe phaseolorum var. *batatatis*	*Diaporthe phaseolorum* (Cke. & Ellis) Sacc. f. sp. *meridionales* Morgan-Jones
Drechslera nodulosa	*Bipolaris nodulosa* (Berk. & M. A. Curtis) Shoemaker
D. oryzae	*B. oryzae* (Breda de Haan) Shoemaker
D. sorokiniana	*B. sorokiniana* (Sacc.) Shoemaker
Epicoccum purpurascens	*Epicoccum nigrum* Link
Fomes annosus	*Heterobasidion annosum* (Fr.:Fr.) Bref.
F. igniarius	*Phellinus igniarius* (L.:Fr.) Quel.
F. pini	*P. pini* (Thore:Fr.) A. Ames
Fusarium nivale	*Microdochium nivale* (Fr.) Samuels & I. C. Hallett
Gerlachia nivalis	*M. nivale*
G. oryzae	*M. oryzae* (Hashioka & Yokogi) Samuels & I. C. Hallett
Gloeosporium musarum	*Colletrotrichum musae* (Berk. & M. A. Curtis) Arx
Helminthosporium avenae	*Drechslera avenae* (Eldam) Scharif
H. gramnium	*D. graminea* (Rabenh.) Shoemaker
H. maydis	*Bipolaris maydis* (Nisikado & Miyake) Shoemaker
H. oryzae	*B. oryzae* (Breda de Haan) Shoemaker
H. sativum	*B. sorokiniana* (Sacc.) Shoemaker
H. sorokinianum	*B. sorokiniana*
H. turcicum	*Exserohilum turcicum* (Pass.) K. J. Leonard & E. G. Suggs
H. victoriae	*Bipolaris victoriae* (F. Meehan & Murphy) Shoemaker
H. zeicola	*B. zeicola* (G. L. Stout) Shoemaker
Kabatiella caulivora	*Aureobasidium caulivora* (Kirchn.) W. B. Cooke
K. lini	*A. lini* (Lafferty) Hermanides-Nijhof
K. zeae	*A. zeae* (Narita & Hiratsuka) J. M. Dingley
Leptosphaeria nodorum	*Phaeosphaeria nodorum* (E. Müller) Hedjaroude
Nakataea irregulare	*Nakataea sigmodea* (Cavara) K. Hara var. *irregulare* Cralley & Tullis
Neovossia barclayana	*Tilletia barclayana* (Bref.) Sacc. & Syd. in Sacc.
N. horrida	*T. barclayana*
Ophiobolus heterostrophus	*Cochliobolus heterostrophus* (Drechs.) Drechs.

Appendix A Revised Scientific Names of Fungal Plant Pathogens (continued)

Previous name	New classification
Peniophora gigantea	*Phanerochaete gigantea* (Fr.:Fr.) S. S. Rattan et al. in S. S. Rattan
Peronospora farinosa f. sp. betae	*Peronospora farinosa* (Fr.:Fr.) Fr.
P. graminicola	*Sclerospora graminicola* (Sacc.) J. Schröt
Peronospora halstedii	*Plasmopara halstedii* (Farl.) Berl. & De Toni in Sacc.
Pestalotia maculans	*Pestalotiopsis maculans* (Corda) Nag Raj
P. palmarum	*P. palmarum* (Cooke) Steyaert
Phytophthora megasperma f. sp. glycinea	*Phytophthora sojae* M. J. Kaufman & J. W. Gerdemann
P. parasitica	*P. nicotianae* Breda de Haan var. *parasitica* (Dastur) G. M. Waterhouse
Polyporus tomentosus	*Inonolus tomentosus* (Fr.:Fr.) S. Teng
Puccinia carthami	*Puccinia calcitrapae* DC. var. *centaureae* (DC.) Cummins
Rhizoctonia bataticola	*Macrophomina phaseolina* (Tassi) Gordanich
Rhynchosporium oryzae	*Microdochium oryzae* (Haskioka & Yokogi) Samuels & I. C. Hallett
Scirrhia acicola	*Mycosphaerella dearnessii* Barr
Scolecotrichum graminis	*Cercosporidium graminis* (Fuckel) Deighton
Sclerospora macrospora	*Sclerophthora macrospora* (Sacc.) Thirumalachar, C. G. Shaw & Narasimhan
S. sorghi	*Peronosclerospora sorghi* (W. Weston & Uppal) C. G. Shaw
Selenophoma donacis	*Pseudoseptoria donacis* (Pass.) Sutton
Septoria avenae	*Stagonospora avenae* (A. B. Frank) Bissett
S. nodorum	*S. nodorum* (Berk.) Castellani & E. G. Germano
Sclerotinia camelliae	*Ciborinia camelliae* L. M. Kohn
S. convoluta	*Botyrotinia convoluta* (Drayton) Whetzel
Selenophoma donacis	*Pseudoseptoria donacis* (Pass.) Sutton
Sphacelotheca cruenta	*Sporisorium cruentum* (Kühn) K. Vánky
S. destruens	*S. destruens* (Schlechtend.) K. Vánky
S. radicinum	*Alternaria radicina* Meier, Drechs, & E. C. Eddy
Tilletia tritici	*Tilletia caries* (DC.) Tul. & C. Tul.
T. foetida	*T. laevis* Kühn in Rabenh.
T. foetens	*T. laevis*
Tolyposporium penicillariae	*Maeziomyces bullatus* (J. Schröt.) K. Vánky
Ustilago avenae	*Ustilago segetum* (Bull.) Roussel. var. *avenae* (Pers.) Brun.
U. hordei	*U. segetum* (Bull.) Roussel. var. *segetum* (Pers.) Brun.
U. kolleri	*U. segetum* var. *segetum*
U. levis	*U. segetum* var. *segetum*
U. nuda	*U. segetum* (Bull.) Roussel var. *tritici* (Pers) Brun.
U. tritici	*U. segetum* var. *tritici*
U. virens	*Ustilaginoidea virens* (Cooke) Takeh.
Whetzelinia sclerotiorum	*Sclerotinia sclerotiorum* (Lib.) de Bary

From Farr, F. F., Bills, G. F., Chamuris, G. P., and Rossman, A. Y., *Fungi on Plants and Plant Products in the United States,* APS Press, St. Paul, MN, 1989; Hansen, E. and Maxwell, D. P., *Mycologia,* 83, 376, 1991; and Duran, R., *Ustilaginales of Mexico,* R. Duran, Publ., Pullman, WA, 331.

Appendix B Anamorphs and Teleomorphs of Some Fungi Causing Plant Disease

Anamorph	Teleomorph
Amphobotrys ricini (Buchwald) Hennebert	*Botryotinia ricini* (Godfrey) Whetzel
Ascochyta pinodes L. K. Jones	*Mycosphaerella pinodes* (Berk & Bloxam) Vestergr.
Asteromella brassicae (Chev.) Boerema & Van Kesteren	*M. brassicicola* (Duby) Lindau in Engl. & Prantl
Bipolaris cynodontis (Marignoni) Shoemaker	*Cochliobolus cynodontis* R. R. Nelson
B. maydis (Nisikado & Miyake) Shoemaker	*C. heterostrophus* (Drechs.) Drechs.
B. oryzae (Breda de Haan) Shoemaker	*C. miyabeanus* (Ito & Kuribayashi) Drechs. ex Dastur
B. setariae (Sawada) Shoemaker	*C. setariae* (Ito & Kuribayashi in Ito) Drechs. ex Dastur
B. sorokiniana (Sacc.) Shoemaker	*C. sativus* (Ito & Kuribayashi) Drechs. ex Dastur
B. victoriae (F. Meehan & Murphy) Shoemaker	*C. victoriae* R. R. Nelson
B. zeicola (G. L. Stout) Shoemaker	*C. carbonum* R. R. Nelson
Cercospora arachidicola S. Hori	*Mycosphaerella arachidis* Deighton
C. janseana (Racib.) O. Const.	*Sphaerulina oryzina* Hara
Chalara quercina B. W. Henry	*Ceratocystis fagacearum* (T. W. Bretz) J. Hunt
Collectotrichum gloeosporioides (Penz.) Penz. & Sacc. in Penz.	*Glomerella cingulata* (Stoneman) Spaul. & H. Schrenk
Curvularia genitculata (Tracy & Earle) Boedijn	*Cochliobolus geniculatus* Nelson
C. lunata (Wakk.) Boedijn	*C. lunatus* R. R. Nelson & Haasis
Cylindrocladium crotalariae (C. A. Loos) D. K. Bell & Sobers	*Calonectria crotalariae* (C. A. Loos) D. K. Bell & Sobers
Drechslera avenae (Eidam) Scharif	*Pyrenophora avenae* Ito & Kuribayashi
D. bromi (Died.) Shoemaker	*P. bromi* (Diad.) Drechs.
D. graminea (Rabenh.) Shoemaker	*P. graminea* Ito & Kuribay
D. teres (Sacc.) Shoemaker	*P. teres* Drechs.
D. tritici-repentis (Died.) Shoemaker	*P. tritici-repentis* (Died.) Drechs.
Ephelis oryzae Syd.	*Balansia oryzae-sativae* Hashioka
Exserohilum turcicum (Pass.) K. J. Leonard & E. G. Suggs	*Setosphaeria turcica* (Luttrell) K. J. Leonard & E. G. Suggs
Fusarium avenaceum (Fr.:Fr.) Sacc.	*Gibberella avenacea* R. J. Cooke
F. graminearum Schwabe	*G. zeae* (Schwein.) Petch
F. lateritium Nees:Fr.	*G. lateritium* (Wallr.) Sacc.
F. moniliforme J. Sheld	*G. fujikuroi* (Sawada) Ito in Ito & K. Kimura
F. subglutinans (Wollenweb. & Reinking) P. E. Nelson, T. A. Toussoun, & Marasas	*G. fujikuroi* var. *subglutinans* (E. Edwards) P. E. Nelson, T. A. Toussoun, & Marasas
Microdochium nivale (Fr.) Samuels & I. C. Hallett	*Monographella nivalis* (Schaffnit) E. Müller
Nigrospora oryzae (Berk. & Broome) Petch	*Khuskia oryzae* H. J. Hudson
Oidium erysiphoides Fr.	*Erysiphe beta* (Varha) Weltz.
Phoma betae A. B. Frank	*Pleospora betae* (Berl.) Nevodovsky
P. lingam (Tode:Fr.) Desmaz.	*Leptosphaeria maculans* (Desmaz.) Ces. & De Not.
P. lycopersici Cooke	*Didymella lycopersici* Kleb.
Phomopsis phaseoli (Desmaz.) Sacc.	*Diaporthe phaseolorum* (Cooke & Ellis) Sacc. var. *caulivora* Athow & Caldwell

Appendix B Anamorphs and Teleomorphs of Some Fungi Causing Plant Disease (continued)

Anamorph	Teleomorph
D. phaseolorum var. *meridionalis* Morgan-Jones	*D. phaseolorum* var. *sojae* (Lehman) Whem.
Pyricularia grisea (Cooke) Sacc.	*Magnaporthe grisea* (T. T. Hebert) Yaegashi & Udagawa
P. oryzae Cavara	*M. grisea* (T. T. Hebert) Yaegashi & Udagawa
Rhizoctonia solani Kühn	*Thanatephorus cucumeris* (A. B. Frank) Donk.
Sclerotium rolfsii Sacc.	*Athelia rolfsii* (Curzi) Tu & Kimbrough
Septoria linicola (Speg.) Garassini	*Mycosphaerella linicola* Naumov
S. tritici Roberge in Desmaz.	*M. graminicola* (Fuckel) J. Schröt. in Cohn
Sphacelia segetum Lev.	*Claviceps purpurea* (Fr.:Fr.) Tul.
Stagonospora avenae (A. B. Frank) Bisselt	*Leptosphaeria avenaria* Weber
S. nodorum (Berk.) Castellani & E. G. Germano	*Phaseosphaeria nodorum* (E. Müller) Hedjaroude
Stemphylium botryosum Wallr.	*Pleospora tarda* E. Simmons
Ustilaginoidea virens (Cooke) Tabah.	*Claviceps oryzae-sativae* Hashioka

From Farr, F. F., Bills, G. F., Chamuris, G. P., and Rossman, A. Y., *Fungi on Plants and Plant Products in the United States*, APS Press, St. Paul, MN, 1989, 1252.

INDEX

A

B

C

D

E

G

H

I

J

K

L

N

O

P

Q

R

S

T

U

V

W

X

Y

Z